Linear Algebra

An Introduction to the Theory and Use of Vectors and Matrices

Linear Algebra

An Introduction to the Theory and Use of Vectors and Matrices ■

Alan Tucker

State University of New York at Stony Brook

Macmillan Publishing Company
New York

Maxwell Macmillan Canada
Toronto

Maxwell Macmillan International
New York   Oxford   Singapore   Sydney

Editor(s):   Robert W. Pirtle
Production Supervisor: Elaine W. Wetterau
Production Manager: Paul Smolenski
Text Designer: Eileen Burke
Cover Designer: Cathleen Norz
Illustrations: York Graphic Services, Inc.

This book was set in 10/12 Times Roman by York Graphic Services, and was printed and bound by R. R. Donnelley & Sons Company— Crawfordsville.
The cover was printed by R. R. Donnelley & Sons Company—Crawfordsville.

Macmillan Publishing Company is part of the Maxwell Communication Group of Companies.

Macmillan Publishing Company
866 Third Avenue, New York, New York 10022

Maxwell Macmillan Canada, Inc.
1200 Eglinton Avenue East
Suite 200
Don Mills, Ontario M3C 3N1

Library of Congress Cataloging in Publication Data

Tucker, Alan,
    Linear algebra : an introduction to the theory and use of vectors
and matrices / Alan Tucker.
        p.   cm.
    Updated ed. of: A unified introduction to linear algebra. © 1988.
    Includes bibliographical references and index.
    ISBN 0-02-421581-3
    1. Algebras, Linear.   I. Tucker, Alan, Unified
introduction to linear algebra.   II. Title.
QA188.T83   1993
512′. 5–dc20
                                                                                    92-23441
                                                                                         CIP

Printing: 2 3 4 5 6 7 8   Year: 4 5 6 7 8 9 0 1 2

Preface

To Students

Linear algebra is an ideal subject for a lower-level college course in mathematics because the theory, computation, and applications are interwoven so beautifully. The theory of linear algebra is powerful, yet easily accessible. Best of all, theory in linear algebra is likable, providing complete, useful answers to almost any question that one can pose. This theory simplifies and explains the workings of linear models and related computations encountered in the past (in solving systems of linear equations, in analytic geometry, and in vector analysis of elementary physics). It provides important answers that go beyond results we could obtain by brute computation. More important, this theory reaches far beyond the initial setting of Euclidean vectors and matrices to give an organizing structure for much of advanced mathematical analysis. This is what mathematics is really about, making things simple and clear. For too many students, mathematics is either a collection of practical techniques, as in calculus, or a collection of formal theory with little relation to applications, as in most mathematics courses after calculus (including many linear algebra courses). A linear algebra course need not be subject to this artificial dichotomy.

Linear models are now used at least as widely as calculus-based models. The world today is commonly thought to consist of large, complex systems with many input and output variables. Linear models are the primary tool for analyzing these systems. A course based on this book (or one like it) should prove to be the most useful college mathematics course most students ever take. With this goal in mind, the material is presented with an eye toward making it easy to remember, not just for the next test but for a lifetime of diverse uses.

Applications of linear algebra are powerful, easily understood, and very diverse. In this book we use examples with concrete applications as much as possible to motivate and illustrate associated theory. Examples include production management, population growth models, Markov chains, computer graphics, economic input–output models, regression and other statistical techniques, linear codes, and more.

The field of linear numerical analysis is very young, having been dependent on digital computers for its development. This field has wrought major changes in what linear algebra theory should be taught in an introductory course. The standout example of such a modern linear algebra text is G. Strang's *Linear Algebra and Its Applications*. A recent NSF-sponsored conference in linear algebra mapped out a more systematic program of what these changes should be—we have incorporated most of that program in this book. Once the theory was needed as an alternative to numerical computation, which was hopelessly difficult. Now theory helps direct and interpret the numerical computation, which computers do for us.

Overview of Text.   Like most introductory linear algebra textbooks, this book develops the concepts of linear algebra initially using concrete settings: vectors in Euclidean space and matrices. Two models introduced in the first section are used over and over again to illustrate many of the new concepts in the first half of the book. In the second half of the book, the focus moves to abstract vector spaces and linear transformations, including inner product spaces and linear operators. The two principal new aspects of this book are (1) a chapter on matrix theory, including a well-motivated early introduction to eigenvalues and eigenvectors; and (2) a ''spiral'' development of central themes of linear algebra, such as orthogonality or solving a system of n equations in n unknowns. Examples illustrating basic concepts early in the book are used simultaneously to raise questions that will be answered in later chapters. Some nonstandard topics toward the end were chosen for the way that they draw on, and reinforce, diverse earlier results.

The book's organization is straightforward: Chapter 1 has basic matrix operations and methods of solving systems of linear equations. Chapter 2 studies properties of vectors. Chapter 3 studies properties of matrices. Chapter 4 develops the theory of vector spaces. Chapter 5 develops the theory of linear transformations. Chapter 6 concludes with more applications.

To Instructors

This book gives students an introduction to the theory, methods, and applications of linear mathematics. It has broader goals than those of most linear algebra texts. Such linear algebra textbooks give their primary attention to the structure of vector spaces, with limited discussion of the structure of matrices and associated linear transformations. Systems of equations arise primarily in the inward-looking process of determining linear dependence or independence of vectors. Matrix multiplication is used to change coordinate bases, and little else. Yet in both applications and many important extensions in higher mathematics, it is the structure of matrices and more general linear operators that is of greater interest. In the larger picture, the importance of vector spaces is the powerful framework they provide for the theory of linear mappings.

While giving a full treatment of abstract vector spaces and linear transformations, we also give substantial attention to the theory of matrices (as recommended in several recent reports about undergraduate linear algebra). We examine a variety of matrix decompositions involving products and sums of special types of matrices. Eigenvectors and eigenvalues are introduced earlier than usual and motivated extensively. Matrix norms, matrix partitioning, and matrix geometric series are developed. These matrix concepts are used to illustrate important theoretical ideas, such as why the dimension of the row space of a matrix equals the dimension of the column space, as well as to analyze applied linear models. The scientific world today uses linear models right and left, and most revolve around matrix multiplication; commonly they involve complex matrix products. For example, the basic projection step in Karmarkar's algorithm for solving linear programs is given by the matrix expression $(\mathbf{I} - \mathbf{A}^\mathrm{T}(\mathbf{A}^\mathrm{T}\mathbf{A})^{-1}\mathbf{A})$.

As mentioned above, a special pedagogical feature of this book is the spiral fashion in which central themes of linear algebra are developed. For example, orthogonality is introduced in Section 2.3 algebraically and geometrically. Illustrations of orthogonality that arose in the first chapter are reviewed: for example, homogeneous systems of equations, rows of $\mathbf{A}^{-1}$ being orthogonal to columns of $\mathbf{A}$, with a note about how easy it is to find the inverse of a matrix with orthogonal columns. Orthogonality is used immediately in the discussion about projections in the next section; the following section on cross products also makes extensive use of orthogonality. Orthogonality is discussed again in Section 3.5, on pseudoinverses. In Chapter 4, orthogonal bases arise in many vector space topics and are extended to function spaces (i.e., Legendre polynomials and Fourier series). In Chapter 5, orthogonality is used in defining new classes of linear transformations and in eigenvector decompositions of symmetric matrices.

In this book we illustrate the methods and theory as often as possible with concrete problems. Some examples are well known, such as Markov chains; others are new to books at the introductory level, such as the use of cross products to determine which faces of a three-dimensional solid are visible to a viewer, and the use of matrix norms to compute the condition number of matrix. Another feature of this book, mentioned earlier, is the use of two models, introduced in the first section, to illustrate new concepts throughout the book.

Especially in the later sections of Chapters 4 and 5 we seek to lay a foundation with students for understanding how the theory of linear analysis will simplify diverse problems in higher mathematics. For example, simple examples of Fourier analysis, eigenfunctions, and spectral decomposition are given, either explicitly or implicitly.

The problem that any linear algebra textbook faces is that linear algebra is a wonderfully multifaceted subject with elegant theory, efficient computational schemes, and diverse, important applications. There is so much to cover and typically only one semester in which to do it. We consciously give the instructor more material than can be covered in a one-semester course. Rather than make the choice for the instructor of a proper mix of vector space theory, matrix theory, and applications, I prefer a book with which an instructor can easily build the mix he or she prefers. The last chapter and later sections in Chapters 2 to 5 can easily be skipped, as desired. *Note that simple matrices, introduced in Section 3.2, are used in Section 4.4 on row and column spaces and in Section 5.4 on eigenvector decompositions of a matrix.*

For the student, the essence of any course should be the homework. This book has a large number of exercises at all levels of difficulty: theoretical exercises, computational exercises, applications. For more information about course outlines, plus suggested homework sets, sample exams, and additional solutions of exercises, see the accompanying *Instructor's Manual*.

At the end of the text is a list of various programming languages and software packages for performing matrix operations. It is recommended that students have access to computers with ready matrix software from the first week.

Acknowledgments

The first people to thank for help with this book are relatives. My father, A. W. Tucker, ignited and nurtured my born-again interest in linear algebra. Notes from the linear algebra course of my brother Tom Tucker at Colgate formed the foundation of the earliest rough drafts. His pedagogical insights from his M.A.A. work on the calculus reform effort were also very instructive. My family, wife Amanda and daughters Lisa and Katie, provided a supportive atmosphere that eased the long hours of writing. The current text is an outgrowth of an earlier more radical book whose development was greatly assisted by collaborative help from Don Small.

Many people have read various versions of the earlier book and this new version, and offered helpful suggestions, including Don Albers, Richard Alo, Tony Ralston, and several anonymous reviewers. Brian Kohn, Tom Hagstrom, and Roger Lautzenheiser class-tested preliminary versions of the original book. Finally, thanks go to Gary Ostedt, Robert Pirtle, and the staff of the editorial and production departments at Macmillan.

A. T.

Organization of Text

As noted in the Preface, this book contains more material than can be covered in one semester and is designed to be used in a variety of courses. The first two chapters (excepting Section 2.5) should always be covered. A vector-space-oriented course would skip most of Chapter 3, covering only eigenvectors (Section 3.3) and parts of 3.1 (which could be incorporated into the presentation of matrix multiplication in Section 1.3), and then would cover all of Chapters 4 and 5 (except possibly Section 5.4). A matrix-oriented course would cover most of Chapter 3, would skim over much of Chapter 4, and cover most of Chapter 5 and selected parts of Chapter 6. Individual instructors have the flexibility to design their own favored mixture of matrix and vector space topics. Further, if they wish, instructors can intermix the material on vector spaces and linear transformations: for example, covering the range and kernel of a linear transformation at the same time as the column space of a matrix.

To assist instructors, the following organization of text table describes dependencies among sections. Where one section is indicated to use briefly material from an earlier section, the earlier section is not required to be covered (it is referenced briefly in a short example or application).

Subject	Builds On	Used In
1.1 Vectors and Matrices		Throughout
1.2 Scalar Product	1.1	Throughout
1.3 Matrix Multiplication	1.2	Throughout
1.4 Gaussian Elimination	1.2	Throughout
1.5 Inverse of a Matrix	1.3, 1.4	Throughout
1.6 Determinant of a Matrix	1.5	2.5, 3.3, 4.2
2.1 Collections of Vectors	1.4	Throughout
2.2 Vector Norms	2.1	3.4, Euclidean norm used throughout
2.3 Angles and Orthogonality	1.5, 2.1	2.4, 2.5, Chapter 4
2.4 Projections and Regression	2.3	3.5, 4.5, 5.3
2.5 Cross Product	1.6, 2.1, 2.3	
3.1 Matrix Algebra	1.3, 2.1	3.2, 3.5, Chapter 6 (4.1, 4.2, 4.5 briefly)
3.2 LU Decomposition Simple Matrices	1.4, 1.6, 3.1	4.4, 5.4
3.3 Eigenvectors and Eigenvalues	1.6	5.3, 5.4, 6.3 (6.1, 5.1 briefly)
3.4 Matrix Norms	1.5, 2.2	6.1
3.5 Pseudoinverse	2.4, 3.1	Optional part of 4.6 (4.5 briefly)
4.1 Subspaces	2.1 (3.3 briefly; optional part uses 3.1)	Chapters 4 and 5
4.2 Linear Independence	1.6, 4.1 (2.3, 3.1 briefly)	Chapters 4 and 5
4.3 Bases	4.2	Chapters 4 and 5
4.4 Vector Spaces of a Matrix	4.3, 3.2	5.2 (4.4 briefly)
4.5 Orthogonal Bases	2.3, 2.4, 4.3, (3.1, 3.5 briefly)	4.6, Chapter 5
4.6 Inner Product Spaces	2.3, 4.5 (optional part uses 3.5)	
5.1 Linear Transformations	4.3 (3.3, 4.5 briefly)	5.2, 5.3
5.2 Range and Kernel	4.4, 4.5, 5.1	
5.3 Diagonalization	2.4, 3.3, 4.3, 5.1 (4.5 briefly)	5.4, 6.3
5.4 Eigenvector Decomposition	3.2, 3.3, 5.3 (4.5 briefly)	
6.1 Iteration and Power Series	3.1, 3.4	6.2
6.2 Markov Chains	1.4, 3.1, 6.1	
6.3 Differential Equations	3.3, 5.3	
6.4 0–1 Matrices	3.1 (4.4 briefly)	

Contents

Matrices   143

Vector Spaces   221

Linear Transformations   299

6 Applications   355

Brief History of Matrices and Linear Algebra   405

References   409

Matrix Algebra Software   413

Solutions to Odd-Numbered Exercises   415

Index   433

Linear Algebra

An Introduction to the Theory and Use of Vectors and Matrices

1 Matrix Operations and Systems of Equations

1.1 Vectors and Matrices

Linear algebra is probably the most widely used mathematical subject in the modern world. In this book we try to show you why. You will see that virtually any question that one can ask about a situation in linear algebra has a complete, satisfying answer. Linear algebra is used in hundreds of diverse applications. It has a powerful, yet accessible body of theory and provides efficient methods for computing numerical results.

First of all, what does one mean by the term *linear algebra*? Or even more simply, what does the term *linear* mean? One good answer is that linear means "like a line." If a variable y is a linear function $f(x)$ of a variable x, the graph of $y = f(x)$ will be a straight line; that is, $f(x)$ has the form $ax + b$, for some constants a and b. Further, in this case y is *linearly dependent* on x. A *linear combination* of variables or *linear expression*, such as $3x + 4y - 6z$, has linear (line-like) terms for each variable. A linear expression cannot have quadratic terms such as x^2 or xy; these terms are *nonlinear*. Linear algebra is the mathematics of linear expressions.

As befits such an important subject, there are many aspects of linear algebra to study, use, and understand.

- There is the notation of vectors, matrices, and more general linear systems, a notation whose power to represent complex problems in mathematics and science concisely is alone worthy of a semester of study.
- There is the theory of linear algebra, the most complete, easiest to comprehend, and most *useful* mathematical theory that you will ever meet.
- There are the computational methods of linear algebra, which are so accurate, efficient, and effective that difficult computational problems in almost every area of the

1

mathematical and physical sciences are solved by recasting them as linear algebra computations.

- There are hundreds of diverse, important applications of linear algebra. Many topics of statistics, economics, management science, and engineering are found to consist largely of linear algebra issues recast in diverse applied settings.

These different aspects of linear algebra interact and help each other in a way that gives the subject an elegance and unity that justifies mathematics' title as the "queen of science."

As in all mathematical subjects, there are two basic types of mathematical building blocks in linear algebra. There are the *objects*—initially, arrays of numbers—and there are *operations* that are performed on these objects—such as adding or multiplying two arrays together. In this section we introduce the basic objects. In subsequent sections in this chapter we present the basic operations.

An ordered list of n numbers is called a ***vector*** or an ***n-vector***. We use lowercase boldfaced letters, such as **v**, to denote vectors. We write v_1 for the first entry in vector **v**, v_2 for the second entry in **v** and v_i for the ith entry in **v**. Examples of vectors are

$$\mathbf{v} = [1, 2, 3, 4] \quad \text{and} \quad \mathbf{c} = \begin{bmatrix} 7 \\ 8 \\ 9 \end{bmatrix}$$

Here **v** is a 4-vector and **c** is a 3-vector. For the **v** and **c** given above, we have $v_2 = 2$ and $c_2 = 8$.

A vector can be treated as a row of numbers or as a column of numbers. For vectors alone, the choice of row or column format is unimportant, but when vectors and matrices are multiplied together, it is very important to distinguish clearly whether a vector is to be treated as a row vector or as a column vector. *The standard convention in this book* (as in most other matrix texts) *is to assume that a vector is a column vector unless explicitly stated otherwise.*

A ***matrix*** is a rectangular array of numbers. We refer to a matrix as an $m \times n$ matrix when the matrix has m rows and n columns, and we use capital boldfaced letters, such as **A**, to denote matrices. (A common handwritten way to indicate a matrix is with a wavy line under the letter, such as $\underset{\sim}{\text{A}}$.) We use the notation a_{ij} to denote the entry in matrix **A** occurring in row i and column j. Examples of matrices are

$$\mathbf{A} = \begin{bmatrix} 4 & 3 & 8 \\ 1 & 2 & 3 \\ 4 & 5 & 6 \end{bmatrix} \quad \text{and} \quad \mathbf{M} = \begin{bmatrix} 9 & 5 & 1 \\ 2 & 7 & 6 \end{bmatrix} \quad \text{where } a_{23} = 3, \quad a_{31} = 4, \quad m_{13} = 1 \quad (1)$$

A column n-vector is also an $n \times 1$ matrix and row n-vector is a $1 \times n$ matrix. Conversely, an $m \times n$ matrix **A** can be thought of as a set of n column vectors (each of length m) or as a set of m row vectors (each of length n). We will use the notation:

$\mathbf{a}_j$ denotes the jth column vector in **A**.
$\mathbf{a}_i'$ denotes the ith row vector in **A**.

(It is common in linear algebra to use the letter i to refer to the number of some row in a matrix and the letter j to refer to the number of some column.) For example, in the matrix $\mathbf{A}$ in (1),

$$\mathbf{a}_2 = \begin{bmatrix} 3 \\ 2 \\ 5 \end{bmatrix} \quad \text{and} \quad \mathbf{a}_1' = [4 \quad 3 \quad 8]$$

Summarizing our matrix notation, we can break down a general matrix $\mathbf{A}$ in the following three ways.

$$\mathbf{A} = \begin{bmatrix} a_{11} & a_{12} & a_{13} & \cdots & a_{1j} & \cdots & a_{1n} \\ a_{21} & a_{22} & a_{23} & \cdots & a_{2j} & \cdots & a_{2n} \\ a_{31} & a_{32} & a_{33} & \cdots & a_{3j} & \cdots & a_{3n} \\ \cdots & \cdots & \cdots & \cdots & \cdots & \cdots & \cdots \\ a_{i1} & a_{i2} & a_{i3} & \cdots & a_{ij} & \cdots & a_{in} \\ \cdots & \cdots & \cdots & \cdots & \cdots & \cdots & \cdots \\ a_{m1} & a_{m2} & a_{m3} & \cdots & a_{mj} & \cdots & a_{mn} \end{bmatrix} \tag{2}$$

$$= \begin{bmatrix} \mathbf{a}_1' \\ \mathbf{a}_2' \\ \mathbf{a}_3' \\ \cdots \\ \mathbf{a}_j' \\ \cdots \\ \mathbf{a}_m' \end{bmatrix} = [\mathbf{a}_1 \quad \mathbf{a}_2 \quad \mathbf{a}_3 \quad \cdots \quad \mathbf{a}_j \quad \cdots \quad \mathbf{a}_n]$$

The following examples will show how vectors and matrices arise naturally in mathematics and applied problems.

EXAMPLE 1.
Geometric View
of Vectors

Vectors can be used to represent points in space. In two-dimensional space, one uses a 2-vector; in three-dimensional space a 3-vector; and in n-dimensional space an n-vector. The point in three-dimensional space with coordinates $x_1 = 2$, $x_2 = 1$, $x_3 = 7$ is written as the vector [2, 1, 7]. Geometrically, one often draws a two- or three-dimensional vector with an arrow going from the origin to the point with the coordinates of the vector. We add vectors by adding the corresponding entries. Geometrically, adding vectors becomes ''adding'' the arrows together. Figure 1.1 displays the addition of [2, 1] and [1, 3] to get the vector [3, 4]. (**Note:** In geometric settings, it is common to write vectors in row format.) ■

EXAMPLE 2.
Oil Refinery
Model

A company runs three oil refineries. Each refinery produces three petroleum-based products: heating oil, diesel oil, and gasoline. Suppose that from one barrel of petroleum, the first refinery produces 4 gallons of heating oil, 2 gallons of diesel oil, and 1

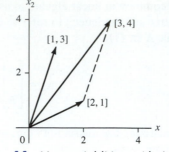

Figure 1.1 Vector Addition with Arrows

gallon of gasoline. The second and third refineries produce different amounts of these three products, as described in the following matrix $\mathbf{A}$.

$$
\mathbf{A} = \begin{array}{c} \text{Heating oil} \\ \text{Diesel oil} \\ \text{Gasoline} \end{array}
\begin{array}{ccc} \text{Refinery 1} & \text{Refinery 2} & \text{Refinery 3} \end{array}
\begin{bmatrix} 4 & 2 & 2 \\ 2 & 5 & 2 \\ 1 & 2.5 & 5 \end{bmatrix} \tag{3}
$$

Each column of $\mathbf{A}$ is a vector of outputs by a refinery. For example, from 1 barrel of oil, refinery 3 produces an output vector $\mathbf{a}_3 = \begin{bmatrix} 2 \\ 2 \\ 5 \end{bmatrix}$. Each row of $\mathbf{A}$ is a vector of amounts produced by different refineries of some product. The row vector for gasoline is $\mathbf{a}_3' = [1, 2.5, 5]$.

Let x_i denote the number of barrels of petroleum used by the ith refinery. Suppose that there is a demand for 600 units of heating oil, 800 units of diesel oil, and 1000 units of gasoline. Then the x_i's need to satisfy the following system of linear equations:

$$
\begin{aligned}
4x_1 + 2x_2 + 2x_3 &= 600 \\
2x_1 + 5x_2 + 2x_3 &= 800 \\
1x_1 + 2.5x_2 + 5x_3 &= 1000
\end{aligned} \tag{4}
$$

or, as a single vector equation in the column vectors of the matrix (we define vector addition formally in the next section),

$$
x_1 \begin{bmatrix} 4 \\ 2 \\ 1 \end{bmatrix} + x_2 \begin{bmatrix} 2 \\ 5 \\ 2.5 \end{bmatrix} + x_3 \begin{bmatrix} 2 \\ 2 \\ 5 \end{bmatrix} = \begin{bmatrix} 600 \\ 800 \\ 1000 \end{bmatrix} \tag{5}
$$

If $\mathbf{b}$ is the column vector of the right-side demands in (5) and $\mathbf{x}$ is a (column) vector of the x_i's, matrix algebra should give us a way to write the system of equations (5) concisely in terms of $\mathbf{A}$, $\mathbf{b}$, and $\mathbf{x}$. Indeed, we learn how to do this in Section 1.2.

Note: We refer to this refinery model repeatedly in this book to illustrate various concepts of linear algebra. It will be our standard example of a system of linear equations. We shall also look at many variations of this model. For example, if the third refinery were closed down, which demand vectors [on the right side of (5)] could the other two refineries still satisfy?

■

**EXAMPLE 3.
Markov Chain
for Weather**

Suppose that we have three states of the weather: sunny, cloudy, or rainy. If it is cloudy today, the probability is $\frac{1}{2}$ that it will be sunny tomorrow, $\frac{1}{4}$ that it will be cloudy tomorrow, and $\frac{1}{4}$ that it will be rainy tomorrow. Other probabilities for tomorrow's weather apply if it is cloudy today or if it is rainy today. It is convenient to display these probabilities in a matrix $\mathbf{A}$.

$$
\begin{array}{c}
\text{Today} \\
\begin{array}{cccc}
\text{Tomorrow} & \text{Sunny} & \text{Cloudy} & \text{Rainy} \\
\end{array} \\
\mathbf{A} = \begin{array}{c}
\text{Sunny} \\
\text{Cloudy} \\
\text{Rainy}
\end{array}
\begin{bmatrix}
\frac{3}{4} & \frac{1}{2} & \frac{1}{4} \\
\frac{1}{8} & \frac{1}{4} & \frac{1}{2} \\
\frac{1}{8} & \frac{1}{4} & \frac{1}{4}
\end{bmatrix}
\end{array}
\qquad (6)
$$

The probabilities in this matrix are called ***transition probabilities***, and the matrix is called a ***transition matrix***. Each column corresponds to the type of weather today. Each row corresponds to the type of weather tomorrow. For example, entry a_{23}, which equals $\frac{1}{2}$, in the cloudy row and rainy column gives the probability that it will be cloudy tomorrow given that it is rainy today. Note that the probabilities in each column of the transition matrix must add up to 1. A convenient way to display the information in a Markov chain is with a transition diagram. The diagram for the weather Markov chain is drawn in Figure 1.2. There is a node for each state and an arrow for each transition probability.

The data in the transition matrix $\mathbf{A}$ are used to compute the probabilities of being sunny or cloudy or rainy tomorrow given the probabilities of being sunny or cloudy or rainy today. Let p_1, p_2, p_3 denote today's probability of sun, clouds, and rain, respectively, and p_1', p_2', p_3' denote tomorrow's probability of sun, clouds, and rain. Then for a given transition matrix $\mathbf{A}$, the following probability formula is used to compute tomorrow's probabilities:

$$
\begin{aligned}
p_1' &= a_{11}p_1 + a_{12}p_2 + a_{13}p_3 \\
p_2' &= a_{21}p_1 + a_{22}p_2 + a_{23}p_3 \\
p_3' &= a_{31}p_1 + a_{32}p_2 + a_{33}p_3
\end{aligned}
\qquad (7)
$$

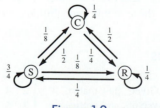

Figure 1.2

For example, in the weather Markov chain if $p_1 = 0$, $p_2 = \frac{1}{2}$, $p_3 = \frac{1}{2}$, tomorrow's distribution p_1', p_2', p_3' is

$$p_1' = \tfrac{3}{4}p_1 + \tfrac{1}{2}p_2 + \tfrac{1}{4}p_3 = 0 + \tfrac{1}{2} \cdot \tfrac{1}{2} + \tfrac{1}{4} \cdot \tfrac{1}{2} = \tfrac{3}{8}$$
$$p_2' = \tfrac{1}{8}p_1 + \tfrac{1}{4}p_2 + \tfrac{1}{2}p_3 = 0 + \tfrac{1}{4} \cdot \tfrac{1}{2} + \tfrac{1}{2} \cdot \tfrac{1}{2} = \tfrac{3}{8} \qquad (8)$$
$$p_3' = \tfrac{1}{8}p_1 + \tfrac{1}{4}p_2 + \tfrac{1}{4}p_3 = 0 + \tfrac{1}{4} \cdot \tfrac{1}{2} + \tfrac{1}{4} \cdot \tfrac{1}{2} = \tfrac{1}{4}$$

We explain the formula for p_1' in (7) and (8) intuitively as follows. We can be in state 1 (sunny) tomorrow either because we are in state 1 today and then stay in state 1—this is the probability $a_{11}p_1$—or because we are in state 2 (cloudy) today and then switch to state 1—this is the probability $a_{12}p_2$—or because we are in state 3 (rainy) today and then switch to state 1—this is the probability $a_{13}p_3$. [To compute the probability of a sequence of two events, such as (1) the probability p_2 of now being in state 2 and (2) the probability a_{12} of switching from state 2 to state 1, we multiply these two probabilities together, to get $a_{12}p_2$.]

We can use the equations in (7) to predict the weather probabilities p_1'', p_2'', p_3'' 2 days ahead using the prediction from (8) for tomorrow's weather chances of $p_1' = \frac{3}{8}$, $p_2' = \frac{3}{8}$, $p_3' = \frac{1}{4}$.

$$p_1'' = \tfrac{3}{4}p_1' + \tfrac{1}{2}p_2' + \tfrac{1}{4}p_3' = \tfrac{3}{4} \cdot \tfrac{3}{8} + \tfrac{1}{2} \cdot \tfrac{3}{8} + \tfrac{1}{4} \cdot \tfrac{1}{4} = \tfrac{34}{64}$$
$$p_2'' = \tfrac{1}{8}p_1' + \tfrac{1}{4}p_2' + \tfrac{1}{2}p_3' = \tfrac{1}{8} \cdot \tfrac{3}{8} + \tfrac{1}{4} \cdot \tfrac{3}{8} + \tfrac{1}{2} \cdot \tfrac{1}{4} = \tfrac{17}{64} \qquad (9)$$
$$p_3'' = \tfrac{1}{8}p_1' + \tfrac{1}{4}p_2' + \tfrac{1}{4}p_3' = \tfrac{1}{8} \cdot \tfrac{3}{8} + \tfrac{1}{4} \cdot \tfrac{3}{8} + \tfrac{1}{4} \cdot \tfrac{1}{4} = \tfrac{13}{64}$$

From probabilities for 2 days hence, we could predict 3 days ahead, and so on. The computations give the following table assuming today's weather probabilities are 0 sunny, $\frac{1}{2}$ cloudy, $\frac{1}{2}$ rainy.

	Sunny	Cloudy	Rainy
Today:	0	$\frac{1}{2}$	$\frac{1}{2}$
1 Day ahead:	$\frac{3}{8}$	$\frac{3}{8}$	$\frac{2}{8}$
2 Days ahead:	$\frac{34}{64}$	$\frac{17}{64}$	$\frac{13}{64}$
3 Days ahead:	$\frac{149}{256}$	$\frac{60}{256}$	$\frac{47}{256}$
5 Days ahead:	$\sim\frac{14}{23}$	$\sim\frac{5}{23}$	$\sim\frac{4}{23}$
10 Days ahead:	$\sim\frac{14}{23}$	$\sim\frac{5}{23}$	$\sim\frac{4}{23}$
100 Days ahead:	$\frac{14}{23}$	$\frac{5}{23}$	$\frac{4}{23}$

(10)

Observe that after several days, the probabilities stabilize at $\frac{14}{23}$, $\frac{5}{23}$, and $\frac{4}{23}$ for sunny, cloudy, and rainy weather, respectively. Later in this chapter we learn how to determine these stable probabilities directly. ■

Section 1.1 Exercises

1. Given the matrix $\mathbf{A} = \begin{bmatrix} 1 & 2 & 3 & 4 \\ 2 & 4 & 6 & 8 \\ 3 & 5 & 7 & 9 \end{bmatrix}$, write out the following row and column vectors and entries.

(a) $\mathbf{a}_1'$ (b) $\mathbf{a}_2$ (c) $\mathbf{a}_3$ (d) a_{22} (e) a_{31}

2. In the matrix of letters $\mathbf{A} = \begin{bmatrix} E & R & S & T & A \\ N & P & O & C & W \\ H & B & U & I & L \\ M & G & Y & F & Y \end{bmatrix}$, spell out the words given by the following sequence of entries.

(a) $a_{31}, a_{11}, a_{35}, a_{22}$

(b) $a_{35}, a_{34}, a_{21}, a_{11}, a_{15}, a_{12}$

(c) $a_{11}, a_{35}, a_{32}, a_{23}, a_{25}$

(d) $a_{25}, a_{15}, a_{14}, a_{24}, a_{31}, a_{23}, a_{33}, a_{14}$

3. Plot the following vectors as arrows as in Figure 1.1 in the x_1–x_2 plane.

(a) $\mathbf{a} = \begin{bmatrix} 0 \\ 3 \end{bmatrix}$ (b) $\mathbf{b} = \begin{bmatrix} 2 \\ 4 \end{bmatrix}$ (c) $\mathbf{c} = \begin{bmatrix} 2 \\ -1 \end{bmatrix}$ (d) $\mathbf{d} = \begin{bmatrix} -1 \\ -4 \end{bmatrix}$ (e) $\mathbf{e} = \begin{bmatrix} -3 \\ 4 \end{bmatrix}$

4. The vectors named in this exercise refer to the vectors in Exercise 3.

(a) Plot the effect of taking vector $\mathbf{a}$ and adding to it vector $\mathbf{c}$.

(b) Plot the effect of taking vector $\mathbf{c}$ and adding to it vector $\mathbf{a}$. Is this the same final vector as obtained in part (a)?

(c) Plot the effect of taking vector $\mathbf{b}$ and adding to it vector $\mathbf{d}$.

(d) Plot the effect of taking vector $\mathbf{c}$ and adding to it vector $\mathbf{e}$.

5. Plot the following vectors as arrows on an x–y–z coordinate grid.

(a) $\begin{bmatrix} 1 \\ 0 \\ 0 \end{bmatrix}$ (b) $\begin{bmatrix} 1 \\ 1 \\ 1 \end{bmatrix}$ (c) $\begin{bmatrix} 2 \\ 4 \\ 1 \end{bmatrix}$ (d) $\begin{bmatrix} 2 \\ -1 \\ 3 \end{bmatrix}$

6. (a) In the matrix $\mathbf{A}$ for the oil refinery model in Example 2, state in words what the numbers in entries a_{23} and entries a_{31} represent. What do the numbers in column $\mathbf{a}_3$ of $\mathbf{A}$ represent?

(b) Suppose that refinery 3 is modernized and its output for each barrel of oil is doubled. What is the new matrix of coefficients?

(c) In the refinery model, suppose that refinery 3 broke down and was out of service. In this case, what is the matrix of coefficients?

7. Consider the following refinery model. There are three refineries, 1, 2, and 3, and from each barrel of crude petroleum, the different refineries produce the following amounts (measured in gallons) of heating oil, diesel oil, and gasoline.

	Refinery 1	Refinery 2	Refinery 3
Heating oil:	6	3	2
Diesel oil:	4	6	3
Gasoline:	3	2	6

Suppose that we have the following demand: 280 gallons of heating oil, 350 gallons of diesel oil, and 350 gallons of gasoline. Write a system of equations whose solution would determine production levels to yield the desired amounts of heating oil, diesel oil, and gasoline. As in Example 2, let x_i be the number of barrels processed by the ith refinery.

8. In Example 2, suppose that refinery 1 processes 15 barrels of petroleum, refinery 2 processes 20 barrels, and refinery 3 processes 60 barrels. With this production schedule, for which product does production deviate the most from the set of demands 600, 800, 1000?

9. The staff dietician at the California Institute of Trigonometry has to make up a meal with 600 calories, 20 grams of protein, and 200 milligrams of vitamin C. There are three food types to choose from: rubbery Jello, dried fish sticks, and mystery meat. They have the following nutritional content per ounce.

	Jello	Fish sticks	Mystery meat
Calories:	10	50	200
Protein:	1	3	0.2
Vitamin C:	30	10	0

Make a mathematical model of the dietician's problem with a system of three linear equations.

10. A furniture manufacturer makes tables, chairs, and sofas. In one month the company has available 300 units of wood, 350 units of labor, and 225 units of upholstery. The manufacturer wants a production schedule for the month that uses all of these resources. The various products require the following amounts of the resources.

	Table	Chair	Sofa
Wood:	4	1	3
Labor:	3	2	5
Upholstery:	2	0	4

Make a mathematical model of this production problem.

11. A company has a budget of $280,000 for computing equipment. There types of equipment are available: microcomputers at $2000 a piece, terminals at $500 a piece, and word processors at $5000 a piece. There should be five times as many terminals as microcomputers and two times as many microcomputers as word processors. Set this problem up as a system of three linear equations.

12. For the weather Markov chain in Example 3, determine the probability distribution for tomorrow's weather, using equations (8), when
    (a) $p_1 = \frac{1}{3}, p_2 = \frac{1}{3}, p_3 = \frac{1}{3}$
    (b) $p_1 = \frac{1}{2}, p_2 = \frac{1}{2}, p_3 = 0$
    (c) $p_1 = \frac{1}{3}, p_2 = 0, p_3 = \frac{2}{3}$
    (d) $p_1 = \frac{14}{23}, p_2 = \frac{5}{23}, p_3 = \frac{4}{23}$
    (e) $p_1 = \frac{1}{2}, p_2 = \frac{1}{4}, p_3 = \frac{1}{4}$

13. Consider the simplified weather Markov chain with two states, sunny and cloudy:

$$\begin{array}{cc} & \begin{array}{cc} \text{Sunny} & \text{Cloudy} \end{array} \\ \begin{array}{c} \text{Sunny} \\ \text{Cloudy} \end{array} & \begin{bmatrix} \frac{1}{2} & \frac{3}{4} \\ \frac{1}{2} & \frac{1}{4} \end{bmatrix} \end{array}$$

In this weather Markov chain, starting with the probability distribution $\begin{bmatrix} 1 \\ 0 \end{bmatrix}$ (a sunny day), compute and plot (in p_1, p_2 coordinates) the distribution over five successive days.

Repeat the process starting with the probability distribution $\begin{bmatrix} 1 \\ 0 \end{bmatrix}$. Can you guess the value of the stable distribution to which your vectors are converging?

14. Consider the following Markov chain model involving the states of mind of Professor Mindthumper. The states are alert (A), hazy (H), and stupor (S). If in state A or H today, the professor has a $\frac{1}{3}$ chance of being in each of the three states tomorrow. If in state S today, tomorrow with probability 1 the professor is still in state S.
    (a) Write the transition matrix **A** for this Markov chain.
    (b) Write out entry a_{23} and column $\mathbf{a}_2$.
    (c) Which pairs of rows and pairs of columns in this Markov chain are the same?

15. The printing press in a newspaper has the following pattern of breakdowns. If it is working today, tomorrow it has a 90% chance of working (and a 10% chance of breaking down). If the press is broken today, it has a 60% chance of working tomorrow (and a 40% chance of being broken again).
    (a) Make a Markov chain for this problem: Give the matrix of transition probabilities and draw the transition diagram.
    (b) If there is a 50–50 chance of the press working today, what are the chances that it will be working tomorrow?
    (c) If the press is working today, what are the chances that it will be working in 2 days?

16. If the local professional basketball team, the Sneakers, wins today's game, they have a $\frac{2}{3}$ chance of winning their next game. If they lose this game, they have a $\frac{1}{2}$ chance of winning their next game.
    (a) Make a Markov chain for this problem: Give the matrix of transition probabilities and draw the transition diagram.
    (b) If there is a 50–50 chance of the Sneakers winning today's game, what are the chances that they win their next game?
    (c) If they won today, what are the chances of their winning the game after next?

17. If the stock market went up today, historical data show that it has a 60% chance of going up tomorrow, a 20% chance of staying the same, and a 20% chance of going down tomorrow. If the market was unchanged today, it has a 20% chance of being unchanged tomorrow, a 40% chance of going up, and a 40% chance of going down tomorrow. If the market goes down today, it has a 20% of going up tomorrow, a 20% chance of being unchanged, and a 60% chance of going down.
    (a) Make a Markov chain for this problem: Give the matrix of transition probabilities and the transition diagram.
    (b) If there is a 30% chance that the market goes up today, a 10% chance that it is unchanged, and a 60% chance that it goes down, what is the probability distribution for the market tomorrow?

18. The following model for learning a concept over a set of lessons identifies four states of learning: I = ignorance, E = exploratory thinking, S = superficial understanding, and M = mastery. If now in state I, after one lesson you have $\frac{1}{2}$ probability of still being in I and $\frac{1}{2}$ probability of being in E. If now in state E, you have $\frac{1}{4}$ probability of being in I, $\frac{1}{2}$ in E, and $\frac{1}{4}$ in S. If now in state S, you have $\frac{1}{4}$ probability of being in E, $\frac{1}{2}$ in S, and $\frac{1}{4}$ in M. If in M, you always stay in M (with probability 1).
    (a) Make a Markov chain model of this learning model.
    (b) If you start in state I, what is your probability distribution after two lessons? After three lessons?

19. (*Computer Project*) Use a computer program to follow the Markov chains in the following exercises for 50 periods by iterating the next-period formula (7) as done in Example 3.
    (a) Exercise 13, starting in the sunny state.
    (b) Exercise 13, starting in the cloudy state.
    (c) Exercise 14, starting in the alert state.
    (d) Exercise 15, starting in the broken state.
    (e) Exercise 16, starting in the win state.
    (f) Exercise 17, starting in the market unchanged state.
    (g) Exercise 18, starting in state I.

1.2 Scalar Product

Scalar Multiplication and Addition of Matrices

The simplest operation in linear algebra is to multiply a vector or matrix by a constant c. This operation is called **scalar multiplication**. The adjective *scalar* is used to refer to activities involving a single number, as opposed to a vector or matrix. Scalar multiplication is performed by multiplying each entry in the vector or matrix by the scalar. For example,

$$\text{if }\ \mathbf{D} = \begin{bmatrix} 2 & 4 & 5 & 1 \\ 3 & 9 & 2 & 5 \\ 1 & 6 & 6 & 2 \end{bmatrix}, \quad \text{then }\ 3\mathbf{D} = \begin{bmatrix} 6 & 12 & 15 & 3 \\ 9 & 27 & 6 & 15 \\ 3 & 18 & 18 & 6 \end{bmatrix}$$

We used scalar multiplication of vectors in Example 2 of Section 1.1 when we rewrote the refinery system of equations

$$\begin{aligned} 4x_1 + \quad 2x_2 + 2x_3 &= \ 600 \\ 2x_1 + \quad 5x_2 + 2x_3 &= \ 800 \\ 1x_1 + 2.5x_2 + 5x_3 &= 1000 \end{aligned} \tag{1}$$

as a vector equation by factoring out the different x_i's and using scalar multiplication of vectors.

$$x_1 \begin{bmatrix} 4 \\ 2 \\ 1 \end{bmatrix} + x_2 \begin{bmatrix} 2 \\ 5 \\ 2.5 \end{bmatrix} + x_3 \begin{bmatrix} 2 \\ 2 \\ 5 \end{bmatrix} = \begin{bmatrix} 600 \\ 800 \\ 1000 \end{bmatrix}$$

The vector version of this system of equations emphasizes that x_1 is the production level for a vector of products from refinery 1, and similarly for x_2 and x_3.

Matrix addition is performed by adding the corresponding entries of matrices. The same method is used to add vectors. For addition to make sense, the two matrices or two vectors must have the same size.

EXAMPLE 1.
Matrices of Test
Scores

Suppose that we are recording the test scores of four students in three subjects. To preserve confidentiality, we will call the students A, B, C, and D, and the subjects 1, 2, and 3. The students have two 1-hour exams and a final exam in each course, each graded out of 10 points. For each of the three tests we form a matrix of test scores with rows for students and columns for subjects. Call the matrices S_1, S_2, and S_3.

$$
\begin{array}{cc}
 & \begin{array}{ccc} 1 & 2 & 3 \end{array} \\
S_1 = \begin{array}{c} A \\ B \\ C \\ D \end{array}
\begin{bmatrix} 6 & 8 & 9 \\ 8 & 5 & 8 \\ 8 & 7 & 8 \\ 4 & 6 & 6 \end{bmatrix}
\end{array}
\qquad
S_2 = \begin{bmatrix} 5 & 9 & 8 \\ 6 & 7 & 9 \\ 7 & 8 & 8 \\ 5 & 6 & 7 \end{bmatrix}
\qquad
S_3 = \begin{bmatrix} 6 & 7 & 9 \\ 8 & 6 & 9 \\ 8 & 7 & 8 \\ 6 & 5 & 6 \end{bmatrix}
$$

Then the matrix **T** of the total course scores of each student in each course (without any weighting to make the final exam count more) is

$$T = S_1 + S_2 + S_3$$

Summing the corresponding entries in S_1, S_2, and S_3, we obtain the matrix **T**:

$$
T = \begin{bmatrix} 6 & 8 & 9 \\ 8 & 5 & 8 \\ 8 & 7 & 8 \\ 4 & 6 & 6 \end{bmatrix}
+ \begin{bmatrix} 5 & 9 & 8 \\ 6 & 7 & 9 \\ 7 & 8 & 8 \\ 5 & 6 & 7 \end{bmatrix}
+ \begin{bmatrix} 6 & 7 & 9 \\ 8 & 6 & 9 \\ 8 & 7 & 8 \\ 6 & 5 & 6 \end{bmatrix}
= \begin{bmatrix} 17 & 24 & 26 \\ 22 & 18 & 26 \\ 23 & 22 & 24 \\ 15 & 17 & 19 \end{bmatrix}
$$

Suppose that the final exam should be weighted twice as much as each 1-hour test. Further, we want the overall course score to be out of 10 points (like each test). That is, the overall course score is a weighted average of the three tests, with the first two tests counting for 25% each and the final exam counting 50%. Then the matrix **C** of weighted course scores is given by the matrix expression

$$C = \tfrac{1}{4}S_1 + \tfrac{1}{4}S_2 + \tfrac{1}{2}S_3 \tag{2}$$

We compute **C** by using the formula in (2) for each entry. For example, the entry c_{12}, student A's weighted overall course score in course 2, is

$$c_{12} = \tfrac{1}{4} \cdot 8 + \tfrac{1}{4} \cdot 9 + \tfrac{1}{2} \cdot 7 = 7\tfrac{3}{4}$$

A computer program to compute all the c_{ij} entries would look as follows:

```
FOR I = 1 TO 4
  FOR J = 1 TO 3
    C(I,J)=.25*S1(I,J)+.25*S2(I,J)+.5*S3(I,J)
  NEXT J
NEXT I
```

Using this program, we obtain **C** (in this matrix, fractions > 0.5 have been rounded up; that is, 3.6 is written as 4):

$$\mathbf{C} = \frac{1}{4}\begin{bmatrix} 6 & 8 & 9 \\ 8 & 5 & 8 \\ 8 & 7 & 8 \\ 4 & 6 & 6 \end{bmatrix} + \frac{1}{4}\begin{bmatrix} 5 & 9 & 8 \\ 6 & 7 & 9 \\ 7 & 8 & 8 \\ 5 & 6 & 7 \end{bmatrix} + \frac{1}{2}\begin{bmatrix} 6 & 7 & 9 \\ 8 & 6 & 9 \\ 8 & 7 & 8 \\ 6 & 5 & 6 \end{bmatrix} = \begin{matrix} \\ A \\ B \\ C \\ D \end{matrix}\begin{matrix} 1 & 2 & 3 \\ \begin{bmatrix} 6 & 8 & 9 \\ 8 & 6 & 9 \\ 8 & 7 & 8 \\ 5 & 6 & 6 \end{bmatrix} \end{matrix} \quad (3)$$

■

Example 1 shows implicitly why matrix addition is defined only for matrices of the same size. We would not want to add the matrices together unless the entries in the matrices matched up in some natural way.

Scalar Product of Vectors

Now we are ready for a form of vector multiplication that is more complex than any sort of multiplication done with single numbers. It is called a ***scalar product***. The following example motivates scalar products. Suppose that we have a vector $\mathbf{p}$ of prices for a set of three vegetables, $\mathbf{p} = [0.80, 1.00, 0.50]$, where the ith entry of $\mathbf{p}$ is the price of the ith vegetable. Suppose that we are also given a vector $\mathbf{d} = [5, 3, 4]$ of the weekly demand in a household for these three vegetables. Then the scalar product of $\mathbf{p}$ and $\mathbf{d}$, written $\mathbf{p} \cdot \mathbf{d}$, equals the cost of the household's weekly demand for these three vegetables. In this case

$$\begin{aligned} \mathbf{p} \cdot \mathbf{d} = [0.80, 1.00, 0.50] \cdot [5, 3, 4] &= 0.80 \cdot 5 + 1.00 \cdot 3 + 0.50 \cdot 4 \\ &= 4 + 3 + 2 = 9 \end{aligned}$$

Definition

Let $\mathbf{a} = [a_1, a_2, \ldots, a_n]$ and $\mathbf{b} = [b_1, b_2, \ldots, b_n]$ be vectors of the same size n. Each vector can be either a row or a column vector. Then the ***scalar product*** $\mathbf{a} \cdot \mathbf{b}$ of $\mathbf{a}$ and $\mathbf{b}$ is a single number (a scalar) equal to the sum of the products $a_i b_i$. That is, $\mathbf{a} \cdot \mathbf{b} = \sum_{i=1}^{n} a_i b_i$.

The scalar product $\mathbf{a} \cdot \mathbf{b}$ makes sense only when $\mathbf{a}$ and $\mathbf{b}$ have the same length.

EXAMPLE 2. Geometric View of Scalar Products

When two vectors point in approximately the same general direction (in their geometric depiction with arrows), their scalar product is positive. When two vectors point in approximately opposite directions, their scalar product is negative. Most interestingly, when two vectors form a right angle, their scalar product will always be zero. This last claim is discussed extensively in Chapter 2. Figure 1.3 illustrates these claims. ■

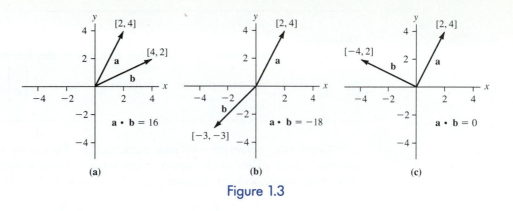

Figure 1.3

EXAMPLE 3.
Oil Refinery
Equations as
Scalar Products

A basic property of scalar products is the fact that the scalar product $\mathbf{c} \cdot \mathbf{d}$ is a linear combination of the entries in each vector. Conversely, *any linear combination of variables or numbers can be expressed as a scalar product*. Consider the first linear equation of the oil refinery model [see (1)].

$$4x_1 + 2x_2 + 2x_3 = 500 \tag{4}$$

The left side of this equation is a linear combination of the variables. If $\mathbf{a} = [4, 2, 2]$ and $\mathbf{x} = \begin{bmatrix} x_1 \\ x_2 \\ x_3 \end{bmatrix}$ (the reason for writing $\mathbf{a}$ as a row vector and $\mathbf{x}$ as a column vector will be explained shortly), the left side of (4) can be written as a scalar product

$$\mathbf{a} \cdot \mathbf{x} = [4, 2, 2] \cdot \begin{bmatrix} x_1 \\ x_2 \\ x_3 \end{bmatrix} = 4x_1 + 2x_2 + 2x_3 \tag{5}$$

Any system of linear equations can be written in terms of a system of scalar products. For example, the left sides of equations (1) of the oil refinery model are

$$
\begin{array}{llll}
\text{Heating oil:} & 4x_1 + & 2x_2 + 2x_3 = [4, 2, 2] \cdot \mathbf{x} = \mathbf{a}_1' \cdot \mathbf{x} \\
\text{Diesel oil:} & 2x_1 + & 5x_2 + 2x_3 = [2, 5, 2] \cdot \mathbf{x} = \mathbf{a}_2' \cdot \mathbf{x} \\
\text{Gasoline:} & 1x_1 + & 2.5x_2 + 5x_3 = [1, 2.5, 5] \cdot \mathbf{x} = \mathbf{a}_3' \cdot \mathbf{x}
\end{array}
\tag{6}
$$

where $\mathbf{a}_i'$ is the ith row of the refinery coefficient matrix.

	Refinery 1	Refinery 2	Refinery 3
Heating oil	4	2	2
$\mathbf{A} =$ Diesel oil	2	5	2
Gasoline	1	2.5	5

∎

We now introduce an important set of special vectors.

Definition

The *coordinate vector* $\mathbf{i}_k$ is a vector with a 1 in the kth position and 0's elsewhere. For example, the coordinate 2-vector $\mathbf{i}_1$ is $\begin{bmatrix} 1 \\ 0 \end{bmatrix}$.

Coordinate vectors are what their name implies. They point in the direction of the coordinate axes. (They can be written as either row or column vectors.) Any 2-vector $\mathbf{v}$ is a linear combination of the coordinate 2-vectors $\mathbf{i}_1$ and $\mathbf{i}_2$. That is,

$$\mathbf{v} = \begin{bmatrix} v_1 \\ v_2 \end{bmatrix} = v_1 \mathbf{i}_1 + v_2 \mathbf{i}_1 = v_1 \begin{bmatrix} 1 \\ 0 \end{bmatrix} + v_2 \begin{bmatrix} 0 \\ 1 \end{bmatrix} \tag{7}$$

Observe also that $\mathbf{v} \cdot \mathbf{i}_2 = \begin{bmatrix} v_1 \\ v_2 \end{bmatrix} \cdot \begin{bmatrix} 0 \\ 1 \end{bmatrix} = v_1 \cdot 0 + v_2 \cdot 1 = v_2$. For example, $\begin{bmatrix} 5 \\ -6 \end{bmatrix} \cdot \begin{bmatrix} 0 \\ 1 \end{bmatrix} = 5 \cdot 0 + (-6) \cdot 1 = -6$. In general, we have the following theorem (whose straightforward proof is left as an exercise).

Theorem 1. The scalar product $\mathbf{v} \cdot \mathbf{i}_k$ of any n-vector $\mathbf{v}$ with the kth coordinate n-vector $\mathbf{i}_k$ is equal to v_k, the kth entry of $\mathbf{v}$.

Matrix–Vector Product

Although we shall encounter many uses of single scalar products in linear algebra, the principal use of scalar products is as a building block for defining the multiplication of a matrix times a vector or a matrix times another matrix.

Definition

The *matrix–vector product* of an $m \times n$ matrix $\mathbf{A}$ and a column n-vector $\mathbf{c}$ is a column vector of scalar products $\mathbf{a}'_i \cdot \mathbf{c}$, of the *rows* $\mathbf{a}'_i$ of $\mathbf{A}$ with $\mathbf{c}$. For example, if $\mathbf{A} = \begin{bmatrix} a_{11} & a_{12} & a_{13} \\ a_{21} & a_{22} & a_{23} \end{bmatrix}$ is a 3×2 matrix and $\mathbf{c} = \begin{bmatrix} c_1 \\ c_2 \\ c_3 \end{bmatrix}$ is a column 3-vector, then

$$\mathbf{Ac} = \begin{bmatrix} \mathbf{a}'_1 \cdot \mathbf{c} \\ \mathbf{a}'_2 \cdot \mathbf{c} \end{bmatrix} = \begin{bmatrix} a_{11}c_1 + a_{12}c_2 + a_{13}c_3 \\ a_{21}c_1 + a_{22}c_2 + a_{23}c_3 \end{bmatrix}$$

In scalar products involving matrices, the first vector in the scalar product must be a row vector and the second vector must be a column vector.

For example, if $\mathbf{A} = \begin{bmatrix} -2 & 1 & 2 \\ 3 & 4 & 5 \end{bmatrix}$ and $\mathbf{c} = \begin{bmatrix} 1 \\ 2 \\ 3 \end{bmatrix}$, then

$$\mathbf{Ac} = \begin{bmatrix} -2 & 1 & 2 \\ 3 & 4 & 5 \end{bmatrix}\begin{bmatrix} 1 \\ 2 \\ 3 \end{bmatrix} = \begin{bmatrix} -2 \cdot 1 + 1 \cdot 2 + 2 \cdot 3 \\ 3 \cdot 1 + 4 \cdot 2 + 5 \cdot 3 \end{bmatrix} = \begin{bmatrix} 6 \\ 29 \end{bmatrix}$$

Remember that the number of columns in $\mathbf{A}$ must equal the length of $\mathbf{c}$.

EXAMPLE 3 (continued). Oil Refinery Model

Returning to the left side of the oil refinery equations (6), we see that if we make a vector of the left sides, we have

$$\begin{bmatrix} 4x_1 + 2x_2 + 2x_3 \\ 2x_1 + 5x_2 + 2x_3 \\ 1x_1 + 2.5x_2 + 5x_3 \end{bmatrix} = \begin{bmatrix} [4, 2, 2] \cdot \mathbf{x} \\ [2, 5, 2] \cdot \mathbf{x} \\ [1, 2.5, 5] \cdot \mathbf{x} \end{bmatrix} = \begin{bmatrix} \mathbf{a}_1' \cdot \mathbf{x} \\ \mathbf{a}_2' \cdot \mathbf{x} \\ \mathbf{a}_3' \cdot \mathbf{x} \end{bmatrix} = \mathbf{Ax} \qquad (8)$$

If we let $\mathbf{b} = \begin{bmatrix} 600 \\ 800 \\ 1000 \end{bmatrix}$ be the column vector of demands for the different products, the system of refinery equations becomes, using (8),

$$\mathbf{Ax} = \mathbf{b}: \quad \begin{bmatrix} 4x_1 + 2x_2 + 2x_3 \\ 2x_1 + 5x_2 + 2x_3 \\ 1x_1 + 2.5x_2 + 5x_3 \end{bmatrix} = \begin{bmatrix} 600 \\ 800 \\ 1000 \end{bmatrix} \qquad (9) \quad ■$$

Thus a system of simultaneous linear equations can be written very compactly in matrix notation as $\mathbf{Ax} = \mathbf{b}$.

EXAMPLE 4. Coordinate Vectors in Matrix–Vector Products

Preceding Theorem 1, we introduced the coordinate vectors $\mathbf{i}_k$, which have a 1 in the kth position and 0's elsewhere. Suppose that we multiply the matrix $\mathbf{A} = \begin{bmatrix} -2 & 1 & 2 \\ 3 & 4 & 5 \end{bmatrix}$ times the first coordinate 3-vector $\mathbf{i}_1$.

$$\mathbf{Ai}_1 = \begin{bmatrix} -2 & 1 & 2 \\ 3 & 4 & 5 \end{bmatrix}\begin{bmatrix} 1 \\ 0 \\ 0 \end{bmatrix} = \begin{bmatrix} -2 \cdot 1 + 1 \cdot 0 + 2 \cdot 0 \\ 3 \cdot 1 + 4 \cdot 0 + 5 \cdot 0 \end{bmatrix} = \begin{bmatrix} -2 \\ 3 \end{bmatrix} \qquad (10)$$

The result is the first column of $\mathbf{A}$. Recall from the definition of matrix–vector products that $\mathbf{Ai}_1$ is a column vector of scalar products $\mathbf{a}_i' \cdot \mathbf{i}_1$ of the rows $\mathbf{a}_i'$ of $\mathbf{A}$ with $\mathbf{i}_1$. By Theorem 1, the scalar product of any vector with $\mathbf{i}_1$ is equal to the first entry of that vector. Thus $\mathbf{Ai}_1$ is a column vector consisting of the first entry in each row of $\mathbf{A}$. The set of first entries in each row yields the first *column* of $\mathbf{A}$. ■

There is another way to view matrix–vector products. In the definition given above, $\mathbf{Ac}$ is defined in terms of the *rows* of $\mathbf{A}$, each of which forms a scalar product with **c**. However, recall that the refinery equations (8) and (9) were expressed at the beginning of this section as a linear combination of the *columns* of the refinery matrix:

$$x_1 \begin{bmatrix} 4 \\ 2 \\ 1 \end{bmatrix} + x_2 \begin{bmatrix} 2 \\ 5 \\ 2.5 \end{bmatrix} + x_3 \begin{bmatrix} 2 \\ 2 \\ 5 \end{bmatrix} = \begin{bmatrix} 600 \\ 800 \\ 1000 \end{bmatrix}$$

The same representation is available for any matrix–vector product. For the product $\mathbf{Ai}_1$ in (10), we have

$$\mathbf{Ai}_1 = 1 \begin{bmatrix} -2 \\ 3 \end{bmatrix} + 0 \begin{bmatrix} 1 \\ 4 \end{bmatrix} + 0 \begin{bmatrix} 2 \\ 5 \end{bmatrix} = \begin{bmatrix} -2 \\ 3 \end{bmatrix}$$

This is much easier than (10). In general, we have:

Column Approach to the Matrix–Vector Product

The product **Ac** can be viewed as a linear combination of the columns of **A**:

$$\mathbf{Ac} = c_1 \mathbf{a}_1 + c_2 \mathbf{a}_2 + \cdots + c_n \mathbf{a}_n$$

where $\mathbf{a}_1, \mathbf{a}_2, \ldots, \mathbf{a}_n$ are the columns of **A** and $c_1, c_2, \ldots, c_n$ are the entries of column vector **c**.

EXAMPLE 5.
Markov Chain for
Weather

In the Markov chain model introduced in Example 3 of Section 1.1, the equations for determining the probabilities p_1', p_2', p_3' of sunny, cloudy, or rainy weather tomorrow given the probabilities p_1, p_2, p_3 of sunny, cloudy, or rainy weather today were

$$p_1' = \tfrac{3}{4}p_1 + \tfrac{1}{2}p_2 + \tfrac{1}{4}p_3$$
$$p_2' = \tfrac{1}{8}p_1 + \tfrac{1}{4}p_2 + \tfrac{1}{2}p_3 \qquad (11)$$
$$p_3' = \tfrac{1}{8}p_1 + \tfrac{1}{4}p_2 + \tfrac{1}{4}p_3$$

If $\mathbf{p} = \begin{bmatrix} p_1 \\ p_2 \\ p_3 \end{bmatrix}$ is the vector of current probabilities, $\mathbf{p}' = \begin{bmatrix} p_1' \\ p_2' \\ p_3' \end{bmatrix}$ is the vector of

tomorrow's probabilities, and **A** is the matrix of transition probabilities:

	Today		
Tomorrow	Sunny	Cloudy	Rainy
Sunny	$\tfrac{3}{4}$	$\tfrac{1}{2}$	$\tfrac{1}{4}$
A = Cloudy	$\tfrac{1}{8}$	$\tfrac{1}{4}$	$\tfrac{1}{2}$
Rainy	$\tfrac{1}{8}$	$\tfrac{1}{4}$	$\tfrac{1}{4}$

then (11) can be written as simply $\mathbf{p}' = \mathbf{Ap}$, since

$$\mathbf{Ap} = \begin{bmatrix} \frac{3}{4} & \frac{1}{2} & \frac{1}{4} \\ \frac{1}{8} & \frac{1}{4} & \frac{1}{2} \\ \frac{1}{8} & \frac{1}{4} & \frac{1}{4} \end{bmatrix} \begin{bmatrix} p_1 \\ p_2 \\ p_3 \end{bmatrix} = \begin{bmatrix} \frac{3}{4}p_1 + \frac{1}{2}p_2 + \frac{1}{4}p_3 \\ \frac{1}{8}p_1 + \frac{1}{4}p_2 + \frac{1}{2}p_3 \\ \frac{1}{8}p_1 + \frac{1}{4}p_2 + \frac{1}{4}p_3 \end{bmatrix} \tag{12}$$

■

EXAMPLE 6.
Geometric View
of Matrix–Vector
Products

In geometric terms, when we premultiply a vector $\mathbf{v}$ by a matrix $\mathbf{A}$, vector $\mathbf{v}$ is transformed by the multiplication into another vector, $\mathbf{w}$. We can view premultiplication by $\mathbf{A}$ as a function: $\mathbf{w} = f(\mathbf{v})$, where $f(\mathbf{v}) = \mathbf{Av}$. This transformation has many important properties which are central to the theory of linear algebra.

Rather than using arrows, this time we will simply represent 2-vectors as points in the plane. Consider the vectors

$$\mathbf{v}_1 = \begin{bmatrix} 0 \\ 0 \end{bmatrix} \quad \mathbf{v}_2 = \begin{bmatrix} 2 \\ 0 \end{bmatrix} \quad \mathbf{v}_3 = \begin{bmatrix} 2 \\ 2 \end{bmatrix} \quad \mathbf{v}_4 = \begin{bmatrix} 0 \\ 2 \end{bmatrix}$$

$$\mathbf{v}_5 = \begin{bmatrix} 1 \\ 0 \end{bmatrix} \quad \mathbf{v}_6 = \begin{bmatrix} 2 \\ 1 \end{bmatrix} \quad \mathbf{v}_7 = \begin{bmatrix} 1 \\ 2 \end{bmatrix} \quad \mathbf{v}_8 = \begin{bmatrix} 0 \\ 1 \end{bmatrix}$$

where $\mathbf{v}_1$, $\mathbf{v}_2$, $\mathbf{v}_3$, $\mathbf{v}_4$ are the corners of a square of side 2 and $\mathbf{v}_5$, $\mathbf{v}_6$, $\mathbf{v}_7$, $\mathbf{v}_8$ are the midpoints of the sides of this square [see Figure 1.4(a)]. If we premultiply these $\mathbf{v}_i$'s by the matrix $\mathbf{A} = \begin{bmatrix} 1 & 1 \\ 1 & -1 \end{bmatrix}$, we obtain the vectors $\mathbf{w}_i = \mathbf{Av}_i$ which form a square on its side, as shown in Figure 1.4(b):

$$\mathbf{w}_1 = \begin{bmatrix} 1 & 1 \\ 1 & -1 \end{bmatrix} \begin{bmatrix} 0 \\ 0 \end{bmatrix} = \begin{bmatrix} 0 \\ 0 \end{bmatrix} \qquad \mathbf{w}_2 = \begin{bmatrix} 1 & 1 \\ 1 & -1 \end{bmatrix} \begin{bmatrix} 2 \\ 0 \end{bmatrix} = \begin{bmatrix} 2 \\ 2 \end{bmatrix}$$

$$\mathbf{w}_3 = \begin{bmatrix} 1 & 1 \\ 1 & -1 \end{bmatrix} \begin{bmatrix} 2 \\ 2 \end{bmatrix} = \begin{bmatrix} 4 \\ 0 \end{bmatrix} \qquad \mathbf{w}_4 = \begin{bmatrix} 1 & 1 \\ 1 & -1 \end{bmatrix} \begin{bmatrix} 0 \\ 2 \end{bmatrix} = \begin{bmatrix} 2 \\ -2 \end{bmatrix}$$

$$\mathbf{w}_5 = \begin{bmatrix} 1 & 1 \\ 1 & -1 \end{bmatrix} \begin{bmatrix} 1 \\ 0 \end{bmatrix} = \begin{bmatrix} 1 \\ 1 \end{bmatrix} \qquad \mathbf{w}_6 = \begin{bmatrix} 1 & 1 \\ 1 & -1 \end{bmatrix} \begin{bmatrix} 2 \\ 1 \end{bmatrix} = \begin{bmatrix} 3 \\ 1 \end{bmatrix} \tag{13}$$

$$\mathbf{w}_7 = \begin{bmatrix} 1 & 1 \\ 1 & -1 \end{bmatrix} \begin{bmatrix} 1 \\ 2 \end{bmatrix} = \begin{bmatrix} 3 \\ -1 \end{bmatrix} \qquad \mathbf{w}_8 = \begin{bmatrix} 1 & 1 \\ 1 & -1 \end{bmatrix} \begin{bmatrix} 0 \\ 1 \end{bmatrix} = \begin{bmatrix} 1 \\ -1 \end{bmatrix}$$

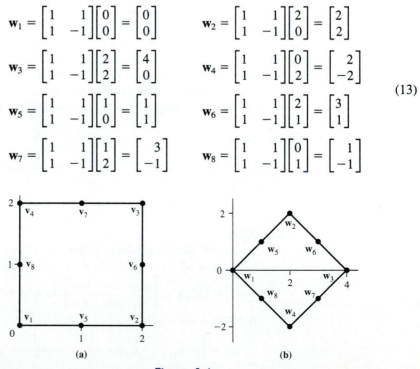

(a)

(b)

Figure 1.4

Note how the midpoint vectors $\mathbf{v}_5, \mathbf{v}_6, \mathbf{v}_7, \mathbf{v}_8$ in the original square are transformed into the midpoint vectors $\mathbf{w}_5, \mathbf{w}_6, \mathbf{w}_7, \mathbf{w}_8$ in the rotated square. ▪

Let us now show that any transformation which takes a 2-vector $\mathbf{v}$ to the 2-vector $\mathbf{w} = \mathbf{A}\mathbf{v}$, for some 2×2 matrix $\mathbf{A}$, maps lines into lines (or points).

Theorem 2. For any 2×2 matrix $\mathbf{A}$, the mapping of 2-space $\mathbf{v} \rightarrow \mathbf{w} = \mathbf{A}\mathbf{v}$ takes lines into lines, except for special cases where it takes a line into a point.

Proof: The line segment L between vectors $\mathbf{v}_1$ and $\mathbf{v}_2$ consists of the set of vectors

$$L = \{\mathbf{u} : \mathbf{u} = c\mathbf{v}_1 + (1 - c)\mathbf{v}_2, \; 0 \le c \le 1\}$$

Let $\mathbf{w}_1 = \mathbf{A}\mathbf{v}_1$ and $\mathbf{w}_2 = \mathbf{A}\mathbf{v}_2$. We need to show that this mapping takes L into the line segment

$$L' = \{\mathbf{y} : \mathbf{y} = c\mathbf{w}_1 + (1 - c)\mathbf{w}_2, \; 0 \le c \le 1\}$$

We shall show that the vector $\mathbf{u} = c\mathbf{v}_1 + (1 - c)\mathbf{v}_2$ is mapped to the vector $\mathbf{y} = c\mathbf{w}_1 + (1 - c)\mathbf{w}_2$, that is, $\mathbf{y} = \mathbf{A}\mathbf{u}$.

$$\mathbf{A}\mathbf{u} = \mathbf{A}(c\mathbf{v}_1 + (1 - c)\mathbf{v}_2) \tag{14a}$$

$$= \mathbf{A}(c\mathbf{v}_1) + \mathbf{A}((1 - c)\mathbf{v}_2) \tag{14b}$$

$$= c\mathbf{A}\mathbf{v}_1 + (1 - c)\mathbf{A}\mathbf{v}_2 = c\mathbf{w}_1 + (1 - c)\mathbf{w}_2 = \mathbf{y} \tag{14c}$$

In the special case where $\mathbf{w}_1 = \mathbf{w}_2$ (this happens occasionally), L' collapses to the point $\mathbf{w}_1$. ▪

In line (14b) we used the distributive law that $\mathbf{A}(\mathbf{b}_1 + \mathbf{b}_2) = \mathbf{A}\mathbf{b}_1 + \mathbf{A}\mathbf{b}_2$ [where $\mathbf{b}_1 = c\mathbf{v}_1$ and $\mathbf{b}_2 = (1 - c)\mathbf{v}_2$], and in (14c) we used the law of scalar factoring $\mathbf{A}(c\mathbf{b}) = c\mathbf{A}\mathbf{b}$. Verification of these laws is given in Section 3.1. (In several places in Chapter 1 we use laws of matrix algebra to illustrate their value; the reader will then have some feeling for these laws when we prove them in Section 3.1.) Because of Theorem 2, the transformation $\mathbf{v} \rightarrow \mathbf{w} = \mathbf{A}\mathbf{v}$ is called a **linear transformation**. The linearity of the mapping $\mathbf{v} \rightarrow \mathbf{w} = \mathbf{A}\mathbf{v}$ is one of the central themes of linear algebra.

Vector–Matrix Product

Multiplication of a vector followed by a matrix is defined in a fashion similar to the product of a matrix followed by a vector.

Definition

The **vector–matrix product** of a row m-vector $\mathbf{c}$ and an $m \times n$ matrix $\mathbf{A}$ is a row vector of scalar products $\mathbf{c} \cdot \mathbf{a}_j$, of $\mathbf{c}$ with the **columns** $\mathbf{a}_j$ of $\mathbf{A}$. For example, if $\mathbf{A} = \begin{bmatrix} a_{11} & a_{12} & a_{13} \\ a_{21} & a_{22} & a_{23} \end{bmatrix}$ is a 3×2 matrix and $\mathbf{c} = [c_1 \quad c_2]$ is a row 2-vector, then

$$\mathbf{c}\mathbf{A} = [\mathbf{c} \cdot \mathbf{a}_1, \quad \mathbf{c} \cdot \mathbf{a}_2,] = [a_{11}c_1 + a_{21}c_2, \quad a_{21}c_1 + a_{22}c_1, \quad a_{31}c_1 + a_{23}c_2]$$

As noted in the definition of a matrix–vector product, for scalar products involving matrices, the first vector in the scalar product must be a row vector and the second vector a column vector. Suppose that $\mathbf{c} = [2, 5]$ and $\mathbf{A} = \begin{bmatrix} 2 & 0 & 1 \\ 4 & 1 & 2 \end{bmatrix}$; then

$$\mathbf{cA} = [2, \quad 5]\begin{bmatrix} 2 & 0 & 1 \\ 4 & 1 & 2 \end{bmatrix} = [\mathbf{c} \cdot \mathbf{a}_1, \quad \mathbf{c} \cdot \mathbf{a}_2, \quad \mathbf{c} \cdot \mathbf{a}_3]$$

$$= [2 \cdot 2 + 5 \cdot 4, \quad 2 \cdot 0 + 5 \cdot 1, \quad 2 \cdot 1 + 5 \cdot 2] = [24, 5, 12]$$

The vector–matrix product with a coordinate vector $\mathbf{i}_k$ is similar to Example 4 except that now the result is a row, instead of a column, of the matrix. That is, for $\mathbf{i}_2 = [0, 1]$ and $\mathbf{A}$ as above,

$$\mathbf{i}_2\mathbf{A} = [0, \quad 1]\begin{bmatrix} 2 & 0 & 1 \\ 4 & 1 & 2 \end{bmatrix} = [0 \cdot 2 + 1 \cdot 4, \quad 0 \cdot 0 + 1 \cdot 1, \quad 0 \cdot 1 + 1 \cdot 2]$$

$$= [4, \quad 1, \quad 2] = \mathbf{a}_2' \tag{15}$$

As with the foregoing column approach to matrix–vector products, there is a row approach to vector–matrix products:

Row Approach to the Vector–Matrix Product

The product $\mathbf{cA}$ can be viewed as a linear combination of the rows of $\mathbf{A}$:

$$\mathbf{cA} = c_1\mathbf{a}_1' + c_2\mathbf{a}_2' + \cdots + c_n\mathbf{a}_n'$$

where $\mathbf{a}_1', \mathbf{a}_2', \ldots, \mathbf{a}_n'$ are the rows of $\mathbf{A}$ and $c_1, c_2, \ldots, c_n$ are the entries of row vector $\mathbf{c}$.

In linear algebra, matrix–vector products $\mathbf{Ac}$ are much more common than vector–matrix products $\mathbf{cA}$. Matrix–vector products arise naturally in systems of equations, such as $\mathbf{Ax} = \mathbf{b}$ in Example 3 [see (8)] and in iterative models such as the Markov chain $\mathbf{p}' = \mathbf{Ap}$ in Example 5 [see (12)].

We end this section with an example in which one multiplies a matrix by vectors both before and after the matrix.

EXAMPLE 7.
Quadratic Forms

Quadratic equations in two variables of the form $ax^2 + bxy + cy^2 = d$ can be expressed and analyzed in terms of matrix expressions called *quadratic forms*. Such quadratic equations form circles, ellipses, and hyperbolas centered at the origin. These equations can be written with matrix algebra in the form $\mathbf{z}^T\mathbf{Az} = d$, where $\mathbf{z} = \begin{bmatrix} x \\ y \end{bmatrix}$ and $\mathbf{z}^T = [x, y]$ ($\mathbf{z}^T$ is called the *transpose* of $\mathbf{z}$) and $\mathbf{A}$ is any 2×2 matrix. For example, if $\mathbf{A} = \begin{bmatrix} 3 & 1 \\ 1 & 3 \end{bmatrix}$, then

$$\mathbf{z}^T\mathbf{A}\mathbf{z} = [x, \quad y]\begin{bmatrix} 3 & 1 \\ 1 & 3 \end{bmatrix}\begin{bmatrix} x \\ y \end{bmatrix} = [3x + y, \ x + 3y]\begin{bmatrix} x \\ y \end{bmatrix}$$

$$= (3x + y)x + (x + 3y)y = 3x^2 + 2xy + 3y^2 \tag{16}$$

Thus, for this $\mathbf{A}$ and a right-side value of 8, $\mathbf{z}^T\mathbf{A}\mathbf{z} = 8$ is $3x^2 + 2xy + 3y^2 = 8$. The graph of this equation is plotted in Figure 1.5. In Chapter 5 linear algebra theory gives us a simple way to determine the axes of symmetry of this ellipse (shown in dashed lines in Figure 1.5). ■

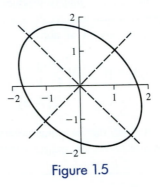

Figure 1.5

Note that in (16), there are two products to compute when multiplying the three objects $\mathbf{z}^T$, $\mathbf{A}$, and $\mathbf{z}$ together. We choose to compute $\mathbf{z}^T\mathbf{A}$ first and then compute the scalar product of the row vector resulting from multiplying $\mathbf{z}^T\mathbf{A}$ by $\mathbf{z}$. The reader should check that the same expression results if one first computes $\mathbf{A}\mathbf{z}$ and then computes the scalar product of $\mathbf{z}^T$ with the column vector resulting from $\mathbf{A}\mathbf{z}$. This computation illustrates the importance of verifying that multiplication in matrix algebra is associative: $(\mathbf{a}\mathbf{B})\mathbf{c} = \mathbf{a}(\mathbf{B}\mathbf{c})$. More generally, we need to know which multiplication laws hold in matrix algebra and which multiplication laws do not hold. For example, $\mathbf{a}\mathbf{B} \neq \mathbf{B}\mathbf{a}$, even when we ignore the fact that $\mathbf{a}\mathbf{B}$ is a row vector and $\mathbf{B}\mathbf{a}$ is a column vector. Concretely,

$$[1, \quad 2]\begin{bmatrix} 1 & 2 \\ 3 & 4 \end{bmatrix} = [7, \quad 10] \quad \text{while} \quad \begin{bmatrix} 1 & 2 \\ 3 & 4 \end{bmatrix}\begin{bmatrix} 1 \\ 2 \end{bmatrix} = \begin{bmatrix} 5 \\ 11 \end{bmatrix}$$

Section 1.2 Exercises

1. Suppose in Example 1 that the final exam counted three times as much as an hour exam so that the weights on the three tests should be $\frac{1}{5}$, $\frac{1}{5}$, $\frac{3}{5}$, respectively. Recompute the course score matrix $\mathbf{C}$ with these weights.

2. Let $\mathbf{A} = \begin{bmatrix} 1 & 2 & 4 \\ 3 & 8 & 6 \end{bmatrix}$ and $\mathbf{B} = \begin{bmatrix} 7 & 9 & 1 \\ 1 & 4 & 3 \end{bmatrix}$. Determine:

   (a) $2\mathbf{A}$ (b) $5\mathbf{B}$ (c) $\mathbf{A} + 2\mathbf{B}$ (d) $2\mathbf{A} - 3\mathbf{B}$

3. Let $A = \begin{bmatrix} 1 & 2 & 3 & 4 \\ 2 & 4 & 6 & 8 \\ 3 & 5 & 7 & 9 \end{bmatrix}$ and $B = \begin{bmatrix} -1 & 0 & 2 & 1 \\ 2 & -1 & -1 & 0 \\ 2 & 0 & 0 & 2 \end{bmatrix}$. Determine:

(a) 3A (b) 2B (c) −3B (d) A + B (e) 2A + 3B (f) 3A − 2B

4. The following matrix **A** gives the price (in cents) of three different kinds of candies bought at three different stores.

	Candy A	Candy B	Candy C
Store A	20	50	50
A = Store B	30	40	60
Store C	20	30	40

(a) Suppose that the price of candy doubles. What will the matrix of candy prices be?
(b) Suppose that the price of candy increases by 50% and there is a tax of 5 cents per candy. What will the matrix of candy prices be now?

5. Let all matrices in this exercise be 4×4. Let **I** denote the 4×4 matrix with 1's in the main diagonal and 0's elsewhere. Let **J** denote the matrix with each entry equal to 1, and

let $A = \begin{bmatrix} 1 & 0 & 1 & 0 \\ 0 & 1 & 0 & 1 \\ 1 & 0 & 1 & 0 \\ 0 & 1 & 0 & 1 \end{bmatrix}$. Express the following matrices in terms of **I**, **J**, and **A**.

(a) $\begin{bmatrix} 6 & 2 & 2 & 2 \\ 2 & 6 & 2 & 2 \\ 2 & 2 & 6 & 2 \\ 2 & 2 & 2 & 6 \end{bmatrix}$ (b) $\begin{bmatrix} 0 & 1 & 0 & 1 \\ 1 & 0 & 1 & 0 \\ 0 & 1 & 0 & 1 \\ 1 & 0 & 1 & 0 \end{bmatrix}$ (c) $\begin{bmatrix} 5 & 3 & 1 & 3 \\ 3 & 5 & 3 & 1 \\ 1 & 3 & 5 & 3 \\ 3 & 1 & 3 & 5 \end{bmatrix}$

6. Let $a = \begin{bmatrix} 2 \\ 1 \end{bmatrix}$, $b = \begin{bmatrix} 2 \\ 3 \end{bmatrix}$, $c = \begin{bmatrix} -2 \\ 4 \end{bmatrix}$, $d = \begin{bmatrix} -3 \\ -2 \end{bmatrix}$. Plot the following pairs of vectors and compute their scalar products.

(a) a, c (b) b, c (c) b, d (d) a, d (e) c, d

7. Let $a = \begin{bmatrix} 1 \\ 2 \\ 3 \end{bmatrix}$, $b = \begin{bmatrix} -1 \\ 3 \\ -1 \end{bmatrix}$, $c = \begin{bmatrix} 2 \\ 5 \\ 8 \end{bmatrix}$. Compute:

(a) $a \cdot b$ (b) $b \cdot c$ (c) $a \cdot (b + c)$ (d) $a \cdot a$

8. Let **a**, **b**, **c** be as in Exercise 7. Let $A = \begin{bmatrix} 1 & 2 & 3 & 4 \\ 2 & 4 & 6 & 8 \\ 3 & 5 & 7 & 9 \end{bmatrix}$, $B = \begin{bmatrix} 1 & 0 & -1 \\ 2 & -2 & 0 \\ 0 & 1 & -1 \end{bmatrix}$,

$C = \begin{bmatrix} 5 & 4 & 1 \\ 1 & 0 & 2 \\ 3 & 2 & 1 \\ 0 & 1 & 3 \end{bmatrix}$. Which of the following matrix calculations are well defined

(the sizes match)? If the computation makes sense, perform it. If needed, **a**, **b**, or **c** may be changed to row vectors.

(a) aA (b) bB (c) cC (d) Aa (e) Bb (f) Cc

9. Calculate the following expressions unless the sizes do not match. The vectors **a**, **b**, **c** are as defined in Exercise 7 and matrices **A**, **B**, **C** are as in Exercise 8. If needed, **a**, **b**, or **c** may be changed to row vectors.
   (a) $(\mathbf{aB}) \cdot \mathbf{c}$ (b) $(\mathbf{a} + \mathbf{c})\mathbf{A}$ (c) $\mathbf{b}(\mathbf{A} + \mathbf{B})$ (d) $\mathbf{Ba} + \mathbf{Ab}$ (e) $(\mathbf{bB}) \cdot (\mathbf{Bb})$

10. Suppose that you want to buy five cantaloupes, four apples, three oranges, and two pineapples. You comparison shop and find that at store A the costs of these four fruits, respectively, are 30 cents, 10 cents, 10 cents, and 75 cents a piece, while at store B the costs are 25 cents, 15 cents, 8 cents, and 80 cents.
    (a) Express the problem of determining the cost of this set of fruit at each store as a matrix–vector product; write out the matrix and vector.
    (b) Compute the costs of the fruits at the two stores.

11. Suppose that you want to have a party catered and will need 10 hero sandwiches, 6 quarts of fruit punch, 3 quarts of potato salad, and 2 plates of hors d'oeuvres. The following matrix gives the costs of these supplies from three different caterers.

	Caterer A	Caterer B	Caterer C
Hero sandwich:	$5	$5	$4
Fruit punch:	$1	$1.50	$0.75
Potato salad:	$0.75	$1	$1
Hors d'oeuvres:	$8	$7	$10

    (a) Express the problem of determining the cost of catering the party by each caterer as a matrix–vector product (be careful whether the vector is first or second in the product).
    (b) Determine the costs of catering with each caterer.

12. Write out the following systems of equations. (Here **x** denotes a column vector of variables $x_1, x_2, \ldots$, where the number of variables equals the number of columns in **A**.)
    (a) $\mathbf{Ax} = \mathbf{b}$ where $\mathbf{A} = \begin{bmatrix} 2 & 5 \\ 4 & 3 \end{bmatrix}$, $\mathbf{b} = \begin{bmatrix} 3 \\ 4 \end{bmatrix}$

    (b) $\mathbf{Ax} = \mathbf{b}$ where $\mathbf{A} = \begin{bmatrix} -1 & 0 \\ 0 & 3 \end{bmatrix}$, $\mathbf{b} = \begin{bmatrix} 0 \\ 3 \end{bmatrix}$

    (c) $\mathbf{Ax} = \mathbf{b}$ where $\mathbf{A} = \begin{bmatrix} 2 & 1 & 0 \\ 5 & -1 & 4 \\ 5 & 2 & 3 \end{bmatrix}$, $\mathbf{b} = \begin{bmatrix} 2 \\ 3 \\ 4 \end{bmatrix}$

    (d) $\mathbf{Ax} = \mathbf{b}$ where $\mathbf{A} = \begin{bmatrix} -1 & 0 & 0 \\ 0 & 0 & 2 \\ 0 & 1 & 0 \end{bmatrix}$, $\mathbf{b} = \begin{bmatrix} 0 \\ 0 \\ 0 \end{bmatrix}$

13. Write the following systems of equations in matrix notation. Define any matrices or vectors you use.

    (a) $3x_1 + 4x_2 = 5$ (b) $2x_1 + x_2 - 2x_3 = 0$ (c) $x_1 = 2x_1 - x_2$
    $\ 2x_1 - 5x_2 = 3$ $\ x_1 + 3x_3 = 3$ $\ x_2 = 3x_1 + 2x_2$
    $\ 3x_1 - x_2 = 5$ $\ x_3 = 4x_1 - 3x_2$

14. Write the following systems of linear inequalities in matrix notation ($\mathbf{Ax} < \mathbf{b}$ means that each entry on the left is less than the corresponding entry on the right.) Define any matrices or vectors you use.

(a) $2x_1 + x_2 < 20$ (b) $x_1 + 2x_2 + 3x_3 > 20$
$\quad\quad x_1 + 3x_2 < 10$ $\quad\quad 2x_1 + x_2 + 2x_3 > 25$
$\quad\quad 4x_1 + 2x_2 < 35$ $\quad\quad 2x_1 - x_2 + 3x_3 > 15$

15. Consider the system of equations

$$2x_1 + 3x_2 - 2x_3 = 5y_1 + 2y_2 - 3y_3 + 200$$
$$x_1 + 4x_2 + 3x_3 = 6y_1 - 4y_2 + 4y_3 - 120$$
$$5x_1 + 2x_2 - x_3 = 2y_1 \quad\quad\quad - 2y_3 + 350$$

(a) Write this system of equations in matrix form. Define the vectors and matrices you introduce.
(b) Rewrite in matrix form with all the variables on the left side (and just scalars on the right).

16. Three different types of computers need varying amounts of four different types of integrated circuits. The following matrix $\mathbf{A}$ gives the number of each circuit needed by each computer.

$$\begin{array}{c} \text{Circuit} \\ \begin{array}{cccc} 1 & 2 & 3 & 4 \end{array} \\ \mathbf{A} = \text{Computer} \begin{array}{c} \text{A} \\ \text{B} \\ \text{C} \end{array} \begin{bmatrix} 2 & 3 & 2 & 1 \\ 5 & 1 & 3 & 2 \\ 3 & 2 & 2 & 2 \end{bmatrix} \end{array}$$

Let $\mathbf{d} = [10, 20, 30]$ be the computer demand vector (how many of each type of computer is needed). Let $\mathbf{p} = \begin{bmatrix} 2 \\ 5 \\ 1 \\ 10 \end{bmatrix}$ be the price vector for the circuits (the cost in dollars of each type of circuit). Write an expression in terms of $\mathbf{A}$, $\mathbf{d}$, $\mathbf{p}$ for the total cost of the circuits needed to produce the set of computers in demand; indicate where the matrix–vector product occurs and where the scalar product occurs. Compute this total cost.

17. Plot of the eight transformed vectors $\mathbf{w}_i = f(\mathbf{v}_i) = \mathbf{Av}_i$ of the eight vectors $\mathbf{v}_1, \mathbf{v}_2, \ldots,$ $\mathbf{v}_8$ in Example 6 for the following matrices. In each case, draw the line segments from $\mathbf{w}_1$ to $\mathbf{w}_2$, from $\mathbf{w}_2$ to $\mathbf{w}_3$, from $\mathbf{w}_3$ to $\mathbf{w}_4$, and from $\mathbf{w}_4$ to $\mathbf{w}_1$, and check that the vectors $\mathbf{w}_5, \mathbf{w}_6,$ $\mathbf{w}_7, \mathbf{w}_8$ lie at the middle of these respective line segments.

(a) $\mathbf{A} = \begin{bmatrix} 2 & 1 \\ 1 & 3 \end{bmatrix}$ (b) $\mathbf{A} = \begin{bmatrix} 0.7 & -0.7 \\ 0.7 & 0.7 \end{bmatrix}$ (c) $\mathbf{A} = \begin{bmatrix} -1 & 1 \\ 0 & -2 \end{bmatrix}$

18. Show that any vector $\mathbf{x} = \begin{bmatrix} x_1 \\ x_2 \end{bmatrix}$ that is a multiple of $\begin{bmatrix} 2 \\ 1 \end{bmatrix}$ (i.e., $\mathbf{x} = c \begin{bmatrix} 2 \\ 1 \end{bmatrix}$ for some c) satisfies the system of equations

$$x_1 - 2x_2 = 0$$
$$-2x_1 + 4x_2 = 0$$

19. Show that for any vector $\mathbf{a}$, $\mathbf{a} \cdot \mathbf{a}$ is the sum of the squares of entries of $\mathbf{a}$.

20. Prove Theorem 1, that the scalar product $\mathbf{v} \cdot \mathbf{i}_k$ of any n-vector $\mathbf{v}$ with the kth coordinate n-vector $\mathbf{i}_k$ is equal to v_k, the kth entry of $\mathbf{v}$.

21. Let $\mathbf{1}$ denote a row n-vector of all 1's. Let $\mathbf{A}$ be the $n \times n$ transition matrix for some Markov chain.
  (a) Show that for each column $\mathbf{a}_j$ of $\mathbf{A}$, $\mathbf{1} \cdot \mathbf{a}_j = 1$.
  (b) Using part (a), show that $\mathbf{1A} = \mathbf{1}$.

22. (a) Let $\mathbf{1}$ be a column 4-vector of 1's and let $\mathbf{A} = \begin{bmatrix} 6 & 2 & 4 & 1 \\ 2 & 6 & 5 & 8 \\ 7 & 2 & 0 & 3 \\ 9 & 1 & 5 & 6 \end{bmatrix}$. Show that $\mathbf{1}^T\mathbf{A1}$

is the sum of all entries in $\mathbf{A}$. Here $\mathbf{1}^T$ denotes the transpose of $\mathbf{1}$, a row 4-vector of 1's.
  (b) Generalizing part (a), show that if $\mathbf{A}$ is an $m \times n$ matrix, $\mathbf{1}$ is a column n-vector and $\mathbf{1}^T$ is a row m-vector, then $\mathbf{1}^T\mathbf{A1}$ equals the sum of the entries of $\mathbf{A}$.
  (c) If $\mathbf{A}$ is an $m \times n$ matrix, give an expression for the average value of the entries of $\mathbf{A}$.

23. Multiply out the quadratic forms $\mathbf{x}^T\mathbf{Ax}$, for the following matrices $\mathbf{A}$. Here $\mathbf{x}$ is a column vector of variables $x_1, x_2, \ldots$, where the number of variables equals the number of columns in $\mathbf{A}$, and $\mathbf{x}^T$ is the transpose of $\mathbf{x}$ (a row vector).

  (a) $\mathbf{A} = \begin{bmatrix} 2 & 4 \\ 4 & 2 \end{bmatrix}$ **(b)** $\mathbf{A} = \begin{bmatrix} 3 & -1 \\ -1 & 3 \end{bmatrix}$ **(c)** $\mathbf{A} = \begin{bmatrix} 2 & 1 & 4 \\ 1 & 3 & -1 \\ 4 & -1 & 5 \end{bmatrix}$

  (d) $\mathbf{A} = \begin{bmatrix} 1 & 0 & -2 \\ 0 & 4 & 1 \\ -2 & 1 & 5 \end{bmatrix}$

24. (a) One can express polynomial multiplication in terms of a matrix–vector product as follows: To multiply the quadratic $2x^2 + 3x + 4$ by $1x^2 - 2x + 5$, we multiply

$$\begin{bmatrix} 2 & 0 & 0 \\ 3 & 2 & 0 \\ 4 & 3 & 2 \\ 0 & 4 & 3 \\ 0 & 0 & 4 \end{bmatrix} \begin{bmatrix} 1 \\ -2 \\ 5 \end{bmatrix}.$$

The resulting vector will give the coefficients in the product. Confirm this.
  (b) For the polynomial multiplication $(x^3 + 2x^2 + 3x + 4)(4x^2 - 3x + 1)$, write the associated matrix–vector product.

25. Write a computer program to add two matrices $\mathbf{A}$ and $\mathbf{B}$ where both are $m \times n$. Assume m, n given and that the entries of the matrices are stored in arrays A(I, J) and B(I, J).

26. Write a computer program to read in scalars r and s and then compute the linear combination $r\mathbf{A} + s\mathbf{B}$ of the $m \times n$ matrices $\mathbf{A}$ and $\mathbf{B}$. Assume m, n given and that the entries of the matrices are stored in arrays A(I, J) and B(I, J).

1.3 Matrix Multiplication

In Section 1.2 we introduced the concept of the scalar product $\mathbf{a} \cdot \mathbf{b}$ of two vectors $\mathbf{a}$, $\mathbf{b}$. By treating a matrix as a sequence of vectors (row vectors or column vectors), we used this scalar product to define the product $\mathbf{Ab}$ of a matrix $\mathbf{A}$ times a vector $\mathbf{b}$ as well as the product $\mathbf{cA}$ of a vector $\mathbf{c}$ times a matrix $\mathbf{A}$. In this section we extend this process the final step to define the product $\mathbf{AB}$ of two matrices.

Matrix Multiplication

Let $\mathbf{A}$ be an $m \times r$ matrix and $\mathbf{B}$ be an $r \times n$ matrix. The number of columns in $\mathbf{A}$ must equal the number of rows in $\mathbf{B}$. The matrix product $\mathbf{AB}$ is an $m \times n$ matrix obtained by forming the scalar product of each row $\mathbf{a}_i'$ in $\mathbf{A}$ with each column $\mathbf{b}_j$ in $\mathbf{B}$. That is, the (i, j)th entry in $\mathbf{AB}$ is $\mathbf{a}_i' \cdot \mathbf{b}_j$.

$$
\mathbf{AB} = \begin{bmatrix} \mathbf{a}_1' \cdot \mathbf{b}_1 & \mathbf{a}_1' \cdot \mathbf{b}_2 & \cdots & \mathbf{a}_1' \cdot \mathbf{b}_n \\ \mathbf{a}_2' \cdot \mathbf{b}_1 & \mathbf{a}_2' \cdot \mathbf{b}_2 & \cdots & \mathbf{a}_2' \cdot \mathbf{b}_n \\ \cdots & \cdots & \cdots & \cdots \\ \cdots & \cdots & \cdots & \cdots \\ \mathbf{a}_m' \cdot \mathbf{b}_1 & \mathbf{a}_m' \cdot \mathbf{b}_2 & \cdots & \mathbf{a}_m' \cdot \mathbf{b}_n \end{bmatrix} \tag{1}
$$

Remember that we multiply the rows of the first matrix times the columns of the second matrix. For example, if $\mathbf{A} = \begin{bmatrix} 1 & 2 \\ -1 & 0 \end{bmatrix}$, $\mathbf{B} = \begin{bmatrix} 4 & 5 & 1 \\ 6 & 7 & 2 \end{bmatrix}$, then

$$
\mathbf{AB} = \begin{bmatrix} 1 & 2 \\ -1 & 0 \end{bmatrix}\begin{bmatrix} 4 & 5 & 1 \\ 6 & 7 & 2 \end{bmatrix} = \begin{bmatrix} 1 \cdot 4 + 2 \cdot 6 & 1 \cdot 5 + 2 \cdot 7 & 1 \cdot 1 + 2 \cdot 2 \\ -1 \cdot 4 + 0 \cdot 6 & -1 \cdot 5 + 0 \cdot 7 & -1 \cdot 1 + 0 \cdot 2 \end{bmatrix}
$$

$$
= \begin{bmatrix} 16 & 19 & 5 \\ -4 & -5 & -1 \end{bmatrix} \tag{2}
$$

There are several ways to interpret matrix multiplication: first, as the scalar product of each row of $\mathbf{A}$ with each column of $\mathbf{B}$, as in (1); and second, we can view it as a sequence of matrix–vector products, that is,

$$
\mathbf{AB} = \mathbf{A}[\mathbf{b}_1, \mathbf{b}_2, \ldots, \mathbf{b}_n] = [\mathbf{Ab}_1, \mathbf{Ab}_2, \ldots, \mathbf{Ab}_n]
$$

For $\mathbf{A}$, $\mathbf{B}$ above in (2), we check that the first column of $\mathbf{AB}$ is the matrix–vector product $\mathbf{Ab}_1$:

$$
\mathbf{Ab}_1 = \begin{bmatrix} 1 & 2 \\ -1 & 0 \end{bmatrix}\begin{bmatrix} 4 \\ 6 \end{bmatrix} = \begin{bmatrix} 1 \cdot 4 + 2 \cdot 6 \\ -1 \cdot 4 + 0 \cdot 6 \end{bmatrix} = \begin{bmatrix} 16 \\ -4 \end{bmatrix}
$$

Finally, we could also view $\mathbf{AB}$ as an extension of the vector–matrix product $\mathbf{a}_i'\mathbf{B}$ to the vector–matrix products of each row of $\mathbf{A}$ with $\mathbf{B}$.

$$
\mathbf{AB} = \begin{bmatrix} \mathbf{a}_1' \\ \mathbf{a}_2' \\ \cdots \\ \cdots \\ \mathbf{a}_m' \end{bmatrix} \mathbf{B} = \begin{bmatrix} \mathbf{a}_1'\mathbf{B} \\ \mathbf{a}_2'\mathbf{B} \\ \cdots \\ \cdots \\ \mathbf{a}_m'\mathbf{B} \end{bmatrix} \tag{3}
$$

For **A**, **B** above, we check that the first row of **AB** is the vector–matrix product $\mathbf{a}'_1\mathbf{B}$:

$$\mathbf{a}'_1\mathbf{B} = [1 \quad 2]\begin{bmatrix} 4 & 5 & 1 \\ 6 & 7 & 2 \end{bmatrix} = [1\cdot 4 + 2\cdot 6 \quad 1\cdot 5 + 2\cdot 7 \quad 1\cdot 1 + 2\cdot 2] = [16 \quad 19 \quad 5]$$

Equivalent Definitions of Matrix Multiplication
1. Entry (i, j) of **AB** is scalar product $\mathbf{a}'_i \cdot \mathbf{b}_j$.
2. Column j of **AB** is matrix–vector product $\mathbf{Ab}_j$.
3. Row i of **AB** is vector–matrix product $\mathbf{a}'_i\mathbf{B}$.

Recall from the column approach to matrix–vector products that $\mathbf{Ab}_j$ is a linear combination of the columns of **A**. Then, by definition 2, each column of **AB** is a linear combinations of the columns of **A**. And by the row approach to vector–matrix products, it follows that each $\mathbf{a}'_i\mathbf{B}$ is a linear combination of the rows of **B**. Then by definition 3, each row of **AB** is a linear combination of the rows of **B**.

Remember that for the matrix product **AB** to make sense, the length of the rows in **A** (= the number of columns in **A**) must equal the length of the columns of **B** (= the number of rows in **B**). Further, if **A** is $m \times r$ and **B** is $r \times n$, **AB** is an $m \times n$ matrix: **AB** has as many rows as **A** and as many columns as **B**.

Note also that *our previous definitions of matrix–vector and vector–matrix products are special cases of matrix multiplication*. In matrix–vector products, the vector was a column vector, which is the same as an $n \times 1$ matrix of one column. If **A** is an $m \times n$ matrix and **b** is an $n \times 1$ matrix (a column n-vector), then as matrix multiplication, **Ab** is an $m \times 1$ matrix (a column m-vector). A similar argument holds for vector–matrix products.

EXAMPLE 1.
Matrix Multiplication Is Not Commutative

Let $\mathbf{A} = \begin{bmatrix} 1 & 1 \\ 3 & 4 \end{bmatrix}$ and $\mathbf{B} = \begin{bmatrix} 1 & -1 \\ 0 & 2 \end{bmatrix}$. Then

$$\mathbf{AB} = \begin{bmatrix} 1\cdot 1 + 1\cdot 0 & 1\cdot(-1) + 1\cdot 2 \\ 3\cdot 1 + 4\cdot 0 & 3\cdot(-1) + 4\cdot 2 \end{bmatrix} = \begin{bmatrix} 1 & 1 \\ 3 & 5 \end{bmatrix}$$

while

$$\mathbf{BA} = \begin{bmatrix} 1\cdot 1 + (-1)\cdot 3 & 1\cdot 1 + (-1)\cdot 4 \\ 0\cdot 1 + 2\cdot 3 & 0\cdot 1 + 2\cdot 4 \end{bmatrix} = \begin{bmatrix} -2 & -3 \\ 6 & 8 \end{bmatrix}$$

Thus $\mathbf{AB} \neq \mathbf{BA}$. ■

EXAMPLE 2.
Special Matrices
That Do Commute

(a) An *identity matrix* **I** is a square matrix that has 1's on the main diagonal and 0's elsewhere.

$$\mathbf{I} = \begin{bmatrix} 1 & 0 & 0 & \cdots & 0 \\ 0 & 1 & 0 & \cdots & 0 \\ 0 & 0 & 1 & \cdots & 0 \\ \cdots & \cdots & \cdots & \cdots & \cdots \\ \cdots & \cdots & \cdots & \cdots & \cdots \\ 0 & 0 & 0 & \cdots & 1 \end{bmatrix}$$

When it is important to emphasize the size of the identity matrix, we write $\mathbf{I}_n$ for the $n \times n$ identity matrix. The kth column or kth row of **I** is the kth coordinate vector $\mathbf{i}_k$ (with a 1 in the kth entry and 0's elsewhere). Coordinate vectors were introduced in Section 1.2.

I is the identity element in matrix multiplication (it plays a role similar to that of the number 1 in scalar multiplication). That is, for any square matrix **A**,

$$\mathbf{IA} = \mathbf{AI} = \mathbf{A} \tag{4}$$

I thus commutes with every square matrix (of its size). For the matrix **A** in Example 1, we have

$$\mathbf{IA} = \begin{bmatrix} 1 & 0 \\ 0 & 1 \end{bmatrix}\begin{bmatrix} 1 & 1 \\ 3 & 4 \end{bmatrix} = \begin{bmatrix} 1 \cdot 1 + 0 \cdot 3 & 1 \cdot 1 + 0 \cdot 4 \\ 0 \cdot 1 + 1 \cdot 3 & 0 \cdot 1 + 1 \cdot 4 \end{bmatrix} = \begin{bmatrix} 1 & 1 \\ 3 & 4 \end{bmatrix}$$

It is left to the reader to check that $\mathbf{AI} = \mathbf{A}$ for this **A**.

(b) Let us use the matrix $\mathbf{A} = \begin{bmatrix} 1 & 1 \\ 3 & 4 \end{bmatrix}$ again, but this time multiply it by $\mathbf{C} = \begin{bmatrix} 4 & -1 \\ -3 & 1 \end{bmatrix}$. We obtain

$$\mathbf{AC} = \begin{bmatrix} 1 & 1 \\ 3 & 4 \end{bmatrix}\begin{bmatrix} 4 & -1 \\ -3 & 1 \end{bmatrix} = \begin{bmatrix} 1 \cdot 4 + 1 \cdot (-3) & 1 \cdot (-1) + 1 \cdot 1 \\ 3 \cdot 4 + 4 \cdot (-3) & 3 \cdot (-1) + 4 \cdot 1 \end{bmatrix} = \begin{bmatrix} 1 & 0 \\ 0 & 1 \end{bmatrix} \tag{5a}$$

and

$$\mathbf{CA} = \begin{bmatrix} 4 & -1 \\ -3 & 1 \end{bmatrix}\begin{bmatrix} 1 & 1 \\ 3 & 4 \end{bmatrix} = \begin{bmatrix} 4 \cdot 1 + (-1) \cdot 3 & 4 \cdot 1 + (-1) \cdot 4 \\ (-3) \cdot 1 + 1 \cdot 3 & (-3) \cdot 1 + 1 \cdot 4 \end{bmatrix} = \begin{bmatrix} 1 & 0 \\ 0 & 1 \end{bmatrix} \tag{5b}$$

So **A** and **C** commute, and their product is our new friend, the identity matrix **I**.

∎

If $ac = 1$, for scalar numbers a and c, then c is called the *inverse* of a, that is, $c = a^{-1}$ (and $a = c^{-1}$). The same term is used in matrix multiplication. If $\mathbf{A}^{-1}$ is the inverse of $\mathbf{A}$, then $\mathbf{A}^{-1}\mathbf{A} = \mathbf{I}$; also, $\mathbf{A}\mathbf{A}^{-1} = \mathbf{I}$. So (5) shows that $\mathbf{C} = \begin{bmatrix} 4 & -1 \\ -3 & 1 \end{bmatrix}$ is the inverse of matrix $\mathbf{A} = \begin{bmatrix} 1 & 1 \\ 3 & 4 \end{bmatrix}$, and we can write $\mathbf{C} = \mathbf{A}^{-1}$. Also, $\mathbf{A} = \mathbf{C}^{-1}$.

A matrix and its inverse always commute, since their product (in either order) is always the identity matrix $\mathbf{I}$. (*Warning:* Not all matrices have inverses.) Inverses are very important in linear algebra. We learn how to compute the inverse $\mathbf{A}^{-1}$ of a matrix $\mathbf{A}$, if the inverse exists, in Section 1.5.

Laws of Matrix Algebra

Having introduced matrix addition and multiplication, we now summarize the laws of matrix algebra that apply to these operations. These laws are the building blocks for most of the theory of matrix algebra developed in this book. With the exception of commutativity, the laws are basically those that applied in single-variable (scalar) algebra learned in high school. However, because the objects here are arrays and the operation of matrix multiplication is much more complicated than scalar multiplication, it is not at all obvious that these matrix laws should be true. It requires some effort to verify them (which we do in Chapter 3). The consequences that follow from these rules yield a theory of matrix algebra that provides elegant, complete answers to virtually any question one can pose about matrices and models based on them.

We now give the four principal laws of matrix algebra. (Other results are presented in Chapter 3.) *We assume here that the matrices in each law have the proper size for the matrix operations of addition and multiplication to make sense.*

Basic Laws of Matrix Algebra

1. *Associative laws.* Matrix addition and multiplication is associative: $(\mathbf{A} + \mathbf{B}) + \mathbf{C} = \mathbf{A} + (\mathbf{B} + \mathbf{C})$ and $(\mathbf{AB})\mathbf{C} = \mathbf{A}(\mathbf{BC})$.
2. *Commutative laws.* Matrix addition is commutative: $\mathbf{A} + \mathbf{B} = \mathbf{B} + \mathbf{A}$. Matrix multiplication is not commutative (except in special cases): $\mathbf{AB} \neq \mathbf{BA}$.
3. *Distributive laws.* $\mathbf{A}(\mathbf{B} + \mathbf{C}) = \mathbf{AB} + \mathbf{AC}$ and $(\mathbf{B} + \mathbf{C})\mathbf{A} = \mathbf{BA} + \mathbf{CA}$.
4. *Law of scalar factoring.* $r(\mathbf{AB}) = (r\mathbf{A})\mathbf{B} = \mathbf{A}(r\mathbf{B})$.

Since a vector is just a $1 \times n$ matrix or an $n \times 1$ matrix, these laws also apply to vectors in expressions containing matrix–vector products. Indeed, in complex matrix expressions involving both matrices and vectors, it is essential always to treat vectors as $1 \times n$ or $n \times 1$ matrices (depending on whether the vector is a row vector or a column vector).

Combining the distributive and scalar factoring laws, we obtain the following fundamental equation of linear algebra for matrix–vector products:

Linearity of Matrix–Vector Products

For any scalar numbers r, q, any $m \times n$ matrix $\mathbf{A}$, and any column n-vectors $\mathbf{b}$, $\mathbf{c}$:

$$\mathbf{A}(r\mathbf{b} + q\mathbf{c}) = r\mathbf{Ab} + q\mathbf{Ac}$$

This result says that multiplying a matrix times a linear combination of vectors is the same as multiplying the matrix times each individual vector and then forming the linear combination of products. This result is also true for vector–matrix products $(r\mathbf{b} + q\mathbf{c})\mathbf{A}$. The linearity of matrix–vector products will be used hundreds of times in this book and is the property that "makes the theory of linear algebra work."

The proof of the linearity of matrix–vector products involves showing that for each i, the ith entry in the vector $\mathbf{A}(r\mathbf{b} + q\mathbf{c})$ equals the ith entry in the vector $r\mathbf{Ab} + q\mathbf{Ac}$. For simplicity, we let $i = 1$ (i.e., we just check the first entry on each side). The first entry in the vector $\mathbf{A}(r\mathbf{b} + q\mathbf{c})$ is the scalar product of the first row $\mathbf{a}_1'$ of $\mathbf{A}$ times $r\mathbf{b} + q\mathbf{c}$:

$$\mathbf{a}_1' \cdot (r\mathbf{b} + q\mathbf{c}) = a_{11}(rb_1 + qc_1) + a_{12}(rb_2 + qc_2) + \cdots + a_{1n}(rb_n + qc_n) \quad (6)$$

The first entry in the vector $r\mathbf{Ab} + q\mathbf{Ac}$ is r times the scalar product of the first row $\mathbf{a}_1'$ times $\mathbf{b}$ plus q times the scalar product of $\mathbf{a}_1'$ times $\mathbf{c}$:

$$\begin{aligned} r\mathbf{a}_1' \cdot \mathbf{b} + q\mathbf{a}_1' \cdot \mathbf{c} &= r\{a_{11}b_1 + a_{12}b_2 + \cdots + a_{1n}b_n\} \\ &+ q\{a_{11}c_1 + a_{12}c_2 + \cdots + a_{1n}c_n\} \end{aligned} \quad (7)$$

Collecting all the terms with r together in (6) and all the terms with q together, we see that the right side of (6) equals the right side of (7). This proves the linearity equation.

Examples of Matrix Multiplication

Now let us examine some uses of matrix multiplication. (By the way, we sympathize with the reader that matrix multiplication can involve a lot of tedious arithmetic. After a little pencil-and-paper practice, it is appropriate to let a computer multiply matrices for you. Indeed, tedious numerical chores in matrix multiplication were a prime motivation for the creation of the digital computer. With three simple loops, the following short computer program performs matrix multiplication.)

```
FOR I=1 TO M
    FOR J=1 TO N
    C(I,J)=0
        FOR K=1 TO R
            C(I,J) = C(I,J)+A(I,K)*B(K,J)
        NEXT K
    NEXT J
NEXT I
```
(8)

As this book develops, we will see dozens of different uses for matrix multiplication. We start with a concrete example in which matrix multiplication is used to compile a table of scalar product computations.

EXAMPLE 3.
Collection of
Computer
Computation
Times

A SUPERDUPER computer requires 3 minutes to do a type 1 job (say, a statistics problem), 4 minutes to do a type 2 job, and 2 minutes to do a type 3 job. The computer has 6 type 1 jobs, 8 type 2 jobs, and 10 type 3 jobs. How long will the computer take to perform all these jobs?

If $\mathbf{t} = [3 \quad 4 \quad 2]$ is the vector of the times to do the various jobs and $\mathbf{n} = \begin{bmatrix} 6 \\ 8 \\ 10 \end{bmatrix}$

is the vector of the numbers of each type of job (the reason $\mathbf{n}$ is written as a column vector will be clear shortly), the total time required will be the value of the scalar product $\mathbf{t} \cdot \mathbf{n}$.

$$\text{Total time} = \mathbf{t} \cdot \mathbf{n} = [3 \quad 4 \quad 2] \cdot \begin{bmatrix} 6 \\ 8 \\ 10 \end{bmatrix}$$

$$= 3 \cdot 6 + 4 \cdot 8 + 2 \cdot 10$$

$$= 18 + 32 + 20 = 70$$

Suppose that instead of one supercomputer we have four supercomputers: SUPERDUPER, WACKO, WHOOPER, ULTIMA. Then instead of a vector of computation times for SUPERDUPER to perform different jobs, we need a matrix $\mathbf{A}$ of computation times for the different supercomputers to perform the different types of jobs.

$$
\begin{array}{c}
\phantom{\mathbf{A} =} \\
\mathbf{A} =
\end{array}
\begin{array}{r}
\text{Type of job} \\
1 \quad 2 \quad 3 \\
\begin{array}{l}
\text{SUPERDUPER} \\
\text{WACKO} \\
\text{WHOOPER} \\
\text{ULTIMA}
\end{array}
\begin{bmatrix} 3 & 4 & 2 \\ 5 & 7 & 3 \\ 1 & 2 & 1 \\ 3 & 3 & 3 \end{bmatrix}
\end{array}
\quad \text{Matrix of times}
$$

To calculate how long it would take each supercomputer to do 6 type 1, 8 type 2, and 10 type 3 jobs, we multiply $\mathbf{A}$ times the column vector $\mathbf{n} = \begin{bmatrix} 6 \\ 8 \\ 10 \end{bmatrix}$:

$$\mathbf{An} = \begin{bmatrix} 3 & 4 & 2 \\ 5 & 7 & 3 \\ 1 & 2 & 1 \\ 3 & 3 & 3 \end{bmatrix} \begin{bmatrix} 6 \\ 8 \\ 10 \end{bmatrix} = \begin{bmatrix} 3 \cdot 6 + 4 \cdot 8 + 2 \cdot 10 \\ 5 \cdot 6 + 7 \cdot 8 + 3 \cdot 10 \\ 1 \cdot 6 + 2 \cdot 8 + 1 \cdot 10 \\ 3 \cdot 6 + 3 \cdot 8 + 3 \cdot 10 \end{bmatrix} = \begin{bmatrix} 70 \\ 116 \\ 32 \\ 72 \end{bmatrix} \quad (9)$$

Now let us do this calculation not just for one set of jobs, but for three sets of jobs.

Set A will be the previous set $\mathbf{n} = \begin{bmatrix} 6 \\ 8 \\ 10 \end{bmatrix}$. Sets B and C will be $\begin{bmatrix} 2 \\ 5 \\ 5 \end{bmatrix}$ and $\begin{bmatrix} 4 \\ 4 \\ 4 \end{bmatrix}$.

Let us calculate the times required to do each set on each computer by expanding the vector $\mathbf{n}$ into a matrix $\mathbf{N}$ of three column vectors.

$$\begin{array}{c} \text{Sets} \\ \begin{array}{ccc} A & B & C \end{array} \end{array}$$

$$\mathbf{N} = \begin{array}{c} \text{Type 1} \\ \text{Type 2} \\ \text{Type 3} \end{array} \begin{bmatrix} 6 & 2 & 4 \\ 8 & 5 & 4 \\ 10 & 5 & 4 \end{bmatrix} \qquad \text{Matrix of jobs}$$

The calculation of $\mathbf{An}$ in (9) required us to multiply each row of $\mathbf{A}$ times the column vector $\mathbf{n}$. Now we need to multiply each row of $\mathbf{A}$ (one for each computer) times each column of $\mathbf{N}$ (one for each set of jobs):

$$\mathbf{AN} = \begin{bmatrix} 3 & 4 & 2 \\ 5 & 7 & 3 \\ 1 & 2 & 1 \\ 3 & 3 & 3 \end{bmatrix} \begin{bmatrix} 6 & 2 & 4 \\ 8 & 5 & 4 \\ 10 & 5 & 4 \end{bmatrix}$$

$$= \begin{bmatrix} 3\cdot6+4\cdot8+2\cdot10 & 3\cdot2+4\cdot5+2\cdot5 & 3\cdot4+4\cdot4+2\cdot4 \\ 5\cdot6+7\cdot8+3\cdot10 & 5\cdot2+7\cdot5+3\cdot5 & 5\cdot4+7\cdot4+3\cdot4 \\ 1\cdot6+2\cdot8+1\cdot10 & 1\cdot2+2\cdot5+1\cdot5 & 1\cdot4+2\cdot4+1\cdot4 \\ 3\cdot6+3\cdot8+3\cdot10 & 3\cdot2+3\cdot5+3\cdot5 & 3\cdot4+3\cdot4+3\cdot4 \end{bmatrix}$$

$$\begin{array}{c} \text{Sets} \\ \begin{array}{ccc} A & B & C \end{array} \end{array}$$

$$= \begin{array}{c} \text{SUPERDUPER} \\ \text{WACKO} \\ \text{WHOOPER} \\ \text{ULTIMA} \end{array} \begin{bmatrix} 70 & 36 & 36 \\ 116 & 60 & 60 \\ 32 & 172 & 16 \\ 72 & 36 & 36 \end{bmatrix} \qquad \begin{array}{c} \text{Matrix of total} \\ \text{computation times} \end{array} \qquad (10)$$

■

Next we consider a more substantial use of matrix multiplication, which greatly expands the power of the Markov chain model introduced in Section 1.1.

EXAMPLE 4.
Powers of Markov
Chain Transition
Matrices

In the Markov chain model for weather used in previous sections, the equations for determining the probabilities p_1', p_2', p_3' of sunny, cloudy, or rainy weather tomorrow given the probabilities p_1, p_2, p_3 of sunny, cloudy, or rainy weather today were

$$\begin{aligned} p_1' &= \tfrac{3}{4}p_1 + \tfrac{1}{2}p_2 + \tfrac{1}{4}p_3 \\ p_2' &= \tfrac{1}{8}p_1 + \tfrac{1}{4}p_2 + \tfrac{1}{2}p_3 \\ p_3' &= \tfrac{1}{8}p_1 + \tfrac{1}{4}p_2 + \tfrac{1}{4}p_3 \end{aligned} \qquad (11)$$

If $\mathbf{p} = \begin{bmatrix} p_1 \\ p_2 \\ p_3 \end{bmatrix}$ is the vector current probabilities, $\mathbf{p}' = \begin{bmatrix} p_1' \\ p_2' \\ p_3' \end{bmatrix}$ is the vector of tomorrow's probabilities and $\mathbf{A}$ is the matrix of transition probabilities:

$$
\begin{array}{cccc}
 & & \text{Today} & \\
\text{Tomorrow} & \text{Sunny} & \text{Cloudy} & \text{Rainy}
\end{array}
$$

$$
\mathbf{A} = \begin{array}{c} \text{Sunny} \\ \text{Cloudy} \\ \text{Rainy} \end{array}
\begin{bmatrix}
\frac{3}{4} & \frac{1}{2} & \frac{1}{4} \\
\frac{1}{8} & \frac{1}{4} & \frac{1}{2} \\
\frac{1}{8} & \frac{1}{4} & \frac{1}{4}
\end{bmatrix}
$$

then (11) can be written as simply $\mathbf{p}' = \mathbf{Ap}$.

$$
\mathbf{p}' = \mathbf{Ap}: \begin{bmatrix} p_1' \\ p_2' \\ p_3' \end{bmatrix} = \begin{bmatrix} \frac{3}{4} & \frac{1}{2} & \frac{1}{4} \\ \frac{1}{8} & \frac{1}{4} & \frac{1}{2} \\ \frac{1}{8} & \frac{1}{4} & \frac{1}{4} \end{bmatrix} \begin{bmatrix} p_1 \\ p_2 \\ p_3 \end{bmatrix} = \begin{bmatrix} \frac{3}{4}p_1 + \frac{1}{2}p_2 + \frac{1}{4}p_3 \\ \frac{1}{8}p_1 + \frac{1}{4}p_2 + \frac{1}{2}p_3 \\ \frac{1}{8}p_1 + \frac{1}{4}p_2 + \frac{1}{4}p_3 \end{bmatrix} \tag{12}
$$

A Markov chain can be followed further into the future. Just as tomorrow's probability vector $\mathbf{p}'$ can be computed by the matrix expression $\mathbf{p}' = \mathbf{Ap}$, so the probability vector $\mathbf{p}''$ for 2 days hence should be given by

$$
\mathbf{p}'' = \mathbf{Ap}' = \mathbf{A(Ap)} = \mathbf{A}^2\mathbf{p} \tag{13}
$$

Equation (13) says to use (11) twice: Compute $\mathbf{p}''$ from $\mathbf{p}'$ using (11) after first computing $\mathbf{p}'$ from $\mathbf{p}$ with (11). An example of computing $\mathbf{p}''$ from $\mathbf{p}'$ was given in Example 3 of Section 1.1. Similarly, the distribution $\mathbf{p}^{(3)}$ 3 days hence should be given by

$$
\mathbf{p}^{(3)} = \mathbf{Ap}'' = \mathbf{A(A}^2\mathbf{p}) = \mathbf{A}^3\mathbf{p} \tag{14}
$$

In (13), we want $\mathbf{A}^2\mathbf{p}$ to be the same as $\mathbf{A(Ap)}$—that is, premultiplying $\mathbf{p}$ by $\mathbf{A}$ twice should be the same as premultiplying $\mathbf{p}$ by $\mathbf{A}^2$; and similarly for $\mathbf{A}^3$. In mathematical terms, we are simply saying that matrix multiplication should be associative— $\mathbf{A(Ap)} = \mathbf{(AA)p}$ and $\mathbf{A(A}^2\mathbf{p}) = \mathbf{(AA}^2)\mathbf{p}$. Fortunately, associativity is one of the laws given above of matrix algebra.

Let us first compute $\mathbf{A}^2$ and $\mathbf{A}^3$ for the weather Markov transition matrix $\mathbf{A}$.

$$
\mathbf{A}^2 = \begin{bmatrix} \frac{3}{4} & \frac{1}{2} & \frac{1}{4} \\ \frac{1}{8} & \frac{1}{4} & \frac{1}{2} \\ \frac{1}{8} & \frac{1}{4} & \frac{1}{4} \end{bmatrix} \begin{bmatrix} \frac{3}{4} & \frac{1}{2} & \frac{1}{4} \\ \frac{1}{8} & \frac{1}{4} & \frac{1}{2} \\ \frac{1}{8} & \frac{1}{4} & \frac{1}{4} \end{bmatrix}
$$

$$
= \begin{bmatrix} \frac{3}{4} \cdot \frac{3}{4} + \frac{1}{2} \cdot \frac{1}{8} + \frac{1}{4} \cdot \frac{1}{8} & \frac{3}{4} \cdot \frac{1}{2} + \frac{1}{2} \cdot \frac{1}{4} + \frac{1}{4} \cdot \frac{1}{4} & \frac{3}{4} \cdot \frac{1}{4} + \frac{1}{2} \cdot \frac{1}{2} + \frac{1}{4} \cdot \frac{1}{4} \\ \frac{1}{8} \cdot \frac{3}{4} + \frac{1}{4} \cdot \frac{1}{8} + \frac{1}{2} \cdot \frac{1}{8} & \frac{1}{8} \cdot \frac{1}{2} + \frac{1}{4} \cdot \frac{1}{4} + \frac{1}{2} \cdot \frac{1}{4} & \frac{1}{8} \cdot \frac{1}{4} + \frac{1}{4} \cdot \frac{1}{2} + \frac{1}{2} \cdot \frac{1}{4} \\ \frac{1}{8} \cdot \frac{3}{4} + \frac{1}{4} \cdot \frac{1}{8} + \frac{1}{4} \cdot \frac{1}{8} & \frac{1}{8} \cdot \frac{1}{2} + \frac{1}{4} \cdot \frac{1}{4} + \frac{1}{4} \cdot \frac{1}{4} & \frac{1}{8} \cdot \frac{1}{4} + \frac{1}{4} \cdot \frac{1}{2} + \frac{1}{4} \cdot \frac{1}{4} \end{bmatrix}
$$

$$
= \begin{bmatrix} \frac{21}{32} & \frac{18}{32} & \frac{16}{32} \\ \frac{6}{32} & \frac{8}{32} & \frac{9}{32} \\ \frac{5}{32} & \frac{6}{32} & \frac{7}{32} \end{bmatrix} \tag{15}
$$

Next we compute $\mathbf{A}^3$.

$$\mathbf{A}^3 = \mathbf{A}\mathbf{A}^2 = \begin{bmatrix} \frac{3}{4} & \frac{1}{2} & \frac{1}{4} \\ \frac{1}{8} & \frac{1}{4} & \frac{1}{2} \\ \frac{1}{8} & \frac{1}{4} & \frac{1}{4} \end{bmatrix} \begin{bmatrix} \frac{21}{32} & \frac{9}{16} & \frac{1}{2} \\ \frac{3}{16} & \frac{1}{4} & \frac{9}{32} \\ \frac{5}{32} & \frac{3}{16} & \frac{7}{32} \end{bmatrix}$$

$$= \begin{bmatrix} \frac{3}{4}\cdot\frac{21}{32}+\frac{1}{2}\cdot\frac{3}{16}+\frac{1}{4}\cdot\frac{5}{32} & \frac{3}{4}\cdot\frac{9}{16}+\frac{1}{2}\cdot\frac{1}{4}+\frac{1}{4}\cdot\frac{3}{16} & \frac{3}{4}\cdot\frac{1}{2}+\frac{1}{2}\cdot\frac{9}{32}+\frac{1}{4}\cdot\frac{7}{32} \\ \frac{1}{8}\cdot\frac{21}{32}+\frac{1}{4}\cdot\frac{3}{16}+\frac{1}{2}\cdot\frac{5}{32} & \frac{1}{8}\cdot\frac{9}{16}+\frac{1}{4}\cdot\frac{1}{4}+\frac{1}{2}\cdot\frac{3}{16} & \frac{1}{8}\cdot\frac{1}{2}+\frac{1}{4}\cdot\frac{9}{32}+\frac{1}{2}\cdot\frac{7}{32} \\ \frac{1}{8}\cdot\frac{21}{32}+\frac{1}{4}\cdot\frac{3}{16}+\frac{1}{4}\cdot\frac{5}{32} & \frac{1}{8}\cdot\frac{9}{16}+\frac{1}{4}\cdot\frac{1}{4}+\frac{1}{4}\cdot\frac{3}{16} & \frac{1}{8}\cdot\frac{1}{2}+\frac{1}{4}\cdot\frac{9}{32}+\frac{1}{4}\cdot\frac{7}{32} \end{bmatrix}$$

$$= \begin{bmatrix} \frac{160}{256} & \frac{152}{256} & \frac{146}{256} \\ \frac{53}{256} & \frac{58}{256} & \frac{62}{256} \\ \frac{43}{256} & \frac{46}{256} & \frac{48}{256} \end{bmatrix} \qquad (16)$$

The entries in $\mathbf{A}^2$ will be transition probabilities for 2 days and the entries in $\mathbf{A}^3$ transition probabilities for 3 days. For example, the value $\frac{6}{32}$ in entry (2, 1) of $\mathbf{A}^2$ means that if we are now in state 1 (sunny), the chance is $\frac{6}{32}$ that in 2 days we will be in state 2 (cloudy). And the value $\frac{53}{256}$ in entry (2, 1) of $\mathbf{A}^3$ tells us that if it is now sunny, the probability is $\frac{53}{256}$ that in 3 days it will be cloudy. The values we obtained in computing $\mathbf{A}^2$ and $\mathbf{A}^3$ look reasonable. In particular, the numbers in each column of $\mathbf{A}^2$ and $\mathbf{A}^3$ sum to 1. Note also that the three columns of $\mathbf{A}^3$ have approximately the same values in the first, second, and third rows. ■

Example 4 illustrates how, with very concise notation, matrix algebra allows us to express quite complex expressions. The computations in (15) and (16) are tedious and much better done by a computer. But if the computations in (15) and (16) seem bad, the reader should consider how much more messy it would be to compute the entries in $\mathbf{p}^{(3)}$ from p_1, p_2, p_3 by resubstituting three times in equations (11). Writing $\mathbf{p}^{(3)} = \mathbf{A}^3\mathbf{p}$ and computing $\mathbf{A}^3$ is much easier. Later in this book, linear algebra theory will allow us to analyze many interesting properties of Markov chains, without the tedium of computing powers of $\mathbf{A}$.

We finish this section with two examples of matrix multiplication in which one matrix operates on the other. A matrix with one 1 in each row and one 1 in each column and other entries all 0 is called a *permutation matrix*.

EXAMPLE 5.
Permuting Rows
and Columns

We can use matrix multiplication to permute the rows or columns of a matrix. Let

$$\mathbf{A} = \begin{bmatrix} 1 & 2 & 3 \\ 4 & 5 & 6 \\ 7 & 8 & 9 \end{bmatrix} \quad \text{and let} \quad \mathbf{P} = \begin{bmatrix} 0 & 1 & 0 \\ 0 & 0 & 1 \\ 1 & 0 & 0 \end{bmatrix}. \quad \mathbf{P} \text{ is a permutation matrix. We}$$

now show why it is called a permutation matrix. Let us compute $\mathbf{PA}$ and $\mathbf{AP}$.

$$\mathbf{PA} = \begin{bmatrix} 0 & 1 & 0 \\ 0 & 0 & 1 \\ 1 & 0 & 0 \end{bmatrix} \begin{bmatrix} 1 & 2 & 3 \\ 4 & 5 & 6 \\ 7 & 8 & 9 \end{bmatrix}$$

$$= \begin{bmatrix} 0 \cdot 1 + 1 \cdot 4 + 0 \cdot 7 & 0 \cdot 2 + 1 \cdot 5 + 0 \cdot 8 & 0 \cdot 3 + 1 \cdot 6 + 0 \cdot 9 \\ 0 \cdot 1 + 0 \cdot 4 + 1 \cdot 7 & 0 \cdot 2 + 0 \cdot 5 + 1 \cdot 8 & 0 \cdot 3 + 0 \cdot 6 + 1 \cdot 9 \\ 1 \cdot 1 + 0 \cdot 4 + 0 \cdot 7 & 1 \cdot 2 + 0 \cdot 5 + 0 \cdot 8 & 1 \cdot 3 + 0 \cdot 6 + 0 \cdot 9 \end{bmatrix} \quad (17)$$

$$= \begin{bmatrix} 4 & 5 & 6 \\ 7 & 8 & 9 \\ 1 & 2 & 3 \end{bmatrix}$$

$$\mathbf{AP} = \begin{bmatrix} 1 & 2 & 3 \\ 4 & 5 & 6 \\ 7 & 8 & 9 \end{bmatrix} \begin{bmatrix} 0 & 1 & 0 \\ 0 & 0 & 1 \\ 1 & 0 & 0 \end{bmatrix}$$

$$= \begin{bmatrix} 1 \cdot 0 + 2 \cdot 0 + 3 \cdot 1 & 1 \cdot 1 + 2 \cdot 0 + 3 \cdot 0 & 1 \cdot 0 + 2 \cdot 1 + 3 \cdot 0 \\ 4 \cdot 0 + 5 \cdot 0 + 6 \cdot 1 & 4 \cdot 1 + 5 \cdot 0 + 6 \cdot 0 & 4 \cdot 0 + 5 \cdot 1 + 6 \cdot 0 \\ 7 \cdot 0 + 8 \cdot 0 + 9 \cdot 1 & 7 \cdot 1 + 8 \cdot 0 + 9 \cdot 0 & 7 \cdot 0 + 8 \cdot 1 + 9 \cdot 0 \end{bmatrix} \quad (18)$$

$$= \begin{bmatrix} 3 & 1 & 2 \\ 6 & 4 & 5 \\ 9 & 7 & 8 \end{bmatrix}$$

Observe in (17) that premultiplying $\mathbf{A}$ by $\mathbf{P}$ permutes the rows of $\mathbf{A}$, moving the first row of $\mathbf{A}$ to the last row of $\mathbf{PA}$, the second row of $\mathbf{A}$ to the first row of $\mathbf{PA}$, and the last row of $\mathbf{A}$ to the second row of $\mathbf{PA}$. This is why $\mathbf{P}$ is called a permutation matrix. Similarly, in (18) we see that postmultiplying $\mathbf{A}$ by $\mathbf{P}$ permutes the columns of $\mathbf{A}$, moving the first column of $\mathbf{A}$ to the second column of $\mathbf{AP}$, and so on.

These results can be explained most easily by referring to the "Equivalent Definitions of Matrix Multiplication" given at the beginning of this section. For (17), use the fact that the ith row of $\mathbf{PA}$ is the ith row $\mathbf{p}_i'$ of $\mathbf{P}$ times $\mathbf{A}$.

$$\mathbf{PA} = \begin{bmatrix} \mathbf{p}_1' \\ \mathbf{p}_2' \\ \mathbf{p}_3' \end{bmatrix} \mathbf{A} = \begin{bmatrix} \mathbf{p}_1'\mathbf{A} \\ \mathbf{p}_2'\mathbf{A} \\ \mathbf{p}_3'\mathbf{A} \end{bmatrix} = \begin{bmatrix} [0 \ \ 1 \ \ 0]\mathbf{A} \\ [0 \ \ 0 \ \ 1]\mathbf{A} \\ [1 \ \ 0 \ \ 0]\mathbf{A} \end{bmatrix} \quad (19)$$

The rows of $\mathbf{P}$ are the coordinate vectors $\mathbf{i}_k$ (with a 1 in the kth entry and 0's elsewhere) introduced in the preceding section and seen in the identity matrix in Example 2. For example, the first row $\mathbf{p}_1'$ equals $\mathbf{i}_2$. Recall from the preceding section that $\mathbf{i}_2\mathbf{A} = \mathbf{a}_2'$, the second row of $\mathbf{A}$. In general, $\mathbf{i}_k\mathbf{A} = \mathbf{a}_k'$.

To understand (18), we recall that the jth column of $\mathbf{AP}$ equals $\mathbf{A}$ times the jth column of $\mathbf{P}$.

$$\mathbf{AP} = \mathbf{A}[\mathbf{p}_1, \mathbf{p}_2, \mathbf{p}_3] = [\mathbf{Ap}_1, \mathbf{Ap}_2, \mathbf{Ap}_3] = \begin{bmatrix} \mathbf{A}\begin{bmatrix} 0 \\ 0 \\ 1 \end{bmatrix}, & \mathbf{A}\begin{bmatrix} 1 \\ 0 \\ 0 \end{bmatrix}, & \mathbf{A}\begin{bmatrix} 0 \\ 1 \\ 0 \end{bmatrix} \end{bmatrix} \quad (20)$$

The columns of $\mathbf{P}$ are also coordinate vectors. From Example 4 of Section 1.2, $\mathbf{Ai}_k = \mathbf{a}_k$, the kth column of $\mathbf{A}$. ▪

EXAMPLE 6.
Other Elementary
Matrices

Continuing Example 5, we now introduce two other types of elementary matrices that correspond to elementary operations on the rows of a matrix A. The operations are:

1. Multiply the ith row of A by a constant r_i.
2. Subtract a multiple r of one row of A from another row.

As an example of operation 1, consider multiplying the first row of a 3×3 matrix A by 4, the second row by $\frac{1}{2}$ (i.e., dividing it by 2), and the third row by $\frac{1}{5}$. We accomplish these changes by putting the multipliers on the main diagonal of a 3×3 matrix D which premultiplies A. So $D = \begin{bmatrix} 4 & 0 & 0 \\ 0 & \frac{1}{2} & 0 \\ 0 & 0 & \frac{1}{5} \end{bmatrix}$. Let's use the same matrix $A = \begin{bmatrix} 1 & 2 & 3 \\ 4 & 5 & 6 \\ 7 & 8 & 9 \end{bmatrix}$ as in Example 5. Then

$$DA = \begin{bmatrix} 4 & 0 & 0 \\ 0 & \frac{1}{2} & 0 \\ 0 & 0 & \frac{1}{5} \end{bmatrix}\begin{bmatrix} 1 & 2 & 3 \\ 4 & 5 & 6 \\ 7 & 8 & 9 \end{bmatrix} = \begin{bmatrix} [4 & 0 & 0]A \\ [0 & \frac{1}{2} & 0]A \\ [0 & 0 & \frac{1}{5}]A \end{bmatrix} = \begin{bmatrix} 4 & 8 & 12 \\ 2 & \frac{5}{2} & 3 \\ \frac{7}{5} & \frac{8}{5} & \frac{9}{5} \end{bmatrix} \quad (21)$$

Recall that $[1, 0, 0]A$, or equivalently, $i_1 A$, equals a_1', the first row of A (as noted in Example 5). Then $[4, 0, 0]A$, or $4i_1 A$, equals $4a_1'$, four times the first row of A; and similarly for the other rows of A. In general, to perform an operation of type 1, we put r_i, the multiplier of the ith row, on the main diagonal in entry (i, i), for each i, with off-diagonal entries all 0.

As an example of operation 2, consider subtracting three times the first row of A from the last. We claim that the desired premultiplying matrix is $S = \begin{bmatrix} 1 & 0 & 0 \\ 0 & 1 & 0 \\ -3 & 0 & 1 \end{bmatrix}$.

We use the same A as before. Then

$$SA = \begin{bmatrix} 1 & 0 & 0 \\ 0 & 1 & 0 \\ -3 & 0 & 1 \end{bmatrix}\begin{bmatrix} 1 & 2 & 3 \\ 4 & 5 & 6 \\ 7 & 8 & 9 \end{bmatrix} = \begin{bmatrix} [1 & 0 & 0]A \\ [0 & 1 & 0]A \\ [-3 & 0 & 1]A \end{bmatrix}$$

$$= \begin{bmatrix} 1 & 2 & 3 \\ 4 & 5 & 6 \\ -3 \cdot 1 + 7 & -3 \cdot 2 + 8 & -3 \cdot 3 + 9 \end{bmatrix} = \begin{bmatrix} 1 & 2 & 3 \\ 4 & 5 & 6 \\ 4 & 2 & 0 \end{bmatrix} \quad (22)$$

Note that by putting D and S after A and multiplying AD and AS, we would perform similar sorts of operations on the columns of A. ■

The row operations of rearranging rows (in Example 5) and multiplying rows by a constant or subtracting a multiple of one row from another row (in Example 6) will be used over and over again in the next section and later as part of Gaussian elimination. We perform these operations without using elementary matrices, but in many applications involving very large matrices (with thousands of rows and columns), it usually is much simpler to work with elementary matrices.

Section 1.3 Exercises

1. Indicate which pairs of the following matrices can be multiplied together and give the size of the resulting product: a 3×7 matrix $\mathbf{A}$, a 2×3 matrix $\mathbf{B}$, a 3×3 matrix $\mathbf{C}$, a 2×2 matrix $\mathbf{D}$, and a 7×2 matrix $\mathbf{E}$.

2. Let $\mathbf{A} = \begin{bmatrix} 1 & 2 \\ 3 & 4 \end{bmatrix}$, $\mathbf{B} = \begin{bmatrix} 1 & -1 & 0 \\ 2 & 10 & -2 \end{bmatrix}$, $\mathbf{C} = \begin{bmatrix} 3 & 1 \\ 2 & 5 \end{bmatrix}$. Compute the following matrix products (if possible).
    (a) $\mathbf{AB}$ (b) $\mathbf{AC}$ (c) $\mathbf{BC}$ (d) $\mathbf{CA}$ (e) $(\mathbf{CA})\mathbf{B}$

3. Let $\mathbf{A} = \begin{bmatrix} 1 & 2 & 3 & 4 \\ 2 & 4 & 6 & 8 \\ 3 & 5 & 7 & 9 \end{bmatrix}$, $\mathbf{B} = \begin{bmatrix} 1 & 0 & -1 \\ 2 & -2 & 0 \\ 0 & 1 & -1 \end{bmatrix}$, and $\mathbf{C} = \begin{bmatrix} 5 & 4 & 1 \\ 1 & 0 & 2 \\ 3 & 2 & 1 \\ 0 & 1 & 3 \end{bmatrix}$. Compute these matrix products (if possible).
    (a) $\mathbf{AB}$ (b) $\mathbf{BA}$ (c) $\mathbf{AC}$ (d) $\mathbf{CA}$ (e) $\mathbf{CB}$

4. Compute just one row or column, as requested, in the following matrix products. ($\mathbf{A}, \mathbf{B}, \mathbf{C}$ are as in Exercise 3.)
    (a) Row 1 in $\mathbf{B}^2$ (b) Column 2 in $\mathbf{AC}$ (c) Column 3 in $\mathbf{CB}$

5. Show that $\mathbf{AB} = \mathbf{BA}$ for the matrices $\mathbf{A} = \begin{bmatrix} 1 & 1 \\ 2 & 3 \end{bmatrix}$ and $\mathbf{B} = \begin{bmatrix} 3 & -1 \\ -2 & 1 \end{bmatrix}$.

6. For $\mathbf{A}, \mathbf{B}, \mathbf{C}$ in Exercise 3, compute entry $(2, 3)$ in $(\mathbf{BA})\mathbf{C}$. Explain first why it is not necessary to multiply out $\mathbf{BA}$ fully to determine entry $(2, 3)$ in $(\mathbf{BA})\mathbf{C}$.

7. Suppose that we are given the following matrices involving the costs of fruits at different stores, the amounts of fruit different people want, and the numbers of different people in different towns.

	Store A	Store B
Apple	$0.10	$0.15
Orange	$0.15	$0.20
Pear	$0.10	$0.10

	Apple	Orange	Pear
Person A	5	10	3
Person B	4	5	5

	Person A	Person B
Town 1	1000	500
Town 2	2000	1000

    (a) Compute a matrix that tells how much each person's fruit purchases cost at each store.
    (b) Compute a matrix that tells how many of each fruit will be purchased in each town.
    (c) Compute a matrix that tells the total cost of everyone's fruit purchases in town 1 and in town 2 when people use store A and when they use store B (a different number for each town and each store).

8. Suppose that we are given the following matrices: Matrix **A** gives the amount of time each of three jobs requires of I/O (input/output), of execution time, and system overhead; matrix **B** gives the charges (per unit of time) of different computer activities under two different charging plans; matrix **C** (actually a vector) tells how many jobs of each type there are; and matrix **D** tells the fraction of the time that each time-charging plan (the columns in matrix **B**) is used each day.

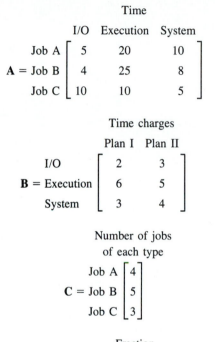

Time

	I/O	Execution	System
Job A	5	20	10
A = Job B	4	25	8
Job C	10	10	5

Time charges

	Plan I	Plan II
I/O	2	3
B = Execution	6	5
System	3	4

Number of jobs
of each type

Job A	4
C = Job B	5
Job C	3

Fraction
of time

Plan I	0.3
D = Plan II	0.7

Find matrix products for the following arrays using **A**, **B**, **C**, and **D**. Compute the numbers in these arrays.
(a) Total cost of each type of job for each charge plan.
(b) Total amount of I/O, execution, and system time for all the jobs. (All jobs are summarized in matrix **C**.)
(c) Total cost of all jobs when run under plan I and under plan II.
(d) Average cost of 1 unit of I/O, of execution, of system time. (Hint: Use matrix **D**.)
(e) Average cost of each type of job (job A, job B, job C).

9. Suppose that we are given the following matrices: Matrix **A** gives the amounts of raw material required to build different products; matrix **B** gives the costs of these raw materials in two different countries; matrix **C** tells how many of the products are needed to build two types of houses; and matrix **D** gives the demand for houses in the two countries.

Raw material

	Wood	Labor	Steel
Item A	5	20	10
A = Item B	4	25	8
Item C	10	10	5

Cost by country

	Spain	Italy
Wood	$2	$3
B = Labor	$6	$5
Steel	$3	$4

Items needed in house

	A	B	C
House I	4	8	3
C = House II	5	5	2

Demand for houses

	House I	House II
Spain	50,000	200,000
D = Italy	80,000	500,000

(a) Compute the first row in the matrix product **AB**.

(b) Which matrix product tells how much of the different items are needed to meet the demand for houses (types I and II combined) in the different countries?

(c) Which matrix product gives the cost of building each type of house in each country?

(d) Which entry in what matrix product would give the total cost of building all homes in Spain?

[Note: If the rows and columns in a matrix **A** must be interchanged in a product, indicate this by using the transpose $\mathbf{A}^T$ of the matrix. (Transposes are formally introduced in Section 3.1.)]

10. Suppose that we are given the following matrices: Matrix **A** gives the number of tradespeople needed each day to build different types of small stores; **B** gives the number of days it takes to build each type of store in each state; matrix **C** gives the cost of tradespeople (per day) of New York and Texas; and matrix **D** gives the number of stores of each type needed in two different shopping centers.

	Carp.	Elect.	Brick.
Store A	5	2	1
A = Store B	4	2	2
Store C	3	1	1

	New York	Texas
Store A	20	15
B = Store B	30	25
Store C	20	20

	New York	Texas
Carpenter	\$100	\$60
C = Electrician	\$80	\$50
Bricklayer	\$80	\$60

	Shopping center I	Shopping center II
Store A	10	5
D = Store B	10	10
Store C	20	20

(a) Compute the first column in **AC**.

(b) Which matrix product tells how many tradespeople per day are needed to build the stores in each type of shopping center? **Don't compute this product.**

(c) Which entry in what matrix product would give the total cost of building three stores, one of each type, in New York? (Total cost covers all the days of construction.)

[Note: If the rows and columns in a matrix **A** must be interchanged in some matrix product, indicate this by using the transpose $\mathbf{A}^T$ of the matrix. (Transposes are formally introduced in Section 3.1.)]

11. Consider a growth model for the numbers of computers (C) and dogs (D) from year to year:

$$D' = 3C + D$$
$$C' = 2C + 2D$$

Let $\mathbf{x} = \begin{bmatrix} C \\ D \end{bmatrix}$ be the initial vector and let $\mathbf{x}^{(k)}$ denote the vector of computers and dogs after k years. Let **A** be the matrix of coefficients in this system. Write an expression for $\mathbf{x}^{(k)}$ in terms of **A** and **x**.

12. Consider a variation on the Markov chain for weather. Now there is just sunny and cloudy weather. The new transition matrix **A** is

	Sunny	Cloudy
Sunny	$\frac{2}{3}$	$\frac{1}{3}$
Cloudy	$\frac{1}{3}$	$\frac{2}{3}$

$\mathbf{A} = $

(a) Compute $\mathbf{A}^2$. What probability does entry (1, 2) in $\mathbf{A}^2$ represent?

(b) Compute $\mathbf{A}^3$. What is the probability if it is sunny today that it will be sunny in 3 days?

(c) Compute $\mathbf{A}^4$. What vector do the columns of $\mathbf{A}^k$, for $k = 2, 3, 4$, seem to be approaching?

13. Consider the following transition matrices for weather Markov chains.

	Sunny	Cloudy	Rainy
Sunny	$\frac{1}{2}$	$\frac{1}{4}$	$\frac{1}{4}$
(1) Cloudy	$\frac{1}{4}$	$\frac{1}{2}$	$\frac{1}{4}$
Rainy	$\frac{1}{4}$	$\frac{1}{4}$	$\frac{1}{2}$

$$
\begin{array}{c@{\quad}ccc}
 & \text{Sunny} & \text{Cloudy} & \text{Rainy} \\
\begin{array}{r}\text{Sunny} \\ (2)\ \text{Cloudy} \\ \text{Rainy}\end{array} &
\left[\begin{array}{ccc}
\frac{1}{2} & \frac{1}{2} & \frac{1}{4} \\
\frac{1}{2} & \frac{1}{4} & \frac{1}{4} \\
0 & \frac{1}{4} & \frac{1}{2}
\end{array}\right]
\end{array}
$$

(a) Compute A^2. What probability does entry $(3, 2)$ in A^2 represent?

(b) Compute A^3. What is the probability if it is sunny today that it will be rainy in 3 days?

14. **(a)** Show that when we multiply $B = \begin{bmatrix} 1 & 0 & -1 \\ 2 & -2 & 0 \\ 0 & 1 & -1 \end{bmatrix}$ and $C = \begin{bmatrix} 5 & 4 & 1 \\ 1 & 0 & 2 \\ 3 & 2 & 1 \\ 0 & 1 & 3 \end{bmatrix}$ times

the 3×3 identity matrix I (see Example 2), the results BI and CI equal B and C, respectively.

(b) Show that if we multiply B by $K = \begin{bmatrix} 1 & 0 & 0 \\ 0 & 2 & 0 \\ 0 & 0 & 3 \end{bmatrix}$, the result BK has the

same first column as B, but the second column is double B's second column, and the third column is triple B's third column. Describe in words what happens when we compute KB.

(c) Suppose that K has k_1 in entry $(1, 1)$, k_2 in $(2, 2)$, and k_3 in $(3, 3)$; and 0's elsewhere. Describe the effect of premultiplying any 3×3 matrix A by K.

15. **(a)** Find the product AB of $A = \begin{bmatrix} 2 & 0 & 0 \\ 0 & 1 & 0 \\ 0 & 0 & 3 \end{bmatrix}$ and $B = \begin{bmatrix} 3 & 0 & 0 \\ 0 & 2 & 0 \\ 0 & 0 & 5 \end{bmatrix}$.

(b) Suppose that a_{11}, a_{22}, a_{33} are the diagonal entries in the diagonal matrix A and b_{11}, b_{22}, b_{33} are the diagonal entries in the diagonal matrix B. Then what are the diagonal entries in AB?

16. **(a)** If we premultiply any 3×3 matrix A by the permutation matrix $P = \begin{bmatrix} 1 & 0 & 0 \\ 0 & 0 & 1 \\ 0 & 1 & 0 \end{bmatrix}$,

show that the resulting matrix PA is just A with rows two and three interchanged. (Hint: As an aid, make up a 3×3 matrix and premultiply by P.)

(b) If we premultiply any 3×3 matrix A by $R = \begin{bmatrix} 0 & 0 & 1 \\ 0 & 2 & 0 \\ 1 & 0 & 0 \end{bmatrix}$, show that the

result RA is a matrix with the rows of A reversed and the values of all entries in row two doubled.

17. **(a)** Show that if we postmultiply any 3×3 matrix A by the permutation matrix P in Exercise 16, the resulting matrix AP is A with columns two and three interchanged.

(b) Show that if we postmultiply A by R in Exercise 16, the resulting matrix AR is A with the order of the columns reversed and the values of all entries in the second column doubled.

18. Describe in words the effect of premultiplying a 3×3 matrix by the following elementary matrices.

(a) $\mathbf{A} = \begin{bmatrix} 2 & 0 & 0 \\ 0 & 1 & 0 \\ 0 & 0 & 3 \end{bmatrix}$ (b) $\mathbf{B} = \begin{bmatrix} 0.5 & 0 & 0 \\ 0 & 4 & 0 \\ 0 & 0 & -1 \end{bmatrix}$

(c) $\mathbf{C} = \begin{bmatrix} 0 & 0 & 1 \\ 1 & 0 & 0 \\ 0 & 1 & 0 \end{bmatrix}$ (d) $\mathbf{D} = \begin{bmatrix} 1 & 0 & 0 \\ 0 & 0 & 1 \\ 0 & 1 & 0 \end{bmatrix}$

(e) $\mathbf{E} = \begin{bmatrix} 1 & 0 & 0 \\ -4 & 1 & 0 \\ 0 & 0 & 1 \end{bmatrix}$ (f) $\mathbf{F} = \begin{bmatrix} 1 & 0 & 0 \\ 0 & 1 & 0 \\ 3 & 0 & 1 \end{bmatrix}$

19. (a) Construct a 3×3 matrix $\mathbf{H}$ which when premultiplying a 3×3 matrix $\mathbf{A}$ has the effect of adding four times the third row of $\mathbf{A}$ to the first row of $\mathbf{A}$.
    (b) Construct a 3×3 matrix $\mathbf{H}$ which when premultiplying a 3×3 matrix $\mathbf{A}$ has the effect of subtracting twice the first row of $\mathbf{A}$ from the second row of $\mathbf{A}$ and also adding three times the first row of $\mathbf{A}$ to the third row of $\mathbf{A}$.

20. (a) Construct a 3×3 matrix $\mathbf{H}$ which when postmultiplying a 3×3 matrix $\mathbf{A}$ has the effect of adding two times the third column of $\mathbf{A}$ to the second column of $\mathbf{A}$.
    (b) Construct a 3×3 matrix $\mathbf{H}$ which when postmultiplying a 3×3 matrix $\mathbf{A}$ has the effect of subtracting four times the first column of $\mathbf{A}$ from the second column of $\mathbf{A}$ and also adding twice the first column of $\mathbf{A}$ to the third column of $\mathbf{A}$.

21. Compute the following matrix product:

$$\begin{bmatrix} 1 & 2 & 1 & 0 \\ 0 & 1 & 3 & 1 \\ 0 & 0 & 1 & 2 \\ 0 & 0 & 0 & 1 \end{bmatrix} \begin{bmatrix} 2 & 0 & 1 & 1 \\ 0 & 1 & 2 & 0 \\ 0 & 0 & 1 & 1 \\ 0 & 0 & 0 & 3 \end{bmatrix}$$

A matrix is called *upper triangular* if the only nonzero entries are on or above the main diagonal. This computation illustrates the fact that the product of two upper triangular matrices is again upper triangular. Give an explanation of why this is true for the product of any two 4×4 upper triangular matrices (or more generally for matrices of any size).

22. Show that in a 2×2 matrix $\begin{bmatrix} a & b \\ c & d \end{bmatrix}$, if the one row is a multiple of the other row, one column is a multiple of the other column.

23. Let $\mathbf{A}$ and $\mathbf{B}$ be 2×2 matrices. If $\mathbf{C} = \mathbf{AB}$ and the second row of $\mathbf{A}$ is three times the first row of $\mathbf{A}$ ($\mathbf{a}_2' = 3\mathbf{a}_1'$), show that the second row of $\mathbf{C}$ is three times the first row of $\mathbf{C}$. (Hint: Compare $c_{11} = \mathbf{a}_1' \cdot \mathbf{b}_1$ with $c_{21} = \mathbf{a}_2' \cdot \mathbf{b}_1$; similarly for c_{12} versus c_{22}.)

24. How many multiplications are required to perform the following matrix operations?
    (a) Square a 10×10 matrix.
    (b) Square a 100×100 matrix.
    (c) Multiply a 20×5 matrix times a 5×20 matrix.
    (d) Multiply a 5×20 matrix times a 20×5 matrix.
    (e) Cube a 10×10 matrix.
    (f) Multiply a 10×10 matrix times itself 10 times.

25. Write a program to multiply a matrix times a vector, take the resulting vector and premultiply it again by the matrix, and so on a specified number of times. (This is the computation needed to follow a Markov chain over many periods.)

26. Write a program to read in two matrices and multiply them together.

27. Write a program to raise a (square) matrix to a specified power.

28. Use a computer program to raise the following Markov transition matrices to the second, fifth, tenth, and twentieth powers. Is there a pattern?
(a) The weather Markov chain
(b) The stock market Markov chain in Exercise 17 of Section 1.1
(c) The learning Markov chain in Exercise 18 of Section 1.1

1.4 Gaussian Elimination

In this section we develop the procedure of Gaussian elimination for solving any system of m linear equations in n variables—to find the unique solution, if one exists, or to find all solutions, or to show that no solution exists. Elimination was devised by Karl Friedrich Gauss around 1820 to solve systems of linear equations that arose in astronomical and land surveying computations. Initially, we assume that the number of equations equals the number of variables—n linear equations in n variables.

Gaussian elimination involves two stages. The first stage transforms the given system of equations, or equivalently, the associated coefficient matrix, into **upper triangular form**, with only 0's below the main diagonal of the matrix, such as

$$
\begin{aligned}
2x_1 + x_2 - x_3 &= 1 \\
x_2 + 4x_3 &= 5 \quad \text{or} \quad \mathbf{A}\mathbf{x} = \mathbf{b} \\
x_3 &= 2
\end{aligned}
$$

$$
\text{where} \quad \mathbf{A} = \begin{bmatrix} 2 & 1 & -1 \\ 0 & 1 & 4 \\ 0 & 0 & 1 \end{bmatrix}, \quad \mathbf{x} = \begin{bmatrix} x_1 \\ x_2 \\ x_3 \end{bmatrix}, \quad \mathbf{b} = \begin{bmatrix} 1 \\ 5 \\ 2 \end{bmatrix}
$$

(1)

The second stage uses back substitution to obtain values for the variables. That is, knowing from the third equation that $x_3 = 2$, we can solve for x_2 in the second equation:

$$x_2 + 4(2) = 5 \quad \text{or} \quad x_2 = -3$$

And now knowing x_2 and x_3, we can solve for x_1 in the first equation:

$$2x_1 + (-3) - 2 = 1 \quad \text{or} \quad x_1 = 3$$

The best way to show how Gaussian elimination works is with some examples. Then we state the procedure in algebraic terms. Since the numbers, not the variables, are what matter in the equations, we often use the coefficient matrix $\mathbf{A}$, with an added column for the right-side vector $\mathbf{b}$. This matrix $[\mathbf{A}|\mathbf{b}]$ of coefficients plus the right-side vector is called the **augmented coefficient matrix**. Initially, we show both the system of equations $\mathbf{A}\mathbf{x} = \mathbf{b}$ and the augmented matrix $[\mathbf{A}|\mathbf{b}]$.

EXAMPLE 1.
Gaussian
Elimination
Example

We start with a simple system of two equations in two unknowns.

$$\begin{array}{ll} \text{(a)} & x + y = 4 \\ \text{(b)} & 2x - y = -1 \end{array} \qquad \begin{bmatrix} 1 & 1 & \bigm| & 4 \\ 2 & -1 & \bigm| & -1 \end{bmatrix}$$

To eliminate the $2x$ term from (b), we subtract 2 times (a) from (b), and obtain the following new second equation:

$$\begin{array}{lll} \text{(b)} & 2x - y = -1 \\ -2\text{(a)} & -(2x + 2y = 8) \\ \hline \text{(b')} = \text{(b)} - 2\text{(a)} & 0 - 3y = -9 \end{array}$$

Our new system of equations is

$$\begin{array}{ll} \text{(a)} & x + y = 4 \\ \text{(b')} & -3y = -9 \end{array} \qquad \begin{bmatrix} 1 & 1 & \bigm| & 4 \\ 0 & -3 & \bigm| & -9 \end{bmatrix}$$

Any solution to (a) and (b) is also a solution to (a) and (b'). Thus we can reverse the step creating (b'). That is, (b') = (b) − 2(a) implies that (b) = (b') + 2(a). Thus (b) is formed from (b') and a multiple of (a), so any solution to (a) and (b') is a solution to (a) and (b). But (b') is trivial to solve, and gives

$$y = 3$$

Substituting $y = 3$ in (a), we have

$$x + 3 = 4 \implies x = 4 - 3 = 1$$

The reader should verify that $x = 1$, $y = 3$ satisfy (a) and (b). *When solving a system of linear equations, one can always verify that the answer is correct by substituting the values for x and y in the original equations and checking for equality.*

▪

In general, Gaussian elimination transforms an augmented coefficient matrix into upper triangular form using the following three row operations.

Elementary Row Operations
1. Multiply or divide a row of an augmented matrix (or an equation) by a nonzero number.
2. Subtract a multiple of one row (or equation) from another row (equation).
3. Interchange two rows (equations).

In Gaussian elimination, we use row operations 1, 2, and 3 repeatedly to reduce a system of linear equations to an upper triangular system, like (1).

Theorem 1. If $Ax = b$ is a system of equations to which elementary row operations were applied to obtain a new system of equations $A'x = b'$, then x^* is a solution to $Ax = b$ if and only if it is also a solution to $A'x = b'$.

Proof: Clearly, applying operations 1 and 3 will not change the solution to a system of equations. Let us look then at the effect of row operation 2. Suppose that a vector x^* satisfies the two equations. Then x^* will also satisfy any equation obtained by subtracting a multiple of the first equation from the second equation. If x^* is still a solution after one elementary row operation, it will continue to be a solution after any number of elementary row operations.

We know that $A'x = b'$ is obtained from $Ax = b$ by elementary row operations. Observe that reversing (undoing) an elementary row operation is also an elementary row operation. If we multiply an equation by the nonzero number r, we reverse this action by dividing by r (or multiplying by $1/r$). If we had subtracted r times the first row from the second row, we reverse by subtracting $-r$ times the first row from the second row. Interchanging the two rows twice puts them in original order. The theorem now follows. ■

Before working more examples of Gaussian elimination, we note that elementary row operations 1 and 2 have associated elementary matrices of type 1 and 2, respectively, introduced in Example 6 at the end of Section 1.3. In Example 1, the elementary row operation of subtracting twice the first equation from the second equation has elementary matrix $S = \begin{bmatrix} 1 & 0 \\ -2 & 1 \end{bmatrix}$. Multiplying S by $A^+ = \begin{bmatrix} 1 & 1 & 4 \\ 2 & -1 & -1 \end{bmatrix}$ yields

$$SA^+ = \begin{bmatrix} 1 & 0 \\ -2 & 1 \end{bmatrix}\begin{bmatrix} 1 & 1 & 4 \\ 2 & -1 & -1 \end{bmatrix}$$

$$= \begin{bmatrix} 1\cdot1+0\cdot2 & 1\cdot1+0-1 & 1\cdot4+0-1 \\ -2\cdot1+1\cdot2 & -2\cdot1+1-1 & -2\cdot4+1-1 \end{bmatrix}\begin{bmatrix} 1 & 1 & 4 \\ 0 & -3 & -9 \end{bmatrix}$$

The third elementary row operation of interchanging rows is a special case of the permutation of rows. In Example 5 of Section 1.3 we showed how premultiplying by a permutation matrix P permutes rows. In the exercises in Section 1.3, readers were asked to verify how elementary matrices perform the elementary row operations used in Gaussian elimination.

**EXAMPLE 2.
Gaussian
Elimination for
Refinery Problem**

Recall the refinery problem introduced in Section 1.1 with three refineries whose production levels had to be chosen to meet the demands for heating oil, diesel oil, and gasoline.

Heating oil:	(a)	$4x_1 + 2x_2 + 2x_3 = 600$	
Diesel oil:	(b)	$2x_1 + 5x_2 + 2x_3 = 800$	
Gasoline:	(c)	$1x_1 + 2.5x_2 + 5x_3 = 1000$	

$$\begin{bmatrix} 4 & 2 & 2 & | & 600 \\ 2 & 5 & 2 & | & 800 \\ 1 & 2.5 & 5 & | & 1000 \end{bmatrix} \quad (2)$$

Use multiples of equation (a) to eliminate x_1 from (b) and (c). First, subtract $\frac{1}{2}$ times (a) from (b) to eliminate the $2x_1$ term from (b) and obtain a new second equation (b′).

$$\begin{array}{rl}
\text{(b)} & 2x_1 + 5x_2 + 2x_3 = 800 \\
-\frac{1}{2}\text{(a)} & \underline{-(2x_1 + 1x_2 + 1x_3 = 300)} \\
\text{(b′)} = \text{(b)} - \frac{1}{2}\text{(a)} & 0 + 4x_2 + 1x_3 = 500
\end{array}$$

In a similar fashion we subtract $\frac{1}{4}$ times (a) from (c) to eliminate the $1x_1$ term from (c) and obtain a new equation (c′):

$$\begin{array}{rl}
\text{(c)} & 1x_1 + 2.5x_2 + 5x_3 = 1000 \\
-\frac{1}{4}\text{(a)} & \underline{-(1x_1 + 0.5x_2 + 0.5x_3 = 150)} \\
\text{(c′)} = \text{(c)} - \frac{1}{4}\text{(a)} & 2x_2 + 4.5x_3 = 850
\end{array}$$

Our new system of equations is now

$$\begin{array}{rl}
\text{(a)} & 4x_1 + 2x_2 + 2x_3 = 600 \\
\text{(b′)} = \text{(b)} - \frac{1}{2}\text{(a)} & 4x_2 + 1x_3 = 500 \\
\text{(c′)} = \text{(c)} - \frac{1}{4}\text{(a)} & 2x_2 + 4.5x_3 = 850
\end{array}
\qquad
\begin{bmatrix}
4 & 2 & 2 & \bigm| & 600 \\
0 & 4 & 1 & \bigm| & 500 \\
0 & 2 & 4.5 & \bigm| & 850
\end{bmatrix}
\qquad (3)$$

Next we use equation (b′) to eliminate the $2x_2$ term from (c′) and obtain a new third equation (c″).

$$\begin{array}{rl}
\text{(c′)} & 2x_2 + 4.5x_3 = 850 \\
-\frac{1}{2}\text{(b′)} & \underline{-(2x_2 + 0.5x_3 = 250)} \\
\text{(c″)} = \text{(c′)} - \frac{1}{2}\text{(b′)} & 4x_3 = 600
\end{array}$$

Our final system of equations is

$$\begin{array}{rl}
\text{(a)} & 4x_1 + 2x_2 + 2x_3 = 600 \\
\text{(b′)} & 4x_2 + 1x_3 = 500 \\
\text{(c″)} = \text{(c′)} - \frac{1}{2}\text{(b′)} & 4x_3 = 600
\end{array}
\qquad
\begin{bmatrix}
4 & 2 & 2 & \bigm| & 600 \\
0 & 4 & 1 & \bigm| & 500 \\
0 & 0 & 4 & \bigm| & 600
\end{bmatrix}
\qquad (4)$$

Any solution to the original system (2) is a solution to the new system (4). Furthermore, by reversing the steps in going from (2) to (4) [so that (2) is formed from linear combinations of the equations in (4)], we see that any solution to (4) is a solution to (2).

System (4) is in upper triangular form and we can solve using back substitution. From (c″) we have

$$x_3 = \tfrac{600}{4} = 150$$

And substituting this value for x_3 in (b′), we have

$$4x_2 + 1(150) = 500 \implies 4x_2 = 500 - 150 \implies x_2 = 87\tfrac{1}{2}$$

Next substituting the values for x_3 and x_2 in (a), we have

$$4x_1 + 2(87\tfrac{1}{2}) + 2(150) = 600 \quad \text{or} \quad x_1 = \frac{600 - 175 - 300}{4} = 31\tfrac{1}{4}$$

So the vector of the production levels of the three refineries is $[31\tfrac{1}{4}, 87\tfrac{1}{2}, 150]$. ■

EXAMPLE 3.
System of
Equations Without
Unique Solution

Suppose that we change the third equation in Example 2.

$$
\begin{array}{ll}
\text{(a) } 4x_1 + 2x_2 + 2x_3 = 600 \\
\text{(b) } 2x_1 + 5x_2 + 2x_3 = 800 \\
\text{(c) } 3x_1 + 3.5x_2 + 2x_3 = 700
\end{array}
\qquad
\left[\begin{array}{ccc|c}
4 & 2 & 2 & 600 \\
2 & 5 & 2 & 800 \\
3 & 3.5 & 2 & 700
\end{array}\right]
\tag{5}
$$

After eliminating x_1 from (b) and (c) as above, we have

$$
\begin{array}{ll}
\text{(a)} & 4x_1 + 2x_2 + 2x_3 = 600 \\
\text{(b')} = \text{(b)} - \tfrac{1}{2}\text{(a)} & 4x_2 + 1x_3 = 500 \\
\text{(c')} = \text{(c)} - \tfrac{3}{4}\text{(a)} & 2x_2 + 0.5x_3 = 250
\end{array}
\qquad
\left[\begin{array}{ccc|c}
4 & 2 & 2 & 600 \\
0 & 4 & 1 & 500 \\
0 & 2 & 0.5 & 250
\end{array}\right]
\tag{6}
$$

Next we subtract $\tfrac{1}{2}$(b') from (c') to eliminate the $2x_2$ term, but this eliminates all of (c').

$$
\begin{array}{ll}
\text{(a)} & 4x_1 + 2x_2 + 2x_3 = 600 \\
\text{(b')} & 4x_2 + 1x_3 = 500 \\
\text{(c'')} = \text{(c')} - \tfrac{1}{2}\text{(b')} & 0 = 0
\end{array}
\qquad
\left[\begin{array}{ccc|c}
4 & 2 & 2 & 600 \\
0 & 4 & 1 & 500 \\
0 & 0 & 0 & 0
\end{array}\right]
\tag{7}
$$

That is, equation (c') is just $\tfrac{1}{2}$(b'). We now have only two equations in three unknowns. This system has an infinite number of solutions, since we can make up any value for x_3 and then use back substitution to determine x_2 and x_1. For example, letting $x_3 = 100$, back substitution yields $x_2 = 100$ and $x_1 = 50$.

Let us reconsider (5) with the third equation having 1000 as the right-side value.

$$
\begin{array}{ll}
\text{(a) } 4x_1 + 2x_2 + 2x_3 = 600 \\
\text{(b) } 2x_1 + 5x_2 + 2x_3 = 800 \\
\text{(c) } 3x_1 + 3.5x_2 + 2x_3 = 1000
\end{array}
\qquad
\left[\begin{array}{ccc|c}
4 & 2 & 2 & 600 \\
2 & 5 & 2 & 800 \\
3 & 3.5 & 2 & 1000
\end{array}\right]
\tag{8}
$$

Eliminating x_1 as before, we get

$$
\begin{array}{ll}
\text{(a) } 4x_1 + 2x_2 + 2x_3 = 600 \\
\text{(b')} 4x_2 + 1x_3 = 500 \\
\text{(c')} 2x_2 + 0.5x_3 = 550
\end{array}
\qquad
\left[\begin{array}{ccc|c}
4 & 2 & 2 & 600 \\
0 & 4 & 1 & 500 \\
0 & 2 & 0.5 & 550
\end{array}\right]
\tag{9}
$$

When we use (b') to eliminate the $2x_2$ term in (c'), we now obtain

$$
\begin{array}{ll}
\text{(a)} & 4x_1 + 2x_2 + 2x_3 = 600 \\
\text{(b')} & 4x_2 + 1x_3 = 500 \\
\text{(c'')} = \text{(c')} - \tfrac{1}{2}\text{(b')} & 0 = 300
\end{array}
\qquad
\left[
\begin{array}{ccc|c}
4 & 2 & 2 & 600 \\
0 & 4 & 1 & 500 \\
0 & 0 & 0 & 300
\end{array}
\right]
\qquad (10)
$$

In (9), (b') and (c') are inconsistent equations, that is, they have no solution, and lead to the impossible equation (c''): $0 = 300$.

By Theorem 1, (8) has no solution if (10) has no solution. ■

Suppose that we have a system of n equations in n unknowns:

$$
\mathbf{Ax} = \mathbf{b}
\qquad
\begin{aligned}
a_{11}x_1 + a_{12}x_2 + \cdots + a_{1n}x_n &= b_1 \\
a_{21}x_1 + a_{22}x_2 + \cdots + a_{2n}x_n &= b_2 \\
&\;\;\vdots \\
a_{i1}x_1 + a_{i2}x_2 + \cdots + a_{in}x_n &= b_i \\
&\;\;\vdots \\
a_{n1}x_1 + a_{n2}x_2 + \cdots + a_{nn}x_n &= b_n
\end{aligned}
\qquad (11)
$$

Since the first equation begins $a_{11}x_1 + \cdots$ and the ith equation begins $a_{i1}x_1 + \cdots$, then multiplying the first equation by $-a_{i1}/a_{11}$ (*assuming that $a_{11} \neq 0$*) will yield a new equation that begins $-a_{i1}x_1 + \cdots$, that is,

$$
-\frac{a_{i1}}{a_{11}}(a_{11}x_1 + a_{12}x_2 + \cdots + a_{1n}x_n = b_1)
$$

equals

$$
-a_{i1}x_1 - \frac{a_{i1}}{a_{11}}a_{12}x_2 - \cdots - \frac{a_{i1}}{a_{11}}a_{1n}x_n = -\frac{a_{i1}}{a_{11}}b_1
\qquad (12)
$$

Subtracting (12) from the ith equation in (11) eliminates x_1 to give

$$
\left(a_{i2} - \frac{a_{i1}}{a_{11}}a_{12}\right)x_2 + \cdots + \left(a_{in} - \frac{a_{i1}}{a_{11}}a_{1n}\right)x_n = b_i - \frac{a_{i1}}{a_{11}}b_1
\qquad (13)
$$

By performing the steps in (12) and (13) for $i = 2, 3, \ldots, n$, we can eliminate the x_1 term from every equation except the first, so that now the second through nth equations will form a system of $n - 1$ equations in $n - 1$ variables. We repeat the elimination process with this $(n - 1) \times (n - 1)$ system, eliminating the x_2 term from the third through nth equations. We continue this method of eliminating variables until we finally have one equation in x_n—which is trivial to solve.

Once x_n is known, we can work backwards to determine the value of x_{n-1}, then of x_{n-2}, and so on, as in the previous examples.

Gaussian Elimination

1. Subtract multiples of the ith equation to eliminate the ith variable from the remaining equations, for $i = 1, \ldots, n - 1$.
2. Solve the resulting upper triangular system of equations using back substitution.

We are assuming here that when it is time to eliminate x_i from equations $i + 1$ through n, the coefficient of x_i in the current ith equation is nonzero. Otherwise, we use elementary row operation 3 to interchange the ith row with a row that does have a nonzero coefficient of x_i. We discuss this case in Example 5 below.

Gauss–Jordan Elimination

We next present a variation on Gaussian elimination, known as *Gauss–Jordan elimination*, that is a little slower but eliminates the need to do back substitution. Gauss–Jordan elimination uses the equation i to eliminate x_i from all other equations before, as well as after, equation i (Gaussian elimination only eliminates x_i in equations after equation i). It also divides equation i by a_{ii} so that the coefficient of x_i in the new equation i is 1.

We use the term *pivot on entry a_{ij}* (the coefficient of x_j in equation i) to denote the process of using equation i to eliminate x_j from all other equations (and making 1 be the new coefficient of x_j in equation i).

EXAMPLE 4.
Gauss–Jordan
Elimination

Let us rework Example 2 using Gauss–Jordan elimination. This time we just show the augmented coefficient matrix $[\mathbf{A}\,|\,\mathbf{b}]$.

$$
\begin{array}{c}
(a) \\
(b) \\
(c)
\end{array}
\left[
\begin{array}{ccc|c}
4 & 2 & 2 & 600 \\
2 & 5 & 2 & 800 \\
1 & 2.5 & 5 & 1000
\end{array}
\right]
\tag{14}
$$

Now we pivot on entry (1, 1).

$$
\begin{array}{c}
(a') = (a)/4 \\
(b') = (b) - \frac{1}{2}(a) \\
(c') = (c) - \frac{1}{4}(a)
\end{array}
\left[
\begin{array}{ccc|c}
1 & \frac{1}{2} & \frac{1}{2} & 150 \\
0 & 4 & 1 & 500 \\
0 & 2 & \frac{9}{2} & 850
\end{array}
\right]
\tag{15}
$$

Next we povot on entry (2, 2).

$$
\begin{array}{c}
(a'') = (a') - \frac{1}{2}(b'') \\
(b'') = (b')/4 \\
(c'') = (c') - 2(b'')
\end{array}
\left[
\begin{array}{ccc|c}
1 & 0 & \frac{3}{8} & 87\frac{1}{2} \\
0 & 1 & \frac{1}{4} & 125 \\
0 & 0 & 4 & 600
\end{array}
\right]
\tag{16}
$$

Finally, we pivot on entry (3, 3).

$$\begin{matrix} (a'') = (a') - \frac{3}{8}(c''') \\ (b''') = (b'') - \frac{1}{4}(c''') \\ (c''') = (c'')/4 \end{matrix} \quad \left[\begin{array}{ccc|c} 1 & 0 & 0 & 31\frac{1}{4} \\ 0 & 1 & 0 & 87\frac{1}{2} \\ 0 & 0 & 1 & 150 \end{array}\right] \qquad (17)$$

Now the upper triangular system of equations corresponding to (17) yields a solution directly, without back substitution.

$$\begin{matrix} x_1 & & & = & 31\frac{1}{4} \\ & x_2 & & = & 87\frac{1}{2} \\ & & x_3 & = & 150 \end{matrix} \qquad (18) \quad ■$$

Note that in (17), the coefficient matrix has been reduced by Gauss–Jordan elimination to the identity matrix $\mathbf{I}$. (Identity matrices were introduced in Example 2 in Section 1.3.)

Note that by interchanging rows (which is elementary row operation 3), one can use any equation to eliminate x_i from the other equations. We illustrate the idea with Gauss–Jordan elimination, but it also applies to Gaussian elimination.

EXAMPLE 5.
Solution with
Interchange of
Rows

Let us repeat Example 4 but with the numbers in equations (b) and (c) changed:

$$\begin{matrix} (a) \\ (b) \\ (c) \end{matrix} \quad \left[\begin{array}{ccc|c} 4 & 2 & 2 & 600 \\ 2 & 1 & 2 & 800 \\ 1 & 2.5 & 2.5 & 550 \end{array}\right] \qquad (19)$$

As above, we want to make entry (1, 1) equal to 1 and the rest of the first-column 0's.

$$\begin{matrix} (a') = (a)/4 \\ (b') = (b) - \frac{1}{2}(a) \\ (c') = (c) - \frac{1}{4}(a) \end{matrix} \quad \left[\begin{array}{ccc|c} 1 & 0.5 & 0.5 & 150 \\ 0 & 0 & 1 & 500 \\ 0 & 2 & 2 & 400 \end{array}\right] \qquad (20)$$

Since entry (2, 2) is 0, we cannot pivot on it. Now we need elementary row operation 3—interchange of two rows. We interchange the second and third rows, (b') and (c').

$$\begin{matrix} (a') \\ (c') \\ (b') \end{matrix} \quad \left[\begin{array}{ccc|c} 1 & 0.5 & 0.5 & 150 \\ 0 & 2 & 2 & 400 \\ 0 & 0 & 1 & 500 \end{array}\right] \qquad (21)$$

We can now pivot on entry (2, 2) in this new matrix [which was entry (3, 2) in (20)].

$$\begin{matrix} (a'') = (a') - \frac{1}{2}(c'') \\ (c'') = (c')/2 \\ (b'') = (b') \end{matrix} \quad \left[\begin{array}{ccc|c} 1 & 0 & 0 & 50 \\ 0 & 1 & 1 & 200 \\ 0 & 0 & 1 & 500 \end{array}\right] \qquad (22)$$

Finally, we pivot on entry $(3, 3)$ in (22) [which was originally entry $(2,3)$].

$$\begin{array}{l} (a''') = (a'') \\ (c''') = (c'') - (b'') \\ (b''') = (b'') \end{array} \quad \left[\begin{array}{ccc|c} 1 & 0 & 0 & 50 \\ 0 & 1 & 0 & -300 \\ 0 & 0 & 1 & 500 \end{array}\right] \qquad (23)$$

We read off the solution, $x_1 = 50$, $x_2 = -300$, $x_3 = 500$.   ■

Notice how the examples we have worked show that there are three possible outcomes when a system of n linear equations in n variables is solved, either a unique solution or an infinite number of solutions, as in (7), or no solution, as in (10).

Alternatives for Solving n Equations in n Variables

Theorem 2.   One of the following three alternatives must result when Gaussian elimination is applied to $\mathbf{Ax} = \mathbf{b}$, a system of n linear equations in n variables.

(i) *Unique solution*. The coefficient matrix $\mathbf{A}$ is reduced by Gaussian elimination to upper triangular form, as in (1), and then a unique solution is obtained by back substitution. Or equivalently, Gauss–Jordan elimination pivots along the main diagonal to reduce the coefficient matrix to an identity matrix.

(ii) *Infinite number of solutions*. In Gaussian elimination, the last rows in the augmented coefficient matrix $[\mathbf{A}\,|\,\mathbf{b}]$ become all 0's, as in (7). If the last k rows are all 0's, then the last k variables (and possibly some other variables) can be given any values and then the remaining variables determined by back substitution.

(iii) *No solution*. In Gaussian elimination, some row in the coefficient matrix becomes all 0's while the associated right-side entry is nonzero, as in (10).

Row Echelon Form

We generalize the previous discussion to a system with m equations in n variables, for any positive integers m and n. We use the same three row operations presented at the start of this section. When the number of rows does not equal the number of columns, it is common in Gaussian elimination to make the first nonzero number in each row be 1 (if the leading nonzero entry in a row is $c \neq 1$, multiply the row by $1/c$). The result of the row operations in Gaussian elimination is now a generalization of upper triangular form, called *row echelon form*.

Row Echelon Form

A matrix is in row echelon form if the following three conditions hold:

1. The first nonzero number in each row is 1, called the *leading 1* of the row.
2. Any row(s) of all 0's appear at the bottom of the matrix.
3. The leading 1 in row i, for $i \geq 2$, is to the right of the leading 1 in earlier rows.

EXAMPLE 6.
Matrices in Row
Echelon Form

The following three matrices are in row echelon form:

$$\begin{bmatrix} 1 & 2 \\ 0 & 1 \end{bmatrix} \quad \begin{bmatrix} 1 & 2 & 3 & 1 \\ 0 & 1 & 0 & 4 \\ 0 & 0 & 0 & 1 \end{bmatrix} \quad \begin{bmatrix} 0 & 1 & 2 & -1 & 6 \\ 0 & 0 & 1 & -3 & 2 \\ 0 & 0 & 0 & 1 & 3 \\ 0 & 0 & 0 & 0 & 0 \end{bmatrix}$$

▪

The upper triangular form of the final augmented coefficient matrices in Gaussian elimination is not row echelon form, since these matrices do not have leading 1's in each row. But they can easily be put in row echelon form by dividing each row by the number on the main diagonal.

The augmented coefficient matrices in Gauss–Jordan elimination are in row echelon form. Their special form is called *reduced row echelon form*.

Reduced Row Echelon Form

A matrix is in reduced row echelon form if it is in row echelon form and:

4. A column that contains a leading 1 for some row has 0's as its other entries.

EXAMPLE 7.
Matrices in
Reduced Row
Echelon Form

The following matrices are in reduced row echelon form.

$$\begin{bmatrix} 1 & 0 \\ 0 & 1 \end{bmatrix} \quad \begin{bmatrix} 1 & 0 & 3 & 0 \\ 0 & 1 & 5 & 0 \\ 0 & 0 & 0 & 1 \end{bmatrix} \quad \begin{bmatrix} 0 & 1 & 0 & 0 & 6 \\ 0 & 0 & 1 & 0 & 2 \\ 0 & 0 & 0 & 1 & 3 \\ 0 & 0 & 0 & 0 & 0 \end{bmatrix}$$

▪

EXAMPLE 8.
Two-Product
Refinery Problem

Suppose in our refinery model that the production of gasoline is not important (there is an excess supply in storage tanks). We are concerned only with the demands for heating oil and diesel oil. The system of equations and augmented coefficient matrix from Example 2 becomes:

$$\text{Heating oil:} \quad \text{(a) } 4x_1 + 2x_2 + 2x_3 = 600 \qquad \begin{bmatrix} 4 & 2 & 2 & | & 600 \\ 2 & 5 & 2 & | & 800 \end{bmatrix} \qquad (24)$$
$$\text{Diesel oil:} \quad \text{(b) } 2x_1 + 5x_2 + 2x_3 = 800$$

Let us put (24) in row echelon form using Gaussian elimination (modified to put leading 1's in each row). We eliminate x_1 from row two (and divide the first row by 4).

$$\begin{aligned}\text{(a') = (a)/4} \qquad & 1x_1 + \tfrac{1}{2}x_2 + \tfrac{1}{2}x_3 = 150 \\ \text{(b') = (b)} - \tfrac{1}{2}\text{(a)} \qquad & 4x_2 + 1x_3 = 500 \end{aligned} \qquad \begin{bmatrix} 1 & \tfrac{1}{2} & \tfrac{1}{2} & | & 150 \\ 0 & 4 & 1 & | & 500 \end{bmatrix}$$

Next we divide row two by 4 to obtain the desired row echelon form.

$$\begin{aligned}\text{(a')} \qquad & 1x_1 + \tfrac{1}{2}x_2 + \tfrac{1}{2}x_3 = 150 \\ \text{(b'') = (b')/4} \qquad & 1x_2 + \tfrac{1}{4}x_3 = 125 \end{aligned} \qquad \begin{bmatrix} 1 & \tfrac{1}{2} & \tfrac{1}{2} & | & 150 \\ 0 & 1 & \tfrac{1}{4} & | & 125 \end{bmatrix} \qquad (25)$$

Variable x_3 is free and can be given any value. One simple approach would be to set it equal to 0. Suppose that we want to set x_3 equal to 60. Back substitution gives

$$x_2 = 125 - \tfrac{1}{4}(60) = 110$$
$$x_1 = 150 - \tfrac{1}{2}(110) - \tfrac{1}{2}(60) = 65$$

From (25) we can obtain the reduced row echelon form by eliminating entry (1, 2) in the augmented coefficient matrix (i.e., the coefficient of x_2 in the first equation).

$$\begin{aligned}\text{(a'') = (a')} - \tfrac{1}{2}\text{(b'')} \quad & 1x_1 \qquad\quad + \tfrac{3}{8}x_3 = 87\tfrac{1}{2} \\ \text{(b'')} \qquad & 1x_2 + \tfrac{1}{4}x_3 = 125 \end{aligned} \qquad \begin{bmatrix} 1 & 0 & \tfrac{3}{8} & | & 87\tfrac{1}{2} \\ 0 & 1 & \tfrac{1}{4} & | & 125 \end{bmatrix} \qquad (26)$$

Bringing x_3 over to the right side of the equations in (26), we have the general solution for (24) in terms of x_3.

$$\begin{aligned} x_1 = \ & 87\tfrac{1}{2} - \tfrac{3}{8}x_3 \\ x_2 = \ & 125 \ - \tfrac{1}{4}x_3 \end{aligned} \quad \text{or} \quad \begin{bmatrix} x_1 \\ x_2 \end{bmatrix} = \begin{bmatrix} 87\tfrac{1}{2} \\ 125 \end{bmatrix} - x_3 \begin{bmatrix} \tfrac{3}{8} \\ \tfrac{1}{4} \end{bmatrix} \qquad (27)$$

■

When there are m equations in n variables with $m < n$, as in (24), we can think of there being $n - m$ additional (unwritten) equations of 0 coefficients on the right sides. Then by Theorem 2, case (i) of a complete upper triangular coefficient matrix cannot occur. Rather, either case (ii) or (iii) in that theorem applies. Thus we have:

Alternatives for Solving m Equations in n Variables ($m < n$)

Theorem 3. One of the following two alternatives results when Gaussian elimination is applied to $\mathbf{Ax} = \mathbf{b}$, a system of m linear equations in n variables, where $m < n$, to reduce the augmented coefficient matrix $[\mathbf{A}\,|\,\mathbf{b}]$ to row echelon form.

(i) *No solution*. When reduced to row echelon form, the leading 1 in some row of $[\mathbf{A}\,|\,\mathbf{b}]$ is $\mathbf{b}$'s column (and the associated equation for this row would be $0 = 1$). Then no solution exists.

(ii) *Infinite number of solutions*. When reduced to row echelon form, no row of $[\mathbf{A}\,|\,\mathbf{b}]$ has a leading 1 in the last column of $\mathbf{A}$. Possibly some of the bottom rows may be all 0's. If there are k rows with leading 1's, the $n - k$ variables corresponding to columns without a leading 1 can be given any values, and then the variables corresponding to columns with leading 1's are determined by back substitution.

The situation where there are more equations than variables is more complicated and must be deferred for now, although we note that the possible outcomes are: a unique solution, an infinite number of solutions, or no solution.

We conclude this section by using the equation-solving tools we have developed to investigate the structure of Markov chains.

**EXAMPLE 9.
Steady State of the
Weather Markov
Chain**

In the weather Markov chain introduced in Section 1.1 with transition matrix,

$$
\mathbf{A} = \begin{array}{c} \text{Tomorrow} \\ \text{Sunny} \\ \text{Cloudy} \\ \text{Rainy} \end{array}
\begin{array}{c} \overset{\text{Today}}{\overset{\text{Sunny}\quad\text{Cloudy}\quad\text{Rainy}}{\left[\begin{array}{ccc} \frac{3}{4} & \frac{1}{2} & \frac{1}{4} \\ \frac{1}{8} & \frac{1}{4} & \frac{1}{2} \\ \frac{1}{8} & \frac{1}{4} & \frac{1}{4} \end{array}\right]}} \end{array}
$$

it was noted that over many periods, the probability distribution stabilized at $\frac{14}{23}$ sunny, $\frac{5}{23}$ cloudy, $\frac{4}{23}$ rainy [see (10) in Section 1.1]. We confirm this fact by showing that if today's distribution $\mathbf{p} = \begin{bmatrix} \frac{14}{23} \\ \frac{5}{23} \\ \frac{4}{23} \end{bmatrix}$, then tomorrow's distribution $\mathbf{p}'(= \mathbf{Ap})$ is also $\begin{bmatrix} \frac{14}{23} \\ \frac{5}{23} \\ \frac{4}{23} \end{bmatrix}$.

$$
\mathbf{p}' = \mathbf{Ap} = \begin{bmatrix} \frac{3}{4} & \frac{1}{2} & \frac{1}{4} \\ \frac{1}{8} & \frac{1}{4} & \frac{1}{2} \\ \frac{1}{8} & \frac{1}{4} & \frac{1}{4} \end{bmatrix} \begin{bmatrix} \frac{14}{23} \\ \frac{5}{23} \\ \frac{4}{23} \end{bmatrix} = \begin{bmatrix} \frac{3}{4}\cdot\frac{14}{23} + \frac{1}{2}\cdot\frac{5}{23} + \frac{1}{4}\cdot\frac{4}{23} \\ \frac{1}{8}\cdot\frac{14}{23} + \frac{1}{4}\cdot\frac{5}{23} + \frac{1}{2}\cdot\frac{4}{23} \\ \frac{1}{8}\cdot\frac{14}{23} + \frac{1}{4}\cdot\frac{5}{23} + \frac{1}{4}\cdot\frac{4}{23} \end{bmatrix} = \begin{bmatrix} \frac{14}{23} \\ \frac{5}{23} \\ \frac{4}{23} \end{bmatrix} \qquad (28)
$$

We call this special vector $\mathbf{p}^* = \begin{bmatrix} \frac{14}{23} \\ \frac{5}{23} \\ \frac{4}{23} \end{bmatrix}$ the **stable distribution** of the Markov chain. We now show how to solve directly for the stable distribution $\mathbf{p}^*$. As checked in (28), $\mathbf{p}^*$ satisfies the matrix equation:

$$\mathbf{p}^* = \mathbf{Ap}^*: \quad \begin{aligned} p_1^* &= \tfrac{3}{4}p_1^* + \tfrac{1}{2}p_2^* + \tfrac{1}{4}p_3^* \\ p_2^* &= \tfrac{1}{8}p_1^* + \tfrac{1}{4}p_2^* + \tfrac{1}{2}p_3^* \\ p_3^* &= \tfrac{1}{8}p_1^* + \tfrac{1}{4}p_2^* + \tfrac{1}{4}p_3^* \end{aligned} \tag{29}$$

Collecting the p^*'s on the left, we get

$$\begin{aligned} \tfrac{1}{4}p_1^* - \tfrac{1}{2}p_2^* - \tfrac{1}{4}p_3^* &= 0 \\ -\tfrac{1}{8}p_1^* + \tfrac{3}{4}p_2^* - \tfrac{1}{2}p_3^* &= 0 \\ -\tfrac{1}{8}p_1^* - \tfrac{1}{4}p_2^* + \tfrac{3}{4}p_3^* &= 0 \end{aligned} \tag{30}$$

Solving by Gaussian elimination, we obtain (using the augmented coefficient matrix)

$$\begin{bmatrix} \tfrac{1}{4} & -\tfrac{1}{2} & -\tfrac{1}{4} & \Big| & 0 \\ -\tfrac{1}{8} & \tfrac{3}{4} & -\tfrac{1}{2} & \Big| & 0 \\ -\tfrac{1}{8} & -\tfrac{1}{4} & \tfrac{3}{4} & \Big| & 0 \end{bmatrix} \Longrightarrow \begin{bmatrix} \tfrac{1}{4} & -\tfrac{1}{2} & -\tfrac{1}{4} & \Big| & 0 \\ 0 & \tfrac{1}{2} & -\tfrac{5}{8} & \Big| & 0 \\ 0 & -\tfrac{1}{2} & \tfrac{5}{8} & \Big| & 0 \end{bmatrix} \Longrightarrow \begin{bmatrix} \tfrac{1}{4} & -\tfrac{1}{2} & -\tfrac{1}{4} & \Big| & 0 \\ 0 & \tfrac{1}{2} & -\tfrac{5}{8} & \Big| & 0 \\ 0 & 0 & 0 & \Big| & 0 \end{bmatrix} \tag{31}$$

In row echelon form, we have

$$\begin{bmatrix} 1 & -2 & -1 & \Big| & 0 \\ 0 & 1 & -\tfrac{5}{4} & \Big| & 0 \\ 0 & 0 & 0 & \Big| & 0 \end{bmatrix} \tag{32a}$$

or

$$\begin{aligned} p_1^* - 2p_2^* - p_3^* &= 0 \\ p_2^* - \tfrac{5}{4}p_3^* &= 0 \end{aligned} \tag{32b}$$

Solving (32b), we get

$$p_2^* = \tfrac{5}{4}p_3^* \quad \text{and} \quad p_1^* = 2p_2^* + p_3^* = 2(\tfrac{5}{4}p_3^*) + p_3^* = \tfrac{7}{2}p_3^* \tag{33}$$

Where is our unique vector of stable probabilities? In (33) we have obtained an infinite number of solutions to (30), one for each value of p_3^*. This difficulty has occurred because we omitted one important fact: that these probabilities must sum to 1,

$$p_1^* + p_2^* + p_3^* = 1 \tag{34}$$

Using (33), we express p_1^* and p_2^* in terms of p_3^* to obtain

$$1 = p_1^* + p_2^* + p_3^* = \tfrac{7}{2}p_3^* + \tfrac{5}{4}p_3^* + p_3^* = \tfrac{23}{4}p_3^*$$

Hence

$$p_3^* = \tfrac{4}{23}$$

and then from (33),

$$p_2^* = \tfrac{5}{23}, \qquad p_1^* = \tfrac{14}{23}$$

These are the probabilities of the stable distribution of the weather Markov chain given above. ■

We could have substituted equation (34) for the last equation in (30), since that last equation is zeroed out during Gaussian elimination. Or we could have added (34) to the three equations in (30) to have a system of four equations in three unknowns. Either of these two new systems could be solved by elimination to determine the stable probabilities.

Systems of equations $\mathbf{Ax} = \mathbf{0}$ with zero right sides, such as (30), arise frequently in linear algebra. They have a special name.

Definition

A system of linear equations $\mathbf{Ax} = \mathbf{0}$ with zero right side is called *homogeneous*.

When solving a homogeneous system, we usually are interested in a nonzero solution, as was the case with system (30). Note that $\mathbf{x} = \mathbf{0}$ is always a solution to any homogeneous system $\mathbf{Ax} = \mathbf{0}$. Thus we need multiple solutions to get a nonzero solution. If $\mathbf{A}$ is an $n \times n$ matrix, then by Theorem 2 there will be multiple solutions if and only if at least one row of $\mathbf{A}$ is zeroed out during elimination. [Note that for homogeneous systems, case (iii) in Theorem 2 cannot happen.] If $\mathbf{A}$ has more columns than rows, then by Theorem 3 there will always be multiple solutions.

Section 1.4 Exercises

1. Solve the following systems of equations by Gaussian elimination.

   (a) $x + y = 5$ (b) $2x - 3y = 4$ (c) $3x - y = 0$
   $\ x - 2y = 4$ $3x + 2y = 5$ $-2x + y = 2$

2. In each of the following sets of three equations, show that the third equation equals the second equation minus some multiple of the first equation: (c) = (b) − r(a) for some r.

   (a) (a) $x + 2y = 4$ (b) (a) $x - y + z = 2$
   (b) $3x + y = 9$ (b) $x + y - z = 3$
   (c) $x - 3y = 1$ (c) $-2x + 4y - 4z = -3$

(c) (a) $2x + y - 2z = -5$
(b) $3x - y + z = 8$
(c) $6x + 0.5y - 2z = 0.5$

3. Solve the following systems of equations using Gaussian elimination.

(a) $2x_1 - 3x_2 + 2x_3 = 0$
$x_1 - x_2 + x_3 = 7$
$-x_1 + 5x_2 + 4x_3 = 4$

(b) $-x_1 - x_2 + x_3 = 2$
$2x_1 + 2x_2 - 4x_3 = -4$
$-x_1 - 2x_2 + 3x_3 = 5$

(c) $-x_1 - 3x_2 + 2x_3 = -2$
$2x_1 + x_2 + 3x_3 = \frac{9}{2}$
$5x_1 + 4x_2 + 6x_3 = 12$

(d) $2x_1 + 4x_2 - 2x_3 = 4$
$x_1 - 2x_2 - 4x_3 = -1$
$-2x_1 - x_2 - 3x_3 = -4$

(e) $x_1 + x_2 + 4x_3 = 4$
$2x_1 + x_2 + 3x_3 = 5$
$5x_1 + 2x_2 + 5x_3 = 11$

(f) $2x_1 - 3x_2 - x_3 = 2$
$3x_1 - 5x_2 - 2x_3 = -1$
$9x_1 + 6x_2 + 4x_3 = 1$

4. Solve the problems in Exercise 3 using Gauss–Jordan elimination.

5. Use Gaussian elimination to solve the following variations on the refinery problem in Example 2. Sometimes the variation will have no solution, sometimes multiple solutions (express such an infinite family of solutions in terms of x_3), sometimes the solution will involve negative numbers (a real-world impossibility).

(a) $20x_1 + 4x_2 + 4x_3 = 500$
$8x_1 + 3x_2 + 5x_3 = 850$
$4x_1 + 5x_2 + 11x_3 = 2050$

(b) $6x_1 + 5x_2 + 6x_3 = 500$
$10x_1 + 10x_2 = 850$
$2x_1 + 12x_3 = 1000$

(c) $6x_1 + 2x_2 + 2x_3 = 500$
$3x_1 + 6x_2 + 3x_3 = 300$
$3x_1 + 2x_2 + 6x_3 = 1000$

(d) $8x_1 + 4x_2 + 3x_3 = 500$
$4x_1 + 8x_2 + 5x_3 = 500$
$12x_2 + 6x_3 = 500$

6. Solve the following systems of equations using Gaussian elimination.

(a) $x_1 + 3x_2 + 2x_3 - x_4 = 7$
$x_1 + x_2 + x_3 + x_4 = 3$
$2x_1 - 2x_2 + x_3 - x_4 = -5$
$x_1 - 3x_2 - x_3 + 2x_4 = -4$

(b) $3x_1 + 2x_2 + x_3 = 3$
$x_1 + x_2 - x_4 = 2$
$2x_1 + x_2 - x_3 + x_4 = -3$
$x_1 + x_2 + x_3 + x_4 = 0$

(c) $x_1 + x_2 - x_3 - x_4 = 4$
$2x_1 + x_4 = 8$
$3x_1 - 2x_4 = 3$
$4x_1 - 2x_2 + x_3 + 3x_4 = 15$

7. Exercise 6(c) can be simplified by first solving the second and third equations for x_1 and x_4, and afterward solving for x_2 and x_3. Solve Exercise 6(c) this way.

8. Determine whether each of the following systems of equations has as unique solution, multiple solutions, or is inconsistent.

(a) $2x - 3y = 6$
$-6x + 9y = 12$

(b) $x_1 + 2x_2 + 3x_3 = 10$
$2x_1 - x_2 + 4x_3 = 20$
$5x_2 + 2x_3 = 0$

(c) $x_1 - x_2 + x_3 = 5$
$x_1 + 3x_2 + 6x_3 = 9$
$-x_1 + 5x_2 + 4x_3 = 10$

(d) $x_1 + x_2 + 2x_3 = 0$
$2x_1 + x_2 - 3x_3 = 0$
$-x_1 + 2x_2 + x_3 = 0$

(e) $x_1 + x_2 + 2x_3 = 3$
$-x_1 - 2x_2 + x_3 = 8$
$x_1 - x_2 + 8x_3 = 25$

(f) $x_1 + 2x_2 + 3x_3 = 5$
$3x_1 - x_2 - 2x_3 = -3$
$-5x_1 + 4x_2 + 10x_3 = 14$

9. Solve each system of equations in Exercise 3 with Gauss–Jordan elimination in which off-diagonal pivots are used—to be exact, pivot on entry $(2, 1)$, then on $(3, 2)$, and finally, on $(1, 3)$.

10. Solve the following systems of equations by Gaussian elimination. When you come to a zero entry on the main diagonal, interchange equations as appropriate.

(a) $\begin{aligned} x_1 + 2x_2 + 3x_3 &= 6 \\ 2x_1 + 4x_2 + 5x_3 &= 12 \\ 2x_1 + 5x_2 - 3x_3 &= 10 \end{aligned}$
(b) $\begin{aligned} x_1 - 3x_2 + x_3 &= 4 \\ -2x_1 + 6x_2 - 2x_3 &= 7 \\ x_1 + x_2 - x_3 &= 3 \end{aligned}$

(c) $\begin{aligned} x_1 - x_2 + x_3 + x_4 &= 6 \\ 2x_1 \quad\quad - x_3 - x_4 &= 5 \\ 2x_1 - 2x_2 \quad\quad + x_4 &= 4 \\ x_2 + x_3 - x_4 &= 3 \end{aligned}$
(d) $\begin{aligned} x_2 + x_3 \quad\quad &= 0 \\ x_3 + x_4 &= 1 \\ x_1 \quad\quad - x_4 &= 2 \\ x_1 + x_2 \quad\quad &= 3 \end{aligned}$

11. The following systems of equations are large but their special tridiagonal form makes them easy to solve. Solve these systems of equations.

(a) $\begin{aligned} x_1 - x_2 \quad\quad\quad\quad\quad &= 2 \\ -x_1 + 2x_2 - x_3 \quad\quad\quad &= 0 \\ -x_2 + 2x_3 - x_4 \quad\quad &= 0 \\ -x_3 + 2x_4 - x_5 &= 0 \\ -x_4 + x_5 &= 2 \end{aligned}$
(b) $\begin{aligned} 2x_1 + x_2 \quad\quad\quad\quad\quad &= 1 \\ x_1 + 2x_2 + x_3 \quad\quad\quad &= 1 \\ x_2 + 2x_3 + x_4 \quad\quad &= 1 \\ x_3 + 2x_4 + x_5 &= 1 \\ x_4 + 2x_5 &= 1 \end{aligned}$

12. Solve the following tridiagonal systems of equations.

(a) $\begin{aligned} x_1 - x_2 \quad\quad\quad\quad\quad\quad\quad &= 1 \\ -x_1 + 2x_2 - x_3 \quad\quad\quad\quad\quad &= 0 \\ -x_2 + 2x_3 - x_4 \quad\quad\quad\quad &= -1 \\ -x_3 + 2x_4 - x_5 \quad\quad &= 1 \\ -x_4 + 2x_5 - x_6 &= 1 \\ -x_5 + 2x_6 &= 2 \end{aligned}$

(b) $\begin{aligned} x_1 + x_2 \quad\quad\quad\quad\quad\quad\quad &= 1 \\ 2x_1 + x_2 + x_3 \quad\quad\quad\quad\quad &= 0 \\ 2x_2 + x_3 + x_4 \quad\quad\quad\quad &= 0 \\ 3x_3 + 2x_4 + x_5 \quad\quad &= 0 \\ 2x_4 + 3x_5 + x_6 &= 0 \\ 2x_5 + 3x_6 &= -4 \end{aligned}$

13. Repeat Exercise 12(a) with the last equation changed to $-x_5 + x_6 = -2$. You now get a set of solutions. Express these solutions in terms of x_6.

14. The staff dietician at the California Institute of Trigonometry has to make up a meal with 600 calories, 20 grams of protein, and 200 milligrams of vitamin C. There are three food types to choose from: rubbery Jello, dried fish sticks, and mystery meat. They have the following nutritional content per ounce.

	Jello	Fish sticks	Mystery meat
Calories:	10	50	200
Protein:	1	3	0.2
Vitamin C:	30	10	0

Set up and solve a system of equations to determine how much of each food should be used.

15. A furniture manufacturer makes tables, chairs, and sofas. In one month the company has available 300 units of wood, 350 units of labor, and 225 units of upholstery. The manufacturer wants a production schedule for the month that uses all of these resources. The various products require the following amounts of the resources.

	Table	Chair	Sofa
Wood:	4	1	3
Labor:	3	2	5
Upholstery:	2	0	4

Set up and solve a system of equations to determine how much of each product should be manufactured.

16. A company has a budget of $280,000 for computing equipment. Three types of equipment are available: microcomputers at $2000 a piece, terminals at $500 a piece, and word processors at $5000 a piece. There should be five times as many terminals as microcomputers and two times as many microcomputers as word processors. Set this problem up as a system of three linear equations and solve to determine how many machines of each type should there be.

17. An investment analyst is trying to find out how much business a secretive TV manufacturer has. The company makes three brands of TV sets: brand A, brand B, brand C. The analyst learns that the manufacturer has ordered from suppliers 450,000 type 1 circuit boards, 300,000 type 2 circuit boards, and 350,000 type 3 circuit boards. Brand A uses 2 type 1 boards, 1 type 2 board, and 2 type 3 boards. Brand B uses 3 type 1 boards, 2 type 2 boards, and 1 type 3 board. Brand C uses 1 board of each type. How many TV sets of each brand are being manufactured?

18. State whether the following matrices are in row echelon form or reduced row echelon form.

(a) $\begin{bmatrix} 1 & 0 & 3 \\ 0 & 0 & 1 \\ 0 & 1 & 4 \end{bmatrix}$ (b) $\begin{bmatrix} 1 & 2 & 3 & 0 \\ 0 & 1 & 1 & 0 \\ 0 & 0 & 1 & 4 \end{bmatrix}$ (c) $\begin{bmatrix} 1 & 0 & 1 & 0 \\ 0 & 1 & 0 & 1 \\ 0 & 0 & 1 & 2 \end{bmatrix}$

(d) $\begin{bmatrix} 1 & 0 & 0 & 4 & 0 \\ 0 & 1 & 0 & 3 & 0 \\ 0 & 0 & 0 & 0 & 1 \end{bmatrix}$ (e) $\begin{bmatrix} 1 & 1 & 0 & 4 & 5 \\ 0 & 0 & 0 & 0 & 0 \\ 0 & 0 & 1 & 3 & 2 \end{bmatrix}$

19. Use elementary row operations to reduce the following matrices to row echelon form.

(a) $\begin{bmatrix} 2 & 3 \\ 1 & 0 \end{bmatrix}$ (b) $\begin{bmatrix} 2 & 0 & 1 \\ 2 & 1 & 0 \\ 1 & 1 & 2 \end{bmatrix}$ (c) $\begin{bmatrix} 1 & 0 & 3 \\ 0 & 0 & 1 \\ 0 & 1 & 4 \end{bmatrix}$ (d) $\begin{bmatrix} 1 & 2 & 3 & 0 \\ 1 & 1 & 1 & 0 \\ 3 & 0 & 1 & 4 \end{bmatrix}$

20. Reduce the following matrices to reduced row echelon form.

(a) $\begin{bmatrix} 1 & 1 \\ 2 & 3 \end{bmatrix}$ (b) $\begin{bmatrix} 2 & 1 & 3 \\ 2 & 2 & 1 \\ 2 & 1 & 3 \end{bmatrix}$ (c) $\begin{bmatrix} 2 & 0 & 1 \\ 0 & 2 & 1 \\ 1 & 1 & 5 \end{bmatrix}$ (d) $\begin{bmatrix} 2 & 1 & 4 & 1 \\ 2 & 1 & 2 & 0 \\ 1 & 0 & 1 & 4 \end{bmatrix}$

21. For the following augmented matrices, state whether the associated system of equations has a unique solution, multiple solutions, or no solution.

(a) $\begin{bmatrix} 3 & 3 & 1 & | & 0 \\ 0 & 1 & 5 & | & 1 \\ 0 & 0 & 1 & | & 2 \end{bmatrix}$ (b) $\begin{bmatrix} 1 & 2 & 2 & | & 3 \\ 0 & 5 & 1 & | & 6 \\ 0 & 0 & 0 & | & 2 \end{bmatrix}$ (c) $\begin{bmatrix} 1 & 2 & 7 & | & 5 \\ 0 & 4 & 2 & | & 1 \\ 0 & 0 & 0 & | & 0 \end{bmatrix}$

(d) $\begin{bmatrix} 4 & 2 & 3 & | & 3 \\ 0 & 4 & 2 & | & 6 \\ 0 & 2 & 1 & | & 3 \end{bmatrix}$ (e) $\begin{bmatrix} 2 & 1 & 3 & | & 0 \\ 0 & 1 & 0 & | & 0 \\ 0 & 0 & 1 & | & 0 \end{bmatrix}$

22. For the following augmented matrices, state whether the associated system of equations has a unique solution, multiple solutions, or no solution.

(a) $\begin{bmatrix} 3 & 3 & 1 & | & 0 \\ 0 & 1 & 5 & | & 1 \end{bmatrix}$ (b) $\begin{bmatrix} 2 & 4 & 5 & 1 & | & 3 \\ 0 & 2 & 1 & 5 & | & 1 \\ 0 & 0 & 3 & 1 & | & 2 \end{bmatrix}$ (c) $\begin{bmatrix} 6 & 3 & 1 & 3 & | & 3 \\ 0 & 0 & 2 & 4 & | & 1 \\ 0 & 0 & 1 & 2 & | & 2 \end{bmatrix}$

23. Find the stable distribution (as done in Example 9) for Markov chains with the following transition matrices.

(a) $\begin{bmatrix} \frac{2}{3} & \frac{1}{3} & \frac{1}{3} \\ 0 & \frac{1}{3} & 0 \\ \frac{1}{3} & \frac{1}{3} & \frac{2}{3} \end{bmatrix}$ (b) $\begin{bmatrix} \frac{1}{2} & \frac{1}{4} & \frac{1}{4} \\ \frac{1}{4} & \frac{1}{2} & \frac{1}{4} \\ \frac{1}{4} & \frac{1}{4} & \frac{1}{2} \end{bmatrix}$ (c) $\begin{bmatrix} 0 & \frac{1}{2} & \frac{1}{2} \\ \frac{1}{2} & 0 & \frac{1}{2} \\ \frac{1}{2} & \frac{1}{2} & 0 \end{bmatrix}$

24. Find the stable distribution (as done in Example 9) for Markov chains with the following transition matrices.

(a) $\begin{bmatrix} \frac{1}{3} & \frac{1}{3} & 0 & 0 & 0 & 0 \\ \frac{2}{3} & \frac{1}{3} & \frac{1}{3} & 0 & 0 & 0 \\ 0 & \frac{1}{3} & \frac{1}{3} & \frac{1}{3} & 0 & 0 \\ 0 & 0 & \frac{1}{3} & \frac{1}{3} & \frac{1}{3} & 0 \\ 0 & 0 & 0 & \frac{1}{3} & \frac{1}{3} & \frac{2}{3} \\ 0 & 0 & 0 & 0 & \frac{1}{3} & \frac{1}{3} \end{bmatrix}$ (b) $\begin{bmatrix} \frac{2}{3} & \frac{2}{3} & 0 & 0 & 0 & 0 \\ \frac{1}{3} & \frac{1}{6} & \frac{2}{3} & 0 & 0 & 0 \\ 0 & \frac{1}{6} & \frac{1}{6} & \frac{2}{3} & 0 & 0 \\ 0 & 0 & \frac{1}{6} & \frac{1}{6} & \frac{2}{3} & 0 \\ 0 & 0 & 0 & \frac{1}{6} & \frac{1}{6} & \frac{2}{3} \\ 0 & 0 & 0 & 0 & \frac{1}{6} & \frac{1}{3} \end{bmatrix}$

(c) $\begin{bmatrix} \frac{2}{3} & \frac{2}{3} & 0 & 0 & 0 & 0 \\ \frac{1}{3} & \frac{1}{6} & \frac{2}{3} & 0 & 0 & 0 \\ 0 & \frac{1}{6} & \frac{1}{6} & \frac{1}{6} & 0 & 0 \\ 0 & 0 & \frac{1}{6} & \frac{1}{6} & \frac{1}{6} & 0 \\ 0 & 0 & 0 & \frac{2}{3} & \frac{1}{6} & \frac{1}{3} \\ 0 & 0 & 0 & 0 & \frac{2}{3} & \frac{2}{3} \end{bmatrix}$ (d) $\begin{bmatrix} 0 & \frac{1}{2} & 0 & 0 & 0 & 0 \\ 1 & 0 & \frac{1}{2} & 0 & 0 & 0 \\ 0 & \frac{1}{2} & 0 & \frac{1}{2} & 0 & 0 \\ 0 & 0 & \frac{1}{2} & 0 & \frac{1}{2} & 0 \\ 0 & 0 & 0 & \frac{1}{2} & 0 & 1 \\ 0 & 0 & 0 & 0 & \frac{1}{2} & 0 \end{bmatrix}$

25. This exercise shows how, instead of interchanging rows, one can pivot (in Gaussian elimination) on entries that do not lie on the main diagonal.

(a) In Example 5, make the second pivot at entry (3, 2) instead of interchanging rows 2 and 3. Where should the third pivot be made? Make it and give the solution.

(b) In Exercise 3(b), make the following sequence of pivots, entry (3, 1), entry (2, 2), then entry (1, 3). What sequence of elementary row operations corresponds to these pivots?

26. For what values of k does the following refinery-type system of equations have a unique solution with all x_i nonnegative?

$$6x_1 + 5x_2 + 3x_3 = 500$$
$$4x_1 + x_2 + 7x_3 = 600$$
$$5x_1 + kx_2 + 5x_3 = 1000$$

27. Write a computer program to perform Gaussian elimination on a system of n equations in n unknowns. (Watch out for 0's on the main diagonal.)

28. Write a computer program to perform Gauss–Jordan elimination on a system of n equations in n unknowns. (Watch out for 0 pivots.)

29. (*Computer Project*) For the following system of equations, pick values for some of the parameters a, b, c, d, e, f, say, a, b, c, d, and then by experimentation determine constraints on the remaining parameters that assure a solution with $x_1 \geq 0$ and $x_2 \geq 0$.

$$ax_1 + bx_2 = e$$
$$cx_1 + dx_2 = f$$

1.5 Inverse of a Matrix

In this section we study a general method for solving a system $\mathbf{Ax} = \mathbf{b}$ of n equations in n variables for any $\mathbf{b}$, instead of for one particular $\mathbf{b}$, as in Section 1.4. Our tool is the (multiplicative) inverse $\mathbf{A}^{-1}$ of a matrix $\mathbf{A}$. As noted in Example 2 of Section 1.3, the inverse $\mathbf{A}^{-1}$ of an $n \times n$ matrix $\mathbf{A}$ is an $n \times n$ matrix with the property

$$\mathbf{AA}^{-1} = \mathbf{I} \quad \text{and} \quad \mathbf{A}^{-1}\mathbf{A} = \mathbf{I}$$

where $\mathbf{I}$ is the $n \times n$ identity matrix (with 1's on the main diagonal and 0's elsewhere). Inverses allow us to "solve" a system of equations symbolically the way one solves the scalar equation $ax = b$ by dividing both sides by a to obtain $x = a^{-1}b$.

Theorem 1. If the $n \times n$ matrix $\mathbf{A}$ has an inverse $\mathbf{A}^{-1}$, the system of equations $\mathbf{Ax} = \mathbf{b}$ has the solution $\mathbf{x} = \mathbf{A}^{-1}\mathbf{b}$.
Proof: As in the scalar case, we divide both sides of $\mathbf{Ax} = \mathbf{b}$ by $\mathbf{A}$, that is, multiply both sides of the matrix equation $\mathbf{Ax} = \mathbf{b}$ by $\mathbf{A}^{-1}$:

$$\mathbf{A}^{-1}(\mathbf{Ax}) = \mathbf{A}^{-1}\mathbf{b} \tag{1}$$

Using the associative law for matrix multiplication and the fact that $\mathbf{A}^{-1}\mathbf{A} = \mathbf{I}$, we can rewrite the left side of (1) as follows:

$$\mathbf{A}^{-1}(\mathbf{Ax}) = (\mathbf{A}^{-1}\mathbf{A})\mathbf{x} = \mathbf{Ix} = \mathbf{x} \tag{2}$$

(Recall that the identity matrix $\mathbf{I}$ times any matrix or vector just equals that matrix or vector; see Example 2 of Section 1.3.) Combining (1) and (2), we have the desired result: $\mathbf{x} = \mathbf{A}^{-1}\mathbf{b}$. ■

Reviewing the proof in Theorem 1, we first multiply both sides of matrix equation $\mathbf{Ax} = \mathbf{b}$ by $\mathbf{A}^{-1}$ and then rewrite the matrix expression $\mathbf{A}^{-1}(\mathbf{Ax})$ using three matrix facts: (1) the associative law for matrix multiplication, (2) $\mathbf{A}^{-1}\mathbf{A} = \mathbf{I}$, and (3) $\mathbf{Ix} = \mathbf{x}$. With these steps we were able to derive a general matrix formula for solving systems of equations. We do not know how to compute inverses, yet we already proved a very important result about inverses. This is the sort of powerful, yet simple-to-obtain result that characterizes linear algebra theory. (The crucial step in the proof is the associative law; given the complicated nature of matrix multiplication, it is far from clear that matrix multiplication should be associative.)

A matrix $\mathbf{A}$ is said to be *invertible* if it has an inverse. In many books, the term *nonsingular* is used instead of invertible (a *singular* matrix has no inverse). Some matrices are invertible and some are not. Much of the theory of linear algebra centers around conditions that will make a matrix invertible.

We will show shortly how to calculate inverses, when they exist. First let us look at three matrices, one with an inverse, one without an inverse, and one with a partial inverse.

EXAMPLE 1.
Matrix with an
Inverse

Matrix $\mathbf{A} = \begin{bmatrix} 3 & 1 \\ 4 & 2 \end{bmatrix}$ has the inverse $\mathbf{A}^{-1} = \begin{bmatrix} 1 & -\frac{1}{2} \\ -2 & \frac{3}{2} \end{bmatrix}$ (For the present, do not worry how this inverse was found.) Multiplying $\mathbf{A}$ times $\mathbf{A}^{-1}$, we have

$$\mathbf{A}\mathbf{A}^{-1} = \begin{bmatrix} 3 & 1 \\ 4 & 2 \end{bmatrix}\begin{bmatrix} 1 & -\frac{1}{2} \\ -2 & \frac{3}{2} \end{bmatrix} = \begin{bmatrix} 3 \cdot 1 + 1 \cdot (-2) & 3 \cdot (-\frac{1}{2}) + 1 \cdot \frac{3}{2} \\ 4 \cdot 1 + 2 \cdot (-2) & 4 \cdot (-\frac{1}{2}) + 2 \cdot \frac{3}{2} \end{bmatrix} = \begin{bmatrix} 1 & 0 \\ 0 & 1 \end{bmatrix}$$

The reader can verify that if $\mathbf{A}^{-1}$ precedes $\mathbf{A}$, again $\mathbf{A}^{-1}\mathbf{A} = \mathbf{I}$. ■

EXAMPLE 2.
Matrices Without
Inverses

(a) We claim that the matrix $\mathbf{B} = \begin{bmatrix} 1 & 4 \\ 2 & 8 \end{bmatrix}$ has no inverse. The key to our claim is the observation that the second row of $\mathbf{B}$ is twice the first row.

$$\mathbf{b}_2' = 2\mathbf{b}_1' \tag{3}$$

where $\mathbf{b}_i'$ denotes the ith row of $\mathbf{B}$. Suppose that $\mathbf{C}$ were the inverse of $\mathbf{B}$, so that $\mathbf{BC} = \mathbf{I} = \begin{bmatrix} 1 & 0 \\ 0 & 1 \end{bmatrix}$. If $\mathbf{c}_1$ and $\mathbf{c}_2$ are the two columns of $\mathbf{C}$, the matrix product $\mathbf{BC}$ is the following collection of scalar products:

$$\mathbf{BC} = \begin{bmatrix} \mathbf{b}_1' \cdot \mathbf{c}_1 & \mathbf{b}_1' \cdot \mathbf{c}_2 \\ \mathbf{b}_2' \cdot \mathbf{c}_1 & \mathbf{b}_2' \cdot \mathbf{c}_2 \end{bmatrix} = \begin{bmatrix} 1 & 0 \\ 0 & 1 \end{bmatrix} \tag{4}$$

From (3), $\mathbf{b}_2' \cdot \mathbf{c}_i = 2\mathbf{b}_1' \cdot \mathbf{c}_i$, for $i = 1, 2$. So row two of $\mathbf{BC}$, $[\mathbf{b}_2' \cdot \mathbf{c}_1, \mathbf{b}_2' \cdot \mathbf{c}_2]$, must be twice row one, $[\mathbf{b}_1' \cdot \mathbf{c}_1, \mathbf{b}_1' \cdot \mathbf{c}_2]$. But the second row of $\mathbf{I}$ is obviously not twice its first row. This contradiction shows that no inverse can exist.

(b) By the definition of an inverse, a matrix must be square to have an inverse. But we claim that the matrix $\mathbf{C} = \begin{bmatrix} -2 & 1 \\ 0 & 1 \\ 2 & 1 \end{bmatrix}$ has a partial (left-sided) inverse

$$\mathbf{C}^+ = \begin{bmatrix} -\frac{1}{4} & 0 & \frac{1}{4} \\ \frac{1}{3} & \frac{1}{3} & \frac{1}{3} \end{bmatrix}:$$

$$\mathbf{C}^+\mathbf{C} = \begin{bmatrix} -\frac{1}{4} & 0 & \frac{1}{4} \\ \frac{1}{3} & \frac{1}{3} & \frac{1}{3} \end{bmatrix} \begin{bmatrix} -2 & 1 \\ 0 & 1 \\ 2 & 1 \end{bmatrix} = \begin{bmatrix} 1 & 0 \\ 0 & 1 \end{bmatrix} = \mathbf{I}$$

On the other hand,

$$\mathbf{C}\mathbf{C}^+ = \begin{bmatrix} -2 & 1 \\ 0 & 1 \\ 2 & 1 \end{bmatrix} \begin{bmatrix} -\frac{1}{4} & 0 & \frac{1}{4} \\ \frac{1}{3} & \frac{1}{3} & \frac{1}{3} \end{bmatrix} = \begin{bmatrix} \frac{5}{6} & \frac{1}{3} & -\frac{1}{6} \\ \frac{1}{3} & \frac{1}{3} & \frac{1}{3} \\ -\frac{1}{6} & \frac{1}{3} & \frac{5}{6} \end{bmatrix}$$

The matrix $\mathbf{C}^+$ is called the *pseudoinverse* of $\mathbf{C}$ and has an important role in many applications (discussed in Section 3.5). We note, however, that for no 3×2 matrix $\mathbf{C}$ can there exist a 2×3 matrix $\mathbf{D}$ such that $\mathbf{CD} = \mathbf{I}$. Facts like this will follow immediately from general theory about linear transformations that will be developed in Chapter 4. ■

The first part of Example 2 shows that if one row of $\mathbf{A}$ is a multiple of another row, no inverse can exist.

If a matrix is square, is a left-sided inverse (e.g., $\mathbf{C}^+$ in Example 2) also a right-sided inverse, and vice versa? The answer is yes, so we can speak of the inverse of a square matrix without worrying about left and right. An argument showing that a left inverse is a right inverse will be given shortly.

Next we ask, if an inverse exists for a matrix $\mathbf{A}$, is it unique? Before we can look for a way to compute the inverse of a matrix, we should know the answer to this question.

Theorem 2. If an $n \times n$ matrix $\mathbf{A}$ has an inverse $\mathbf{A}^{-1}$, this inverse must be unique.
Proof: Suppose that $\mathbf{Q}$ is another matrix with the property that $\mathbf{AQ} = \mathbf{QA} = \mathbf{I}$. Then by the associative law for matrix algebra, we can compute the triple product $\mathbf{QAA}^{-1}$ in two ways:

$$\mathbf{QAA}^{-1} = \mathbf{Q}(\mathbf{AA}^{-1}) = \mathbf{QI} = \mathbf{Q} \quad \text{and} \quad \mathbf{QAA}^{-1} = (\mathbf{QA})\mathbf{A}^{-1} = \mathbf{IA}^{-1} = \mathbf{A}^{-1}$$

Thus we conclude that $\mathbf{Q} = \mathbf{A}^{-1}$. ■

Note the critical role again played by the associative law in the proof of Theorem 2. The following example shows how to compute the inverse of a 2×2 matrix.

EXAMPLE 3.
Computing Inverse
of a 2 × 2 Matrix

Consider the 2 × 2 matrix **A** and its (unknown) inverse **X**:

$$\mathbf{A} = \begin{bmatrix} 3 & 1 \\ 4 & 2 \end{bmatrix} \qquad \mathbf{X} = \begin{bmatrix} x_{11} & x_{12} \\ x_{21} & x_{22} \end{bmatrix}$$

We require that **AX** = **I**:

$$\mathbf{AX} = \begin{bmatrix} 3 & 1 \\ 4 & 2 \end{bmatrix}\begin{bmatrix} x_{11} & x_{12} \\ x_{21} & x_{22} \end{bmatrix} = \begin{bmatrix} 1 & 0 \\ 0 & 1 \end{bmatrix} = \mathbf{I} \tag{5}$$

We determine **X** (= $\mathbf{A}^{-1}$) one column at a time from (5). We find the first column $\mathbf{x}_1 = \begin{bmatrix} x_{11} \\ x_{21} \end{bmatrix}$ of matrix **X** by setting **Ax₁**—the first column in the product **AX** of (5)—equal to the $\mathbf{i}_1$, the first column of **I**:

$$\mathbf{Ax}_1 = \mathbf{i}_1\colon \quad \begin{bmatrix} 3 & 1 \\ 4 & 2 \end{bmatrix}\begin{bmatrix} x_{11} \\ x_{21} \end{bmatrix} = \begin{bmatrix} 1 \\ 0 \end{bmatrix}$$

or

$$\begin{aligned} 3x_{11} + x_{21} &= 1 \\ 4x_{11} + 2x_{21} &= 0 \end{aligned} \tag{6a}$$

Similarly, the second column of (5) yields the system

$$\begin{aligned} 3x_{12} + x_{22} &= 0 \\ 4x_{12} + 2x_{22} &= 1 \end{aligned} \tag{6b}$$

Using Gauss–Jordan elimination on the augmented coefficient matrix for (6a), we obtain

$$\begin{bmatrix} 3 & 1 & | & 1 \\ 4 & 2 & | & 0 \end{bmatrix} \Longrightarrow \begin{bmatrix} 1 & \frac{1}{3} & | & \frac{1}{3} \\ 0 & \frac{2}{3} & | & -\frac{4}{3} \end{bmatrix} \Longrightarrow \begin{bmatrix} 1 & 0 & | & 1 \\ 0 & 1 & | & -2 \end{bmatrix} \tag{7a}$$

so $x_{11} = 1$, $x_{21} = -2$. For (6b), we obtain

$$\begin{bmatrix} 3 & 1 & | & 0 \\ 4 & 2 & | & 1 \end{bmatrix} \Longrightarrow \begin{bmatrix} 1 & \frac{1}{3} & | & 0 \\ 0 & \frac{2}{3} & | & 1 \end{bmatrix} \Longrightarrow \begin{bmatrix} 1 & 0 & | & -\frac{1}{2} \\ 0 & 1 & | & \frac{3}{2} \end{bmatrix} \tag{7b}$$

so $x_{12} = -\frac{1}{2}$, $x_{22} = \frac{3}{2}$. Substituting these values for x_{ij} back into **X** (= $\mathbf{A}^{-1}$), we have

$$\mathbf{X} = \begin{bmatrix} 1 & -\frac{1}{2} \\ -2 & \frac{3}{2} \end{bmatrix}. \qquad ∎$$

The method in Example 3 can be extended to determine the inverse, when it exists, for a matrix of any size. The right-side vectors in (6a) and (6b) will become the columns of the appropriate-sized identity matrix $\mathbf{I}$, column vectors with a 1 in one position and 0's in all the other positions. The column vectors $\mathbf{i}_1, \mathbf{i}_2, \mathbf{i}_3, \ldots$ of $\mathbf{I}$ are coordinate vectors.

Theorem 3. Let $\mathbf{A}$ be an $n \times n$ matrix and $\mathbf{i}_1, \mathbf{i}_2, \mathbf{i}_3, \ldots$ be the coordinate n-vectors (the columns of the $n \times n$ identity matrix $\mathbf{I}$). Let the n-vectors $\mathbf{x}_1, \mathbf{x}_2, \mathbf{x}_3, \ldots,$ $\mathbf{x}_n$ be the solutions to

$$\mathbf{A}\mathbf{x}_j = \mathbf{i}_j$$

for $j = 1, 2, \ldots, n$. Then the $n \times n$ matrix $\mathbf{X}$ with column vectors $\mathbf{x}_j$ is the inverse of $\mathbf{A}$.

$$\mathbf{A}^{-1} = \mathbf{X} = [\mathbf{x}_1, \mathbf{x}_2, \ldots, \mathbf{x}_n] \tag{8}$$

Note: *If the systems* $\mathbf{A}\mathbf{x}_j = \mathbf{i}_j$ *do not all have solutions, then* $\mathbf{A}$ *does not have an inverse.*

Recall that when we solve a system of equations by elimination, the right sides play a passive role. That is, using a different right side $\mathbf{b}$ does not change any of the calculations involving the coefficients. It affects only the final values that appear on the right side. Thus when we performed Gauss–Jordan elimination on the coefficient matrix in (7a) and (7b) of Example 3, we could have simultaneously applied the elimination steps to an augmented coefficient matrix $[\mathbf{A} \quad \mathbf{I}]$ that contained both right-side vectors. The computations would be

$$\begin{bmatrix} 3 & 1 & | & 1 & 0 \\ 4 & 2 & | & 0 & 1 \end{bmatrix} \Longrightarrow \begin{bmatrix} 1 & \frac{1}{3} & | & \frac{1}{3} & 0 \\ 0 & \frac{2}{3} & | & -\frac{4}{3} & 1 \end{bmatrix} \Longrightarrow \begin{bmatrix} 1 & 0 & | & 1 & -\frac{1}{2} \\ 0 & 1 & | & -2 & \frac{3}{2} \end{bmatrix} \tag{9}$$

So starting with $[\mathbf{A} \quad \mathbf{I}]$, we apply Gauss–Jordan elimination to put the matrix in reduced row echelon form $[\mathbf{I} \quad \mathbf{A}^{-1}]$.

Corollary A. The right inverse $\mathbf{A}^{-1}$ of $\mathbf{A}$, given by (8), is also a left inverse of $\mathbf{A}$.
Proof: The elimination process from $[\mathbf{A} \quad \mathbf{I}]$ to $[\mathbf{I} \quad \mathbf{A}^{-1}]$ is reversible (see proof of Theorem 1 of Section 1.4). But by switching $[\mathbf{I} \quad \mathbf{A}^{-1}]$ to $[\mathbf{A}^{-1} \quad \mathbf{I}]$ (rearranging columns), the reversal takes $[\mathbf{A}^{-1} \quad \mathbf{I}]$ to $[\mathbf{I} \quad \mathbf{A}]$, so $\mathbf{A}$ is the right inverse of $\mathbf{A}^{-1}$. For example, applying this reversal to the elimination process in (9), we have (reading from right to left)

$$[\mathbf{I} \quad \mathbf{A}]\begin{bmatrix} 1 & 0 & | & 3 & 1 \\ 0 & 1 & | & 4 & 2 \end{bmatrix} \Longleftarrow \begin{bmatrix} \frac{1}{3} & 0 & | & 1 & \frac{1}{3} \\ -\frac{4}{3} & 1 & | & 0 & \frac{2}{3} \end{bmatrix} \Longleftarrow \begin{bmatrix} 1 & -\frac{1}{2} & | & 1 & 0 \\ -2 & \frac{3}{2} & | & 0 & 1 \end{bmatrix} = [\mathbf{A}^{-1} \quad \mathbf{I}] \tag{9a}$$

If $\mathbf{A}$ is the right inverse of $\mathbf{A}^{-1}$, that is, $\mathbf{A}^{-1}\mathbf{A} = \mathbf{I}$, then $\mathbf{A}^{-1}$ is the left inverse of $\mathbf{A}$. ■

Corollary B. If the $n \times n$ matrix $\mathbf{A}$ is invertible, then $\mathbf{A}$ is the product of elementary matrices.

Proof: Recall that premultiplying a matrix by an elementary matrix is equivalent to performing an elementary row operation on the matrix (see the discussion preceding Example 2 in Section 1.4). If $\mathbf{A}$ is invertible, it has an inverse $\mathbf{A}^{-1}$, and Gauss–Jordan elimination will reduce $[\mathbf{A}^{-1} \quad \mathbf{I}]$ to $[\mathbf{I} \quad \mathbf{A}]$. The sequence of elementary row operations in this Gauss–Jordan elimination is equivalent to successively premultiplying $[\mathbf{A}^{-1} \quad \mathbf{I}]$ by a set of elementary matrices. The product $\mathbf{E}^*$ of these elementary matrices satisfies the equation $\mathbf{E}^*[\mathbf{A}^{-1} \quad \mathbf{I}] = [\mathbf{I} \quad \mathbf{A}]$. Thus $\mathbf{E}^*\mathbf{A}^{-1} = \mathbf{I}$, so $\mathbf{E}^* = \mathbf{A}$, since $\mathbf{A}$ is the unique inverse of $\mathbf{A}^{-1}$. ■

For example, in (9a) going from the right (augmented) matrix to the middle matrix is accomplished by premultiplying first by the elementary matrix $\mathbf{E}_1 = \begin{bmatrix} 1 & \frac{1}{3} \\ 0 & 1 \end{bmatrix}$ to add $\frac{1}{3}$ of the second row to the first row and then by premultiplying by $\mathbf{E}_2 = \begin{bmatrix} 1 & 0 \\ 0 & \frac{2}{3} \end{bmatrix}$ to multiply the second row by $\frac{2}{3}$. Going from the middle matrix to the left matrix is accomplished by premultiplying first by $\mathbf{E}_3 = \begin{bmatrix} 1 & 0 \\ 4 & 1 \end{bmatrix}$ to add 4 times the first row to the second row and then by premultiplying by $\mathbf{E}_4 = \begin{bmatrix} 3 & 0 \\ 0 & 1 \end{bmatrix}$ to multiply the first row by 3. Then by the argument in the proof of Corollary B, $\mathbf{A} = \mathbf{E}_4\mathbf{E}_3\mathbf{E}_2\mathbf{E}_1$ (the reader can verify this). Theorem 3 gives us the following procedure for computing inverses.

Computation of the Inverse of an $n \times n$ Matrix A

Pivot on entries $(1, 1), (2, 2), \ldots, (n, n)$ in the augmented matrix $[\mathbf{A} \quad \mathbf{I}]$. The resulting reduced row echelon form will be $[\mathbf{I} \quad \mathbf{A}^{-1}]$.

EXAMPLE 4.
Inverse of a 3 × 3 Matrix

Let $\mathbf{A} = \begin{bmatrix} 1 & 0 & 2 \\ 2 & 4 & 2 \\ 1 & 2 & 6 \end{bmatrix}$. Then we compute the inverse by pivoting along the main diagonal of the augmented matrix $[\mathbf{A} \quad \mathbf{I}]$.

$$[\mathbf{A} \quad \mathbf{I}] = \begin{matrix} (a) \\ (b) \\ (c) \end{matrix} \left[\begin{array}{ccc|ccc} 1 & 0 & 2 & 1 & 0 & 0 \\ 2 & 4 & 2 & 0 & 1 & 0 \\ 1 & 2 & 6 & 0 & 0 & 1 \end{array}\right]$$

$$
\begin{array}{l}
(a') = (a) \\
(b') = (b') - 2(a) \\
(c') = (c) - (a)
\end{array}
\quad
\left[\begin{array}{ccc|ccc}
1 & 0 & 2 & 1 & 0 & 0 \\
0 & 4 & -2 & -2 & 1 & 0 \\
0 & 2 & 4 & -1 & 0 & 1
\end{array}\right]
$$

$$
\begin{array}{l}
(a'') = (a') \\
(b'') = (b)/4 \\
(c'') = (c') - 2(b'')
\end{array}
\quad
\left[\begin{array}{ccc|ccc}
1 & 0 & 2 & 1 & 0 & 0 \\
0 & 1 & -\frac{1}{2} & -\frac{1}{2} & \frac{1}{4} & 0 \\
0 & 0 & 5 & 0 & -\frac{1}{2} & 1
\end{array}\right]
$$

$$
\begin{array}{l}
(a''') = (a'') - 2(c''') \\
(b''') = (b'') + \frac{1}{2}(c''') \\
(c''') = (c'')/5
\end{array}
\quad
\left[\begin{array}{ccc|ccc}
1 & 0 & 0 & 1 & \frac{1}{5} & -\frac{2}{5} \\
0 & 1 & 0 & -\frac{1}{2} & \frac{1}{5} & \frac{1}{10} \\
0 & 0 & 1 & 0 & -\frac{1}{10} & \frac{1}{5}
\end{array}\right]
$$

(10)

Thus

$$
\mathbf{A}^{-1} =
\begin{bmatrix}
1 & \frac{1}{5} & -\frac{2}{5} \\
-\frac{1}{2} & \frac{1}{5} & \frac{1}{10} \\
0 & -\frac{1}{10} & \frac{1}{5}
\end{bmatrix}
$$

If the entry (3, 3) in **A** were 1 instead of 6, yielding $\mathbf{A'} = \begin{bmatrix} 1 & 0 & 2 \\ 2 & 4 & 2 \\ 1 & 2 & 1 \end{bmatrix}$, then

after the first two pivots in (10), we would get

$$
\left[\begin{array}{ccc|ccc}
1 & 0 & 2 & 1 & 0 & 0 \\
0 & 1 & -\frac{1}{2} & -\frac{1}{2} & \frac{1}{4} & 0 \\
0 & 0 & 0 & 0 & -\frac{1}{2} & 1
\end{array}\right]
$$

The 0's on the left side of the last row indicate that Gauss–Jordan elimination fails and that **A'** has no inverse. ■

EXAMPLE 5.
Use of Inverse in
Multiple Right-
Hand Sides

In Example 4 of Section 1.4, we solved the refinery system of equations by Gauss–Jordan elimination. Let us use the same sequence of pivots with the augmented matrix [**A   I**] to compute the inverse.

$$
\begin{array}{l}
(a) \\
(b) \\
(c)
\end{array}
\left[\begin{array}{ccc|ccc}
4 & 2 & 2 & 1 & 0 & 0 \\
2 & 5 & 2 & 0 & 1 & 0 \\
1 & 2.5 & 5 & 0 & 0 & 1
\end{array}\right]
$$

$$
\begin{array}{l}
(a') = \tfrac{1}{4}(a) \\
(b') = (b) - \tfrac{1}{2}(a) \\
(c') = (c) - \tfrac{1}{4}(a)
\end{array}
\left[
\begin{array}{ccc|ccc}
1 & \tfrac{1}{2} & \tfrac{1}{2} & \tfrac{1}{4} & 0 & 0 \\
0 & 4 & 1 & -\tfrac{1}{2} & 1 & 0 \\
0 & 2 & 4\tfrac{1}{2} & -\tfrac{1}{4} & 0 & 1
\end{array}
\right]
$$

$$
\begin{array}{l}
(a'') = (a') - \tfrac{1}{2}(b) \\
(b'') = (b')/4 \\
(c'') = (c') - 2(b'')
\end{array}
\left[
\begin{array}{ccc|ccc}
1 & 0 & \tfrac{3}{8} & \tfrac{5}{16} & -\tfrac{1}{8} & 0 \\
0 & 1 & \tfrac{1}{4} & -\tfrac{1}{8} & \tfrac{1}{4} & 0 \\
0 & 0 & 4 & 0 & -\tfrac{1}{2} & 1
\end{array}
\right]
$$

$$
\begin{array}{l}
(a''') = (a'') - \tfrac{3}{8}(c''') \\
(b''') = (b'') - \tfrac{1}{4}(c''') \\
(c''') = (c'')/4
\end{array}
\left[
\begin{array}{ccc|ccc}
1 & 0 & 0 & \tfrac{5}{16} & -\tfrac{5}{64} & -\tfrac{3}{32} \\
0 & 1 & 0 & -\tfrac{1}{8} & \tfrac{9}{32} & -\tfrac{1}{16} \\
0 & 0 & 1 & 0 & -\tfrac{1}{8} & \tfrac{1}{4}
\end{array}
\right]
$$

(11)

The inverse is thus

$$
\mathbf{A}^{-1} =
\begin{bmatrix}
\tfrac{5}{16} & -\tfrac{5}{64} & -\tfrac{3}{32} \\
-\tfrac{1}{8} & \tfrac{9}{32} & -\tfrac{1}{16} \\
0 & -\tfrac{1}{8} & \tfrac{1}{4}
\end{bmatrix}
\tag{12}
$$

If we were given a new right-hand-side vector for the refinery system, say $\mathbf{b}^* = \begin{bmatrix} 200 \\ 200 \\ 100 \end{bmatrix}$, the new solution can be obtained by computing $\mathbf{x}^* = \mathbf{A}^{-1}\mathbf{b}^*$.

$$
\begin{aligned}
\mathbf{x}^* = \mathbf{A}^{-1}\mathbf{b}^* &=
\begin{bmatrix}
\tfrac{5}{16} & -\tfrac{5}{64} & -\tfrac{3}{32} \\
-\tfrac{1}{8} & \tfrac{9}{32} & -\tfrac{1}{16} \\
0 & -\tfrac{1}{8} & \tfrac{1}{4}
\end{bmatrix}
\begin{bmatrix} 200 \\ 200 \\ 100 \end{bmatrix} \\
&=
\begin{bmatrix}
\tfrac{5}{16}\cdot 200 - \tfrac{5}{64}\cdot 200 - \tfrac{3}{32}\cdot 100 \\
-\tfrac{1}{8}\cdot 200 + \tfrac{9}{32}\cdot 200 - \tfrac{1}{16}\cdot 100 \\
0\cdot 200 - \tfrac{1}{8}\cdot 200 + \tfrac{1}{4}\cdot 100
\end{bmatrix}
=
\begin{bmatrix} 37\tfrac{1}{2} \\ 25 \\ 0 \end{bmatrix}
\end{aligned}
\tag{13}
$$

Next consider what happens to a solution $\mathbf{x}$ when we change the right-side $\mathbf{b}$ a little, say, we increase the first component (heating oil) by 1 gallon. So $\mathbf{b}$ changes to $\mathbf{b} + \Delta\mathbf{b}$, where $\Delta\mathbf{b} = \begin{bmatrix} 1 \\ 0 \\ 0 \end{bmatrix}$. Then we claim that the solution will change by

$$
\Delta\mathbf{x} = \mathbf{A}^{-1}\,\Delta\mathbf{b} = \mathbf{A}^{-1}\begin{bmatrix} 1 \\ 0 \\ 0 \end{bmatrix} = (\mathbf{A}^{-1})_1; \quad \text{in this problem,} \quad (\mathbf{A}^{-1})_1 = \begin{bmatrix} \tfrac{5}{16} \\ -\tfrac{1}{8} \\ 0 \end{bmatrix}
\tag{14}
$$

where $(\mathbf{A}^{-1})_1$ denotes the first column of $\mathbf{A}^{-1}$. Recall that by Example 4 of Section 1.2, for any matrix $\mathbf{C}$, $\mathbf{C}i_1 = \mathbf{c}_1$ (the first column of $\mathbf{C}$). To verify that (14) is the change in the solution, we observe that $\mathbf{x}$ $(= \mathbf{A}^{-1}\mathbf{b})$ changes to

$$\mathbf{A}^{-1}(\mathbf{b} + \Delta\mathbf{b}) = \text{(by linearity) } \mathbf{A}^{-1}\mathbf{b} + \mathbf{A}^{-1}\Delta\mathbf{b} = \mathbf{x} + \Delta\mathbf{x} \tag{15}$$

Thus the change $\Delta\mathbf{x}$ in the solution to produce 1 more unit of the first product is the first column of $\mathbf{A}$, $(\mathbf{A}^{-1})_1$. Similarly, the second and third columns of $\mathbf{A}^{-1}$ tell how the solution will change if we need 1 more unit of the second or third product. In sum, the columns of $\mathbf{A}^{-1}$ show us how the solution $\mathbf{x}$ changes when the right-side vector $\mathbf{b}$ changes.

As a specific example, let us consider how our solution $\mathbf{x}^*$ in (13) changes when we change from

$$\mathbf{b}^* = \begin{bmatrix} 200 \\ 200 \\ 100 \end{bmatrix} \text{ to } \mathbf{b}^\circ = \begin{bmatrix} 400 \\ 200 \\ 100 \end{bmatrix}$$

We take the solution

$$\mathbf{x}^* = \mathbf{A}^{-1}\mathbf{b}^* = \begin{bmatrix} 37.5 \\ 25 \\ 0 \end{bmatrix}$$

computed in (13) for $\mathbf{b}^*$ and change it by $\Delta\mathbf{x} = \mathbf{A}^{-1}\Delta\mathbf{b}$, where $\Delta\mathbf{b} = \mathbf{b}^\circ - \mathbf{b}^*$

$= \begin{bmatrix} 200 \\ 0 \\ 0 \end{bmatrix}$. By (14), $\Delta\mathbf{x}$ will be $200(\mathbf{A}^{-1})_1$, 200 times the first column of $\mathbf{A}^{-1}$.

$$\mathbf{x}^\circ = \mathbf{A}^{-1}\begin{bmatrix} 400 \\ 200 \\ 100 \end{bmatrix} = \mathbf{A}^{-1}\begin{bmatrix} 200 \\ 200 \\ 100 \end{bmatrix} + \mathbf{A}^{-1}\begin{bmatrix} 200 \\ 0 \\ 0 \end{bmatrix} = \begin{bmatrix} 37.5 \\ 25 \\ 0 \end{bmatrix} + 200\begin{bmatrix} -\frac{5}{16} \\ -\frac{1}{8} \\ 0 \end{bmatrix}$$

$$= \begin{bmatrix} 37.5 \\ 25 \\ 0 \end{bmatrix} + \begin{bmatrix} 62.5 \\ -25 \\ 0 \end{bmatrix} = \begin{bmatrix} 100 \\ 0 \\ 0 \end{bmatrix} \tag{16}$$

Observe how the linearity of the matrix–vector products, in this case, $\mathbf{A}^{-1}(\mathbf{b} + \Delta\mathbf{b}) = \mathbf{A}^{-1}\mathbf{b} + \mathbf{A}^{-1}\Delta\mathbf{b}$, makes it easy to adjust a solution for changes in the right side. The analysis of the effects of small changes in demand is of crucial importance in many economic applications. This subject is called *sensitivity analysis* in economics textbooks.

We now develop some theory about inverses.

Theorem 4   (Fundamental Theorem for Solving Ax = b).

The following three statements are equivalent for any $n \times n$ matrix $\mathbf{A}$.
  (i) For some particular vector $\mathbf{b}$, the system of equations $\mathbf{Ax} = \mathbf{b}$ has a unique solution.
  (ii) For all $\mathbf{b}$, the system of equations $\mathbf{Ax} = \mathbf{b}$ always has a unique solution.
  (iii) $\mathbf{A}$ has an inverse.

Proof: (i) $\Rightarrow$ (ii). As noted in Theorem 2 of Section 1.4, when Gaussian elimination produces a unique solution, the outcome depends only on the coefficients of $\mathbf{A}$, not on $\mathbf{b}$. Thus if Gaussian elimination produces a unique solution for a particular $\mathbf{b}$, it will produce a unique solution for all $\mathbf{b}$.

(ii) $\Rightarrow$ (iii). By Theorem 3, the inverse is found by solving $\mathbf{Ax} = \mathbf{i}_j$, for $j = 1, 2, \ldots, n$.

(iii) $\Rightarrow$ (i). If $\mathbf{A}^{-1}$ exists, then by Theorems 1 and 2, there is a unique solution of $\mathbf{x} = \mathbf{A}^{-1}\mathbf{b}$. ■

We have the following useful corollary.

Corollary. If for some $\mathbf{b}$, the system of equations $\mathbf{Ax} = \mathbf{b}$ has two solutions, $\mathbf{A}$ is not invertible.

Proof: If (i) in Theorem 4 is false, equivalent statement (iii) must also be false. ■

Note that if a matrix $\mathbf{A}$ does *not* have an inverse, Gauss–Jordan elimination will fail to reduce $[\mathbf{A} \quad \mathbf{I}]$ to $[\mathbf{I} \quad \mathbf{A}^{-1}]$. This is the fastest way in general to determine if $\mathbf{A}^{-1}$ exists.

Let us now present some of the laws governing matrix inverses.

Theorem 5. Let $\mathbf{A}$ be an $n \times n$ invertible matrix. Then
  (i) $\mathbf{A}^{-1}$ is invertible and $(\mathbf{A}^{-1})^{-1} = \mathbf{A}$.
 (ii) $(r\mathbf{A})^{-1} = \frac{1}{r}\mathbf{A}^{-1}$.
(iii) $\mathbf{A}^k$ is invertible and $(\mathbf{A}^k)^{-1} = (\mathbf{A}^{-1})^k$.
 (iv) If $\mathbf{B}$ is an $n \times n$ invertible matrix, then $\mathbf{AB}$ is invertible and $(\mathbf{AB})^{-1} = \mathbf{B}^{-1}\mathbf{A}^{-1}$.

Proof: (i) Since $\mathbf{AA}^{-1} = \mathbf{I} = \mathbf{A}^{-1}\mathbf{A}$, then $\mathbf{A}$ is the inverse of $\mathbf{A}^{-1}$.

(ii) We verify that $\frac{1}{r}\mathbf{A}^{-1}$ is the inverse of $r\mathbf{A}$: $r\mathbf{A}(\frac{1}{r}\mathbf{A}^{-1}) = r\frac{1}{r}(\mathbf{AA}^{-1}) = \mathbf{AA}^{-1} = \mathbf{I}$.

(iii) Since $\mathbf{A}^n = \mathbf{AA} \ldots$ (*n* times) $\ldots \mathbf{A}$, and $(\mathbf{A}^{-1})^n = \mathbf{A}^{-1}\mathbf{A}^{-1}\cdots\mathbf{A}^{-1}$, then

$$\mathbf{A}^n(\mathbf{A}^{-1})^n = (\mathbf{AA}\cdots\mathbf{A})(\mathbf{A}^{-1}\mathbf{A}^{-1}\cdots\mathbf{A}^{-1}) = \mathbf{A}\cdot\mathbf{A}\cdots\mathbf{AA}^{-1}\mathbf{A}^{-1}\cdots\mathbf{A}^{-1} = \mathbf{I}$$

(iv) We verify that $\mathbf{B}^{-1}\mathbf{A}^{-1}$ is the inverse of $\mathbf{AB}$: $(\mathbf{AB})(\mathbf{B}^{-1}\mathbf{A}^{-1}) = \mathbf{A}(\mathbf{BB}^{-1})\mathbf{A}^{-1} = \mathbf{AA}^{-1} = \mathbf{I}$. ■

If we have to solve a system of equations $\mathbf{Ax} = \mathbf{b}$ for many different right-hand sides, it is useful to know $\mathbf{A}^{-1}$. For each new $\mathbf{b}^*$, we find the solution of $\mathbf{Ax} = \mathbf{b}^*$ as $\mathbf{x} = \mathbf{A}^{-1}\mathbf{b}^*$. However, it is only fair to note that the inverse is very rarely used in large real-world problems, because it requires more computation than Gaussian elimination, thereby offering more chances for roundoff errors to occur.

We conclude this section with the two examples applying the inverse in two linear models, one old and one new.

EXAMPLE 6.
Reversing a
Markov Chain

The transition matrix **A** in a Markov chain is used to compute the probability distribution **p′** in the next period from the current probability distribution **p** according to the matrix equation

$$\mathbf{p'} = \mathbf{Ap} \tag{17}$$

Suppose that we want to run the Markov chain backwards—earlier in time—so that the relation in (17) becomes reversed and **p′** is used to determine **p**. Then the new transition matrix should be $\mathbf{A}^{-1}$, since solving (17) for **p** yields $\mathbf{p} = \mathbf{A}^{-1}\mathbf{p'}$.

We now try to invert the weather Markov chain introduced in Section 1.1. Starting with the transition matrix augmented by the identity matrix, we reduce this augmented matrix to row echelon form.

$$
\left[\begin{array}{ccc|ccc}
\frac{3}{4} & \frac{1}{2} & \frac{1}{4} & 1 & 0 & 0 \\
\frac{1}{8} & \frac{1}{4} & \frac{1}{2} & 0 & 1 & 0 \\
\frac{1}{8} & \frac{1}{4} & \frac{1}{4} & 0 & 0 & 1
\end{array}\right]
\Longrightarrow
\left[\begin{array}{ccc|ccc}
1 & \frac{2}{3} & \frac{1}{3} & \frac{4}{3} & 0 & 0 \\
0 & \frac{1}{6} & \frac{11}{24} & -\frac{1}{6} & 1 & 0 \\
0 & \frac{1}{6} & \frac{5}{24} & -\frac{1}{6} & 0 & 1
\end{array}\right]
$$

$$
\Longrightarrow
\left[\begin{array}{ccc|ccc}
1 & 0 & \frac{1}{4} & 2 & -4 & 0 \\
0 & 1 & \frac{11}{4} & -1 & 6 & 0 \\
0 & 1 & -\frac{1}{4} & 0 & -1 & 1
\end{array}\right]
\Longrightarrow
\left[\begin{array}{ccc|ccc}
1 & 0 & 0 & 2 & 2 & -6 \\
0 & 1 & 0 & 1 & -5 & 11 \\
0 & 0 & 1 & 0 & 4 & -4
\end{array}\right] \tag{18}
$$

$$
\Longrightarrow \mathbf{A}^{-1} = \begin{bmatrix}
2 & 2 & -6 \\
-1 & -5 & 11 \\
0 & 4 & -4
\end{bmatrix}
$$

The inverse exists, but it has negative entries. This can lead to some paradoxical situations. For example, if we knew that tomorrow there will be a 50–50 chance of rain or sun, so that $\mathbf{p'} = \begin{bmatrix} \frac{1}{2} \\ 0 \\ \frac{1}{2} \end{bmatrix}$, then today's vector **p** is

$$
\mathbf{p} = \mathbf{A}^{-1}\mathbf{p'} = \begin{bmatrix}
2 & 2 & -6 \\
-1 & -5 & 11 \\
0 & 4 & -4
\end{bmatrix}\begin{bmatrix} \frac{1}{2} \\ 0 \\ \frac{1}{2} \end{bmatrix} = \begin{bmatrix} -2 \\ 5 \\ -2 \end{bmatrix}.
$$

But the probabilities in **p** are nonsense. Why? The answer is that with this weather Markov chain, it must be impossible for there to be $\frac{1}{2}$ probability of sunny and $\frac{1}{2}$ probability of rainy weather tomorrow. That is, there is no possible probability distribution **p** for today that would give that distribution for tomorrow. ■

EXAMPLE 7.
Encoding and
Decoding
Alphabetic
Messages
(Optional)

A common approach to coding an alphabetic message is to treat each letter in the message as a number between 1 and 26: A ↔ 1, B ↔ 2, C ↔ 3, . . . , Z ↔ 26. So MAYDAY would be the numeric sequence 13, 1, 25, 4, 1, 25. Then we apply some algebraic formula to encode each letter (number) as some other code letter (number). To ensure that the result of some calculation is a number between 1 and 26, one usually assumes that all arithmetic is done mod 26. As an example, suppose that letter L_x is encoded as code letter C_x using the formula $C_x = 7L_x + 6$. Then the letter M (13) is encoded

$$C_M = 7M + 6 = 7 \cdot 13 + 6 = 97 \equiv 19 \quad (\text{mod } 26) \longleftrightarrow S$$

(since: $97 = 3 \cdot 26 + 19$).

Such linear encoding schemes are easy to break. An encoding scheme that is simple to use but fairly hard to break is to group letters into pairs L_1, L_2 and encode them into a pair of code letters C_1, C_2 with two linear equations, such as

$$C_1 \equiv 9L_1 + 17L_2 \quad (\text{mod } 26) \qquad C_2 \equiv 7L_1 + 2L_2 \quad (\text{mod } 26) \tag{19}$$

If $L_1 = E \ (= 5)$ and $L_2 = C \ (= 3)$, then (19) would encode them as

$$C_1 = 9 \cdot 5 + 17 \cdot 3 = 96 \equiv 18 \quad (\text{mod } 26) \longleftrightarrow R$$
$$C_2 = 7 \cdot 5 + \ 2 \cdot 3 = 41 \equiv 15 \quad (\text{mod } 26) \longleftrightarrow O$$

In matrix form, with $\mathbf{c} = \begin{bmatrix} C_1 \\ C_2 \end{bmatrix}$, $\mathbf{l} = \begin{bmatrix} L_1 \\ L_2 \end{bmatrix}$, and $\mathbf{E} = \begin{bmatrix} 9 & 17 \\ 7 & 2 \end{bmatrix}$, (19) becomes

$$\mathbf{c} \equiv \mathbf{El} \quad (\text{mod } 26) \tag{20}$$

The person who receives the coded pair $\mathbf{c}$ will decode $\mathbf{c}$ back into the original message pair $\mathbf{l}$ by using the inverse of $\mathbf{E}$:

$$\mathbf{l} \equiv \mathbf{E}^{-1}\mathbf{c} \tag{21}$$

We compute the inverse by reducing $[\mathbf{E} | \mathbf{I}]$ to reduced row echelon form, except that now all computations must be performed mod 26. Subtraction mod 26 is performed by adding; for example, $-1 \equiv 25$. Division mod 26 is performed by multiplication; for example, $\frac{1}{9} = 3$ since $9 \cdot 3 = 27 \equiv 1$.

$$\begin{bmatrix} 9 & 17 & | & 1 & 0 \\ 7 & 2 & | & 0 & 1 \end{bmatrix} \Longrightarrow \begin{bmatrix} 1 & 25 & | & 3 & 0 \\ 0 & 9 & | & 5 & 1 \end{bmatrix} \quad (\text{mod } 26)$$

$$\Longrightarrow \begin{bmatrix} 1 & 0 & | & 18 & 3 \\ 0 & 1 & | & 15 & 3 \end{bmatrix} \quad (\text{mod } 26)$$

Thus

$$\mathbf{E}^{-1} = \begin{bmatrix} 18 & 3 \\ 15 & 3 \end{bmatrix} \quad (\text{mod } 26)$$

The decoding equations are

$$L_1 \equiv 18C_1 + 3C_2 \quad (\text{mod } 26) \qquad L_2 \equiv 15C_1 + 3C_2 \quad (\text{mod } 26) \tag{22}$$

For example, the pair R, O (= 18, 15) is decoded using (22) as

$$L_1 = 18 \cdot 18 + 3 \cdot 15 = 324 + 45 = 369 \equiv 5 \pmod{26} \longleftrightarrow E$$
$$L_2 = 15 \cdot 18 + 3 \cdot 15 = 270 + 45 = 315 \equiv 3 \pmod{26} \longleftrightarrow C$$

So R, O decode back to the original pair E, C, as required. ■

Section 1.5 Exercises

1. Verify for the matrix A in Example 1 that $A^{-1}A = I$.

2. Write the system of equations that entries in the inverse of the following matrices must satisfy. Then find inverses (as in Example 3) or show none can exist (following the reasoning in Example 2).

(a) $\begin{bmatrix} 1 & 1 \\ 1 & 0 \end{bmatrix}$ (b) $\begin{bmatrix} 0 & 1 \\ 1 & 0 \end{bmatrix}$ (c) $\begin{bmatrix} 2 & 1 \\ 7 & 4 \end{bmatrix}$ (d) $\begin{bmatrix} -1 & 3 \\ 2 & -6 \end{bmatrix}$

(e) $\begin{bmatrix} 1 & 2 & 1 \\ 2 & 4 & 2 \\ 2 & 5 & 1 \end{bmatrix}$ (f) $\begin{bmatrix} 1 & 1 & 0 \\ 0 & 1 & 1 \\ 1 & 2 & 1 \end{bmatrix}$ (g) $\begin{bmatrix} 3 & 2 & 1 \\ 1 & 1 & 1 \\ 7 & 6 & 5 \end{bmatrix}$

3. (a) Write out the system of equations that the first column of the inverse of A must satisfy, where $A = \begin{bmatrix} 1 & 0 & 2 \\ 0 & 1 & 3 \\ 1 & 0 & 4 \end{bmatrix}$.

   (b) Determine the first column of A^{-1} using Theorem 3.

4. This exercise gives a "picture" of how when two columns of A are almost the same, the inverse of A almost does not exist. For the following matrices A, solve the system $A \begin{bmatrix} u_1 \\ u_2 \end{bmatrix} = \begin{bmatrix} 1 \\ 0 \end{bmatrix}$. Then plot $u_1\mathbf{a}_1$ and $u_2\mathbf{a}_2$ in a two-dimensional coordinate system and show geometrically how the sum of vectors $u_1\mathbf{a}_1$ and $u_2\mathbf{a}_2$ is $\begin{bmatrix} 1 \\ 0 \end{bmatrix}$.

(a) $\begin{bmatrix} 2 & 3 \\ 1 & 2 \end{bmatrix}$ (b) $\begin{bmatrix} 2 & 3 \\ 2 & 2 \end{bmatrix}$ (c) $\begin{bmatrix} 8 & 10 \\ 7 & 7 \end{bmatrix}$ (d) $\begin{bmatrix} 8 & 9 \\ 7 & 7 \end{bmatrix}$

5. Use Gauss–Jordan elimination to find the inverse of the following matrices.

(a) $\begin{bmatrix} 2 & -3 & 2 \\ 1 & -1 & 1 \\ -1 & 5 & 4 \end{bmatrix}$ (b) $\begin{bmatrix} -1 & -1 & 1 \\ 2 & 2 & -4 \\ 1 & -2 & 3 \end{bmatrix}$ (c) $\begin{bmatrix} -1 & -3 & 2 \\ 2 & 1 & 3 \\ 5 & 4 & 6 \end{bmatrix}$

(d) $\begin{bmatrix} 2 & 4 & -2 \\ 1 & -2 & -4 \\ -2 & -1 & -3 \end{bmatrix}$ (e) $\begin{bmatrix} 1 & 1 & 4 \\ 2 & 1 & 3 \\ 5 & 2 & 5 \end{bmatrix}$ (f) $\begin{bmatrix} 2 & -3 & -1 \\ 3 & -5 & -2 \\ 9 & 6 & 4 \end{bmatrix}$

6. For each matrix A in Exercise 5, solve $A\mathbf{x} = \mathbf{b}$, where $\mathbf{b} = \begin{bmatrix} 10 \\ 10 \\ 10 \end{bmatrix}$.

7. For each matrix **A** in Exercise 5, how much will the solution of **Ax** = **b** change if **b** is changed as follows?

(a) From the vector $\begin{bmatrix} b_1 \\ b_2 \\ b_3 \end{bmatrix}$ to the vector $\begin{bmatrix} b_1 \\ b_2 + 1 \\ b_3 \end{bmatrix}$

(b) From the vector $\begin{bmatrix} b_1 \\ b_2 \\ b_3 \end{bmatrix}$ to the vector $\begin{bmatrix} b_1 \\ b_2 \\ b_3 - 2 \end{bmatrix}$

(c) From the vector $\begin{bmatrix} b_1 \\ b_2 \\ b_3 \end{bmatrix}$ to the vector $\begin{bmatrix} b_1 \\ b_2 + 1 \\ b_3 - 1 \end{bmatrix}$

8. Use Gauss–Jordan elimination to find the inverse of the following matrices.

(a) $\begin{bmatrix} 1 & 1 & -1 & -1 \\ 2 & 0 & 0 & 1 \\ 3 & 0 & 0 & -2 \\ 4 & -2 & 1 & 3 \end{bmatrix}$ (b) $\begin{bmatrix} 3 & 2 & 1 & 0 \\ 1 & 1 & 0 & -1 \\ 2 & 1 & -1 & 1 \\ 1 & 1 & 1 & 1 \end{bmatrix}$

(c) $\begin{bmatrix} 1 & 3 & 2 & -1 \\ 1 & 1 & 1 & 1 \\ 2 & -2 & 1 & -1 \\ 1 & -3 & -1 & 2 \end{bmatrix}$

9. (a) Find the inverse of the following tridiagonal matrix. (*Tridiagonal* means that all non-zero entries are on the main diagonal or in entries adjacent to the main diagonal.)

$$\begin{bmatrix} 1 & -1 & 0 & 0 & 0 \\ -1 & 2 & -1 & 0 & 0 \\ 0 & -1 & 2 & -1 & 0 \\ 0 & 0 & -1 & 2 & -1 \\ 0 & 0 & 0 & -1 & 2 \end{bmatrix}$$

Note that the inverse is not tridiagonal or full of zeros.

(b) Change entry (1, 1) from a 1 to a 2 and repeat part (a). Does this small change affect the inverse substantially?

10. Let **A** be an 8 × 8 matrix. How many multiplications (approximately) are required to perform each of the following operations?
(a) Compute $\mathbf{A}^2$.
(b) Solve **Ax** = **b**.
(c) Compute $\mathbf{A}^{-1}$.

11. (Continuation of Exercise 14 in Section 1.4) The staff dietician at the California Institute of Trigonometry has to make up a meal with 600 calories, 20 grams of protein, and 200 milligrams of vitamin C. There are three food types to choose from: rubbery Jello, dried fish sticks, and mystery meat. They have the following nutritional content per ounce.

	Jello	Fish sticks	Mystery meat
Calories:	10	50	200
Protein:	1	3	0.2
Vitamin C:	30	10	0

(a) Find the inverse of this data matrix and use it to compute the amount of Jello, fish sticks, and mystery meat required.

(b) If the protein requirement is increased by 4, how will this change the number of units of Jello in the meal?

(c) If the vitamin C requirement is decreased by k milligrams, how much will this change the number of fish sticks in a meal?

12. (Continuation of Exercise 15 in Section 1.4) A furniture manufacturer makes tables, chairs, and sofas. In one month the company has available 300 units of wood, 350 units of labor, and 225 units of upholstery. The manufacturer wants a production schedule for the month that uses all of these resources. The various products require the following amounts of the resources.

	Table	Chair	Sofa
Wood:	4	1	3
Labor:	3	2	5
Upholstery:	2	0	4

(a) Find the inverse of this data matrix and use it to determine how much of each product should be manufactured.

(b) If the amount of wood is increased by 30 units, how will this change the number of sofas produced?

(c) If the amount of labor is decreased by k, how will this change your answer in part (a)?

13. (Continuation of Exercise 17 in Section 1.4) An investment analyst is trying to find out how much business a secretive TV manufacturer has. The company makes three brands of TV sets: brand A, brand B, brand C. The analyst learns that the manufacturer has ordered from suppliers 450,000 type 1 circuit boards, 300,000 type 2 circuit boards, and 350,000 type 3 circuit boards. Brand A uses 2 type 1 boards, 1 type 2 board, and 2 type 3 boards. Brand B uses 3 type 1 boards, 2 type 2 boards, and 1 type 3 board. Brand C uses 1 board of each type.

(a) Set up this problem as a system $Ax = b$. Find the inverse of A and use it to determine how many TV sets of each brand are being manufactured.

(b) If the number of type 2 boards used is increased by 100,000, how will this change your answer in part (a)?

(c) If the number of type 1 boards is decreased by $10,000k$, how much will this change your answer in part (a)?

14. Why must a matrix be square if it has an inverse?

15. Show, by the reasoning in Example 2, that if a matrix has a row (or column) that is all 0's, then the matrix cannot have an inverse.

16. **(a)** Following the reasoning in Example 2, show that $A = \begin{bmatrix} 1 & 2 & 3 \\ 1 & 4 & 5 \\ 1 & 6 & 7 \end{bmatrix}$ cannot have

an inverse because the third column is the sum of the other two columns.

(b) Generalize the argument in part (a) to show that if one column (row) is a linear combination of two other (rows), $a_i = c a_k + d a_h$, the matrix cannot have an inverse.

17. Find the inverse of the diagonal matrix $A = \begin{bmatrix} a_{11} & 0 & 0 \\ 0 & a_{22} & 0 \\ 0 & 0 & a_{33} \end{bmatrix}$. (Hint: The inverse

is also a diagonal matrix.)

18. **(a)** Use the following fact: The inverse of an upper triangular matrix (if the inverse exists) is itself upper triangular, to determine what the main diagonal entries must be

in the inverse of the upper triangular matrix $\begin{bmatrix} 2 & 3 & 4 \\ 0 & 4 & 2 \\ 0 & 0 & 5 \end{bmatrix}$. (Do not use

Gauss–Jordan elimination.)

(b) Using the main diagonal entries in the inverse from part (a) and the fact that the inverse is upper triangular, determine the other entires in the inverse.

19. **(a)** Find A^{-1} for the upper triangular matrix $A = \begin{bmatrix} 1 & 2 & 3 & 4 \\ 0 & 1 & 1 & 2 \\ 0 & 0 & 2 & 4 \\ 0 & 0 & 0 & 1 \end{bmatrix}$ using back

substitution, or equivalently, pivoting on the augmented matrix [A I], except start with entry (4, 4), then entry (3, 3), then entry (2, 2), and then entry (1, 1).

(b) Generalize the computation in part (a) to show that the inverse of an upper triangular matrix is upper triangular.

20. **(a)** Determine the inverse of the matrix $\begin{bmatrix} 1 & 0 & 0 & 0 \\ 0 & 1 & 0 & 0 \\ a & 0 & 1 & 0 \\ 0 & 0 & 0 & 1 \end{bmatrix}$. (Hint: The inverse has

a simple form; try trial-and-error guesswork.)

(b) Generalize your result in part (a) to give the inverse of an $n \times n$ matrix with 1's on the main diagonal and 0's elsewhere except one position, entry (i, j), $i \neq j$, whose value is a.

21. Prove that if A and B are $n \times n$ matrices and if AB is invertible, A and B are both invertible. [Hint: Write A^{-1} as a product of $(AB)^{-1}$ and B.]

22. Which of the following conditions guarantees that a square matrix A has an inverse; which guarantees that it does not have an inverse? Explain the reason for your answer briefly.
    (a) $Ax = 0$ has only the solution $x = 0$. (**0** is a vector of all zeros.)
    (b) $Ax = b$ has a unique solution for some **b**.
    (c) $Ax = b'$ has two solutions for some other **b'**?
    (d) **b** equals a column of A.

23. Which of the following conditions guarantees that a matrix A has an inverse; which guarantees it does not have an inverse? Explain the reason for your answer briefly.
    (a) A has twice as many rows as columns.

(b) A is a 4×4 Markov chain matrix.

(c) The first row of A is twice the last row.

(d) The system of equations $Ax = b$ can be solved for any b.

24. Suppose that the inverse of A equals $\begin{bmatrix} 2 & 5 \\ 1 & 3 \end{bmatrix}$.

(a) What is the inverse of 3A?

(b) What is the inverse of A^3?

(c) What is A?

25. Suppose that the inverse of A equals $\begin{bmatrix} 2 & 0 & 1 \\ 2 & 1 & 0 \\ 0 & 0 & 5 \end{bmatrix}$.

(a) What is the inverse of $\frac{1}{3}A$?

(b) What is the inverse of A^2?

(c) What is A?

26. (a) Find the inverse of the transition matrix A for a weather Markov chain (like the one introduced in Section 1.1), where there are two states and $A = \begin{bmatrix} \frac{3}{4} & \frac{1}{2} \\ \frac{1}{4} & \frac{1}{2} \end{bmatrix}$.

(b) If p is today's weather probability distribution, $p° = A^{-1}p$ is yesterday's probability distribution. Find yesterday's weather probability distribution if today's weather probability distribution p is

(i) $\begin{bmatrix} \frac{1}{2} \\ \frac{1}{2} \end{bmatrix}$ (ii) $\begin{bmatrix} \frac{2}{3} \\ \frac{1}{3} \end{bmatrix}$ (iii) $\begin{bmatrix} 0 \\ 1 \end{bmatrix}$

(c) Use a computer program to determine the weather probability distribution 20 days ago for the current distributions in part (b).

27. Reverse the following Markov chains. Then find the "distribution" in the previous period if the current distribution is $\begin{bmatrix} 0.5 \\ 0 \\ 0.5 \end{bmatrix}$. Is this distribution really a probability distribution?

(a) $\begin{bmatrix} 0.5 & 0 & 0 \\ 0.5 & 1 & 0.5 \\ 0 & 0 & 0.5 \end{bmatrix}$ (b) $\begin{bmatrix} 0.5 & 0.25 & 0 \\ 0.5 & 0.5 & 0.5 \\ 0 & 0.25 & 0.5 \end{bmatrix}$ (c) $\begin{bmatrix} 0.4 & 0.3 & 0.3 \\ 0.3 & 0.4 & 0.3 \\ 0.3 & 0.3 & 0.4 \end{bmatrix}$

28. (a) There is just one special type of $n \times n$ Markov transition matrix A* (for each n) that has an inverse $A*^{-1}$ such that if p is a probability distribution, $p° = A*^{-1}p$ is always a probability distribution. Describe A*.

(b) Give an informal argument why no other such reversible Markov chain can exist.

29. (a) Find the system of equations for decoding the following encoding schemes.

(i) $C_1 = 3L_1 + 5L_2$
$C_2 = 5L_1 + 8L_2$

(ii) $C_1 = 11L_1 + 6L_2$
$C_2 = 8L_1 + 5L_2$

(iii) $C_1 = 2L_1 + 3L_2$
$C_2 = 7L_1 + 5L_2$

(Hint: The inverse of 7 is -11; the inverse of -11 is 7.)

(b) Decode the coded pair EF in each of these schemes.

30. Find a 2×2 matrix $\mathbf{A}$, $\mathbf{A} \neq \mathbf{I}$, such that $\mathbf{A} = \mathbf{A}^{-1}$. (That is, $\mathbf{A}^2 = \mathbf{I}$.)

31. (*Computer Project*) For one or more of the Markov transition matrices in Exercise 27, explore the behavior of running the Markov chain in reverse starting with different initial distributions $\mathbf{p}$. For each Markov chain, describe those $\mathbf{p}$'s for which $\mathbf{A}^{-1}\mathbf{p}$ is a probability distribution. Determine this set of $\mathbf{p}$'s by numerical experimentation.

1.6   Determinant of a Matrix

When mathematicians first looked systematically at solving systems of linear equations, they wanted to get a formula for their solution, in the spirit of the solution to a quadratic equation $ax^2 + bx + c = 0$ given by the quadratic formula

$$x = \frac{-b \pm \sqrt{b^2 - 4ac}}{2a}$$

What grew out of this objective was the theory of determinants and Cramer's rule (given below) which does provide the desired "formula" for solving a system of linear equations. The easier approach of reducing the system of equations to upper triangular form was not developed by Gauss until a century later.

Consider the following general system of two linear equations in two variables:

$$\mathbf{Ax = b}: \quad \begin{array}{l} a_{11}x_1 + a_{12}x_2 = b_1 \\ a_{21}x_1 + a_{22}x_2 = b_2 \end{array} \tag{1}$$

We shall develop a general formula for solving (1) for x_1 and x_2. Multiplying the first equation by a_{22} and the second by a_{12} and then subtracting, we obtain

$$\begin{array}{l} a_{11}a_{22}x_1 + a_{12}a_{22}x_2 = a_{22}b_1 \\ \underline{-(a_{12}a_{21}x_1 + a_{12}a_{22}x_2 = a_{12}b_2)} \\ (a_{11}a_{22} - a_{12}a_{21})x_1 = a_{22}b_1 - a_{12}b_2 \end{array}$$

Solving for x_1, we have

$$x_1 = \frac{a_{22}b_1 - a_{12}b_2}{a_{11}a_{22} - a_{12}a_{21}} \tag{2}$$

Substituting (2) in the first equation of (1) and simplifying, we obtain

$$x_2 = \frac{a_{11}b_2 - a_{21}b_1}{a_{11}a_{22} - a_{12}a_{21}} \tag{3}$$

**EXAMPLE 1.
Solving a 2 × 2
System by
Formula**

Consider the following system of two linear equations in two variables:

$$\begin{array}{l} 2x_1 - 3x_2 = 4 \\ x_1 + 2x_2 = 9 \end{array} \tag{4}$$

Let us use formulas (2) and (3) to solve for x_1 and x_2.

$$x_1 = \frac{a_{22}b_1 - a_{12}b_2}{a_{11}a_{22} - a_{12}a_{21}} = \frac{2 \cdot 4 - (-3) \cdot 9}{2 \cdot 2 - (-3) \cdot 1} = \frac{35}{7} = 5$$

$$x_2 = \frac{a_{11}b_2 - a_{21}b_1}{a_{11}a_{22} - a_{12}a_{21}} = \frac{2 \cdot 9 - 1 \cdot 4}{2 \cdot 2 - (-3) \cdot 1} = \frac{14}{7} = 2$$

(5) ■

It is possible to extend these formulas to obtain expressions for the solutions to three linear equations in three variables and more generally to n linear equations in n variables. However, these expressions become huge, and evaluating them normally takes far longer than solving by Gaussian elimination.

This common denominator, $a_{11}a_{22} - a_{12}a_{21}$, in (2) and (3) is called the determinant of system (1).

Definition

The ***determinant***, det($\mathbf{A}$), of a 2×2 matrix $\mathbf{A} = \begin{bmatrix} a_{11} & a_{12} \\ a_{21} & a_{22} \end{bmatrix}$ is defined to be

$$\det(\mathbf{A}) = a_{11}a_{22} - a_{12}a_{21}$$

The determinant of $\mathbf{A} = \begin{bmatrix} a_{11} & a_{12} \\ a_{21} & a_{22} \end{bmatrix}$ is often written $\begin{vmatrix} a_{11} & a_{12} \\ a_{21} & a_{22} \end{vmatrix}$. In the 2×2 case, det($\mathbf{A}$) is simply the product of the two main diagonal entries minus the product of the two off-diagonal entries. A complex recursive formula for determinants of larger matrices will be given shortly. As in the case of 2×2 matrices, the determinant of any square matrix $\mathbf{A}$ will turn out to be the denominator in the algebraic expressions for the solution of the system $\mathbf{Ax} = \mathbf{b}$.

We can write the numerators in formulas (2) and (3) for x_1 and x_2 as determinants of the matrices obtained by replacing the first and second columns, respectively, of $\mathbf{A}$ by the right-side vector $\mathbf{b}$. That is, let

$$\mathbf{A}_1 = \begin{bmatrix} b_1 & a_{12} \\ b_2 & a_{22} \end{bmatrix} \quad \text{and} \quad \mathbf{A}_2 = \begin{bmatrix} a_{11} & b_1 \\ a_{21} & b_2 \end{bmatrix}$$

(6)

Then

$$\det(\mathbf{A}_1) = \begin{vmatrix} b_1 & a_{12} \\ b_2 & a_{22} \end{vmatrix} = b_1 a_{22} - a_{12} b_2$$

$$\det(\mathbf{A}_2) = \begin{vmatrix} a_{11} & b_1 \\ a_{21} & b_2 \end{vmatrix} = a_{11} b_2 - b_1 a_{21}$$

(7)

The expressions in (7) are exactly the numerators in (2) and (3). So using det($\mathbf{A}_1$) and det($\mathbf{A}_2$), our formulas for x_1 and x_2 are

$$x_1 = \frac{\det(\mathbf{A}_1)}{\det(\mathbf{A})} \quad \text{and} \quad x_2 = \frac{\det(\mathbf{A}_2)}{\det(\mathbf{A})}$$

The numerators in systems of n linear equations in n variables turn out to have the same form. That is, if we define $\mathbf{A}_i$ to be the matrix obtained from $\mathbf{A}$ by replacing the ith column $\mathbf{a}_i$ of $\mathbf{A}$ by the right-hand-side vector $\mathbf{b}$,

$$\mathbf{A}_i = [\mathbf{a}_1, \mathbf{a}_2, \ldots, \mathbf{a}_{i-1}, \mathbf{b}, \mathbf{a}_{i+1}, \ldots, \mathbf{a}_n] \tag{8}$$

the solution to $\mathbf{A}\mathbf{x} = \mathbf{b}$ is given by the following determinant-based formula.

Theorem 1 (Cramer's Rule)

Let $\mathbf{A}\mathbf{x} = \mathbf{b}$ be a system of n linear equations in n variables. If $\det(\mathbf{A}) \neq 0$, the solution is given by $x_i = \det(\mathbf{A}_{ii})/\det(\mathbf{A})$, $i = 1, 2, \ldots, n$.

We shall not prove Cramer's rule. (The interested reader should refer to *Elementary Linear Algebra*, 6th ed., by H. Anton, Wiley, New York, 1991.)

Applying Cramer's rule to the system of equations in Example 1,

$$2x_1 - 3x_2 = 4$$
$$x_1 + 2x_2 = 9$$

we obtain

$$x_1 = \frac{\begin{vmatrix} 4 & -3 \\ 9 & 2 \end{vmatrix}}{\begin{vmatrix} 2 & -3 \\ 1 & 2 \end{vmatrix}} = \frac{4 \cdot 2 - (-3) \cdot 9}{2 \cdot 2 - (-3) \cdot 1} = \frac{35}{7} = 5$$

$$x_2 = \frac{\begin{vmatrix} 2 & 4 \\ 1 & 9 \end{vmatrix}}{\begin{vmatrix} 2 & -3 \\ 1 & 2 \end{vmatrix}} = \frac{2 \cdot 9 - 4 \cdot 1}{2 \cdot 2 - (-3) \cdot 1} = \frac{14}{7} = 2$$

which is the same solution we obtained earlier in (5).

Cramer's rule also yields an easy-to-remember formula for the entries in the inverse $\mathbf{A}^{-1}$ of a 2×2 matrix $\mathbf{A}$. Recall that since $\mathbf{A}\mathbf{A}^{-1} = \mathbf{I}$, the first column $\mathbf{a}_1^*$ of the inverse satisfies $\mathbf{A}\mathbf{a}_1^* = \begin{bmatrix} 1 \\ 0 \end{bmatrix}$ and the second column $\mathbf{a}_2^*$ of the inverse satisfies $\mathbf{A}\mathbf{a}_2^* = \begin{bmatrix} 0 \\ 1 \end{bmatrix}$. For any 2×2 matrix $\mathbf{A}$, the system of equations $\mathbf{A}\mathbf{a}_1^* = \begin{bmatrix} 1 \\ 0 \end{bmatrix}$ has the solution by Cramer's rule:

$$a_{11}^* = \frac{\begin{vmatrix} 1 & a_{12} \\ 0 & a_{22} \end{vmatrix}}{\det(\mathbf{A})} = \frac{a_{22}}{\det(\mathbf{A})} \qquad a_{21}^* = \frac{\begin{vmatrix} a_{11} & 1 \\ a_{21} & 0 \end{vmatrix}}{\det(\mathbf{A})} = \frac{-a_{21}}{\det(\mathbf{A})} \tag{9}$$

The simple form of the numerator comes from having a right-side vector of $\begin{bmatrix} 1 \\ 0 \end{bmatrix}$.

The same simplification occurs in solving $\mathbf{Aa}_2^* = \begin{bmatrix} 0 \\ 1 \end{bmatrix}$ by Cramer's rule, yielding

$$a_{12}^* = \frac{-a_{12}}{\det(\mathbf{A})}, \qquad a_{22}^* = \frac{a_{11}}{\det(\mathbf{A})}$$

These single-number numerators lead to the following general formula for the inverse of a 2×2 matrix.

Formula for Inverse of a 2 × 2 Matrix

If $\mathbf{A} = \begin{bmatrix} a_{11} & a_{12} \\ a_{21} & a_{22} \end{bmatrix}$, then $\mathbf{A}^{-1} = \dfrac{1}{\det(\mathbf{A})} \begin{bmatrix} a_{22} & -a_{12} \\ -a_{21} & a_{11} \end{bmatrix}$. $\tag{10}$

In words, the inverse of a 2×2 matrix $\mathbf{A}$ is obtained as follows: Divide all entries of $\mathbf{A}$ by the determinant, then interchange the two diagonal entries and change the sign of the two off-diagonal entries. No such simple construction of the inverse is possible for larger matrices.

EXAMPLE 2.
Inverse of a 2 × 2
Matrix

Let us use (10) to obtain the inverse of the 2×2 matrix $\mathbf{A} = \begin{bmatrix} 3 & 1 \\ 4 & 2 \end{bmatrix}$. (This matrix's inverse was computed by elimination in Example 3 of Section 1.5.) We first compute $\det(\mathbf{A}) = 3 \cdot 2 - 1 \cdot 4 = 2$. Then, by (10),

$$\mathbf{A}^{-1} = \frac{1}{2} \begin{bmatrix} 2 & -1 \\ -4 & 3 \end{bmatrix} = \begin{bmatrix} 1 & -\frac{1}{2} \\ -2 & \frac{3}{2} \end{bmatrix} \tag{11}$$

■

We now present a general recursive formula for the determinant of an $n \times n$ matrix. We need to introduce two terms. The (i, j)-**minor** $\mathbf{A}_{ij}$ of a matrix $\mathbf{A}$ is a submatrix of $\mathbf{A}$ obtained by deleting the ith row and jth column of $\mathbf{A}$. The (i, j)-**cofactor** C_{ij} of $\mathbf{A}$ is the determinant of the minor $\mathbf{A}_{ij}$ multiplied by $(-1)^{i+j}$:

$$C_{ij} = (-1)^{i+j}\det(\mathbf{A}_{ij}) \tag{12}$$

> **Definition**
>
> The ***determinant*** of an $n \times n$ matrix $\mathbf{A}$ is defined to be
>
> $$\det(\mathbf{A}) = a_{11}C_{11} + a_{12}C_{12} + a_{13}C_{13} + \cdots + a_{1n}C_{1n} \qquad (13)$$

Actually, in formula (13), the terms a_{11}, a_{12}, a_{13}, . . . , a_{1n} (and associated cofactors) can be replaced by any row (or column), so that our definition generalizes to

$$\det(\mathbf{A}) = a_{i1}C_{i1} + a_{i2}C_{i2} + a_{i3}C_{i3} + \cdots + a_{in}C_{in} \qquad \text{for some } i, \ 1 \le i \le n \quad (14)$$

In the case of the determinant of a 3×3 matrix $\mathbf{A}$, the cofactors involve 2×2 determinants which we know how to compute. Thus we have

$$\det(\mathbf{A}) = \begin{vmatrix} a_{11} & a_{12} & a_{13} \\ a_{21} & a_{22} & a_{23} \\ a_{31} & a_{32} & a_{33} \end{vmatrix} = a_{11}\begin{vmatrix} a_{22} & a_{23} \\ a_{32} & a_{33} \end{vmatrix} - a_{12}\begin{vmatrix} a_{21} & a_{23} \\ a_{31} & a_{33} \end{vmatrix} + a_{13}\begin{vmatrix} a_{21} & a_{22} \\ a_{31} & a_{32} \end{vmatrix} \qquad (15)$$

$$= a_{11}(a_{22}a_{33} - a_{23}a_{32}) - a_{12}(a_{21}a_{33} - a_{23}a_{31}) + a_{13}(a_{21}a_{32} - a_{22}a_{31})$$

**EXAMPLE 3.
Determinant of a
3 × 3 Matrix**

Using (15), we can compute the following 3×3 determinant:

$$\begin{vmatrix} 1 & 2 & 3 \\ 4 & 5 & 6 \\ 7 & 8 & 0 \end{vmatrix} = 1\begin{vmatrix} 5 & 6 \\ 8 & 0 \end{vmatrix} - 2\begin{vmatrix} 4 & 6 \\ 7 & 0 \end{vmatrix} + 3\begin{vmatrix} 4 & 5 \\ 7 & 8 \end{vmatrix}$$

$$= 1(5 \cdot 0 - 6 \cdot 8) - 2(4 \cdot 0 - 6 \cdot 7) + 3(4 \cdot 8 - 5 \cdot 7)$$

$$= 1(-48) - 2(-42) + 3(-3) = -48 + 84 - 9$$

$$= 27 \qquad\qquad ■$$

Knowing how to compute the determinant of a 3×3 matrix allows us to compute the determinant of a 4×4 matrix, since the right side in formula (13) involves 3×3 cofactors. In general, with enough stamina (or better, a computer program) we can compute the determinant of a matrix of any size. A simpler way to compute large determinants is given below. One immediate consequence of the definition in (13) and the fact that it applies to any row or column is that if a row or column of $\mathbf{A}$ is all 0's, then $\det(\mathbf{A}) = 0$.

Lemma. If a square matrix $\mathbf{A}$ has a row or column of 0's, then $\det(\mathbf{A}) = 0$.

For the determinant of a 3×3 matrix $\mathbf{A}$, there is a simple picture that can be used in which one multiplies the numbers lying on the six "diagonals" in the augmented

3×5 array shown below. The products marked by solid lines have plus signs and the products marked by dashed lines have minus signs.

$$\det(\mathbf{A}) = \begin{vmatrix} a_{11} & a_{12} & a_{13} \\ a_{21} & a_{22} & a_{23} \\ a_{31} & a_{32} & a_{33} \end{vmatrix} \begin{matrix} a_{11} & a_{12} \\ a_{21} & a_{22} \\ a_{31} & a_{32} \end{matrix} = a_{11}a_{22}a_{33} + a_{12}a_{23}a_{31} + a_{13}a_{21}a_{32} \\ - a_{13}a_{22}a_{31} - a_{11}a_{23}a_{32} - a_{12}a_{21}a_{33} \qquad (16)$$

EXAMPLE 4.
Solving the
Refinery
Equations by
Cramer's Rule

In Section 1.1 we presented a system of equations for controlling the production of three oil refineries. Let us solve this system with Cramer's rule.

$$\begin{array}{llll} \text{Heating oil:} & 4x_1 + & 2x_2 + 2x_3 = & 600 \\ \text{Diesel oil:} & 2x_1 + & 5x_2 + 2x_3 = & 800 \\ \text{Gasoline:} & 1x_1 + & 2.5x_2 + 5x_3 = & 1000 \end{array}$$

The determinant of this coefficient matrix is, by (15),

$$\det(\mathbf{A}) = \begin{vmatrix} 4 & 2 & 2 \\ 2 & 5 & 2 \\ 1 & 2.5 & 5 \end{vmatrix} = 4 \begin{vmatrix} 5 & 2 \\ 2.5 & 5 \end{vmatrix} - 2 \begin{vmatrix} 2 & 2 \\ 1 & 5 \end{vmatrix} + 2 \begin{vmatrix} 2 & 5 \\ 1 & 2.5 \end{vmatrix}$$

$$= 4 \cdot (5 \cdot 5 - 2 \cdot 2.5) - 2(2 \cdot 5 - 2 \cdot 1) + 2(2 \cdot 2.5 - 5 \cdot 1) \qquad (17)$$

$$= 4(20) - 2(8) + 2(0) = 64$$

To determine x_1 using Cramer's rule, we need $\det(\mathbf{A}_1)$. Recall that $\mathbf{A}_1$ is obtained from $\mathbf{A}$ by replacing the first column of $\mathbf{A}$ by the numbers on the right side of the equations. This time we use (16) to compute the $\det(\mathbf{A}_1)$.

$$\det(\mathbf{A}_1) = \begin{vmatrix} 600 & 2 & 2 \\ 800 & 5 & 2 \\ 1000 & 2.5 & 5 \end{vmatrix} = 600 \cdot 5 \cdot 5 + 2 \cdot 2 \cdot 1000 + 2 \cdot 800 \cdot 2.5 - 2 \cdot 5 \cdot 1000 \\ - 2 \cdot 800 \cdot 5 - 600 \cdot 2 \cdot 2.5$$

$$= 15{,}000 + 4000 + 4000 - 10{,}000 - 8000 - 3000$$

$$= 2000 \qquad (18)$$

We thus have

$$x_1 = \frac{\det(\mathbf{A}_1)}{\det(\mathbf{A})} = \frac{2000}{64} = 31\tfrac{1}{4}$$

In a similar fashion we use (16) to compute $\det(\mathbf{A}_2)$:

$$\det(\mathbf{A}_2) = \begin{vmatrix} 4 & 600 & 2 \\ 2 & 800 & 2 \\ 1 & 1000 & 5 \end{vmatrix} = 4 \cdot 800 \cdot 5 + 600 \cdot 2 \cdot 1 + 2 \cdot 2 \cdot 1000 - 2 \cdot 800 \cdot 1 \\ - 2 \cdot 1000 \cdot 4 - 5 \cdot 2 \cdot 600$$

$$= 16{,}000 + 1200 + 4000 - 1600 - 8000 - 6000 \qquad (19)$$

$$= 5600$$

and thus

$$x_2 = \frac{\det(\mathbf{A}_2)}{\det(\mathbf{A})} = \frac{5600}{64} = 87\tfrac{1}{2}$$

It is left as an exercise for the reader to compute $\det(\mathbf{A}_3)$ and determine x_3. ■

Properties of Determinants

We now develop a little of the theory of determinants. Most introductory linear algebra books devote a whole chapter to determinants. (For example, see *Elementary Linear Algebra*, 6th ed., by H. Anton, Wiley, New York, 1991.) Although determinants used to be considered an important subject, modern approaches to linear algebra and its applications have allocated a greatly diminished role to determinants. In particular, determinants have essentially no role in contemporary numerical methods for solving systems of linear equations. The principal uses of determinants in this book are finding eigenvalues (in Section 3.3) and computing cross products (in Section 2.5).

We start by giving another equivalent definition of the computation of the determinant.

> **Alternative Definition of a Determinant.** The *determinant* of an $n \times n$ matrix $\mathbf{A}$ is the sum of all possible products of n entries, each from a different row and different column (there is a technical rule of signs for determining whether the product gets a plus or minus sign).

The reader should check that our formulas for 2×2 and 3×3 determinants involved all products of this sort. A counting argument shows that there are $n!$ $[= n(n-1)(n-2)\cdots 3 \cdot 2 \cdot 1]$ such products in an $n \times n$ determinant. For example, a 10×10 determinant has $10! = 3,628,800$ products. However, the determinant formula in (13) reduces the work some by recursively factoring out common terms.

Fortunately, a simple shortcut exists for computing determinants that uses Gaussian elimination. As a first step, we note that the alternative definition of a determinant shows that it is easy to compute the determinant of an upper triangular matrix. Recall that a square matrix is *upper triangular* if all entries below the main diagonal are zero, such as

$$\mathbf{A} = \begin{bmatrix} 2 & 4 & 1 \\ 0 & 1 & 7 \\ 0 & 0 & 2 \end{bmatrix} \tag{20}$$

The final matrix produced by Gaussian elimination is an upper triangular matrix. That is, Gaussian elimination uses elementary row operations to convert any square matrix into an upper triangular matrix. A *lower triangular* matrix has 0's above the main diagonal.

Theorem 2. If $\mathbf{A}$ is an upper or lower triangular matrix, $\det(\mathbf{A}) = a_{11}a_{22}\cdots a_{nn}$, the product of entries on the main diagonal.

Proof: Apply our alternative definition of a determinant to an upper triangular matrix. Except for the product of main diagonal entries, any other product of n entries, each in a different row and column, will have to contain a 0 entry below (or above, in the case of a lower triangular matrix) the main diagonal, so all such other products are 0. The sign of this product of main diagonal entries is positive (according to the technical rule for signs). ■

Theorem 2 can also be derived (with a longer argument) from the original recursive formula (13) for a determinant. By Theorem 2, $\det(\mathbf{A}) = 2 \cdot 1 \cdot 2 = 4$ for the upper triangular matrix $\mathbf{A}$ in (20). A special upper (and lower) triangular matrix is an identity matrix $\mathbf{I}$, which has 1's on the main diagonal and 0's elsewhere. By Theorem 2, $\det(\mathbf{I}) = 1$, irrespective of the size of $\mathbf{I}$. Next we give some properties of determinants.

Theorem 3. Let $\mathbf{A}$ be a square matrix.
   (i) If two rows of $\mathbf{A}$ are interchanged to form a new matrix $\mathbf{A}'$, $\det(\mathbf{A}') = -\det(\mathbf{A})$.
  (ii) If two rows of $\mathbf{A}$ are equal, $\det(\mathbf{A}) = 0$.
 (iii) If $\mathbf{A}'$ is obtained from $\mathbf{A}$ by multiplying each entry in the kth row by a scalar r,

$$\det(\mathbf{A}') = r \cdot det(\mathbf{A})$$

  (iv) If one row of $\mathbf{A}$ is a multiple of another row, $\det(\mathbf{A}) = 0$.
   (v) Suppose that $\mathbf{A}$ and $\mathbf{A}'$ are the same except they have different values in row k. Let $\mathbf{A}''$ be the same as $\mathbf{A}$ except its row k is the sum of row k in $\mathbf{A}$ and row k in $\mathbf{A}'$. Then

$$\det(\mathbf{A}'') = \det(\mathbf{A}) + \det(\mathbf{A}')$$

  (vi) If $\mathbf{A}'$ is the matrix formed by subtracting a multiple of one row of $\mathbf{A}$ from another row of $\mathbf{A}$,

$$\det(\mathbf{A}') = \det(\mathbf{A})$$

Proof: (i) The proof of (i) is not hard but involves a technical argument using the cofactor definition (13) of $\det(\mathbf{A})$ and is omitted.

(ii) Let $\mathbf{A}'$ be obtained from $\mathbf{A}$ by interchanging the two equal rows of $\mathbf{A}$. This may seem silly since the rows are the same, so $\mathbf{A}' = \mathbf{A}$. But by (i),

$$\det(\mathbf{A}') = -\det(\mathbf{A})$$

Since $\mathbf{A}' = \mathbf{A}$, obviously

$$\det(\mathbf{A}') = \det(\mathbf{A})$$

The only way both equations can be true is if $\det(\mathbf{A}) = 0$.

(iii) In the alternative definition of a determinant, every product of one entry from each row and column will have one entry from row k, so each such product in $\mathbf{A}'$ will be r times the value of the same product in $\mathbf{A}$. Hence $\det(\mathbf{A}') = r \cdot \det(\mathbf{A})$.

(iv) Suppose that row k is r times row h. Multiplying row k by $1/r$ yields a matrix $\mathbf{A}'$ in which rows k and h are equal. By (iii), $\det(\mathbf{A}') = (1/r)\det(\mathbf{A})$. By (ii), $\det(\mathbf{A}') = 0$. Hence $\det(\mathbf{A}) = 0$.

(v) In the alternative definition of a determinant, every product in $\mathbf{A}''$ of one entry from each row and column will equal the sum of the corresponding products in $\mathbf{A}$ and $\mathbf{A}'$. Hence $\det(\mathbf{A}'') = \det(\mathbf{A}) + \det(\mathbf{A}')$.

(vi) Suppose that $\mathbf{A}'$ is formed by subtracting r times row h from row k. Let $\mathbf{A}^\circ$ be formed from $\mathbf{A}$ by replacing the row k by $-r$ times row h. Then $\mathbf{A}' = \mathbf{A} + \mathbf{A}^\circ$. By (iv), $\det(\mathbf{A}^\circ) = 0$. By (v), $\det(\mathbf{A}') = \det(\mathbf{A}) + \det(\mathbf{A}^\circ) = \det(\mathbf{A}) + 0 = \det(\mathbf{A})$. ■

Theorem 4. Let $\mathbf{A}$ be an $n \times n$ matrix and let $\mathbf{U}$ be the upper triangular matrix resulting when Gaussian elimination is applied to $\mathbf{A}$. Assuming no row interchanges during elimination, then $\det(\mathbf{A}) = u_{11}u_{22} \cdots u_{nn}$, the product of the main diagonal entries of $\mathbf{U}$.

Proof: Gaussian elimination involves repeatedly subtracting a multiple of one row from another row. By Theorem 3(vi), this step does not change the value of the determinant. Therefore, $\det(\mathbf{A}) = \det(\mathbf{U})$, and, by Theorem 2, $\det(\mathbf{U}) = u_{11}u_{22} \cdots u_{nn}$. [If there were a need to interchange pairs of rows during elimination, each such interchange would change the sign of the determinant, according to Theorem 3(i).] ■

Theorem 4 provides us with a simple way to determine the determinant of a square matrix of any size. This is the fastest way to compute a determinant of matrix of size 3 or larger.

EXAMPLE 5.
Determinant of the
Refinery Problem
Revisited

Let us use Theorem 4 to obtain the determinant of the coefficient matrix in the oil refinery problem of Example 4. The coefficient matrix is reduced by Gaussian elimination as follows (see Example 2 in Section 1.4).

$$\mathbf{A} = \begin{bmatrix} 4 & 2 & 2 \\ 2 & 5 & 2 \\ 1 & 2.5 & 5 \end{bmatrix} \implies \begin{bmatrix} 4 & 2 & 2 \\ 0 & 4 & 1 \\ 0 & 2 & 4.5 \end{bmatrix} \implies \begin{bmatrix} 4 & 2 & 2 \\ 0 & 4 & 1 \\ 0 & 0 & 4 \end{bmatrix} = \mathbf{U}$$

Then $\det(\mathbf{A}) = u_{11}u_{22}u_{33} = 4 \cdot 4 \cdot 4 = 64$. This is the same result we got in Example 4. ■

We next state the product rule for determinants. (See Exercise 24 for the proof.)

Theorem 5. The determinant of a matrix product is the product of the determinants:

$$\det(\mathbf{AB}) = \det(\mathbf{A}) \det(\mathbf{B})$$

EXAMPLE 6.
Determinant of
the Product of
Matrices

Take the refinery matrix $\mathbf{A}$ (from Example 5) and multiply it by $\mathbf{B} = \begin{bmatrix} 1 & 2 & 3 \\ 2 & 4 & 6 \\ 0 & 1 & 2 \end{bmatrix}$.

The product $\mathbf{C}$ is computed to be

$$\mathbf{C} = \mathbf{AB} = \begin{bmatrix} 4 & 2 & 2 \\ 2 & 5 & 2 \\ 1 & 2.5 & 5 \end{bmatrix} \begin{bmatrix} 1 & 2 & 3 \\ 2 & 4 & 6 \\ 0 & 1 & 2 \end{bmatrix} = \begin{bmatrix} 8 & 18 & 28 \\ 12 & 26 & 40 \\ 6 & 17 & 28 \end{bmatrix}$$

Since the second row of $\mathbf{B}$ equals twice the first row of $\mathbf{B}$, $\det(\mathbf{B}) = 0$ by Theorem 3(iv). By Theorem 5, $\det(\mathbf{C}) = \det(\mathbf{A}) \det(\mathbf{B}) = 64 \cdot 0 = 0$. We confirm this result by computing the determinant of $\mathbf{C}$ using Theorem 4. That is, we apply Gaussian elimination to $\mathbf{C}$.

$$\mathbf{C} = \begin{bmatrix} 8 & 18 & 28 \\ 12 & 26 & 40 \\ 6 & 17 & 28 \end{bmatrix} \Longrightarrow \begin{bmatrix} 8 & 18 & 28 \\ 0 & -1 & -2 \\ 0 & 3.5 & 7 \end{bmatrix} \Longrightarrow \begin{bmatrix} 8 & 18 & 28 \\ 0 & -1 & -2 \\ 0 & 0 & 0 \end{bmatrix} = \mathbf{U} \quad (21)$$

By Theorem 4, $\det(\mathbf{C})$ equals the product of the diagonal elements in $\mathbf{U}$. Since the last diagonal entry is 0, the product of diagonal elements is 0, as expected. ■

Now we present two theorems that link determinants with inverses and unique solutions of a system of equations.

Theorem 6. Let $\mathbf{A}$ be a square matrix and suppose that the inverse $\mathbf{A}^{-1}$ exists. Then:
(i) $\det(\mathbf{A}) \neq 0$.
(ii) $\det(\mathbf{A}^{-1}) = 1/\det(\mathbf{A})$.
Proof: Since $\mathbf{AA}^{-1} = \mathbf{I}$, and, as noted earlier, $\det(\mathbf{I}) = 1$, then by Theorem 5, we have

$$\det(\mathbf{A}) \det(\mathbf{A}^{-1}) = \det(\mathbf{I}) = 1 \qquad (22)$$

If the product of two real numbers is 1, both numbers must be nonzero. Thus part (i) follows. Dividing both sides of (22) by $\det(\mathbf{A})$ gives part (ii). ■

EXAMPLE 7.
Determinant of
Inverse

The matrix $\mathbf{A} = \begin{bmatrix} 3 & 1 \\ 4 & 2 \end{bmatrix}$ has inverse $\mathbf{A}^{-1} = \begin{bmatrix} 1 & -\frac{1}{2} \\ -2 & \frac{3}{2} \end{bmatrix}$ (see Example 2).

Det$(\mathbf{A}) \doteq 3 \cdot 2 - 1 \cdot 4 = 2$, and $\det(\mathbf{A}^{-1}) = 1 \cdot \frac{3}{2} - (-2)(-\frac{1}{2}) = \frac{1}{2}$. Thus $\det(\mathbf{A}^{-1}) = 1/\det(\mathbf{A})$, as required. ■

In Theorem 4 of Section 1.5, we gave some equivalent conditions that guaranteed a system $\mathbf{Ax} = \mathbf{b}$ of n equations in n unknowns had a solution. Now we can add another condition, that the determinant is nonzero.

Theorem 7 (Expanded Fundamental Theorem for Solving Ax = b)

The following four statements are equivalent for any $n \times n$ matrix $\mathbf{A}$.

(i) For some particular vector $\mathbf{b}$, the system of equations $\mathbf{Ax} = \mathbf{b}$ has a unique solution.

(ii) For all $\mathbf{b}$, the system of equations $\mathbf{Ax} = \mathbf{b}$ always has a unique solution.

(iii) $\mathbf{A}$ has an inverse.

(iv) $\det(\mathbf{A}) \neq 0$.

Proof: In Section 1.5 we proved the equivalence of parts (i), (ii), and (iii). By Theorem 6, (iii) implies (iv), and Theorem 1 (Cramer's rule) says that (iv) implies (i). ■

EXAMPLE 6 (continued). Application of Theorem 7

In Example 6, we multiplied the refinery matrix $\mathbf{A} = \begin{bmatrix} 4 & 2 & 2 \\ 2 & 5 & 2 \\ 1 & 2.5 & 5 \end{bmatrix}$ times

$\mathbf{B} = \begin{bmatrix} 1 & 2 & 3 \\ 2 & 4 & 6 \\ 0 & 1 & 2 \end{bmatrix}$ to get matrix $\mathbf{C} = \mathbf{AB} = \begin{bmatrix} 8 & 18 & 28 \\ 12 & 26 & 40 \\ 6 & 17 & 28 \end{bmatrix}$. We know from

earlier work that the refinery problem $\mathbf{Ax} = \mathbf{b}$ always has a unique solution. Suppose that we want to know if $\mathbf{Cx} = \mathbf{b}$ also always has a unique solution. Since $\det(\mathbf{B}) = 0$ (see Example 6), then by Theorem 5, $\det(\mathbf{C}) = \det(\mathbf{A}) \det(\mathbf{B})$ is also 0, and, by Theorem 7, $\mathbf{Cx} = \mathbf{b}$ never has a unique solution. ■

Matrices Whose Determinant Is 0

The determinant has a role like the expression $b^2 - 4ac$ under the square root sign in the quadratic formula $x = (-b \pm \sqrt{b^2 - 4ac})/2a$ for the solution of $ax^2 + bx + c$ ($b^2 - 4ac$ is called the *discriminant* of the quadratic equation). The discriminant must be *nonnegative* for a quadratic equation to have real solutions (otherwise, x is imaginary). In the case of a system of equations $\mathbf{Ax} = \mathbf{b}$, the determinant $\det(\mathbf{A})$ must be *nonzero* for a well-defined solution $x_i = \det(\mathbf{A}_i)/\det(\mathbf{A})$ in Cramer's rule. If $\det(\mathbf{A}) = 0$, Cramer's rule involves division by zero and cannot be used.

Let us consider things the other way around. In Section 1.4 on Gaussian elimination, we saw examples where a system $\mathbf{Ax} = \mathbf{b}$ of three equations in three variables had an infinite number of solutions or had no solution. In such systems $\det(\mathbf{A})$ must be zero, since, as just noted, when $\det(\mathbf{A}) \neq 0$, Cramer's rule gives a unique solution.

Although systems of equations without unique solutions must have determinants equal to 0, Cramer's rule does not tell us exactly what happens in $\mathbf{Ax} = \mathbf{b}$ if $\det(\mathbf{A}) = 0$. There might be no solution or there might be an infinite number of solutions. To get a better understanding of what might happen when $\det(\mathbf{A}) = 0$, let us return to Cramer's rule.

We take its formula for x_i and multiply both sides by $\det(\mathbf{A})$:

$$x_i = \frac{\det(\mathbf{A}_i)}{\det(\mathbf{A})} \implies \det(\mathbf{A}) \cdot x_i = \det(\mathbf{A}_i) \tag{23}$$

The derivation of Cramer's rule involves getting the expression on the right and then dividing by $\det(\mathbf{A})$. For example, in the 2×2 case at the start of this section, we obtained the equation

$$(a_{11}a_{22} - a_{12}a_{21})x_1 = a_{22}b_1 - a_{12}b_2 \tag{24}$$

and, from it, obtained Cramer's rule for x_1 for 2×2 matrices [see (2)]. In (24), suppose that $(a_{11}a_{22} - a_{12}a_{21}) = 0$. If $a_{22}b_1 - a_{12}b_2 \neq 0$, no value of x_1 can satisfy (24). But if $a_{22}b_1 - a_{12}b_2 = 0$, any value of x_1 satisfies (24). More generally, in the equation on the right in (23), if $\det(\mathbf{A}) = 0$ but $\det(\mathbf{A}_i) \neq 0$, there is no possible value of x_i that satisfies the equation. But if $\det(\mathbf{A}_i) = 0$ as well as $\det(\mathbf{A}) = 0$, then any value of x_i satisfies $\det(\mathbf{A}) \cdot x_i = \det(\mathbf{A}_i)$. We illustrate this situation with a system of equations arising from the weather Markov chain.

EXAMPLE 8.
System of
Equations with
Multiple Solutions

In Example 9 of Section 1.4, we solved the following system of equations to determine the stable probability vector (which stays the same from one period to the next) for our weather Markov chain model.

$$\mathbf{p}^* = \mathbf{A}\mathbf{p}^*: \quad \begin{aligned} p_1^* &= \tfrac{3}{4}p_1^* + \tfrac{1}{2}p_2^* + \tfrac{1}{4}p_3^* \\ p_2^* &= \tfrac{1}{8}p_1^* + \tfrac{1}{4}p_2^* + \tfrac{1}{2}p_3^* \\ p_3^* &= \tfrac{1}{8}p_1^* + \tfrac{1}{4}p_2^* + \tfrac{1}{4}p_3^* \end{aligned} \tag{25}$$

Let us now try to re-solve (25) using Cramer's rule. First we collect the p^*'s on the left to obtain

$$\begin{aligned} \tfrac{1}{4}p_1^* - \tfrac{1}{2}p_2^* - \tfrac{1}{4}p_3^* &= 0 \\ -\tfrac{1}{8}p_1^* + \tfrac{3}{4}p_2^* - \tfrac{1}{2}p_3^* &= 0 \\ -\tfrac{1}{8}p_1^* - \tfrac{1}{4}p_2^* + \tfrac{3}{4}p_3^* &= 0 \end{aligned} \tag{26}$$

Next we compute the determinant of the coefficient matrix $\mathbf{A}$ of (26).

$$\det(\mathbf{A}) = \begin{vmatrix} \tfrac{1}{4} & -\tfrac{1}{2} & -\tfrac{1}{4} \\ -\tfrac{1}{8} & \tfrac{3}{4} & -\tfrac{1}{2} \\ -\tfrac{1}{8} & -\tfrac{1}{4} & \tfrac{3}{4} \end{vmatrix} = \frac{1}{4}\begin{vmatrix} \tfrac{3}{4} & -\tfrac{1}{2} \\ -\tfrac{1}{4} & \tfrac{3}{4} \end{vmatrix} + \frac{1}{2}\begin{vmatrix} -\tfrac{1}{8} & -\tfrac{1}{2} \\ -\tfrac{1}{8} & \tfrac{3}{4} \end{vmatrix} - \frac{1}{4}\begin{vmatrix} -\tfrac{1}{8} & \tfrac{3}{4} \\ -\tfrac{1}{8} & -\tfrac{1}{4} \end{vmatrix}$$

$$= \frac{1}{4}\left(\frac{7}{16}\right) + \frac{1}{2}\left(-\frac{5}{32}\right) - \frac{1}{4}\left(\frac{2}{16}\right) = \frac{7 - 5 - 2}{64} = 0 \tag{27}$$

Observe that since the right-side vector of (26) is all 0's, the matrices $\mathbf{A}_1$, $\mathbf{A}_2$, $\mathbf{A}_3$ will each have a column of 0's. This means that their determinants will be zero (see the lemma earlier in this section). So Cramer's rule gives $p_i^* = 0/0$ and we are stuck. In a certain sense, we need Cramer's rule to fail, since $p_1^* = p_2^* = p_3^* = 0$ is always a solution to (26), so if Cramer's rule gave us a unique solution, it would have to be $p_1^* = p_2^* = p_3^* = 0$. But we are looking for a stable probability vector, not the zero vector.

Recall that in Section 1.4 we solved (26) by Gaussian elimination and got an infinite set of nonzero solutions (to get the unique stable probability vector, we had to add the probability constraint: $p_1^* + p_2^* + p_3^* = 1$). ■

In Section 1.4 we called a system of equations $\mathbf{Ax} = \mathbf{b}$ *homogeneous* if $\mathbf{b} = \mathbf{0}$. Thus the system (26) is homogeneous. As in Example 8, it is common with homogeneous systems to be seeking another (more interesting) solution other than the zero solution. In order to have another solution—that is, the solution should not be unique—*the determinant must be* 0. Recall from Theorem 2 in Section 1.4 that if $\mathbf{Ax} = \mathbf{0}$ does not have a unique solution, it either has no solution or an infinite number of solutions. However, a homogeneous system cannot have no solution, since $\mathbf{x} = \mathbf{0}$ is always a solution. Thus when $\det(\mathbf{A}) = 0$, $\mathbf{Ax} = \mathbf{0}$ must have an infinite number of solutions.

Theorem 8. Let $\mathbf{Ax} = \mathbf{0}$ be a homogeneous system of n linear equations in n variables.
  (i) If $\det(\mathbf{A}) \neq 0$, the system has the unique solution $\mathbf{x} = \mathbf{0}$.
  (ii) If $\det(\mathbf{A}) = 0$, the system has an infinite number of solutions.

Section 1.6 Exercises

1. Compute the determinant of the following matrices.

(a) $\begin{vmatrix} 1 & 3 \\ 5 & -2 \end{vmatrix}$ (b) $\begin{vmatrix} 2 & 4 \\ -3 & -6 \end{vmatrix}$ (c) $\begin{vmatrix} 3 & 2 \\ 1 & 0 \end{vmatrix}$

2. Find the (unique) solution to the following systems of equations, if possible, using Cramer's rule.

(a) $\begin{aligned} x + y &= 34 \\ 2x - y &= 30 \end{aligned}$ (b) $\begin{aligned} 2x - 3y &= 5 \\ -4x + 6y &= 10 \end{aligned}$ (c) $\begin{aligned} 3x + y &= 7 \\ 2x - 2y &= 7 \end{aligned}$

3. Consider the two-refinery production of diesel oil and gasoline. The second refinery has not been built, but when it is built it will produce twice as much gas as diesel oil from each barrel of crude oil. We have

$$\text{Diesel oil: } 10x_1 + ax_2 = D$$
$$\text{Gasoline: } 5x_1 + 2ax_2 = G$$

where a is to be determined, D is demand for diesel oil, G demand for gasoline (and x_i is the number of barrels of crude oil processed by refinery i, $i = 1, 2$). Solve this system of equations to determine x_1 and x_2 in terms of a, D, G using Cramer's rule.

4. Consider the two-refinery production of diesel oil and gasoline. The second refinery has not been built, but when it is built it will produce 15 gallons of gasoline and k gallons of diesel oil from each barrel of crude oil. We have

$$\text{Diesel oil: } 10x_1 + kx_2 = D$$
$$\text{Gasoline: } 5x_1 + 15x_2 = G$$

where k is to be determined, D is the demand for diesel oil, G is the demand for gasoline (and x_i is the number of barrels of crude oil processed by refinery i, $i = 1, 2$). Solve this system of equations to determine x_1 and x_2 in terms of k, D, G using Cramer's rule. What values of k yield a nonunique solution? In practical terms, what does this nonuniqueness mean?

5. Which of the following systems of equations have nonzero solutions? If the solution is not unique, give the set of all possible solutions.

(a) $3x + 4y = 0$ (b) $4x - y = 0$ (c) $2x - 6y = 0$
    $6x + 2y = 0$ $4x - y = 0$ $-x + 3y = 0$

6. When the right-hand side is nonzero and the determinant is 0, there may be no solution to the system of equations. Which of the following systems of equations have no solution?

(a) $3x + 2y = 2$ (b) $2x - 3y = 2$ (c) $2x - 6y = 4$
    $6x + 4y = 2$ $2x - 3y = 2$ $-x + 3y = -2$

7. Compute the following determinants.

(a) $\begin{vmatrix} 1 & 1 & 1 \\ 1 & 2 & 1 \\ 3 & 1 & 1 \end{vmatrix}$ (b) $\begin{vmatrix} 2 & 0 & -1 \\ 0 & 1 & 3 \\ -4 & 0 & 2 \end{vmatrix}$ (c) $\begin{vmatrix} 0 & 1 & 0 \\ 1 & 2 & 1 \\ 2 & 5 & 2 \end{vmatrix}$

(d) $\begin{vmatrix} 1 & 0 & 1 \\ 0 & 2 & 2 \\ 1 & 2 & 3 \end{vmatrix}$ (e) $\begin{vmatrix} 2 & 1 & 0 \\ 0 & 0 & 2 \\ 2 & 2 & 2 \end{vmatrix}$ (f) $\begin{vmatrix} \frac{1}{6} & \frac{1}{7} & \frac{1}{8} \\ \frac{1}{7} & \frac{1}{8} & \frac{1}{9} \\ \frac{1}{8} & \frac{1}{9} & \frac{1}{10} \end{vmatrix}$

8. In which of the matrices in Exercise 7 is one row or column a multiple of another?

9. Use Cramer's rule to solve for x_3 in Example 4.

10. Solve the following systems of equations using Cramer's rule.

(a) $2x - y + 2z = 4$ (b) $x + y = 3$ (c) $-x + 3y - z = 4$
    $x + 3z = 6$ $2x + 3y + z = 9$ $2x - y = 6$
    $2y - z = 1$ $-x - y - z = -4$ $x + z = 3$

11. If you double the first row in the following system, show using Cramer's rule that the solution does not change.

$$ax + by = e$$
$$cx + dy = f$$

12. If you double the first column in the following system, show using Cramer's rule that the value of x is half as large and the value of y is unchanged.

$$ax + by = e$$
$$cx + dy = f$$

13. From the computational definition for a determinant, deduce that for any square matrix $\mathbf{A}$, its transpose $\mathbf{A}^T$, obtained by interchanging rows and columns, has the same determinant as $\mathbf{A}$ (actually, the details of whether the sign of the determinant is unchanged were not specified). Thus $\mathbf{A}^T\mathbf{x} = \mathbf{b}$ has a unique solution if and only if $\mathbf{A}\mathbf{x} = \mathbf{b}$ does.

14. Compute the determinant of the following matrices.

(a) $\begin{bmatrix} 2 & 3 & 7 & 8 \\ 0 & 3 & 9 & 1 \\ 0 & 0 & 1 & 5 \\ 0 & 0 & 0 & 4 \end{bmatrix}$ (b) $\begin{bmatrix} 3 & 0 & 0 & 0 \\ 2 & 6 & 0 & 0 \\ 4 & 5 & 1 & 0 \\ 3 & 4 & 5 & 4 \end{bmatrix}$ (c) $\begin{bmatrix} 0 & 2 & 0 & 0 & 0 \\ 0 & 0 & 3 & 0 & 0 \\ 0 & 0 & 0 & 1 & 0 \\ 0 & 0 & 0 & 0 & 2 \\ 2 & 0 & 0 & 0 & 0 \end{bmatrix}$

15. Use Gaussian elimination and Theorem 4 to compute the determinant of the following matrices.

(a) $\begin{bmatrix} 2 & 0 & -1 \\ 0 & 1 & 3 \\ 2 & 0 & -1 \end{bmatrix}$ (b) $\begin{bmatrix} 1 & 1 & 1 \\ 1 & 2 & 1 \\ 3 & 1 & 1 \end{bmatrix}$ (c) $\begin{bmatrix} 3 & 2 & 1 & 0 \\ 1 & 1 & 0 & -1 \\ 2 & 1 & -1 & 1 \\ 1 & 1 & 1 & 1 \end{bmatrix}$

16. For what value(s) of k do the following matrices fail to have an inverse?

(a) $\begin{bmatrix} 2 & k \\ 3 & 6 \end{bmatrix}$ (b) $\begin{bmatrix} 1 & k \\ k & 9 \end{bmatrix}$ (c) $\begin{bmatrix} 2 & 1 & 3 \\ 1 & 0 & k \\ 0 & 2 & 1 \end{bmatrix}$

17. If $\mathbf{A}$ is a 3×3 matrix with $\det(\mathbf{A}) = 3$, determine:
(a) $\det(\mathbf{A}^2)$ (b) $\det(\mathbf{A}^4)$ (c) $\det(\mathbf{A}^{-1})$ (d) $\det(3\mathbf{A})$

18. If $\mathbf{A}, \mathbf{B}, \mathbf{C}$ are 4×4 matrices such that $\mathbf{C} = \mathbf{AB}$, $\det(\mathbf{A}) = 3$, and $\det(\mathbf{C}) = 6$, determine:
(a) $\det(\mathbf{B})$ (b) $\det(4\mathbf{B})$ (c) $\det(\mathbf{ABC})$ (d) $\det(\mathbf{B}^{-1}\mathbf{A})$

19. If $\mathbf{A}$ and $\mathbf{B}$ are square matrices and $\mathbf{A}$ is invertible, show that $\det(\mathbf{B}) = \det(\mathbf{A}^{-1}\mathbf{BA})$.

20. Let $\mathbf{A}$ be an $n \times n$ matrix. Which of the following statements guarantee that $\mathbf{Ax} = \mathbf{b}$ has a unique solution? Which guarantee that $\mathbf{Ax} = \mathbf{b}$ does not have a unique solution?
(a) $\det(\mathbf{A}) = 2$.
(b) One row of $\mathbf{A}$ is a multiple of another row of $\mathbf{A}$.
(c) Two rows of $\mathbf{A}$ are equal.
(d) $\det(\mathbf{A}^{-1}) = 5$.
(e) $\mathbf{A} = \mathbf{BC}$ and $\det(\mathbf{B}) = \det(\mathbf{C}) = 2$.

21. Which of the following statements guarantees that the homogeneous system $\mathbf{Ax} = \mathbf{0}$, where $\mathbf{A}$ is a square matrix, has only $\mathbf{x} = \mathbf{0}$ as a solution? Which guarantee that $\mathbf{Ax} = \mathbf{0}$ has an infinite number of solutions?
(a) $\det(\mathbf{A}) = 2$.
(b) $\det(\mathbf{A}^{-1}) = 2$.
(c) Two rows of $\mathbf{A}$ are equal.
(d) $\mathbf{A}$ is not invertible.
(e) $\mathbf{Ax} = \mathbf{b}$ has a unique solution, for some $\mathbf{b}$.

22. Let $\mathbf{A} = \begin{bmatrix} a_{11} & a_{12} \\ a_{21} & a_{22} \end{bmatrix}$ and $\mathbf{B} = \begin{bmatrix} b_{11} & b_{12} \\ b_{21} & b_{22} \end{bmatrix}$ be arbitrary 2×2 matrices. Symbolically multiply out $\mathbf{AB}$ and determine $\det(\mathbf{AB})$. Then show that $\det(\mathbf{AB}) = \det(\mathbf{A}) \det(\mathbf{B})$.

23. In the accompanying figure, the area of the triangle ABC can be expressed as

$$\text{area } ABC = \text{area } ABB'A' + \text{area } BCC'B' - \text{area } ACC'A' \qquad (*)$$

Using $(*)$ and the fact that the area of a trapezoid is $\frac{1}{2}$ of the distance between the parallel sides times the sum of the lengths of the parallel sides, show that

$$\text{area } ABC = \frac{1}{2} \begin{vmatrix} x_1 & y_1 & 1 \\ x_2 & y_2 & 1 \\ x_3 & y_3 & 1 \end{vmatrix}$$

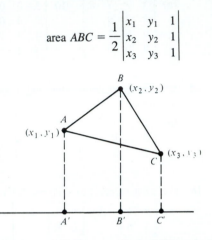

24. This exercise develops the proof of Theorem 5, $\det(\mathbf{AB}) = \det(\mathbf{A})\det(\mathbf{B})$.

(a) *Case A*: $\mathbf{A}$ is not invertible and, by Theorem 7, has determinant 0. Then we must show that $\det(\mathbf{AB}) = 0$. Using Theorem 7, restate this problem as: If $\mathbf{A}$ is not invertible, $\mathbf{AB}$ is not invertible. (*Hint:* Use Exercise 21 in Section 1.5.)

(b) *Case B*: $\mathbf{A}$ is invertible. Use Corollary B of Theorem 3 in Section 1.5, which says that an invertible matrix can be written as a product of elementary matrices. So $\mathbf{A} = \mathbf{E}_k \cdots \mathbf{E}_2 \mathbf{E}_1$. Show that for any elementary matrix $\mathbf{E}$, $\det(\mathbf{EC}) = \det(\mathbf{E})\det(\mathbf{C})$. (See the end of Section 1.3 for a description of elementary matrices.) Now apply $\det(\mathbf{EC}) = \det(\mathbf{E})\det(\mathbf{C})$ repeatedly to show that:

$$(*) \ \det(\mathbf{AC}) = \det(\mathbf{E}_k \cdots \mathbf{E}_2 \mathbf{E}_1 \mathbf{C}) = \det(\mathbf{E}_k) \cdots \det(\mathbf{E}_2)\det(\mathbf{E}_1)\det(\mathbf{C})$$

Then in $(*)$, set $\mathbf{C} = \mathbf{I}$ to get $\det(\mathbf{A}) = \det(\mathbf{E}_k) \cdots \det(\mathbf{E}_2)\det(\mathbf{E}_1)$, and next set $\mathbf{C} = \mathbf{B}$ to get $\det(\mathbf{AB}) = \det(\mathbf{A})\det(\mathbf{B})$.

2 Vectors

2.1 Collections of Vectors

Geometry of Vectors

In this chapter we examine mathematical properties of individual vectors and collections of vectors. Vectors have a geometric and an algebraic interpretation. In the first chapter we used vectors in an algebraic setting. There, vectors were row vectors or column vectors that arose in matrix models, were manipulated by matrix algebra, and were evaluated by matrix computations. The geometric approach to vectors is much older and grew out of physics problems in which one must consider the direction and magnitude of a moving object or a force applied to the object.

In this chapter we restrict ourselves to vectors in two and three dimensions. We start with two dimensions. The name mathematicians give to the two-dimensional space of real numbers is $\mathbf{R}^2$. The standard way to give a geometric representation of a vector is with a pointed line segment starting at the origin of $\mathbf{R}^2$ and extending to a point with the coordinates given by the numbers in the vector. Figure 2.1(a) displays

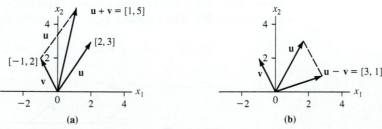

Figure 2.1

the two vectors $\mathbf{u} = [2, 3]$, $\mathbf{v} = [-1, 2]$. The first entry of each vector is the x_1-coordinate of the vector, the distance of the endpoint of the vector from the origin in the direction of the x_1-axis. The second entry is the x_2-coordinate of the vector, the distance of the endpoint from the origin in the direction of the x_2-axis.

It is common when discussing vectors in geometric settings, as in most of this chapter, to write *vectors as row vectors*. Note that when matrices are not present, the distinction between row and column vectors is unimportant.

If the two vectors $\mathbf{u} = [2, 3]$, $\mathbf{v} = [-1, 2]$ in Figure 2.1(a) represent two forces acting on an object, the net force from $\mathbf{u}$ and $\mathbf{v}$ is, not surprisingly, $\mathbf{u} + \mathbf{v} = [2 + (-1), 3 + 2] = [1, 5]$. Visually, we depict the addition of two vectors in Figure 2.1(a) by displacing $\mathbf{u}$ so that the head of $\mathbf{u}$ is at the tail of $\mathbf{v}$ ($\mathbf{u}$ still points in the same direction and has the same length). We can also form the difference $\mathbf{u} - \mathbf{v}$ of two vectors, $\mathbf{u} - \mathbf{v} = [2 - (-1), 3 - 2] = [3, 1]$ [see Figure 2.1(b)].

**EXAMPLE 1.
Airplane Flight
Direction**

A turboprop airplane is flying due east at 200 miles per hour. It meets with a crosswind coming from the southeast with a westerly component (the negative of east direction) of 40 miles per hour and northerly component of 30 miles per hour. Figure 2.2 depicts the two velocity vectors (thick vector lines) acting on the plane. To find the net velocity of the airplane, we add the two vectors to get the (lighter) vector in Figure 2.2. Their sum is $[200, 0] + [-40, 30] = [160, 30]$. (We assume that the first component of a vector is the easterly speed and the second component the northerly speed.) ■

In $\mathbf{R}^3$, the three-dimensional space of real numbers, depictions of vectors are a little bit more difficult to draw on a page. Otherwise, the concepts and visualizations are the same as in $\mathbf{R}^2$. Now the three entries in a vector are the values of the three coordinates of the vector's endpoint. These coordinates represent distances along the x_1-, x_2-, and x_3-axes (with the x_1-axis pointing "out" of the page toward the reader, the x_2-axis pointing to the right, and the x_3-axis pointing up; see Figure 2.3).

**EXAMPLE 2.
Vectors in $\mathbf{R}^3$**

The vectors $\mathbf{u} = [3, 0, 3]$ and $\mathbf{v} = [-2, 2, 0]$ are depicted in Figure 2.3. Their sum $\mathbf{w} = \mathbf{u} + \mathbf{v}$ is also shown in Figure 2.3, where $\mathbf{w} = \mathbf{u} + \mathbf{v} = [3, 0, 3] + [-2, 2, 0] = [1, 2, 3]$. ■

Lines

In linear algebra it is more common to work with collections of vectors than individual vectors. There are several natural types of collections of vectors. Initially, we consider collections in $\mathbf{R}^2$. The simplest type is the collection of multiples of a vector $\mathbf{d}$,

$$L(\mathbf{d}) = \{r\mathbf{d} : r \text{ a scalar}\} = \{[rd_1, rd_2], \text{ for } r \text{ a scalar}\} \tag{1}$$

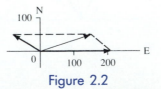

Figure 2.2

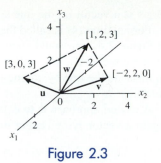

Figure 2.3

Another type is the set of vectors whose entries satisfy some equation, that is, the set of vectors $\mathbf{x}$ such that $a_1 x_1 + a_2 x_2 = b$, or $\mathbf{a} \cdot \mathbf{x} = b$, for some vector $\mathbf{a}$ and some scalar b. For example, consider the following two sets of vectors that form lines

$$\mathrm{L}_1 = \{[x_1, x_2]: 4x_1 + 5x_2 = 0\} \quad \text{and} \quad \mathrm{L}_2 = \{[x_1, x_2]: 3x_1 + 4x_2 = 1\} \tag{2}$$

L_1 and L_2 are drawn in Figure 2.4. (Notice that they intersect at a single vector $[-5, 4]$.)

Observe that L_1 equals the set $\mathrm{L}(\mathbf{d})$ of multiples of the vector $\mathbf{d} = [5, -4]$. (The reader should check this; see Figure 2.4.) Any line through the origin in $\mathbf{R}^2$ will consist of multiples of some vector. To see this, we note that when a line is written as $x_2 = cx_1 + d$, it is being defined by its slope c and its x_2-intercept. If the x_2-intercept is 0, the equation can be written simply as

$$x_2 = cx_1 \quad \text{or} \quad \frac{x_2}{x_1} = c \tag{3}$$

The multiples $\mathbf{x} = r\mathbf{d}$ of a vector $\mathbf{d} = [d_1, d_2]$ clearly satisfy (3) when $c = d_2/d_1$, for if $[x_1, x_2] = r[d_1, d_2]$, then $x_2/x_1 = rd_2/rd_1 = d_2/d_1 = c$.

A line that does not pass through the origin can be defined in a fashion similar to $\mathrm{L}(\mathbf{d})$. There are two ways to do this. One way is to write the line as

$$\mathbf{v} + \mathrm{L}(\mathbf{d}) = \{\mathbf{v} + r\mathbf{d} : r \text{ a scalar}\} = \{[v_1 + rd_1, v_2 + rd_2], \text{ for } r \text{ a scalar}\} \tag{4}$$

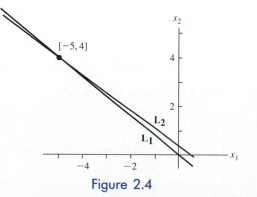

Figure 2.4

where $\mathbf{v}$ is a vector on the line and $\mathbf{d}$ is a vector pointing in the same direction as the line. Formulation (4) is called the **parametric form of a line**. For the line $3x_1 + 4x_2 = 1$ in Figure 2.4, one could choose the vector $\mathbf{v} = [0, \frac{1}{4}]$ and $\mathbf{d} = [4, -3]$ (see Figure 2.4). Here we take a line $L(\mathbf{d})$ through the origin that is parallel to the given line and then translate the line by adding the vector $\mathbf{v}$ to it [i.e., adding $\mathbf{v}$ to each of the vectors of $L(\mathbf{d})$].

A related approach is to define the line in terms of the difference $\mathbf{x} - \mathbf{v}$, where $\mathbf{v}$ is again a vector lying on the line. Define the set of vectors

$$L(\mathbf{d}, \mathbf{v}) = \{\mathbf{x} : \mathbf{x} - \mathbf{v} = r\mathbf{d}, \text{ for some scalar } r\} \tag{5}$$

Since $\mathbf{x} - \mathbf{v} = r\mathbf{d}$ in $L(\mathbf{d}, \mathbf{v})$, then $\mathbf{x} = \mathbf{v} + r\mathbf{d}$. Then (5) can be rewritten as $L(\mathbf{d}, \mathbf{v}) = \{[v_1 + rd_1, v_2 + rd_2], \text{ for } r \text{ a scalar}\}$. That is, $L(\mathbf{d}, \mathbf{v}) = \mathbf{v} + L(\mathbf{d})$.

In definition (5) we observe that if $x_1 = v_1 + rd_1$, then $r = (x_1 - v_1)/d_1$. Also, $r = (x_2 - v_2)/d_2$. Setting these two expressions for r equal to one another, we obtain

$$\frac{x_1 - v_1}{d_1} = \frac{x_2 - v_2}{d_2} \tag{6}$$

Formulation (6) is called the **symmetric form of a line**. *Note: The symmetric form of a line cannot be used if any of the components of $\mathbf{d}$ are zero.*

The parametric and symmetric forms for defining a line in $\mathbf{R}^2$ extend to lines in $\mathbf{R}^3$. The parametric form in 3-space becomes

$$\mathbf{v} + L(\mathbf{d}) = \{[v_1 + rd_1, v_2 + rd_2, v_3 + rd_3], \text{ for } r \text{ a scalar}\} \tag{7}$$

and the symmetric form becomes

$$\frac{x_1 - v_1}{d_1} = \frac{x_2 - v_2}{d_2} = \frac{x_3 - v_3}{d_3} \tag{8}$$

We summarize the forms for describing a line in $\mathbf{R}^2$ and $\mathbf{R}^3$.

Three Forms of a Line in $\mathbf{R}^2$

1. *Standard equation.* $a_1x_1 + a_2x_2 = b$, for given $\mathbf{a}$ and b.
2. *Parametric form.* $\mathbf{v} + L(\mathbf{d}) = \{[v_1 + rd_1, v_2 + rd_2], \text{ for } r \text{ a scalar}\}$.
3. *Symmetric form.* $\dfrac{x_1 - v_1}{d_1} = \dfrac{x_2 - v_2}{d_2}$, for given $\mathbf{v}$ and $\mathbf{d}$.

Two Forms of a Line in $\mathbf{R}^3$

1. *Parametric form.* $\mathbf{v} + L(\mathbf{d}) = \{[v_1 + rd_1, v_2 + rd_2, v_3 + rd_3], \text{ for } r \text{ a scalar}\}$.
2. *Symmetric form.* $\dfrac{x_1 - v_1}{d_1} = \dfrac{x_2 - v_2}{d_2} = \dfrac{x_3 - v_3}{d_3}$.

We will learn another way to determine the equation of a line in $\mathbf{R}^2$ in Section 2.3.

Suppose that we are given two vectors $\mathbf{u} = [1, 6]$ and $\mathbf{v} = [-1, 2]$ on a line and wish to find the parametric form or the symmetric form of the line through $\mathbf{u}$ and $\mathbf{v}$ or the standard equation $a_1x_1 + a_2x_2 = b$ of the line. The difference vector $\mathbf{u} - \mathbf{v}$ provides the needed vector $\mathbf{d}$ pointing in the direction of the line.

$$\mathbf{d} = \mathbf{u} - \mathbf{v} = [1, 6] - [-1, 2] = [2, 4]$$

Then the parametric form is $\mathbf{v} + L(\mathbf{d}) = \{[-1 + 2r, 2 + 4r],$ for r a scalar$\}$ and the symmetric form is

$$\frac{x_1 + 1}{2} = \frac{x_2 - 2}{4}$$

Note that it would have been more natural to choose $\mathbf{d} = [1, 2]$ (factoring out common multiples).

To find the equation $a_1x_1 + a_2x_2 = b$ of the line through $\mathbf{u}$ and $\mathbf{v}$, we first rewrite the equation as $a_1x_1 + a_2x_2 - b = 0$. Since $\mathbf{u} = [1, 6]$ and $\mathbf{v} = [-1, 2]$ must satisfy this equation, we obtain the equations

$$a_1 + 6a_2 - b = 0$$
$$-a_1 + 2a_2 - b = 0$$

Reducing the associated augmented matrix to reduced row echelon form, we have

$$\begin{bmatrix} 1 & 6 & -1 & \Big| & 0 \\ -1 & 2 & -1 & \Big| & 0 \end{bmatrix} \Longrightarrow \begin{bmatrix} 1 & 6 & -1 & \Big| & 0 \\ 0 & 8 & -2 & \Big| & 0 \end{bmatrix} \Longrightarrow \begin{bmatrix} 1 & 0 & \frac{1}{2} & \Big| & 0 \\ 0 & 1 & -\frac{1}{4} & \Big| & 0 \end{bmatrix} \tag{9}$$

So (9) yields $a_1 + \frac{1}{2}b = 0$, $a_2 - \frac{1}{4}b = 0$ or $a_1 = -\frac{1}{2}b$, $a_2 = \frac{1}{4}b$. Setting $b = 4$, we get $a_1 = -2$, $a_2 = 1$ and the desired equation is $-2x_1 + x_2 = 4$. Picking a different value for b just multiplies the values of a_1, a_2, and b by a constant, but does not change the line defined by the equation.

EXAMPLE 3.
Line in $\mathbf{R}^2$
Determined by
Two Vectors

Given vectors $\mathbf{u} = [4, -1]$ and $\mathbf{v} = [2, 2]$, find the parametric form and the symmetric forms of the line through $\mathbf{u}$ and $\mathbf{v}$ and its standard equation $a_1x_1 + a_2x_2 = b$. As above, we obtain a vector $\mathbf{d}$ pointing in the direction of the line from the difference of the given vectors

$$\mathbf{d} = \mathbf{u} - \mathbf{v} = [4, -1] - [2, 2] = [2, -3]$$

Then the parametric form of the line through $\mathbf{u}$ and $\mathbf{v}$ is $\mathbf{v} + L(\mathbf{d}) = \{[2 + 2r, 2 - 3r],$ for r a scalar$\}$ and the symmetric form is

$$\frac{x_1 - 2}{2} = \frac{x_2 - 2}{-3}$$

Since $\mathbf{u}$ and $\mathbf{v}$ must satisfy the equation $a_1x_1 + a_2x_2 - b = 0$, we have $4a_1 - a_2 - b = 0$ and $2a_1 + 2a_2 - b = 0$. Reducing the associated augmented matrix to reduced row echelon form, we have

$$\begin{bmatrix} 4 & -1 & -1 & | & 0 \\ 2 & 2 & -1 & | & 0 \end{bmatrix} \Longrightarrow \begin{bmatrix} 1 & -\frac{1}{4} & -\frac{1}{4} & | & 0 \\ 0 & \frac{5}{2} & -\frac{1}{2} & | & 0 \end{bmatrix} \Longrightarrow \begin{bmatrix} 1 & 0 & -\frac{3}{10} & | & 0 \\ 0 & 1 & -\frac{1}{5} & | & 0 \end{bmatrix}$$

So $a_1 - \frac{3}{10}b = 0$, $a_2 - \frac{1}{5}b = 0$ or $a_1 = \frac{3}{10}b$, $a_2 = \frac{1}{5}b$. Setting $b = 10$, we get $a_1 = 3$, $a_2 = 2$ and the desired standard equation is $3x_1 + 2x_2 = 10$. ■

EXAMPLE 4.
Line in $\mathbf{R}^3$
Determined by
Two Vectors

Given vectors $\mathbf{u} = [2, -1, 3]$ and $\mathbf{v} = [0, 2, -1]$, find the parametric form and the symmetric form of the line through $\mathbf{u}$ and $\mathbf{v}$. (*Note:* There is no standard equation for a line in $\mathbf{R}^3$.) As above, we obtain a vector $\mathbf{d}$ pointing in the direction of the line from the difference of the given vectors

$$\mathbf{d} = \mathbf{u} - \mathbf{v} = [2, -1, 3] - [0, 2, -1] = [2, -3, 4]$$

Then the parametric form of the line through $\mathbf{u}$ and $\mathbf{v}$ is $\mathbf{v} + L(\mathbf{d}) = \{[2r, 2 - 3r, -1 + 4r]$, for r a scalar and the symmetric form is

$$\frac{x_1}{2} = \frac{x_2 - 2}{-3} = \frac{x_3 + 1}{4}$$

■

We can also define a collection of vectors by a system of equations. For example, given a matrix $\mathbf{A}$ and a vector $\mathbf{b}$, we can define the set of all vectors $\mathbf{x}$ for which $\mathbf{Ax} = \mathbf{b}$. This set may be a plane, a line, or some other geometric form, or there may be no solution.

EXAMPLE 5.
Solution Set for
Two-Product
Refinery Problem

Suppose that in our oil refinery model, introduced in Section 1.1, the production of gasoline is not important (there is an excess supply in storage tanks). We are only concerned with the demands for heating oil and diesel oil. The system of equations reduces to

$$\text{Heating oil: } 4x_1 + 2x_2 + 2x_3 = 600$$
$$\text{Diesel oil: } \quad 2x_1 + 5x_2 + 2x_3 = 800 \tag{10}$$

Each equation in (10) is satisfied by a set of vectors that form a two-dimensional plane, since for every x_1 and x_2, there is a vector $[x_1, x_2, x_3]$ with $x_3 = (600 - 4x_1 - 2x_2)/2$ satisfying the heating oil equation; similarly for the diesel oil equation. We reduce the augmented coefficient matrix to reduced row echelon form.

$$\begin{bmatrix} 4 & 2 & 2 & | & 600 \\ 2 & 5 & 2 & | & 800 \end{bmatrix} \Longrightarrow \begin{bmatrix} 1 & \frac{1}{2} & \frac{1}{2} & | & 150 \\ 0 & 4 & 1 & | & 500 \end{bmatrix} \Longrightarrow \begin{bmatrix} 1 & 0 & \frac{3}{8} & | & 87\frac{1}{2} \\ 0 & 1 & \frac{1}{4} & | & 125 \end{bmatrix}$$

Expressing x_1 and x_2 in terms of x_3, the reduced row echelon form gives

$$x_1 = 87\tfrac{1}{2} - \tfrac{3}{8}x_3 \quad \text{and} \quad x_2 = 125 - \tfrac{1}{4}x_3$$

In a parametric form, the solutions are the set of vectors $[87\tfrac{1}{2} - \tfrac{3}{8}x_3,\ 125 - \tfrac{1}{4}x_3,\ x_3]$, which can be written as $\mathbf{v} + L(\mathbf{d})$ with $\mathbf{v} = [87\tfrac{1}{2},\ 125,\ 0]$ and $\mathbf{d} = [-\tfrac{3}{8},\ -\tfrac{1}{4},\ 1]$ (see Figure 2.5). ■

In problems such as Example 5 with a set of vectors satisfying some constraints, there is often a cost function that one seeks to minimize, such as $35x_1 + 25x_2 + 40x_3$. Typically, one also usually requires that all $x_i \geq 0$. The mathematical subject of linear programming is concerned with optimizing a linear function over a collection of vectors defined by a system of linear equations.

Planes

Planes in $\mathbf{R}^3$ can be defined parametrically and with equations analogously to how lines are defined in $\mathbf{R}^2$ and $\mathbf{R}^3$, except that a plane is a two-dimensional structure, so its parametric form requires multiples of two vectors. We cannot use the symmetric form because there is no single vector that points in the direction of a plane. We start with planes that pass through the origin. For example, if $\mathbf{d}_1 = [3, 4, 1]$ and $\mathbf{d}_2 = [2, 5, 0]$, we can define the plane P:

$$P(\mathbf{d}_1, \mathbf{d}_2) = \{\mathbf{x} : \mathbf{x} = r_1\mathbf{d}_1 + r_2\mathbf{d}_2, \text{ for scalars } r_1, r_2\} \tag{11}$$

As with lines, the general parametric form of a plane is $\mathbf{v} + P(\mathbf{d}_1, \mathbf{d}_2)$, where $\mathbf{v}$ is a vector lying in the plane.

A plane can also be defined by an equation of the form $a_1x_1 + a_2x_2 + a_3x_3 = b$. Such an equation does define a plane since for any values chosen for x_1 and x_2, there is a vector satisfying this equation with $x_3 = (a_1x_1 + a_2x_2)/a_3$ (assuming that $a_3 \neq 0$).

Suppose that we are given three vectors $\mathbf{u}$, $\mathbf{v}$, $\mathbf{w}$ and want to find the parametric form and equation of the plane. We assume here that these three vectors do not lie on a line. (If they do lie on a line, the parametric form and equation form would yield that line.) We proceed in the same fashion as for determining lines. To find the direction vectors $\mathbf{d}_1$ and $\mathbf{d}_2$, we take the differences of the given vectors: $\mathbf{d}_1 = \mathbf{u} - \mathbf{v}$ and $\mathbf{d}_2 = \mathbf{u} - \mathbf{w}$. (Any two differences could be used.) To find the equation of the line, we

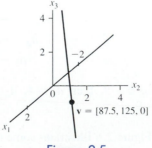

Figure 2.5

substitute each of the three vectors into the equation for the line to get a system of three equations which we solve to determine the parameters in the equation. The following example illustrates these computations.

EXAMPLE 6.
Plane Determined
by Three Vectors

Given the vectors $\mathbf{u} = [1, -1, 2]$, $\mathbf{v} = [1, 3, 4]$, and $\mathbf{w} = [3, 1, -1]$, find the parametric form of the plane and the equation $a_1x_1 + a_2x_2 + a_3x_3 = b$ of the plane.
 For our direction vectors $\mathbf{d}_1$ and $\mathbf{d}_2$ we use

$$\mathbf{d}_1 = \mathbf{u} - \mathbf{v} = [1, -1, 2] - [1, 3, 4] = [0, -4, -2]$$
$$\mathbf{d}_2 = \mathbf{u} - \mathbf{w} = [1, -1, 2] - [3, 1, -1] = [-2, -2, 3]$$

Now the parametric form is $\mathbf{v} + P(\mathbf{d}_1, \mathbf{d}_2) = \{\mathbf{v} + r_1\mathbf{d}_1 + r_2\mathbf{d}_2 : \text{for scalars } r_1, r_2\}$.
 Since $\mathbf{u} = [1, -1, 2]$, $\mathbf{v} = [1, 3, 4]$, and $\mathbf{w} = [3, 1, -1]$ must satisfy the equation of the plane, we have

$$1a_1 - 1a_2 + 2a_3 - b = 0$$
$$1a_1 + 3a_2 + 4a_3 - b = 0 \tag{12}$$
$$3a_1 + 1a_2 - 1a_3 - b = 0$$

This time, for variety, we will use Gauss–Jordan elimination to reduce the system of equations rather than the augmented coefficient matrix. Eliminating in the first column, we get

$$1a_1 - 1a_2 + 2a_3 - b = 0$$
$$4a_2 + 2a_3 = 0$$
$$4a_2 - 7a_3 + 2b = 0$$

Eliminating coefficients in the second column (and making the coefficient of a_2 in the second equation 1), we get

$$a_1 \quad + \tfrac{5}{2}a_3 - b = 0$$
$$a_2 + \tfrac{1}{2}a_3 = 0$$
$$- 9a_3 + 2b = 0$$

Finally, eliminating in the third column gives

$$a_1 \qquad - \tfrac{4}{9}b = 0 \qquad\qquad a_1 = \tfrac{4}{9}b$$
$$a_2 \quad + \tfrac{1}{9}b = 0 \qquad \text{or} \qquad a_2 = -\tfrac{1}{9}b \tag{13}$$
$$a_3 - \tfrac{2}{9}b = 0 \qquad\qquad a_3 = \tfrac{2}{9}b$$

Setting $b = 9$, then $a_1 = 4$, $a_2 = -1$, and $a_3 = 2$ and the equation is $4x_1 - x_2 + 2x_3 = 9$. ■

 Figure 2.6 illustrates some of the possible geometric situations arising when one tries to solve a system of three equations in three variables. Each equation defines a

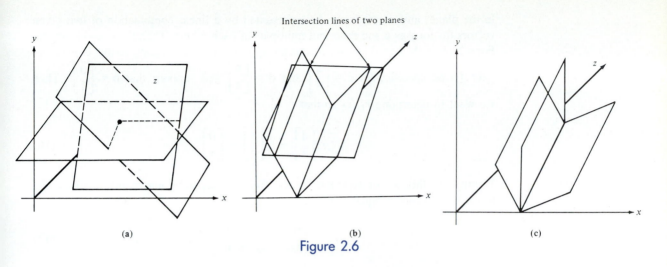

Figure 2.6

plane. Figure 2.6(a) shows the case where the three planes intersect at a point. Figure 2.6(b) shows the case where the intersection of each pair of planes forms a line (as was illustrated in Example 5), but the three lines are parallel and never intersect, so that the planes have no common point and the associated system of three equations in three variables has no solution. Figure 2.6(c) shows the case where the three planes intersect in a common line. (For concrete examples, see Example 3 in Section 1.4, where we presented two systems of equations, one with no solution and one with a line of solutions.)

Row and Column Geometry of Solving Two Equations

At the start of the discussion about lines, we presented two lines in standard equation form

$$L_1 = \{[x_1, x_2] : 4x_1 + 5x_2 = 0\} \quad \text{and} \quad L_2 = \{[x_1, x_2] : 3x_1 + 4x_2 = 1\}$$

The vector where these two lines intersect, that is, the vector that lies on both lines, is the solution to the system of equations

$$\begin{aligned} 4x_1 + 5x_2 &= 0 \\ 3x_1 + 4x_2 &= 1 \end{aligned} \tag{14}$$

This is the geometric way of picturing the solution of two equations in two unknowns (see Figure 2.4). Note that if the two equations describe parallel lines, there will be no vector satisfying both equations (unless the two equations describe the same line). We call (14) the *row* approach to solving a system of equations.

Suppose that we have two vectors **c** and **d** and we are interested in the set Q of vectors generated by linear combinations of **c** and **d**.

$$Q = \{\mathbf{b} : \mathbf{b} = r_1\mathbf{c} + r_2\mathbf{d}, \text{ for some scalars } r_1, r_2\} \tag{15}$$

In the plane, any vector **b** can be represented by a linear combination of two given vectors (as long as **c** and **d** are not multiples of each other). Then the question arises: For a 2-vector **b**, how do we determine the appropriate scalars r_1, r_2, so that $\mathbf{b} = r_1\mathbf{c} + r_2\mathbf{d}$? To be specific, let $\mathbf{c} = \begin{bmatrix} 4 \\ 3 \end{bmatrix}$ and $\mathbf{d} = \begin{bmatrix} 5 \\ 4 \end{bmatrix}$ and suppose that $\mathbf{b} = \begin{bmatrix} 0 \\ 1 \end{bmatrix}$. Then we want to determine r_1, r_2 so that

$$r_1\begin{bmatrix} 4 \\ 3 \end{bmatrix} + r_2\begin{bmatrix} 5 \\ 4 \end{bmatrix} = \begin{bmatrix} 0 \\ 1 \end{bmatrix} \tag{16}$$

Writing out this vector equation for the first and second components, we have the following system of equations:

$$\begin{aligned} 4r_1 + 5r_2 &= 0 \\ 3r_1 + 4r_2 &= 1 \end{aligned} \tag{17}$$

The solution is found to be $r_1 = -5$ and $r_2 = 4$. Figure 2.7 gives the geometric picture of how $-5\begin{bmatrix} 4 \\ 3 \end{bmatrix} + 4\begin{bmatrix} 5 \\ 4 \end{bmatrix}$ equals $\begin{bmatrix} 0 \\ 1 \end{bmatrix}$. This is the *column* approach to solving a system of equations. Notice that (14) and (17) are the same system of equations.

The row and column approach just described applies to vectors of any size.

Solution of a System of Equations

1. *Row problem*. Find a vector **x** that simultaneously satisfies each of the row equations $\mathbf{a}'_i \cdot \mathbf{x} = b_i$ in the system of equations $\mathbf{Ax} = \mathbf{b}$.
2. *Column problem*. Find a linear combination of the columns $\mathbf{a}_i$ of **A** that yield **b**; that is, find r_i's so that $r_1\mathbf{a}_1 + r_2\mathbf{a}_2 + \cdots + r_n\mathbf{a}_n = \mathbf{b}$.

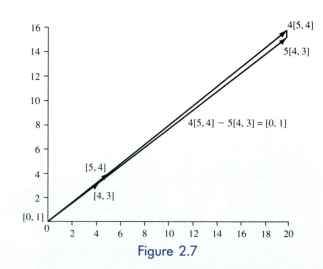

Figure 2.7

Laws of Vector Algebra

We conclude this section with a summary of the laws for vector algebra and scalar products. There are two special vectors that we shall encounter frequently. The *zero vector* **0** has all entries equal to 0. The *ones vector* **1** has all entries equal to 1. There is one zero vector and one ones vector for 2-vectors, for 3-vectors, and so on.

The *negative*, or *additive inverse*, $-\mathbf{v}$ of a vector **v** is obtained by multiplying the entries of **v** by -1. Intuitively, *vector subtraction is accomplished by reversing the direction of the vector*. For example, if $\mathbf{v} = [2, 3]$, then $-\mathbf{v} = [-2, -3]$.

We now state the principal laws of vector arithmetic and prove them by algebraic arguments.

Laws of Vector Algebra and Scalar Products

Theorem 1. The following properties are valid in vector arithmetic in $\mathbf{R}^2$ or $\mathbf{R}^3$ and actually in $\mathbf{R}^n$ for any positive integer n (assuming that all vectors in each equation have the same number of components).

(i) $\mathbf{u} + \mathbf{v} = \mathbf{v} + \mathbf{u}$.

(ii) $(\mathbf{u} + \mathbf{v}) + \mathbf{w} = \mathbf{u} + (\mathbf{v} + \mathbf{w})$.

(iii) $\mathbf{u} + \mathbf{0} = \mathbf{u} = \mathbf{0} + \mathbf{u}$.

(iv) $\mathbf{u} + (-\mathbf{u}) = \mathbf{u} - \mathbf{u} = \mathbf{0}$.

(v) $\mathbf{u} \cdot \mathbf{v} = \mathbf{v} \cdot \mathbf{u}$.

(vi) $\mathbf{u} \cdot (\mathbf{v} + \mathbf{w}) = \mathbf{u} \cdot \mathbf{v} + \mathbf{u} \cdot \mathbf{w}$ and $(\mathbf{v} + \mathbf{w}) \cdot \mathbf{u} = \mathbf{v} \cdot \mathbf{u} + \mathbf{w} \cdot \mathbf{u}$.

(vii) $\mathbf{u} \cdot (r\mathbf{v}) = (r\mathbf{u}) \cdot \mathbf{v} = r(\mathbf{u} \cdot \mathbf{v})$, for any real number r.

(viii) $\mathbf{u} \cdot \mathbf{0} = \mathbf{0} \cdot \mathbf{u} = 0$.

(ix) $\mathbf{1} \cdot \mathbf{u} = \mathbf{u} \cdot \mathbf{1} = \Sigma\, u_i$ (the sum of the components of **u**).

(x) $\mathbf{u} \cdot \mathbf{u} = \Sigma\, u_i^2$ (the sum of the squares of the components of **u**).

In addition, scalars r, s, t, 1, and 0 obey the following laws:

(xi) $r(s\mathbf{u}) = (rs)\mathbf{u}$.

(xii) $r(\mathbf{u} + \mathbf{v}) = r\mathbf{u} + r\mathbf{v}$ and $(r + s)\mathbf{u} = r\mathbf{u} + s\mathbf{u}$.

(xiii) $1\mathbf{u} = \mathbf{u}$ and $0\mathbf{u} = \mathbf{0}$.

Proof: We shall give proofs for vectors with two components. Unless stated, the extension to three or more components follows by the same reasoning. The first four properties hold for vector addition because they hold for scalar (single-number) addition, and vector addition is just a collection of scalar additions, one for each component.

(i) Commutativity of vector addition follows from the commutativity of scalar addition:

$$\mathbf{u} + \mathbf{v} = [u_1 + v_1, u_2 + v_2] = [v_1 + u_1, v_2 + u_2] = \mathbf{v} + \mathbf{u}$$

(ii) Associativity of vector addition follows from the associativity of scalar addition:

$$(\mathbf{u} + \mathbf{v}) + \mathbf{w} = [u_1 + v_1, u_2 + v_2] + [w_1, w_2]$$
$$= [u_1 + v_1 + w_1, u_2 + v_2 + w_2]$$
$$= [u_1, u_2] + [v_1 + w_1, v_2 + w_2] = \mathbf{u} + (\mathbf{v} + \mathbf{w})$$

(iii) and (iv) Proofs are similar to (i) and (ii) and left to the reader.
(v) This property follows from the commutativity of scalar multiplication:

$$\mathbf{u} \cdot \mathbf{v} = u_1 v_1 + u_2 v_2 = v_1 u_1 + v_2 u_2 = \mathbf{v} \cdot \mathbf{u}$$

(vi) Distributivity of scalar products follows from the distributivity of scalar arithmetic:

$$\mathbf{u} \cdot (\mathbf{v} + \mathbf{w}) = u_1(v_1 + w_1) + u_2(v_2 + w_2)$$
$$= (u_1 v_1 + u_1 w_1) + (u_2 v_2 + u_2 w_2)$$
$$= (u_1 v_1 + u_2 v_2) + (u_1 w_1 + u_2 w_2) = \mathbf{u} \cdot \mathbf{v} + \mathbf{u} \cdot \mathbf{w}$$

The case of $(\mathbf{v} + \mathbf{w}) \cdot \mathbf{u} = \mathbf{v} \cdot \mathbf{u} + \mathbf{w} \cdot \mathbf{u}$ proceeds in the same way.
(vii) The proof is similar to (v) and (vi) and is left to the reader.
(viii) Follows immediately from the scalar rule that any number times zero is zero.
(ix) By (v), we know that $\mathbf{1} \cdot \mathbf{u} = \mathbf{u} \cdot \mathbf{1}$. And $\mathbf{1} \cdot \mathbf{u} = 1u_1 + 1u_2 = u_1 + u_2$.
(x) The proof is straightforward.
(xi) If $\mathbf{u}$ is the 2-vector $[u_1, u_2]$, then

$$r(s\mathbf{u}) = r([su_1, su_2]) = [r(su_1), r(su_2)] \quad \text{while} \quad (rs)\mathbf{u} = [(rs)u_1, (rs)u_2]$$

The equality of corresponding components follows from the associative law for scalar multiplication.
(xii) Similar to (xi) except that now the distributive law for scalars is used.
(xiii) Left as an exercise. ■

**EXAMPLE 7.
Simplifying a
Scalar Product
Expression**

Consider the following scalar product expression involving n-vectors $\mathbf{u}$ and $\mathbf{v}$.

$$(\mathbf{u} + \mathbf{v}) \cdot (\mathbf{u} - \mathbf{v})$$

Let us use the laws of scalar products in Theorem 2 to expand and simplify this expression.

$$(\mathbf{u} + \mathbf{v}) \cdot (\mathbf{u} - \mathbf{v}) = \mathbf{u} \cdot (\mathbf{u} - \mathbf{v}) \quad + \mathbf{v} \cdot (\mathbf{u} - \mathbf{v}) \qquad [\text{by (vi)}]$$
$$= \mathbf{u} \cdot \mathbf{u} - \mathbf{u} \cdot \mathbf{v} + \mathbf{v} \cdot \mathbf{u} - \mathbf{v} \cdot \mathbf{v} \qquad [\text{by (vi)}]$$
$$= \mathbf{u} \cdot \mathbf{u} - \mathbf{v} \cdot \mathbf{v} + \mathbf{v} \cdot \mathbf{u} - \mathbf{u} \cdot \mathbf{v} \qquad (18)$$
$$= \mathbf{u} \cdot \mathbf{u} - \mathbf{v} \cdot \mathbf{v} + \mathbf{v} \cdot \mathbf{u} - \mathbf{v} \cdot \mathbf{u} \qquad [\text{by (v)}]$$
$$= \mathbf{u} \cdot \mathbf{u} - \mathbf{v} \cdot \mathbf{v}$$

■

We close this section by noting that there are some reasonable-looking properties of vector arithmetic that are *not* true. For example,

$$(\mathbf{u} \cdot \mathbf{v})\mathbf{w} \neq \mathbf{u}(\mathbf{v} \cdot \mathbf{w}) \tag{19}$$

The parentheses make this look like some sort of associative law, but the problem is that the two vectors within a set of parentheses yield a single (scalar) number when their scalar product is computed, not a vector. The left side of (19) is a scalar multiple of $\mathbf{w}$, where the scalar is equal to $\mathbf{u} \cdot \mathbf{v}$, while the right side of (19) is a scalar multiple of $\mathbf{u}$, where the scalar now is equal to $\mathbf{v} \cdot \mathbf{w}$. A multiple of vector $\mathbf{w}$ will not normally be equal to a multiple of another vector. However, if $\mathbf{u}$ is a multiple of $\mathbf{w}$ or if one of the three vectors equals the zero vector $\mathbf{0}$, the two sides will be equal.

Section 2.1 Exercises

1. Plot the following vectors in the x_1–x_2 plane (drawing the vectors as arrows from the origin to the point with the given coordinates).

   (a) $\mathbf{a} = \begin{bmatrix} 2 \\ 0 \end{bmatrix}$ (b) $\mathbf{b} = \begin{bmatrix} 2 \\ 3 \end{bmatrix}$ (c) $\mathbf{c} = \begin{bmatrix} 2 \\ -3 \end{bmatrix}$ (d) $\mathbf{d} = \begin{bmatrix} -3 \\ -2 \end{bmatrix}$ (e) $\mathbf{e} = \begin{bmatrix} -2 \\ 5 \end{bmatrix}$

2. The vectors named in this exercise refer to the vectors in Exercise 1.
   (a) Plot the effect of adding vector $\mathbf{a}$ to vector $\mathbf{c}$.
   (b) Plot the effect of adding vector $\mathbf{c}$ and to vector $\mathbf{a}$. Is this the same final vector as obtained in part (a)?
   (c) Plot the effect of adding vector $\mathbf{b}$ to vector $\mathbf{d}$.
   (d) Plot the effect of subtracting vector $\mathbf{b}$ from vector $\mathbf{e}$.
   (e) Plot the effect of subtracting vector $\mathbf{c}$ from vector $\mathbf{d}$.

3. Plot the following vectors as arrows in x_1–x_2–x_3 coordinates (as in Figure 2.3).

   (a) $\begin{bmatrix} 2 \\ 0 \\ 0 \end{bmatrix}$ (b) $\begin{bmatrix} 1 \\ 2 \\ 1 \end{bmatrix}$ (c) $\begin{bmatrix} 2 \\ -2 \\ 1 \end{bmatrix}$ (d) $\begin{bmatrix} -2 \\ -1 \\ -4 \end{bmatrix}$

4. A ship is traveling 15 kilometers per hour due west and an ocean current is moving the water with a westerly component of 5 kilometers per hour and a southerly component of 10 kilometers per hour. Plot the vectors of the two forces acting on the ship and determine the net velocity vector of the ship.

5. A person is paddling a canoe 5 kilometers per hour due east on a lake. A motor mounted at an angle on the rear of the canoe is propelling the canoe with components of 5 kilometers per hour north and 5 kilometers per hour east. There is also a current moving the water 3 kilometers per hour south. Plot the vectors of the three forces acting on the canoe and determine the net velocity vector of the canoe.

6. Plot the following lines, given in parametric form.
   (a) $[1 + r, 2 + 2r]$ (b) $[3, 1 + 2r]$ (c) $[-1 - r, 3 + 2r]$
   (d) $[-3 + 3r, -1 - 2r]$ (e) $[-4 + 2r, 3r]$

7. Plot the following lines, given in symmetric form.

   (a) $\dfrac{x_1 - 2}{2} = \dfrac{x_2 - 1}{3}$ (b) $\dfrac{x_1 - 1}{3} = \dfrac{x_2 + 1}{-1}$ (c) $\dfrac{x_1 + 3}{2} = x_2$

   (d) $x_1 - 2 = \dfrac{x_2 - 2}{-3}$ (e) $\dfrac{x_1 + 1}{-1} = \dfrac{x_2 + 4}{-2}$

8. Find the equation of the line in symmetric form through the vector **v** and pointing in the direction of vector **d**.
   (a) $\mathbf{v} = [1, 3]$, $\mathbf{d} = [1, 2]$ (b) $\mathbf{v} = [0, 2]$, $\mathbf{d} = [-3, 2]$
   (c) $\mathbf{v} = [-2, -1]$, $\mathbf{d} = [-1, -2]$ (d) $\mathbf{v} = [3, 1]$, $\mathbf{d} = [0, 0]$

9. For each of the following pairs of vectors, find the line going through the two vectors. Describe the line in parametric form.
   (a) $[1, 2]$, $[2, 3]$ (b) $[-1, 3]$, $[2, 1]$ (c) $[3, -2]$, $[3, 3]$ (d) $[-2, -1]$, $[-1, 3]$
   (e) $[4, 0]$, $[-1, -2]$ (f) $[1, 2]$, $[1, 2]$

10. For each of the pairs of vectors in Exercise 9, give the equation in symmetric form of the line going through the two vectors.

11. For each of the pairs of vectors in Exercise 9, give the standard equation $a_1 x_1 + a_2 x_2 = b$ of the line going through the two vectors.

12. For each of the following pairs of vectors, find the line going through the two vectors. Describe the line in parametric form.
    (a) $[1, 2, 1]$, $[2, 0, 3]$ (b) $[-1, 2, 3]$, $[0, 3, 1]$ (c) $[-3, 3, -2]$, $[3, 3, -1]$
    (d) $[0, -2, -1]$, $[3, 4, -1]$ (e) $[4, -1, -2]$, $[-1, -2, -3]$
    (f) $[1, 4, 3]$, $[1, 4, 3]$

13. For each of the pairs of vectors in Exercise 12, give the equation in symmetric form of the line going through the two vectors.

14. Determine the line in parametric form satisfying the following pairs of linear equations.

    (a) $\begin{aligned} x_1 + 2x_2 + x_3 &= 3 \\ 2x_1 + x_2 + x_3 &= 6 \end{aligned}$ (b) $\begin{aligned} 2x_1 + 4x_2 + x_3 &= 4 \\ x_1 + x_2 + x_3 &= 1 \end{aligned}$ (c) $\begin{aligned} x_1 + 2x_2 + x_3 &= 1 \\ 3x_1 + 3x_2 + 2x_3 &= 4 \end{aligned}$

    (d) $\begin{aligned} 2x_1 + x_2 + 3x_3 &= 3 \\ 2x_1 + x_2 + x_3 &= 6 \end{aligned}$ (e) $\begin{aligned} 3x_1 + 2x_2 + 4x_3 &= 3 \\ 2x_1 + x_2 + 2x_3 &= 2 \end{aligned}$ (f) $\begin{aligned} 2x_1 + 6x_2 + 4x_3 &= 6 \\ x_1 + 3x_2 + 2x_3 &= 3 \end{aligned}$

15. Determine the plane in parametric form that passes through the following sets of three vectors.
    (a) $[1, 1, 2]$, $[0, 2, 3]$, $[4, 1, 2]$ (b) $[3, 0, 1]$, $[5, 2, -2]$, $[-2, 3, 1]$
    (c) $[2, 0, 0]$, $[0, 1, 0]$, $[0, 0, 3]$ (d) $[-2, -4, 3]$, $[2, -5, -1]$, $[1, 6, -2]$

16. Determine the standard equation of the plane that passes through the sets of three vectors in Exercise 15.

17. Find a linear combination $r_1 \mathbf{u} + s_2 \mathbf{v}$ of the vectors **u** and **v** that equals vector **b**. Plot $r_1 \mathbf{u}$ and $s_2 \mathbf{v}$ and their sum (which equals **b**).
    (a) $\mathbf{u} = [2, 1]$, $\mathbf{v} = [1, -1]$, $\mathbf{b} = [6, 0]$ (b) $\mathbf{u} = [1, 3]$, $\mathbf{v} = [1, 1]$, $\mathbf{b} = [1, 7]$
    (c) $\mathbf{u} = [1, 2]$, $\mathbf{v} = [-1, -1]$, $\mathbf{b} = [3, 0]$ (d) $\mathbf{u} = [2, 3]$, $\mathbf{v} = [3, -2]$, $\mathbf{b} = [10, 2]$
    (e) $\mathbf{u} = [4, 5]$, $\mathbf{v} = [5, 6]$, $\mathbf{b} = [0, 1]$ (f) $\mathbf{u} = [10, 11]$, $\mathbf{v} = [9, 10]$, $\mathbf{b} = [2, 0]$

18. Simplify the following vector expressions.
    (a) $(\mathbf{u} + \mathbf{v}) - (\mathbf{v} + \mathbf{w} + \mathbf{u})$ (b) $(\mathbf{u} + \mathbf{u} + \mathbf{u}) \cdot (\mathbf{u} + \mathbf{u})$
    (c) $(\mathbf{u} + \mathbf{v}) \cdot \mathbf{w} + (\mathbf{w} + \mathbf{v}) \cdot \mathbf{u} - \mathbf{v} \cdot (\mathbf{u} + \mathbf{w})$ (d) $\mathbf{u} \cdot \mathbf{u} + \mathbf{v} \cdot \mathbf{v} - (\mathbf{u} - \mathbf{v}) \cdot (\mathbf{u} - \mathbf{v})$
    (e) $(\mathbf{u} + \mathbf{v}) \cdot (\mathbf{u} + \mathbf{v}) - (\mathbf{u} - \mathbf{v}) \cdot (\mathbf{u} - \mathbf{v})$ (f) $(\mathbf{u} + \mathbf{v}) \cdot (\mathbf{u} - \mathbf{w}) - (\mathbf{u} - \mathbf{v}) \cdot (\mathbf{u} + \mathbf{w})$
    (g) $(1 \cdot \mathbf{u} + 1 \cdot \mathbf{v} - \mathbf{w} \cdot 1)1 + 1(\mathbf{u} \cdot 1 + \mathbf{w} \cdot 1 - \mathbf{v} \cdot 1)$
    (h) $(\mathbf{u} \cdot \mathbf{v} + \mathbf{v} \cdot \mathbf{v} - \mathbf{w} \cdot \mathbf{u})(\mathbf{v} \cdot \mathbf{w}) + (\mathbf{u} - \mathbf{v}) \cdot (\mathbf{w} + \mathbf{u})(\mathbf{v} \cdot \mathbf{w})$

19. Prove part (iii) of Theorem 1.

20. Prove part (iv) of Theorem 1.

21. Prove the second half of part (vi) of Theorem 1.

22. Prove part (vii) of Theorem 1.

23. Prove that $(\mathbf{u} \cdot \mathbf{v})\mathbf{w} = \mathbf{u}(\mathbf{v} \cdot \mathbf{w})$ if and only if at least one of the following three properties hold.
    (i) $\mathbf{u}$ is a multiple of $\mathbf{w}$.
    (ii) One of $\mathbf{u}$, $\mathbf{v}$, or $\mathbf{w}$ is the zero vector $\mathbf{0}$.
    (iii) $\mathbf{u} \cdot \mathbf{v} = \mathbf{v} \cdot \mathbf{w} = 0$.

2.2   Vector Norms

Geometric Norm

Physical vectors, such as wind, have a direction and a magnitude. In two dimensions, the direction can be described by the angle the vector makes with respect to the x_1-axis (easterly direction). Indeed, this angle, along with the magnitude of a vector, forms the polar coordinates of a vector. In higher dimensions, the direction is described by the angles formed with respect to the various axes. In this section we consider the magnitude of a vector. We shall show how to compute the angle between two vectors in the next section.

The formal term used in mathematics for the magnitude of an object is *norm*. There are several norms that can be used to measure the magnitude of a vector. In geometric settings, the standard norm is the Euclidean distance that students learn in high school geometry.

Definition

The **Euclidean norm** $|\mathbf{v}|_e$ of a vector $\mathbf{v} = [v_1, v_2, \ldots, v_n]$ is the square root of the sum of the squares of its coordinates.

$$|\mathbf{v}|_e = \sqrt{v_1^2 + v_2^2 + \cdots + v_n^2}$$

EXAMPLE 1.
Euclidean Norm of a Vector

If $\mathbf{v} = [3, -4]$, then $|\mathbf{v}|_e = \sqrt{3^2 + (-4)^2} = \sqrt{9 + 16} = \sqrt{25} = 5$.

If $\mathbf{v} = [2, 3]$, then $|\mathbf{v}|_e = \sqrt{2^2 + 3^2} = \sqrt{4 + 9} = \sqrt{13} \approx 3.6$.

If $\mathbf{v} = [2, 3, 5]$ $|\mathbf{v}|_e = \sqrt{2^2 + 3^2 + 5^2} = \sqrt{4 + 9 + 25} = \sqrt{38} \approx 6.2$.   ■

EXAMPLE 2.
Distance to the
Corner of a Box

Consider a box with one corner at the origin and neighboring edges given by the vectors $\mathbf{u} = [2, 0, 0]$, $\mathbf{v} = [0, 1, 0]$, and $\mathbf{w} = [0, 0, 3]$ as shown in Figure 2.8. What is the length from the origin to the farthest corner, $[2, 1, 3]$? The length of the vector $[2, 1, 3]$ (from the origin to the corner point $[2, 1, 3]$) is $\sqrt{2^2 + 1^2 + 3^2} = \sqrt{14}$.

■

In $\mathbf{R}^2$, a vector of unit length that forms an angle of θ with the x-axis is given by $\mathbf{v}_\theta = [\cos\theta, \sin\theta]$. We check that the length of $\mathbf{v}_\theta$ is 1:

$$|\mathbf{v}_\theta|_e = \sqrt{(\cos\theta)^2 + (\sin\theta)^2} = \sqrt{1} = 1$$

It is common in mathematical settings to give the direction of a vector in terms of a unit-length vector pointing in the appropriate direction. For vector $\mathbf{v}$, $(1/|\mathbf{v}|_e)\mathbf{v}$ is the desired vector of magnitude one pointing in the same direction as $\mathbf{v}$.

EXAMPLE 3.
Representing a
Vector by a
Magnitude and
Direction

Consider the vector $\mathbf{v} = [2, 2]$. Let us represent $\mathbf{v}$ as a magnitude and a direction (a unit-length vector pointing in the direction of $\mathbf{v}$). The magnitude (length) of $\mathbf{v}$ is

$$|\mathbf{v}|_e = \sqrt{2^2 + 2^2} = \sqrt{4 + 4} = \sqrt{8} = 2\sqrt{2}$$

To get a vector of unit length, we divide $\mathbf{v}$ by its length $|\mathbf{v}|_e$:

$$\left(\frac{1}{|\mathbf{v}|_e}\right)\mathbf{v} = \frac{1}{2\sqrt{2}}[2, 2] = \left[\frac{1}{\sqrt{2}}, \frac{1}{\sqrt{2}}\right] = [\cos 45°, \sin 45°]$$

Or in polar coordinates, $[2, 2]$ becomes $(2\sqrt{2}, 45°)$.

■

Although the norm is defined for a single vector, one of its most common uses is to measure the distance between two vectors.

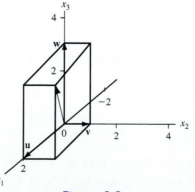

Figure 2.8

Distance Between Two Vectors

The Euclidean distance between two vectors **u** and **v** is given by the Euclidean norm of their difference $|\mathbf{u} - \mathbf{v}|_e$.

EXAMPLE 4.
Distance Between
Pairs of Vectors

(a) The distance between $\mathbf{u} = [1, -2]$ and $\mathbf{v} = [4, 2]$ is the Euclidean norm of $\mathbf{u} - \mathbf{v} = [1 - 4, -2 - 2] = [-3, -4]$ (see Figure 2.9(a)). Then

$$|\mathbf{u} - \mathbf{v}|_e = \sqrt{(-3)^2 + (-4)^2} = \sqrt{9 + 16} = \sqrt{25} = 5$$

(b) The distance between $\mathbf{u} = [1, 3, 4]$ and $\mathbf{v} = [4, 0, 2]$ is the Euclidean norm of $\mathbf{u} - \mathbf{v} = [1 - 4, 3 - 0, 4 - 2] = [-3, 3, 2]$ (see Figure 2.9(b)). Then

$$|\mathbf{u} - \mathbf{v}|_e = \sqrt{(-3)^2 + 3^2 + 2^2} = \sqrt{9 + 9 + 4} = \sqrt{22} \approx 4.7. \qquad ■$$

Recall that $\mathbf{u} \cdot \mathbf{u}$ equals $\Sigma\, u_i^2$, that is, the sum of the squares of the components of **u**. [This fact was Theorem 2(x) of Section 2.1.] Using this fact, we can write the formula for the Euclidean norm in vector notation, that is, without reference to individual components.

Theorem 1
　　(i) The Euclidean norm $|\mathbf{v}|_e$ equals $\sqrt{\mathbf{v} \cdot \mathbf{v}}$.
　　(ii) A vector **v** is a unit vector in the Euclidean norm ($|\mathbf{v}|_e = 1$) if and only if $\mathbf{v} \cdot \mathbf{v} = 1$.
There are three basic mathematical properties of the Euclidean norm (every norm in mathematics must satisfy these three properties). All other properties of a norm can be deduced from these three. The third property is called the ***triangle inequality***.

Theorem 2. The Euclidean norm has the following three properties:
　　(i) For any vector **v**, $|\mathbf{v}|_e \geq 0$; further, $|\mathbf{v}|_e = 0$ if and only if $\mathbf{v} = \mathbf{0}$.
　　(ii) For any vector **v** and any scalar r, $|r\mathbf{v}|_e = |r| \cdot |\mathbf{v}|_e$.
　　(iii) For any two vectors **u**, **v** (of the same dimension): $|\mathbf{u} + \mathbf{v}|_e \leq |\mathbf{u}|_e + |\mathbf{v}|_e$.

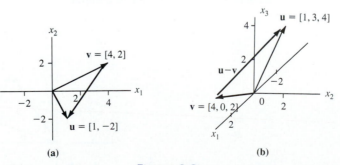

(a)　　　　　　　　　　　　(b)

Figure 2.9

Proof: (i) It is obvious that $|\mathbf{v}|_e \geq 0$. Note that $|\mathbf{v}|_e = 0$ if and only if $(|\mathbf{v}|_e)^2 = 0$. But $(|\mathbf{v}|_e)^2 = v_1^2 + v_2^2 + \cdots + v_n^2 = 0$ only when each $v_i = 0$, that is, when $\mathbf{v} = \mathbf{0}$.

(ii) One can prove (ii) directly from the definition of $|\mathbf{v}|_e$. Instead, we shall prove (ii) without considering individual components of $\mathbf{v}$. By part (i) of Theorem 1, $|\mathbf{v}|_e = \sqrt{\mathbf{v} \cdot \mathbf{v}}$. Then

$$|r\mathbf{v}|_e = \sqrt{(r\mathbf{v}) \cdot (r\mathbf{v})} = \sqrt{r^2(\mathbf{v} \cdot \mathbf{v})} = |r|\sqrt{\mathbf{v} \cdot \mathbf{v}} = |r| \cdot |\mathbf{v}|_e \tag{1}$$

The second equation in (1) uses property (vii)—$\mathbf{u} \cdot (r\mathbf{v}) = (r\mathbf{u}) \cdot \mathbf{v} = r(\mathbf{u} \cdot \mathbf{v})$—in Theorem 1 of Section 2.1.

(iii) Observe that geometrically this result is simply a mathematical form of the well-known saying, "The shortest distance between two points is a straight line." The result is called the triangle inequality because when we geometrically add $\mathbf{u}$ and $\mathbf{v}$ (as shown in Figure 2.10), $\mathbf{u}$ and $\mathbf{v}$ form two sides of a triangle and $\mathbf{u} + \mathbf{v}$ is the third side. Thus this result follows immediately from the geometric fact that the sum of lengths of any two sides of a triangle must exceed the length of the third side. A proof of this fact follows from the law of cosines of trigonometry: $|\mathbf{a} - \mathbf{b}|_e^2 = |\mathbf{a}|_e^2 + |\mathbf{b}|_e^2 - 2|\mathbf{a}|_e|\mathbf{b}|_e \cos \theta$. Letting $\mathbf{a} = \mathbf{u}$ and $\mathbf{b} = -\mathbf{v}$ (so $-\mathbf{b} = \mathbf{v}$), the law of cosines gives

$$\begin{aligned} |\mathbf{u} + \mathbf{v}|_e^2 &= |\mathbf{u}|_e^2 + |\mathbf{v}|_e^2 - 2|\mathbf{u}|_e|\mathbf{v}|_e \cos \theta \\ &\leq |\mathbf{u}|_e^2 + |\mathbf{v}|_e^2 + 2|\mathbf{u}|_e|\mathbf{v}|_e = (|\mathbf{u}|_e + |\mathbf{v}|_e)^2 \end{aligned} \tag{2}$$

The inequality follows from the fact that $\cos \theta \geq -1$ and thus $-2|\mathbf{u}|_e|\mathbf{v}|_e \cos \theta \leq 2|\mathbf{u}|_e|\mathbf{v}|_e$. Now taking square roots in (2), we obtain the triangle inequality. ■

Nongeometric Norms

We next present two other vector norms, the sum norm and the max norm. These two norms are numerical measures of the size of a vector that lack the geometric meaning of the Euclidean norm, but they are very useful in nongeometric situations and have the advantage of being easier to compute.

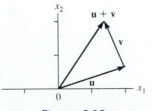

Figure 2.10

Definition

The **sum norm** $|\mathbf{v}|_s$ of a vector $\mathbf{v} = [v_1, v_2, \ldots, v_n]$ is the sum of the absolute values of its entries.

$$|\mathbf{v}|_s = \Sigma|v_i| = |v_1| + |v_2| + \cdots + |v_n|$$

Definition

The **max norm** $|\mathbf{v}|_m$ of a vector $\mathbf{v} = [v_1, v_2, \ldots, v_n]$ is the largest number among the absolute values of its entries.

$$|\mathbf{v}|_m = \max\{|v_i|\} = \max\{|v_1|, |v_2|, \ldots, |v_n|\}$$

Note that the sum norm would not work if it were just the sum of the entries (without absolute values), since then a vector of large size, such as $[1000, -1001]$, could have a very small norm (and possibly a negative norm). Clearly, the sum and max norms are far easier to compute than the Euclidean norm. Observe that a unit coordinate vector, such as $[1, 0, 0]$, has a norm of 1 in all three of the norms.

While the Euclidean norm $|\cdot|_e$ is the correct norm for geometric interpretations of vectors, the sum and max norms are appropriate norms for many nongeometric interpretations of vectors. The following examples of vectors that arose in matrix models in Chapter 1 illustrate this claim.

**EXAMPLE 5.
Sum Norm in the
Markov Chain
Weather Model**

In the Markov chain for weather introduced in Section 1.1, the equations for determining the probabilities p_1', p_2', p_3' of sunny, cloudy, or rainy weather tomorrow given the probabilities p_1, p_2, p_3 of sunny, cloudy, or rainy weather today were

$$p_1' = \tfrac{3}{4}p_1 + \tfrac{1}{2}p_2 + \tfrac{1}{4}p_3$$
$$p_2' = \tfrac{1}{8}p_1 + \tfrac{1}{4}p_2 + \tfrac{1}{2}p_3 \tag{3}$$
$$p_3' = \tfrac{1}{8}p_1 + \tfrac{1}{4}p_2 + \tfrac{1}{4}p_3$$

If $\mathbf{p} = \begin{bmatrix} p_1 \\ p_2 \\ p_3 \end{bmatrix}$ is the vector of current probabilities, $\mathbf{p}' = \begin{bmatrix} p_1' \\ p_2' \\ p_3' \end{bmatrix}$ is the vector of tomorrow's probabilities, and $\mathbf{A}$ is the matrix of transition probabilities:

	Today		
Tomorrow	Sunny	Cloudy	Rainy
$\mathbf{A} = $ Sunny	$\tfrac{3}{4}$	$\tfrac{1}{2}$	$\tfrac{1}{4}$
Cloudy	$\tfrac{1}{8}$	$\tfrac{1}{4}$	$\tfrac{1}{2}$
Rainy	$\tfrac{1}{8}$	$\tfrac{1}{4}$	$\tfrac{1}{4}$

then (3) can be written as simply $\mathbf{p}' = \mathbf{Ap}$.

The probabilities in $\mathbf{p}$ and $\mathbf{p}'$ must both add up to 1. We express this fact by the statements

$$|\mathbf{p}|_s \ (= p_1 + p_2 + p_3) = 1 \quad \text{and} \quad |\mathbf{p}'|_s \ (= p_1' + p_2' + p_3') = 1$$

We omitted the absolute value signs around the p's since they are known to be positive.

Also, the entries in each column in $\mathbf{A}$ must sum to 1 (recall that the probabilities in, say, column one tell the chances of sunny, cloudy, or rainy weather tomorrow if it is sunny today). Thus, for each column $\mathbf{a}_i$ of $\mathbf{A}$, $|\mathbf{a}_i|_s = 1$. (*Hint of things to come*: In Section 3.4, we define norms for matrices and the sum norm of a probability transition matrix, such as $\mathbf{A}$, will be 1.) ■

We note that the Euclidean norm for a probability vector $\mathbf{p}$ will be some value less than or equal to 1, with no useful interpretation.

EXAMPLE 6.
Sum and Max
Norms in Refinery
Model

In the oil refinery production model introduced in Section 1.1, we had the following system of equations (where x_i is the number of barrels of crude oil processed in the ith refinery):

$$\begin{array}{lrl}
\text{Heating oil:} & 4x_1 + 2x_2 + 2x_3 = & 600 \\
\text{Diesel oil:} & 2x_1 + 5x_2 + 2x_3 = & 800 \quad \text{or} \quad \mathbf{Ax} = \mathbf{b} \quad (4) \\
\text{Gasoline:} & 1x_1 + 2.5x_2 + 5x_3 = & 1000
\end{array}$$

where

$$\mathbf{A} = \begin{array}{l} \\ \text{Heating oil} \\ \text{Diesel oil} \\ \text{Gasoline} \end{array} \begin{array}{ccc} \text{Refinery 1} & \text{Refinery 2} & \text{Refinery 3} \\ \left[\begin{array}{ccc} 4 & 2 & 2 \\ 2 & 5 & 2 \\ 1 & 2.5 & 5 \end{array} \right] \end{array} \quad \text{and} \quad \mathbf{b} = \begin{bmatrix} 600 \\ 800 \\ 1000 \end{bmatrix}$$

The sum norm of $\mathbf{b}$ is the total amount of petroleum products that must be produced: $|\mathbf{b}|_s = 600 + 800 + 1000 = 2400$. The sum norm of a column of $\mathbf{A}$ is the total amount of petroleum products generated by one of the refineries from one barrel of crude oil. For example, $|\mathbf{a}_3|_s = 2 + 2 + 5 = 9$. And $|\mathbf{x}|_s$ is the total number of barrels of crude oil processed by all refineries combined in the solution to the production problem in (4).

The max norm of $\mathbf{b}$ is the largest amount of any petroleum product that must be produced—$|\mathbf{b}|_m = \max\{600, 800, 1000\} = 1000$. The max norm of the third column, $|\mathbf{a}_3|_m = \max\{2, 2, 5\} = 5$, is the largest amount of any petroleum product produced by the third refinery from one barrel of crude oil. And $|\mathbf{x}|_m$ is the largest entry in the solution to (4). ■

We now prove that the sum norm and max norm are indeed norms in the formal mathematical sense.

Theorem 3. The sum norm and max norm have the following three properties:
    (i) For any vector $\mathbf{v}$, $|\mathbf{v}| \geq 0$; further $|\mathbf{v}| = 0$ if and only if $\mathbf{v} = \mathbf{0}$.
    (ii) For any vector $\mathbf{v}$ and any scalar r, $|r\mathbf{v}| = r|\mathbf{v}|$.
    (iii) For any two vectors $\mathbf{u}$, $\mathbf{v}$ (of the same dimension): $|\mathbf{u} + \mathbf{v}| \leq |\mathbf{u}| + |\mathbf{v}|$.
    Proof: (i) and (ii) are both obvious from the definitions of the sum norm and max norm.

      (iii) We prove the triangle inequality of the sum norm and leave the max norm as an exercise for the reader. To keep the notation simple, we consider just 2-vectors. Our result follows from the following inequality, which is true for any two (scalar) numbers: $|a + b| \leq |a| + |b|$.

$$|\mathbf{u} + \mathbf{v}|_s = |u_1 + v_1| + |u_2 + v_2| \leq |u_1| + |v_1| + |u_2| + |v_2|$$
$$= (|u_1| + |u_2|) + (|v_1| + |v_2|) = |\mathbf{u}|_s + |\mathbf{v}|_s \tag{5}$$
■

    As noted earlier, all mathematical properties of a norm follow from the three properties in Theorem 3.

    Most readers are accustomed to having only one way to perform a mathematical calculation. Linear algebra, and much of higher mathematics, frequently has many equally valid choices of which formula to use, of how to "solve" a problem, or of how to give a geometric interpretation to a problem. Here we have given three ways to measure the size of a vector. In Section 2.1 we saw that solving a system of equations could be viewed as finding a vector that is the intersection point of the planes defined by each *row* equation, or as finding the right linear combination of *columns* that equals the right-side vector. This variety of possible mathematical viewpoints is one of the aspects of linear algebra that makes it such a powerful theory.

Section 2.2 Exercises

1. Give the Euclidean norm of the following vectors.
    (a) [1, 1, 1]   **(b)** [3, 0, 0]   **(c)** [−1, 1, 4]   **(d)** [−1, 4, 3]   **(e)** [4, 4, 4, 4]

2. Represent the following vectors by a magnitude and a direction (as in Example 3).
    (a) [3, 4]   **(b)** $[6, -\frac{5}{2}]$ **(c)** [−2, −2]

3. Determine the Euclidean distance between the following pairs of vectors.
    (a) [2, 4], [6, 1]   **(b)** [1, 2], [3, 1]   **(c)** [2, 5, 1], [4, 1, 3]
    (d) [3, −2, −3], [−1, 6, −5]

4. Give the sum and max norm of the following vectors.
    (a) [1, 1, 1]   **(b)** [3, 0, 0]   **(c)** [−1, 1, 4]   **(d)** [−1, 4, 3]   **(e)** [4, 4, 4, 4]

5. The distance between two vectors $\mathbf{a}$, $\mathbf{b}$ is defined to be the norm of their difference $|\mathbf{a} - \mathbf{b}|$.
    (a) What is the distance between the two vectors [2, 5, 7] and [3, −1, 4] if we use the sum norm and max norm?
    (b) Explain in words what the distance between two vectors in the sum norm measures.
    (c) Explain in words what the distance between two vectors in the max norm measures.

6. For $\mathbf{A} = \begin{bmatrix} 3 & 1 \\ 2 & 2 \end{bmatrix}$, determine for each of the following vectors the ratio $|\mathbf{A}\mathbf{x}|_s / |\mathbf{x}|_s$.

    (a) $\mathbf{x} = \begin{bmatrix} 1 \\ 1 \end{bmatrix}$ **(b)** $\mathbf{x} = \begin{bmatrix} -1 \\ 2 \end{bmatrix}$ **(c)** $\mathbf{x} = \begin{bmatrix} 1 \\ 0 \end{bmatrix}$

7. **(a)** Describe those vectors for which the Euclidean norm and max norm are equal. Explain the reason for your answer.
   (b) Describe those vectors for which the sum norm and max norm are equal. Explain the reason for your answer.
   (c) Describe those vectors for which the Euclidean norm, sum norm, and max norm are all equal. Why must these be the only vectors with this property?

8. Show that any two vectors $\mathbf{a}$ and $\mathbf{b}$, $|\mathbf{a} + \mathbf{b}|_e = |\mathbf{a}|_e + |\mathbf{b}|_e$ if and only if $\mathbf{b}$ is a positive multiple of $\mathbf{a}$. (Hint: See the proof of Theorem 2(iii).)

9. Under what conditions on $\mathbf{a}$ and $\mathbf{b}$ is it true that $|\mathbf{a} + \mathbf{b}|_s = |\mathbf{a}|_s + |\mathbf{b}|_s$?

10. Under what conditions on $\mathbf{a}$ and $\mathbf{b}$ is it true that $|\mathbf{a} + \mathbf{b}|_m = |\mathbf{a}|_m + |\mathbf{b}|_m$?

11. We prove that $|\mathbf{a}|_m \leq |\mathbf{a}|_e \leq |\mathbf{a}|_s$.
    (a) Show that the max norm of a vector $\mathbf{a}$ is always less than or equal to the Euclidean norm of $\mathbf{a}$.
    (b) Show that the Euclidean norm of a vector $\mathbf{a}$ is always less than or equal to the sum norm of $\mathbf{a}$.

12. Let $\mathbf{a}$ be an n-vector. Show that $|\mathbf{a}|_s \leq n|\mathbf{a}|_m$.

13. Show that $|\mathbf{a} \cdot \mathbf{b}| \leq |\mathbf{a}|_s \cdot |\mathbf{b}|_m$.

14. Show that $|\mathbf{a} + \mathbf{b}|_m \leq |\mathbf{a}|_m + |\mathbf{b}|_m$.

15. Let $\mathbf{a}$ be an n-vector. Show that $|\mathbf{a}|_s \leq \sqrt{n}|\mathbf{a}|_e$. (Hint: Square both sides and prove that $2a_i a_j \leq a_i^2 + a_j^2$.)

2.3 Angles and Orthogonality

In this section we develop and apply a formula for the angle between two vectors. We are particularly interested in vectors that are perpendicular to each other. We use the Euclidean norm $|\mathbf{a}| = \sqrt{a_1^2 + a_2^2 + \cdots + a_n^2}$ (we omit the subscript e in writing the norm $|\cdot|_e$), because we will be thinking geometrically. Norms will be the geometric lengths of vectors.

Our initial goal is a formula for the cosine of the angle between two vectors $\mathbf{a}$ and $\mathbf{b}$. This is the angle formed at the origin by the line segments to the two vector points. (The angle is measured in the plane formed by the two line segments; see Figure 2.11.) The formula is simple and its proof is fairly short and not hard to follow. However, the proof uses an indirect algebraic argument that gives one no feel for why the formula is true.

Cosine of Angle Between Two Vectors

Theorem 1. The cosine of the angle θ between two nonzero vectors $\mathbf{a}$, $\mathbf{b}$ is

$$\cos \theta = \frac{\mathbf{a} \cdot \mathbf{b}}{|\mathbf{a}||\mathbf{b}|} \tag{1}$$

If $\mathbf{a}$ and $\mathbf{b}$ are unit-length vectors, (1) becomes $\cos \theta = \mathbf{a} \cdot \mathbf{b}$.

Proof: Our proof involves expressing the square of the distance between **a** and **b**, $|\mathbf{a} - \mathbf{b}|^2$, in two very different ways. The result will follow immediately when we set the two expressions for $|\mathbf{a} - \mathbf{b}|^2$ equal to one another. The first way we express $|\mathbf{a} - \mathbf{b}|^2$ uses the law of cosines from trigonometry:

$$|\mathbf{a} - \mathbf{b}|^2 = |\mathbf{a}|^2 + |\mathbf{b}|^2 - 2|\mathbf{a}||\mathbf{b}| \cos \theta \qquad (2)$$

For the second way, we recall that the square of the Euclidean norm $|\mathbf{c}|^2$ of any vector **c** is simply $\mathbf{c} \cdot \mathbf{c}$ (this fact was part of Theorem 1 in Section 2.2). Letting $\mathbf{c} = \mathbf{a} - \mathbf{b}$, we obtain (using the distributive law for scalar products from Theorem 1 of Section 2.1)

$$|\mathbf{a} - \mathbf{b}|^2 = (\mathbf{a} - \mathbf{b}) \cdot (\mathbf{a} - \mathbf{b}) = \mathbf{a} \cdot \mathbf{a} + \mathbf{b} \cdot \mathbf{b} - 2\mathbf{a} \cdot \mathbf{b}$$
$$= |\mathbf{a}|^2 + |\mathbf{b}|^2 - 2\mathbf{a} \cdot \mathbf{b} \qquad (3)$$

The right sides of (3) and (2) must be equal. Moreover, the right side of (3) is the same as the right side of (2) except for the last terms. So these last terms must be equal:

$$-2|\mathbf{a}||\mathbf{b}| \cos \theta = -2\mathbf{a} \cdot \mathbf{b}$$

Solving for $\cos \theta$ yields Theorem 1. [Wasn't that a sneaky way to verify (1)!]

■

EXAMPLE 1.
Examples of
Angles Between
Vectors

(a) If $\mathbf{a} = [1, 0, 0]$ and $\mathbf{b} = [0, 1, 0]$, then $\cos \theta = \mathbf{a} \cdot \mathbf{b} = 0$ and we conclude that **a** and **b** form a 90° angle.

(b) If $\mathbf{a} = [3, 4]$ and $\mathbf{b} = [12, 5]$, then

$$|\mathbf{a}| = 5(= \sqrt{9 + 16}) \qquad \text{and} \qquad |\mathbf{b}| = 13(= \sqrt{144 + 25})$$

So $\cos \theta = \mathbf{a} \cdot \mathbf{b}/|\mathbf{a}||\mathbf{b}| = (3 \cdot 12 + 4 \cdot 5)/5 \cdot 13 = 56/65 \approx 0.86$. The angle with a cosine of 0.86 is 30.7° [see Figure 2.11(a)].

(c) If $\mathbf{a} = [0.6, 0.8]$, with $|\mathbf{a}| = 1$ and $\mathbf{b} = [1, 0]$, then $\cos \theta = \mathbf{a} \cdot \mathbf{b} = 0.6 \cdot 1 + 0.8 \cdot 0 = 0.6$—this is just the first coordinate [see Figure 2.11(b)].

(d) If $\mathbf{a} = [\cos \theta, \sin \theta]$ and $\mathbf{b} = [1, 0]$, with $|\mathbf{a}| = (\cos \theta)^2 + (\sin \theta)^2 = 1$, then $\cos \theta = \mathbf{a} \cdot \mathbf{b} = \cos \theta \cdot 1 + \sin \theta \cdot 0 = \cos \theta$. ■

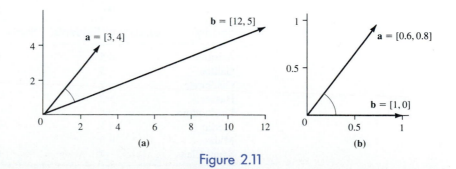

Figure 2.11

Before pursuing the mathematical uses of Theorem 1, we note that it has an interesting application in statistics. The cosine of the angle $\theta(\mathbf{x}, \mathbf{y})$ between two vectors $\mathbf{x}$ and $\mathbf{y}$ tells us if the vectors are close together [when $\cos \theta(\mathbf{x}, \mathbf{y})$ is near 1], or opposites of one another [when $\cos \theta(\mathbf{x}, \mathbf{y})$ is near -1], or are unrelated, that is, close to perpendicular [when $\cos \theta(\mathbf{x}, \mathbf{y})$ is near 0].

Suppose that $\mathbf{x}$ and $\mathbf{y}$ are vectors of data from an experiment, say $\mathbf{x}$ is the vector of scores of 10 students on a math test and $\mathbf{y}$ is the vector of scores of the same 10 students on a language test. Then $\cos \theta(\mathbf{x}, \mathbf{y})$ tells us how closely related these two sets of data are and helps us predict future relations between math and language scores. If $\cos \theta(\mathbf{x}, \mathbf{y})$ is 0.8, performance on these two tests is closely related and we can view a student's score on one test as a reasonably good predictor of how he or she will do on the other test. If $\cos \theta(\mathbf{x}, \mathbf{y})$ is -0.7, the score vectors point in almost opposite directions and a high score on one test is very likely to produce a below-average score on the other test. If $\cos \theta(\mathbf{x}, \mathbf{y}) = 0$ (the vectors $\mathbf{x}, \mathbf{y}$ are at right angles), performance on one test tells us nothing about the likely performance on the other test. (In statistics, one says that the two data variables are uncorrelated.)

Definition. Let $\mathbf{x} = [x_1, x_2, \ldots, x_n]$ and $\mathbf{y} = [y_1, y_2, \ldots, y_n]$ be two sets of observations with the property that the average x-value and the average y-value are each 0. Then the *correlation coefficient* $\text{Cor}(\mathbf{x}, \mathbf{y})$ of $\mathbf{x}$ and $\mathbf{y}$ is defined to be $\cos \theta(\mathbf{x}, \mathbf{y})$.

$$\text{Cor}(\mathbf{x}, \mathbf{y}) = \frac{\mathbf{x} \cdot \mathbf{y}}{|\mathbf{x}||\mathbf{y}|} = \frac{\Sigma \, x_i y_i}{\sqrt{\Sigma \, x_i^2}\sqrt{\Sigma \, y_i^2}} \tag{4}$$

Recall that the average x-value is $\bar{x} = (1/n) \, \Sigma \, x_i$. If $\bar{x} \neq 0$, we can subtract $\bar{x}$ from each x_i to get a revised vector that does have an average value of 0; similarly for y-values. We need an average value of 0 so that the opposite of a high score (a positive value) will be a low score (a negative value). This way the terms $x_i y_i$ in (4) for pairs of oppositely correlated entries x_i, y_i will be negative (when $x_i y_i$ is the product of a positive and a negative number), leading to a negative correlation.

**EXAMPLE 2.
Correlation
Coefficient**

Suppose that we ask the eight faculty members of the Podunk University Alchemy Department to rate the quality of their graduate students and we poll the students to get a rating of the quality of each of the eight faculty. The results of our experiment are presented in the following table (where we have processed the data to make the average value 0 in each category).

Faculty	x_i Quality of Students	y_i Student Rating
Aristotle	+5	+2
Galileo	−5	−7
Goldbrick	−2	0
Hasbeen	+3	−1
Leadbottom	−4	−5
Merlin	+5	+3
Midas	+5	0
Santa Claus	−7	+8

Applying formula (4) to this data, we obtain

$$\text{Cor}(\mathbf{x}, \mathbf{y}) = \frac{\Sigma\, x_i y_i}{\sqrt{\Sigma\, x_i^2}\sqrt{\Sigma\, y_i^2}} = \frac{5 \cdot 2 + (-5)(-7) + \cdots + (-7)8}{\sqrt{178} \cdot \sqrt{152}}$$

$$= \frac{21}{(13.3)(12.3)} \approx 0.1$$

Looking back at the data in the table, we are a little surprised to see such a low correlation since the numbers in the two columns correspond fairly well for most faculty with the glaring exception of Santa Claus. Statisticians would call Santa Claus's data pair $(-7, 8)$ an *outlier*, an observation that fits poorly with the rest of the data. (A little investigating reveals that Santa Claus is a terrible teacher but is still well liked because he gives the students lots of candy every December.)

Let us throw out Santa Claus's numbers and recompute the correlation coefficient. This requires us to adjust the data so that the averages in each column are again 0. The new numbers are

Faculty	Quality of Students	Student Rating
Aristotle	+4	+3
Galileo	−6	−6
Goldbrick	−3	+1
Hasbeen	+2	0
Leadbottom	−5	−4
Merlin	+4	+4
Midas	+4	+1

$$\text{Cor}(\mathbf{x}, \mathbf{y}) = \frac{85}{\sqrt{122} \cdot \sqrt{79}} \approx \frac{85}{(11)(8.9)} \approx 0.9$$

a high degree of correlation. ■

The concept of a correlation coefficient gives a good example of the power of applying geometric interpretations to vectors that arise from nongeometric settings.

Orthogonal Vectors

The most interesting angle between two vectors is a right angle (90°), that is, when the two vectors are perpendicular to one another. There is a special mathematical name for perpendicular vectors.

Definition

Two nonzero vectors are called **orthogonal** when the angle between them is 90°.

Since the cosine of 90° is 0, then from Theorem 1 we have

Theorem 2. Two nonzero vectors **a**, **b** are orthogonal if and only if $\mathbf{a} \cdot \mathbf{b} = 0$.

Note that in Theorem 2, we can omit the denominator in the formula for the cosine because the fraction is zero if and only if its numerator is zero. (Recall that the denominator cannot be 0 since both vectors are nonzero and hence have positive norms.)

We now examine orthogonality in some topics that we encountered earlier—homogeneous systems of equations, the definition of a line and plane, and matrix inverses.

**EXAMPLE 3.
Homogeneous
Systems of
Equations and
Orthogonality**

As noted at the end of Section 1.4, there are many applications where we get homogeneous systems of linear equations, that is, systems $\mathbf{Ax} = \mathbf{0}$ with right-hand sides all 0's. In Section 1.4 a homogeneous system arose in determining the stable probability distribution of the weather Markov chain. Now we can give a geometric interpretation to homogeneous systems.

The set of solutions **x** *to a homogeneous system of linear equations* $\mathbf{Ax} = \mathbf{0}$
are those vectors **x** *that are orthogonal to each of the row vectors* $\mathbf{a}'_i$ *of* **A**. ■

**EXAMPLE 4.
Inverse Matrices
and Orthogonality**

When we multiply two matrices **A** and **B** together, entry (i, j) in the product **AB** is the scalar product of the ith row $\mathbf{a}'_i$ of **A** and the jth column $\mathbf{b}_j$ of **B**. In the case of a matrix **A** and its inverse $\mathbf{A}^{-1}$, their product $\mathbf{A}^{-1}\mathbf{A}$ equals the identity matrix **I**, in which all entries are 0 except on the main diagonal. Thus the ith row of $\mathbf{A}^{-1}$ has a scalar product of 0 with all columns of **A** except the ith column, and hence is orthogonal to all columns of **A** except the jth column. Since $\mathbf{AA}^{-1}$ also equals **I**, the jth column of $\mathbf{A}^{-1}$ is orthogonal to all rows of **A** except the jth row. As an example, the 2×2 matrix $\mathbf{A} = \begin{bmatrix} 3 & 1 \\ 4 & 2 \end{bmatrix}$ has the inverse $\mathbf{A}^{-1} = \begin{bmatrix} 1 & -\frac{1}{2} \\ -2 & \frac{3}{2} \end{bmatrix}$. Figure 2.12 plots the columns of **A** and the rows of $\mathbf{A}^{-1}$. ■

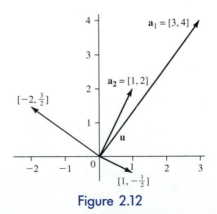

Figure 2.12

Because of the central role of orthogonality in matrix inverses, *if a matrix A has columns that are mutually orthogonal, its inverse A^{-1} is very easy to determine* (with no elimination computation needed). Let **A** have orthogonal columns. Form a matrix **B** whose ith row $\mathbf{b}_i'$ is a vector equal to the ith column $\mathbf{a}_i$ of **A**. Then **BA** will have 0's everywhere except on the main diagonal, since entry (i, j) of **BA** is $\mathbf{b}_i' \cdot \mathbf{a}_j = \mathbf{a}_i \cdot \mathbf{a}_j = 0$ (by the orthogonality of columns $\mathbf{a}_i, \mathbf{a}_j$). So **BA** is almost **I**. Next look at diagonal entry (i, i) of **BA**: $\mathbf{b}_i' \cdot \mathbf{a}_i = \mathbf{a}_i \cdot \mathbf{a}_i \ (= |\mathbf{a}_i|^2)$. To make this entry equal to 1, we modify **B** by dividing the ith row of **B** by $|\mathbf{a}_i|^2$. Call this matrix **C**, whose ith row $\mathbf{c}_i'$ is equal to the vector $(1/|\mathbf{a}_i|^2)\mathbf{a}_i$. **CA** has 0's off the main diagonal and entry (i, i) will be $\mathbf{c}_i' \cdot \mathbf{a}_i = (1/|\mathbf{a}_i|^2)\mathbf{a}_i \cdot \mathbf{a}_i = 1$. Thus $\mathbf{C} = \mathbf{A}^{-1}$.

Theorem 3 (Inverse of Matrix with Orthogonal Columns)
Let **A** be a square matrix with orthogonal columns $\mathbf{a}_j$. Then the ith row of $\mathbf{A}^{-1}$ equals the vector $(1/|\mathbf{a}_i|^2)\mathbf{a}_i$.

**EXAMPLE 5.
Inverse of Matrix
with Orthogonal
Columns**

The matrix

$$A = \begin{bmatrix} 2 & 1 & -2 \\ 1 & 2 & 2 \\ 2 & -2 & 1 \end{bmatrix}$$

has orthogonal columns. Then

$$A^{-1} = \begin{bmatrix} \frac{2}{9} & \frac{1}{9} & \frac{2}{9} \\ \frac{1}{9} & \frac{2}{9} & -\frac{2}{9} \\ -\frac{2}{9} & \frac{2}{9} & \frac{1}{9} \end{bmatrix}$$

by Theorem 3. For example, the second row of $\mathbf{A}^{-1}$ is the vector formed by taking the second column $\mathbf{a}_2$ of **A** and dividing its entries by $|\mathbf{a}_2|^2 = \mathbf{a}_2 \cdot \mathbf{a}_2 = 1 \cdot 1 + 2 \cdot 2 + (-2) \cdot (-2) = 1 + 4 + 4 = 9$. ■

A vector **a** is said to be *normal* to a line in $\mathbf{R}^2$ (or a plane in $\mathbf{R}^3$) if the line segment from the origin to the vector **a** is perpendicular to the line (or plane). See Figure 2.13, where **a** is normal to vector **b** and line segment L. Suppose that **u** and **v** are two vectors on a line L in $\mathbf{R}^2$. If **a** is a vector normal to L, then **a** is orthogonal to the vector $\mathbf{u} - \mathbf{v}$, that is, $\mathbf{a} \cdot (\mathbf{u} - \mathbf{v}) = 0$. This fact can be used to develop another form for representing

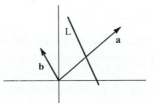

Figure 2.13   **b** and **c** are normal to **a**.

lines. Given a vector $\mathbf{v}$ on a line L in $\mathbf{R}^2$ and a vector $\mathbf{a}$ normal to L, the ***normal form*** of a line L is written

$$\mathbf{a} \cdot (\mathbf{x} - \mathbf{v}) = 0 \quad \text{or} \quad a_1(x_1 - v_1) + a_2(x_2 - v_2) = 0 \tag{5}$$

This result generalizes naturally to planes in $\mathbf{R}^3$. Let $\mathbf{u}$ and $\mathbf{v}$ be any two vectors on a plane P in $\mathbf{R}^3$ and $\mathbf{a}$ be a vector normal to P. Then $\mathbf{a}$ will be orthogonal to the vector $\mathbf{u} - \mathbf{v}$, that is, $\mathbf{a} \cdot (\mathbf{u} - \mathbf{v}) = 0$. Given a vector $\mathbf{v}$ on a plane P in $\mathbf{R}^3$ and a vector $\mathbf{a}$ normal to P, the *normal form* of plane P is written

$$\mathbf{a} \cdot (\mathbf{x} - \mathbf{v}) = 0 \quad \text{or} \quad a_1(x_1 - v_1) + a_2(x_2 - v_2) + a_3(x_3 - v_3) = 0 \tag{6}$$

As an exercise in vector algebra, let us verify that a 2-vector $\mathbf{x}$ satisfying the standard equation for a line in $\mathbf{R}^2$, $a_1x_1 + a_2x_2 = c$—or, equivalently, $\mathbf{a} \cdot \mathbf{x} = c$—satisfies the normal form, $\mathbf{a} \cdot (\mathbf{x} - \mathbf{v}) = 0$. Using the distributive law for scalar products, we have

$$\mathbf{a} \cdot (\mathbf{x} - \mathbf{v}) = \mathbf{a} \cdot \mathbf{x} - \mathbf{a} \cdot \mathbf{v} = c - c = 0$$

The analysis is similar for vectors on a plane in $\mathbf{R}^3$. As a consequence of this vector algebra exercise, we have seen that the components of the normal vector $\mathbf{a}$ are exactly the coefficients in the standard equation for the line or plane. Summarizing, we have

---

Theorem 4 (Normal Form of a Line or Plane)

(i) Given a vector $\mathbf{v}$ on a line in $\mathbf{R}^2$ and a vector $\mathbf{a}$ normal to the line, the normal form of the line is

$$\mathbf{a} \cdot (\mathbf{x} - \mathbf{v}) = 0 \quad \text{or} \quad a_1(x_1 - v_1) + a_2(x_2 - v_2) = 0$$

(ii) Given a vector $\mathbf{v}$ on a plane in $\mathbf{R}^3$ and a vector $\mathbf{a}$ normal to the plane, the normal form of the plane is

$$\mathbf{a} \cdot (\mathbf{x} - \mathbf{v}) = 0 \quad \text{or} \quad a_1(x_1 - v_1) + a_2(x_2 - v_2) + a_3(x_3 - v_3) = 0$$

(iii) The components of $\mathbf{a}$ are the coefficients in the standard equation for the line or plane, and the constant c equals $\mathbf{a} \cdot \mathbf{v}$.

EXAMPLE 6.
Equation of a
Line in $\mathbf{R}^2$

Find the normal form and the standard equation of a line in $\mathbf{R}^2$ through the vector $\mathbf{v} = [2, 3]$ and orthogonal to the vector $\mathbf{a} = [1, 4]$. The normal form is $\mathbf{a} \cdot (\mathbf{x} - \mathbf{v}) = 0$ or $1(x_1 - 2) + 4(x_2 - 3) = 0$. The standard equation of this line will be $\mathbf{a} \cdot \mathbf{x} = c$ or $1x_1 + 4x_2 = c$. As noted above, $c = \mathbf{a} \cdot \mathbf{v} = 1 \cdot 2 + 4 \cdot 3 = 14$. So the desired equation is $x_1 + 4x_2 = 14$. ■

EXAMPLE 7.
Equation of a
Plane in R^3

Find the normal form and standard equation of the plane in R^3 containing the vector $v = [4, -1, 3]$ and orthogonal to the vector $a = [2, 5, 1]$. The normal form is $a \cdot (x - v) = 0$ or $2(x_1 - 4) + 5(x_2 + 1) + (x_3 - 3) = 0$. The standard equation of the plane will be $a \cdot x = c$ or $2x_1 + 5x_2 + 1x_3 = c$. The value of c is $2 \cdot 4 + 5 \cdot (-1) + 1 \cdot 3 = 6$ and the desired equation is $2x_1 + 5x_2 + 1x_3 = 6$. ▪

These examples illustrate the pervasive role of orthogonality in a variety of linear algebra settings that would seem to have no role for geometry. Linear algebra is full of unexpected, but useful, geometric interpretations of vector and matrix expressions.

Section 2.3 Exercises

1. Compute the cosine of the angle, and determine the angle, made by the following pairs of vectors.
   (a) [1, 0], [1, 1] (b) [3, 4], [-3, 4] (c) [1, 2], [3, 1] (d) [1, 0, 1], [0, 1, 0]
   (e) [1, 1, 1], [1, -1, 2] (f) [1, 1, 1], [2, -1, 3]

2. If a triangle has corners at the following three points, find the cosine of each of the angles of the triangle.
   (a) [1, 1], [3, 4], and [5, 2] (b) [3, 0], [-2, 1], and [5, 6]
   (c) [2, 0, 1], [3, -1, 1], and [4, 2, 0]

3. Determine which of the following pairs of vectors are parallel (the angle between them is 0° or 180°), orthogonal, or neither.
   (a) [2, 3], [6, 9] (b) [6, 4], [-2, 3] (c) [3, 2], [3, 2] (d) [1, 0, 2], [0, 4, 0]
   (e) [3, -2, 4], [-6, 4, 8] (f) [1, 2, -3], [2, -1, 3]

4. Suppose that $u = [2, 5]$ and $v = [1, d]$. Determine d so that u and v are orthogonal.

5. Consider the following data of student performance, where the x-value is a scaled score (to have average value of 0) of high school grades, the y-value is a scaled score of SAT scores, and the z-value is an unscaled score of college grades.

	Student				
	A	B	C	D	E
x:	-4	-2	0	2	4
y:	2	-1	-2	-1	2
z:	3	6	7	7	6

Compute the correlation coefficient between the vectors of x and y values and between the vectors of x and z values.

6. The following data show scores that three students received on a battery of six different tests.

	Gerry	Jimmie	Ronnie
General IQ:	12	20	10
Math:	8	22	4
Reading:	16	14	10
Running:	24	16	12
Speaking:	12	10	30
Watching:	12	14	8

Compute the correlation coefficient between:
(a) Gerry and Jimmie
(b) Gerry and Ronnie
(c) Jimmie and Ronnie
(Hint: Remember first to subtract the average value from each number.)

7. (a) Compute the correlation coefficient of the following readings from seven students of their IQs and scores at Zaxxon.

Student:	A	B	C	D	E	F	G
IQ:	120	130	105	90	125	120	110
Zaxxon:	11,000	7000	10,000	12,000	8000	100,000	8000

    (b) Delete student F and recompute the correlation coefficient. (Hint: Remember first to subtract the average value from each number.)

8. In the homogeneous systems of equations $\mathbf{Ax} = \mathbf{0}$, for the following matrices $\mathbf{A}$, find a nonzero solution $\mathbf{x}^*$ and check graphically that $\mathbf{x}^*$ is orthogonal to the first row of $\mathbf{A}$.

(a) $\begin{bmatrix} 1 & 2 \\ 2 & 4 \end{bmatrix}$ (b) $\begin{bmatrix} 1 & -2 \\ -3 & 6 \end{bmatrix}$ (c) $\begin{bmatrix} 1 & 3 \\ -3 & -9 \end{bmatrix}$

9. For each of the following 2×2 matrices $\mathbf{A}$, find the inverse $\mathbf{A}^{-1}$ and confirm graphically that the ith column of $\mathbf{A}$ is orthogonal to the jth row of $\mathbf{A}^{-1}$ for $i \neq j$.

(a) $\begin{bmatrix} 1 & 2 \\ 2 & 3 \end{bmatrix}$ (b) $\begin{bmatrix} 0 & 1 \\ 1 & 2 \end{bmatrix}$ (c) $\begin{bmatrix} 2 & -1 \\ -4 & 3 \end{bmatrix}$

10. For the following matrices $\mathbf{A}$ with orthogonal columns, determine the inverse $\mathbf{A}^{-1}$ using Theorem 3.

(a) $\begin{bmatrix} 1 & -4 \\ 2 & 2 \end{bmatrix}$ (b) $\begin{bmatrix} -6 & 4 \\ 3 & 8 \end{bmatrix}$ (c) $\begin{bmatrix} -1 & 4 & -1 \\ 2 & 1 & -2 \\ 1 & 2 & 3 \end{bmatrix}$ (d) $\begin{bmatrix} 2 & -3 & 6 \\ -6 & 2 & 3 \\ 3 & 6 & 2 \end{bmatrix}$

11. Show that for any positive real numbers r, s, the vectors $\mathbf{u} = [r, s]$ and $\mathbf{v} = [s, -r]$ are orthogonal.

12. Give a 3-vector orthogonal to $[r, s, t]$, where r, s, t are real numbers.

13. Find the normal form and standard equation of the line through vector $\mathbf{v}$ and orthogonal to vector $\mathbf{a}$.
(a) $\mathbf{v} = [2, 3]$, $\mathbf{a} = [1, 1]$ (b) $\mathbf{v} = [1, -4]$, $\mathbf{a} = [2, -1]$ (c) $\mathbf{v} = [4, 1]$, $\mathbf{a} = [0, 3]$
(d) $\mathbf{v} = [-2, 3]$, $\mathbf{a} = [3, 4]$

14. Find the normal form and standard equation of the plane through the vector **v** and orthogonal to the vector **a**.

   (a) **v** = [1, 0, 2], **a** = [1, −2, 2]   **(b)** **v** = [−1, 1, 4], **a** = [0, 3, 1]
   (c) **v** = [2, 3, −1], **a** = [0, 1, 3]   **(d)** **v** = [5, 2, 1], **a** = [0, −2, 0]

15. **(a)** Given vectors **u** and **v** in $\mathbf{R}^2$ or $\mathbf{R}^3$, with $|\mathbf{u}|_e = |\mathbf{v}|_e = 1$. Show that $\mathbf{w} = \frac{1}{2}(\mathbf{u} + \mathbf{v})$ bisects the angle between **u** and **v**. (Hint: It is sufficient to show that the angle between **u** and **w** equals the angle between **w** and **u**.)

   (b) Generalize part (a) to vectors **u** and **v** in $\mathbf{R}^2$ or $\mathbf{R}^3$, with $|\mathbf{u}|_e = r$ and $|\mathbf{v}|_e = s$. Show that $\mathbf{w} = [1/(r + s)](s\mathbf{u} + r\mathbf{v})$ bisects the angle between **u** and **v**.

2.4 Projections and Regression

In this section we consider a vector closely related to the angle between two vectors. The *projection* **p** of vector **b** onto vector **a** is a multiple of **a**, that is, $\mathbf{p} = q\mathbf{a}$, for some scalar q, and is defined as follows. Form a right triangle with one corner at the origin, the hypotenuse being vector **b** (i.e., the line segment from the origin to point **b**), the projection **p** being the side adjacent to the origin and the third side being **b** − **p** [see Figure 2.14(a)]. The following picture can be associated in $\mathbf{R}^2$ with Figure 2.14(a). Suppose that there is a line L through the origin and **a** and there is a powerful light infinitely far away from the origin in a direction orthogonal to the line. Also assume that **b** is on the same side of the line L as the light. Then the shadow on line L made by **b** is the projection of **b** onto **a**. The length of the projection of **b** onto **a** can be interpreted as the value of **b**'s coordinate along the a-axis, if a coordinate system were used in which **a** was one of the axes [see Figure 2.14(b)].

Observe from these figures that the projection $q\mathbf{a}$ of **b** onto **a** is *the multiple of* **a** *that is closest to* **b**. That is, along the line formed by all multiples $r\mathbf{a}$ of **a**, the projection $q\mathbf{a}$ is the vector that minimizes the distance $|\mathbf{b} - r\mathbf{a}|$. (See Exercise 17 for a formal proof.)

When we compute the scalar product of any 2-vector $\mathbf{b} = [b_1, b_2]$ with the coordinate vector $\mathbf{i}_1 = [1, 0]$, we obtain $\mathbf{b} \cdot \mathbf{i}_1 = [b_1, b_2] \cdot [1, 0] = b_1$, simply the first coordinate of **b**, but b_1 is also the length of projection of **b** onto $\mathbf{i}_1$. Moreover, for the unit-length vector $\mathbf{b} = [\cos \theta, \sin \theta]$ (**b** is unit length since $|\mathbf{b}| = \sqrt{(\cos \theta)^2 + (\sin \theta)^2} = 1$), $\mathbf{b} \cdot \mathbf{i}_1 = \cos \theta \cdot 1 + \sin \theta \cdot 0 = \cos \theta$. So for unit-length vectors, the length of their projection onto $\mathbf{i}_1$ is the same as the cosine of the angle they make with $\mathbf{i}_1$. Extending this observation, we shall see shortly that the formulas for the

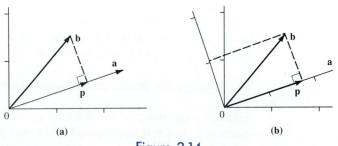

(a)   (b)

Figure 2.14

length of a projection of **b** onto **a** and for the cosine of the angle between **a** and **b** are very similar.

We summarize the key properties we shall need of the projection **p** of a vector **b** onto a vector **a**.

Property 1. The projection **p** is a multiple of **a**: $\mathbf{p} = q\mathbf{a}$, for some scalar q.

Property 2. The projection **p** is orthogonal to the vector $\mathbf{b} - \mathbf{p}$.

Property 3. The projection **p** is the multiple of **a** closest to **b**.

By property 1, the projection of **b** onto **a** is totally characterized by the scalar q. We now give a simple formula for q.

Theorem 1

The projection $q\mathbf{a}$ of vector **b** onto vector **a** is given by the formula

$$q = \frac{\mathbf{a} \cdot \mathbf{b}}{\mathbf{a} \cdot \mathbf{a}} \tag{1}$$

When **a** is a unit-length vector, $q = \mathbf{a} \cdot \mathbf{b}$.

Proof: Combining properties 1 and 2, the projection **p**, which equals $q\mathbf{a}$, is orthogonal to $\mathbf{b} - \mathbf{p}$, that is, to $\mathbf{b} - q\mathbf{a}$. If $q\mathbf{a}$ is orthogonal to $\mathbf{b} - q\mathbf{a}$, then **a** itself is also orthogonal to $\mathbf{b} - q\mathbf{a}$. The orthogonality of **a** and $\mathbf{b} - q\mathbf{a}$ means that $(\mathbf{b} - q\mathbf{a}) \cdot \mathbf{a} = 0$ (by Theorem 2 of Section 2.3). A little vector algebra then yields

$$(\mathbf{b} - q\mathbf{a}) \cdot \mathbf{a} = 0 \Rightarrow \mathbf{b} \cdot \mathbf{a} - q\mathbf{a} \cdot \mathbf{a} = 0 \tag{2}$$
$$\Rightarrow \mathbf{b} \cdot \mathbf{a} = q\mathbf{a} \cdot \mathbf{a} \Rightarrow q = \frac{\mathbf{a} \cdot \mathbf{b}}{\mathbf{a} \cdot \mathbf{a}}$$
■

Corollary. The Euclidean length of the projection $q\mathbf{a}$ of vector **b** onto vector **a** is $|(\mathbf{a} \cdot \mathbf{b}/|\mathbf{a}|)|$.
Proof: We use the scalar factoring property of norms: $|r\mathbf{a}| = |r| \cdot |\mathbf{a}|$. The length of the projection $q\mathbf{a}$ is then $|q| \cdot |\mathbf{a}|$. Since $\mathbf{a} \cdot \mathbf{a} = |\mathbf{a}|^2$, we have $|q| \cdot |\mathbf{a}| = |(\mathbf{a} \cdot \mathbf{b}/|\mathbf{a}|^2)||\mathbf{a}| = |(\mathbf{a} \cdot \mathbf{b}/|\mathbf{a}|)|$, as claimed. ■

The number q, the projection coefficient of **b** onto **a**, is often called the **a-*coordinate*** of **b**.

Since the projection **p** of vector **b** onto vector **a** is orthogonal to $\mathbf{b} - \mathbf{p}$ (property 2 of projections), then **b** is decomposed into a part **p** that is a multiple of **a** and a part $\mathbf{b} - \mathbf{p}$ orthogonal to **a**.

EXAMPLE 1.
Projections and
Orthogonal
Decompositions

(a) Find the projection $\mathbf{p}$ of $\mathbf{b} = [-1, 3]$ onto $\mathbf{a} = \mathbf{i}_2 = [0, 1]$ and use it to decompose $\mathbf{b}$ into orthogonal parts, one part parallel to $\mathbf{a}$ and one part orthogonal to $\mathbf{a}$. This projection of $\mathbf{b}$ is $q\mathbf{i}_2$, where

$$q = \frac{\mathbf{a} \cdot \mathbf{b}}{\mathbf{a} \cdot \mathbf{a}} = \frac{0 \cdot (-1) + 1 \cdot 3}{0 \cdot 0 + 1 \cdot 1} = \frac{3}{1} = 3$$

So $\mathbf{p} = 3\mathbf{i}_2 = [0, 3]$, and the orthogonal part is $\mathbf{b} - \mathbf{p} = [-1, 3] - [0, 3] = [-1, 0]$ [see Figure 2.15(a)].

(b) Find the projection $\mathbf{p}$ of $\mathbf{b} = [-1, 3]$ onto $\mathbf{a} = [2, 4]$ and use it to decompose $\mathbf{b}$ into orthogonal parts. This projection is $q\mathbf{a}$, where

$$q = \frac{\mathbf{a} \cdot \mathbf{b}}{\mathbf{a} \cdot \mathbf{a}} = \frac{2 \cdot (-1) + 4 \cdot 3}{2 \cdot 2 + 4 \cdot 4} = \frac{10}{20} = \frac{1}{2}$$

So $\mathbf{p} = q\mathbf{a} = 0.5[2, 4] = [1, 2]$ and the orthogonal part is $\mathbf{b} - \mathbf{p} = [-1, 3] - [1, 2] = [-2, 1]$ [see Figure 2.15(b)].

(c) Find the projection $\mathbf{p}$ of $\mathbf{b} = [2, -4]$ onto $\mathbf{a} = [2, 1]$ and use it to decompose $\mathbf{b}$ into orthogonal parts. Now

$$q = \frac{\mathbf{a} \cdot \mathbf{b}}{\mathbf{a} \cdot \mathbf{a}} = \frac{2 \cdot 2 + 1 \cdot (-4)}{2 \cdot 2 + 1 \cdot 1} = \frac{0}{5}$$

Then $\mathbf{a}$ and $\mathbf{b}$ are orthogonal, so the projection $\mathbf{p}$ is the zero vector $0\mathbf{a} = [0, 0]$, and the orthogonal part is $\mathbf{b} - \mathbf{p} = \mathbf{b} = [2, -4]$.

(d) Find the projection $\mathbf{p}$ of $\mathbf{b} = [-1, 5, 3]$ onto $\mathbf{a} = [-2, 0, 2]$ and use it to decompose $\mathbf{b}$ into orthogonal parts. Here

$$q = \frac{\mathbf{a} \cdot \mathbf{b}}{\mathbf{a} \cdot \mathbf{a}} = \frac{(-2) \cdot (-1) + 0 \cdot 5 + 2 \cdot 3}{(-2) \cdot (-2) + 0 \cdot 0 + 2 \cdot 2} = \frac{8}{8} = 1$$

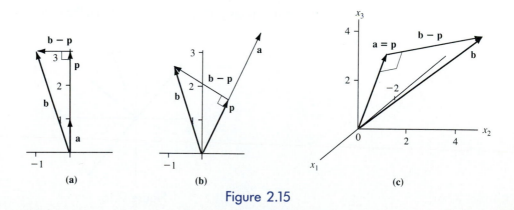

Figure 2.15

So the projection is $\mathbf{p} = 1\mathbf{a} = 1[-2, 0, 2] = [-2, 0, 2]$, and the orthogonal part is $\mathbf{b} - \mathbf{p} = [-1, 5, 3] - [-2, 0, 2] = [1, 5, 1]$ (see Figure 2.15(c)). ■

Theorem 3 in Section 2.3 discussed the form of the inverse of a matrix $\mathbf{A}$ whose columns are orthogonal. The key fact was that the ith row of $\mathbf{A}^{-1}$ equals the vector $(1/|\mathbf{a}_i|^2)\mathbf{a}_i$ (the ith column of $\mathbf{A}$ divided the square of the column's norm). Suppose that for such an $\mathbf{A}$, we wanted to solve the system of equations $\mathbf{Ax} = \mathbf{b}$. Then $\mathbf{x} = \mathbf{A}^{-1}\mathbf{b}$, and the ith component $x_i = \mathbf{A}^{-1}\mathbf{b}$ is the scalar product of the ith row of $\mathbf{A}^{-1}$ times $\mathbf{b}$. That is,

$$x_i = \left(\frac{1}{|\mathbf{a}_i|^2}\right)\mathbf{a}_i \cdot \mathbf{b} = \frac{\mathbf{a}_i \cdot \mathbf{b}}{\mathbf{a}_i \cdot \mathbf{a}_i} \tag{3}$$

Thus x_i is just the projection coefficient of $\mathbf{b}$ onto $\mathbf{a}_i$. Small world!

Theorem 2 (Solving Ax = b When A Has Orthogonal Columns)
The solution of the system of equations $\mathbf{Ax} = \mathbf{b}$, where $\mathbf{A}$ is a square matrix with orthogonal columns, is the lengths of the projection of $\mathbf{b}$ onto the columns of $\mathbf{A}$, as given in (3).

**EXAMPLE 2.
Solving Ax = b for a Matrix with Orthogonal Columns**

Consider $\mathbf{Ax} = \mathbf{b}$, where $\mathbf{A} = \begin{bmatrix} 3 & 4 \\ -4 & 3 \end{bmatrix}$ has orthogonal columns and $\mathbf{b} = \begin{bmatrix} 3 \\ 6 \end{bmatrix}$. As a linear combination of columns problem (see the middle of Section 2.1, we have

$$x_1 \begin{bmatrix} 3 \\ -4 \end{bmatrix} + x_2 \begin{bmatrix} 4 \\ 3 \end{bmatrix} = \begin{bmatrix} 3 \\ 6 \end{bmatrix} \tag{4}$$

Figure 2.16 shows the situation graphically. Since $\mathbf{A}$'s columns $\mathbf{a}_1 = \begin{bmatrix} 3 \\ -4 \end{bmatrix}$ and $\mathbf{a}_2 = \begin{bmatrix} 4 \\ 3 \end{bmatrix}$ are orthogonal, in the decomposition of $\mathbf{b}$ into its projection $\mathbf{p}$ onto $\mathbf{a}_1$ and an orthogonal part $\mathbf{b} - \mathbf{p}$ (as performed in Example 1), the orthogonal part $\mathbf{b} - \mathbf{p}$ will be

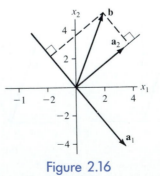

Figure 2.16

a multiple of $\mathbf{a}_2$ (all vectors orthogonal to $\mathbf{a}_1$ in $\mathbf{R}^2$ are multiples of one another). So $\mathbf{b} - \mathbf{p}$ will be a projection of $\mathbf{b}$ on $\mathbf{a}_2$. Thus we have argued that (4) can be solved by letting x_1 be the projection coefficient of $\mathbf{b}$ onto $\mathbf{a}_1$ and x_2 be the projection coefficient of $\mathbf{b}$ onto $\mathbf{a}_2$. Using (3), we have

$$x_1 = \frac{\mathbf{a}_1 \cdot \mathbf{b}}{\mathbf{a}_1 \cdot \mathbf{a}_1} = \frac{3 \cdot 3 + (-4) \cdot 6}{3 \cdot 3 + (-4) \cdot (-4)} = -\frac{15}{25} = -\frac{3}{5}$$

$$x_2 = \frac{\mathbf{a}_2 \cdot \mathbf{b}}{\mathbf{a}_2 \cdot \mathbf{a}_2} = \frac{4 \cdot 3 + 3 \cdot 6}{4 \cdot 4 + 3 \cdot 3} = \frac{30}{25} = \frac{6}{5}$$

(5) ■

**EXAMPLE 3.
Solving Ax = b for
a 3 × 3 Matrix
with Orthogonal
Columns**

Consider the matrix

$$\mathbf{A} = \begin{bmatrix} 2 & 1 & -2 \\ 1 & 2 & 2 \\ 2 & -2 & 1 \end{bmatrix}$$

with orthogonal columns. Let $\mathbf{b} = \begin{bmatrix} 1 \\ 4 \\ 6 \end{bmatrix}$. Then the solution of $\mathbf{Ax} = \mathbf{b}$ is, by (3),

$$x_1 = \frac{\mathbf{a}_1 \cdot \mathbf{b}}{\mathbf{a}_1 \cdot \mathbf{a}_1} = \frac{2 \cdot 1 + 1 \cdot 4 + 2 \cdot 6}{2 \cdot 2 + 1 \cdot 1 + 2 \cdot 2} = \frac{18}{9} = 2$$

$$x_2 = \frac{\mathbf{a}_2 \cdot \mathbf{b}}{\mathbf{a}_2 \cdot \mathbf{a}_2} = \frac{1 \cdot 1 + 2 \cdot 4 + (-2) \cdot 6}{1 \cdot 1 + 2 \cdot 2 + (-2) \cdot (-2)} = \frac{-3}{9} = -\frac{1}{3}$$

$$x_3 = \frac{\mathbf{a}_3 \cdot \mathbf{b}}{\mathbf{a}_3 \cdot \mathbf{a}_3} = \frac{(-2) \cdot 1 + 2 \cdot 4 + 1 \cdot 6}{(-2) \cdot (-2) + 2 \cdot 2 + 1 \cdot 1} = \frac{12}{9} = \frac{4}{3}$$

(6)

The inverse $\mathbf{A}^{-1}$ of $\mathbf{A}$ is given in Example 5 of Section 2.3. The reader can check that computing the solution $\mathbf{x}$ by $\mathbf{x} = \mathbf{A}^{-1}\mathbf{b}$ yields the same values as in (6). ■

**EXAMPLE 4.
Projections in
Physics**

A common vector problem in physics is to have a force applied to an object that is free to move only along a line. Suppose that a round ball with a hole through it is mounted on a string and the string passes through the origin at an angle of $30°$, that is, in the direction of the unit-length vector $\mathbf{u} = [\sqrt{3}/2, \frac{1}{2}]$. (The reader should check that $|\mathbf{u}| = 1$.) A force with a strength of 4 units in the x_1 direction and a strength of 8 units in the x_2 direction is applied to the ball. What is the strength of this force in the direction of the string vector $\mathbf{u}$? (See Figure 2.17.)

We need to find the length of the projection of the force vector $\mathbf{f} = [4, 8]$ onto $\mathbf{u}$. We compute

$$q = \frac{\mathbf{f} \cdot \mathbf{u}}{\mathbf{u} \cdot \mathbf{u}} = \frac{4 \cdot \sqrt{3}/2 + 8 \cdot \frac{1}{2}}{1} = 2\sqrt{3} + 4 \approx 7.5$$

(7)

So the strength of the force along $\mathbf{u}$ is about 7.5 units. Note that in (7), we do not multiply out the scalar product $\mathbf{u} \cdot \mathbf{u}$ in the denominator because we know that $|\mathbf{u}|^2$ $(= \mathbf{u} \cdot \mathbf{u}) = 1$. ■

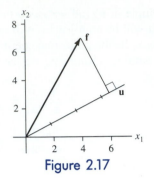

Figure 2.17

EXAMPLE 5.
Distance from a
Point to a Line

Find the distance from the vector $\mathbf{v} = [7, 3]$ to the line $3x_1 + x_2 = 9$.

We seek the length of the vector $\mathbf{t}$ that goes from the closest point on the line to $\mathbf{v}$. See Figure 2.18. Clearly, $\mathbf{t}$ will be orthogonal to the line. To find $\mathbf{t}$, form the difference vector $\mathbf{v} - \mathbf{w}$ from some vector $\mathbf{w}$ on the line to $\mathbf{v}$ and project $\mathbf{v} - \mathbf{w}$ onto the vector $\mathbf{a} = [3, 1]$, which (by Theorem 4 of Section 2.3) is orthogonal to our line $\mathbf{a} \cdot \mathbf{x} = 9$. The projection of $\mathbf{v} - \mathbf{w}$ onto $\mathbf{a}$ will be $\mathbf{t}$ (see Figure 2.18). To find a vector $\mathbf{w}$ on the line $3x_1 + x_2 = 9$, we can set $w_2 = 0$. Then $w_1 = 3$ (so that $3w_1 + w_2 = 9$). If $\mathbf{w} = [3, 0]$, then $\mathbf{v} - \mathbf{w} = [7, 3] - [3, 0] = [4, 3]$. By Theorem 1, the projection of $\mathbf{v} - \mathbf{w}$ onto $\mathbf{a}$ is $q\mathbf{a}$, where

$$q = \frac{(\mathbf{v} - \mathbf{w}) \cdot \mathbf{a}}{\mathbf{a} \cdot \mathbf{a}} = \frac{4 \cdot 3 + 3 \cdot 1}{3 \cdot 3 + 1 \cdot 1} = \frac{15}{10} = \frac{3}{2} \tag{8}$$

So the projection $\mathbf{t}$ is $\frac{3}{2}\mathbf{a} = [\frac{9}{2}, \frac{3}{2}]$. The required distance is the length of $\mathbf{t}$, $|\mathbf{t}| = \sqrt{(\frac{9}{2})^2 + (\frac{3}{2})^2} = \sqrt{\frac{90}{4}} = \frac{3}{2}\sqrt{10}$. ■

We next apply the concept of projections to a very important problem in statistics, one seemingly unrelated to projections and angles. The problem is statistical regression. In regression, one is given a set of data points (x_i, y_i) and one seeks to fit a line $\hat{y} = cx + d$ as closely as possible to a set of data points. The point $(x_i, \hat{y}_i)$, where $\hat{y}_i = cx_i + d$, would be the regression estimate for (x_i, y_i). The name **regression**,

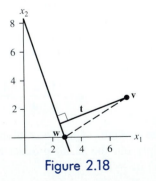

Figure 2.18

which means movement back to a less developed state, comes from the idea that our model recaptures a simple relationship between the x_i and the y_i which randomness has obscured (the variables regress to a linear relationship).

For the present we consider the simple case where the line has the form $\hat{y} = cx$.

EXAMPLE 6.
Simple Regression
Problem

Suppose that we want to fit the three points (0, 1), (2, 1), and (4, 4) to a line of the form

$$\hat{y} = cx \qquad (9)$$

(see Figure 2.19). The x-value might represent the number of semesters of college mathematics a student has taken and the y-value the student's score on some test. There are thousands of other settings that might give rise to these values. The estimate (9) would help us predict the y-values for other x-values (e.g., predict how other students might do on the test based on the amount of math they have taken).

The three points in this problem are readily seen not to lie on a common line, much less a line through the origin. [Any line of the form (9) passes through the origin.] So we have to find a choice of c that gives the best possible fit, that is, a line $\hat{y} = cx$ passing as close to these three points as possible.

What do "best possible fit" and "as close as possible" mean? The most common approach used in such problems is to minimize $\Sigma\,(\hat{y}_i - y_i)^2$, the sum of the squares of the errors. The error at a point (x_i, y_i) will be $|\hat{y}_i - y_i| = |cx_i - y_i|$, the absolute difference between the value cx_i predicted by (9) and the true value y_i. (The absolute value is needed so that a "negative" error and a "positive" error cannot offset each other.) However, absolute values are not easy to use in mathematical equations. Taking the squares of differences yields positive numbers without using absolute values. There is also a geometric reason we shall give shortly for using squares.

For the points (0, 1), (2, 1), (4, 4), the expression $\Sigma\,(\hat{y}_i - y_i)^2$ for the sum of squares of the errors (SSE) is

$$
\begin{aligned}
\text{SSE} &= (0c - 1)^2 + (2c - 1)^2 + (4c - 4)^2 \\
&= 1 + (4c^2 - 4c + 1) + (16c^2 - 32c + 16) \qquad (10) \\
&= 20c^2 - 36c + 18
\end{aligned}
$$

We could use calculus to find the value of c that minimizes the expression in (10). However, there is a geometric interpretation of the sum of squares of errors that will

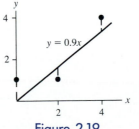

Figure 2.19

allow us to convert the minimization of (10) into a projection problem. Let $\mathbf{x}$ be the vector of our x-values and let $\mathbf{y}$ be the vector of our corresponding y-values. Then $\mathbf{x} = \begin{bmatrix} 0 \\ 2 \\ 4 \end{bmatrix}$ and $\mathbf{y} = \begin{bmatrix} 1 \\ 1 \\ 4 \end{bmatrix}$. Furthermore, let $\hat{\mathbf{y}}$ be the vector of the estimates for $\mathbf{y}$. Equation (9) can now be rewritten

$$\hat{\mathbf{y}} = c\mathbf{x} \tag{11}$$

That is, the estimates $\hat{\mathbf{y}} = \begin{bmatrix} \hat{y}_1 \\ \hat{y}_2 \\ \hat{y}_3 \end{bmatrix}$ from (9) will be c times the x-values $\begin{bmatrix} 0 \\ 2 \\ 4 \end{bmatrix}$.

Think of $\mathbf{x}$, $\mathbf{y}$, and $\hat{\mathbf{y}}$ as vectors in three-dimensional space, where $\hat{\mathbf{y}}$ is a multiple of $\mathbf{x}$. Then the obvious strategy for approximating $\mathbf{y}$ by $\hat{\mathbf{y}}$ is to pick the value of c that makes $c\mathbf{x} (= \hat{\mathbf{y}})$ as geometrically close as possible to $\mathbf{y}$ (see Figure 2.20). That is, we want to minimize the distance $|c\mathbf{x} - \mathbf{y}|$ in three-dimensional space between $c\mathbf{x}$ and $\mathbf{y}$. This distance between

$$c\mathbf{x} = \begin{bmatrix} 0c \\ 2c \\ 4c \end{bmatrix} \quad \text{and} \quad \mathbf{y} = \begin{bmatrix} 1 \\ 1 \\ 4 \end{bmatrix}$$

is simply

$$|c\mathbf{x} - \mathbf{y}| = \sqrt{(0c - 1)^2 + (2c - 1)^2 + (4c - 4)^2} \tag{12}$$

Comparing (10) and (12) we see that $|c\mathbf{x} - \mathbf{y}|$ is the square root of SSE. So minimizing SSE is equivalent to minimizing the distance $|c\mathbf{x} - \mathbf{y}|$.

But the multiple of $\mathbf{x}$ that is closest to $\mathbf{y}$ is the projection of $\mathbf{y}$ onto $\mathbf{x}$ (property 3 of projections). Then by Theorem 1,

$$c = \frac{\mathbf{y} \cdot \mathbf{x}}{\mathbf{x} \cdot \mathbf{x}} = \frac{1 \cdot 0 + 1 \cdot 2 + 4 \cdot 4}{0 \cdot 0 + 2 \cdot 2 + 4 \cdot 4} = \frac{18}{20} = 0.9$$

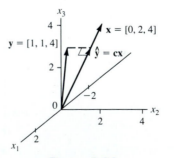

Figure 2.20

Then the regression line is $\hat{y} = 0.9x$, and our estimates $\hat{y}_i$ are

$$\hat{y}_1 = 0.9x_1 = 0.9 \cdot 0 = 0$$
$$\hat{y}_2 = 0.9x_2 = 0.9 \cdot 2 = 1.8$$
$$\hat{y}_3 = 0.9x_3 = 0.9 \cdot 4 = 3.6$$

(see Figure 2.19). ■

Section 2.4 Exercises

1. Determine the projection coefficient of **b** onto **a**.
   (a) **a** = [3, 4], **b** = [5, 1]   (b) **a** = [3, 1], **b** = [−3, 5]   (c) **a** = [0, 1], **b** = [4, 0]
   (d) **a** = [1, 3], **b** = [−6, 2]   (e) **a** = [−3, −4], **b** = [3, 7]   (f) **a** = [5, 2], **b** = [5, 2]

2. Determine the projection of **b** onto **a**.
   (a) **a** = [2, 2, 1], **b** = [3, 4, 1]   (b) **a** = [3, 2, 1], **b** = [1, −3, 3]
   (c) **a** = [3, −4, 2], **b** = [0, 6, −2]   (d) **a** = [3, 0, 4], **b** = [0, 2, 1]
   (e) **a** = [−2, 3, 1], **b** = [6, 3, 3]   (f) **a** = [−2, −4, −5], **b** = [1, 2, 4]

3. Decompose each vector **b** in Exercise 1 into two orthogonal components, one component parallel to **a** and one component orthogonal to **a**.

4. Decompose each vector **b** in Exercise 2 into two orthogonal components, one component parallel to **a** and one component orthogonal to **a**.

5. Use Theorem 2 to solve the system of linear equations **Ax** = **b**, where the columns of **A** are orthogonal. Plot the columns of **A** and **b**, and the projections of **b** onto the columns of **A**.

   (a) $\mathbf{A} = \begin{bmatrix} 1 & -4 \\ 2 & 2 \end{bmatrix}$, $\mathbf{b} = \begin{bmatrix} 4 \\ 2 \end{bmatrix}$ (b) $\mathbf{A} = \begin{bmatrix} -6 & 4 \\ 3 & 8 \end{bmatrix}$, $\mathbf{b} = \begin{bmatrix} 7 \\ 2 \end{bmatrix}$

   (c) $\mathbf{A} = \begin{bmatrix} -3 & 3 \\ 1 & 9 \end{bmatrix}$, $\mathbf{b} = \begin{bmatrix} -5 \\ 1 \end{bmatrix}$

6. Use Theorem 2 to solve the system of linear equations **Ax** = **b**, where the columns of **A** are orthogonal.

   (a) $\mathbf{A} = \begin{bmatrix} -1 & 4 & -1 \\ 2 & 1 & -2 \\ 1 & 2 & 3 \end{bmatrix}$, $\mathbf{b} = \begin{bmatrix} 4 \\ 1 \\ 2 \end{bmatrix}$ (b) $\mathbf{A} = \begin{bmatrix} 2 & -3 & 6 \\ -6 & 2 & 3 \\ 3 & 6 & 2 \end{bmatrix}$, $\mathbf{b} = \begin{bmatrix} -2 \\ 5 \\ -1 \end{bmatrix}$

7. Suppose that a ball is moving along a string in the direction **u** = [0.6, −0.8] away from the origin in the x_1–x_2 plane. Find the strength of the following force vectors **f** in the direction of the string vector **u**?
   (a) **f** = [3, 0]   (b) **f** = [3, 6]   (c) **f** = [4, 3]   (d) **f** = [2, 3]

8. Suppose that a train is moving along a track in a northeast direction—the unit vector in this direction is [$1/\sqrt{2}$, $1/\sqrt{2}$]. For the following direction and velocity of a wind, what is the component of this wind in the direction the train is moving?
   (a) 20 kilometers per hour due east (a vector of [20, 0])
   (b) 10 kilometers per hour due north (a vector of [0, 10])

(c) 12 kilometers per hour to the east and 8 kilometers per hour to the north
(d) 6 kilometers per hour to the west and 10 kilometers per hour to the south
(e) 20 kilometers per hour due southeast

9. Find the distance from the following vectors to the line $x_1 + 2x_2 = 8$.
   (a) [5, 3] **(b)** [−2, 4] **(c)** [4, 2]

10. Find the distance from the following vectors to the line $5x_1 + 3x_2 = 11$.
   (a) [2, 3] **(b)** [0, 1] **(c)** [−2, 0]

11. Find the distance from the following vectors to the plane $3x_1 + 2x_2 + x_3 = 12$.
   (a) [0, 0, 1] **(b)** [3, 2, 1] **(c)** [4, 1, 2]

12. Seven students earned the following scores on a test after studying for different numbers of weeks.

Student:	A	B	C	D	E	F	G
Weeks of study:	0	1	2	3	4	5	6
Test score:	3	4	7	6	10	6	10

Fit these data with a regression model of the form $\hat{y} = qx$, where x is the number of weeks studied and y is the test score. Plot the observed scores and the predicted scores.

13. The following data give the number of accidents bus drivers had in one year as a function of the number of years on the job.

Years on job:	2	4	6	8	10	12
Accidents:	10	8	3	8	4	5

Fit these data with a regression model of the form $\hat{y} = qx$, where x is the number of years experience and y is the number of bus accidents. Plot the observed numbers of accidents and the predicted numbers. How good is the fit?

14. The following data give the number of fish caught by a fisher on successive weekends during a fishing season.

Weekend number:	1	2	3	4	5	6	7
Fish caught:	4	9	13	14	18	22	28

Fit these data with a regression model of the form $\hat{y} = qx$, where x is the number of the weekend and y is the number of fish caught.

15. Consider the following data, which are believed to obey (approximately) a relation of the form $\hat{y} = qx^2$.

x-value:	1	2	3	4	5	6	7
y-value:	0.5	1	2	3.5	7	9	12

Perform a transformation $y = f(y)$ on y so that the transformed model is $\hat{y}' = q'x$. Then determine q', reverse the transformation to determine q, and plot the curve $\hat{y} = qx^2$.

16. Consider the following data, which are believed to obey a square root law $\hat{y} = q\sqrt{x}$.

$$\begin{array}{lccccccc} \text{Age:} & 10 & 15 & 20 & 25 & 30 & 35 & 40 \\ \text{Strength:} & 7 & 12 & 13 & 16 & 17 & 19 & 20 \end{array}$$

Perform a transformation $y' = f(y)$ on y so that the transformed model is $\hat{y}' = q'x$. Then determine q', reverse the transformation to determine q, and plot the curve $\hat{y} = q\sqrt{x}$.

17. Prove that if $\mathbf{p} = q\mathbf{a}$ is the projection of vector $\mathbf{b}$ onto vector $\mathbf{a}$, then $\mathbf{p}$ is the multiple of $\mathbf{a}$ closest to $\mathbf{b}$. That is, $s = q$ minimizes the expression $|\mathbf{b} - s\mathbf{a}|$, where $q = \mathbf{a} \cdot \mathbf{b}/\mathbf{a} \cdot \mathbf{a}$. [Hint: Let $\mathbf{r} = c\mathbf{a}$ be any multiple of $\mathbf{a}$ and form the right triangle with corners at $\mathbf{r}$, $\mathbf{b}$, and $\mathbf{p}$. (The corner at $\mathbf{p}$ is the right angle.)]

2.5 Cross Product

In this section we discuss a binary operation, the cross product, on two vectors whose result is a vector. Recall that the scalar product $\mathbf{a} \cdot \mathbf{b}$ yields a scalar. The cross product originally arose in physical science problems, where, for example, it gives the direction of the flux in electromagnetic fields. Today, cross products play an important role in a wider range of applications, including computer graphics.

Definition

If $\mathbf{u} = [u_1, u_2, u_3]$ and $\mathbf{v} = [v_1, v_2, v_3]$ are vectors in $\mathbf{R}^3$, the *cross product* $\mathbf{u} \times \mathbf{v}$ is the 3-vector defined by

$$\mathbf{u} \times \mathbf{v} = [u_2v_3 - u_3v_2, \; u_3v_1 - u_1v_3, \; u_1v_2 - u_2v_1]$$

Note that the cross product can be defined in terms of determinants. That is,

$$\mathbf{u} \times \mathbf{v} = \left[\begin{vmatrix} u_2 & u_3 \\ v_2 & v_3 \end{vmatrix}, \; -\begin{vmatrix} u_1 & u_3 \\ v_1 & v_3 \end{vmatrix}, \; \begin{vmatrix} u_1 & u_2 \\ v_1 & v_2 \end{vmatrix} \right] \tag{1}$$

Using the coordinate 3-vectors $\mathbf{i}_1 = [1, 0, 0]$, $\mathbf{i}_2 = [0, 1, 0]$, $\mathbf{i}_3 = [0, 0, 1]$, (1) becomes

$$\mathbf{u} \times \mathbf{v} = \begin{vmatrix} u_2 & u_3 \\ v_2 & v_3 \end{vmatrix} \mathbf{i}_1 - \begin{vmatrix} u_1 & u_3 \\ v_1 & v_3 \end{vmatrix} \mathbf{i}_2 + \begin{vmatrix} u_1 & u_2 \\ v_1 & v_2 \end{vmatrix} \mathbf{i}_3 \tag{2a}$$

An easy-to-remember way to write (2a) is obtained by using the cofactor formulation of determinants [expression (13) in Section 1.6]:

$$\mathbf{u} \times \mathbf{v} = \begin{vmatrix} \mathbf{i}_1 & \mathbf{i}_2 & \mathbf{i}_3 \\ u_1 & u_2 & u_3 \\ v_1 & v_2 & v_3 \end{vmatrix} = \begin{vmatrix} u_2 & u_3 \\ v_2 & v_3 \end{vmatrix} \mathbf{i}_1 - \begin{vmatrix} u_1 & u_3 \\ v_1 & v_3 \end{vmatrix} \mathbf{i}_2 + \begin{vmatrix} u_1 & u_2 \\ v_1 & v_2 \end{vmatrix} \mathbf{i}_3 \tag{2b}$$

EXAMPLE 1.
Cross Products

(a) For $\mathbf{u} = [2, 3, 4]$ and $\mathbf{v} = [3, -1, 0]$, the cross product is

$$\mathbf{u} \times \mathbf{v} = [3 \cdot 0 - 4 \cdot (-1), 4 \cdot 3 - 2 \cdot 0, 2 \cdot (-1) - 3 \cdot 3] = [4, 12, -11]$$

(b) For $\mathbf{i}_1 = [1, 0, 0]$ and $\mathbf{i}_2 = [0, 1, 0]$, the cross product is

$$\mathbf{u} \times \mathbf{v} = [0 \cdot 0 - 0 \cdot 1, 0 \cdot 0 - 1 \cdot 0, 1 \cdot 1 - 0 \cdot 0] = [0, 0, 1] = \mathbf{i}_3$$

(c) For $\mathbf{u} = [2, 3, 4]$ and $\mathbf{v} = 2\mathbf{u} = [4, 6, 8]$, the cross product is

$$\mathbf{u} \times \mathbf{v} = [3 \cdot 8 - 4 \cdot 6, 4 \cdot 4 - 2 \cdot 8, 2 \cdot 6 - 3 \cdot 4] = [0, 0, 0] = \mathbf{0} \qquad ■$$

Extending part (b) of Example 1, one can check that

$$\begin{array}{lll} \mathbf{i}_1 \times \mathbf{i}_2 = \mathbf{i}_3 & \mathbf{i}_2 \times \mathbf{i}_3 = \mathbf{i}_1 & \mathbf{i}_3 \times \mathbf{i}_1 = \mathbf{i}_2 \\ \mathbf{i}_2 \times \mathbf{i}_1 = -\mathbf{i}_3 & \mathbf{i}_3 \times \mathbf{i}_2 = -\mathbf{i}_1 & \mathbf{i}_1 \times \mathbf{i}_3 = -\mathbf{i}_2 \\ \mathbf{i}_1 \times \mathbf{i}_1 = \mathbf{0} & \mathbf{i}_2 \times \mathbf{i}_2 = \mathbf{0} & \mathbf{i}_3 \times \mathbf{i}_3 = \mathbf{0} \end{array} \qquad (3)$$

The normal development of new concepts in this book has been to give motivating examples and problems first and then develop the theorems. In this section we give the theorems first, because these theorems describe interesting properties of cross products that are not apparent in numerical examples. For starters, we present the following theorem that summarizes the basic algebraic properties of the cross product. Most important of these is the fact that the cross product of a vector with itself yields the zero vector $\mathbf{0}$.

Theorem 1. If $\mathbf{u}$, $\mathbf{v}$, and $\mathbf{w}$ are 3-vectors, the following properties are true.
   (i) $\mathbf{u} \times \mathbf{v} = -\mathbf{v} \times \mathbf{u}$.
   (ii) $\mathbf{u} \times (\mathbf{v} + \mathbf{w}) = \mathbf{u} \times \mathbf{v} + \mathbf{u} \times \mathbf{w}$ and $(\mathbf{u} + \mathbf{v}) \times \mathbf{w} = \mathbf{u} \times \mathbf{w} + \mathbf{v} \times \mathbf{w}$.
   (iii) $k(\mathbf{u} \times \mathbf{v}) = (k\mathbf{u}) \times \mathbf{v} = \mathbf{u} \times (k\mathbf{v})$.
   (iv) $\mathbf{u} \times \mathbf{0} = \mathbf{0} \times \mathbf{u} = \mathbf{0}$ and $\mathbf{u} \times \mathbf{u} = \mathbf{0}$.
   Proof: (i) By inspection we see that $\mathbf{v} \times \mathbf{u} = [v_2 u_3 - v_3 u_2, v_3 u_1 - v_1 u_3, v_1 u_2 - v_2 u_1]$ is the entry-by-entry negation of $\mathbf{u} \times \mathbf{v} = [u_2 v_3 - u_3 v_2, u_3 v_1 - u_1 v_3, u_1 v_2 - u_2 v_1]$. The property also follows from the property of determinants that interchanging the rows in the determinant in (2b) reverses its sign [Theorem 3(i) in Section 1.6].
   (ii) Using (2b), we have

$$\mathbf{u} \times (\mathbf{v} + \mathbf{w}) = \begin{vmatrix} \mathbf{i}_1 & \mathbf{i}_2 & \mathbf{i}_3 \\ u_1 & u_2 & u_3 \\ v_1 + w_1 & v_2 + w_2 & v_3 + w_3 \end{vmatrix}$$

$$= \begin{vmatrix} \mathbf{i}_1 & \mathbf{i}_2 & \mathbf{i}_3 \\ u_1 & u_2 & u_3 \\ v_1 & v_2 & v_3 \end{vmatrix} + \begin{vmatrix} \mathbf{i}_1 & \mathbf{i}_2 & \mathbf{i}_3 \\ u_1 & u_2 & u_3 \\ w_1 & w_2 & w_3 \end{vmatrix} = (\mathbf{u} \times \mathbf{v}) + (\mathbf{u} \times \mathbf{w})$$

where the equality between determinants is Theorem 3(v) in Section 1.6. A similar argument proves the other distributive law for cross products.

(iii) Proof is obvious.

(iv) Proof of $\mathbf{u} \times \mathbf{0} = \mathbf{0} \times \mathbf{u} = \mathbf{0}$ is obvious. The first entry of $\mathbf{u} \times \mathbf{u}$ is $u_2 u_3 - u_3 u_2 = 0$ and similarly for the other entries of $\mathbf{u} \times \mathbf{u}$. ■

Note that, except in special cases, $\mathbf{u} \times (\mathbf{v} \times \mathbf{w}) \neq (\mathbf{u} \times \mathbf{v}) \times \mathbf{w}$.

One of the interesting properties of the cross product $\mathbf{u} \times \mathbf{v}$ is that it is orthogonal to both $\mathbf{u}$ and $\mathbf{v}$. Thus if P is the plane of vectors formed by all linear combinations of the 3-vectors $\mathbf{u}$ and $\mathbf{v}$ (assuming that $\mathbf{u}$ is not a multiple of $\mathbf{v}$), the vector $\mathbf{u} \times \mathbf{v}$ is normal to P, that is, orthogonal to P. Before proving this and some related results, we first note a simple way to express the scalar product–cross product $\mathbf{u} \cdot (\mathbf{v} \times \mathbf{w})$.

Lemma.
$$\mathbf{u} \cdot (\mathbf{v} \times \mathbf{w}) = \begin{vmatrix} u_1 & u_2 & u_3 \\ v_1 & v_2 & v_3 \\ w_1 & w_2 & w_3 \end{vmatrix}.$$

Proof: By (2b), $\mathbf{v} \times \mathbf{w} = \left[\begin{vmatrix} v_2 & v_3 \\ w_2 & w_3 \end{vmatrix}, \ - \begin{vmatrix} v_1 & v_3 \\ w_1 & w_3 \end{vmatrix}, \ \begin{vmatrix} v_1 & v_2 \\ w_1 & w_2 \end{vmatrix} \right]$, so

$$\mathbf{u} \cdot (\mathbf{v} \times \mathbf{w}) = u_1 \begin{vmatrix} v_2 & v_3 \\ w_2 & w_3 \end{vmatrix} - u_2 \begin{vmatrix} v_1 & v_3 \\ w_1 & w_3 \end{vmatrix} + u_3 \begin{vmatrix} v_1 & v_2 \\ w_1 & w_2 \end{vmatrix} \tag{4}$$

But the right side of (4) is the cofactor-based definition of the determinant

$$\begin{vmatrix} u_1 & u_2 & u_3 \\ v_1 & v_2 & v_3 \\ w_1 & w_2 & w_3 \end{vmatrix}$$
■

Theorem 2. If $\mathbf{u}$ and $\mathbf{v}$ are 3-vectors, the following properties are true.

(i) $\mathbf{u} \cdot (\mathbf{u} \times \mathbf{v}) = 0$.

(ii) $\mathbf{v} \cdot (\mathbf{u} \times \mathbf{v}) = 0$.

(iii) If $\mathbf{w}$ is a linear combination of $\mathbf{u}$ and $\mathbf{v}$, that is, $\mathbf{w} = r_1\mathbf{u} + r_2\mathbf{v}$, for some scalars r_1, r_2, then $\mathbf{w}$ is orthogonal to $\mathbf{u} \times \mathbf{v}$.

Proof: (i) By the lemma, $\mathbf{u} \cdot (\mathbf{u} \times \mathbf{v})$ is a determinant in which the first row equals the second row. By Theorem 3(ii) in Section 1.6, any determinant with two equal rows is 0.

(ii) The proof is the same as in (i).

(iii) We must show that $\mathbf{w} \cdot \mathbf{u} \times \mathbf{v} = 0$, or equivalently, $(a\mathbf{u} + b\mathbf{v}) \cdot (\mathbf{u} \times \mathbf{v}) = 0$. Using (i), (ii), and the linearity of the scalar product (Theorem 1 in Section 2.1), we have

$$(a\mathbf{u} + b\mathbf{v}) \cdot (\mathbf{u} \times \mathbf{v}) = a\mathbf{u} \cdot (\mathbf{u} \times \mathbf{v}) + b\mathbf{v} \cdot (\mathbf{u} \times \mathbf{v})$$
$$= a\{\mathbf{u} \cdot (\mathbf{u} \times \mathbf{v})\} + b\{\mathbf{v} \cdot (\mathbf{u} \times \mathbf{v})\} \tag{5}$$
$$= a\{ \quad 0 \quad \} + b\{ \quad 0 \quad \} = 0$$

Corollary. If **a**, **b**, **c** are vectors in a plane P in $\mathbf{R}^3$, then $(\mathbf{b} - \mathbf{a}) \times (\mathbf{c} - \mathbf{a})$ is normal to P.

Proof: Any vector **d** in P can be written in parametric form as $\mathbf{d} = \mathbf{a} + r_1(\mathbf{b} - \mathbf{a}) + r_2(\mathbf{c} - \mathbf{a})$, for some scalars r_1, r_2 (see Section 2.1). To show that $(\mathbf{b} - \mathbf{a}) \times (\mathbf{c} - \mathbf{a})$ is normal to P means showing that $(\mathbf{b} - \mathbf{a}) \times (\mathbf{c} - \mathbf{a})$ is orthogonal to $\mathbf{d} - \mathbf{a}$, for any such parametrically defined **d**. Letting $\mathbf{u} = \mathbf{b} - \mathbf{a}$, $\mathbf{v} = \mathbf{c} - \mathbf{a}$, and $\mathbf{w} = \mathbf{d} - \mathbf{a}$, then

$$\mathbf{w} = \mathbf{d} - \mathbf{a} = \{\mathbf{a} + r_1(\mathbf{b} - \mathbf{a}) + r_2(\mathbf{c} - \mathbf{a})\} - \mathbf{a}$$
$$= r_1(\mathbf{b} - \mathbf{a}) + r_2(\mathbf{c} - \mathbf{a}) = r_1(\mathbf{u}) + r_2(\mathbf{v})$$

Thus **w** is orthogonal to $\mathbf{u} \times \mathbf{v}$ by part (iii) of Theorem 2.   ■

Notice how we are making unexpected, but very helpful, use of the theory of determinants here.

There is one little geometric detail that is left hanging by Theorem 2. A 3-vector orthogonal (perpendicular) to a plane can point in two possible directions away from the plane. There is a simple geometric rule, called the **right-hand rule**, that tells us in which direction the cross-product vector points. If your right hand is cupped as shown in Figure 2.21 with your fingers bending in the direction from **u** toward **v**, the upward direction of a pointing thumb is the direction of $\mathbf{u} \times \mathbf{v}$.

EXAMPLE 2.
Alternative Way
to Define a Plane
Through Three
Vectors

Find the equation $a_1x_1 + a_2x_2 + a_3x_3 = b$, or $\mathbf{a} \cdot \mathbf{x} = b$ of a plane P containing the vectors $\mathbf{r} = [1, -1, 2]$, $\mathbf{s} = [1, 3, 4]$, and $\mathbf{t} = [3, 1, -1]$. In Section 2.3 we learned that **a** is a vector normal to the plane P. [In that section we developed the related normal form $\mathbf{a} \cdot (\mathbf{x} - \mathbf{v}) = 0$ of the equation of a plane.] With Theorem 2, we can determine a normal vector **a** from the given three vectors in the plane. The differences $\mathbf{s} - \mathbf{r}$ and $\mathbf{t} - \mathbf{r}$ are two vectors pointing in directions lying in our plane P. Then the cross product $(\mathbf{s} - \mathbf{r}) \times (\mathbf{t} - \mathbf{r})$, by Theorem 2, will be normal to P.

$$(\mathbf{s} - \mathbf{r}) \times (\mathbf{t} - \mathbf{r}) = [0, 4, 2] \times [2, 2, -3]$$
$$= [4 \cdot (-3) - 2 \cdot 2, 2 \cdot 2 - 0 \cdot (-3), 0 \cdot 2 - 4 \cdot 2] \qquad (6)$$
$$= [-16, 4, -8]$$

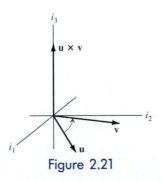

Figure 2.21

Factoring out the largest common denominator, we can use the normal vector $\mathbf{a} = [4, -1, 2]$. So the desired equation is $4x_1 - x_2 + 2x_3 = b$. We determine b by using one of the given vectors which must lie on the plane, say $\mathbf{r}: \mathbf{a} \cdot \mathbf{r} = 4(1) - (-1) + 2(2) = 9$. Then $b = 9$. ▪

Recall that the scalar product is associated with the cosine of the angle θ between two vectors $\mathbf{u}$, $\mathbf{v}$, namely $\mathbf{u} \cdot \mathbf{v} = |\mathbf{u}||\mathbf{v}| \cos \theta$ (Theorem 1 in Section 2.3) or $\cos \theta = \mathbf{u} \cdot \mathbf{v}/|\mathbf{u}||\mathbf{v}|$. In a similar fashion, the cross product is related to the sine of the angle between two vectors. As in Section 2.4, the vector norm here is the Euclidean norm.

Theorem 3 (Lagrange's Identity). If $\mathbf{u}$ and $\mathbf{v}$ are 3-vectors, then

$$|\mathbf{u} \times \mathbf{v}|^2 = |\mathbf{u}|^2|\mathbf{v}|^2 - (\mathbf{u} \cdot \mathbf{v})^2 \tag{7}$$

And from this it follows that

$$|\mathbf{u} \times \mathbf{v}| = |\mathbf{u}||\mathbf{v}| \sin \theta \tag{8}$$

Proof: The left side of (7) is

$$|\mathbf{u} \times \mathbf{v}|^2 = (\mathbf{u} \times \mathbf{v}) \cdot (\mathbf{u} \times \mathbf{v})$$
$$= (u_2v_3 - u_3v_2)^2 + (u_3v_1 - u_1v_3)^2 + (u_1v_2 - u_2v_1)^2$$

and the right side of (7) is

$$|\mathbf{u}|^2|\mathbf{v}|^2 - (\mathbf{u} \cdot \mathbf{v})^2 = (\mathbf{u} \cdot \mathbf{u})^2(\mathbf{v} \cdot \mathbf{v})^2 - (\mathbf{u} \cdot \mathbf{v})^2$$
$$= (u_1^2 + u_2^2 + u_3^2)(v_1^2 + v_2^2 + v_3^2) + (u_1v_1 + u_2v_2 + u_3v_3)^2$$

It is left to the reader to do the algebra of showing that these two expressions are equal.

To prove (8), we take (7) and use the fact that $\mathbf{u} \cdot \mathbf{v} = |\mathbf{u}||\mathbf{v}| \cos \theta (0 \le \theta \le \pi)$ plus the familiar trigonometric identity $\sin^2\theta + \cos^2\theta = 1$.

$$|\mathbf{u} \times \mathbf{v}|^2 = |\mathbf{u}|^2|\mathbf{v}|^2 - (\mathbf{u} \cdot \mathbf{v})^2 = |\mathbf{u}|^2|\mathbf{v}|^2 - (|\mathbf{u}||\mathbf{v}| \cos \theta)^2$$
$$= |\mathbf{u}|^2|\mathbf{v}|^2(1 - \cos^2\theta) = |\mathbf{u}|^2|\mathbf{v}|^2(\sin^2\theta) \tag{9}$$

Taking square roots in (9), we have the desired result. ▪

The formula $|\mathbf{u} \times \mathbf{v}| = |\mathbf{u}||\mathbf{v}| \sin \theta$ has a variety of interesting consequences. The first one is that we now are assured that the *cross-product vector* $\mathbf{u} \times \mathbf{v}$ is independent of the particular coordinate system we are using. The Euclidean length of a vector will be the same in any coordinate system. So will the angle between vectors $\mathbf{u}$ and $\mathbf{v}$. Thus the quantity $|\mathbf{u}||\mathbf{v}| \sin \theta$ is independent of our coordinate system. Similarly, the direction of $\mathbf{u} \times \mathbf{v}$ in $\mathbf{R}^3$, the direction orthogonal to $\mathbf{u}$ and $\mathbf{v}$, is also independent of the coordinate system we use.

The expression $|\mathbf{u}||\mathbf{v}| \sin \theta$ has an important geometric interpretation. It happens to be equal to the area of a parallelogram with sides $\mathbf{u}$ and $\mathbf{v}$ (see Figure 2.22). Recall

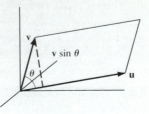

Figure 2.22

that the area of a parallelogram equals the base times the altitude. As shown in Figure 2.22, if $\mathbf{u}$ is treated as the base, the altitude of the parallelogram is $|\mathbf{v}|\sin\theta$. Then

$$\text{Area} = (\text{base})(\text{altitude}) = (|\mathbf{u}|)(|\mathbf{v}|\sin\theta) \tag{10}$$

By (8), this expression for the area is equal to $|\mathbf{u} \times \mathbf{v}|$. We have proved the following theorem.

Theorem 4.   The area of the parallelogram generated by the 3-vectors $\mathbf{u}$ and $\mathbf{v}$ is $|\mathbf{u} \times \mathbf{v}|$.

EXAMPLE 3.
Area of a
Parallelogram

(a) Find the area of the parallelogram generated by the vectors $\mathbf{u} = [2, 3, 4]$ and $\mathbf{v} = [3, -1, 0]$. By Theorem 4, we compute

$$\mathbf{u} \times \mathbf{v} = [3 \cdot 0 - 4 \cdot (-1), 4 \cdot 3 - 2 \cdot 0, 2 \cdot (-1) - 3 \cdot 3] = [4, 12, -11]$$

Then $|\mathbf{u} \times \mathbf{v}| = \sqrt{4^2 + 12^2 + (-11)^2} = \sqrt{16 + 144 + 121} = \sqrt{281} \approx 16.8$.

(b) Find the area of parallelogram generated by the vectors $\mathbf{u} = [3, 1]$ and $\mathbf{v} = [2, 5]$. The cross product is defined for 3-vectors. To make $\mathbf{u}$ and $\mathbf{v}$ 3-vectors, we simply add a third component and set it equal to 0. Thus we work with $\mathbf{u} = [3, 1, 0]$ and $\mathbf{v} = [2, 5, 0]$. Using Theorem 4, we compute

$$\mathbf{u} \times \mathbf{v} = [1 \cdot 0 - 0 \cdot 5, 0 \cdot 2 - 3 \cdot 0, 3 \cdot 5 - 1 \cdot 2] = [0, 0, 13]$$

and $|\mathbf{u} \times \mathbf{v}| = 13$.   ■

Note in part (b) of Example 3 how the cross product simplified when the $\mathbf{u}$ and $\mathbf{v}$ lie in $\mathbf{R}^2$ (with their third components 0). Since $\mathbf{u}$ and $\mathbf{v}$ lie in $\mathbf{R}^2$, this cross product $\mathbf{u} \times \mathbf{v}$ is orthogonal to $\mathbf{u}$ and $\mathbf{v}$, by Theorem 2, and so only the third component of $\mathbf{u} \times \mathbf{v}$ is nonzero. From formula (1) defining cross products, we see that $\mathbf{u} \times \mathbf{v} = [0, 0, u_1v_2 - u_2v_1]$. So the area of the parallelogram generated by $\mathbf{u}$ and $\mathbf{v}$ is simply the absolute value of $u_1v_2 - u_2v_1$, which is the determinant whose rows (or columns) are $\mathbf{u}$ and $\mathbf{v}$. Summarizing, we have

Theorem 5.   The area of a parallelogram generated by vectors $\mathbf{u}$ and $\mathbf{v}$ in $\mathbf{R}^2$ is the absolute value of $\begin{vmatrix} u_1 & u_2 \\ v_1 & v_2 \end{vmatrix}$.

A further link between determinants and the area $a(\mathbf{u}, \mathbf{v})$ of the parallelogram generated by $\mathbf{u}$ and $\mathbf{v}$ is as follows. If we multiply $\mathbf{u}$ and $\mathbf{v}$ by a 2×2 matrix $\mathbf{A}$ to obtain vectors $\mathbf{u}' = \mathbf{Au}$ and $\mathbf{v}' = \mathbf{Av}$, and if $a(\mathbf{u}', \mathbf{v}')$ denotes the area of parallelogram generated $\mathbf{u}'$ and $\mathbf{v}'$, then $a(\mathbf{u}', \mathbf{v}') = \det(\mathbf{A})a(\mathbf{u}, \mathbf{v})$. This interesting result (which is closely related to the role of the Jacobian in integral calculus) is proved in the exercises.

Now we demonstrate the useful role that cross products play in computer graphics. Recall from Section 2.3 that the scalar product $\mathbf{a} \cdot \mathbf{b}$ is positive when $\mathbf{a}$ and $\mathbf{b}$ form an angle of less than 90° and $\mathbf{a} \cdot \mathbf{b}$ is negative when $\mathbf{a}$ and $\mathbf{b}$ form an angle of greater than 90° (and $\mathbf{a} \cdot \mathbf{b} = 0$ when $\mathbf{a}$ and $\mathbf{b}$ form an angle of 90°).

EXAMPLE 4.
Left Turns and
Right Turns in the
Plane

Suppose that we are generating a figure in $\mathbf{R}^2$ that requires us to travel in the direction of vector $\mathbf{a}_1$ for a certain distance, then travel in direction $\mathbf{a}_2$ for a certain distance, then in direction $\mathbf{a}_3$, and so on. When we switch from direction $\mathbf{a}_i$ to direction $\mathbf{a}_{i+1}$, it is often important to know whether this change of direction constitutes a left turn or a right turn. The cross product and associated right-hand rule (see Figure 2.21) provide a simple way to test whether we turned to the left or right.

As in part (b) of Example 3, we treat the directions $\mathbf{a}_i$ as 3-vectors by adding a third component, which is 0. Assume that the unit vectors $\mathbf{i}_1$, $\mathbf{i}_2$, $\mathbf{i}_3$ are oriented as shown in Figure 2.21. When the turn from direction $\mathbf{u}$ to direction $\mathbf{v}$ is a left turn, then $\mathbf{u} \times \mathbf{v}$ will point in the direction of $\mathbf{i}_3$, and $\mathbf{i}_3 \cdot (\mathbf{u} \times \mathbf{v})$ will be positive (see Figure 2.21). When the turn from direction $\mathbf{u}$ to direction $\mathbf{v}$ is a right turn, then $\mathbf{u} \times \mathbf{v}$ will point in the direction of $-\mathbf{i}_3$, and $\mathbf{i}_3 \cdot (\mathbf{u} \times \mathbf{v})$ will be negative. Note that the computation of $\mathbf{i}_3 \cdot (\mathbf{u} \times \mathbf{v})$ only requires computing the third component of $\mathbf{u} \times \mathbf{v}$, since $\mathbf{i}_3 = [0, 0, 1]$. Actually, as noted above, since $\mathbf{u}$ and $\mathbf{v}$ are really 2-vectors, only the third component of $\mathbf{u} \times \mathbf{v}$ is nonzero anyway. (This third component is the determinant in Theorem 5 whose absolute value is the area of the parallelogram generated by $\mathbf{u}$ and $\mathbf{v}$.)

For example, suppose that we were going in direction $\mathbf{u} = [1, 2]$, then turned to direction $\mathbf{v} = [-2, 3]$ and then turned to direction $\mathbf{w} = [-1, -2]$ (see Figure 2.23). At the first turn we have

$$\begin{vmatrix} u_1 & u_2 \\ v_1 & v_2 \end{vmatrix} = 1 \cdot 3 - 2 \cdot (-2) = 7$$

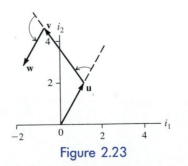

Figure 2.23

The number is positive, so the first turn is to the left. Next we compute

$$\begin{vmatrix} v_1 & v_2 \\ w_1 & w_2 \end{vmatrix} = (-2) \cdot (-2) - 3 \cdot (-1) = 7$$

The number is positive, so the next turn is also to the left.   ■

EXAMPLE 5.
Hidden Surfaces

An important problem in computer graphics is determining which surfaces of a three-dimensional object are visible to a viewer looking at the object from a given location. Consider the distorted tetrahedron (consisting of four triangular faces) in Figure 2.24 with vertices

$$\mathbf{a} = [1, 1, 0] \qquad \mathbf{b} = [3, 2, 0] \qquad \mathbf{c} = [2, 3, 1] \qquad \mathbf{d} = [1, 2, 1]$$

Let us determine whether the face F defined by vertices $\mathbf{a}$, $\mathbf{b}$, $\mathbf{c}$ is visible to a viewer looking at the solid from location $\mathbf{v} = [5, 3, 4]$.

If the fourth corner $\mathbf{d}$ of the tetrahedron (and hence the faces that have $\mathbf{d}$ as a corner) are on the same side of F formed by $\mathbf{a}$, $\mathbf{b}$, $\mathbf{c}$ as the viewer at $\mathbf{v}$ is, the viewer cannot see F. That is, F will be on the far side of the solid from $\mathbf{v}$. If $\mathbf{d}$ and $\mathbf{v}$ are on opposite sides of F, the viewer can see F. To determine the relative sides of $\mathbf{d}$ and $\mathbf{v}$ with respect to face F, consider vectors $\mathbf{d} - \mathbf{a}$ and $\mathbf{v} - \mathbf{a}$, which point away from F toward $\mathbf{d}$ and $\mathbf{v}$, respectively. We also use the vector $\mathbf{n} = (\mathbf{b} - \mathbf{a}) \times (\mathbf{c} - \mathbf{a})$, which, by the corollary of Theorem 2, is normal to face F. (But we do not know whether $\mathbf{n}$ points outward from F away from the solid or inward from F toward the interior of the solid.) We use $\mathbf{n}$ simply as a reference direction. If the signs of scalar products $(\mathbf{d} - \mathbf{a}) \cdot \mathbf{n}$ and $(\mathbf{v} - \mathbf{a}) \cdot \mathbf{n}$ are the same, then $\mathbf{d}$ and $\mathbf{v}$ are on the same side of F, and the viewer cannot see F. If the signs of $(\mathbf{d} - \mathbf{a}) \cdot \mathbf{n}$ and $(\mathbf{v} - \mathbf{a}) \cdot \mathbf{n}$ are different, the viewer at $\mathbf{v}$ is on the opposite side from $\mathbf{d}$ and can see face F.

First we determine

$$\mathbf{b} - \mathbf{a} = [3, 2, 0] - [1, 1, 0] = [2, 1, 0] \qquad \mathbf{c} - \mathbf{a} = [2, 3, 1] - [1, 1, 0] = [1, 2, 1]$$
$$\mathbf{d} - \mathbf{a} = [1, 2, 1] - [1, 1, 0] = [0, 1, 1] \qquad \mathbf{v} - \mathbf{a} = [5, 3, 4] - [1, 1, 0] = [4, 2, 4]$$

Recall from the lemma that $(\mathbf{d} - \mathbf{a}) \cdot \mathbf{n} = (\mathbf{d} - \mathbf{a}) \cdot \{(\mathbf{b} - \mathbf{a}) \times (\mathbf{c} - \mathbf{a})\}$ is the determinant with $\mathbf{d} - \mathbf{a}$ as first row, $\mathbf{b} - \mathbf{a}$ the second row, and $\mathbf{c} - \mathbf{a}$ the third row; similarly for $(\mathbf{v} - \mathbf{a}) \cdot \mathbf{n} = (\mathbf{v} - \mathbf{a}) \cdot \{(\mathbf{b} - \mathbf{a}) \times (\mathbf{c} - \mathbf{a})\}$.

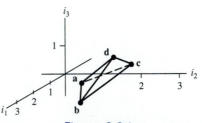

Figure 2.24

$$(\mathbf{d} - \mathbf{a}) \cdot \mathbf{n} = (\mathbf{d} - \mathbf{a}) \cdot \{(\mathbf{b} - \mathbf{a}) \times (\mathbf{c} - \mathbf{a})\} = \begin{vmatrix} 0 & 1 & 1 \\ 2 & 1 & 0 \\ 1 & 2 & 1 \end{vmatrix}$$

$$= 0 + 0 + 4 - 1 - 0 - 2 = 1$$

$$(\mathbf{v} - \mathbf{a}) \cdot \mathbf{n} = (\mathbf{v} - \mathbf{a}) \cdot \{(\mathbf{b} - \mathbf{a}) \times (\mathbf{c} - \mathbf{a})\} = \begin{vmatrix} 4 & 2 & 4 \\ 2 & 1 & 0 \\ 1 & 2 & 1 \end{vmatrix}$$

$$= 4 + 0 + 16 - 4 - 0 - 4 = 12$$

The signs are the same, so face F is hidden from a viewer at [5, 3, 4]. ■

Section 2.5 Exercises

1. Determine the cross product $\mathbf{u} \times \mathbf{v}$ of the following pairs of vectors.
   (a) $\mathbf{u} = [1, 2, 5]$, $\mathbf{v} = [-3, 4, 0]$ (b) $\mathbf{u} = [3, 1, 6]$, $\mathbf{v} = [6, -1, 2]$
   (c) $\mathbf{u} = [0, 1, 6]$, $\mathbf{v} = [0, 4, 5]$ (d) $\mathbf{u} = [4, -3, -1]$, $\mathbf{v} = [-3, -2, -4]$
   (e) $\mathbf{u} = [1, 0, 5]$, $\mathbf{v} = [0, 4, 0]$ (f) $\mathbf{u} = [1, 2, 4]$, $\mathbf{v} = [2, 4, 8]$

2. Determine the cross product $\mathbf{u} \times \mathbf{v}$ of the following pairs of vectors.
   (a) $\mathbf{u} = [a, a, a]$, $\mathbf{v} = [a, -a, 0]$ (b) $\mathbf{u} = [a, -a, -a]$, $\mathbf{v} = [0, a, -a]$
   (c) $\mathbf{u} = [a, b, a]$, $\mathbf{v} = [b, a, b]$ (d) $\mathbf{u} = [a, b, 0]$, $\mathbf{v} = [0, b, a]$

3. Let $\mathbf{u} = [1, 3, 2]$, $\mathbf{v} = [5, 0, -2]$, and $\mathbf{w} = [2, -6, 3]$. Determine the value of the following expressions.
   (a) $(\mathbf{u} \times \mathbf{v}) - (\mathbf{u} \times (-\mathbf{v}))$ (b) $\mathbf{u} \cdot (\mathbf{v} \times \mathbf{w})$ (c) $(\mathbf{u} + \mathbf{v} + \mathbf{w}) \times (\mathbf{u} + \mathbf{v} + \mathbf{w})$
   (d) $(\mathbf{u} \times \mathbf{w}) \times (\mathbf{u} \times \mathbf{v})$ (e) $(\mathbf{u} \times \mathbf{v}) \times (\mathbf{v} \times \mathbf{u})$ (f) $(\mathbf{u} \times \mathbf{v}) \cdot (\mathbf{u} \times \mathbf{w}) \cdot (\mathbf{w} \times \mathbf{u})$

4. Use the cross product to find a unit vector orthogonal to the following pairs of vectors.
   (a) $\mathbf{u} = [1, 2, 5]$, $\mathbf{v} = [-3, 4, 0]$ (b) $\mathbf{u} = [3, 1, 6]$, $\mathbf{v} = [6, -1, 2]$
   (c) $\mathbf{u} = [2, 1, -1]$, $\mathbf{v} = [2, -2, 0]$ (d) $\mathbf{u} = [2, -6, -2]$, $\mathbf{v} = [-1, 5, -8]$

5. Verify the first two parts of Theorem 2 for $\mathbf{u} = [5, 3, 1]$ and $\mathbf{v} = [-4, 0, -2]$.

6. If $\mathbf{u} = [3, 1, -4]$ and $\mathbf{v} = [6, 2, 5]$, determine the set of vectors $\mathbf{z}$ such that $\mathbf{u} \times \mathbf{z} = \mathbf{v}$. (Hint: Use the definition of the cross product to define a system of linear equations that the entries of $\mathbf{z}$ must satisfy.)

7. For any 3-vectors $\mathbf{u}$, $\mathbf{v}$, and $\mathbf{w}$, show that $(\mathbf{u} \times \mathbf{v}) \cdot \mathbf{w} = \mathbf{u} \cdot (\mathbf{v} \times \mathbf{w})$.

8. Use the method in Example 2 to find the equation of the plane containing the following three vectors.
   (a) $\mathbf{u} = [1, 1, 2]$, $\mathbf{v} = [0, 2, 3]$, $\mathbf{w} = [4, 1, 2]$
   (b) $\mathbf{u} = [3, 0, 1]$, $\mathbf{v} = [5, -2, 2]$, $\mathbf{w} = [-2, 3, 1]$
   (c) $\mathbf{u} = [2, 0, 0]$, $\mathbf{v} = [0, 1, 0]$, $\mathbf{w} = [0, 0, 3]$
   (d) $\mathbf{u} = [-2, -4, 3]$, $\mathbf{v} = [2, -5, -1]$, $\mathbf{w} = [1, 6, -2]$

9. Find the area of the parallelograms generated by the following pairs of 2-vectors.
   (a) $\mathbf{u} = [1, 4]$, $\mathbf{v} = [0, 3]$ (b) $\mathbf{u} = [0, -3]$, $\mathbf{v} = [3, -4]$ (c) $\mathbf{u} = [2, 2]$, $\mathbf{v} = [5, 2]$
   (d) $\mathbf{u} = [1, 4]$, $\mathbf{v} = [-2, -8]$

10. Find the area of the parallelograms generated by the following pairs of 3-vectors.
    (a) $\mathbf{u} = [2, 1, 4]$, $\mathbf{v} = [0, 5, 3]$ (b) $\mathbf{u} = [3, 0, -3]$, $\mathbf{v} = [8, -3, -4]$
    (c) $\mathbf{u} = [6, 1, 2]$, $\mathbf{v} = [8, 2, 0]$ (d) $\mathbf{u} = [1, -2, 3]$, $\mathbf{v} = [-3, 6, -9]$

11. Let $\mathbf{u} = [u_1, u_2, u_3]$, $\mathbf{v} = [v_1, v_2, v_3]$, and $\mathbf{w} = [w_1, w_2, w_3]$ be three 3-vectors that do not lie in a plane. Show that the volume of the parallelepiped generated by $\mathbf{u}$, $\mathbf{v}$, and $\mathbf{w}$ (i.e., the parallelepiped with a corner at the origin and the three edges going out from the origin formed by $\mathbf{u}$, $\mathbf{v}$, and $\mathbf{w}$) is given by the absolute value of

$$\begin{vmatrix} u_1 & u_2 & u_3 \\ v_1 & v_2 & v_3 \\ w_1 & w_2 & w_3 \end{vmatrix}$$

[Hint: (1) Use Theorem 3; and (2) show that the base of the parallelepiped is $|\mathbf{u} \times \mathbf{v}|$.]

12. Using the result in Exercise 11, find the volumes of the parallelepipeds generated by:
(a) $\mathbf{u} = [1, 3, 2]$, $\mathbf{v} = [4, 1, 0]$, $\mathbf{w} = [5, 3, -1]$
(b) $\mathbf{u} = [-3, 1, 5]$, $\mathbf{v} = [1, 2, -2]$, $\mathbf{w} = [1, 0, 2]$
(c) $\mathbf{u} = [8, 0, 1]$, $\mathbf{v} = [4, 1, 0]$, $\mathbf{w} = [0, 2, -1]$

13. Starting from the origin, we go in a direction specified by 2-vector $\mathbf{a}_1$. From the end of vector $\mathbf{a}_1$, we next go in a direction specified by the 2-vector $\mathbf{a}_2$, and then in a direction specified by 2-vector $\mathbf{a}_3$. As in Example 4, tell whether we turn to the left or to the right in changing direction from $\mathbf{a}_1$ to $\mathbf{a}_2$ and in changing direction from $\mathbf{a}_2$ to $\mathbf{a}_3$.
(a) $\mathbf{a}_1 = [2, 3]$, $\mathbf{a}_2 = [-3, -3]$, $\mathbf{a}_3 = [4, 5]$
(b) $\mathbf{a}_1 = [-2, 1]$, $\mathbf{a}_2 = [4, -3]$, $\mathbf{a}_3 = [-6, 5]$
(c) $\mathbf{a}_1 = [2, -5]$, $\mathbf{a}_2 = [-5, 12]$, $\mathbf{a}_3 = [3, -7]$

14. A person looks at a tetrahedron with corners $\mathbf{a}$, $\mathbf{b}$, $\mathbf{c}$, $\mathbf{d}$ from position $\mathbf{v}$. Is the face determined by corners $\mathbf{a}$, $\mathbf{b}$, $\mathbf{c}$ visible to the person at $\mathbf{v}$, for the following vectors $\mathbf{a}$, $\mathbf{b}$, $\mathbf{c}$, $\mathbf{d}$, $\mathbf{v}$?
(a) $\mathbf{a} = [1, 2, 3]$, $\mathbf{b} = [3, 7, 1]$, $\mathbf{c} = [2, 0, 0]$, $\mathbf{d} = [2, 3, 6]$, and $\mathbf{v} = [6, 6, 6]$
(b) $\mathbf{a} = [-4, 1, 3]$, $\mathbf{b} = [2, -1, 1]$, $\mathbf{c} = [0, 2, 5]$, $\mathbf{d} = [1, -5, 0]$, and $\mathbf{v} = [6, 6, 6]$
(c) $\mathbf{a} = [0, 3, 1]$, $\mathbf{b} = [2, -3, 1]$, $\mathbf{c} = [5, 1, 0]$, $\mathbf{d} = [-3, 1, 6]$, and $\mathbf{v} = [2, -4, -3]$

15. A person looks at a tetrahedron with corners $\mathbf{a}$, $\mathbf{b}$, $\mathbf{c}$, $\mathbf{d}$ from position $\mathbf{v}$. Are the corners $\mathbf{a}$ and $\mathbf{b}$ visible to the person at $\mathbf{v}$, for the following vectors $\mathbf{a}$, $\mathbf{b}$, $\mathbf{c}$, $\mathbf{d}$, $\mathbf{v}$?
(a) $\mathbf{a} = [1, 2, 3]$, $\mathbf{b} = [3, 7, 1]$, $\mathbf{c} = [2, 0, 0]$, $\mathbf{d} = [2, 3, 6]$, and $\mathbf{v} = [6, 6, 6]$
(b) $\mathbf{a} = [-4, 1, 3]$, $\mathbf{b} = [2, -1, 1]$, $\mathbf{c} = [0, 2, 5]$, $\mathbf{d} = [1, -5, 0]$, and $\mathbf{v} = [6, 6, 6]$

16. Let $\mathbf{u}' = A\mathbf{u}$ and $\mathbf{v}' = A\mathbf{v}$ for $A = \begin{bmatrix} 3 & 2 \\ 2 & 3 \end{bmatrix}$. For each $\mathbf{u}$ and $\mathbf{v}$, determine the area of the parallelograms generated by $\mathbf{u}$, $\mathbf{v}$ and by $\mathbf{u}'$, $\mathbf{v}'$, and in each case check that the area of the second parallelogram is five times the area of the first parallelogram. [Note: $\det(A) = 5$.]

(a) $\mathbf{u} = \begin{bmatrix} 2 \\ 1 \end{bmatrix}$, $\mathbf{v} = \begin{bmatrix} 1 \\ 4 \end{bmatrix}$ (b) $\mathbf{u} = \begin{bmatrix} 0 \\ 4 \end{bmatrix}$, $\mathbf{v} = \begin{bmatrix} -1 \\ 3 \end{bmatrix}$ (c) $\mathbf{u} = \begin{bmatrix} 2 \\ 3 \end{bmatrix}$, $\mathbf{v} = \begin{bmatrix} 3 \\ -2 \end{bmatrix}$

17. Let $\mathbf{u} = \begin{bmatrix} u_1 \\ u_2 \end{bmatrix}$ and $\mathbf{v} = \begin{bmatrix} v_1 \\ v_2 \end{bmatrix}$. Then, let $A = \begin{bmatrix} a_{11} & a_{12} \\ a_{21} & a_{22} \end{bmatrix}$, $\mathbf{u}' = \begin{bmatrix} u_1' \\ u_2' \end{bmatrix} = A\mathbf{u}$, and $\mathbf{v}' = \begin{bmatrix} v_1' \\ v_2' \end{bmatrix} = A\mathbf{v}$. If $a(\mathbf{x}, \mathbf{y})$ denotes the area of the parallelogram generated by 2-vectors $\mathbf{x}$ and $\mathbf{y}$, show that $a(\mathbf{u}', \mathbf{v}') = \det(A)a(\mathbf{u}, \mathbf{v})$.

18. The triple cross product of the 3-vectors $\mathbf{u}$, $\mathbf{v}$, $\mathbf{w}$ is defined to be $\mathbf{u} \times (\mathbf{v} \times \mathbf{w})$.
(a) Show that $\mathbf{u} \times (\mathbf{v} \times \mathbf{w})$ lies in the plane determined by $\mathbf{v}$ and $\mathbf{w}$.
(b) Show that $\mathbf{u} \times (\mathbf{v} \times \mathbf{w}) = (\mathbf{u} \cdot \mathbf{w})\mathbf{v} - (\mathbf{u} \cdot \mathbf{v})\mathbf{w}$. (Hint: First prove the result for the three cases where $\mathbf{w}$ is a unit coordinate vector.)

3 Matrices

3.1 Matrix Algebra

In this chapter we examine the theory, applications, and computations of matrices in greater depth. This section is devoted to matrix addition and multiplication. We verify the rules of matrix algebra, introduce matrix transposes and block matrices, and apply these new concepts to a problem involving graphs and 0–1 matrices.

We repeat the basic laws of matrix algebra from Section 1.3.

Basic Laws of Matrix Algebra
1. *Associative laws*. Matrix addition is associative: $(\mathbf{A} + \mathbf{B}) + \mathbf{C} = \mathbf{A} + (\mathbf{B} + \mathbf{C})$. Matrix multiplication is associative: $(\mathbf{AB})\mathbf{C} = \mathbf{A}(\mathbf{BC})$.
2. *Commutative laws*. Matrix addition is commutative: $\mathbf{A} + \mathbf{B} = \mathbf{B} + \mathbf{A}$. Matrix multiplication is not commutative (except in special cases): $\mathbf{AB} \neq \mathbf{BA}$.
3. *Distributive laws*. $\mathbf{A}(\mathbf{B} + \mathbf{C}) = \mathbf{AB} + \mathbf{AC}$ and $(\mathbf{B} + \mathbf{C})\mathbf{A} = \mathbf{BA} + \mathbf{CA}$.
4. *Law of scalar factoring*. $r(\mathbf{AB}) = (r\mathbf{A})\mathbf{B} = \mathbf{A}(r\mathbf{B})$.

Note that a vector is a matrix with just one column (or one row). Thus these matrix laws apply to vectors. The vector versions of the laws (except for the associativity and noncommutivity of matrix multiplication) were given at the end of the Section 2.1.

EXAMPLE 1.
Examples of Laws
of Matrix Algebra

Let $\mathbf{A} = \begin{bmatrix} 1 & 2 \\ 3 & 4 \end{bmatrix}$, $\mathbf{B} = \begin{bmatrix} 2 & 0 \\ 1 & 2 \end{bmatrix}$, and $\mathbf{C} = \begin{bmatrix} 6 & 1 \\ 0 & 4 \end{bmatrix}$. Then

$$\mathbf{AB} = \begin{bmatrix} 1 & 2 \\ 3 & 4 \end{bmatrix}\begin{bmatrix} 2 & 0 \\ 1 & 2 \end{bmatrix} = \begin{bmatrix} 4 & 4 \\ 10 & 8 \end{bmatrix}$$

and $(\mathbf{AB})\mathbf{C} = \left(\begin{bmatrix} 4 & 4 \\ 10 & 8 \end{bmatrix}\right)\begin{bmatrix} 6 & 1 \\ 0 & 4 \end{bmatrix} = \begin{bmatrix} 24 & 20 \\ 60 & 42 \end{bmatrix}$, while $\mathbf{BC} = \begin{bmatrix} 2 & 0 \\ 1 & 2 \end{bmatrix}\begin{bmatrix} 6 & 1 \\ 0 & 4 \end{bmatrix} =$

$\begin{bmatrix} 12 & 2 \\ 6 & 9 \end{bmatrix}$ and $\mathbf{A}(\mathbf{BC}) = \begin{bmatrix} 1 & 2 \\ 3 & 4 \end{bmatrix}\left(\begin{bmatrix} 12 & 2 \\ 6 & 9 \end{bmatrix}\right) = \begin{bmatrix} 24 & 20 \\ 60 & 42 \end{bmatrix}$. Thus

$$(\mathbf{AB})\mathbf{C} = \mathbf{A}(\mathbf{BC}) = \begin{bmatrix} 24 & 20 \\ 60 & 42 \end{bmatrix}$$

Also, $\mathbf{A}(\mathbf{B} + \mathbf{C}) = \begin{bmatrix} 1 & 2 \\ 3 & 4 \end{bmatrix}\left(\begin{bmatrix} 2 & 0 \\ 1 & 2 \end{bmatrix} + \begin{bmatrix} 6 & 1 \\ 0 & 4 \end{bmatrix}\right) = \begin{bmatrix} 1 & 2 \\ 3 & 4 \end{bmatrix}\begin{bmatrix} 8 & 1 \\ 1 & 6 \end{bmatrix} = \begin{bmatrix} 10 & 13 \\ 28 & 27 \end{bmatrix}$

while $\mathbf{AB} = \begin{bmatrix} 4 & 4 \\ 10 & 8 \end{bmatrix}$ and $\mathbf{AC} = \begin{bmatrix} 1 & 2 \\ 3 & 4 \end{bmatrix}\begin{bmatrix} 6 & 1 \\ 0 & 4 \end{bmatrix} = \begin{bmatrix} 6 & 9 \\ 18 & 19 \end{bmatrix}$, so that

$$\mathbf{AB} + \mathbf{AC} = \begin{bmatrix} 4 & 4 \\ 10 & 8 \end{bmatrix} + \begin{bmatrix} 6 & 9 \\ 18 & 19 \end{bmatrix} = \begin{bmatrix} 10 & 13 \\ 28 & 27 \end{bmatrix}$$ ■

We shall now verify these laws. The additive laws of associativity and commutativity are straightforward, since they follow directly from similar laws for scalar arithmetic.

Distributive Law. The distributive law for matrix multiplication can be derived from the distributive law for scalar products that was proved in Section 2.1. We verify matrix equations by showing that an individual entry (i, j) of the expression on the left side has the same value as entry (i, j) of the expression on the right side. To prove $\mathbf{A}(\mathbf{B} + \mathbf{C}) = \mathbf{AB} + \mathbf{AC}$, we must show that the (i, j)th entry of the product matrix $\mathbf{A}(\mathbf{B} + \mathbf{C})$ equals the (i, j)th entry of the sum of products $\mathbf{AB} + \mathbf{AC}$. This entry in $\mathbf{A}(\mathbf{B} + \mathbf{C})$ is $\mathbf{a}_i' \cdot (\mathbf{b}_j + \mathbf{c}_j)$, the scalar product of $\mathbf{a}_i'$, the ith row of $\mathbf{A}$, times the jth column $\mathbf{b}_j + \mathbf{c}_j$ of $\mathbf{B} + \mathbf{C}$. Entry (i, j) of $\mathbf{AB} + \mathbf{AC}$ is $\mathbf{a}_i' \cdot \mathbf{b}_j + \mathbf{a}_i' \cdot \mathbf{c}_j$. Showing that $\mathbf{a}_i' \cdot (\mathbf{b}_j + \mathbf{c}_j) = \mathbf{a}_i' \cdot \mathbf{b}_j + \mathbf{a}_i' \cdot \mathbf{c}_j$ follows from the distributive law for scalar products which was part (vi) of Theorem 1 in Section 2.1. Similar reasoning verifies that $(\mathbf{B} + \mathbf{C})\mathbf{A} = \mathbf{BA} + \mathbf{CA}$.

Associative Law. The associative law for matrix multiplication, $(\mathbf{AB})\mathbf{C} = \mathbf{A}(\mathbf{BC})$, was used in many critical places in Chapter 1, as, for example, in studying the weather Markov chain, where the vector of probabilities after two periods $\mathbf{p}^{(2)} = \mathbf{A}(\mathbf{Ap})$ was rewritten as $(\mathbf{A}^2)\mathbf{p}$. Entry (i, j) of $(\mathbf{AB})\mathbf{C}$ is the scalar product of the ith row of $\mathbf{AB}$ times the jth column $\mathbf{c}_j$ of $\mathbf{C}$. Remember from "Equivalent Definitions of Matrix Multiplication" near the beginning of Section 1.3 that the ith row of $\mathbf{AB}$ is $\mathbf{a}_i'\mathbf{B}$ (where $\mathbf{a}_i'$ denotes the ith row of $\mathbf{A}$). So entry (i, j) of $(\mathbf{AB})\mathbf{C}$ is $(\mathbf{a}_i'\mathbf{B}) \cdot \mathbf{c}_j$. Similarly, entry (i, j) of $\mathbf{A}(\mathbf{BC})$ is the scalar product of the ith row $\mathbf{a}_i'$ of $\mathbf{A}$ times the jth column of $\mathbf{BC}$, which equals $\mathbf{Bc}_j$, and thus equals $\mathbf{a}'_i \cdot (\mathbf{Bc}_j)$. The associative law for matrix multiplication will follow if we can show that

$$\mathbf{a}'_i \mathbf{B} \cdot \mathbf{c}_j = \mathbf{a}'_i \cdot \mathbf{Bc}_j \tag{1}$$

The row vector $\mathbf{a}'_i \mathbf{B}$ equals

$$[\mathbf{a}'_i \mathbf{b}_1, \mathbf{a}'_i \mathbf{b}_2, \ldots, \mathbf{a}'_i \mathbf{b}_n] = \left[\sum_k a_{ik} b_{k1}, \sum_k a_{ik} b_{k2}, \ldots, \sum_k a_{ik} b_{kn} \right]$$

Thus

$$\mathbf{a}'_i \mathbf{B} \cdot \mathbf{c}_j = \left(\sum_k a_{ik} b_{k1} \right) c_{1j} + \left(\sum_k a_{ik} b_{k2} \right) c_{2j} + \cdots + \left(\sum_k a_{ik} b_{kn} \right) c_{nj} \tag{2}$$

$$= \sum_h \sum_k a_{ik} b_{kh} c_{hj}$$

The double summation is quite a mess! Next we observe that the column vector $\mathbf{Bc}_j$ has a kth entry of $\mathbf{b}'_k \cdot \mathbf{c}_j$, which equals $\sum_h b_{kh} c_{hj}$. Then the scalar product $\mathbf{a}'_i \cdot (\mathbf{Bc}_j)$ equals

$$\mathbf{a}'_i \cdot (\mathbf{Bc}_j) = a_{i1} \left(\sum_h b_{1h} c_{hj} \right) + a_{i2} \left(\sum_h b_{2h} c_{hj} \right) + \cdots + a_{in} \left(\sum_h b_{nh} c_{hj} \right) \tag{3}$$

$$= \sum_k \sum_h a_{ik} b_{kh} c_{hj}$$

The right side of (3) equals the right side of (2). The associated law is verified.

Scalar factoring for matrix multiplication follows from scalar factoring for scalar products (proved in Theorem 1 in Section 2.1); details are left to the exercises.

We now repeat previously stated results about the products of a matrix with the identity matrix and special 0–1 vectors. In products of vectors and matrices, it is essential to know whether a vector is a column vector or a row vector. As noted at the beginning of this book, *unless stated otherwise, we always assume that vectors are column vectors.*

The kth coordinate vector $\mathbf{i}_k$ has a 1 in the kth entry and 0's elsewhere. The one's vector $\mathbf{1}$ has a 1 in every entry. The identity matrix $\mathbf{I}$ is a square matrix with 1's on the main diagonal and 0's elsewhere. The coordinate vectors, the 1's vector, and the identity matrix can have any size. We assume that $\mathbf{i}_k$ and $\mathbf{1}$ are column vectors. To define corresponding row vectors, we introduce the notion of the *transpose of a col-*

umn vector. The transpose $\mathbf{1}^T$ of the column 3-vector $\mathbf{1} = \begin{bmatrix} 1 \\ 1 \\ 1 \end{bmatrix}$ is the row vector $\mathbf{1}^T = [1, 1, 1]$.

Coordinate vectors $\mathbf{i}_k$: $\mathbf{Ai}_k = \mathbf{a}_k$ (*k*th column of $\mathbf{A}$)
$\mathbf{i}_k^T\mathbf{A} = \mathbf{a}_k'$ (*k*th row of $\mathbf{A}$)

One's vector $\mathbf{1}$: $\mathbf{A1} = $ a column vector of the row sums of $\mathbf{A}$
$\mathbf{1}^T\mathbf{A} = $ a row vector of the column sums of $\mathbf{A}$

Identity matrix $\mathbf{I}$: $\mathbf{AI} = \mathbf{IA} = \mathbf{A}$

EXAMPLE 2.
Example of
Coordinate and
One's Vectors

Let $\quad \mathbf{A} = \begin{bmatrix} 1 & 0 & 2 \\ 3 & 1 & 5 \\ 7 & 3 & 1 \end{bmatrix}$ $\quad$ Then $\quad \mathbf{Ai}_2 = \begin{bmatrix} 0 \\ 1 \\ 3 \end{bmatrix}$, $\quad \mathbf{i}_3^T\mathbf{A} = [7, 3, 1]$, $\quad \mathbf{A1} = \begin{bmatrix} 3 \\ 9 \\ 11 \end{bmatrix}$, $\quad$ and $\mathbf{1A} = [11, 4, 8]$. ∎

We illustrate the 1's vector property and some other laws of matrix algebra in the following example about Markov chain matrices.

EXAMPLE 3.
Matrix Algebra in
Markov Chains

We showed in Section 1.2 that the transition equations for a Markov chain can be written in matrix notation as

$$\mathbf{p}' = \mathbf{Ap}$$

where $\mathbf{A}$ is the Markov transition matrix, $\mathbf{p}$ is the (column) vector of the current probability distribution, and $\mathbf{p}'$ is the (column) vector of the next-period probability distribution. For the weather Markov chain, the system of transition equations $\mathbf{p}' = \mathbf{Ap}$ is

$$\begin{aligned} p_1' &= \tfrac{3}{4}p_1 + \tfrac{1}{2}p_2 + \tfrac{1}{4}p_3 \\ p_2' &= \tfrac{1}{8}p_1 + \tfrac{1}{4}p_2 + \tfrac{1}{2}p_3 \\ p_3' &= \tfrac{1}{8}p_1 + \tfrac{1}{4}p_2 + \tfrac{1}{4}p_3 \end{aligned} \tag{4}$$

The next-period calculations represented by (4) can be repeated to find the distribution $\mathbf{p}''$ after two periods. (A numerical example of $\mathbf{p}''$ was given in Example 3 of Section 1.1.) Using the associative law, one has

$$\mathbf{p}'' = \mathbf{A}(\mathbf{Ap}) = \mathbf{A}^2\mathbf{p}$$

and after n periods, the distribution vector $\mathbf{p}^{(n)}$ is

$$\mathbf{p}^{(n)} = \mathbf{A}^n\mathbf{p}$$

Recall that the current probabilities p_i, the entries in $\mathbf{p}$, must sum to 1. We can express this fact with the 1's vector $\mathbf{1}$ as

$$\mathbf{1} \cdot \mathbf{p} = 1: \quad p_1 + p_2 + \cdots + p_n = 1 \tag{5}$$

The entries in the columns $\mathbf{a}_i$ in $\mathbf{A}$ must also sum to 1.

$$\mathbf{1} \cdot \mathbf{a}_j = 1: \quad a_{1j} + a_{2j} + \cdots + a_{nj} = 1 \tag{6}$$

Combining all the columns together into $\mathbf{A}$, we see that (6) yields (where $\mathbf{1}$ is now a *row* vector)

$$\begin{aligned}
\mathbf{1A} &= [\mathbf{1} \cdot \mathbf{a}_1, \mathbf{1} \cdot \mathbf{a}_2, \ldots, \mathbf{1} \cdot \mathbf{a}_n] \\
&= [1, 1, \cdot, 1] = \mathbf{1}
\end{aligned} \tag{7}$$

Matrix algebra allows us first to state concisely that a particular column sum is 1 and then also allows us to state the fact for all columns at once as $\mathbf{1A} = \mathbf{1}$.

Equations $\mathbf{p}' = \mathbf{Ap}$, $\mathbf{1} \cdot \mathbf{p} = 1$, and $\mathbf{1A} = \mathbf{1}$ can be used to show that in the next period distribution vector $\mathbf{p}'$, the entries p_i' also sum to 1. That is, we want to prove

$$\begin{aligned}
\mathbf{1} \cdot \mathbf{p}' \stackrel{?}{=} 1: \quad \mathbf{1} \cdot \mathbf{p}' &= \mathbf{1} \cdot (\mathbf{Ap}) && \text{since } \mathbf{p}' = \mathbf{Ap} \\
&= (\mathbf{1A}) \cdot \mathbf{p} && \text{by the associative law} \\
&= \mathbf{1} \cdot \mathbf{p} && \text{since } \mathbf{1A} = \mathbf{1} \\
&= 1 && \text{since } \mathbf{1} \cdot \mathbf{p} = 1
\end{aligned} \tag{8}$$

This argument can be repeated to show that the entries sum to 1 in $\mathbf{p}''$, the distribution vector after two periods, and more generally in $\mathbf{p}^{(n)}$. ■

Transpose of a Matrix and Symmetric Matrices

The operation of transposing a matrix has many theoretical and practical uses. In this book, its primary use is with pseudoinverses (in Section 3.5) and inverses of matrices with orthogonal columns (in Sections 4.5, 5.3, and 5.4).

The *transpose* of a matrix $\mathbf{A}$, written $\mathbf{A}^T$, is the matrix obtained from $\mathbf{A}$ by interchanging rows and columns. Another way to think of it is, flipping $\mathbf{A}$'s entries around the main diagonal. For example, if

$$\mathbf{A} = \begin{bmatrix} 1 & 2 \\ 4 & 5 \\ 7 & 8 \end{bmatrix}, \quad \text{then} \quad \mathbf{A}^T = \begin{bmatrix} 1 & 4 & 7 \\ 2 & 5 & 8 \end{bmatrix}$$

The transpose of a column vector was defined above. We can more generally define the transpose of any vector by considering a row vector as a $1 \times n$ matrix and a column vector an $m \times 1$ matrix. So the transpose of a row vector is a column vector and the transpose of a column vector is a row vector. For example, if $\mathbf{a} = [1, 2, 5]$,

$$\text{then } \mathbf{a}^\mathrm{T} = \begin{bmatrix} 1 \\ 2 \\ 5 \end{bmatrix}.$$

Theorem 1. Transposes have the following properties.
(i) $\mathbf{A}^\mathrm{T} + \mathbf{B}^\mathrm{T} = (\mathbf{A} + \mathbf{B})^\mathrm{T}$.
(ii) $(\mathbf{AB})^\mathrm{T} = \mathbf{B}^\mathrm{T}\mathbf{A}^\mathrm{T}$ and $(\mathbf{Ab})^\mathrm{T} = \mathbf{b}^\mathrm{T}\mathbf{A}^\mathrm{T}$.
(iii) $(\mathbf{A}^\mathrm{T})^\mathrm{T} = \mathbf{A}$.
The order of multiplying $\mathbf{A}$ and $\mathbf{B}$ must be reversed on the left side in (ii) because transposing reverses the roles of rows and columns: If $\mathbf{A}$ is $m \times r$ and $\mathbf{B}$ is $r \times n$, then $\mathbf{A}^\mathrm{T}$ is $r \times m$ and $\mathbf{B}^\mathrm{T}$ is $n \times r$.
Proof: Part (i) follows immediately from the definition of matrix addition. Part (iii) is obvious, since interchanging rows and columns twice gets you back to where you started. To prove part (ii), we must show for any i, j, that entry (i, j) in $(\mathbf{AB})^\mathrm{T}$ equals entry (i, j) in $\mathbf{B}^\mathrm{T}\mathbf{A}^\mathrm{T}$. Entry (i, j) in $\mathbf{B}^\mathrm{T}\mathbf{A}^\mathrm{T}$ equals the scalar product of the ith row of $\mathbf{B}^\mathrm{T}$ ($=$ ith column of $\mathbf{B}$) times the jth column of $\mathbf{A}^\mathrm{T}$ ($=$ jth row of $\mathbf{A}$). Thus entry (i, j) of $\mathbf{B}^\mathrm{T}\mathbf{A}^\mathrm{T} = \mathbf{b}_i \cdot \mathbf{a}_j'$. On the other hand, entry (j, i) in $\mathbf{AB} = \mathbf{a}_j' \cdot \mathbf{b}_i$. Since $\mathbf{b}_i \cdot \mathbf{a}_j' = \mathbf{a}_j' \cdot \mathbf{b}_i$, (ii) is proved. The second equation in (ii) is just the special case where $\mathbf{B}$ is an $n \times 1$ matrix—a column vector $\mathbf{b}$. ■

Corollary
(i) $\mathbf{b} \cdot \mathbf{c} = \mathbf{b}^\mathrm{T}\mathbf{c}$, for any two column n-vectors $\mathbf{b}, \mathbf{c}$.
(ii) $(\mathbf{Ax}) \cdot (\mathbf{Ax}) = \mathbf{x}^\mathrm{T}\mathbf{A}^\mathrm{T}\mathbf{Ax}$.
Proof: (i) For the scalar product of two vectors, we do not worry whether the vectors are row or column vectors. However, in matrix multiplication, we require scalar products to multiply a row vector times a column vector. To make $\mathbf{b}$ into a row vector, we take its transpose. Then $\mathbf{b}^\mathrm{T}\mathbf{c}$ is a valid matrix product (a $1 \times n$ matrix times an $n \times 1$ matrix).

(ii) Let $\mathbf{b} = \mathbf{c} = \mathbf{Ax}$. Applying part (i) and Theorem 1(ii), we get $(\mathbf{Ax}) \cdot (\mathbf{Ax}) = (\mathbf{Ax})^\mathrm{T}(\mathbf{Ax}) = (\mathbf{x}^\mathrm{T}\mathbf{A}^\mathrm{T})(\mathbf{Ax}) = \mathbf{x}^\mathrm{T}\mathbf{A}^\mathrm{T}\mathbf{Ax}$. ■

One of the most useful properties that a matrix can have is symmetry. A matrix $\mathbf{A}$ is *symmetric* if $\mathbf{A} = \mathbf{A}^\mathrm{T}$. A symmetric matrix must be a square matrix. Later in this book, we shall see that symmetric matrices have many nice theoretical and computational properties. If $\mathbf{A}$ is a symmetric matrix, all the information in the matrix is contained on or above the main diagonal.

A familiar example of a symmetric matrix is a mileage chart on a road map. Because of its symmetric structure, only the upper (or lower) triangular portion of this table is usually given. Symmetric matrices are very common in physical science applications.

There is a useful symmetric matrix associated with any unsymmetric matrix $\mathbf{A}$. It is the matrix $\mathbf{A}^\mathrm{T}\mathbf{A}$. Entry (i, j) in $\mathbf{A}^\mathrm{T}\mathbf{A}$ will be the scalar product $\mathbf{a}_i \cdot \mathbf{a}_j$ of the ith and jth columns of $\mathbf{A}$ (since the ith row of $\mathbf{A}^T$ equals the ith column of $\mathbf{A}$). Since $\mathbf{a}_i \cdot \mathbf{a}_j = \mathbf{a}_j \cdot \mathbf{a}_i$, then entry (i, j) and entry (j,i) in $\mathbf{A}^T\mathbf{A}$ are the same. So $\mathbf{A}^T\mathbf{A}$ will be symmetric. (One computes scalar products of pairs of rows in the related symmetric matrix $\mathbf{A}\mathbf{A}^T$.)

There are many problems where one wants to measure in some informal way how similar various pairs of columns are in a matrix. Scalar products of the columns, as computed in $\mathbf{A}^T\mathbf{A}$, provide one good measure.

EXAMPLE 4.
Scalar Products of
Columns as a
Similarity Measure

Suppose that five students A, B, C, D, E have been asked to rate six subjects—linguistics, mathematics, necromancy, optometry, philosophy, and quantum mechanics—as subjects they like (rating = 1) or as subjects they do not like (rating = -1). The following rating matrix $\mathbf{R}$ was obtained.

$$
\mathbf{R} = \begin{array}{r}
\\
\text{Linguistics} \\
\text{Mathematics} \\
\text{Necromancy} \\
\text{Optometry} \\
\text{Philosophy} \\
\text{Quantum mechanics}
\end{array}
\begin{array}{c}
\begin{array}{rrrrr} A & B & C & D & E \end{array} \\
\left[\begin{array}{rrrrr}
1 & -1 & 1 & -1 & 1 \\
1 & 1 & -1 & -1 & 1 \\
1 & -1 & 1 & -1 & -1 \\
1 & 1 & -1 & -1 & 1 \\
-1 & -1 & -1 & -1 & -1 \\
1 & -1 & 1 & -1 & 1
\end{array} \right]
\end{array}
\tag{9}
$$

To measure the similarity of interests among students, we want to use the scalar product of pairs of columns. Observe that the scalar product will be positive if two students' ratings tend to agree and will be negative if they tend to disagree. (Note the similarity of these scalar products with the correlation coefficient introduced in Section 2.3.) To get these scalar products, we simply compute $\mathbf{R}^T\mathbf{R}$. Since $\mathbf{R}^T\mathbf{R}$ is symmetric, we only need to compute the entries on or above the main diagonal.

This computation yields

$$
\mathbf{R}^T\mathbf{R} = \begin{array}{r}
A \\
B \\
C \\
D \\
E
\end{array}
\begin{array}{c}
\begin{array}{rrrrr} A & B & C & D & E \end{array} \\
\left[\begin{array}{rrrrr}
6 & 0 & 2 & -4 & 4 \\
 & 6 & -4 & 2 & 2 \\
 & & 6 & 0 & 0 \\
 & & & 6 & -2 \\
 & & & & 6
\end{array} \right]
\end{array}
\tag{10}
$$

Computing $\mathbf{R}\mathbf{R}^T$ would yield scalar products of pairs of rows in $\mathbf{R}$. These products would measure how similar different pairs of subjects are perceived to be (i.e., a large positive number would mean that most students give the same rating to the two subjects). ■

The matrix $\mathbf{A}^T\mathbf{A}$ arises in a surprising number of different settings in linear algebra. It is central to least squares solutions (discussed in Section 3.5), arises in matrix norms (Section 3.4), and yields a simple computer-based test for linear independence

(Section 4.2). The following theorem will prove to be very important in future uses of $A^T A$. It converts the question of the uniqueness of solutions for an $m \times n$ matrix to the question of uniqueness of solutions for a square $n \times n$ matrix. [Recall that there is a nice theory about uniqueness of solutions for square matrices: A unique solution to $Ax = 0$ is equivalent to $\det(A) \neq 0$ and the invertibility of A.]

Theorem 2. Let A be an $m \times n$ matrix A $(m > n)$ with associated $n \times n$ matrix $A^T A$. Then $(A^T A)x = 0$ has the unique solution $x = 0$ if and only if $Ax = 0$ has the unique solution $x = 0$.

Proof: If $Ax = 0$ has two solutions, say $Ax' = 0$ and $Ax'' = 0$, we get that $A^T Ax = 0$ has two solutions: $A^T Ax' = A^T(Ax') = 0$ and $A^T Ax'' = A^T(Ax'') = 0$.

Suppose next that $Ax = 0$ has the unique solution $x = 0$. If $A^T Ax^* = 0$, then obviously $x^{*T}A^T Ax^* = x^{*T}(A^T Ax^*) = 0$ (here 0 is a scalar). By part (ii) of the Corollary of Theorem 1, $x^{*T}A^T Ax^* = (Ax^*) \cdot (Ax^*)$. Then

$$|Ax^*|^2 = (Ax^*) \cdot (Ax^*) = x^{*T}A^T Ax^* = 0$$

If $|Ax^*|^2 = 0$, then $|Ax^*| = 0$, which implies that $Ax^* = 0$ (since only the vector 0 has norm 0). Since we are assuming that $Ax = 0$ has the unique solution $x = 0$, then $x^* = 0$. ■

0–1 Matrices and Graphs

We now introduce a new application of matrix multiplication. The application, involving graphs and symmetric matrices, provides a good setting to review the tabular interpretation of matrix multiplication presented in Section 1.3 and also a good example of partitioned matrices, a topic introduced shortly.

A ***graph*** $G = (N, E)$ consists of a set N of nodes and a collection E of edges that are pairs of nodes. There is a natural way to "draw" a graph. We make a point for each node and then draw lines linking each pair of nodes that forms an edge. For example, the graph G that has the node set $N = \{a, b, c, d, e\}$ and the edge set $E = \{(a, b), (a, c), (a, d), (b, c), (b, d), (d, e), (e, e)\}$ is drawn in Figure 3.1. An edge may link a node with itself, as at node e in Figure 3.1. Such an edge is called a *loop*.

A flowchart for a computer program is a form of graph. The data structures that are used to organize complex sets of data are graphs. Organizational charts, electrical circuits, telephone networks, and road maps are other examples of graphs. Industry spends billions of dollars every year working on applied graph problems.

One common type of question asked about graphs concerns paths. A ***path*** is a sequence of nodes with edges linking consecutive nodes. The length of a path is the number of edges on it. For example, in Figure 3.1, (a, b, d, e) is a path of length 3 between a and e. A single edge is a path of length 1. We may want to find the shortest path between two nodes, or determine whether or not any path exists between a given pair of nodes. Finding paths through graphs arises when one wants to route a set of telephone calls through a network between prescribed cities without exceeding the capacity of any edge. The question of whether a path exists between two nodes arises over and over again in studying the effect on networks of random disruption, say, due to lightning. For example, in a given 1000-edge network one might want to know the probability that if five randomly chosen edges are destroyed, the network will become disconnected.

The purpose of mentioning all these graph problems is to motivate the importance of having good methods to represent and manipulate graphs in a computer. One common way to represent graphs is with a 0–1 matrix.

Definition

The *adjacency matrix* $\mathbf{A}(G)$ *of a graph* G tells which pairs of nodes are adjacent (i.e., which pairs form edges). Entry $a_{ij} = 1$ if there is an edge linking the ith and jth nodes; otherwise, $a_{ij} = 0$. Note that $a_{ii} = 0$ unless there is a loop at the ith node.

The adjacency matrix $\mathbf{A}(G)$ of the graph G in Figure 3.1 is

$$
\mathbf{A}(G) = \begin{array}{c} \\ a \\ b \\ c \\ d \\ e \end{array}
\begin{array}{c} \begin{array}{ccccc} a & b & c & d & e \end{array} \\
\left[\begin{array}{ccccc}
0 & 1 & 1 & 1 & 0 \\
1 & 0 & 1 & 1 & 0 \\
1 & 1 & 0 & 0 & 0 \\
1 & 1 & 0 & 0 & 1 \\
0 & 0 & 0 & 1 & 1
\end{array}\right] \end{array}
\tag{11}
$$

Matrix $\mathbf{A}(G)$ is symmetric, since adjacency is a symmetric relationship.

We claim that the ith and jth nodes, for $i \neq j$, can be joined by a path of length 2 if and only if entry (i, j) in $\mathbf{A}^2(G)$, the square of the adjacency matrix $\mathbf{A}(G)$, is positive. First let us compute $\mathbf{A}^2(G)$ for the graph in Figure 3.1. To do this, we must find the scalar product of each row of $\mathbf{A}(G)$ with each column of $\mathbf{A}(G)$. Since $\mathbf{A}(G)$ is symmetric, this is equivalent to finding the scalar product of each row with every other row. Consider the scalar product of a's row and b's row in $\mathbf{A}(G)$:

$$
\begin{array}{c} a \\ b \\ \\ \end{array}
\left[\begin{array}{ccccc}
0 & 1 & 1 & 1 & 0 \\
1 & 0 & 1 & 1 & 0 \\
\cdot & \cdot & \cdot & \cdot & \cdot \\
\cdot & \cdot & \cdot & \cdot & \cdot
\end{array}\right]
\tag{12}
$$

$$[0, 1, 1, 1, 0] \cdot [1, 0, 1, 1, 0] = 0 \cdot 1 + 1 \cdot 0 + 1 \cdot 1 + 1 \cdot 1 + 0 \cdot 0 = 2$$

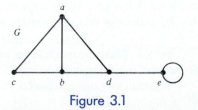

Figure 3.1

The product of two entries in this scalar product will be 1 if and only if the two entries are both 1. Thus the value of the scalar product is simply the number of positions where the two vectors both have a 1.

We now interpret the computation of the scalar product (12) in terms of adjacencies in the graph. In (12), when rows a and b have a 1 in q's column, a and b are both adjacent to q. From (12), we see that a and b are both adjacent to nodes c and d. Then (a, c, b) and (a, d, b) are paths of length 2 between a and b. In general, when two nodes n_i and n_j are adjacent to a common node n_k, then (n_i, n_k, n_j) will be a path of length 2 between n_i and n_j. This proves that the (i, j)th entry in $\mathbf{A}^2(G)$ equals the number of paths of length 2 between the ith and jth nodes.

EXAMPLE 5. **Paths in Graphs**	Find which pairs of different nodes in the graph in Figure 3.1 are joined by a path of length 2. By the preceding discussion, we can answer this question by computing $\mathbf{A}^2(G)$.

$$\mathbf{A}^2(G) = \begin{array}{c} \\ a \\ b \\ c \\ d \\ e \end{array} \begin{array}{c} \begin{array}{ccccc} a & b & c & d & e \end{array} \\ \left[\begin{array}{ccccc} 3 & 2 & 1 & 1 & 1 \\ 2 & 3 & 1 & 1 & 1 \\ 1 & 1 & 2 & 2 & 0 \\ 1 & 1 & 2 & 3 & 1 \\ 1 & 1 & 0 & 1 & 2 \end{array} \right] \end{array} \qquad (13)$$

The positive off-diagonal entries tell us which pairs of different nodes are joined by a path of length 2. The answer from (13) is all pairs of different nodes except c, e.

■

This reasoning can be extended to higher powers of $\mathbf{A}(G)$. The entries of $\mathbf{A}^3(G)$ tell how many paths of length 3 join different pairs of nodes. And for any positive integer m, the entries of $\mathbf{A}^m(G)$ tell how many paths of length m join different pairs of nodes. Illustrative examples and mathematical verification of this property of $\mathbf{A}^m(G)$ are left to the exercises.

A graph G is **connected** if every pair of nodes in G is joined by a path. Using powers of $\mathbf{A}(G)$ we can determine whether or not a graph is connected. If G has n nodes, any path between two nodes in G has length $\leq n-1$. So G is connected when there exist paths of length $\leq n-1$ between all pairs of nodes. To determine whether all such paths exist, we compute $\mathbf{A}^2, \mathbf{A}^3, \ldots, \mathbf{A}^{n-1}(G)$ and check for each (i, j) pair, $i \neq j$, whether entry (i, j) is positive for some power of $\mathbf{A}$. For example, the graph G in Figure 3.1 is seen to be connected; all entries but (c, e) and (e, c) are positive in $\mathbf{A}^2(G)$, and these two zero entries become positive in $\mathbf{A}^3(G)$.

Note: There are much faster methods to check for connectedness of a graph that do not use matrix multiplication.

Summarizing our discussion of paths and connectedness, we have

Theorem 3 (Graphs and Matrix Multiplication)

(i) Let $\mathbf{A}(G)$ be the adjacency matrix of graph G. Then the entry (i, j) in $\mathbf{A}^2(G)$ tells how many paths of length 2 join node i with node j, and, more generally, entry (i, j) in $\mathbf{A}^m(G)$ tells how many paths of length m join node i with node j.

(ii) Let G be an n-vertex graph. G is connected if and only if for each (i, j) pair, $i \neq j$, entry (i, j) is positive in some power $\mathbf{A}^k$, $k = 1, 2, \ldots,$ $n - 1$.

Partitioning Matrices

Any vector $\mathbf{a}$ can be partitioned into two or more subvectors such as

$$\mathbf{a} = [\mathbf{a}_1, \mathbf{a}_2, \mathbf{a}_3] \tag{14}$$

For example, if $\mathbf{a} = [1, 2, 3, 0, 0, 0, 1, 2, 3]$ and if $\mathbf{a}^* = [1, 2, 3]$ and $\mathbf{0}_3$ is the 3-entry zero row vector, we can write $\mathbf{a}$ as $[\mathbf{a}^*, \mathbf{0}_3, \mathbf{a}^*]$. A matrix $\mathbf{A}$ can be partitioned into submatrices, such as

$$\mathbf{A} = [\mathbf{A}_1 \quad \mathbf{A}_2] \tag{15}$$

or

$$\mathbf{A} = \begin{bmatrix} \mathbf{A}_{11} & \mathbf{A}_{12} \\ \mathbf{A}_{21} & \mathbf{A}_{22} \end{bmatrix} \tag{16}$$

or

$$\mathbf{A} = \left[\begin{array}{cc|c} & \mathbf{A}^* & \mathbf{A}_1 \\ & & \mathbf{A}_2 \\ \hline \mathbf{A}_1' & \mathbf{A}_2' & \mathbf{A}_3 \end{array} \right] \tag{17}$$

For example, we might partition

$$\left[\begin{array}{cccc|cc} 1 & 2 & 3 & 4 & 1 & 1 \\ 2 & 3 & 4 & 5 & 1 & 1 \\ 3 & 4 & 5 & 6 & 0 & 0 \\ 4 & 5 & 6 & 7 & 0 & 0 \\ \hline 1 & 1 & 0 & 0 & 1 & 1 \\ 1 & 1 & 0 & 0 & 1 & 1 \end{array} \right] = \left[\begin{array}{cc|c} & & \mathbf{A}_1 \\ \mathbf{A}^* & & \\ & & \mathbf{A}_0 \\ \hline \mathbf{A}_1 & \mathbf{A}_0 & \mathbf{A}_1 \end{array} \right]$$

where $\mathbf{A}_1$ is a 2×2 matrix of 1's and $\mathbf{A}_0$ is a 2×2 matrix of 0's.

The partition of a matrix **A** will typically correspond to different components of the underlying model. A partition of **A** in the form of (16) would arise in a Markov chain transition matrix if the states divide in some natural way into two groups, group S_1 and group S_2. The partition in (15) arises naturally if the columns of **A** represent two different types of variables. The structure of a graph can provide a natural way to partition its adjacency matrix.

**EXAMPLE 6.
Partitioning the
Adjacency Matrix
of a Graph**

The graph G in Figure 3.2 has the following adjacency matrix:

$$\mathbf{A}(G) = \begin{array}{c} \\ a \\ b \\ c \\ d \\ e \\ f \\ g \\ h \end{array} \begin{array}{c} \begin{array}{cccccccc} a & b & c & d & e & f & g & h \end{array} \\ \left[\begin{array}{cccccccc} 0 & 1 & 1 & 0 & 1 & 0 & 0 & 0 \\ 1 & 0 & 1 & 1 & 0 & 1 & 0 & 0 \\ 1 & 1 & 0 & 1 & 0 & 0 & 1 & 0 \\ 0 & 1 & 1 & 0 & 0 & 0 & 0 & 1 \\ 1 & 0 & 0 & 0 & 0 & 1 & 1 & 0 \\ 0 & 1 & 0 & 0 & 1 & 0 & 1 & 1 \\ 0 & 0 & 1 & 0 & 1 & 1 & 0 & 1 \\ 0 & 0 & 0 & 1 & 0 & 1 & 1 & 0 \end{array} \right] \end{array} \tag{18}$$

$\mathbf{A}(G)$ has the nice partitioned form

$$\mathbf{A}(G) = \begin{bmatrix} \mathbf{A}^* & \mathbf{I} \\ \mathbf{I} & \mathbf{A}^* \end{bmatrix} \text{ with } \mathbf{A}^* = \begin{bmatrix} 0 & 1 & 1 & 0 \\ 1 & 0 & 1 & 1 \\ 1 & 1 & 0 & 1 \\ 0 & 1 & 1 & 0 \end{bmatrix} \text{ and } \mathbf{I} \text{ the } 4 \times 4 \text{ identity matrix} \tag{19}$$

■

Partitioning is very useful in matrix multiplication, because we can treat the submatrices like scalar entries. For example, if

$$\mathbf{A} = \begin{bmatrix} \mathbf{A}_{11} & \mathbf{A}_{12} \\ \mathbf{A}_{21} & \mathbf{A}_{22} \end{bmatrix} \quad \text{and} \quad \mathbf{B} = \begin{bmatrix} \mathbf{B}_{11} & \mathbf{B}_{12} \\ \mathbf{B}_{21} & \mathbf{B}_{22} \end{bmatrix} \tag{20}$$

then

$$\mathbf{AB} = \begin{bmatrix} \mathbf{A}_{11}\mathbf{B}_{11} + \mathbf{A}_{12}\mathbf{B}_{21} & \mathbf{A}_{11}\mathbf{B}_{12} + \mathbf{A}_{12}\mathbf{B}_{22} \\ \mathbf{A}_{21}\mathbf{B}_{11} + \mathbf{A}_{22}\mathbf{B}_{21} & \mathbf{A}_{21}\mathbf{B}_{12} + \mathbf{A}_{22}\mathbf{B}_{22} \end{bmatrix} \tag{21}$$

Verification of (21) is left as an exercise. Of course, (21) requires that the number of columns in the **A** submatrices equal the number of rows in the appropriate **B** submatrices. The situation with partitioning is similar to all the matrix algebra rules presented earlier in this section. Unless it is expressly prohibited, anything you would like to be true about partitioning probably is true.

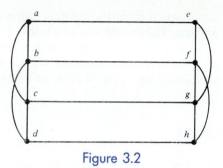

Figure 3.2

If some of the submatrices of **A** and **B** have nice forms (e.g., **O** or **I**) or if several submatrices are the same, the amount of work needed to compute the matrix product **AB** is greatly reduced by using (21).

EXAMPLE 6 (continued). Squaring a Partitioned Adjacency Matrix

Let us use (21) to compute the square of the partitioned adjacency matrix **A**(*G*) given in (19).

$$\mathbf{A}(G)^2 = \begin{bmatrix} \mathbf{A}^*\mathbf{A}^* + \mathbf{I}\mathbf{I} & \mathbf{A}^*\mathbf{I} + \mathbf{I}\mathbf{A}^* \\ \mathbf{A}^*\mathbf{I} + \mathbf{I}\mathbf{A}^* & \mathbf{A}^*\mathbf{A}^* + \mathbf{I}\mathbf{I} \end{bmatrix} = \begin{bmatrix} \mathbf{A}^{*2} + \mathbf{I} & 2\mathbf{A}^* \\ 2\mathbf{A}^* & \mathbf{A}^{*2} + \mathbf{I} \end{bmatrix} \quad (22)$$

Computing $\mathbf{A}^{*2}$ can be done by inspection faster than entering the numbers in a computer program. It just involves counting how many 1-entries each pair of rows in **A*** have in common (as explained earlier in this section).

$$\mathbf{A}^{*2} = \begin{bmatrix} 2 & 1 & 1 & 2 \\ 1 & 3 & 2 & 1 \\ 1 & 2 & 3 & 1 \\ 2 & 1 & 1 & 2 \end{bmatrix} \quad \text{and} \quad \mathbf{A}^{*2} + \mathbf{I} = \begin{bmatrix} 3 & 1 & 1 & 2 \\ 1 & 4 & 2 & 1 \\ 1 & 2 & 4 & 1 \\ 2 & 1 & 1 & 3 \end{bmatrix} \quad (23)$$

Inserting $\mathbf{A}^{*2} + \mathbf{I}$ and $2\mathbf{A}^*$ into the partition product in (22), we obtain

$$\mathbf{A}(G)^2 = \begin{array}{c} \\ a \\ b \\ c \\ d \\ e \\ f \\ g \\ h \end{array} \begin{array}{cccccccc} a & b & c & d & e & f & g & h \end{array} \left[\begin{array}{cccc|cccc} 3 & 1 & 1 & 2 & 0 & 2 & 2 & 0 \\ 1 & 4 & 2 & 1 & 2 & 0 & 2 & 2 \\ 1 & 2 & 4 & 1 & 2 & 2 & 0 & 2 \\ 2 & 1 & 1 & 3 & 0 & 2 & 2 & 0 \\ \hline 0 & 2 & 2 & 0 & 3 & 1 & 1 & 2 \\ 2 & 0 & 2 & 2 & 1 & 4 & 2 & 1 \\ 2 & 2 & 0 & 2 & 1 & 2 & 4 & 1 \\ 0 & 2 & 2 & 0 & 2 & 1 & 1 & 3 \end{array} \right] \quad (24)$$

Using partitioning, the only matrix product we had to calculate was $\mathbf{A}^{*2}$, a 4×4 problem. Halving the size of a matrix reduces the computation in a matrix product by a factor of 8, since $\mathbf{A}^2$ requires 8×8 scalar products of 8-entry vectors—a total of $8 \cdot 8 \cdot 8 = 512$ multiplications—while $\mathbf{A}^{*2}$ requires 4×4 scalar products of 4-entry vectors—a total of $4 \cdot 4 \cdot 4 = 64$ multiplications. ■

Section 3.1 Exercises

1. Evaluate the following products involving the 1's (column) vector $\mathbf{1}$, the identity matrix $\mathbf{I}$, and the kth coordinate (column) vector $\mathbf{i}_k = [0, 0, \ldots, 1, \ldots, 0, 0]^T$. Assume that all have size n.

   (a) $\mathbf{I}^2$ (b) $\mathbf{1} \cdot \mathbf{1}$ (c) $\mathbf{I1}$ (d) $\mathbf{Ii}_k$ (e) $\mathbf{1} \cdot \mathbf{i}_k$ (f) $\mathbf{i}_k \cdot \mathbf{i}_k$ (g) $\mathbf{i}_k \cdot \mathbf{i}_j (k \neq j)$ (h) $\mathbf{1}^T \mathbf{I1}$
   (i) $\mathbf{j}_k^T \mathbf{Ii}_j (k \neq j)$

2. (a) Write the system of equations in matrix notation in the form $\mathbf{Ax} = \mathbf{b}$. (Define $\mathbf{A}$, $\mathbf{x}$, and $\mathbf{b}$.)

$$3x_1 + 5x_2 + 7x_3 = 8$$
$$2x_1 - x_2 + x_3 = 4$$
$$x_1 + 6x_2 - 2x_3 = 6$$

   (b) Rewrite the matrix equation in part (a) to reflect the operation of bringing the right sides over to the left sides (so that the right side is now 0's).

3. (a) Write the system of equations in matrix notation [see Exercise 2(a)].

$$2x_1 - 3x_2 = x_1$$
$$5x_1 + 4x_2 = x_2$$

   (b) Rewrite the matrix equation in part (a) to reflect the operation of bringing the right-side variables over to the left side. Your new equation should be of the form $\mathbf{Qx} = \mathbf{0}$, where $\mathbf{Q}$ is a matrix expression involving $\mathbf{I}$, the identity matrix.

4. Consider the system of equations

$$2x_1 + 3x_2 - 2x_3 = 5y_1 + 2y_2 - 3y_3 + 200$$
$$x_1 + 4x_2 + 3x_3 = 6y_1 - 4y_2 + 4y_3 - 120$$
$$5x_1 + 2x_2 - x_3 = 2y_1 \qquad - 2y_3 + 350$$

   (a) Write this system of equations in matrix notation. (Define the vectors and matrices that you introduce.)
   (b) Rewrite in matrix notation with all the variables on the left side and just numbers on the right.
   (c) Rewrite in matrix notation so that x_1 is the only term on the left in the first equation, x_2 is the only term on the left in the second equation, and x_3 is the only term on the left in the third equation.

5. Let **A** and **B** be 2×2 matrices and **x**, **y**, **z** be 2-vectors such that $\mathbf{Ax} = \mathbf{By} = \mathbf{1}\left(= \begin{bmatrix} 1 \\ 1 \end{bmatrix}\right)$, $\mathbf{Ay} = \begin{bmatrix} 1 \\ 0 \end{bmatrix}$, $\mathbf{Bx} = \begin{bmatrix} 0 \\ 1 \end{bmatrix}$. Determine **z** when:

(a) $\mathbf{z} = \mathbf{A}(2\mathbf{x} - \mathbf{y})$ (b) $\mathbf{z} = (\mathbf{A} - \mathbf{B})\mathbf{x}$ (c) $\mathbf{z} = (\mathbf{A} + \mathbf{B})\mathbf{x} - 2(\mathbf{A} + \mathbf{B})\mathbf{y}$ (d) $\mathbf{z} = (3\mathbf{A} + \mathbf{B})(\mathbf{x} + \mathbf{y})$ (e) $\mathbf{z} = [(\mathbf{A} - \mathbf{B})\mathbf{y}] \cdot [(\mathbf{A} + 3\mathbf{B})(\mathbf{x} - \mathbf{y})]\mathbf{1}$

6. (a) Consider the following system of equations for the growth of rabbits and foxes from year to year:

$$R' = 1.5R - 0.2F + 100$$
$$F' = 0.3R + 0.9F + 50$$

Write this system in matrix notation, with $\mathbf{p} = \begin{bmatrix} R \\ F \end{bmatrix}$ and $\mathbf{p}' = \begin{bmatrix} R' \\ F' \end{bmatrix}$.

(b) Write a matrix equation for $\mathbf{p}''$, the vector of the number of rabbits and foxes after 2 years.

(c) Write a matrix equation for $\mathbf{p}^{(3)}$, the vector of the number of rabbits and foxes after 3 years.

(d) Using summation notation (Σ), write a matrix equation for $\mathbf{p}^{(n)}$, the vector of the number of rabbits and foxes after n years.

7. Verify that $\mathbf{bI} = \mathbf{b}$ for any $\mathbf{b} = [b_1, b_2, \ldots, b_n]$.

8. Verify that $\mathbf{IB} = \mathbf{B}$ for a 3×3 matrix $\mathbf{B} = \begin{bmatrix} b_{11} & b_{12} & b_{13} \\ b_{21} & b_{22} & b_{23} \\ b_{31} & b_{32} & b_{33} \end{bmatrix}$.

9. Let $\mathbf{i}_k$ be the kth coordinate vector $\mathbf{i}_k = [0, 0, \ldots, 1, \ldots, 0, 0]^T$. What is the value of $\mathbf{1}^T \mathbf{Ai}_k$?

10. (a) Show that **B1** yields a vector whose ith position is the sum of the entries in the ith row.

(b) Show that $\mathbf{1}^T\mathbf{B1}$ equals the sum of all the entries in **B**.

11. Show that if the second row of **A** is all 0's, the second row in the product **AB** (if defined) is all 0's.

12. Show that if $\mathbf{x}^\circ$ is a solution to $\mathbf{Ax} = \mathbf{b}$, $r\mathbf{x}^\circ$ is a solution to $\mathbf{Ax} = r\mathbf{b}$.

13. Show that if $\mathbf{x}^\circ$ is a solution to $\mathbf{Ax} = \mathbf{b}$ and if $\mathbf{x}^*$ is a solution to the homogeneous system $\mathbf{Ax} = \mathbf{0}$, then $\mathbf{x}^\circ + \mathbf{x}^*$ is also a solution to $\mathbf{Ax} = \mathbf{b}$.

14. Let **A** and **B** be $n \times n$ matrices. If $\mathbf{C} = \mathbf{AB}$ and each entry in the second column of **B** is five times the corresponding entry in the first column of **B** ($\mathbf{b}_2 = 5\mathbf{b}_1$), show that the second column of **C** is five times the first column of **C**.

15. Given a linear model of the form $\mathbf{x}' = \mathbf{Ax} + \mathbf{b}$, let us expand the $n \times n$ matrix **A** into an $(n + 1) \times (n + 1)$ matrix **A*** by including **b**, row vector **0** of 0's, and a 1 in entry $(n + 1, n + 1)$ so that **A*** has the form $\mathbf{A}^* = \begin{bmatrix} \mathbf{A} & \mathbf{b} \\ \mathbf{0} & 1 \end{bmatrix}$. We should also add to **x** an $(n + 1)$st entry equal to 1; call the new vector **x***, and now our linear model has the form $\mathbf{x}^{*\prime} = \mathbf{A}^*\mathbf{x}^*$. Give the new **A*** for the following linear models.

(a) $x_1' = 3x_1 + 2x_2 + 10$
$\quad\ x_2' = 4x_1 - 5x_2 + \ \ 8$

(b) $x_1' = \ \ x_1 + 2x_2 + 5x_3 + 20$
$\quad\ x_2' = 2x_1 - \ \ x_2 - 2x_3 - 10$
$\quad\ x_3' = 3x_1 + 4x_2 + 6x_3 + 30$

16. Show that if $\mathbf{A}$ is the transition matrix of a Markov chain with five states, then $\mathbf{1A1} = 5$.

17. Extend the reasoning in Example 3 to show that $\mathbf{1} \cdot \mathbf{p}'' = 1$, where $\mathbf{p}''$ is the distribution after two periods.

18. **(a)** Write a matrix equation involving the 1's vector $\mathbf{1}$ to express the fact that for any Markov transition matrix $\mathbf{A}$, the column sums in $\mathbf{A}^2$ equal 1.
    (b) Through repeated use of equation (7), verify the equation you wrote in part (a).
    (c) Prove that the column sums in $\mathbf{A}^3$ equal 1 using matrix algebra. [Follow the reasoning in parts (a) and (b).]
    (d) Use induction to prove that column sums in $\mathbf{A}^n$ equal 1.

19. Prove that $(\mathbf{B} + \mathbf{C})\mathbf{A} = \mathbf{BA} + \mathbf{CA}$ by mimicking the argument in the text used to show $\mathbf{A}(\mathbf{B} + \mathbf{C}) = \mathbf{AB} + \mathbf{AC}$.

20. Give an example involving three 2-vectors to show that the equality $(\mathbf{a} \cdot \mathbf{b})\mathbf{c} = \mathbf{a}(\mathbf{b} \cdot \mathbf{c})$ makes no sense.

21. Show that the identity $\mathbf{A}(\mathbf{b} \cdot \mathbf{c}) = (\mathbf{Ab}) \cdot \mathbf{c}$ makes no sense by testing the identity with any 2×2 matrix $\mathbf{A}$, 2-vector $\mathbf{b}$, and 2-vector $\mathbf{c}$. What goes wrong?

22. Show that the identity $\mathbf{A}(\mathbf{bC})^{\mathrm{T}} = (\mathbf{Ab}^{\mathrm{T}})\mathbf{C}$ makes no sense by making up matrices $\mathbf{A}$, $\mathbf{C}$ and row vector $\mathbf{b}$ with $\mathbf{A}$ 3×2 so that the matrix expression $\mathbf{A}(\mathbf{bC})^{\mathrm{T}}$ makes sense (the sizes fit together properly), and then show that the sizes are wrong for $(\mathbf{Ab}^{\mathrm{T}})\mathbf{C}$.

23. **(a)** Why is the following identity false: $(\mathbf{AB})^2 = \mathbf{A}^2\mathbf{B}^2$? What expression is $(\mathbf{AB})^2$ actually equal to?
    (b) Can you find two nonzero matrices $\mathbf{A}$, $\mathbf{B}$ for which $(\mathbf{AB})^2 = \mathbf{A}^2\mathbf{B}^2$?

24. Why is the following identity false: $(\mathbf{A} + \mathbf{B})^2 = \mathbf{A}^2 + 2\mathbf{AB} + \mathbf{B}^2$? What is $(\mathbf{A} + \mathbf{B})^2$ actually equal to?

25. Let $\mathbf{x}$ be a column n-vector such that $\mathbf{x} \cdot \mathbf{x} = 1$. Define $\mathbf{xx}^{\mathrm{T}}$ to be the $n \times n$ matrix resulting from the matrix product of the $n \times 1$ matrix $\mathbf{x}$ times the $1 \times n$ matrix $\mathbf{x}^{\mathrm{T}}$. Show that the $n \times n$ matrix $\mathbf{H} = \mathbf{I} - \mathbf{xx}^{\mathrm{T}}$ has the following three properties.
    (a) $\mathbf{H}$ is symmetric. **(b)** $\mathbf{H}^2 = \mathbf{I}$. **(c)** $\mathbf{H}^{\mathrm{T}}\mathbf{H} = \mathbf{I}$.

26. Compute $\mathbf{RR}^{\mathrm{T}}$ in Example 4 to find a measure of how much different students share common views of their subjects.

27. The faculties in the four divisions of the College of Arts and Sciences at Wayward Univ. (Nat. Sci./Math., Bio. Sci., Arts&Hum., Soc. Sci.) have taken stands for or against the following five issues.

	NS/M	Bio	A&H	SS
Wayward needs to change its name:	No	Yes	Yes	Yes
Wayward U. is a friendly campus:	Yes	Yes	No	No
CompSci 112 is too hard:	No	No	Yes	Yes
The Alfred E. Neuman dorm is ugly:	No	Yes	Yes	No
Wayward athletes should perform better:	No	No	No	Yes

Compute a matrix of similarities between the divisions. (Remember that, by symmetry, you only have to compute the entries on or above the main diagonal.)

28. Verify that the inverse of A^T is $(A^{-1})^T$. [Hint: Use the multiplication rule for transposes, $(CD)^T = D^T C^T$.]

29. Show that if A is symmetric, then A^2 is symmetric.

30. Draw the graphs with the following adjacency matrices.

(a) $\begin{bmatrix} 0 & 0 & 0 & 1 \\ 0 & 0 & 1 & 0 \\ 0 & 1 & 0 & 0 \\ 1 & 0 & 0 & 0 \end{bmatrix}$
(b) $\begin{bmatrix} 0 & 1 & 1 & 1 \\ 1 & 0 & 0 & 1 \\ 1 & 0 & 0 & 1 \\ 1 & 1 & 1 & 0 \end{bmatrix}$
(c) $\begin{bmatrix} 0 & 0 & 1 & 1 & 0 & 0 \\ 0 & 1 & 1 & 0 & 1 & 1 \\ 1 & 1 & 0 & 0 & 0 & 1 \\ 1 & 0 & 0 & 1 & 1 & 1 \\ 0 & 1 & 0 & 1 & 0 & 0 \\ 0 & 1 & 1 & 1 & 0 & 1 \end{bmatrix}$

31. Write the adjacency matrices for the following graphs.

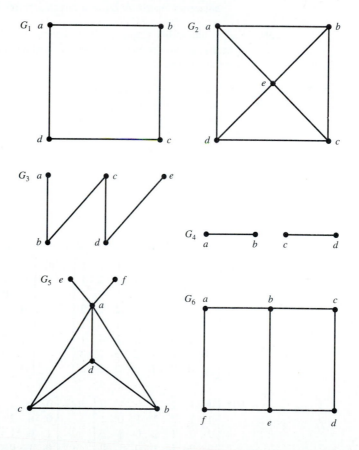

32. Compute the square of the adjacency matrix for the following graphs from Exercise 31.
(a) G_1 (b) G_2 (c) G_3 (d) G_4 (e) G_5 (f) G_6
Use your answer to tell how many paths of length 2 there are in each graph between vertex a and vertex d.

33. Compute the cube of the adjacency matrix for the following graphs from Exercise 31.
    (a) G_1 (b) G_2 (c) G_3 (d) G_4
    (Note: You may want to use a computer program to do this computation.) Use your answer here along with that in Exercise 32 to tell if all vertices in the graph are joined by a path of length ≤ 3.

34. Use your calculation in Exercise 32 to show that G_1 and G_2 are connected.

35. In a directed graph, each edge is directed from the first node to the second node in the edge pair. For example, edge (a, d) goes from a to d. In the adjacency matrix of a directed graph, edge (a, d) causes the entry in column a and row d to be $+1$ and the entry in row a and column d to remain 0. Direct the edges in the graphs G_1, G_2 in Exercise 31 from the earlier node to the later node according to the alphabetical order of the nodes; for example, edge (a, c) goes from a to c.
    (a) Write the adjacency matrix for directed graphs G_1 and G_2 and compute the square of each.
    (b) What interpretation, if any, can be given to entry (i, j) in the square $\mathbf{A}^2(D)$ of the adjacency matrix $\mathbf{A}(D)$ of a directed graph D?

36. For an undirected graph G, show that the result of the matrix–vector product $\mathbf{A}(G)\mathbf{1}$ (where $\mathbf{1}$ is a vector of 1's) is a vector in which the ith position tells how many nodes are adjacent to node i.

37. (a) Explain in words why, if entry (i, j) in $\mathbf{A}^2(G)$ is the number of paths of length 2 between nodes i and j in graph G, entry (i, j) in $\mathbf{A}^3(G)$ is the number of paths of length 3 between nodes i and j in G.
    (b) (Advanced) Extend the argument in part (a) by induction to show that entry (i, j) in $\mathbf{A}^k(G)$ is the number of paths of length k between nodes i and j in G.

38. (a) Suppose that we redefined the adjacency matrix $\mathbf{A}(G)$ so that the diagonal entries (i, i) were all 1. Now what is the interpretation of entry (i, j) being positive in $\mathbf{A}^2(G)$? In $\mathbf{A}^k(G)$?
    (b) Compute $\mathbf{A}^2(G)$ for this redefined adjacency matrix for graph G_1 in Exercise 31.

39. Partition the following matrices into appropriate submatrices.

(a)
$$\begin{bmatrix} 2 & 2 & 2 & 2 & 1 & 1 & 1 & 1 \\ 2 & 2 & 2 & 2 & 1 & 1 & 1 & 1 \\ 1 & 1 & 1 & 1 & 2 & 2 & 2 & 2 \\ 1 & 1 & 1 & 1 & 2 & 2 & 2 & 2 \\ 2 & 2 & 2 & 2 & 1 & 1 & 1 & 1 \\ 2 & 2 & 2 & 2 & 1 & 1 & 1 & 1 \\ 1 & 1 & 1 & 1 & 2 & 2 & 2 & 2 \\ 1 & 1 & 1 & 1 & 2 & 2 & 2 & 2 \end{bmatrix}$$

(b)
$$\begin{bmatrix} 1 & 0 & 1 & 0 & 0 & 1 & 0 & 1 \\ 0 & 1 & 0 & 1 & 1 & 0 & 1 & 0 \\ 1 & 0 & 1 & 0 & 0 & 1 & 0 & 1 \\ 0 & 1 & 0 & 1 & 1 & 0 & 1 & 0 \\ 2 & 0 & 2 & 0 & 0 & 1 & 0 & 1 \\ 0 & 2 & 0 & 2 & 1 & 0 & 1 & 0 \\ 2 & 0 & 2 & 0 & 0 & 1 & 0 & 1 \\ 0 & 2 & 0 & 2 & 1 & 0 & 1 & 0 \end{bmatrix}$$

(c)
$$\begin{bmatrix} 1 & 1 & 2 & 2 & 0 & 1 & 0 \\ 1 & 1 & 2 & 2 & 1 & 0 & 1 \\ 2 & 2 & 1 & 1 & 0 & 1 & 0 \\ 2 & 2 & 1 & 1 & 0 & 1 & 0 \\ 1 & 1 & 1 & 1 & 1 & 0 & 1 \\ 1 & 1 & 1 & 1 & 0 & 1 & 0 \end{bmatrix}$$

(d)
$$\begin{bmatrix} 0 & 0 & 2 & 2 & 3 & 0 & 3 & 0 \\ 0 & 0 & 2 & 2 & 0 & 3 & 0 & 3 \\ 1 & 0 & 1 & 0 & 3 & 0 & 3 & 0 \\ 0 & 1 & 0 & 1 & 0 & 3 & 0 & 3 \\ 1 & 0 & 1 & 0 & 0 & 0 & 0 & 0 \\ 0 & 1 & 0 & 1 & 0 & 0 & 0 & 0 \end{bmatrix}$$

40. Write the adjacency matrices of the following graphs and define a partitioned form of the matrices.

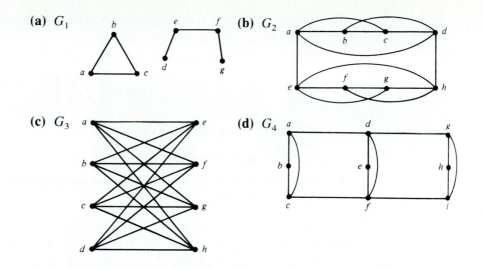

(a) G_1

(b) G_2

(c) G_3

(d) G_4

41. Consider the following Markov chain model. In the maze below, a person in any given room has equal chances of leaving by any door out of the room (but never remains in a room). Write the Markov transition matrix for this maze. Write the matrix in partitioned form.

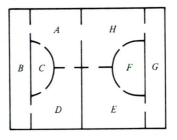

42. Determine the square of the matrix in part (b) of Exercise 39.

43. Determine $\mathbf{AA}^T$ for the matrix in part (a) of Exercise 39.

44. **(a)** Partition the matrix

$$\mathbf{A} = \begin{bmatrix} 1 & 1 & 0 & 0 & 0 & 0 & 0 & 0 & 0 \\ 1 & 0 & 1 & 0 & 0 & 0 & 0 & 0 & 0 \\ 0 & 1 & 1 & 0 & 0 & 0 & 0 & 0 & 0 \\ 0 & 0 & 0 & 2 & 2 & 0 & 0 & 0 & 0 \\ 0 & 0 & 0 & 2 & 0 & 2 & 0 & 0 & 0 \\ 0 & 0 & 0 & 0 & 2 & 2 & 0 & 0 & 0 \\ 0 & 0 & 0 & 0 & 0 & 0 & 3 & 3 & 0 \\ 0 & 0 & 0 & 0 & 0 & 0 & 3 & 0 & 3 \\ 0 & 0 & 0 & 0 & 0 & 0 & 0 & 3 & 3 \end{bmatrix}$$

in terms of the matrix $\mathbf{B} = \begin{bmatrix} 1 & 1 & 0 \\ 1 & 0 & 1 \\ 0 & 1 & 1 \end{bmatrix}$ and the 3 × 3 zero matrix $\mathbf{O}$.

(b) Write $\mathbf{A}^2$ and $\mathbf{A}^3$ in partitioned form in terms of $\mathbf{B}$ and $\mathbf{O}$.

(c) Write out $\mathbf{A}^2$ entry by entry.

(d) Compute $\mathbf{vA}^2$ where $\mathbf{v} = [1, -1, 1, 0, 1, 0, 1, -1, 1]$.

45. Compute the square of each adjacency matrix in Exercise 40 using the partitioned form of the matrix.

46. Suppose that the adjacency matrix $\mathbf{A}(G)$ of a graph G has the partitioned form

$$\mathbf{A}(G) = \begin{bmatrix} \mathbf{J}_3 & \mathbf{O} \\ \mathbf{O} & \mathbf{J}_3 \end{bmatrix}$$

where $\mathbf{J}_3$ is a 3×3 matrix with each entry 1.
(a) Draw G.
(b) Write out all the entries in $\mathbf{A}^3(G)$.

47. Determine the partitioned form of $\mathbf{A}^3(G)$ in Example 6 [in terms of $\mathbf{A}^*$ and $\mathbf{I}$, just as $\mathbf{A}^2(G)$ is expressed in (22)].

48. **(a)** Let $\mathbf{A} = \begin{bmatrix} 1 & 2 & 0 \\ -1 & 3 & 1 \\ 0 & 1 & 2 \end{bmatrix}$ and $\mathbf{B} = \begin{bmatrix} 1 & 2 \\ 3 & 4 \\ 1 & 0 \end{bmatrix}$ Partition $\mathbf{A}$ into three 3×1 submatrices and $\mathbf{B}$ into three 1×2 submatrices and use this partition to compute the product $\mathbf{AB}$. Compute the matrix product $\mathbf{AB}$ the normal way and compare the arithmetic in the two methods.

(b) Extend part (a) to show that in any matrix product $\mathbf{AB}$, the $m \times r$ matrix $\mathbf{A}$ can be partitioned into r submatrices each consisting of one of $\mathbf{A}$'s columns and the $r \times n$ matrix $\mathbf{B}$ partitioned into r submatrices each consisting of one of $\mathbf{B}$'s rows. Explain in words the effect of this partitioned product. [Generalize the comparison you made in part (a).]

49. Give an example to show that the product of two 2×2 symmetric matrices need not be symmetric.

3.2 LU Decomposition and Simple Matrices

In this section we return to Gaussian elimination and present two important matrix decompositions associated with the elimination process. We use the refinery example introduced in Chapter 1 to illustrate these new ideas. Initially, we repeat Gaussian elimination for that system of equations.

EXAMPLE 1.
Solving Refinery
Problem by
Gaussian
Elimination

Recall that there are three refineries that produce given amounts of petroleum products from a barrel of crude oil. The variable x_i is the amount of crude oil processed at the ith refinery. The right-side numbers are the demands for the different products.

$$\begin{array}{lll} \text{Heating oil:} & \text{(a) } 4x_1 + 2x_2 + 2x_3 = 600 \\ \text{Diesel oil:} & \text{(b) } 2x_1 + 5x_2 + 2x_3 = 800 \\ \text{Gasoline:} & \text{(c) } 1x_1 + 2.5x_2 + 5x_3 = 1000 \end{array} \qquad \begin{bmatrix} 4 & 2 & 2 & | & 600 \\ 2 & 5 & 2 & | & 800 \\ 1 & 2.5 & 5 & | & 1000 \end{bmatrix} \qquad (1)$$

We subtract multiples of the first row (equation) from the other rows (equations) to eliminate the coefficients of x_1 from the second and third rows. Our new system of equations is

$$
\begin{array}{lll}
\text{(a)} & 4x_1 + 2x_2 + 2x_3 = 600 \\
\text{(b')} = \text{(b)} - \tfrac{1}{2}\text{(a)} & 4x_2 + 1x_3 = 500 \\
\text{(c')} = \text{(c)} - \tfrac{1}{4}\text{(a)} & 2x_2 + 4.5x_3 = 850
\end{array}
\qquad
\begin{bmatrix}
4 & 2 & 2 & \Big| & 600 \\
0 & 4 & 1 & \Big| & 500 \\
0 & 2 & 4.5 & \Big| & 850
\end{bmatrix}
\qquad (2)
$$

Next we perform the same elimination step in the second row. The goal is to reduce the original coefficient matrix to an upper triangular matrix. Our final system of equations is

$$
\begin{array}{lll}
\text{(a)} & 4x_1 + 2x_2 + 2x_3 = 600 \\
\text{(b')} & 4x_2 + 1x_3 = 500 \\
\text{(c'')} = \text{(c')} - \tfrac{1}{2}\text{(b')} & 4x_3 = 600
\end{array}
\qquad
\begin{bmatrix}
4 & 2 & 2 & \Big| & 600 \\
0 & 4 & 1 & \Big| & 500 \\
0 & 0 & 4 & \Big| & 600
\end{bmatrix}
\qquad (3)
$$

Next we use substitution to solve (3). That is, from (3)(c''), we see that $x_3 = 600/4 = 150$. Substituting this value for x_3 in (b') we can solve for x_2, and then substituting the values of x_2 and x_3 in (a) we solve for x_1 (details of the back substitution were given in Example 2 of Section 1.4), obtaining $x_1 = 31\tfrac{1}{4}$, $x_2 = 87\tfrac{1}{2}$, $x_3 = 150$. ■

We now form two matrices from the elimination process. If $\mathbf{A}$ is the initial $m \times n$ matrix as in (1), let $\mathbf{U}$ denote the upper triangular $m \times n$ matrix of coefficients in the final system (3), and let $\mathbf{L}$ be the $m \times m$ matrix of multipliers l_{ij} telling how many times equation j is subtracted from equation i. For reasons to be explained shortly, we set $l_{ii} = 1$.

EXAMPLE 2.
L and U Matrices for Re-solving Refinery Equations

The final system (3) has coefficient matrix

$$
\mathbf{U} = \begin{bmatrix}
4 & 2 & 2 \\
0 & 4 & 1 \\
0 & 0 & 4
\end{bmatrix}
\qquad (4)
$$

Looking at (2), we see that x_1 was eliminated from equations (b) and (c) by subtracting $\tfrac{1}{2}$ times (a) from (b) and $\tfrac{1}{4}$ times (a) from (c). Thus $l_{21} = \tfrac{1}{2}$ and $l_{31} = \tfrac{1}{4}$. Next we eliminated x_2 from the last equation by subtracting $\tfrac{1}{2}$ times (b') from (c'). Thus $l_{32} = \tfrac{1}{2}$. Putting 1's on the main diagonal, we have

$$
\mathbf{L} = \begin{bmatrix}
1 & 0 & 0 \\
\tfrac{1}{2} & 1 & 0 \\
\tfrac{1}{4} & \tfrac{1}{2} & 1
\end{bmatrix}
\qquad (5)
$$

To solve the refinery problem (1) for another right-side vector $\mathbf{b}^*$, simply perform the elimination steps on $\mathbf{b}^*$ specified by $\mathbf{L}$ to get the final right-side vector $\mathbf{b}^\circ$ and then solve the reduced system $\mathbf{Ux} = \mathbf{b}^\circ$ by back substitution.

For example, suppose that $\mathbf{b}^* = \begin{bmatrix} 400 \\ 600 \\ 700 \end{bmatrix}$. Then upon repeating the elimination steps (using $\mathbf{L}$) on the new right sides, we have

$$
\begin{aligned}
&\text{(a)} & &= 400 \\
&\text{(b)} & &= 600 \\
&\text{(c)} & &= 700 \\[6pt]
&\text{(a)} & &= 400 \\
&\text{(b')} = \text{(b)} - \tfrac{1}{2}\text{(a)} & &= 400 \\
&\text{(c')} = \text{(c)} - \tfrac{1}{4}\text{(a)} & &= 600 \\[6pt]
&\text{(a)} & &= 400 \\
&\text{(b')} & &= 400 \\
&\text{(c'')} = \text{(c')} - \tfrac{1}{2}\text{(b')} & &= 400
\end{aligned}
$$

The new reduced system (using $\mathbf{U}$) is

$$
\begin{aligned}
&\text{(a)} & 4x_1 + 2x_2 + 2x_3 &= 400 \\
&\text{(b')} & 4x_2 + 1x_3 &= 400 \\
&\text{(c'')} & 4x_3 &= 400
\end{aligned} \tag{6}
$$

Using back substitution, we find that

$$
x_3 = \frac{400}{4} = 100
$$

then

$$
x_2 = \frac{400 - 1(100)}{4} = 75
$$

and finally,

$$
x_1 = \frac{400 - 2(75) - 2(100)}{4} = \frac{50}{4} = 12\tfrac{1}{2}
$$

▪

EXAMPLE 3.
Gaussian
Elimination and L
and U

Consider the following augmented coefficient matrix to which we apply Gaussian elimination.

$$
\begin{aligned}
&\text{(a)} \\
&\text{(b)} \\
&\text{(c)}
\end{aligned}
\left[
\begin{array}{rrrr|r}
-2 & 2 & 0 & 4 & 100 \\
1 & 5 & 2 & 2 & 100 \\
6 & 3 & 3 & -6 & -75
\end{array}
\right] \tag{7}
$$

First we use multiples of equation (a) to eliminate the x_1 term from equations (b) and (c).

$$
\begin{array}{l}
\text{(a)} \\
\text{(b')} = \text{(b)} + \tfrac{1}{2}\text{(a)} \\
\text{(c')} = \text{(c)} + 3\text{(a)}
\end{array}
\left[
\begin{array}{cccc|c}
-2 & 2 & 0 & 4 & 100 \\
0 & 6 & 2 & 4 & 150 \\
0 & 9 & 3 & 6 & 225
\end{array}
\right]
\tag{8}
$$

Next we use multiples of equation (b') to eliminate the x_2 term from equation (c').

$$
\begin{array}{l}
\text{(a)} \\
\text{(b')} \\
\text{(c'')} = \text{(c')} - \tfrac{3}{2}\text{(b')}
\end{array}
\left[
\begin{array}{cccc|c}
-2 & 2 & 0 & 4 & 100 \\
0 & 6 & 2 & 3 & 150 \\
0 & 0 & 0 & 0 & 0
\end{array}
\right]
\tag{9}
$$

There are multiple solutions possible now by choosing any value we want for x_4 and x_3, and then using back substitution to obtain x_2 and x_1. **L** and **U** are seen to be

$$
\mathbf{L} =
\begin{bmatrix}
1 & 0 & 0 \\
-\tfrac{1}{2} & 1 & 0 \\
-3 & \tfrac{3}{2} & 1
\end{bmatrix}
\qquad
\mathbf{U} =
\begin{bmatrix}
-2 & 2 & 0 & 4 \\
0 & 6 & 2 & 3 \\
0 & 0 & 0 & 0
\end{bmatrix}
\tag{10}
$$

If we had started with a right-side vector of $\begin{bmatrix} 40 \\ 60 \\ 90 \end{bmatrix}$ instead of $\begin{bmatrix} 100 \\ 100 \\ -75 \end{bmatrix}$, then

using the first column of **L**, the right side in (8) would be $\begin{bmatrix} 40 \\ 60 + \tfrac{1}{2}\cdot 40 \\ 90 + 3\cdot 40 \end{bmatrix} = \begin{bmatrix} 40 \\ 80 \\ 210 \end{bmatrix}$.

And from the second column of **L**, we see that the right side of (9) would be

$\begin{bmatrix} 40 \\ 80 \\ 210 - \tfrac{3}{8}\cdot 80 \end{bmatrix} = \begin{bmatrix} 40 \\ 80 \\ 90 \end{bmatrix}$. Again, we solve by back substitution after choosing

any desired values for x_4 and x_3. ■

Note that, ignoring the 1's on **L**'s main diagonal, the data in **L** and **U** can be stored together in one square matrix.

We now state a remarkable theorem.

Theorem 1 (LU Decomposition)

Let **A** be a $m \times n$ matrix. Let **U** denote the upper triangular $m \times n$ matrix of coefficients in the reduced matrix after Gaussian elimination. Let **L** be the lower triangular $m \times m$ matrix of multipliers l_{ij} telling how many times equation j is subtracted from equation i, with $l_{ii} = 1$. Then

$$
\mathbf{A} = \mathbf{LU}
\tag{11}
$$

Note: We are assuming that **A**'s rows are arranged so that no 0's occur on the main diagonal during elimination. If the order of the rows of **A** needs to be permuted, we premultiply **A** by a permutation matrix **P** (see Example 5 of Section 1.3) and (11) becomes **PA** = **LU**.

Let us check (11) for **L** and **U** in Example 2. We want to multiply:

$$\mathbf{LU} = \begin{bmatrix} 1 & 0 & 0 \\ \frac{1}{2} & 1 & 0 \\ \frac{1}{4} & \frac{1}{2} & 1 \end{bmatrix} \begin{bmatrix} 4 & 2 & 2 \\ 0 & 4 & 1 \\ 0 & 0 & 4 \end{bmatrix} \tag{12}$$

Let us compute **LU** by the row definition of matrix multiplication, which says that the ith row in **LU** equals $\mathbf{l}_i'\mathbf{U}$ (where $\mathbf{l}_i'$ is the ith row of **L**). The first row of **LU** in (12) is

$$\mathbf{l}_1'\mathbf{U} = \begin{bmatrix} 1 & 0 & 0 \end{bmatrix} \begin{bmatrix} 4 & 2 & 2 \\ 0 & 4 & 1 \\ 0 & 0 & 4 \end{bmatrix} = \begin{bmatrix} 4 & 2 & 2 \end{bmatrix} \tag{13a}$$

The second row is

$$\mathbf{l}_2'\mathbf{U} = \begin{bmatrix} \frac{1}{2} & 1 & 0 \end{bmatrix} \begin{bmatrix} 4 & 2 & 2 \\ 0 & 4 & 1 \\ 0 & 0 & 4 \end{bmatrix} = \tfrac{1}{2}\begin{bmatrix} 4 & 2 & 2 \end{bmatrix} + 1\begin{bmatrix} 0 & 4 & 1 \end{bmatrix} = \begin{bmatrix} 2 & 5 & 2 \end{bmatrix} \tag{13b}$$

The third row is

$$\mathbf{l}_3'\mathbf{U} = \begin{bmatrix} \frac{1}{4} & \frac{1}{2} & 1 \end{bmatrix} \begin{bmatrix} 4 & 2 & 2 \\ 0 & 4 & 1 \\ 0 & 0 & 4 \end{bmatrix} = \tfrac{1}{4}\begin{bmatrix} 4 & 2 & 2 \end{bmatrix} + \tfrac{1}{2}\begin{bmatrix} 0 & 4 & 1 \end{bmatrix} + 1\begin{bmatrix} 0 & 0 & 4 \end{bmatrix} \\ = \begin{bmatrix} 1 & 2.5 & 5 \end{bmatrix} \tag{13c}$$

Putting the three rows of **LU** computed in (13a)–(13c) together, we have **A**.

Theorem 1 can be proved by generalizing the computation done in (13a)–(13c). In the elimination process, we are forming new equations as linear combinations of the original equations. Conversely, the original equations are linear combinations of the final reduced equations. The latter property is exactly what the computations in (13a)–(13c) illustrate. For example, (13b) shows that $\mathbf{a}_2'$, the second row of **A**, is the following linear combination of **U**'s rows: $\mathbf{a}_2' = \tfrac{1}{2}\mathbf{u}_1' + \mathbf{u}_2'$.

We shall present a different proof of the **LU** decomposition later in this section. A third way to approach the **LU** decomposition involves elementary matrices. Recall from the beginning of Section 1.4 that each elementary row operation used in elimination has an elementary matrix **E** such that premultiplying **A** by **E** "performs" the row operation. If $\mathbf{E}_1$, $\mathbf{E}_2$, $\mathbf{E}_3$ are elementary row operations performed to get equations (2)(b'), (2)(c'), and (3)(c''), respectively, in Example 1, in reducing the refinery matrix **A** to **U**, then $\mathbf{E}_3\mathbf{E}_2\mathbf{E}_1\mathbf{A} = \mathbf{U}$, implying that $\mathbf{A} = (\mathbf{E}_3\mathbf{E}_2\mathbf{E}_1)^{-1}\mathbf{U}$. Indeed, one can show that $\mathbf{L} = (\mathbf{E}_3\mathbf{E}_2\mathbf{E}_1)^{-1}$. See Exercise 17 for further details.

The **LU** decomposition of a square matrix has many important uses. One consequence is a simple proof of the Gaussian elimination–based formula for the determinant, originally proved in Section 1.6.

Theorem 2. For any square matrix $\mathbf{A}$,

$$\det(\mathbf{A}) = u_{11} \cdot u_{22} \cdots \cdots u_{nn} \tag{14}$$

That is, $\det(\mathbf{A})$ equals the product of the main diagonal entries in $\mathbf{U}$, where $\mathbf{U}$ is the reduced-system matrix in the decomposition $\mathbf{A} = \mathbf{LU}$.

Proof: Since $\mathbf{A} = \mathbf{LU}$, then $\det(\mathbf{A}) = \det(\mathbf{L})\det(\mathbf{U})$, by the product rule for determinants (Theorem 5 of Section 1.6). Recall that the determinant of a lower or upper triangular matrix (such as $\mathbf{L}$ or $\mathbf{U}$) is just the product of the main diagonal entries (Theorem 2 of Section 1.6). Then $\det(\mathbf{L}) = 1$. So $\det(\mathbf{A}) = \det(\mathbf{U}) = $ product of $\mathbf{U}$'s main diagonal entries. ■

In Theorem 2, if the rows were interchanged during elimination (to avoid zero pivots), each interchange changes the sign of $\det(\mathbf{A})$ (see Theorem 4 in Section 1.6).

Using (14), we can determine the value of the determinant of the coefficient matrix (1) for the refinery problem. The three diagonal entries in $\mathbf{U}$ for this matrix are $u_{11} = u_{22} = u_{33} = 4$ [see (4)], and thus the determinant equals $u_{11}u_{22}u_{33} = 4 \cdot 4 \cdot 4 = 64$.

Simple Matrices

A **simple matrix** $\mathbf{K}$ is an $m \times n$ matrix formed by matrix multiplication $\mathbf{c} * \mathbf{d}$ of a *column m*-vector $\mathbf{c}$ times *row n*-vector $\mathbf{d}$ in which $\mathbf{c}$ is treated as an $m \times 1$ matrix and $\mathbf{d}$ as a $1 \times n$ matrix (we use the new symbol $*$ to emphasize that this is *not* a scalar product nor a cross product of vectors). Thus $\mathbf{K}$ has the form

$$\mathbf{K} = \mathbf{c} * \mathbf{d} = \begin{bmatrix} c_1 \\ c_2 \\ \vdots \\ c_m \end{bmatrix} [d_1, d_2, \ldots, d_n] = \begin{bmatrix} c_1d_1 & c_1d_2 & \ldots & c_1d_n \\ c_2d_1 & c_2d_2 & \ldots & c_2d_n \\ \ldots & \ldots & \ldots & \ldots \\ \ldots & \ldots & \ldots & \ldots \\ c_md_1 & c_md_2 & \ldots & c_md_n \end{bmatrix} \tag{15}$$

We also refer to this product $\mathbf{c} * \mathbf{d}$ as an **outer product** of vectors. All rows in a simple matrix are multiples of each other, and similarly for all columns. If we perform Gaussian elimination on a simple matrix, the first step of the elimination process will convert all other rows to rows of 0's. (Verification is left as an exercise.)

As an example, the following outer product of vectors yields the simple matrix:

$$\begin{bmatrix} 3 \\ -1 \end{bmatrix} * [1, 2, 3] = \begin{bmatrix} 3 \cdot 1 & 3 \cdot 2 & 3 \cdot 3 \\ -1 \cdot 1 & -1 \cdot 2 & -1 \cdot 3 \end{bmatrix} = \begin{bmatrix} 3 & 6 & 9 \\ -1 & -2 & -3 \end{bmatrix}$$

Using simple matrices, we can give a new way to interpret matrix multiplication.

Theorem 3. Let $\mathbf{C}$ be an $m \times r$ matrix with columns $\mathbf{c}_j$ and $\mathbf{D}$ be an $r \times n$ matrix with rows $\mathbf{d}'_i$. Then the matrix product $\mathbf{CD}$ can be decomposed into a sum of the simple matrices $\mathbf{c}_j * \mathbf{d}'_i$ of the column vectors of $\mathbf{C}$ times the row vectors of $\mathbf{D}$.

$$\mathbf{CD} = \mathbf{c}_1 * \mathbf{d}'_1 + \mathbf{c}_2 * \mathbf{d}'_2 + \cdots + \mathbf{c}_r * \mathbf{d}'_r \tag{16}$$

Proof: We verify (16) by performing matrix multiplication on partitioned matrices. We partition $\mathbf{C}$ into r $m \times 1$ matrices, the $\mathbf{c}_i$, and partition $\mathbf{D}$ into r $1 \times n$ matrices, the $\mathbf{d}'_j$. Thus

$$\mathbf{CD} = [\mathbf{c}_1 \quad \mathbf{c}_2 \quad \cdots \quad \mathbf{c}_r] \begin{bmatrix} \mathbf{d}_1 \\ \mathbf{d}_2 \\ \cdots \\ \cdots \\ \mathbf{d}_r \end{bmatrix} = [\mathbf{c}_1 * \mathbf{d}'_1 + \mathbf{c}_2 * \mathbf{d}'_2 + \cdots + \mathbf{c} * \mathbf{d}'_r] \quad ■$$

We illustrate this theorem with the following product of two matrices.

EXAMPLE 4. Decomposition of Matrix Multiplication into a Sum of Simple Matrices

Let

$$\mathbf{C} = \begin{bmatrix} 1 & 2 & 3 \\ 4 & 5 & 6 \end{bmatrix} \quad \text{and} \quad \mathbf{D} = \begin{bmatrix} 11 & 12 & 13 \\ 14 & 15 & 16 \\ 17 & 18 & 19 \end{bmatrix}$$

Standard matrix multiplication of $\mathbf{CD}$ yields (17a), which we rewrite as follows.

$$\mathbf{CD} = \begin{bmatrix} 1 \cdot 11 + 2 \cdot 14 + 3 \cdot 17 & 1 \cdot 12 + 2 \cdot 15 + 3 \cdot 18 & 1 \cdot 13 + 2 \cdot 16 + 3 \cdot 19 \\ 4 \cdot 11 + 5 \cdot 14 + 6 \cdot 17 & 4 \cdot 12 + 5 \cdot 15 + 6 \cdot 18 & 4 \cdot 13 + 5 \cdot 16 + 6 \cdot 19 \end{bmatrix} \quad (17a)$$

$$= \begin{bmatrix} 1 \cdot 11 & 1 \cdot 12 & 1 \cdot 13 \\ 4 \cdot 11 & 4 \cdot 12 & 4 \cdot 13 \end{bmatrix} + \begin{bmatrix} 2 \cdot 14 & 2 \cdot 15 & 2 \cdot 16 \\ 5 \cdot 14 & 5 \cdot 15 & 5 \cdot 16 \end{bmatrix} + \begin{bmatrix} 3 \cdot 17 & 3 \cdot 18 & 3 \cdot 19 \\ 6 \cdot 17 & 6 \cdot 18 & 6 \cdot 19 \end{bmatrix} \quad (17b)$$

$$= \begin{bmatrix} 1 \\ 4 \end{bmatrix} * [11, 12, 13] + \begin{bmatrix} 2 \\ 5 \end{bmatrix} * [14, 15, 16] + \begin{bmatrix} 3 \\ 6 \end{bmatrix} * [17, 18, 19] \quad (17c)$$

$$= \qquad \mathbf{c}_1 * \mathbf{d}'_1 \qquad + \qquad \mathbf{c}_2 * \mathbf{d}'_2 \qquad + \qquad \mathbf{c}_3 * \mathbf{d}'_3$$

The first simple matrix $\mathbf{c}_1 * \mathbf{d}'_1$ contains the first term of each scalar product in the entries of $\mathbf{CD}$ in (17a); and similarly for the second and third simple matrices. ■

We will now show how an $m \times n$ matrix $\mathbf{A}$ can be decomposed into a sum of simple matrices in a manner that mimics the action of Gaussian elimination. Instead of summing certain simple matrices together to get $\mathbf{A}$, we successively subtract the simple matrices from $\mathbf{A}$ to eliminate all entries in $\mathbf{A}$ (to reduce $\mathbf{A}$ to the zero matrix).

Our strategy will be to form a simple matrix $\mathbf{S}_1 = \mathbf{l}_1 * \mathbf{u}'_1$ whose first column equals $\mathbf{a}_1$ (the first column of $\mathbf{A}$) and whose first row equals $\mathbf{a}'_1$ (the first row of $\mathbf{A}$). Then $\mathbf{A} - \mathbf{S}_1$ will have 0's in its first row and column. We will form $\mathbf{S}_2$ to remove the second row and column of $\mathbf{A}$; possibly we will zero out additional rows and columns in the process. We continue similarly with $\mathbf{S}_3$, and so on, until all rows and columns of $\mathbf{A}$ are reduced to 0's. Let $\mathbf{u}'_1 = \mathbf{a}'_1$ and let

$$\mathbf{l}_1 = \left(\frac{1}{a_{11}} \right) \mathbf{a}_1 = \begin{bmatrix} \dfrac{a_{11}}{a_{11}} \\ \dfrac{a_{21}}{a_{11}} \\ \dfrac{a_{31}}{a_{11}} \end{bmatrix}$$

(actually, the first entry of l_1 is 1, since $a_{11}/a_{11} = 1$). Then

$$l_1 * u_1' = \begin{bmatrix} 1 \\ \dfrac{a_{21}}{a_{11}} \\ \dfrac{a_{31}}{a_{11}} \end{bmatrix} * [a_{11}, a_{12}, a_{13}] = \begin{bmatrix} a_{11} & a_{12} & a_{13} \\ a_{21} & \cdots & \cdots \\ a_{31} & \cdots & \cdots \end{bmatrix} \tag{18}$$

(Note that a_{11}, and other diagonal entries, used in elimination are assumed to be nonzero.)

So $A - S_1 (= A - l_1 * u_1')$ deletes the first row and column of A. For the refinery

matrix $A = \begin{bmatrix} 4 & 2 & 2 \\ 2 & 5 & 2 \\ 1 & 2.5 & 5 \end{bmatrix}$, we have

$$l_1 = \tfrac{1}{4} a_1 = \begin{bmatrix} 1 \\ \tfrac{1}{2} \\ \tfrac{1}{4} \end{bmatrix} \quad \text{and} \quad u_1' = a_1' = [4, 2, 2]$$

Then

$$A - S_1 = A - l_1 * u_1' = \begin{bmatrix} 4 & 2 & 2 \\ 2 & 5 & 2 \\ 1 & 2.5 & 5 \end{bmatrix} - \begin{bmatrix} 1 \\ \tfrac{1}{2} \\ \tfrac{1}{4} \end{bmatrix} \times [4, 2, 2]$$

$$= \begin{bmatrix} 4 & 2 & 2 \\ 2 & 5 & 2 \\ 1 & 2.5 & 5 \end{bmatrix} - \begin{bmatrix} 4 & 2 & 2 \\ 2 & 1 & 1 \\ 1 & 0.5 & 0.5 \end{bmatrix} = \begin{bmatrix} 0 & 0 & 0 \\ 0 & 4 & 1 \\ 0 & 2 & 4.5 \end{bmatrix} \tag{19}$$

Observe that l_1 is the vector of elimination multipliers (ignoring the first entry of 1) used to eliminate x_1 from other equations in Gaussian elimination (see Example 2). Indeed, l_1 is exactly the first column in the matrix L in the LU decomposition. Similarly, u_1' is the first row of U (which equals the first row of A). When we subtract from A the simple matrix $l_1 * u_1'$, we obtain the coefficient matrix (ignoring the first row of 0's) for the remaining $n - 1$ equations of $Ax = b$ after x_1 is eliminated in the first stage of Gaussian elimination. In the case of the refinery problem, compare (19) with (2).

Continuing this process, we let $S_2 = l_2 * u_2'$. Subtracting S_2 from $A - S_1$ will have the effect of the second stage of elimination—zeroing out the second column of A—plus zeroing out the second row of A. For the refinery problem, we have from L and U in Example 2, $l_2 = \begin{bmatrix} 0 \\ 1 \\ \tfrac{1}{2} \end{bmatrix}$ and $u_2' = [0, 4, 1]$. Then

$$A - S_1 - S_2 = (A - S_1) - l_2 * u_2' = \begin{bmatrix} 0 & 0 & 0 \\ 0 & 4 & 1 \\ 0 & 2 & 4.5 \end{bmatrix} - \begin{bmatrix} 0 \\ 1 \\ \tfrac{1}{2} \end{bmatrix} \times [0, 4, 1]$$

$$= \begin{bmatrix} 0 & 0 & 0 \\ 0 & 4 & 1 \\ 0 & 2 & 4.5 \end{bmatrix} - \begin{bmatrix} 0 & 0 & 0 \\ 0 & 4 & 1 \\ 0 & 2 & 0.5 \end{bmatrix} = \begin{bmatrix} 0 & 0 & 0 \\ 0 & 0 & 0 \\ 0 & 0 & 4 \end{bmatrix} \tag{20}$$

The other S_i are defined and perform similarly. Thus in each round of Gaussian elimination, if we zero out the ith row after we eliminate nonzero entries in the ith column, *we can use simple matrices to represent Gaussian elimination.*

Theorem 4. Gaussian elimination can be viewed as a decomposition of **A** into a sum of simple matrices:

$$\mathbf{A} = \mathbf{l}_1 * \mathbf{u}_1' + \mathbf{l}_2 * \mathbf{u}_2' + \cdots + \mathbf{l}_k * \mathbf{u}_k' \tag{21}$$

where $\mathbf{l}_i$ is the ith column of **L** (the matrix of elimination multipliers), $\mathbf{u}_i'$ is the ith row of **U** (the reduced matrix in Gaussian elimination), and k is the number of nonzero rows of **U** (or equivalently, the number of pivots performed during elimination).

From Theorem 3 it follows that if **A** equals the sum of simple matrices $\mathbf{l}_i * \mathbf{u}_i'$, then **A = LU**—we have proved Theorem 1, the validity of the **LU** decomposition.

**EXAMPLE 5.
Refinery Matrix
Expressed As a
Sum of Simple
Matrices**

Earlier we gave the **LU** decomposition of our refinery matrix

$$\mathbf{A} = \mathbf{LU}: \begin{bmatrix} 4 & 2 & 2 \\ 2 & 5 & 2 \\ 1 & 2.5 & 5 \end{bmatrix} = \begin{bmatrix} 1 & 0 & 0 \\ \frac{1}{2} & 1 & 0 \\ \frac{1}{4} & \frac{1}{2} & 1 \end{bmatrix} \begin{bmatrix} 4 & 2 & 2 \\ 0 & 4 & 1 \\ 0 & 0 & 4 \end{bmatrix}$$

Instead of subtracting simple matrices from **A**, we use (21) to write **A** as

$$\mathbf{A} = \quad \mathbf{l}_1 * \mathbf{u}_1' \quad + \quad \mathbf{l}_2 * \mathbf{u}_2' \quad + \quad \mathbf{l}_3 * \mathbf{u}_3'$$

$$= \begin{bmatrix} 1 \\ \frac{1}{2} \\ \frac{1}{4} \end{bmatrix} * [4, 2, 2] + \begin{bmatrix} 1 \\ 1 \\ \frac{1}{2} \end{bmatrix} * [0, 4, 1] + \begin{bmatrix} 0 \\ 0 \\ 1 \end{bmatrix} * [0, 0, 4]$$

$$= \begin{bmatrix} 4 & 2 & 2 \\ 2 & 1 & 1 \\ 1 & 0.5 & 0.5 \end{bmatrix} + \begin{bmatrix} 0 & 0 & 0 \\ 0 & 4 & 1 \\ 0 & 2 & 0.5 \end{bmatrix} + \begin{bmatrix} 0 & 0 & 0 \\ 0 & 0 & 0 \\ 0 & 0 & 4 \end{bmatrix} \tag{22}$$

The reader should check that this set of three simple matrices adds up to **A**. ■

**EXAMPLE 6.
A Nonsquare
Matrix Expressed
As a Sum of
Simple Matrices**

In Example 3 we obtained **L** and **U** for the 3×4 matrix

$$\mathbf{A} = \begin{bmatrix} -2 & 2 & 0 & 4 \\ 1 & 5 & 2 & 2 \\ 6 & 3 & 3 & -6 \end{bmatrix} \quad \mathbf{L} = \begin{bmatrix} 1 & 0 & 0 \\ -\frac{1}{2} & 1 & 0 \\ -3 & \frac{3}{2} & 1 \end{bmatrix} \quad \mathbf{U} = \begin{bmatrix} -2 & 2 & 0 & 4 \\ 0 & 6 & 2 & 4 \\ 0 & 0 & 0 & 0 \end{bmatrix} \tag{23}$$

By (21), we can write $\mathbf{A}$ as (we omit the third simple matrix since $\mathbf{u}_3' = \mathbf{0}$)

$$\mathbf{A} = \mathbf{l}_1 * \mathbf{u}_1' + \mathbf{l}_2 * \mathbf{u}_2'$$

$$= \begin{bmatrix} 1 \\ -\frac{1}{2} \\ -3 \end{bmatrix} * [-2,\ 2,\ 0,\ 4] \quad + \begin{bmatrix} 0 \\ 1 \\ \frac{3}{2} \end{bmatrix} * [0,\ 6,\ 2,\ 4]$$

$$= \begin{bmatrix} -2 & 2 & 0 & 4 \\ 1 & -1 & 0 & -2 \\ 6 & -6 & 0 & -12 \end{bmatrix} + \begin{bmatrix} 0 & 0 & 0 & 0 \\ 0 & 6 & 2 & 4 \\ 0 & 9 & 3 & 6 \end{bmatrix} \tag{24}$$

The reader should check that this set of two simple matrices adds up to $\mathbf{A}$. ■

Section 3.2 Exercises

1. Find the **LU** decomposition for the coefficient matrix in each of the following systems of equations.

   (a) $\begin{aligned} 2x_1 - 3x_2 + 2x_3 &= 0 \\ x_1 - x_2 + x_3 &= 7 \\ -x_1 + 5x_2 + 4x_3 &= 4 \end{aligned}$ (b) $\begin{aligned} -x_1 - x_2 + x_3 &= 2 \\ x_1 - 2x_2 + 3x_3 &= -4 \\ 2x_1 + 2x_2 - 4x_3 &= 5 \end{aligned}$

   (c) $\begin{aligned} -x_1 - 3x_2 + 2x_3 &= -2 \\ 2x_1 + x_2 + 3x_3 &= \tfrac{9}{2} \\ 5x_1 + 4x_2 + 6x_3 &= 12 \end{aligned}$ (d) $\begin{aligned} 2x_1 + 4x_2 - 2x_3 &= 4 \\ x_1 - 2x_2 - 4x_3 &= -1 \\ -2x_1 - x_2 - 3x_3 &= -4 \end{aligned}$

   (e) $\begin{aligned} x_1 + x_2 + 4x_3 &= 4 \\ 2x_1 + x_2 + 3x_3 &= 5 \\ 5x_1 + 2x_2 + 5x_3 &= 11 \end{aligned}$ (f) $\begin{aligned} 2x_1 - 3x_2 - x_3 &= 2 \\ 3x_1 - 5x_2 - 2x_3 &= -1 \\ 9x_1 + 6x_2 + 4x_3 &= 1 \end{aligned}$

2. Find the **LU** decomposition for the following matrices.

   (a) $\begin{bmatrix} 1 & 2 & 3 \\ 3 & 4 & 5 \\ 7 & 8 & 9 \end{bmatrix}$ (b) $\begin{bmatrix} 2 & 1 & 5 \\ 0 & 3 & -2 \\ 6 & 0 & 3 \end{bmatrix}$ (c) $\begin{bmatrix} 1 & 4 & 6 \\ 0 & 1 & -3 \\ 4 & -2 & 3 \end{bmatrix}$ (d) $\begin{bmatrix} 2 & 3 & -1 \\ 2 & 0 & -3 \\ 1 & 3 & 0 \end{bmatrix}$

3. Multiply **L** times **U** to show that the product is **A** for each coefficient matrix **A** in Exercise 2.

4. Find the determinant of each matrix in Exercise 1 using Theorem 2.

5. Solve each system in Exercise 1 with the new right-hand-side vector $\begin{bmatrix} 10 \\ 5 \\ 10 \end{bmatrix}$ using the numbers in the **L** and **U** matrices you found in Exercise 1.

6. For the right-side vector $\mathbf{b} = \begin{bmatrix} 1 \\ 2 \\ 3 \end{bmatrix}$, solve the system of equations $\mathbf{Ax} = \mathbf{b}$ where instead of **A**, you are given the **LU** decomposition of **A**.

   (a) $\mathbf{L} = \begin{bmatrix} 1 & 0 & 0 \\ 1 & 1 & 0 \\ 2 & 3 & 1 \end{bmatrix}$, $\mathbf{U} = \begin{bmatrix} 2 & 1 & 1 \\ 0 & 3 & 2 \\ 0 & 0 & -2 \end{bmatrix}$ (b) $\mathbf{L} = \begin{bmatrix} 1 & 0 & 0 \\ -2 & 1 & 0 \\ 4 & -1 & 1 \end{bmatrix}$, $\mathbf{U} = \begin{bmatrix} 1 & -2 & 2 \\ 0 & 5 & 2 \\ 0 & 0 & 2 \end{bmatrix}$

(c) $L = \begin{bmatrix} 1 & 0 & 0 \\ 3 & 1 & 0 \\ -1 & 5 & 1 \end{bmatrix}$, $U = \begin{bmatrix} 5 & 0 & -2 \\ 0 & 1 & -3 \\ 0 & 0 & 4 \end{bmatrix}$

(d) $L = \begin{bmatrix} 1 & 0 & 0 \\ 0 & 1 & 0 \\ 3 & 2 & 1 \end{bmatrix}$, $U = \begin{bmatrix} -1 & 2 & -4 \\ 0 & -1 & 0 \\ 0 & 0 & 3 \end{bmatrix}$

7. Given the **LU** decomposition of an $n \times n$ matrix **A**, how many multiplications are required to compute **A** as the matrix product **LU** (allowing for known 0's in **L** and **U**)?

8. For an arbitrary 2×2 system of equations

$$ax + by = r$$
$$cx + dy = s$$

   (a) Determine the **LU** decomposition of the coefficient matrix **A**.
   (b) Verify that **L** times **U** equals **A**.

9. Use Theorem 2 to show that if one row of **A** is a multiple of another row, then $\det(A) = 0$.

10. Consider the following 3×3 matrix whose entries are functions. Find the **LU** decomposition of this matrix and find its determinant.

$$\begin{bmatrix} 3x & 6x^2 & e^x \\ x^2 & 3x^3 & xe^x \\ 6 & 3x & e^x/x \end{bmatrix}$$

11. Let $\mathbf{a} = [1, 2, 3]$, $\mathbf{b} = [2, 0]$, $\mathbf{c} = \begin{bmatrix} 3 \\ 1 \end{bmatrix}$, $\mathbf{d} = \begin{bmatrix} -1 \\ 2 \\ 1 \end{bmatrix}$. Compute the following simple matrices.
   (a) $\mathbf{c} * \mathbf{a}$ **(b)** $\mathbf{d} * \mathbf{a}$ **(c)** $\mathbf{c} * \mathbf{b}$ **(d)** $\mathbf{d} * \mathbf{b}$

12. Show that each of the following matrices is a simple matrix by giving the column and row vectors whose matrix product equals the matrix.
   (a) $\begin{bmatrix} 1 & 2 \\ 2 & 4 \end{bmatrix}$ **(b)** $\begin{bmatrix} 2 & -3 \\ -6 & 9 \end{bmatrix}$ **(c)** $\begin{bmatrix} 6 & -4 \\ -9 & 6 \end{bmatrix}$

   (d) $\begin{bmatrix} 12 & -6 & 9 \\ 8 & -4 & 6 \\ 4 & -2 & 3 \end{bmatrix}$ **(e)** $\begin{bmatrix} 4 & 8 & -6 \\ -2 & -4 & 3 \\ 6 & 12 & -9 \end{bmatrix}$ **(f)** $\begin{bmatrix} 6 & 3 & 0 \\ 2 & 1 & 0 \\ -4 & -2 & 0 \end{bmatrix}$

13. Write each of the following matrices as the sum of two simple matrices.
   (a) $\begin{bmatrix} 1 & 2 \\ 2 & 3 \end{bmatrix}$ **(b)** $\begin{bmatrix} 4 & 0 \\ 2 & 4 \end{bmatrix}$ **(c)** $\begin{bmatrix} 2 & -2 \\ 5 & 1 \end{bmatrix}$ **(d)** $\begin{bmatrix} 0 & 2 \\ 5 & 0 \end{bmatrix}$

14. Write each of the following matrices as the sum of as few simple matrices as possible.
   (a) $\begin{bmatrix} 1 & 2 & 3 \\ 3 & 4 & 5 \\ 7 & 8 & 9 \end{bmatrix}$ **(b)** $\begin{bmatrix} 2 & 1 & 5 \\ 0 & 3 & -2 \\ 6 & 0 & 3 \end{bmatrix}$ **(c)** $\begin{bmatrix} 1 & 2 & 9 \\ 0 & -1 & -3 \\ 4 & 6 & 6 \end{bmatrix}$ **(d)** $\begin{bmatrix} 2 & 3 & -1 \\ 2 & 0 & -3 \\ 1 & 3 & 0 \end{bmatrix}$

15. Write each of the following $m \times n$ matrices as the sum of as few simple matrices as possible.

(a) $\begin{bmatrix} 1 & 2 \\ 3 & 4 \\ 5 & 6 \end{bmatrix}$ (b) $\begin{bmatrix} 2 & 5 & 2 \\ 1 & 3 & 6 \end{bmatrix}$ (c) $\begin{bmatrix} 1 & 5 & 7 & 2 \\ 2 & 1 & 0 & 3 \end{bmatrix}$ (d) $\begin{bmatrix} 2 & -6 & 4 \\ -1 & 3 & -2 \end{bmatrix}$

16. Prove that if $\mathbf{A} = \mathbf{c} * \mathbf{d}$ is a simple matrix and we perform Gaussian (or Gauss–Jordan) elimination on $\mathbf{A}$, then in the first step we will make all other rows of $\mathbf{A}$ (except the first row) into rows of 0's.

17. (a) Construct elementary matrices $\mathbf{E}_1$, $\mathbf{E}_2$, $\mathbf{E}_3$ corresponding to the elementary row operations performed in Example 1 to get equations (2)(b′), (2)(c′), and (3)(c″), respectively.

(b) If $\mathbf{E}$ is a 3×3 matrix with 1's on the main diagonal and all other entries 0 except one off-diagonal entry (i, j), $i \neq j$, with value a, show that $\mathbf{E}^{-1}$ has the same form as $\mathbf{E}$ except that entry (i, j) is $-a$.

(c) Using part (b), find the inverses of the elementary matrices $\mathbf{E}_1$, $\mathbf{E}_2$, $\mathbf{E}_3$ found in part (a) and then verify that $(\mathbf{E}_3\mathbf{E}_2\mathbf{E}_1)^{-1}$ equals $\mathbf{L}$ [in (5)]. Recall that $(\mathbf{E}_3\mathbf{E}_2\mathbf{E}_1)^{-1} = \mathbf{E}_1^{-1}\mathbf{E}_2^{-1}\mathbf{E}_3^{-1}$.

(d) Show that if $\mathbf{E}_1$, $\mathbf{E}_2$, $\mathbf{E}_3$ are the elementary matrices corresponding to the elementary row operations to zero out entries $(2, 1)$, $(3, 1)$, and $(3, 2)$, respectively, during Gaussian elimination on an arbitrary 3×3 matrix $\mathbf{A}$, then $(\mathbf{E}_3\mathbf{E}_2\mathbf{E}_1)^{-1} = \mathbf{L}$, where $\mathbf{A} = \mathbf{LU}$. (You may assume that no row interchanges are needed.)

3.3 Eigenvectors and Eigenvalues

When $\mathbf{A}$ is a square matrix, multiplying a vector by $\mathbf{A}$ occasionally has exactly the same effect as multiplying it by a simple scalar λ. That is, for some vectors $\mathbf{u}$, $\mathbf{Au}$ equals $\lambda\mathbf{u}$. While such an occurrence might seem unlikely, it is actually common in many matrix models. For example, in the weather Markov chain introduced in Section 1.1, as we experimented with computing successive probability distributions over many periods, we saw that eventually the probability vector converged to a stable distribution $\mathbf{p}*$ which stayed the same from one period to the next, that is, $\mathbf{p}* = \mathbf{Ap}*$. So for this $\mathbf{p}*$, computing $\mathbf{Ap}*$ is the same as multiplying $\mathbf{p}*$ by 1.

When $\mathbf{Au} = \lambda\mathbf{u}$ (where $\mathbf{u} \neq \mathbf{0}$), the vector $\mathbf{u}$ is called an *eigenvector* of $\mathbf{A}$ (*eigen* is the German word for "proper") and the scalar λ is called an *eigenvalue* of $\mathbf{A}$. In this section we introduce eigenvalues and eigenvectors and try to give the reader some sense of their great usefulness.

EXAMPLE 1.
Eigenvalues in a
Growth Model

Consider a fictitious growth model for computers and dogs given in (1). The terms C and D represent the current numbers of computers and dogs, respectively, in our university, and C' and D' stand for the numbers of computers and dogs, respectively, one year from now.

$$C' = 3C + D$$
$$D' = 2C + 2D \tag{1}$$

If initially we had $C = 1$, $D = 1$, then we compute $C' = 3(1) + 1(1) = 4$, $D' = 2(1) + 2(1) = 4$. Letting $C = 4$, $D = 4$, we obtain $C' = 3(4) + 1(4) = 16$, $D' = (4) + 2(4) = 16$. Whenever $\begin{bmatrix} C \\ D \end{bmatrix} = \begin{bmatrix} a \\ a \end{bmatrix}$, then $\begin{bmatrix} C' \\ D' \end{bmatrix} = \begin{bmatrix} 4a \\ 4a \end{bmatrix}$. So 4 is an eigenvalue of $\mathbf{A}$, the coefficient matrix in (1), and any vector of the form $\begin{bmatrix} a \\ a \end{bmatrix}$ is an eigenvector of $\mathbf{A}$. Looking at powers of $\mathbf{A}$, we have

$$\mathbf{A}^2 \begin{bmatrix} a \\ a \end{bmatrix} = \mathbf{A}\left(\mathbf{A}\begin{bmatrix} a \\ a \end{bmatrix} \right) = \mathbf{A}\begin{bmatrix} 4a \\ 4a \end{bmatrix} = \begin{bmatrix} 16a \\ 16a \end{bmatrix}$$

and in general

$$\mathbf{A}^k \begin{bmatrix} a \\ a \end{bmatrix} = 4^k \begin{bmatrix} a \\ a \end{bmatrix} \tag{2}$$

Note that if initially we had the (nonsense) vector $\begin{bmatrix} C \\ D \end{bmatrix} = \begin{bmatrix} 1 \\ -2 \end{bmatrix}$, then $\begin{bmatrix} C' \\ D' \end{bmatrix} = \begin{bmatrix} 1 \\ -2 \end{bmatrix}$. Hence 1 is also an eigenvalue of $\mathbf{A}$ with associated eigenvector $\begin{bmatrix} 1 \\ -2 \end{bmatrix}$, or any multiple of $\begin{bmatrix} 1 \\ -2 \end{bmatrix}$. ■

EXAMPLE 2.
Stable Probability
Vector for
Weather Markov
Chain

In Section 1.1 we introduced the weather Markov chain with transition matrix,

Today

$$\mathbf{A} = \begin{array}{c} \text{Tomorrow} \\ \text{Sunny} \\ \text{Cloudy} \\ \text{Rainy} \end{array} \begin{array}{ccc} \text{Sunny} & \text{Cloudy} & \text{Rainy} \\ \begin{bmatrix} \frac{3}{4} & \frac{1}{2} & \frac{1}{4} \\ \frac{1}{8} & \frac{1}{4} & \frac{1}{2} \\ \frac{1}{8} & \frac{1}{4} & \frac{1}{4} \end{bmatrix} \end{array}$$

We showed in Section 1.4 that $\mathbf{p}^* = \begin{bmatrix} \frac{14}{23} \\ \frac{5}{23} \\ \frac{4}{23} \end{bmatrix}$ is a stable probability distribution for this transition matrix $\mathbf{A}$. That is,

$$\mathbf{A}\mathbf{p}^* = \mathbf{p}^*: \begin{bmatrix} \frac{3}{4} & \frac{1}{2} & \frac{1}{4} \\ \frac{1}{8} & \frac{1}{4} & \frac{1}{2} \\ \frac{1}{8} & \frac{1}{4} & \frac{1}{4} \end{bmatrix} \begin{bmatrix} \frac{14}{23} \\ \frac{5}{23} \\ \frac{4}{23} \end{bmatrix} = \begin{bmatrix} \frac{3}{4} \cdot \frac{14}{23} + \frac{1}{2} \cdot \frac{5}{23} + \frac{1}{4} \cdot \frac{4}{23} \\ \frac{1}{8} \cdot \frac{14}{23} + \frac{1}{4} \cdot \frac{5}{23} + \frac{1}{2} \cdot \frac{4}{23} \\ \frac{1}{8} \cdot \frac{14}{23} + \frac{1}{4} \cdot \frac{5}{23} + \frac{1}{4} \cdot \frac{4}{23} \end{bmatrix} = \begin{bmatrix} \frac{14}{23} \\ \frac{5}{23} \\ \frac{4}{23} \end{bmatrix}$$

Thus $\mathbf{A}\mathbf{p}^* = 1\mathbf{p}^*$, and $\mathbf{p}^*$ is an eigenvector of $\mathbf{A}$ with eigenvalue 1. ■

This property of matrix multiplication, acting like scalar multiplication for certain vectors, happens for all matrices. It is the key to understanding the behavior of many linear models. An $n \times n$ matrix usually has n eigenvalues, each with an infinite collection of eigenvectors.

Observe that if e is an eigenvector of A with $Ae = \lambda e$, then any multiple re of e is also an eigenvector, since $A(re) = r(Ae) = r(\lambda e) = \lambda(re)$.

The following example shows how eigenvectors provide a simplifying way to carry out matrix–vector computations.

**EXAMPLE 3.
Eigenvectors As a
Coordinate System**

The computer/dog growth model from Example 1 has the form

$$c' = Ac, \quad \text{where} \quad A = \begin{bmatrix} 3 & 1 \\ 2 & 2 \end{bmatrix}, \quad c = \begin{bmatrix} C \\ D \end{bmatrix}, \quad c' = \begin{bmatrix} C' \\ D' \end{bmatrix}$$

In Example 1 we saw that the two eigenvalues and associated eigenvectors of this matrix A are $\lambda_1 = 4$ with $e_1 = \begin{bmatrix} 1 \\ 1 \end{bmatrix}$ and $\lambda_2 = 1$ with $e_2 = \begin{bmatrix} 1 \\ -2 \end{bmatrix}$.

Suppose that we want to determine the effects of this growth model over 20 periods with the starting vector $c = \begin{bmatrix} 1 \\ 7 \end{bmatrix}$. Using the fact that it is much easier to compute $A^{20}e_1$ and $A^{20}e_2$ [see (2)] than $A^{20}c$, let us express c as a linear combination of e_1 and e_2.

$$c = ae_1 + be_2: \begin{bmatrix} 1 \\ 7 \end{bmatrix} = a\begin{bmatrix} 1 \\ 1 \end{bmatrix} + b\begin{bmatrix} 1 \\ -2 \end{bmatrix} \quad \text{or} \quad \begin{aligned} 1 &= 1a + 1b \\ 7 &= 1a - 2b \end{aligned} \tag{3}$$

The system (3) can be solved by elimination to yield $a = 3$, $b = -2$. So $c = 3e_1 - 2e_2$.

Using the linearity of matrix–vector products, we can write

$$\begin{aligned} Ac = A(3e_1 - 2e_2) &= 3(Ae_1) - 2(Ae_2) \\ &= 3(4e_1) - 2(1e_2) \quad \text{(since } e_1, e_2 \text{ are eigenvectors)} \\ &= 12e_1 - 2e_2 \end{aligned} \tag{4}$$

and for 20 periods, we have

$$\begin{aligned} A^{20}c = A^{20}(3e_1 - 2e_2) &= 3(A^{20}e_1) - 2(A^{20}e_2) \\ &= 3(4^{20}e_1) - 2(1^{20}e_2) \end{aligned} \tag{5}$$

$$= 3 \cdot 4^{20}\begin{bmatrix} 1 \\ 1 \end{bmatrix} - 2\begin{bmatrix} 1 \\ -2 \end{bmatrix} = \begin{bmatrix} 3 \cdot 4^{20} \\ 3 \cdot 4^{20} \end{bmatrix} - \begin{bmatrix} 2 \\ -4 \end{bmatrix}$$

Note how the eigenvector with the larger eigenvalue swamps the other eigenvector. The relative effect of the other eigenvector is so small that it can be neglected. So after n periods we have

$$\mathbf{A}^n\mathbf{c} \approx \mathbf{A}^n(3\mathbf{e}_1) = 3 \cdot 4^n\mathbf{e}_1 = 3 \cdot 4^n\begin{bmatrix} 1 \\ 1 \end{bmatrix} = \begin{bmatrix} 3 \cdot 4^n \\ 3 \cdot 4^n \end{bmatrix} \tag{6}$$

This is much easier than multiplying $\mathbf{A}^n\mathbf{c}$ out directly for various n. ■

As shown in Example 3, one important use of eigenvectors is to simplify computations of the form $\mathbf{A}^n\mathbf{c}$. We express $\mathbf{c}$ as a linear combination of eigenvectors, $\mathbf{c} = a\mathbf{e}_1 + b\mathbf{e}_2$. Then the messy matrix calculation $\mathbf{A}^n\mathbf{c}$ can be rewritten as the much easier $a\mathbf{A}^n\mathbf{e}_1 + b\mathbf{A}^n\mathbf{e}_2$. We give two more examples of this process.

EXAMPLE 4.
Expressing a
Vector in Terms
of Eigenvectors

Consider the following model for growth of cats (C) and dinosaurs (D).

$$\mathbf{c}' = \mathbf{Ac}, \quad \text{where} \quad \mathbf{A} = \begin{bmatrix} 1 & 2 \\ 3 & 2 \end{bmatrix}, \mathbf{c} = \begin{bmatrix} C \\ D \end{bmatrix}, \mathbf{c}' = \begin{bmatrix} C' \\ D' \end{bmatrix}$$

We are given the two eigenvalues and associated eigenvectors of this matrix $\mathbf{A}$: $\lambda_1 = 4$ with $\mathbf{e}_1 = \begin{bmatrix} 2 \\ 3 \end{bmatrix}$ and $\lambda_2 = -1$ with $\mathbf{e}_2 = \begin{bmatrix} 1 \\ -1 \end{bmatrix}$.

Suppose that we want to determine the effects of this growth model after 20 periods with the starting vector $\mathbf{c} = \begin{bmatrix} 6 \\ 4 \end{bmatrix}$. We first express $\mathbf{c}$ as a linear combination of $\mathbf{e}_1$ and $\mathbf{e}_2$.

$$\mathbf{c} = a\mathbf{e}_1 + b\mathbf{e}_2: \quad \begin{bmatrix} 6 \\ 4 \end{bmatrix} = a\begin{bmatrix} 2 \\ 3 \end{bmatrix} + b\begin{bmatrix} 1 \\ -1 \end{bmatrix} \quad \text{or} \quad \begin{bmatrix} 2 & 1 \\ 3 & -1 \end{bmatrix}\begin{bmatrix} a \\ b \end{bmatrix} = \begin{bmatrix} 6 \\ 4 \end{bmatrix} \tag{7}$$

We can solve (7) by Gaussian elimination or Gauss–Jordan elimination, or by computing the inverse of $\begin{bmatrix} 2 & 1 \\ 3 & -1 \end{bmatrix}$. Gauss–Jordan elimination on the augmented coefficient matrix yields

$$\begin{bmatrix} 2 & 1 & | & 6 \\ 3 & -1 & | & 4 \end{bmatrix} \Longrightarrow \begin{bmatrix} 1 & \frac{1}{2} & | & 3 \\ 0 & -\frac{5}{2} & | & -5 \end{bmatrix} \Longrightarrow \begin{bmatrix} 1 & 0 & | & 2 \\ 0 & 1 & | & 2 \end{bmatrix} \tag{8}$$

Thus $\mathbf{c} = 2\mathbf{e}_1 + 2\mathbf{e}_2$; that is, $\begin{bmatrix} 6 \\ 4 \end{bmatrix} = 2\begin{bmatrix} 2 \\ 3 \end{bmatrix} + 2\begin{bmatrix} 1 \\ -1 \end{bmatrix}$.

To find the population vector after 20 periods, we need to compute $\mathbf{A}^{20}\mathbf{c}$.

$$\mathbf{A}^{20}\mathbf{c} = \mathbf{A}^{20}(2\mathbf{e}_1 + 2\mathbf{e}_2) = 2(\mathbf{A}^{20}\mathbf{e}_1) + 2(\mathbf{A}^{20}\mathbf{e}_2)$$

$$= 2(4^{20}\mathbf{e}_1) + 2((-1)^{20}\mathbf{e}_2) \qquad [\text{note that } (-1^{20}) = 1] \qquad (9)$$

$$= 2 \cdot 4^{20}\begin{bmatrix} 2 \\ 3 \end{bmatrix} + 2\begin{bmatrix} 1 \\ -1 \end{bmatrix} \approx 2 \cdot 4^{20}\begin{bmatrix} 2 \\ 3 \end{bmatrix} \qquad ■$$

EXAMPLE 5.
Eigenvalues and
Eigenvectors of a
3 × 3 Matrix

Consider the following model for growth of canaries (C), dogs (D), and elephants (E).

$$\mathbf{c}' = \mathbf{Ac}, \quad \text{where} \quad \mathbf{A} = \begin{bmatrix} 5 & 4 & 2 \\ 4 & 5 & 2 \\ 2 & 2 & 2 \end{bmatrix}, \quad \mathbf{c} = \begin{bmatrix} C \\ D \\ E \end{bmatrix}, \quad \mathbf{c}' = \begin{bmatrix} C' \\ D' \\ E' \end{bmatrix}$$

We are given the three eigenvalues and associated eigenvectors of this matrix $\mathbf{A}$:

$$\lambda_1 = 10 \text{ with } \mathbf{e}_1 = \begin{bmatrix} 2 \\ 2 \\ 1 \end{bmatrix}, \ \lambda_2 = 1 \text{ with } \mathbf{e}_2 = \begin{bmatrix} 1 \\ 0 \\ -2 \end{bmatrix}, \text{ and } \lambda_3 = 1 \text{ with } \mathbf{e}_3 = \begin{bmatrix} 0 \\ 1 \\ -2 \end{bmatrix}.$$

Note that $\lambda_2 = \lambda_3$.

Suppose now we want to determine the effects of this growth model after 10 periods with the starting vector $\mathbf{c} = \begin{bmatrix} 6 \\ 9 \\ 6 \end{bmatrix}$. We first express $\mathbf{c}$ as a linear combination of $\mathbf{e}_1$, $\mathbf{e}_2$, and $\mathbf{e}_3$.

$$\mathbf{c} = a\mathbf{e}_1 + b\mathbf{e}_2 + c\mathbf{e}_3: \begin{bmatrix} 6 \\ 9 \\ 6 \end{bmatrix} = a\begin{bmatrix} 2 \\ 2 \\ 1 \end{bmatrix} + b\begin{bmatrix} 1 \\ 0 \\ -2 \end{bmatrix} + c\begin{bmatrix} 0 \\ 1 \\ -2 \end{bmatrix}$$

$$\text{or} \quad \begin{bmatrix} 2 & 1 & 0 \\ 2 & 0 & 1 \\ 1 & -2 & -2 \end{bmatrix}\begin{bmatrix} a \\ b \\ c \end{bmatrix} = \begin{bmatrix} 6 \\ 9 \\ 6 \end{bmatrix} \tag{10}$$

Gaussian elimination on the augmented coefficient matrix yields

$$\begin{bmatrix} 2 & 1 & 0 & | & 6 \\ 2 & 0 & 1 & | & 9 \\ 1 & -2 & -2 & | & 6 \end{bmatrix} \Longrightarrow \begin{bmatrix} 2 & 1 & 0 & | & 6 \\ 0 & -1 & 1 & | & 3 \\ 0 & -\frac{5}{2} & -2 & | & 3 \end{bmatrix} \Longrightarrow \begin{bmatrix} 2 & 1 & 0 & | & 6 \\ 0 & -1 & 1 & | & 3 \\ 0 & 0 & -\frac{9}{2} & | & -\frac{9}{2} \end{bmatrix}$$
$$(11)$$

So $c = 1$ and back substitution yields

$$b = \frac{3-1}{-1} = -2 \quad \text{and} \quad a = \frac{6-1(-2)}{2} = 4$$

Then $\mathbf{c} = 4\mathbf{e}_1 - 2\mathbf{e}_2 + \mathbf{e}_3$, or

$$\begin{bmatrix} 6 \\ 9 \\ 6 \end{bmatrix} = 4 \begin{bmatrix} 2 \\ 2 \\ 1 \end{bmatrix} - 2 \begin{bmatrix} 1 \\ 0 \\ -2 \end{bmatrix} + 1 \begin{bmatrix} 0 \\ 1 \\ -2 \end{bmatrix}$$

To find the population vector after 10 periods, we need to compute $\mathbf{A}^{10}\mathbf{c}$.

$$\begin{aligned} \mathbf{A}^{10}\mathbf{c} &= \mathbf{A}^{10}(4\mathbf{e}_1 - 2\mathbf{e}_2 + \mathbf{e}_3) \\ &= 4(\mathbf{A}^{10}\mathbf{e}_1) - 2(\mathbf{A}^{10}\mathbf{e}_2) + \mathbf{A}^{10}\mathbf{e}_3 \\ &= 4(10^{10}\mathbf{e}_1) - 2(1^{10}\mathbf{e}_2) + 1(1^{10}\mathbf{e}_3) \end{aligned} \tag{12}$$

$$= 4 \cdot 10^{10} \begin{bmatrix} 2 \\ 2 \\ 1 \end{bmatrix} - 2 \begin{bmatrix} 1 \\ 0 \\ -2 \end{bmatrix} + \begin{bmatrix} 0 \\ 1 \\ -2 \end{bmatrix} \approx 4 \cdot 10^{10} \begin{bmatrix} 2 \\ 2 \\ 1 \end{bmatrix} \qquad ■$$

Let us generalize the pattern after many periods in the previous examples. Suppose that we are given eigenvectors $\mathbf{e}_1$, $\mathbf{e}_2$ of a 2×2 matrix $\mathbf{A}$ with eigenvalues λ_1 and λ_2, respectively, where $\lambda_1 > \lambda_2$. We express the vector $\mathbf{c}$ as a linear combination of $\mathbf{e}_1$ and $\mathbf{e}_2$, that is, $\mathbf{c} = a\mathbf{e}_1 + b\mathbf{e}_2$. Then by the linearity of the matrix–vector products, $\mathbf{Ac}$ and $\mathbf{A}^2\mathbf{c}$ can be calculated as

$$\begin{aligned} \mathbf{Ac} &= \mathbf{A}(a\mathbf{e}_1 + b\mathbf{e}_2) = a\mathbf{A}\mathbf{e}_1 + b\mathbf{A}\mathbf{e}_2 = a\lambda_1\mathbf{e}_1 + b\lambda_2\mathbf{e}_2 \\ \mathbf{A}^2\mathbf{c} &= \mathbf{A}^2(a\mathbf{e}_1 + b\mathbf{e}_2) = a\mathbf{A}^2\mathbf{e}_1 + b\mathbf{A}^2\mathbf{e}_2 = a\lambda_1^2\mathbf{e}_1 + b\lambda_2^2\mathbf{e}_2 \end{aligned} \tag{13}$$

and more generally,

$$\mathbf{A}^n\mathbf{c} = \mathbf{A}^n(a\mathbf{e}_1 + b\mathbf{e}_2) = a\mathbf{A}^n\mathbf{e}_1 + b\mathbf{A}^n\mathbf{e}_2 = a\lambda_1^n\mathbf{e}_1 + b\lambda_2^n\mathbf{e}_2 \tag{14}$$

As noted in Example 5, for large n, λ_1^n will be much larger than λ_2^n, since $\lambda_1 > \lambda_2$, and so we have

$$\mathbf{A}^n\mathbf{c} \approx a\lambda_1^n\mathbf{e}_1 \tag{15}$$

To give a geometric picture of the convergence of vectors toward multiples of $\lambda_1^n\mathbf{e}_1$ in Example 3, Figure 3.3 shows how $\mathbf{A}$, $\mathbf{A}^2$, and $\mathbf{A}^3$ map the positive quadrant of the plane toward vectors close to $\mathbf{e}_1$. (While eigenvectors might have seemed a bit mysterious at first, readers hopefully now realize that eigenvectors are our friends.)

Eigenvectors clearly provide a very simple way to follow growth models over many periods. In many other types of problems in linear algebra, such as systems of linear differential equations, eigenvalues and eigenvectors provide the simplest way to obtain solutions. A note of warning: Some $n \times n$ matrices do not have n different eigenvectors (see Example 4 in Section 5.3) and hence the preceding method does not apply to them.

We call an eigenvalue λ^* of the square matrix $\mathbf{A}$ the *dominant eigenvalue* of $\mathbf{A}$ if λ^* is larger in absolute value than any other eigenvalue of $\mathbf{A}$. An eigenvector $\mathbf{e}^*$

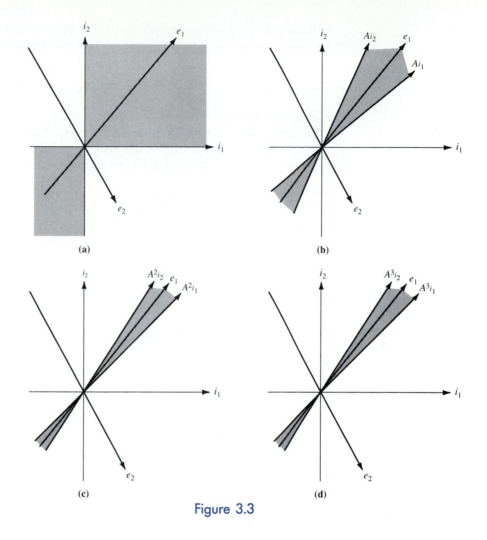

Figure 3.3

associated with the dominant eigenvalue of **A** is called a ***dominant eigenvector*** of **A**. Note that not all square matrices have dominant eigenvalues, for it is possible to have two eigenvalues of largest (absolute) size. For example, the identity matrix **I** has all eigenvalues of size 1 (and all vectors are eigenvectors). Example 9 at the end of this section gives other examples of matrices without dominant eigenvalues. However, most square matrices have a dominant eigenvalue and dominant eigenvector, and the nice behavior we saw in the previous examples is what normally happens.

Principle for Long-Term Behavior of $A^n c$

Let square matrix **A** have a dominant eigenvector **e***. Then for any vector **c**, the expression $A^n c$ normally approaches a multiple of **e*** as n becomes large.

Example 9 also gives an example of a 3×3 matrix with a dominant eigenvector $\mathbf{e}^*$ for which, for some $\mathbf{c}$, $\mathbf{A}^n\mathbf{c}$ does not approach a multiple of $\mathbf{e}^*$. The problem is that $\mathbf{c}$ might be a linear combination of eigenvectors associated with smaller eigenvalues.

Corollary to Principle

Let square matrix $\mathbf{A}$ have a dominant eigenvector $\mathbf{e}^*$ and assume that none of the coordinate vectors $\mathbf{i}_1, \mathbf{i}_2, \ldots$ are eigenvectors of $\mathbf{A}$. Then the columns of $\mathbf{A}^n$ normally approach multiples of $\mathbf{e}^*$ as n becomes large.

Proof: Recall that $\mathbf{i}_k$ is the vector with 1 in the kth position and 0's elsewhere. By the principle above, $\mathbf{A}^n\mathbf{i}_k$ normally approaches a multiple of $\mathbf{e}^*$ as n becomes large. But $\mathbf{A}^n\mathbf{i}_k$ equals the kth column of $\mathbf{A}^n$ (refer to the beginning of Section 3.1). ■

As an example of this corollary, for

$$\mathbf{A} = \begin{bmatrix} 3 & 1 \\ 2 & 2 \end{bmatrix} \quad \text{with} \quad \lambda_1 = 4 \quad \text{and} \quad \mathbf{e}_1 = \begin{bmatrix} 1 \\ 1 \end{bmatrix}$$

we have

$$\mathbf{A}^2 = \begin{bmatrix} 11 & 5 \\ 10 & 6 \end{bmatrix} \quad \mathbf{A}^4 = \begin{bmatrix} 171 & 85 \\ 170 & 86 \end{bmatrix} \quad \mathbf{A}^8 = \begin{bmatrix} 43{,}691 & 21{,}845 \\ 43{,}690 & 21{,}846 \end{bmatrix}$$

If $\mathbf{A}$ is a symmetric matrix ($a_{ij} = a_{ji}$), $\mathbf{A}$ can be decomposed into a sum of simple matrices generated by its eigenvectors. The details are presented in Section 5.4. We present an example of this result here to illustrate how symmetric matrices can be expressed as a linear combination of their eigenvectors.

**EXAMPLE 6.
Eigenvector
Decomposition of
a Symmetric
Matrix**

Suppose we know that the symmetric matrix $\mathbf{A} = \begin{bmatrix} 3 & 2 & 2 \\ 2 & 2 & 0 \\ 2 & 0 & 4 \end{bmatrix}$ has eigenvalues

$\lambda_1 = 6$, $\lambda_2 = 3$, $\lambda_3 = 0$ with associated eigenvectors $\mathbf{e}_1 = \begin{bmatrix} \frac{2}{3} \\ \frac{1}{3} \\ \frac{2}{3} \end{bmatrix}$, $\mathbf{e}_2 = \begin{bmatrix} \frac{1}{3} \\ \frac{2}{3} \\ -\frac{2}{3} \end{bmatrix}$,

$\mathbf{e}_3 = \begin{bmatrix} -\frac{2}{3} \\ \frac{2}{3} \\ \frac{1}{3} \end{bmatrix}$, where the $\mathbf{e}_i$ have been chosen so that $|\mathbf{e}_i|_e = 1$ (i.e., $\mathbf{e}_i \cdot \mathbf{e}_i = 1$).

Simple matrices were introduced in the preceding section. Recall that the simple matrix $\mathbf{a} * \mathbf{b}$ generated by the outer product of *column* vector $\mathbf{a}$ and *row* vector $\mathbf{b}$ has entry (i, j) equal to $a_i b_j$ (we perform matrix multiplication of an $m \times 1$ matrix times a $1 \times n$ matrix). From Theorem 2 in Section 5.4,

$$\mathbf{A} = \lambda_1 \mathbf{e}_1 * \mathbf{e}_1^T + \lambda_2 \mathbf{e}_2 * \mathbf{e}_2^T + \lambda_3 \mathbf{e}_3 * \mathbf{e}_3^T$$

$$= 6 \begin{bmatrix} \frac{2}{3} \\ \frac{1}{3} \\ \frac{2}{3} \end{bmatrix} * [\frac{2}{3}, \frac{1}{3}, \frac{2}{3}] + 3 \begin{bmatrix} \frac{1}{3} \\ \frac{2}{3} \\ -\frac{2}{3} \end{bmatrix} * [\frac{1}{3}, \frac{2}{3}, -\frac{2}{3}] + 0 \begin{bmatrix} -\frac{2}{3} \\ \frac{2}{3} \\ \frac{1}{3} \end{bmatrix} * [-\frac{2}{3}, \frac{2}{3}, \frac{1}{3}] \qquad (16)$$

$$= 6 \begin{bmatrix} \frac{4}{9} & \frac{2}{9} & \frac{4}{9} \\ \frac{2}{9} & \frac{1}{9} & \frac{2}{9} \\ \frac{4}{9} & \frac{2}{9} & \frac{4}{9} \end{bmatrix} + 3 \begin{bmatrix} \frac{1}{9} & \frac{2}{9} & -\frac{2}{9} \\ \frac{2}{9} & \frac{4}{9} & -\frac{4}{9} \\ -\frac{2}{9} & -\frac{4}{9} & \frac{4}{9} \end{bmatrix} = \begin{bmatrix} 3 & 2 & 2 \\ 2 & 2 & 0 \\ 2 & 0 & 4 \end{bmatrix}$$

The third simple matrix in (16) is omitted on the last line because it is multiplied by the eigenvalue 0 and so is the zero matrix. ■

The determination of the eigenvalues and eigenvectors of a square matrix is one of the most carefully studied computational problems in all of mathematics. For 2×2 matrices, the theory of determinants (introduced in Section 1.6) yields a simple computational procedure for computing eigenvalues, which we will present shortly. For most larger matrices, the problem is more complicated. We present a method for any symmetric matrix in Section 5.4.

We first note how eigenvalues and eigenvectors of $\mathbf{A}$ are related to eigenvalues and eigenvectors of $\mathbf{A}^n$ and $\mathbf{A}^{-1}$.

Theorem 1. Let $\mathbf{A}$ be a square matrix and let λ be an eigenvalue of $\mathbf{A}$ with associated eigenvector $\mathbf{e}$.
  (i) Then λ^n is an eigenvalue of $\mathbf{A}^n$ with associated eigenvector $\mathbf{e}$.
  (ii) If $\mathbf{A}$ is invertible, then $1/\lambda$ is an eigenvector of $\mathbf{A}^{-1}$ with associated eigenvector $\mathbf{e}$.

Proof: (i) Follows immediately from the preceding discussion; see (2).
  (ii) We know that $\mathbf{A}^{-1}\mathbf{A} = \mathbf{I}$. Then

$$\mathbf{A}^{-1}\mathbf{A}\mathbf{e} = \mathbf{I}\mathbf{e} = \mathbf{e} \qquad (17)$$

Since $\mathbf{A}\mathbf{e} = \lambda\mathbf{e}$, then we also have

$$\mathbf{A}^{-1}\mathbf{A}\mathbf{e} = \mathbf{A}^{-1}(\lambda\mathbf{e}) = \lambda(\mathbf{A}^{-1}\mathbf{e}) \qquad (18)$$

Equating the right sides of (17) and (18), we have $\mathbf{e} = \lambda(\mathbf{A}^{-1}\mathbf{e})$ or $(1/\lambda)\mathbf{e} = \mathbf{A}^{-1}\mathbf{e}$. ■

Note that there is a relation between eigenvalues and the existence of an inverse. If 0 is an eigenvalue of $\mathbf{A}$, there are multiple solutions to $\mathbf{A}\mathbf{x} = \mathbf{0}$ ($= 0\mathbf{x}$). But by Theorem 4 of Section 1.5, $\mathbf{A}$ has an inverse if and only if there is a unique solution to $\mathbf{A}\mathbf{x} = \mathbf{0}$. We have proved:

Theorem 2. $\mathbf{A}$ has an inverse if and only if 0 is not an eigenvalue of $\mathbf{A}$.

Determining the Eigenvalues of a 2 × 2 Matrix

The defining equation for an eigenvalue λ of a square matrix $\mathbf{A}$ and its eigenvector $\mathbf{e}$ is

$$\mathbf{Ae} = \lambda\mathbf{e} \quad \text{or} \quad \mathbf{Ae} - \lambda\mathbf{e} = 0 \tag{19}$$

Writing $\lambda\mathbf{e}$ as $\lambda\mathbf{Ie}$, we have

$$(\mathbf{A} - \lambda\mathbf{I})\mathbf{e} = 0 \tag{20}$$

Once an eigenvalue λ is found, we can obtain an associated eigenvector $\mathbf{e}$ by solving the system of equations (20). More important, we can use (20) to determine the eigenvalues themselves. The key to finding the eigenvalues is to remember that each eigenvalue has an infinite set of associated eigenvectors, for if $\mathbf{Ae} = \lambda\mathbf{e}$, then for any scalar r, $\mathbf{A}(r\mathbf{e}) = r\mathbf{Ae} = \lambda r\mathbf{e}$. Thus when λ is an eigenvalue, the system (20) has an infinite number of solutions.

Let us think back to what we learned in Chapter 1 about systems of equations with an infinite number of solutions. In particular, recall Theorem 8 in Section 1.6, which said that (20) has only one solution, namely $\mathbf{e} = \mathbf{0}$ (the zero vector), if $\det(\mathbf{A} - \lambda\mathbf{I}) \neq 0$, while (20) has an infinite number of solutions if $\det(\mathbf{A} - \lambda\mathbf{I}) = 0$. Thus, eigenvalues must be values λ that make $\det(\mathbf{A} - \lambda\mathbf{I}) = 0$. For any $n \times n$ matrix $\mathbf{A}$, $\det(\mathbf{A} - \lambda\mathbf{I})$ will be a polynomial of degree n in λ, called the ***characteristic polynomial*** of $\mathbf{A}$. A polynomial of degree n can have up to n zeros. These zeros of the characteristic polynomial of $\mathbf{A}$ are, by Theorem 8 of Section 1.6, the only possible choices for eigenvalues of $\mathbf{A}$.

Theorem 3. The values λ that make $\det(\mathbf{A} - \lambda\mathbf{I}) = 0$ are the set of eigenvalues of $\mathbf{A}$. The associated eigenvector(s) for λ are the nonzero solutions to $(\mathbf{A} - \lambda\mathbf{I})\mathbf{x} = \mathbf{0}$.

Finding the zeros of polynomials generally requires some numerical iterative scheme, but for degree 2 polynomials the quadratic formula can be used. Note that to find a dominant eigenvector and associated dominant eigenvalue (assuming that there is a unique dominant eigenvalue), there is an iterative method available from the "principle for long-term behavior of $\mathbf{A}^n\mathbf{c}$": For any vector $\mathbf{c}$ not equal to an eigenvector of $\mathbf{A}$, as n becomes large, the expression $\mathbf{A}^n\mathbf{c}$ normally approaches a multiple of the dominant eigenvector $\mathbf{e}^*$; and from $\mathbf{e}^*$, the associated eigenvalue λ^* is obtained (since $\mathbf{Ae}^* = \lambda^*\mathbf{e}^*$).

EXAMPLE 7.
Determining
Eigenvalues and
Eigenvectors

Consider the system of computer/dog growth equations again.

$$C' = 3C + D \quad \text{or} \quad \mathbf{c}' = \mathbf{Ac}, \quad \text{where} \quad \mathbf{A} = \begin{bmatrix} 3 & 1 \\ 2 & 2 \end{bmatrix}$$
$$D' = 2C + 2D$$

Earlier the eigenvalues and eigenvectors for $\mathbf{A}$ were given to us. Now we calculate them by Theorem 3. We must find the zeros of the characteristic polynomial $\det(\mathbf{A} - \lambda\mathbf{I})$.

$$\det(\mathbf{A} - \lambda\mathbf{I}) = \begin{vmatrix} 3 - \lambda & 1 \\ 2 & 2 - \lambda \end{vmatrix} = (3 - \lambda)(2 - \lambda) - 1 \cdot 2 = (6 - 5\lambda + \lambda^2) - 2$$
$$= 4 - 5\lambda + \lambda^2 = (4 - \lambda)(1 - \lambda)$$

$$(21)$$

The zeros of $\det(\mathbf{A} - \lambda\mathbf{I}) = (4 - \lambda)(1 - \lambda)$ are thus 4 and 1.

To find an eigenvector $\mathbf{e}$ for the eigenvalue 4, we must solve the system $\mathbf{Ae} = 4\mathbf{e}$ or, as in (20), $(\mathbf{A} - 4\mathbf{I})\mathbf{e} = 0$, where

$$\mathbf{A} - 4\mathbf{I} = \begin{bmatrix} 3 & 1 \\ 2 & 2 \end{bmatrix} - \begin{bmatrix} 4 & 0 \\ 0 & 4 \end{bmatrix} = \begin{bmatrix} -1 & 1 \\ 2 & -2 \end{bmatrix}$$

Writing out $(\mathbf{A} - 4\mathbf{I})\mathbf{e} = 0$, we have

$$\begin{aligned} -e_1 + e_2 &= 0 \implies e_1 = e_2 \\ 2e_1 - 2e_2 &= 0 \end{aligned}$$

$$(22)$$

The second equation in (22) here is just -2 times the first equation (so it is superfluous). Then $\mathbf{e}$ is an eigenvector if $e_1 = e_2$, or equivalently, if $\mathbf{e}$ is a multiple of $\begin{bmatrix} 1 \\ 1 \end{bmatrix}$.

It is left as an exercise for the reader to verify that $\mathbf{e}' = \begin{bmatrix} 1 \\ -2 \end{bmatrix}$ is an eigenvector for $\lambda = 1$ by showing that this $\mathbf{e}'$ is a solution to $(\mathbf{A} - \mathbf{I})\mathbf{e}' = 0$. ■

We next completely analyze a linear growth model using our new knowledge of eigenvectors and eigenvalues.

EXAMPLE 8.
Eigenvalues and
Eigenvectors for
Rabbit/Fox
Population Model

We consider a simple ecological model for survival of rabbits and foxes. Suppose that the current number of rabbits (R) naturally grows by 10% a year in the absence of foxes. So next year's number of rabbits (R') follows the growth law: $R' = 1.1R$. Also suppose that the current number of foxes (F) decreases by 15% a year in the absence of rabbits, so that $F' = 0.85F$. However, when foxes and rabbits are in the same habitat, the foxes eat the rabbits, decreasing the number of rabbits and allowing the foxes to increase. The model we propose is as follows.

$$\begin{aligned} R' &= 1.1R - 0.15F \\ F' &= 0.1R + 0.85F \end{aligned} \quad \text{or} \quad \mathbf{r}' = \mathbf{Ar}: \quad \begin{bmatrix} R' \\ F' \end{bmatrix} = \begin{bmatrix} 1.1 & -0.15 \\ 0.1 & 0.85 \end{bmatrix} \begin{bmatrix} R \\ F \end{bmatrix} \quad (23)$$

First, to get some numerical picture of the model, we use (23) to compute the following table of rabbit and fox populations over many periods when we start with the sample population vector $\begin{bmatrix} R \\ F \end{bmatrix} = \begin{bmatrix} 10 \\ 8 \end{bmatrix}$.

$$
\begin{array}{lll}
\text{0 Months:} & \text{10 rabbits,} & \text{8 foxes} \\
\text{1 Month:} & \text{9.8 rabbits,} & \text{7.8 foxes} \\
\text{2 Months:} & \text{9.6 rabbits,} & \text{7.6 foxes} \\
\text{3 Months:} & \text{9.4 rabbits,} & \text{7.4 foxes} \\
\vdots & & \\
\text{10 Months:} & \text{8.4 rabbits,} & \text{6.4 foxes} \\
\vdots & & \\
\text{20 Months:} & \text{7.4 rabbits,} & \text{5.4 foxes} \\
\vdots & & \\
\text{50 Months:} & \text{6.3 rabbits,} & \text{4.3 foxes} \\
\vdots & & \\
\text{100 Months:} & \text{6.02 rabbits,} & \text{4.02 foxes}
\end{array}
\tag{24}
$$

The graph in Figure 3.4 plots the trajectory for the population vectors in (24). They converge toward the vector $\begin{bmatrix} 6 \\ 4 \end{bmatrix}$ and lie on a line with a slope of 1. The graph also shows the trajectories given by this model for other starting vectors. All trajectories move along a line with slope 1 and converge toward some vector that lies on the line $F = \frac{2}{3}R$. Let us try to obtain a mathematical analysis of the behavior shown in (24) and Figure 3.4. As in Example 3, we can simplify the computation of $\mathbf{A}^n\mathbf{r}$ by finding the eigenvalues λ_1, λ_2 and eigenvectors $\mathbf{e}$, $\mathbf{e}'$ for this matrix and using them to write our starting population vector

$$
\mathbf{r} = \begin{bmatrix} 10 \\ 8 \end{bmatrix}
$$

in terms of $\mathbf{e}$ and $\mathbf{e}'$, $\mathbf{r} = a\mathbf{e} + b\mathbf{e}'$.

We first compute $\det(\mathbf{A} - \lambda\mathbf{I})$, the characteristic polynomial of $\mathbf{A}$.

$$
\det(\mathbf{A} - \lambda\mathbf{I}) = \begin{bmatrix} 1.1 - \lambda & -0.15 \\ 0.1 & 0.85 - \lambda \end{bmatrix} = (1.1 - \lambda)(0.85 - \lambda) - 0.1(-0.15)
$$
$$
= \lambda^2 - 1.95\lambda + 0.95 = (\lambda - 1)(\lambda - 0.95)
\tag{25}
$$

Thus the zeros of this polynomial are the eigenvalues $\lambda = 1$ and $\lambda = 0.95$.

To find an eigenvector $\mathbf{e}$ associated with $\lambda = 1$, we solve $(\mathbf{A} - \mathbf{I})\mathbf{e} = \mathbf{0}$, where

$$
\mathbf{A} - \mathbf{I} = \begin{bmatrix} 1.1 & -0.15 \\ 0.1 & 0.85 \end{bmatrix} - \begin{bmatrix} 1 & 0 \\ 0 & 1 \end{bmatrix} = \begin{bmatrix} 0.1 & -0.15 \\ 0.1 & -0.15 \end{bmatrix}.
$$

$$
(A - I)e = 0: \quad \begin{array}{l} 0.1e_1 - 0.15e_2 = 0 \\ 0.1e_1 - 0.15e_2 = 0 \end{array} \implies e_1 = \tfrac{3}{2}e_2 \implies \mathbf{e} = \begin{bmatrix} 3 \\ 2 \end{bmatrix}
$$

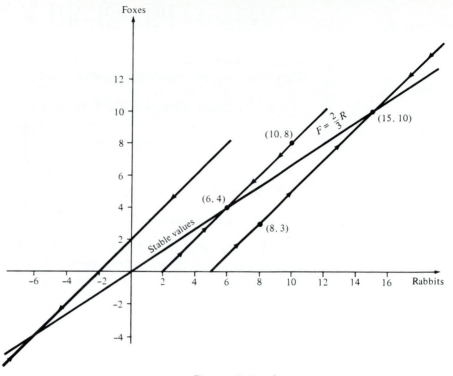

Figure 3.4

Thus the nonzero multiples of

$$\mathbf{e} = \begin{bmatrix} 3 \\ 2 \end{bmatrix}$$

are eigenvectors associated with $\lambda = 1$. Note that an eigenvector associated with an eigenvalue of 1 will be a "stable" population vector that stays the same period after period, that is, $\mathbf{e} = \mathbf{Ae}$.

Next we determine the eigenvector of $\lambda = 0.95$. We solve $(\mathbf{A} - 0.95\mathbf{I})\mathbf{e}'$, where

$$\mathbf{A} - 0.95\mathbf{I} = \begin{bmatrix} 1.1 & -0.15 \\ 0.1 & 0.85 \end{bmatrix} - \begin{bmatrix} 0.95 & 0 \\ 0 & 0.95 \end{bmatrix} = \begin{bmatrix} 0.15 & -0.15 \\ 0.1 & -0.1 \end{bmatrix}$$

$$(\mathbf{A} - 0.95\mathbf{I})\mathbf{e}' = \mathbf{0}: \quad \begin{matrix} 0.15e_1' - 0.15e_2' = 0 \\ 0.1e_1' - 0.1e_2' = 0 \end{matrix} \implies e_1' = e_2' \implies \mathbf{e}' = \begin{bmatrix} 1 \\ 1 \end{bmatrix}$$

The nonzero multiples of $\mathbf{e}' = \begin{bmatrix} 1 \\ 1 \end{bmatrix}$ are eigenvectors associated with $\lambda = 0.95$.

We now express the starting vector $\mathbf{r} = \begin{bmatrix} 10 \\ 8 \end{bmatrix}$ used in (24) in terms of eigenvectors $\mathbf{e} = \begin{bmatrix} 3 \\ 2 \end{bmatrix}$ and $\mathbf{e}' = \begin{bmatrix} 1 \\ 1 \end{bmatrix}$:

$$\mathbf{r} = a\mathbf{e} + b\mathbf{e}': \begin{bmatrix} 10 \\ 8 \end{bmatrix} = a\begin{bmatrix} 3 \\ 2 \end{bmatrix} + b\begin{bmatrix} 1 \\ 1 \end{bmatrix} \quad \text{or} \quad \begin{matrix} 3a + b = 10 \\ 2a + b = 8 \end{matrix} \tag{26}$$

By elimination, we find that $a = 2$, $b = 4$, so $\mathbf{r} = 2\mathbf{e} + 4\mathbf{e}'$.

Then using (26) to compute the population sizes in (24) gives

$$\mathbf{r}^{(n)} = \mathbf{A}^n\mathbf{r} = 2\mathbf{A}^n\mathbf{e} + 4\mathbf{A}^n\mathbf{e}' = 2(1^n\mathbf{e}) + 4(0.95^n\mathbf{e}')$$

$$= 2\begin{bmatrix} 3 \\ 2 \end{bmatrix} + 4 \cdot 0.95^n\begin{bmatrix} 1 \\ 1 \end{bmatrix} = \begin{bmatrix} 6 \\ 4 \end{bmatrix} + 0.95^n\begin{bmatrix} 4 \\ 4 \end{bmatrix} \tag{27}$$

So $\mathbf{r}^{(n)}$ is composed of a stable population term $\begin{bmatrix} 6 \\ 4 \end{bmatrix}$ and a second term of $0.95^n\begin{bmatrix} 4 \\ 4 \end{bmatrix}$ that slowly decays away. (The vectors in the second term lie on a line with slope $4/4 = 1$.) With (27), the behavior in table (24) is completely explained!

Generalizing the calculation in (27), we find that if the starting vector $\mathbf{r}$ has the eigenvector representation $\mathbf{r} = a\mathbf{e} + b\mathbf{e}'$, then

$$\mathbf{r}^{(n)} = \mathbf{A}^n\mathbf{r} = a\mathbf{A}^n\mathbf{e} + b\mathbf{A}^n\mathbf{e}' = a\mathbf{e} + b(0.95^n\mathbf{e}')$$

$$= \begin{bmatrix} 3a \\ 2a \end{bmatrix} + 0.95^n\begin{bmatrix} b \\ b \end{bmatrix} \tag{28}$$

And the long-term stable population is $\begin{bmatrix} 3a \\ 2a \end{bmatrix}$. The critical number is a. To find a for any starting vector $\mathbf{r} = \begin{bmatrix} R \\ F \end{bmatrix}$, we substitute the general vector $\begin{bmatrix} R \\ F \end{bmatrix}$ for $\begin{bmatrix} 10 \\ 8 \end{bmatrix}$ in (26) and solve by elimination or Cramer's rule. Using Cramer's rule, we have

$$a = \frac{\det(\mathbf{A}_1)}{\det(\mathbf{A})} = \frac{\begin{vmatrix} R & 1 \\ F & 1 \end{vmatrix}}{\begin{vmatrix} 3 & 1 \\ 2 & 1 \end{vmatrix}} = \frac{R - F}{1} = R - F \qquad ■$$

We conclude this section with an example of some matrices without unique dominant eigenvalues.

**EXAMPLE 9.
Nonunique
Eigenvalues**

Consider the following matrices, which each have two eigenvalues of the same absolute value.

$$\mathbf{A} = \begin{bmatrix} 0 & 1 \\ 1 & 0 \end{bmatrix} \qquad \mathbf{B} = \begin{bmatrix} 0 & 1 & 0 \\ 1 & 0 & 0 \\ 0 & 0 & 3 \end{bmatrix} \qquad \mathbf{C} = \begin{bmatrix} \cos 45° & \sin 45° \\ -\sin 45° & \cos 45° \end{bmatrix}$$

Observe what happens when we use as a starting vector i_1, the first coordinate vector.

$$Ai_1 = \begin{bmatrix} 0 & 1 \\ 1 & 0 \end{bmatrix}\begin{bmatrix} 1 \\ 0 \end{bmatrix} = \begin{bmatrix} 0 \\ 1 \end{bmatrix} = i_2 \qquad A^2 i_1 = \begin{bmatrix} 1 & 0 \\ 0 & 1 \end{bmatrix}\begin{bmatrix} 1 \\ 0 \end{bmatrix} = \begin{bmatrix} 1 \\ 0 \end{bmatrix} = i_1$$

We have a cycling phenomenon here. Odd powers of A yield i_2 and even powers of A yield i_1. The characteristic polynomial is $\det(A - \lambda I) = \lambda^2 - 1 = (\lambda - 1)(\lambda + 1)$, so the eigenvalues are 1 and -1 (of the same absolute value). Then the associated eigenvectors are $e_1 = \begin{bmatrix} 1 \\ 1 \end{bmatrix}$ and $e_2 = \begin{bmatrix} 1 \\ -1 \end{bmatrix}$ and $i_1 = \frac{1}{2}e_1 + \frac{1}{2}e_2$. If we had used B with i_1 (now i_1 would be a 3-vector), we would get the same cycling between i_1 and i_2, and never reach the dominant eigenvector of i_3, associated with the dominant eigenvalue of 3.

Next we consider C.

$$Ci_1 = \begin{bmatrix} \cos 45° & \sin 45° \\ -\sin 45° & \cos 45° \end{bmatrix}\begin{bmatrix} 1 \\ 0 \end{bmatrix} = \begin{bmatrix} \cos 45° \\ \sin 45° \end{bmatrix}$$

$$C^2 i_1 = \begin{bmatrix} \cos 90° & \sin 90° \\ -\sin 90° & \cos 90° \end{bmatrix}\begin{bmatrix} 1 \\ 0 \end{bmatrix} = \begin{bmatrix} \cos 90° \\ \sin 90° \end{bmatrix} = \begin{bmatrix} 0 \\ 1 \end{bmatrix}$$

Each time we multiply by C, we rotate 45° farther around the origin. The characteristic polynomial of C is [using the fact that $(\cos 45°)^2 + (\sin 45°)^2 = 1$]:

$$\det(C - \lambda I) = \lambda^2 - 2 \cos 45°\lambda + 1$$
$$= (\lambda - \cos 45° + i \sin 45°)(\lambda - \cos 45° - i \sin 45°)$$

C has imaginary roots $\cos 45° + i \sin 45°$ and $\cos 45° - i \sin 45°$ (i is the imaginary number equal to $\sqrt{-1}$). Recall that imaginary roots to real-valued polynomials always come in "conjugate" pairs (with the same absolute value). So if the characteristic polynomial has an imaginary root for the largest eigenvalue, standard iteration methods will not work. The eigenvectors are also complex-valued. (For details, see the exercises.)

■

Section 3.3 Exercises

1. (a) The matrix $\begin{bmatrix} 1 & 1 \\ 0 & 2 \end{bmatrix}$ has eigenvectors $e_1 = \begin{bmatrix} 1 \\ 1 \end{bmatrix}$ and $e_2 = \begin{bmatrix} 1 \\ 0 \end{bmatrix}$. What are the corresponding eigenvalues for these eigenvectors?

   (b) The matrix $\begin{bmatrix} 1 & -2 \\ -2 & 1 \end{bmatrix}$ has eigenvectors $e_1 = \begin{bmatrix} 1 \\ 1 \end{bmatrix}$ and $e_2 = \begin{bmatrix} 1 \\ -1 \end{bmatrix}$. What are the corresponding eigenvalues for these eigenvectors?

   (c) The matrix $\begin{bmatrix} 1 & 6 \\ -2 & -6 \end{bmatrix}$ has eigenvectors $e_1 = \begin{bmatrix} -2 \\ 1 \end{bmatrix}$ and $e_2 = \begin{bmatrix} -3 \\ 2 \end{bmatrix}$. What are the corresponding eigenvalues for these eigenvectors?

(d) The matrix $\begin{bmatrix} -4 & 4 & 4 \\ -1 & 1 & 2 \\ -3 & 2 & 4 \end{bmatrix}$ has eigenvectors $\mathbf{e}_1 = \begin{bmatrix} 2 \\ 0 \\ 1 \end{bmatrix}$, $\mathbf{e}_2 = \begin{bmatrix} 6 \\ 4 \\ 5 \end{bmatrix}$, and $\mathbf{e}_3 = \begin{bmatrix} 4 \\ 3 \\ 2 \end{bmatrix}$.

What are the corresponding eigenvalues for these eigenvectors?

2. Verify for each Markov transition matrix $\mathbf{A}$ that the given vector is a stable probability vector.

(a) $\mathbf{A} = \begin{bmatrix} \frac{1}{3} & \frac{2}{3} \\ \frac{2}{3} & \frac{1}{3} \end{bmatrix}$, $\mathbf{p} = \begin{bmatrix} \frac{1}{2} \\ \frac{1}{2} \end{bmatrix}$ **(b)** $\mathbf{A} = \begin{bmatrix} \frac{1}{4} & \frac{1}{2} \\ \frac{3}{4} & \frac{1}{2} \end{bmatrix}$, $\mathbf{p} = \begin{bmatrix} \frac{2}{5} \\ \frac{3}{5} \end{bmatrix}$

(c) $\mathbf{A} = \begin{bmatrix} \frac{1}{3} & \frac{1}{2} & 0 \\ \frac{2}{3} & 0 & \frac{2}{3} \\ 0 & \frac{1}{2} & \frac{1}{3} \end{bmatrix}$, $\mathbf{p} = \begin{bmatrix} \frac{3}{10} \\ \frac{4}{10} \\ \frac{3}{10} \end{bmatrix}$

3. The matrix $\begin{bmatrix} 2 & 5 \\ 6 & 1 \end{bmatrix}$ has eigenvalue $\lambda_1 = 7$ with eigenvector $\mathbf{e}_1 = \begin{bmatrix} 1 \\ 1 \end{bmatrix}$ and eigenvalue $\lambda_2 = -4$ with eigenvector $\mathbf{e}_2 = \begin{bmatrix} -5 \\ 6 \end{bmatrix}$.

(a) We want to compute $\mathbf{A}^3\mathbf{v}$ where $\mathbf{v} = \begin{bmatrix} -2 \\ 9 \end{bmatrix}$. Writing $\mathbf{v}$ as $\mathbf{v} = 3\mathbf{e}_1 + \mathbf{e}_2$, compute $\mathbf{A}^3\mathbf{v}$ indirectly as in Example 3.

(b) Give an approximate formula for $\mathbf{A}^n\mathbf{v}$.

(c) Determine a and b so that the vector $\mathbf{v} = \begin{bmatrix} 2 \\ 13 \end{bmatrix}$ can be written as $\mathbf{v} = a\mathbf{e}_1 + b\mathbf{e}_2$, and use this representation of $\mathbf{v}$ to compute $\mathbf{A}^3\mathbf{v}$.

4. The matrix $\begin{bmatrix} 1 & 1 \\ 0 & 2 \end{bmatrix}$ has eigenvectors $\mathbf{e}_1 = \begin{bmatrix} 1 \\ 1 \end{bmatrix}$ and $\mathbf{e}_2 = \begin{bmatrix} 1 \\ 0 \end{bmatrix}$.

(a) We want to compute $\mathbf{A}^4\mathbf{v}$ where $\mathbf{v} = \begin{bmatrix} 3 \\ 1 \end{bmatrix}$. Writing $\mathbf{v}$ as $\mathbf{v} = \mathbf{e}_1 + 2\mathbf{e}_2$, compute $\mathbf{A}^4\mathbf{v}$ indirectly as in Example 3.

(b) Give an approximate formula for $\mathbf{A}^n\mathbf{v}$.

(c) Determine a and b so that the vector $\mathbf{v} = \begin{bmatrix} 6 \\ 9 \end{bmatrix}$ can be written as $\mathbf{v} = a\mathbf{e}_1 + b\mathbf{e}_2$, and use this representation of $\mathbf{v}$ to compute $\mathbf{A}^5\mathbf{v}$.

5. **(a)** Determine the eigenvalues of each of the following matrices. [Parts (vi) to (ix) require computer assistance.]

(i) $\begin{bmatrix} 4 & 0 \\ 2 & 2 \end{bmatrix}$ **(ii)** $\begin{bmatrix} 1 & 2 \\ 5 & 4 \end{bmatrix}$ **(iii)** $\begin{bmatrix} 2 & 1 \\ 2 & 3 \end{bmatrix}$ **(iv)** $\begin{bmatrix} 4 & -1 \\ 1 & 2 \end{bmatrix}$ **(v)** $\begin{bmatrix} 2 & 5 \\ 1 & -2 \end{bmatrix}$

(vi) $\begin{bmatrix} 0 & 1 & 1 \\ 1 & 0 & 1 \\ 1 & 1 & 0 \end{bmatrix}$ **(vii)** $\begin{bmatrix} 3 & 2 & 4 \\ 2 & 0 & 2 \\ 4 & 2 & 3 \end{bmatrix}$ **(viii)** $\begin{bmatrix} 1 & 1 & 2 \\ 1 & 3 & 1 \\ 2 & 1 & 1 \end{bmatrix}$

(ix) $\begin{bmatrix} 3 & 2 & 0 & 1 \\ 1 & 4 & 0 & 1 \\ 1 & -2 & 2 & -3 \\ -1 & 2 & 0 & 5 \end{bmatrix}$

(b) Determine an eigenvector associated with the largest eigenvalue for the matrices in part (a).

6. The matrix $\mathbf{B} = \begin{bmatrix} \frac{2}{3} & -\frac{1}{3} \\ -\frac{1}{3} & \frac{2}{3} \end{bmatrix}$ is the inverse of $\mathbf{A} = \begin{bmatrix} 2 & 1 \\ 1 & 2 \end{bmatrix}$.

   (a) Verify that $\mathbf{e}_1 = \begin{bmatrix} 1 \\ 1 \end{bmatrix}$ and $\mathbf{e}_2 = \begin{bmatrix} 1 \\ -1 \end{bmatrix}$ are eigenvectors of both $\mathbf{A}$ and $\mathbf{B}$.

   (b) Determine the eigenvalues of $\mathbf{A}$ and $\mathbf{B}$.

7. (a) For the following rabbit/fox models, determine both eigenvalues.

   (i) $R' = 1.1R - 0.3F$ (ii) $R' = 1.3R - 0.2F$ (iii) $R' = 1.1R + 0.1F$
   $F' = 0.2R + 0.4F$ $F' = 0.15R + 0.9F$ $F' = 0.2R + 1.1F$

   (b) Determine an eigenvector $\mathbf{e}$ associated with the largest eigenvalue in each system in part (a).

   (c) Determine the other eigenvector $\mathbf{e}'$ (associated with the smaller eigenvalue) for each system in part (a).

   (d) If the initial population is $\mathbf{x} = \begin{bmatrix} 10 \\ 10 \end{bmatrix}$, express $\mathbf{x}^{(k)}$ as a linear combination of $\mathbf{e}$ and $\mathbf{e}'$ as in equation (26), for each system in part (a). Use this expression to describe in words the behavior of this model over time.

8. The following growth model for elephants (E) and mice (M) predicts population changes from decade to decade.

$$E' = 3E + M$$
$$M' = 2E + 4M$$

   (a) Determine the eigenvalues and associated eigenvectors for this system.

   (b) Suppose that initially we have $\mathbf{p} = \begin{bmatrix} E \\ M \end{bmatrix} = \begin{bmatrix} 5 \\ 5 \end{bmatrix}$. Write $\mathbf{p}$ as a linear combination of the eigenvectors.

   (c) Use the information in part (b) to determine an approximate value for the population sizes in eight decades.

9. The following growth model for computer science teachers (T), computer operators (O), and computer programmers (P) predicts population changes from decade to decade.

$$T' = T - O$$
$$O' = -T + 2O - P$$
$$P' = - O + P$$

   (a) Determine the eigenvalues and associated eigenvectors for this system.

   (b) Suppose that initially we have $\mathbf{p} = \begin{bmatrix} T \\ O \\ P \end{bmatrix} = \begin{bmatrix} 50 \\ 0 \\ 10 \end{bmatrix}$. Write $\mathbf{p}$ as a linear combination of the eigenvectors.

   (c) Use the information in part (b) to determine an approximate value for the population sizes in 12 decades.

10. (a) Let $\mathbf{A}$ be an $n \times n$ matrix with n distinct eigenvalues $|\lambda_1| > |\lambda_2| > \cdots > |\lambda_n|$, with associated eigenvectors $\mathbf{e}_i$. Let n-vector $\mathbf{v}$ be expressed as the linear combination of eigenvectors

$$\mathbf{v} = a_1\mathbf{e}_1 + a_2\mathbf{e}_2 + a_3\mathbf{e}_3 + \cdots + a_n\mathbf{e}_n$$

Then show that

$$\mathbf{A}^k\mathbf{v} = \lambda_1^k\left\{a_1\mathbf{e}_1 + \left(\frac{\lambda_2}{\lambda_1}\right)^k a_2\mathbf{e}_2 + \left(\frac{\lambda_3}{\lambda_1}\right)^k a_3\mathbf{e}_3 + \cdots + \left(\frac{\lambda_n}{\lambda_1}\right)^k a_n\mathbf{e}_n\right\}$$

(b) Use part (a) to show that $\mathbf{A}^k\mathbf{v} \to a_1\lambda_1^k\mathbf{e}_1$ as $k \to \infty$.

11. Verify that the constant term in the characteristic polynomial is det(A).

12. Show that if **A** is a 2×2 matrix with all entries positive and det(A) > 0, the two eigenvalues of **A** are both positive, real numbers. (**Hint:** Use the quadratic formula.)

13. Show that the product of the eigenvalues of a 2×2 matrix **A** equals det(A). [**Hint:** See Exercise 11 and show that the product of the eigenvalues is the constant term in the characteristic polynomial det(**A** − λ**I**).] Note that this result is true for matrices of any size.

14. Show that the sum of the eigenvalues of a 2×2 matrix **A** equals the sum of the main diagonal entries of **A**. [**Hint:** show these quantites are both the coefficient of $-\lambda$ in the characteristic polynomial det(**A** − λ**I**).] Note that this result is true for matrices of any size.

15. Show that if **A** is a 2×2 symmetric matrix of the form $\begin{bmatrix} a & b \\ b & a \end{bmatrix}$, the eigenvalues of

A are $a + b$ and $a - b$. Verify that $\begin{bmatrix} 1 \\ 1 \end{bmatrix}$ and $\begin{bmatrix} 1 \\ -1 \end{bmatrix}$ are the associated eigenvectors.

16. **(a)** For a matrix $\mathbf{A} = \begin{bmatrix} a & q \\ 0 & b \end{bmatrix}$, show that the eigenvalues are a and b.

   (b) Generalize the result in part (a) to show that in any upper triangular matrix, the eigenvalues are just the entries on the main diagonal. (**Hint:** See Theorem 2 in Section 1.6.)

17. This exercise illustrates a famous result in linear algebra known as the *Cayley–Hamilton theorem*, which says that a square matrix **A** satisfies its characteristic equation, det(**A** − λ**I**) = 0.

   (a) Let $\mathbf{A} = \begin{bmatrix} 2 & 1 \\ 1 & 2 \end{bmatrix}$. So det(**A** − λ**I**) = $(2 - \lambda)(2 - \lambda) - 1 \cdot 1 = \lambda^2 - 4\lambda + 3$, and the characteristic equation of **A** is $\lambda^2 - 4\lambda + 3 = 0$. Verify that **A** satisfies its characteristic equation; that is, show that $\mathbf{A}^2 - 4\mathbf{A} + 3\mathbf{I} = \mathbf{O}$.

   (b) The characteristic equation can be factored to $(\lambda - 3)(\lambda - 1) = 0$. Check that (**A** − 3**I**)(**A** − **I**) = **O**.

   (c) Following the same steps as in part (a), check that the matrix $\mathbf{A} = \begin{bmatrix} 3 & 1 \\ 2 & 2 \end{bmatrix}$ for the computer/dog model satisfies its characteristic equation.

18. **(a)** Determine the dominant eigenvalue and an associated eigenvector for the following system of equations by computing successive $\begin{bmatrix} x' \\ y' \end{bmatrix}$ pairs for many periods. Use $\begin{bmatrix} 1 \\ 0 \end{bmatrix}$ as your starting vector.

$$x' = 0.707x - 0.707y$$
$$y' = 0.707x + 0.707y$$

(b) Plot the successive iterates on graph paper. Try other starting vectors. State in words the effect of x–y-coordinates of this linear model.

(c) Solve the characteristic equation $\det(\mathbf{A} - \lambda\mathbf{I}) = 0$ to determine the eigenvalues for this matrix of coefficients. You are finding out that imaginary eigenvalues correspond to rotations. Note that $0.707 = \sin 45° = \cos 45°$.

19. **(a)** Determine the dominant eigenvalue and an associated eigenvector for the following system of equations by computing successive $\begin{bmatrix} x' \\ y' \end{bmatrix}$ pairs for many periods. Use $\begin{bmatrix} 1 \\ 1 \end{bmatrix}$ as your starting vector.

$$x' = 2x - y$$
$$y' = 3x - 2y$$

(b) Repeat the iteration starting with $\begin{bmatrix} 1 \\ 3 \end{bmatrix}$.

(c) Solve the characteristic equation $\det(\mathbf{A} - \lambda\mathbf{I}) = 0$ to determine the eigenvalues for this matrix of coefficients. Does this give you any hints about what is wrong in the iteration procedure in part (b)?

20. (*Computer Project*) Growth models for a species that is split into three age groups, baby (*b*), youth (*y*), adult (*a*), have the form $\mathbf{p}' = A\mathbf{p}$, where $\mathbf{p} = \begin{bmatrix} b \\ y \\ a \end{bmatrix}$, $\mathbf{p}' = \begin{bmatrix} b' \\ y' \\ a' \end{bmatrix}$,

$A = \begin{bmatrix} 0 & 0 & c \\ s_1 & 0 & 0 \\ 0 & s_2 & s_3 \end{bmatrix}$, and $c =$ number of children born each period by an adult, $s_1 =$ probability of a baby surviving one period to become a youth, $s_2 =$ probability of a youth surviving one period to become an adult, and $s_3 =$ probability of an adult surviving the current period. A growth model of this form is a called a *Leslie model*. Explore some Leslie growth models for different values of c, s_1, s_2, s_3 of your choice. Use software packages to determine the eigenvectors and eigenvalues. Initially set $s_3 = 0$.

21. (*Computer Project*)
(a) Use computer software to determine the eigenvalues and eigenvectors, with entries of the dominant eigenvector chosen to sum to 1 (recall that the dominant eigenvector is the stable probability distribution; see Example 2) for the weather Markov chain introduced in Section 1.1. Express the starting vector $\begin{bmatrix} \frac{1}{2} \\ \frac{1}{2} \\ 0 \end{bmatrix}$ as a linear combination of these eigenvectors.

(b) Repeat part (a) for some of the Markov transition matrices in Exercise 23 in Section 1.4. What common patterns do you see?

3.4   Norm and Condition Number of a Matrix

In Section 2.2 we presented three norms for a vector: the standard Euclidean distance norm and two nongeometric norms, the sum norm and the max norm. In this section we discuss norms for square matrices. We then use them to develop the condition number, a measure of how "well behaved" solutions of $Ax = b$ are. First we review the three vector norms.

$$\text{Euclidean norm: } |a|_e = \sqrt{a_1^2 + a_2^2 + \cdots + a_n^2}$$
$$\text{Sum norm: } |a|_s = \Sigma |a_i| \tag{1}$$
$$\text{Max norm: } |a|_m = \max\{|a_i|\}$$

If $a = [-1, 2, 2]$, then $|a|_e = \sqrt{1^2 + 2^2 + 2^2} = \sqrt{9} = 3$, $|a|_s = 1 + 2 + 2 = 5$, and $|a|_m = \max\{1, 2, 2\} = 2$. The coordinate vectors $i_1, i_2, \ldots, i_n$, which have all entries 0's except for a 1 in the position indicated by the subscript, have a value of 1 in all three norms.

The norm $\|A\|$ of a square matrix A is a bound on the magnifying effect A has when it multiplies some vector. We define $\|A\|$ to be the (smallest) bound so that

$$|Ax| \le \|A\| \cdot |x| \qquad \text{for all } x \tag{2}$$

Thus

$$\|A\| = \max_{x \ne 0} \frac{|Ax|}{|x|} \tag{3}$$

There is an immediate extension of (2) to powers of A.

$$|A^k x| \le \|A\|^k \cdot |x| \tag{4}$$

Since the magnifying effect depends on how the size of the vector is measured in (3), for each of our three vector norms we get a corresponding matrix norm: a Euclidean matrix norm $\|A\|_e$, a sum matrix norm $\|A\|_s$, and a max matrix norm $\|A\|_m$. Each of these three matrix norms has its own special properties.

From Section 3.3 we saw that when we represent a vector x as a linear combination of eigenvectors, the maximum growth in Ax occurs in the dominant eigenvector. So it would seem that the $\|A\|$ should be the dominant eigenvalue. For symmetric matrices, the Euclidean norm of A is indeed equal to (the absolute value of) the largest eigenvalue of A.

As with the Euclidean norm for vectors, the Euclidean norm for square matrices is the most commonly used matrix norm in linear algebra and has the best theoretical properties. For nonsymmetric matrices, the Euclidean norm is a more complicated number that comes from the largest eigenvalue of a related symmetric matrix. For matrices larger than 2×2, the largest eigenvalue, and hence the Euclidean norm of a matrix, requires a computer to calculate. On the other hand, the sum and max norms are easy to determine by hand. It is also much easier to prove properties of the sum and max norms (at this point we lack several theoretical results needed to work with the Euclidean norm of a matrix). For this reason we consider just the sum and max norm in the rest of this section.

Theorem 1. The sum matrix norm $\|A\|_s$ equals the largest column sum of A (in absolute value). The max matrix norm $\|A\|_m$ equals the largest row sum of A (in absolute value).

(i) $\|\mathbf{A}\|_s = \max_j \{|\mathbf{a}_j|_s\}$

(ii) $\|\mathbf{A}\|_m = \max_i \{|\mathbf{a}_i'|_s\}$

The proof of (i) and (ii) is given shortly. First let us illustrate these formulas.

EXAMPLE 1.
Sum and Max
Norms of a Matrix

Let us use Theorem 1 to determine the sum and max norms of $\mathbf{A} = \begin{bmatrix} 1 & 2 & 3 \\ 4 & 5 & 6 \\ 7 & 8 & 9 \end{bmatrix}$. The last column has the largest column sum, so the sum matrix norm of $\mathbf{A}$ is

$$\|\mathbf{A}\|_s = |\mathbf{a}_3|_s = 3 + 6 + 9 = 18$$

The last row has the largest row sum, so the max matrix norm of $\mathbf{A}$ is

$$\|\mathbf{A}\|_m = |\mathbf{a}_3'|_s = 7 + 8 + 9 = 24$$

Let us see how these norms bound the magnifying effect of multiplying a vector $\mathbf{x}$ by $\mathbf{A}$. Since $|\mathbf{A}\mathbf{x}| \leq \|\mathbf{A}\| \cdot |\mathbf{x}|$, using the sum and max norms we have

$$|\mathbf{A}\mathbf{x}|_s \leq 18|\mathbf{x}|_s \qquad |\mathbf{A}\mathbf{x}|_m \leq 24|\mathbf{x}|_m \tag{5}$$

We claim that we can attain the sum norm bound in (5) using $\mathbf{i}_3 = \begin{bmatrix} 0 \\ 0 \\ 1 \end{bmatrix}$, where $|\mathbf{i}_3|_s = 1$.

$$|\mathbf{A}\mathbf{i}_3|_s \leq \|\mathbf{A}\|_s |\mathbf{i}_3|_s = 18 \cdot 1 = 18$$

$|\mathbf{A}\mathbf{i}_3|_s$ equals this bound:

$$\mathbf{A}\mathbf{i}_3 = \begin{bmatrix} 1 & 2 & 3 \\ 4 & 5 & 6 \\ 7 & 8 & 9 \end{bmatrix} \begin{bmatrix} 0 \\ 0 \\ 1 \end{bmatrix} = \begin{bmatrix} 3 \\ 6 \\ 9 \end{bmatrix}, \quad \text{so} \quad |\mathbf{A}\mathbf{i}_3|_s = \left| \begin{bmatrix} 3 \\ 6 \\ 9 \end{bmatrix} \right|_s = 18 \tag{6}$$

The reader should check that we attain the max norm bound of 24 with

$$\mathbf{x} = \mathbf{1} = \begin{bmatrix} 1 \\ 1 \\ 1 \end{bmatrix}. \qquad ■$$

Our use of a coordinate vector and the 1's vector to achieve the sum and max norm bounds, respectively, in (5) always works for positive matrices.

Theorem 2. Let $\mathbf{A}$ be a matrix with nonnegative entries.
(i) *Sum norm.* If the kth column of $\mathbf{A}$ has the largest sum, then coordinate vector $\mathbf{i}_k$ achieves the sum norm bound: $|\mathbf{A}\mathbf{i}_k|_s = \|\mathbf{A}\|_s |\mathbf{i}_k|_s$.

(ii) *Max norm*. The 1's vector **1** achieves the max norm bound: $|A1|_m = \|A\|_m|1|_m$.

Proof of (i) *in Theorems* 1 *and* 2: First we note that in the definition (3) of the sum norm $\|A\|_s = \max(|Ax|_s/|x|_s)$, it is sufficient to consider only vectors x with $|x|_s = 1$. For if $|x|_s = k$, $y = (1/k)x$ has sum norm 1 and

$$\frac{|Ay|_s}{|y|_s} = \frac{1/k|Ax|_s}{1/k|x|_s} = \frac{|Ax|_s}{|x|_s}$$

For concreteness, we work with matrix A in Example 1. We want to find a vector x with $|x|_s = 1$ that maximizes $|Ax|_s \, (= |Ax|_s/|x|_s)$. Let us write the matrix–vector product Ax in the following vector form:

$$Ax = \begin{bmatrix} 1 & 2 & 3 \\ 4 & 5 & 6 \\ 7 & 8 & 9 \end{bmatrix}\begin{bmatrix} x_1 \\ x_2 \\ x_3 \end{bmatrix} = x_1\begin{bmatrix} 1 \\ 4 \\ 7 \end{bmatrix} + x_2\begin{bmatrix} 2 \\ 5 \\ 8 \end{bmatrix} + x_3\begin{bmatrix} 3 \\ 6 \\ 9 \end{bmatrix} \tag{7}$$

Since $|x_1| + |x_2| + |x_3| = 1$, we must pick the x_i's to make the sum norm of a linear combination of column vectors in (7) as large as possible (in absolute value). Clearly, (7) is maximized when the x_i associated with the largest column—column 3—is 1 (and the other x_j's are 0). Thus the maximizing x is $i_3 = \begin{bmatrix} 0 \\ 0 \\ 1 \end{bmatrix}$

and the sum norm of (7) with i_3 is $|a_3|_s$, the sum of the third column's entries. This argument is valid for any matrix, no matter what the signs of its entries.

Proof of (ii) *in Theorems* 1 *and* 2: As in the proof of (i), to find $\|A\|_m = \max(|Ax|_m/|x|_m)$, it suffices to find the x with $|x|_m = 1$ that maximizes $|Ax|_m$. The max norm of Ax is $|Ax|_m = \max\{|a_i' \cdot x|\}$, with $|x_j| \le 1$. Assuming that the entries of A are nonnegative, all the $a_i' \cdot x$ are maximized (and hence $\max\{|a_i' \cdot x|\}$ is maximized) by making each $x_j = 1$. Thus $x = 1$ and $a_i' \cdot x = a_i' \cdot 1 = |a_i'|_s$. Then $\|A\|_m = \max\{|a_i' \cdot 1|\} = \max\{|a_i'|_s\}$, as claimed. ■

A simple alteration of the 1's vector is required to achieve the max norm when A has negative entries (see Exercise 14).

| **EXAMPLE 2.**
Norm Bound on
Growth in a
Population Model | Consider the growth model from Section 3.3 for the numbers of computers and dogs in successive years. Let C be the number of computers this year and C' the number next year. Similarly, let D be the number of dogs this year and D' the number next year. |

$$\begin{array}{l} C' = 3C + D \\ D' = 2C + 2D \end{array} \quad \text{or} \quad c' = Ac, \quad \text{where} \quad A = \begin{bmatrix} 3 & 1 \\ 2 & 2 \end{bmatrix}, \quad c = \begin{bmatrix} C \\ D \end{bmatrix}, \quad c = \begin{bmatrix} C' \\ D' \end{bmatrix} \tag{8}$$

The sum norm $\|A\|_s$ of the coefficient matrix A is 5 (= the sum of coefficients in the first column). The max norm $\|A\|_m$ of the coefficient matrix A is 4 (= the sum of coefficients in either row).

By (2), $|\mathbf{c}'| = |\mathbf{Ac}| \le \|\mathbf{A}\| \cdot |\mathbf{c}|$. With the numerical values of the sum and max norms of $\mathbf{A}$, we have

$$|\mathbf{c}'|_s \le 5|\mathbf{c}|_s \quad \text{and} \quad |\mathbf{c}'|_m \le 4|\mathbf{c}|_m \tag{9}$$

Suppose that we start with $C = 20$ and $D = 10$. Then $|\mathbf{c}|_s = 30$ and $|\mathbf{c}|_m = 20$. So (9) yields the sum and max norm bounds

$$|\mathbf{c}'|_s \le 5|\mathbf{c}|_s = 5 \cdot 30 = 150 \quad \text{and} \quad |\mathbf{c}'|_m \le 4|\mathbf{c}|_m = 4 \cdot 20 = 80$$

From (8) we compute $C' = 3(20) + 10 = 70$, $D' = 2(20) + 2(10) = 60$, so $|\mathbf{c}'|_s = 130$ and $|\mathbf{c}'|_m = 70$. Thus for $\mathbf{c} = \begin{bmatrix} 20 \\ 10 \end{bmatrix}$, the sum norm bound and the max norm bound are both decent estimates.

Bound (4) can be used for the population $\mathbf{c}^{(k)}$ after k periods:

$$|\mathbf{c}^{(k)}|_s = |\mathbf{A}^k \mathbf{c}|_s \le \|\mathbf{A}\|^k |\mathbf{c}|_s = 5^k \cdot 30 \tag{10}$$

■

**EXAMPLE 3.
Sum Norm of a
Markov Transition
Matrix**

Since the sum norm equals the largest column sum and the entries in every column $\mathbf{a}_i$ sum to 1 in a Markov transition matrix, it follows that a Markov transition matrix $\mathbf{A}$ has sum norm $\|\mathbf{A}\|_s = 1$. Because powers of a transition matrix also have column sums of 1, all powers of $\mathbf{A}$ have sum norm of 1.

Any probability vector $\mathbf{p}$ achieves the sum norm bound:

$$|\mathbf{Ap}|_s = \|\mathbf{A}\|_s |\mathbf{p}|_s \quad \text{or} \quad |\mathbf{Ap}|_s = |\mathbf{p}|_s \quad (\text{since } \|\mathbf{A}\|_s = 1)$$

since the vectors $\mathbf{p}$ and $\mathbf{Ap}$ ($= \mathbf{p}'$) are probability vectors and thus $|\mathbf{p}|_s = |\mathbf{p}'|_s = 1$. ■

Theorem 3. Let $\mathbf{A}$ and $\mathbf{B}$ be $n \times n$ matrices. Then any matrix norm has the following properties.

  (i) $\|r\mathbf{A}\| = |r| \cdot \|\mathbf{A}\|$
  (ii) $\|\mathbf{AB}\| \le \|\mathbf{A}\| \cdot \|\mathbf{B}\|$
  (iii) $\|\mathbf{A}^k\| \le (\|\mathbf{A}\|)^k$
  (iv) $\|\mathbf{A} + \mathbf{B}\| \le \|\mathbf{A}\| + \|\mathbf{B}\|$

Proof: (i) By scalar factoring for matrix products and for vector norms, we know that $|(r\mathbf{A})\mathbf{x}| = |r| \cdot |\mathbf{Ax}|$. Whatever the value of $|\mathbf{Ax}|/|\mathbf{x}|$ is, then $|r\mathbf{Ax}|/|\mathbf{x}| = |r| \cdot |\mathbf{Ax}|/|\mathbf{x}|$, and $\max(|r\mathbf{Ax}|/|\mathbf{x}|) = |r|\{\max(|\mathbf{Ax}|/|\mathbf{x}|)\}$. These are the maximum ratios defining the matrix norm. Then $\|r\mathbf{A}\| = |r| \cdot \|\mathbf{A}\|$.

(ii) Repeated use of $|\mathbf{Dx}| \le \|\mathbf{D}\| \cdot |\mathbf{x}|$ gives

$$|(\mathbf{AB})\mathbf{x}| = |\mathbf{A}(\mathbf{Bx})| \le \|\mathbf{A}\|(|\mathbf{Bx}|) \le \|\mathbf{A}\| \cdot (\|\mathbf{B}\| \cdot |\mathbf{x}|) \tag{11}$$

Assuming that $\mathbf{x} \ne 0$, we can divide by $|\mathbf{x}|$ in (11) to obtain

$$\frac{|(\mathbf{AB})\mathbf{x}|}{|\mathbf{x}|} \le \|\mathbf{A}\| \cdot \|\mathbf{B}\| \tag{12}$$

In particular, the inequality in (12) holds for the **x** that maximizes the ratio on the left. That maximum ratio is $\|AB\|$. Thus (ii) is proved.

(iii) This result follows from repeated use of (ii) when $B = A$.

(iv) The addition inequality is hard to prove in general. For the sum norm and the max norm, the result is fairly simple when we think how these norms are defined: The sum matrix norm $\|A\|_s$ equals the largest column sum of A (in absolute value), and the max matrix norm $\|A\|_m$ equals the largest row sum of A (in absolute value). Clearly, the largest (absolute value) column sum of $A + B$ cannot exceed the largest column sum of A plus the largest column sum of B; and similarly for row sums. ■

One of the most important uses of norms is to determine error bounds.

EXAMPLE 4.
Use of Matrix
Norm in Error
Bounds

Consider again our growth model for the numbers of computers C and dogs D in successive years:

$$\begin{array}{l} C' = 3C + D \\ D' = 2C + 2D \end{array} \quad \text{or} \quad \mathbf{c}' = \mathbf{Ac}, \quad \text{where} \quad A = \begin{bmatrix} 3 & 1 \\ 2 & 2 \end{bmatrix}, \quad \mathbf{c} = \begin{bmatrix} C \\ D \end{bmatrix}, \quad \mathbf{c}' = \begin{bmatrix} C' \\ D' \end{bmatrix}$$

The sum norm $\|A\|_s = 5$. The max norm $\|A\|_m = 4$.

Suppose that there is an error in determining **c** and we mistakenly use the initial vector **b**, where $\mathbf{b} = \mathbf{c} + \mathbf{e}$. Here **e** is the vector of errors. Then the error one year later is

$$\begin{aligned} \mathbf{b}' - \mathbf{c}' &= \mathbf{Ab} - \mathbf{Ac} \\ &= \mathbf{A}(\mathbf{b} - \mathbf{c}) = \mathbf{Ae} \end{aligned} \tag{13}$$

Taking (sum) norm bounds, we have the error bound

$$|\mathbf{b}' - \mathbf{c}'|_s \le \|A\|_s |\mathbf{e}|_s = 5|\mathbf{e}|_s \tag{14}$$

If we know that the sum of the errors is no more than $\frac{1}{2}$, then the sum of our errors after one year is at most $5|\mathbf{e}|_s = 5 \cdot \frac{1}{2} = \frac{5}{2}$.

Following the model for n years, we let $\mathbf{c}^{(n)}$ denote the numbers of computers and dogs after n years. Then

$$\mathbf{c}^{(n)} = \mathbf{A}^n \mathbf{c}$$

and after n years, the error is

$$\begin{aligned} \mathbf{b}^{(n)} - \mathbf{c}^{(n)} &= \mathbf{A}^n \mathbf{b} - \mathbf{A}^n \mathbf{c} \\ &= \mathbf{A}^n(\mathbf{b} - \mathbf{c}) \\ &= \mathbf{A}^n \mathbf{e} \end{aligned} \tag{15}$$

Taking norms, we have the error bound

$$|\mathbf{b}^{(n)} - \mathbf{c}^{(n)}|_s \le \|A^n\|_s |\mathbf{e}|_s \le \|A\|_s^n |\mathbf{e}|_s = 5^n |\mathbf{e}|_s \tag{16}$$

For example, suppose that $\mathbf{c} = \begin{bmatrix} 20 \\ 10 \end{bmatrix}$ but $\mathbf{b} = \begin{bmatrix} 22 \\ 11 \end{bmatrix}$, so $\mathbf{e} = \begin{bmatrix} 2 \\ 1 \end{bmatrix}$ with $|\mathbf{e}|_s = 3$. Then computing $\mathbf{c}'$ and $\mathbf{b}'$, we find that

$$\mathbf{c}' = \mathbf{Ac} = \begin{bmatrix} 3 & 1 \\ 2 & 2 \end{bmatrix}\begin{bmatrix} 20 \\ 10 \end{bmatrix} = \begin{bmatrix} 70 \\ 60 \end{bmatrix}$$

and

$$\mathbf{b}' = \mathbf{Ab} = \begin{bmatrix} 3 & 1 \\ 2 & 2 \end{bmatrix}\begin{bmatrix} 22 \\ 11 \end{bmatrix} = \begin{bmatrix} 77 \\ 66 \end{bmatrix}$$

So $\mathbf{b}' - \mathbf{c}' = \begin{bmatrix} 7 \\ 6 \end{bmatrix}$ with $|\mathbf{b}' - \mathbf{c}'|_s = 13$. The sum norm bound on this error is, from (14),

$$|\mathbf{b}' - \mathbf{c}'|_s \le \|\mathbf{A}\|_s|\mathbf{e}|_s = 5 \cdot 3 = 15 \tag{17}$$

This bound of 15 compares well with the observed error of 13. Using the max norm yields

$$|\mathbf{b}' - \mathbf{c}'|_m \le \|\mathbf{A}\|_m|\mathbf{e}|_m = 4 \cdot 2 = 8 \tag{18}$$

This max bound of 8 compares well with the observed max error of 7. If we had iterated n times, the error between $\mathbf{b}^{(n)} = \mathbf{A}^n\mathbf{b}$ and $\mathbf{c}^{(n)} = \mathbf{A}^n\mathbf{c}$ would be bounded, using (16), by $\|\mathbf{A}\|_s^n|\mathbf{e}|_s = 5^n \cdot 3$. ■

Condition Number of a Matrix

Now we turn to the study of systems of linear equations whose solution is very sensitive to small changes in the coefficients. Such systems are called ***ill-conditioned***. Consider the following system of equations.

$$\begin{aligned} x_1 + kx_2 &= 1 \\ x_1 + x_2 &= 4 \end{aligned} \tag{19}$$

When k becomes close to 1, these two equations represent almost parallel lines. A small change in the value of k near 1 has a large effect on where these lines will intersect. To see this, note that the determinant of the coefficient matrix in (19) is $1 - k$, and Cramer's rule gives

$$x_1 = \frac{1 - 4k}{1 - k} \quad \text{and} \quad x_2 = \frac{-3}{1 - k} \tag{20}$$

As $k \to 1$, $1 - k \to 0$. So in (20) we are almost dividing by 0 and problems will abound. For example, if $k = 0.99$, then while the coefficients in (19) are all about 1, the solution is $x_1 = -296$ and $x_2 = 300$. And if $k = 0.99999$, then

$$x_1 = -299,996 \quad \text{and} \quad x_2 = 300,000 \tag{21}$$

[Note that to be precise, the critical issue here is not the size of the det$(\mathbf{A})$ in $\mathbf{Ax} = \mathbf{b}$ but the size of det$(\mathbf{A})$ *relative* to the size of the numerator in Cramer's rule.]

A related problem arises with the inverse of a square matrix $\mathbf{A}$. If det$(\mathbf{A}) \neq 0$, the inverse exists. However, as det$(\mathbf{A})$ approaches 0, say, because a variable like k in (19) is changing, the inverse starts to "blow up." For example, the inverse of the coefficient matrix in (19) is

$$\mathbf{A}^{-1} = \begin{bmatrix} \dfrac{1}{1-k} & \dfrac{-k}{1-k} \\ \dfrac{-1}{1-k} & \dfrac{1}{1-k} \end{bmatrix} \tag{22}$$

When $k = 0.99999$, the entries in (22) will be around 100,000.

Imagine a problem where the coefficients come from observations. A small change in the reading of a coefficient could have a drastic effect on the solution of the system of equations, if the system is poorly behaved as in (19).

The reason for the poor behavior is found by considering the geometry of the problem. Looking at (19) as a linear combination of vectors, we have

$$x_1 \begin{bmatrix} 1 \\ 1 \end{bmatrix} + x_2 \begin{bmatrix} k \\ 1 \end{bmatrix} = \begin{bmatrix} 1 \\ 4 \end{bmatrix} \tag{23}$$

As $k \to 1$, the two vectors on the left side in (23) become the same and linear combinations of them will only produce vectors of the form $\begin{bmatrix} b \\ b \end{bmatrix}$. When $\begin{bmatrix} 1 \\ 1 \end{bmatrix}$ and $\begin{bmatrix} k \\ 1 \end{bmatrix}$ are almost the same, then to find the right values for x_1 and x_2 in (23) takes a "trick," as in (21), involving the difference of huge multiples of the two vectors. In (23) it is

$$-299,996 \begin{bmatrix} 1 \\ 1 \end{bmatrix} + 300,000 \begin{bmatrix} 0.99999 \\ 1 \end{bmatrix} = \begin{bmatrix} 1 \\ 4 \end{bmatrix} \tag{24}$$

A more rigorous analysis of errors requires use of the norm of the coefficient matrix $\mathbf{A}$. For easy reference, we repeat the key inequality satisfied by matrix norms: For any vector $\mathbf{x}$,

$$\|\mathbf{Ax}\| \leq \|\mathbf{A}\| \cdot |\mathbf{x}| \tag{25}$$

Suppose that $\mathbf{E}$ represents a matrix of errors (either in recording data or roundoff errors); the true matrix $\mathbf{A}$ has been altered to become the matrix $\mathbf{A} + \mathbf{E}$. Then let us see how changing $\mathbf{A}$ to $\mathbf{A} + \mathbf{E}$ changes the solution to the matrix equation $\mathbf{Ax} = \mathbf{b}$. Suppose that $\mathbf{x}$ is the solution to the correct equation and $\mathbf{x} + \mathbf{e}$ represents the solution to the altered equation. Thus we have

$$\mathbf{Ax} = \mathbf{b} \quad \text{and} \quad (\mathbf{A} + \mathbf{E})(\mathbf{x} + \mathbf{e}) = \mathbf{b} \tag{26}$$

We will now derive a bound on the relative size of $\mathbf{e}$ in terms of the relative size of $\mathbf{E}$ and $\mathbf{A}$.

Theorem 4. Let $\mathbf{A}$ be a square, invertible matrix. Suppose that instead of solving $\mathbf{Ax} = \mathbf{b}$, we introduce an error changing $\mathbf{A}$ to $\mathbf{A} + \mathbf{E}$ and solve the system $(\mathbf{A} + \mathbf{E})(\mathbf{x} + \mathbf{e}) = \mathbf{b}$. Then

$$\frac{|\mathbf{e}|}{|\mathbf{x} + \mathbf{e}|} \leq c(\mathbf{A})\frac{\|\mathbf{E}\|}{\|\mathbf{A}\|} \tag{27}$$

where $c(\mathbf{A}) = \|\mathbf{A}^{-1}\| \cdot \|\mathbf{A}\|$ is called the **condition number** of $\mathbf{A}$.

We derive (27) by subtracting the first equation in (26) from the second to obtain

$$\{(\mathbf{A} + \mathbf{E})(\mathbf{x} + \mathbf{e})\} - \mathbf{Ax} = \mathbf{b} - \mathbf{b} = \mathbf{0} \quad \text{or} \quad \{(\mathbf{Ax} + \mathbf{Ae} + \mathbf{E}(\mathbf{x} + \mathbf{e})\} - \mathbf{Ax} = \mathbf{0} \tag{28}$$

Canceling the $\mathbf{Ax}$ terms, we have

$$\mathbf{Ae} + \mathbf{E}(\mathbf{x} + \mathbf{e}) = \mathbf{0} \tag{29}$$

or

$$\mathbf{Ae} = -\mathbf{E}(\mathbf{x} + \mathbf{e}) \tag{30}$$

We must assume that $\mathbf{A}$ is invertible (otherwise, the solution does not exist or is not unique). Then we can solve (30) for $\mathbf{e}$.

$$\mathbf{e} = -\mathbf{A}^{-1}\mathbf{E}(\mathbf{x} + \mathbf{e}) \tag{31}$$

Taking norms on both sides of (31) and applying the norm inequality (25) on the right side, we have

$$|\mathbf{e}| = |\mathbf{A}^{-1}\mathbf{E}(\mathbf{x} + \mathbf{e})| \leq \|\mathbf{A}^{-1}\mathbf{E}\| \cdot |\mathbf{x} + \mathbf{e}| \tag{32}$$

From Theorem 3(ii) we have

$$\|\mathbf{A}^{-1}\mathbf{E}\| \leq \|\mathbf{A}^{-1}\| \cdot \|\mathbf{E}\| \tag{33}$$

Substituting (33) into (32) gives

$$|\mathbf{e}| \cdot |\mathbf{x} + \mathbf{e}| \leq \|\mathbf{A}^{-1}\| \cdot \|\mathbf{E}\| \cdot |\mathbf{x} + \mathbf{e}| \tag{34}$$

Dividing both sides of (34) by $|\mathbf{x} + \mathbf{e}|$ yields the bound we were seeking.

$$\frac{|\mathbf{e}|}{|\mathbf{x} + \mathbf{e}|} \leq \|\mathbf{A}^{-1}\| \cdot \|\mathbf{E}\| \tag{35}$$

Equation (35) can be rewritten as

$$\frac{|\mathbf{e}|}{|\mathbf{x} + \mathbf{e}|} \leq (\|\mathbf{A}^{-1}\| \cdot \|\mathbf{A}\|)\frac{\|\mathbf{E}\|}{\|\mathbf{A}\|} \tag{36}$$

So the condition number $c(\mathbf{A})$ in (27) turns out to be $c(\mathbf{A}) = \|\mathbf{A}^{-1}\| \cdot \|\mathbf{A}\|$.

A small condition number means that the matrix is well behaved and yields stable computations during elimination, since by (27) small relative errors in **A** will only produce small errors in the solution vector.

The condition number can also be shown to bound the effects on **x** of an error **e***
in the right-side vector **b**. Again, if **x** + **e** is the solution to the erroneous system
A(**x** + **e**) = **b** + **e***, then (see Exercise 20 for proof)

$$\frac{|\mathbf{e}|}{|\mathbf{x}|} \le c(\mathbf{A})\frac{|\mathbf{e}^*|}{|\mathbf{b}|} \tag{37}$$

The exact value of the condition number is dependent on which matrix norm we use. For the sum and max norm, the condition number rarely is less than the number of rows in the matrix. The best possible condition number is for the identity matrix **I** whose inverse is also **I**. The norm of **I** is 1 in all matrix norms (see Exercise 15), so $c(\mathbf{I}) = \|\mathbf{I}\| \cdot \|\mathbf{I}\| = 1$.

**EXAMPLE 5.
Moderately Well-
Conditioned
System**

Consider the refinery system of equations introduced in Section 1.1 whose coefficient matrix is

$$\mathbf{A} = \begin{bmatrix} 4 & 2 & 2 \\ 2 & 5 & 2 \\ 1 & 2.5 & 5 \end{bmatrix}$$

In Example 5 of Section 1.5, we computed its inverse to be

$$\mathbf{A}^{-1} = \begin{bmatrix} \frac{5}{16} & -\frac{5}{64} & -\frac{3}{32} \\ -\frac{1}{8} & \frac{9}{32} & -\frac{1}{16} \\ 0 & -\frac{1}{8} & \frac{1}{4} \end{bmatrix}$$

Taking the maximum (absolute value) of the column sums, we have $\|\mathbf{A}\|_s = 9.5$ (second column) and $\|\mathbf{A}^{-1}\|_s = \frac{31}{64}$ (second column). Thus the condition number of **A** is

$$c(\mathbf{A}) = \|\mathbf{A}\|_s \|\mathbf{A}^{-1}\|_s = 9.5 \cdot \frac{31}{64} \approx 4.6$$

This is a reasonably well conditioned matrix. For the demand vector $\mathbf{b} = \begin{bmatrix} 600 \\ 800 \\ 1000 \end{bmatrix}$
used in this model, we found (in Section 1.4) that the solution **x** of **Ax** = **b** was

$$x_1 = 31\tfrac{1}{4} \qquad x_2 = 87\tfrac{1}{2} \qquad x_3 = 150$$

If we change entry (3, 3) of **A** from 5 to 4 to get **A**′, then the error matrix **E** [a matrix of all 0's except entry (3, 3) is 1] has $\|\mathbf{E}\|_s = 1$. The error bound (27) gives

$$\frac{|\mathbf{e}|_s}{|\mathbf{x} + \mathbf{e}|_s} \le c(\mathbf{A})\frac{\|\mathbf{E}\|_s}{\|\mathbf{A}\|_s} = 4.6\frac{\|\mathbf{E}\|_s}{\|\mathbf{A}\|_s} = 4.6 \cdot \frac{1}{9.5} \approx 48\% \tag{38}$$

Thus a 1/9.5 or 10% change in the norm of **A** can yield up to a 48% error in the norm of the solution. Solving $\mathbf{A'x'} = \mathbf{b}$ for the same **b**, one obtains

$$x_1' = 12\tfrac{1}{2} \qquad x_2' = 75 \qquad x_3' = 200 \tag{39}$$

Writing $\mathbf{x'}$ as $\mathbf{x'} = \mathbf{x} + \mathbf{e}$ yields

$$\mathbf{e} = \mathbf{x'} - \mathbf{x} = \begin{bmatrix} 12.5 \\ 75 \\ 200 \end{bmatrix} - \begin{bmatrix} 31.25 \\ 87.5 \\ 150 \end{bmatrix} \approx \begin{bmatrix} -18.75 \\ -12.5 \\ 50 \end{bmatrix}$$

So $|\mathbf{e}|_s \approx 81$. From (39), $|\mathbf{x} + \mathbf{e}|_s = |\mathbf{x'}|_s \approx 287$. Thus $|\mathbf{e}|_s/|\mathbf{x} + \mathbf{e}|_s \approx \frac{81}{287} = 28\%$, which is about $\frac{3}{5}$ of the theoretical maximum error of 48%. ■

EXAMPLE 6.
Matrix with Small
Condition Number

Consider the matrix **A** with inverse $\mathbf{A}^{-1}$.

$$\mathbf{A} = \begin{bmatrix} 3 & 4 \\ -4 & 3 \end{bmatrix} \quad \text{and} \quad \mathbf{A}^{-1} = \begin{bmatrix} -\frac{3}{25} & -\frac{4}{25} \\ \frac{4}{25} & \frac{3}{25} \end{bmatrix}$$

$\|\mathbf{A}\|_s = 7$ and $\|\mathbf{A}^{-1}\|_s = \frac{7}{25}$ and $c(\mathbf{A}) = 7 \cdot \frac{7}{25} = \frac{49}{25} \approx 2$. ■

EXAMPLE 7.
Poorly
Conditioned
Matrix

Consider the system of equations at the beginning of our discussion of the condition number.

$$\begin{aligned} x_1 + kx_2 &= 1 \\ x_1 + x_2 &= 4 \end{aligned} \tag{40}$$

The sum norm of the coefficient matrix **A** in (40) is 2 (assuming that $0 \le k \le 1$). The inverse of the coefficient matrix is

$$\mathbf{A}^{-1} = \begin{bmatrix} \dfrac{1}{1-k} & \dfrac{-k}{1-k} \\ \dfrac{-1}{1-k} & \dfrac{k}{1-k} \end{bmatrix}$$

Then $\|\mathbf{A}^{-1}\|_s = 2/(1-k)$ (taking the absolute value of entries in the first column of $\mathbf{A}^{-1}$) and $c(\mathbf{A}) = 4/(1-k)$. For $k = 0.9$, $c(\mathbf{A}) \approx 40$ and for $k = 0.99$, $c(\mathbf{A}) \approx 400$. These are terribly conditioned matrices. ■

EXAMPLE 8.
Extremely Ill-
Conditioned
Matrix

Consider the following system of equations:

$$\begin{bmatrix} \frac{1}{3} & \frac{1}{4} & \frac{1}{5} \\ \frac{1}{4} & \frac{1}{5} & \frac{1}{6} \\ \frac{1}{5} & \frac{1}{6} & \frac{1}{7} \end{bmatrix} \begin{bmatrix} x \\ y \\ z \end{bmatrix} = \begin{bmatrix} 0 \\ 1 \\ 0 \end{bmatrix} \tag{41}$$

The condition number of the matrix in (41) is over 5000! The system's solution is $x = -900$, $y = 2880$, $z = -2100$.

Matrices with coefficients of the form $1/k$, where k increases by 1 as the column or row number increases, are famously ill-conditioned. Such matrices are called **Hilbert matrices**. Unfortunately, such matrices arise in some important applications. (A Hilbert matrix arises in Section 4.6 with a condition number of about 5,000,000.) ■

In the exercises, the reader has the chance to examine how the solution to $\mathbf{Ax} = \mathbf{b}$ is affected by small changes in a coefficient of $\mathbf{A}$, for a variety of well-conditioned and ill-conditioned matrices $\mathbf{A}$.

In closing we observe that the condition number of an $n \times n$ matrix $\mathbf{A}$ gives us a more sensitive indicator of whether the system of linear equations $\mathbf{Ax} = \mathbf{b}$ has a solution. We learned in Chapter 1 that for a square matrix $\mathbf{A}$, $\mathbf{Ax} = \mathbf{b}$ always has a (unique) solution if $\mathbf{A}$ has an inverse, or equivalently, if $\det(\mathbf{A}) \neq 0$. The existence of an inverse $\mathbf{A}^{-1}$ is like having a well-defined value for the inverse $1/x$ of a scalar x. Technically, $1/x$ is only undefined for $x = 0$, but informally we sense that for x very close to 0, $1/x$ is going to assume wildly large values that probably would not make sense in real-world situations.

The same is true for the inverse matrix $\mathbf{A}^{-1}$. Before it ceases officially to exist, it starts to "blow up" (when, say, one row is almost a multiple of another row). In practical settings where round off and other errors are present, an inverse not only needs to exist to be useful, but it must come from a well-conditioned matrix with a reasonable condition number.

Principle. The condition number $c(\mathbf{A})$ of a square matrix $\mathbf{A}$ can be viewed akin to the value of the inverse $1/x$ of a scalar x. Just as when x is near 0, $1/x$ is still defined but has wildly large values, so a matrix can be close to having no inverse, when $c(\mathbf{A})$ becomes very large, in which case entries of $\mathbf{A}^{-1}$, and also solutions of $\mathbf{Ax} = \mathbf{b}$, may be unreliable.

Finally, we note that even a well-conditioned problem can yield an unreliable solution if a poor algorithm is used to solve it. A computational algorithm is called **unstable** if it yields a "solution" that is not close to the true solution, nor close to any solution that might result from small changes in the data (coefficients). Finding efficient, stable algorithms is the heart of numerical analysis, a major discipline in applied mathematics.

Section 3.4 Exercises

1. Give the sum and max norms of the following matrices.

(a) $\begin{bmatrix} 1 & 4 \\ 5 & 3 \end{bmatrix}$ (b) $\begin{bmatrix} 0 & 3 \\ -5 & 3 \end{bmatrix}$ (c) $\begin{bmatrix} 8 & 1 \\ 0 & 2 \end{bmatrix}$ (d) $\begin{bmatrix} -2 & -1 \\ -5 & 1 \end{bmatrix}$

(e) $\begin{bmatrix} 1 & 2 & 2 \\ 6 & 1 & 3 \\ 5 & 1 & 2 \end{bmatrix}$ (f) $\begin{bmatrix} -5 & 4 & 6 \\ 8 & 0 & 2 \\ -6 & 7 & 7 \end{bmatrix}$ (g) $\begin{bmatrix} -3 & -4 & -6 \\ -1 & 0 & -3 \\ -4 & 2 & -5 \end{bmatrix}$ (h) $\begin{bmatrix} 0 & 0 & 0 \\ 0 & 0 & 0 \\ 0 & 2 & 0 \end{bmatrix}$

2. (a) For each of the matrices in Exercise 1, give the vector $\mathbf{x}^*$ such that $|\mathbf{A}\mathbf{x}^*|_s = \|\mathbf{A}\|_s \cdot |\mathbf{x}^*|_s$.
    (b) For each of the matrices in Exercise 1, give the vector $\mathbf{x}^*$ such that $|\mathbf{A}\mathbf{x}^*|_m = \|\mathbf{A}\|_m \cdot |\mathbf{x}^*|_m$.

3. (a) What is the sum norm of the matrix $\mathbf{A} = \begin{bmatrix} 2 & 4 & -5 \\ -3 & 3 & 3 \\ 4 & 1 & -1 \end{bmatrix}$?
    (b) If $\mathbf{v}$ is a vector with sum norm of 3, give an upper bound on the sum norm of $\mathbf{A}\mathbf{v}$.
    (c) Give a vector with sum norm of 3 for which the bound in part (b) is achieved.
    (d) If $\mathbf{w}$ is a vector with sum norm of 5, give an upper bound on the sum norm of $\mathbf{A}^2\mathbf{w}$.

4. (a) What is the max norm of the matrix $\mathbf{A} = \begin{bmatrix} 1 & 3 & 2 \\ 2 & 1 & 3 \\ 1 & 1 & 1 \end{bmatrix}$?
    (b) If $\mathbf{v}$ is a vector with max norm of 4, give an upper bound on the max norm of $\mathbf{A}\mathbf{v}$.
    (c) Give a vector with max norm of 4 for which the bound in part (b) is achieved.
    (d) If $\mathbf{w}$ is a vector with max norm of 6, give an upper bound on the max norm of $\mathbf{A}^3\mathbf{w}$.

5. In the computer/dog model in Example 2, give a bound on the size of $\mathbf{p}^{(5)} = \mathbf{A}^5\mathbf{p}$ in the sum and max norms when $\mathbf{p} = \begin{bmatrix} 100 \\ 100 \end{bmatrix}$.

6. In Example 4, suppose we assume that $\mathbf{c} = \begin{bmatrix} 15 \\ 5 \end{bmatrix}$ when the correct value is actually $\begin{bmatrix} 14 \\ 7 \end{bmatrix}$. What is the maximum size that the error could be after 3 years (using the sum norm)?

7. In Example 4 we discussed the absolute size of errors, but it is often more interesting to consider the relative size of errors. The relative error in $\mathbf{b}$ is $|\mathbf{b} - \mathbf{c}|/|\mathbf{c}|$ if $\mathbf{b}$ is used when $\mathbf{c}$ really should be used. (Use the sum norm.)
    (a) If

    $$\begin{aligned} R' &= R + F \\ F' &= 3R - 4F \end{aligned} \quad \text{and} \quad \mathbf{p} = \begin{bmatrix} R \\ F \end{bmatrix}$$

    is set equal to $\begin{bmatrix} 3 \\ 1 \end{bmatrix}$ when it really should have been $\begin{bmatrix} 2 \\ 2 \end{bmatrix}$, then what is the relative error in $\mathbf{p}$ and what is the relative error in $\mathbf{p}'$?
    (b) If

    $$\begin{aligned} R' &= 2R + 3F \\ F' &= -5R - 7F \end{aligned} \quad \text{and} \quad \mathbf{p} = \begin{bmatrix} R \\ F \end{bmatrix}$$

    is set equal to $\begin{bmatrix} 5 \\ 5 \end{bmatrix}$ when it really should have been $\begin{bmatrix} 6 \\ 4 \end{bmatrix}$, what is the relative error in $\mathbf{p}$ and what is the relative error in $\mathbf{p}'$?

8. Explain why the sum norm and max norm of a symmetric matrix are the same (symmetric means $a_{ij} = a_{ji}$).

9. Let $\mathbf{A}^T$ be the transpose of $\mathbf{A}$ ($\mathbf{A}^T$ is obtained from $\mathbf{A}$ by interchanging rows and columns). Show that $\|\mathbf{A}^T\|_m = \|\mathbf{A}\|_s$ and $\|\mathbf{A}^T\|_s = \|\mathbf{A}\|_m$.

10. If **A** is a matrix that is all 0's except one entry that has value a, show that $\|\mathbf{A}\|_s = \|\mathbf{A}\|_m = |a|$.

11. **(a)** Give the adjacency matrix $\mathbf{A}(G)$ for the accompanying graph.

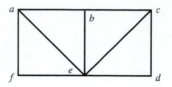

(b) What is the sum norm of $\mathbf{A}(G)$?
(c) Explain in words an interpretation that can be given to the sum norm of the adjacency matrix of a graph.

12. **(a)** In choosing the 2-vector **v** with max norm 1 so that $|\mathbf{Av}|_m = \|\mathbf{A}\|_m |\mathbf{v}|_m$, explain how to modify **v** if the 2×2 matrix **A** has negative entries. Explicitly, consider the case of
$$\mathbf{A} = \begin{bmatrix} 3 & -5 \\ -2 & 7 \end{bmatrix}.$$
(b) Generalize part (a) to $n \times n$ matrices.

13. The Euclidean norm $\|\mathbf{A}\|_e$ of **A** satisfies $|\mathbf{Ax}|_e \le \|\mathbf{A}\|_e \cdot |\mathbf{x}|_e$, where $|\cdot|_e$ denotes the Euclidean distance norm of a vector. If **A** is a symmetric matrix, it can be proved that $\|\mathbf{A}\|_e$ equals the largest eigenvalue of **A** (in absolute value). Compute the Euclidean norm of the following matrices and compare this value with the sum norm and max norm of these matrices.

(a) $\begin{bmatrix} 2 & -1 \\ -1 & 2 \end{bmatrix}$ **(b)** $\begin{bmatrix} -5 & 2 \\ 2 & 1 \end{bmatrix}$ **(c)** $\begin{bmatrix} 3 & 5 \\ 5 & 0 \end{bmatrix}$ **(d)** $\begin{bmatrix} 0 & 3 \\ 3 & 0 \end{bmatrix}$

14. The Euclidean norm (see Exercise 13) of a nonsymmetric matrix **A** is equal to the square root of the largest eigenvalue (in absolute value) of the symmetric matrix $\mathbf{A}^T\mathbf{A}$ (where $\mathbf{A}^T$ is the transpose of **A**). Compute the Euclidean norm of the following matrices and compare this value with the sum norm and max norm of these matrices.

(a) $\begin{bmatrix} 1 & 2 \\ 3 & 0 \end{bmatrix}$ **(b)** $\begin{bmatrix} 3 & 0 \\ 1 & 1 \end{bmatrix}$ **(c)** $\begin{bmatrix} -1 & 1 \\ 2 & 4 \end{bmatrix}$ **(d)** $\begin{bmatrix} 0 & 4 \\ 3 & 0 \end{bmatrix}$

15. What is the condition number of the $n \times n$ identity matrix?

16. Determine the condition number of the following matrices. Comment on whether or not small errors in data of each matrix **A** can result in large errors in the solution of $\mathbf{Ax} = \mathbf{b}$.

(a) $\begin{bmatrix} 1 & 3 \\ 2 & 4 \end{bmatrix}$ **(b)** $\begin{bmatrix} 2 & -3 \\ -2 & 1 \end{bmatrix}$ **(c)** $\begin{bmatrix} 1 & -3 \\ -2 & 6 \end{bmatrix}$ **(d)** $\begin{bmatrix} \frac{1}{3} & \frac{1}{4} & \frac{1}{5} \\ \frac{1}{4} & \frac{1}{5} & \frac{1}{6} \\ \frac{1}{5} & \frac{1}{6} & \frac{1}{7} \end{bmatrix}$

17. **(a)** In the refinery problem in Example 5, if entry $(1, 1)$ is changed from 4 to 2 (yielding matrix **A'**), how large a relative error in the solution to $\mathbf{A'x} = \mathbf{b}$ is possible?
(b) Solve the system $\mathbf{A'x} = \mathbf{b}$ for the **A'** in part (a) and compare the actual relative error with the relative error bound given in part (a).

18. **(a)** Compute the condition number of $\mathbf{A} = \begin{bmatrix} 1 & 1 & 1 \\ 2 & 4 & -1 \\ 1 & -1 & 2 \end{bmatrix}$.

(b) Solve $\mathbf{Ax} = \mathbf{b}$ where $\mathbf{b} = \begin{bmatrix} 1 \\ 2 \\ 3 \end{bmatrix}$.

(c) Suppose that we change entry $(2, 1)$ from 2 to 1 to get a new matrix $\mathbf{A}'$. How large a relative change in solution of $\mathbf{Ax} = \mathbf{b}$ is possible with this change in $\mathbf{A}$? (Use the condition number estimate in Theorem 4.)

(d) Solve $\mathbf{A}'\mathbf{x} = \mathbf{b}$ for this new $\mathbf{A}'$, and compare the observed relative change to the one predicted in part (c).

(e) The large condition number of $\mathbf{A}$ in part (a) means that this matrix is close to being noninvertible (i.e., that some combination of two rows of $\mathbf{A}$ almost equals a third row). Show that one-third of the sum of two of the rows almost equals the other row.

19. Answer parts (a) and (b) using equation (37) in the text.

(a) If $\mathbf{b} = \begin{bmatrix} 2 \\ 1 \end{bmatrix}$ is changed to $\begin{bmatrix} 2 \\ 2 \end{bmatrix}$, how large a change can this yield in the solution to $\mathbf{Ax} = \mathbf{b}$ for $\mathbf{A}$ in Exercise 16(a)? Find the actual relative change.

(b) If $\mathbf{b} = \begin{bmatrix} 1 \\ 2 \\ 3 \end{bmatrix}$ is changed to $\begin{bmatrix} 2 \\ 2 \\ 3 \end{bmatrix}$, how large a change can this yield in the solution to $\mathbf{Ax} = \mathbf{b}$ for $\mathbf{A}$ in Exercise 18? Find the actual relative change.

20. Derive the bound $|\mathbf{e}|/|\mathbf{x}| \le c(\mathbf{A})\{|\mathbf{e}*|/|\mathbf{b}|\}$ in equation (37) by following the reasoning in equations (30) and (32) to obtain $|\mathbf{e}| \le \|\mathbf{A}^{-1}\| \cdot |\mathbf{e}*|$; then divide by $|\mathbf{x}|$ ($\ge |\mathbf{b}|/\|\mathbf{A}\|$).

21. (Continuation of Exercise 14 in Section 1.4 and Exercise 11 in Section 1.5) The staff dietician at the California Institute of Trigonometry has to make up a meal with 600 calories, 20 grams of protein, and 200 milligrams of vitamin C. There are three food types to choose from: rubbery Jello, dried fish sticks, and mystery meat. They have the following nutritional content per ounce.

	Jello	Fish sticks	Mystery meat
Calories:	10	50	200
Protein:	1	3	0.2
Vitamin C:	30	10	0

If there is at most a 5% error in (sum) norm of any column in this matrix of data $\mathbf{A}$, how large a relative error can occur in solving the dietician's problem?

22. (Continuation of Exercise 15 in Section 1.4 and Exercise 12 in Section 1.5) A furniture manufacturer makes tables, chairs, and sofas. In one month the company has available 300 units of wood, 350 units of labor, and 225 units of upholstery. The manufacturer wants a production schedule for the month in which all of these resources are used. The various products require the following amounts of the resources.

	Table	Chair	Sofa
Wood:	4	1	3
Labor:	3	2	5
Upholstery:	2	0	4

If the amount of wood needed to make a table is accidentally entered as 3 instead of 4, how large a relative error in the solution to this production problem is possible?

23. (Continuation of Exercise 17 in Section 1.4 and Exercise 13 in Section 1.5) An investment analyst is trying to find out how much business a secretive TV manufacturer has. The company makes three brands of TV sets: brand A, brand B, and brand C. The analyst learns that the manufacturer has ordered from suppliers 450,000 type 1 circuit boards, 300,000 type 2 circuit boards, and 350,000 type 3 circuit boards. Brand A uses 2 type 1 boards, 1 type 2 board, and 2 type 3 boards. Brand B uses 3 type 1 boards, 2 type 2 boards, and 1 type 3 board. Brand C uses 1 board of each type. If there is a mistake in getting the type 1 circuit board orders and the analyst thinks 350,000 boards were ordered instead of 450,000 boards, how large a relative error is possible in the solution to this TV production problem?

24. Show that for any invertible matrix $\mathbf{A}$ with condition number $c(\mathbf{A})$ (using the sum norm), $c(\mathbf{A}) \geq 1$. (Hint: Use the fact that $\|\mathbf{AB}\| \leq \|\mathbf{A}\| \cdot \|\mathbf{B}\|$.)

3.5   Pseudoinverse of a Matrix

In this section we present a method for obtaining an approximate solution that can be used on an $m \times n$ system $\mathbf{Ax} = \mathbf{b}$ that has no true solution. We seek an approximate "solution" $\mathbf{w}$ that gives a vector $\mathbf{p} = \mathbf{Aw}$ which is as close as possible to $\mathbf{b}$. In the following discussion we use the Euclidean norm $|\mathbf{a}| = \sqrt{a_1^2 + a_2^2 + \cdots + a_n^2}$ because we will be treating the vectors $\mathbf{p}$ and $\mathbf{b}$ as points in Euclidean space and seeking to minimize the size of the error $|\mathbf{b} - \mathbf{p}|$.

EXAMPLE 1.
Refinery Problem
Revisited

Recall our refinery model with three refineries producing three petroleum-based products: heating oil, diesel oil, and gasoline.

$$
\begin{array}{llll}
\text{Heating oil:} & 4x_1 + & 2x_2 + 2x_3 = & 600 \\
\text{Diesel oil:} & 2x_1 + & 5x_2 + 2x_3 = & 800 \\
\text{Gasoline:} & 1x_1 + & 2.5x_2 + 5x_3 = & 1000
\end{array}
\tag{1}
$$

Suppose that the second refinery is out of service. We still want to attempt to produce the same amounts of these products. That is, we want to satisfy the system (as best we can)

$$
\begin{array}{ll}
\text{Heating oil:} & 4x_1 + 2x_3 = 600 \\
\text{Diesel oil:} & 2x_1 + 2x_3 = 800 \\
\text{Gasoline:} & 1x_1 + 5x_3 = 1000
\end{array}
\quad \text{or} \quad
x_1 \begin{bmatrix} 4 \\ 2 \\ 1 \end{bmatrix} + x_3 \begin{bmatrix} 2 \\ 2 \\ 5 \end{bmatrix} = \begin{bmatrix} 600 \\ 800 \\ 1000 \end{bmatrix}
\tag{2}
$$

In Section 1.4 we solved (1) by Gaussian elimination, obtaining $\mathbf{x} = \begin{bmatrix} 31.25 \\ 87.5 \\ 150 \end{bmatrix}$.

Since this solution is unique and involves a nonzero value for x_2, we will not be able to solve (2) exactly.

Let $\mathbf{A}$ be the matrix of coefficients in (2) and let $\mathbf{b}$ be the right-side vector. We seek an approximate solution $\mathbf{w}$ to $\mathbf{Ax} = \mathbf{b}$ which minimizes $|\mathbf{b} - \mathbf{p}|$, where $\mathbf{p} = \mathbf{Aw}$.

That is, we want a vector $\mathbf{w}$ so that $\mathbf{Aw} = w_1 \begin{bmatrix} 4 \\ 2 \\ 1 \end{bmatrix} + w_3 \begin{bmatrix} 2 \\ 2 \\ 5 \end{bmatrix}$ is as close as possible to $\begin{bmatrix} 600 \\ 800 \\ 1000 \end{bmatrix}$. ▪

This type of approximate solution is called a ***least squares solution*** because minimizing $|\mathbf{b} - \mathbf{p}| = \sqrt{\Sigma \, (b_i - p_i)^2}$ involves minimizing a sum of squares. (The square root is minimized by minimizing the sum of squares expression under the square root.) For such approximate solutions, we assume that $\mathbf{A}$ has more rows than columns; otherwise, a more sophisticated theory is needed.

Recall that we encountered least squares solutions in regression at the end of Section 2.4. We review the regression problem presented in Section 2.4.

EXAMPLE 2.
Simple Linear
Regression

We wanted to fit the three x–y points $(0, 1)$, $(2, 1)$, $(4, 4)$ to a line of the form $\hat{y} = qx$. Here $\hat{y}$ is the approximating y-value. The x-value might be the number of college math courses taken and the y-value a score on some test. The ideal is that $y = qx$. For these three points, this ideal equation yields the system

$$\begin{array}{l} 0q = 1 \\ 2q = 1 \\ 4q = 4 \end{array} \quad \text{or} \quad q\mathbf{x} = \mathbf{y}, \quad \text{where} \quad \mathbf{x} = \begin{bmatrix} 0 \\ 2 \\ 4 \end{bmatrix}, \quad \mathbf{y} = \begin{bmatrix} 1 \\ 1 \\ 4 \end{bmatrix} \qquad (3)$$

Figure 3.5 shows the points to be estimated by this line, and Figure 3.6 shows the vectors $\mathbf{y}$ and $q\mathbf{x}$ in 3-space. Our goal is to find q so that the estimates $\hat{y}_i = qx_i$ in (3) are as close as possible to the true y_i. In vector form we want the multiple $q\mathbf{x}$ of $\mathbf{x}$ that is as close as possible to $\mathbf{y}$. In Section 2.4 we noted that such a closest multiple of $\mathbf{x}$ to $\mathbf{y}$ was simply the projection of $\mathbf{y}$ onto $\mathbf{x}$ (see Figure 3.6).

The formula for q in the projection $\mathbf{p} = q\mathbf{x}$ of $\mathbf{y}$ onto $\mathbf{x}$, developed in Section 2.4, is

$$q = \frac{\Sigma \, x_i y_i}{\Sigma \, x_i^2} = \frac{\mathbf{x} \cdot \mathbf{y}}{\mathbf{x} \cdot \mathbf{x}} \qquad (4)$$

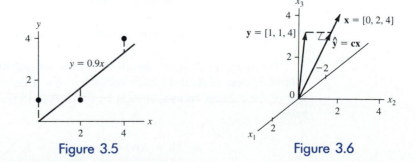

Figure 3.5   Figure 3.6

For the values of $\mathbf{x}$ and $\mathbf{y}$ in this example, we have

$$q = \frac{0 \cdot 1 + 2 \cdot 1 + 4 \cdot 4}{0 \cdot 0 + 2 \cdot 2 + 4 \cdot 4} = \frac{18}{20} = 0.9$$

Thus the equation of our line is $\hat{y} = 0.9x$. The regression estimates $\hat{y}_i$ for the y_i-values are $\hat{y}_1 = 0.9(0) = 0$, $\hat{y}_2 = 0.9(2) = 1.8$, and $\hat{y}_3 = 0.9(4) = 3.6$. ■

We seek to extend the situation in Example 2, finding an approximate solution to a system of equations (3) involving one variable, to the more general situation of finding a vector $\mathbf{w}$ that is the approximate solution to the system $\mathbf{Ax} = \mathbf{b}$. We choose $\mathbf{w}$ so that the vector $\mathbf{p} = \mathbf{Aw}$ is as close as possible to the given vector $\mathbf{b}$. Generalizing Example 2, $\mathbf{p}$ will be the projection of $\mathbf{b}$ onto the collection of vectors

$$R(\mathbf{A}) = \{\mathbf{d} : \mathbf{Ax} = \mathbf{d}, \text{ for some } \mathbf{x}\} \tag{5}$$

The collection $R(\mathbf{A})$ is called the **range** of $\mathbf{A}$. We shall study the range of a matrix extensively in Section 5.2. To determine $\mathbf{w}$ and $\mathbf{p}$, we use the following property of the projection $\mathbf{p}$ of $\mathbf{b}$ onto the range of $\mathbf{A}$:

Projection Property. The error vector $\mathbf{b} - \mathbf{p}$ is orthogonal to vectors in the range of $\mathbf{A}$.

This is the critical property of projections that we used repeatedly in Section 2.4. Recall that if $\mathbf{a}$ and $\mathbf{b}$ are orthogonal, then $\mathbf{a} \cdot \mathbf{b} = 0$.

We review the argument, based on this projection property, used in Section 2.4 to derive the projection formula (4) for q. The error vector in this case is $\mathbf{y} - q\mathbf{x}$, and the range of the system (3) is simply all multiples of $\mathbf{x}$. Since the error vector $\mathbf{y} - q\mathbf{x}$ is orthogonal to $\mathbf{x}$, we have

$$\mathbf{x} \cdot (\mathbf{y} - q\mathbf{x}) = 0 \quad \text{or} \quad \mathbf{x} \cdot \mathbf{y} - q\mathbf{x} \cdot \mathbf{x} = 0 \tag{6}$$

Solving for q, we obtain the regression formula (4).

Next consider the general case where we want an approximate solution $\mathbf{w}$ to $\mathbf{Ax} = \mathbf{b}$ for some $m \times n$ matrix $\mathbf{A}$ and some m-vector $\mathbf{b}$. The error vector $\mathbf{b} - \mathbf{p} = \mathbf{b} - \mathbf{Aw}$ should be orthogonal to every vector in the range of $\mathbf{A}$. Recall that $\mathbf{Aw} = \mathbf{p}$ can be written in terms of the n column vectors $\mathbf{a}_j$ of $\mathbf{A}$ as

$$w_1\mathbf{a}_1 + w_2\mathbf{a}_2 + \cdots + w_n\mathbf{a}_n = \mathbf{p} \tag{7}$$

The range of $\mathbf{A}$ is the set of all vectors formed by linear combinations of the column vectors of $\mathbf{A}$. If the error vector $\mathbf{b} - \mathbf{Aw}$ is orthogonal to any linear combination of the column vectors, it must be orthogonal to these column vectors $\mathbf{a}_j$. So we have

$$\mathbf{a}_j \cdot (\mathbf{b} - \mathbf{Aw}) = 0 \quad \text{for } j = 1, 2, \ldots, n \tag{8}$$

The n scalar products in (8) constitute a system of n linear equations—enough equations to allow us to determine $\mathbf{w}$. If we make a matrix $\mathbf{A}^*$ whose rows are the *columns* of $\mathbf{A}$, we can collect the n equations in (8) together into a matrix equation.

$$\mathbf{A}^*(\mathbf{b} - \mathbf{Aw}) = \mathbf{0} \tag{9}$$

The scalar products of the rows of $\mathbf{A}^*$ times the vector $\mathbf{b} - \mathbf{Aw}$ give the equations in (8).

Observe that matrix $\mathbf{A}^*$ is simply $\mathbf{A}^T$, the transpose of $\mathbf{A}$ (whose rows are the columns of $\mathbf{A}$). So (9) can be written

$$\mathbf{A}^T(\mathbf{b} - \mathbf{Aw}) = \mathbf{0} \quad \text{or} \quad \mathbf{A}^T\mathbf{Aw} = \mathbf{A}^T\mathbf{b} \tag{10}$$

Assuming that the matrix $\mathbf{A}^T\mathbf{A}$ is invertible, we can solve (10) for $\mathbf{w}$ to obtain

$$\mathbf{w} = (\mathbf{A}^T\mathbf{A})^{-1}\mathbf{A}^T\mathbf{b} \tag{11}$$

The right side of (11) is a pretty messy expression. Since the rows of $\mathbf{A}^T$ are the columns of $\mathbf{A}$, entry (i, j) in the matrix product $\mathbf{A}^T\mathbf{A}$ is the scalar product $\mathbf{a}_i \cdot \mathbf{a}_j$ of the ith column of $\mathbf{A}$ times the jth column of $\mathbf{A}$. When $\mathbf{A}$ consists of a single column $\mathbf{x}$, as in the regression model $q\mathbf{x} = \mathbf{y}$, then (11) reduces to $q = (\mathbf{x} \cdot \mathbf{x})^{-1}\mathbf{x} \cdot \mathbf{y}$ or $q = \mathbf{x} \cdot \mathbf{y}/\mathbf{x} \cdot \mathbf{x}$—the formula we obtained above in (4).

We call the product of matrices on the right in (11),

$$\mathbf{A}^+ = (\mathbf{A}^T\mathbf{A})^{-1}\mathbf{A}^T \tag{12}$$

the *pseudoinverse* of $\mathbf{A}$ (the term *generalized inverse* is also used). If $\mathbf{A}$ is an $m \times n$ matrix, $\mathbf{A}^+$ will be an $n \times m$ matrix.

Theorem 1

The least squares solution $\mathbf{w}$ to $\mathbf{Ax} = \mathbf{b}$ is $\mathbf{w} = \mathbf{A}^+\mathbf{b}$, where $\mathbf{A}^+ = (\mathbf{A}^T\mathbf{A})^{-1}\mathbf{A}^T$. The projection of $\mathbf{b}$ onto the range of $\mathbf{A}$ is $\mathbf{p} = \mathbf{Aw} = \mathbf{AA}^+\mathbf{b}$. Also, $\mathbf{A}^+$ is the left inverse of $\mathbf{A}$.

As noted above, *for the pseudoinverse to exist, we must assume that $\mathbf{A}^T\mathbf{A}$ is invertible*. The invertibility of $\mathbf{A}^T\mathbf{A}$ is discussed at the end of this section.

The last sentence of the theorem is easily verified: $\mathbf{A}^+\mathbf{A} = [(\mathbf{A}^T\mathbf{A})^{-1}\mathbf{A}^T]\mathbf{A} = (\mathbf{A}^T\mathbf{A})^{-1}(\mathbf{A}^T\mathbf{A}) = \mathbf{I}$, since we are multiplying $\mathbf{A}^T\mathbf{A}$ times its inverse.

Note: *The identity, $\mathbf{A}^+\mathbf{A} = \mathbf{I}$, can be used to check that one has computed $\mathbf{A}^+$ correctly.*

If $\mathbf{A}$ is an invertible $n \times n$ matrix, the pseudoinverse $\mathbf{A}^+$ equals the regular inverse $\mathbf{A}^{-1}$ (see Exercise 17). If $\mathbf{b}$ happens to lie in the range of $\mathbf{A}$, then $\mathbf{Aw}$ will be the exact solution; that is, $\mathbf{Aw}$ equals $\mathbf{b}$.

While (12) is complex, the fact that such a matrix $\mathbf{A}^+$ exists at all is impressive.

EXAMPLE 3.
Least Squares
Solution to
Refinery Problem

Let us find the least squares solution to the system of equations we had in Example 1, where the first and third refineries alone had to try to satisfy the demand vector.

$$4x_1 + 2x_3 = 600$$
$$2x_1 + 2x_3 = 800 \qquad (13)$$
$$1x_1 + 5x_3 = 1000$$

If $\mathbf{A}$ is the coefficient matrix in (13), we compute $\mathbf{A}^T\mathbf{A}$ and $(\mathbf{A}^T\mathbf{A})^{-1}$ to be [recall that entry (i, j) in $\mathbf{A}^T\mathbf{A}$ is the scalar product of columns i and j of $\mathbf{A}$]

$$\mathbf{A}^T\mathbf{A} = \begin{bmatrix} 21 & 17 \\ 17 & 33 \end{bmatrix} \quad \text{and} \quad (\mathbf{A}^T\mathbf{A})^{-1} = \frac{1}{404}\begin{bmatrix} 33 & -17 \\ -17 & 21 \end{bmatrix} = \begin{bmatrix} 0.0817 & -0.0421 \\ -0.0421 & 0.0520 \end{bmatrix}$$

The pseudoinverse $\mathbf{A}^+$ of $\mathbf{A}$ is

$$\mathbf{A}^+ = (\mathbf{A}^T\mathbf{A})^{-1}\mathbf{A}^T = \begin{bmatrix} 0.0817 & -0.0421 \\ -0.0421 & 0.0520 \end{bmatrix}\begin{bmatrix} 4 & 2 & 1 \\ 2 & 2 & 5 \end{bmatrix}$$

$$= \begin{bmatrix} 0.2426 & 0.0792 & -0.1287 \\ -0.0644 & 0.0198 & 0.2178 \end{bmatrix} \qquad (14)$$

With (14), we can now find the least squares solution $\mathbf{w}$ to (13):

$$\mathbf{w} = \mathbf{A}^+\mathbf{b} = \begin{bmatrix} 0.2426 & 0.0792 & -0.1287 \\ -0.0644 & 0.0198 & 0.2178 \end{bmatrix}\begin{bmatrix} 600 \\ 800 \\ 1000 \end{bmatrix} \approx \begin{bmatrix} 80.2 \\ 195.0 \end{bmatrix} \qquad (15)$$

This solution produces the following approximating output vector:

$$\mathbf{Aw} = \begin{bmatrix} 711 \\ 551 \\ 1055 \end{bmatrix}$$

with an error vector of

$$\mathbf{b} - \mathbf{Aw} = \begin{bmatrix} 600 \\ 800 \\ 1000 \end{bmatrix} - \begin{bmatrix} 711 \\ 551 \\ 1055 \end{bmatrix} = \begin{bmatrix} -111 \\ 249 \\ -55 \end{bmatrix}$$

This is a so-so approximation. A little thought shows that using just the first and third refineries (columns) forces us to produce more of the first product (heating oil) than of the second product (diesel oil). Given that shortcoming, the least squares solution in (15) is not bad. ■

Let us return to regression. The regression model of fitting a line $\hat{y} = qx$ to a set of points (x_i, y_i) can be extended to more general regression models. We start with a modest extension to the line $\hat{y} = qx + r$.

EXAMPLE 4.
Regression Model
$\hat{y} = qx + r$

Let us use the pseudoinverse to solve the regression problem with points $(0, 1)$, $(2, 1)$, $(4, 4)$ (see Figure 3.5) and the model $\hat{y} = qx + r$. The formal system of equations is $q\mathbf{x} + r\mathbf{1} = \mathbf{y}$:

$$
\begin{aligned}
0q + r &= 1 \\
2q + r &= 1 \quad \text{or} \quad \mathbf{X}\mathbf{q} = \mathbf{y}, \quad \text{where} \quad \mathbf{X} = \begin{bmatrix} 0 & 1 \\ 2 & 1 \\ 4 & 1 \end{bmatrix}, \quad \mathbf{y} = \begin{bmatrix} 1 \\ 1 \\ 4 \end{bmatrix}, \quad \mathbf{q} = \begin{bmatrix} q \\ r \end{bmatrix} \quad (16) \\
4q + r &= 4
\end{aligned}
$$

Then

$$
\mathbf{X}^T\mathbf{X} = \begin{bmatrix} 20 & 6 \\ 6 & 3 \end{bmatrix}, \quad (\mathbf{X}^T\mathbf{X})^{-1} = \begin{bmatrix} \frac{1}{8} & -\frac{1}{4} \\ -\frac{1}{4} & \frac{5}{6} \end{bmatrix}
$$

We compute the pseudoinverse

$$
\mathbf{X}^+ = (\mathbf{X}^T\mathbf{X})^{-1}\mathbf{X}^T = \begin{bmatrix} \frac{1}{8} & -\frac{1}{4} \\ -\frac{1}{4} & \frac{5}{6} \end{bmatrix} \begin{bmatrix} 0 & 2 & 4 \\ 1 & 1 & 1 \end{bmatrix} = \begin{bmatrix} -\frac{1}{4} & 0 & \frac{1}{4} \\ \frac{5}{6} & \frac{1}{3} & -\frac{1}{6} \end{bmatrix} \quad (17)
$$

Then

$$
\mathbf{q} = \begin{bmatrix} q \\ r \end{bmatrix} = \mathbf{X}^+\mathbf{y} = \begin{bmatrix} -\frac{1}{4} & 0 & \frac{1}{4} \\ \frac{5}{6} & \frac{1}{3} & -\frac{1}{6} \end{bmatrix} \begin{bmatrix} 1 \\ 1 \\ 4 \end{bmatrix} = \begin{bmatrix} \frac{3}{4} \\ \frac{1}{2} \end{bmatrix}
$$

So $q = 0.75$, $r = 0.5$. Our regression estimates for the y-values are given by $\mathbf{X}\mathbf{q} = \begin{bmatrix} 0.5 \\ 2 \\ 3.5 \end{bmatrix}$. Note that this is not much better than the regression estimates $\begin{bmatrix} 0 \\ 1.8 \\ 3.6 \end{bmatrix}$ we got in Example 2 with the simpler model $\hat{y} = qx$. In this case, the extra parameter r did not help much. ■

The **general regression model** has the form (where y depends on many input values)

$$
\hat{y} = q_1 x^{(1)} + q_2 x^{(2)} + \cdots + q_n x^{(n)} + r \quad (18)
$$

Note here that the coefficient of the last variable r is always 1. Applying Theorem 1 to the general regression model, we obtain

Corollary. Consider the regression model $\hat{y} = q_1 x^{(1)} + q_2 x^{(2)} + \cdots + q_n x^{(n)} + r$ with associated matrix equation

$$\mathbf{y} = \mathbf{X}\mathbf{q} \tag{19}$$

where $\mathbf{y}$ is the column vector of y-value observations, $\mathbf{X}$ is the matrix whose jth column is the set of $x^{(j)}$-values and whose last column is the 1's vector, and $\mathbf{q} = [q_1, q_2, \ldots, q_n, r]^T$. Then the regression model parameters $\mathbf{q}$ are given by

$$\mathbf{q} = (\mathbf{X}^T\mathbf{X})^{-1}\mathbf{X}^T\mathbf{y} \tag{20}$$

EXAMPLE 5.
Least Squares
Polynomial Fitting

Suppose that we want to try to fit a quadratic curve, instead of a straight line, through the set of points $(0, 7)$, $(1, 5)$, $(2, 4)$, $(3, 4)$, $(4, 8)$, $(5, 12)$ using a least squares approximation. Our model is

$$\hat{y} = ax^2 + bx + c \tag{21}$$

We will treat x^2 as a separate variable, say $z = x^2$, so that a linear regression model can be used:

$$\hat{y} = az + bx + c \tag{22}$$

For the given set of points, our system of equations is $\mathbf{X}\mathbf{q} = \mathbf{y}$, where

$$\mathbf{X} = \begin{bmatrix} 0 & 0 & 1 \\ 1 & 1 & 1 \\ 4 & 2 & 1 \\ 9 & 3 & 1 \\ 16 & 4 & 1 \\ 25 & 5 & 1 \end{bmatrix} \quad \text{with} \quad \mathbf{y} = \begin{bmatrix} 7 \\ 5 \\ 4 \\ 4 \\ 8 \\ 12 \end{bmatrix} \quad \text{and} \quad \mathbf{q} = \begin{bmatrix} a \\ b \\ c \end{bmatrix} \tag{23}$$

Using a computer program, we obtain

$$\mathbf{X}^T\mathbf{X} = \begin{bmatrix} 979 & 225 & 55 \\ 225 & 55 & 15 \\ 55 & 15 & 6 \end{bmatrix}$$

and

$$\mathbf{X}^+ = (\mathbf{X}^T\mathbf{X})^{-1}\mathbf{X}^T = \frac{1}{56}\begin{bmatrix} 5 & -1 & -4 & -4 & -1 & 5 \\ -33 & 0.2 & 18.4 & 21.6 & 9.8 & -17 \\ 46 & 18 & 0 & -8 & -6 & 6 \end{bmatrix}$$

and hence

$$\mathbf{q} = \mathbf{X}^+\mathbf{y} \approx \begin{bmatrix} 0.893 \\ -3.493 \\ 7.214 \end{bmatrix} \quad \text{with } \hat{y}\text{-estimates} \quad \hat{\mathbf{y}} = \mathbf{X}\mathbf{q} \approx \begin{bmatrix} 7.2 \\ 4.6 \\ 3.8 \\ 4.8 \\ 7.5 \\ 12.1 \end{bmatrix} \qquad (24)$$

Our quadratic estimate is thus $\hat{y} \approx 0.89x^2 - 3.49x + 7.21$. Although the estimated y-values work out closely to the observed y-values, a word of warning is important. This is a very poorly conditioned problem—the columns of $\mathbf{X}$, and hence of $\mathbf{X}^T\mathbf{X}$, are all fairly similar. In fact, the condition number of the matrix $\mathbf{X}^T\mathbf{X}$ is 2000 (in the sum norm)! A small change in the data could produce a large change in $\mathbf{X}^+$ and our answer. ■

To compute the pseudoinverse $(\mathbf{A}^T\mathbf{A})^{-1}\mathbf{A}^T$, we need to know that the matrix $\mathbf{A}^T\mathbf{A}$ is invertible. Although it is not true that $\mathbf{A}^T\mathbf{A}$ is always invertible, in practical problems in regression and elsewhere it is virtually certain that $\mathbf{A}^T\mathbf{A}$ will be invertible. At the end of this section we prove a theorem justifying the claim that $\mathbf{A}^T\mathbf{A}$ is usually invertible.

Pseudoinverses of Matrices with Orthogonal Columns

There is an important special case in which computation of the pseudoinverse becomes very easy. This is when the columns $\mathbf{a}_i$ of the matrix $\mathbf{A}$ are orthogonal. Then the scalar product of different columns $\mathbf{a}_i \cdot \mathbf{a}_j$ will equal 0. Since entry (i, j) in $\mathbf{A}^T\mathbf{A}$ is exactly this scalar product, $\mathbf{A}^T\mathbf{A}$ will be all 0's except on the main diagonal. This simple form of $\mathbf{A}^T\mathbf{A}$ leads to a simple form for $(\mathbf{A}^T\mathbf{A})^{-1}$ and $\mathbf{A}^+$. (Recall that in Section 2.3, Theorem 3, we saw that the inverse $\mathbf{A}^{-1}$ of $\mathbf{A}$ also has a simple form when $\mathbf{A}$'s columns are orthogonal.)

EXAMPLE 6.
Regression with
Orthogonal
Columns

We shall repeat the analysis of the regression problem in Example 4 with points $(0, 1)$, $(2, 1)$, $(4, 4)$ and model $\hat{y} = qx + r$, but we shall shift the x-values so that the new average x-value is 0. The average x-value for the original data is $(0 + 2 + 4)/3 = 2$. If we subtract 2 from each x-value to obtain points $(-2, 1)$, $(0, 1)$, $(2, 4)$, the new average x-value is 0. (Subtracting the average value always makes the new average be 0.)

Let us repeat the pseudoinverse computations of Example 4 for these new points in the regression model $q\mathbf{x} + r\mathbf{1} = \mathbf{y}$:

$$\begin{aligned} -2q + r &= 1 \\ 0q + r &= 1 \\ 2q + r &= 4 \end{aligned} \quad \text{or} \quad \mathbf{X}\mathbf{q} = \mathbf{y}, \quad \text{where} \quad \mathbf{X} = \begin{bmatrix} -2 & 1 \\ 0 & 1 \\ 2 & 1 \end{bmatrix}, \quad \mathbf{y} = \begin{bmatrix} 1 \\ 1 \\ 4 \end{bmatrix}, \quad \mathbf{q} = \begin{bmatrix} q \\ r \end{bmatrix}$$

$$(25)$$

Observe that the two columns of **X** are now orthogonal. The scalar product of the two columns $\mathbf{x} \cdot \mathbf{1} = \Sigma\, x_i$ must be 0, since the average x-value $[= (1/m)\, \Sigma\, x_i]$ was designed to be 0. Then

$$\mathbf{X}^\mathrm{T}\mathbf{X} = \begin{bmatrix} 8 & 0 \\ 0 & 3 \end{bmatrix} \qquad (\mathbf{X}^\mathrm{T}\mathbf{X})^{-1} = \begin{bmatrix} \frac{1}{8} & 0 \\ 0 & \frac{1}{3} \end{bmatrix} \tag{26}$$

The inverse $(\mathbf{X}^\mathrm{T}\mathbf{X})^{-1}$ in (26) can be computed by the determinant formula, but there is a simpler formula for computing the inverse of a diagonal matrix **D** (with 0's everywhere off the main diagonal): replace each diagonal entry with its inverse, as in (26).

The two diagonal entries in $\mathbf{X}^\mathrm{T}\mathbf{X}$ are, in symbolic terms, $\mathbf{x} \cdot \mathbf{x}$ and $\mathbf{1} \cdot \mathbf{1}$. The latter scalar product $\mathbf{1} \cdot \mathbf{1}$ equals m (number of points in the regression problem). Thus, when the average x-value is 0, $(\mathbf{X}^\mathrm{T}\mathbf{X})^{-1}$ has the simple form

$$(\mathbf{X}^\mathrm{T}\mathbf{X})^{-1} = \begin{bmatrix} \dfrac{1}{\mathbf{x} \cdot \mathbf{x}} & 0 \\ 0 & \dfrac{1}{m} \end{bmatrix} \tag{27}$$

The pseudoinverse $\mathbf{X}^+$ is now

$$\mathbf{X}^+ = (\mathbf{X}^\mathrm{T}\mathbf{X})^{-1}\mathbf{X}^\mathrm{T} = \begin{bmatrix} \frac{1}{8} & 0 \\ 0 & \frac{1}{3} \end{bmatrix} \begin{bmatrix} -2 & 0 & 2 \\ 1 & 1 & 1 \end{bmatrix} = \begin{bmatrix} -\frac{1}{4} & 0 & \frac{1}{4} \\ \frac{1}{3} & \frac{1}{3} & \frac{1}{3} \end{bmatrix} \tag{28}$$

When we premultiply any matrix **B** by a diagonal matrix **D**, then **D** has the effect of multiplying the ith row of **B** by the ith diagonal entry of **D**, as in (28) (see Example 6 of Section 1.3). Looking at the values of the diagonal entries in $(\mathbf{X}^\mathrm{T}\mathbf{X})^{-1}$ [see (27)], we see that $\mathbf{X}^+$ is simply the transpose of **X** with the first column **x** of **X** divided by its sum of squares $(\mathbf{x} \cdot \mathbf{x})$ and the second column **1** divided by m (the number of points). That is, the symbolic form of $\mathbf{X}^+$ when columns are orthogonal is

$$\mathbf{X}^+ = (\mathbf{X}^\mathrm{T}\mathbf{X})^{-1}\mathbf{X}^\mathrm{T} = \begin{bmatrix} \dfrac{1}{\mathbf{x} \cdot \mathbf{x}} & 0 \\ 0 & \dfrac{1}{m} \end{bmatrix} \begin{bmatrix} \mathbf{x}^\mathrm{T} \\ \mathbf{1} \end{bmatrix} = \begin{bmatrix} \left(\dfrac{1}{\mathbf{x} \cdot \mathbf{x}}\right)\mathbf{x}^\mathrm{T} \\ \dfrac{1}{m}\mathbf{1} \end{bmatrix} \tag{29}$$

Solving our regression problem from (28), we obtain

$$\mathbf{q} = \begin{bmatrix} q \\ r \end{bmatrix} = \mathbf{X}^+\mathbf{y} = \begin{bmatrix} -\frac{1}{4} & 0 & \frac{1}{4} \\ \frac{1}{3} & \frac{1}{3} & \frac{1}{3} \end{bmatrix} \begin{bmatrix} 1 \\ 1 \\ 4 \end{bmatrix} = \begin{bmatrix} \frac{3}{4} \\ 2 \end{bmatrix}$$

Observe that q is the scalar product of the first row of $\mathbf{X}^+$ with **y**. But we just noted that the first row of $\mathbf{X}^+$ is simply $\mathbf{x}^\mathrm{T}$ divided by the number $\mathbf{x} \cdot \mathbf{x}$ [see (29)].

Similarly, r equals the scalar product of the second row of $\mathbf{X}^+$, which is just $(1/m)\mathbf{1}$, times $\mathbf{y}$. Thus we have the simple formulas

$$q = \frac{\mathbf{x} \cdot \mathbf{y}}{\mathbf{x} \cdot \mathbf{x}} \qquad r = \frac{\mathbf{1} \cdot \mathbf{y}}{\mathbf{1} \cdot \mathbf{1}} \quad \left(= \frac{1}{m} \Sigma\, y_i = \text{average } y\text{-value} \right) \qquad (30)$$

■

The formulas for q and r are simply those for the lengths of the projections of $\mathbf{y}$ onto $\mathbf{x}$ and $\mathbf{1}$, respectively [see (4)]. The nice results obtained in Example 6 will be true for the pseudoinverse of any matrix with orthogonal columns.

Theorem 2 (Pseudoinverse with Orthogonal Columns)

If the $m \times n$ matrix $\mathbf{A}$ $(m > n)$ has orthogonal columns, the pseudoinverse $\mathbf{A}^+$ is obtained by dividing each column of $\mathbf{A}$ by the sum of the squares of the column's entries and then taking the transpose of the resulting matrix: The ith row of $\mathbf{A}^+$ is $\mathbf{a}_i^T/(\mathbf{a}_i \cdot \mathbf{a}_i)$. Further, the least squares solution $\mathbf{w} = \mathbf{A}^+\mathbf{b}$ is just the projection of $\mathbf{b}$ onto the columns of $\mathbf{A}$: $w_i = \mathbf{a}_i \cdot \mathbf{b}/\mathbf{a}_i \cdot \mathbf{a}_i$.

Recall that in Theorem 3 of Section 2.3, we obtained exactly the same situation for the regular inverse of a square matrix $\mathbf{A}$ with orthogonal columns. In that case, the ith row of $\mathbf{A}^{-1}$ is the ith column $\mathbf{a}_i$ of $\mathbf{A}$ divided by its sum of squares $\mathbf{a}_i \cdot \mathbf{a}_i$. And in Theorem 2 of Section 2.4 we saw that the solution $\mathbf{x}$ of $\mathbf{Ax} = \mathbf{b}$ consisted of the lengths of the projections of $\mathbf{b}$ onto the columns of $\mathbf{A}$: $x_i = \mathbf{a}_i \cdot \mathbf{b}/(\mathbf{a}_i \cdot \mathbf{a}_i)$.

Suppose that we have a regression model with several input variables, such as

$$\hat{y} = q_1 u + q_2 v + q_3 x + r \qquad (31)$$

and suppose that the vectors $\mathbf{u}$, $\mathbf{v}$, $\mathbf{x}$, $\mathbf{1}$ (of the u-values, v-values, x-values, and 1's vector) are orthogonal. Then Theorem 3 tells us, generalizing (30), that the regression parameters are the projections of $\mathbf{y}$ onto $\mathbf{u}$, $\mathbf{v}$, $\mathbf{x}$, and $\mathbf{1}$:

$$q_1 = \frac{\mathbf{u} \cdot \mathbf{y}}{\mathbf{u} \cdot \mathbf{u}} \qquad q_2 = \frac{\mathbf{v} \cdot \mathbf{y}}{\mathbf{v} \cdot \mathbf{v}} \qquad q_3 = \frac{\mathbf{x} \cdot \mathbf{y}}{\mathbf{x} \cdot \mathbf{x}} \qquad r = \frac{\mathbf{1} \cdot \mathbf{y}}{\mathbf{1} \cdot \mathbf{1}} = \frac{\Sigma\, y_i}{m} \qquad (32)$$

But what chance is there that the $\mathbf{u}$, $\mathbf{v}$, $\mathbf{x}$, and $\mathbf{1}$ vectors will be orthogonal? The answer is often up to the person who collects the data. If the u-, v-, and x-values measure settings of control knobs on a complex machine and the y-value measures the task performed by the machine, a researcher who knows about Theorem 3 could pick settings to make the vectors $\mathbf{u}$, $\mathbf{v}$, $\mathbf{x}$, and $\mathbf{1}$ orthogonal. This is a problem in the field of statistics called *design of experiments*.

EXAMPLE 7.
Experiment with
Orthogonal
Columns

Suppose that there are two knobs with readings u and v on a thrill machine which gives a thrill measured by the variable y. Suppose that a set of five experiments is run in which the vector of u- and v-values are orthogonal to each other and to the vector of 1's. Let us fit the data

$$
\begin{array}{rrrrrr}
u: & -4 & -2 & 0 & 2 & 4 \\
v: & 2 & -1 & -2 & -1 & 2 \\
y: & 3 & 6 & 7 & 7 & 6
\end{array}
$$

to the regression model $\hat{y} = q_1 u + q_2 v + r$.

By the mutual orthogonality of the vectors $\mathbf{u}$, $\mathbf{v}$, and $\mathbf{1}$, we may use Theorem 2, or for regression, the formulas in (32).

$$
q_1 = \frac{\mathbf{u} \cdot \mathbf{y}}{\mathbf{u} \cdot \mathbf{u}} = \frac{(-4) \cdot 3 + (-2) \cdot 6 + 0 \cdot 7 + 2 \cdot 7 + 4 \cdot 6}{(-4)^2 + (-2)^2 + 0^2 + 2^2 + 4^2} = \frac{14}{40} = 0.35
$$

$$
q_2 = \frac{\mathbf{v} \cdot \mathbf{y}}{\mathbf{v} \cdot \mathbf{v}} = \frac{2 \cdot 3 + (-1) \cdot 6 + (-2) \cdot 7 + (-1) \cdot 7 + 2 \cdot 6}{2^2 + (-1)^2 + (-2)^2 + (-1)^2 + 2^2} = \frac{-9}{14} \approx -0.64
$$

$$
r = \frac{\mathbf{1} \cdot \mathbf{y}}{\mathbf{1} \cdot \mathbf{1}} = \frac{\Sigma\, y_i}{m} = \frac{3 + 6 + 7 + 7 + 6}{5} = \frac{29}{5} = 5.8
$$

So the regression model is $\hat{y} = 0.35u - 0.64v + 5.8$. ■

We conclude this section with a theorem that tells when an $m \times n$ matrix $\mathbf{A}$ $(m > n)$ has a pseudoinverse.

Theorem 3. Let $\mathbf{A}$ be an $m \times n$ matrix $\mathbf{A}$ $(m > n)$. The $n \times n$ matrix $\mathbf{A}^T\mathbf{A}$ is invertible—and thus the pseudoinverse $\mathbf{A}^+$ exists—if and only if $\mathbf{A}\mathbf{x} = \mathbf{0}$ has the unique solution $\mathbf{x} = \mathbf{0}$.

Proof: We want to know when the $n \times n$ matrix $\mathbf{A}^T\mathbf{A}$ is invertible. By Theorem 7 of Section 1.6, a square matrix $\mathbf{C}$ has an inverse if and only if $\mathbf{C}\mathbf{x} = \mathbf{b}$ has a unique solution for any given $\mathbf{b}$. Choosing $\mathbf{b} = \mathbf{0}$, we must then show that $\mathbf{A}^T\mathbf{A}\mathbf{x} = \mathbf{0}$ has a unique solution—this unique solution will be $\mathbf{x} = \mathbf{0}$—if and only if $\mathbf{A}\mathbf{x} = \mathbf{0}$ has the unique solution $\mathbf{x} = \mathbf{0}$. But this equivalence is exactly the result of Theorem 2 in Section 3.1 (in the discussion about $\mathbf{A}^T\mathbf{A}$). ■

Section 3.5 Exercises

1. Compute the pseudoinverse of the following matrices.

(a) $\begin{bmatrix} 1 \\ 2 \end{bmatrix}$ (b) $\begin{bmatrix} 1 \\ 2 \\ 3 \end{bmatrix}$ (c) $\begin{bmatrix} 1 & 0 \\ 2 & -1 \\ 1 & 1 \end{bmatrix}$ (d) $\begin{bmatrix} 4 & -1 \\ 2 & 2 \\ 1 & 0 \end{bmatrix}$

(e) $\begin{bmatrix} 2 & 0 & 1 \\ 1 & 1 & 0 \\ 0 & -2 & 1 \\ -1 & 1 & 2 \end{bmatrix}$ **(f)** $\begin{bmatrix} 0 & 2 & -1 \\ 1 & -1 & 3 \\ 2 & 1 & 0 \\ -1 & 4 & 1 \end{bmatrix}$

2. Determine the condition number of the matrix (A^TA) in Example 3. Is this matrix poorly conditioned?

3. Seven students earned the following scores on a test after studying the subject matter different numbers of weeks.

Student:	A	B	C	D	E	F	G
Length of study (x_i):	0	1	2	3	4	5	6
Test score (y_i):	3	4	7	6	10	6	10

Fit these data with a regression model of the form $\hat{y} = qx + r$. Determine q and r by computing the pseudoinverse of X, the matrix whose first column is the vector of x_i's and whose second column is a 1's vector. Plot the observed scores and the predicted scores.

4. The following data indicate the numbers of accidents bus drivers had in one year as a function of the numbers of years on the job.

Years on job (x_i):	2	4	6	8	10	12
Accidents (y_i):	10	8	3	8	4	5

   (a) Fit these data with a regression model of the form $\hat{y} = qx + r$. Determine q and r by computing the pseudoinverse of X, the matrix whose first column is the vector of x_i's and whose second column is a 1's vector.
   (b) What is the condition number of the matrix (X^TX)? Is the problem poorly conditioned?
   (c) Repeat the calculations in part (a) by first shifting the x-values to make the average x-value be 0 (see Example 6).

5. The following data show the GPA and the job salary (five years after graduation) of six mathematics majors from Podunk U.

GPA:	2.3	3.1	2.7	3.4	3.7	2.8
Salary:	25,000	38,000	28,000	35,000	30,000	32,000

   (a) Fit these data with a regression model of the form $\hat{y} = qx + r$ using pseudoinverses.
   (b) What is the condition number of the matrix (X^TX)? Is the problem poorly conditioned?
   (c) Repeat the calculations in part (a) by first shifting the x-values to make the average x-value be 0 (see Example 6).

6. Compute the pseudoinverse, and then solve, the refinery problem in Example 1 when refinery 1 is shut down. (The other two refineries operate.)

7. **(a)** Compute the pseudoinverse, and then solve, the refinery problem in Example 1 when refinery 3 is shut down. (The other two refineries operate.)
   (b) Compute the error vector $e = b - Aw$ for part (a). Compute the angle between the error vector e in part (a) and the solution vector Aw. It should be about $90°$. Is it?
   (c) Which refinery closing, of the three refineries, has the smallest error vector (in the sum norm)? This assumes that you have done Exercise 6.

8. In each case, find the linear combination of the first two vectors that is as close as possible (in the least squares sense) to the third vector.

(a) $\begin{bmatrix} 1 \\ 2 \\ 1 \end{bmatrix}, \begin{bmatrix} 2 \\ 0 \\ -1 \end{bmatrix}; \begin{bmatrix} 3 \\ -1 \\ 0 \end{bmatrix}$ (b) $\begin{bmatrix} 1 \\ 0 \\ 1 \end{bmatrix}, \begin{bmatrix} 0 \\ 1 \\ 1 \end{bmatrix}; \begin{bmatrix} 0 \\ 0 \\ 5 \end{bmatrix}$ (c) $\begin{bmatrix} 0 \\ -2 \\ 3 \end{bmatrix}, \begin{bmatrix} 1 \\ 1 \\ 1 \end{bmatrix}; \begin{bmatrix} 1 \\ -5 \\ 10 \end{bmatrix}$

(d) $\begin{bmatrix} 2 \\ 0 \\ 1 \end{bmatrix}, \begin{bmatrix} -1 \\ 0 \\ 1 \end{bmatrix}; \begin{bmatrix} 4 \\ 3 \\ 2 \end{bmatrix}$ (e) $\begin{bmatrix} 0 \\ 1 \\ 1 \\ 0 \end{bmatrix}, \begin{bmatrix} 1 \\ -1 \\ -1 \\ 1 \end{bmatrix}; \begin{bmatrix} 2 \\ 0 \\ 2 \\ 0 \end{bmatrix}$

9. (a) Factory A produces 30 cars, 40 light trucks, and 20 heavy trucks per day, while factory B produces 60 cars, 20 light trucks, and 20 heavy trucks a day. If the monthly demand is 1000 cars, 500 light trucks, and 400 heavy trucks, what is the least squares solution (days of production for each factory)?

(b) If the monthly demand increased by 10 cars, how much longer would factory A have to work each month? (Hint: See Example 5 in Section 1.5.)

(c) If the monthly demand increased by 10 light trucks and 5 heavy trucks, how much longer would factory B have to work each month?

10. (a) Bureaucratic office A produces 40 new regulations, inspects 90 defective appliances, and approves 300 applications a week. Bureaucratic office B produces 80 new regulations, inspects 40 defective appliances, and approves 200 applications a week. How many weeks would each office have to work to best approximate (in the least squares sense) a demand of producing 1000 new regulations, inspecting 700 defective appliances, and approving 2000 applications?

(b) What is the condition number of the matrix (A^TA) in the pseudoinverse computations? Is this problem poorly conditioned?

(c) If the demand for new regulations increased by 10, how much longer would office A have to work? (Hint: See Example 5 in Section 1.5.)

11. Consider the regression problem in which high school GPA and total SAT score (verbal plus math) are used to predict a person's college GPA:

$$\text{GPA college} = q_1(\text{GPA hi sch}) + q_2(\text{total SAT}/1000) + r$$

Suppose that the data for five people are as follows:

	GPAcol	GPAhi	SAT
A:	2.8	3.0	1.05
B:	3.0	2.8	1.15
C:	3.6	3.8	1.30
D:	3.2	3.6	1.00
E:	3.8	3.4	1.35

(a) Compute the pseudoinverse $(X^TX)^{-1}X^T$. In the process, determine the condition number of (X^TX). Is this problem poorly conditioned?

(b) Then determine q_1, q_2, and r.

(c) Determine the error vector $\mathbf{e}$ (differences between true GPA–college and estimated GPA–college). Is it orthogonal to the estimated GPA–college vector?

12. In Example 5, re-solve the quadratic least squares approximation problem for the following data points. Note that for parts (a), (b), and (c) the x-values are the same, so the pseudoinverse will be the same. (Just use the $\mathbf{X}^+$ in the text.)
    (a) Same as in Example 5 except that the fourth point is (3, 5)
    (b) Same as in Example 5 except that the first point is (0, 9)
    (c) (0, 7), (1, 5), (2, 7), (3, 9), (4, 13), (5, 12)
    (d) (0, 2), (1, 4), (2, 10)
    (e) (−2, 7), (−1, 5), (0, 4), (1, 4), (2, 8), (3, 12)

13. Fit a cubic polynomial to the following data points using the same idea as in the quadratic fit in Example 5: (−1, −2), (0, 3), (1, 2), (2, 8), (3, 12), (4, 100). What is the condition number of $(\mathbf{X}^T\mathbf{X})$?

14. Consider the regression model $\hat{z} = qx + ry + s$ for the following data, where the x-value is a scaled score (to have average value of 0) of high school grades, the y-value is a scaled score of SAT scores, and the z-value is an unscaled score of college grades.

Student:	A	B	C	D	E
x:	−4	−2	0	2	4
y:	2	−1	−2	−1	2
z:	3	6	7	7	6

    Determine q, r, and s. Note that the $\mathbf{x}$, $\mathbf{y}$, and $\mathbf{1}$ vectors (in the regression matrix equation $\mathbf{z} = q\mathbf{x} + r\mathbf{y} + s\mathbf{1}$) are orthogonal.

15. Suppose that there are two dials, A and B, on a machine that produces steel. We want to find out how settings a_i and b_i of the two dials affect the quality c_i of the steel. We use a regression model $\hat{c} = pa + qb + r$. For each of the following vectors $\mathbf{a}$ of settings for dial A, find a vector $\mathbf{b}$ of settings of dial B that is orthogonal to the dial A vector and also orthogonal to the 1's vector.

    (a) $\begin{bmatrix} 2 \\ 1 \\ 0 \\ -1 \\ -2 \end{bmatrix}$ (b) $\begin{bmatrix} -4 \\ -1 \\ 0 \\ 2 \\ 3 \end{bmatrix}$ (c) $\begin{bmatrix} 2 \\ 6 \\ 1 \\ -4 \\ -3 \\ -2 \end{bmatrix}$

16. Consider the regression model $y_i = qx_i$, $i = 1, 2, \ldots, n$. Compute the pseudoinverse for this regression problem and solve for q (in terms of the x_i, y_i values). As a matrix system $\mathbf{Xq} = \mathbf{y}$, the matrix $\mathbf{X}$ is the $n \times 1$ column vector of x-values and $\mathbf{y}$ is the vector of y-values. Your answer should agree with the formula for q in equation (4).

17. Show that the pseudoinverse $\mathbf{A}^+$ equals the true inverse $\mathbf{A}^{-1}$ if the $n \times n$ matrix $\mathbf{A}$ is invertible.

4 Vector Spaces

4.1 Subspaces and Spanning Sets

In this chapter we generalize the properties of vectors used in previous chapters. In the process we develop a general theory about vectors that extends much of the familiar structure of $\mathbf{R}^n$ to collections of continuous functions and of other mathematical objects that one would not think of as vectors. Simultaneously, our theory will give us powerful insights into the structure of $\mathbf{R}^n$ and subsets of vectors in $\mathbf{R}^n$. These insights are appealing in their own right and can be used to better understand solutions to linear equations and to simplify the analysis of linear models in hundreds of diverse applied settings.

As an illustration of what we seek to achieve in this chapter, consider the problem of generalizing the structure of vectors in $\mathbf{R}^n$ to collections of functions. One can imagine how a wide class of functions might be represented as linear combinations of some set of special functions, the way the vector $[3, 2, -5]$ is a linear combination $3\mathbf{i}_1 + 2\mathbf{i}_2 - 5\mathbf{i}_3$ of the coordinate vectors $\mathbf{i}_1 = [1, 0, 0]$, $\mathbf{i}_2 = [0, 1, 0]$, $\mathbf{i}_3 = [0, 0, 1]$. For example, polynomials of degree 2, such as $5x^2 + 3x - 2$, can be represented as linear combinations of the functions x^2, x, 1 (where 1 is the constant function that is 1 for all values of x). What are the critical laws of vectors in $\mathbf{R}^n$ that we need to require of functions in order to obtain the same nice properties of $\mathbf{R}^n$?

The following definition captures the important properties of most collections of vectors that we have encountered.

Definition

A *vector space* is any nonempty collection V of elements $\mathbf{v}$ called *vectors* such that:

1. *Laws for addition.* The binary operation of addition $+$ is defined for any pair of vectors on V. There is a zero element $\mathbf{0}$ in V and a negation operation. Addition, negation, and $\mathbf{0}$ obey the standard laws of addition:
   (a) $\mathbf{u} + \mathbf{v} = \mathbf{v} + \mathbf{u}$
   (b) $(\mathbf{u} + \mathbf{v}) + \mathbf{w} = \mathbf{u} + (\mathbf{v} + \mathbf{w})$
   (c) $\mathbf{v} + \mathbf{0} = \mathbf{v}$
   (d) $\mathbf{v} + (-\mathbf{v}) = \mathbf{0}$
2. *Closure.* One can construct vectors in V by the following operations:
   (e) For $\mathbf{v} \in V$ and scalar r, $\mathbf{w} = r\mathbf{v}$ is a vector in V.
   (f) For $\mathbf{u}, \mathbf{v} \in V$ and scalars r_1, r_2, $\mathbf{w} = r_1\mathbf{u} + r_2\mathbf{v}$ is a vector in V.
3. *Scalar factoring.* The laws of scalar factoring apply:
   (g) $r(\mathbf{u} + \mathbf{v}) = r\mathbf{u} + r\mathbf{v}$
   (h) $(r + s)\mathbf{v} = r\mathbf{v} + s\mathbf{v}$
   (i) $r(s\mathbf{v}) = (rs)\mathbf{v}$
   (j) $1\mathbf{v} = \mathbf{v}$ and $0\mathbf{v} = \mathbf{0}$

Observe that since vector addition is commutative, $\mathbf{v} + \mathbf{0} = \mathbf{v}$ implies that $\mathbf{0} + \mathbf{v} = \mathbf{v}$; similarly, $\mathbf{v} + (-\mathbf{v}) = \mathbf{0}$ implies that $(-\mathbf{v}) + \mathbf{v} = \mathbf{0}$.

Parts 1 and 3 in the definition above were proved to hold for $\mathbf{R}^n$ in Theorem 1 (laws of vector algebra) in Section 2.1.

A mathematical shorthand for (e) is to say that *V is closed under scalar multiplication*. Similarly, a mathematical shorthand for part (f) of the definition is to say that *V is closed under linear combinations*. The scalar numbers used here are all assumed to be *real numbers*. In some settings, the scalars are complex numbers. Formally, the scalars can be any collection that forms an algebraic field.

Note that (e) can be subsumed as a special case of (f). If we let $\mathbf{v}_2$ in (f) equal $\mathbf{0}$, then (f) reduces to (e): for $\mathbf{v}_1 \in V$ and scalar r_1, $\mathbf{v}' = r_1\mathbf{v}_1$ is a vector in V. Also note that if $r_1 = r_2 = 1$ in (f), then (f) says simply that any sum of two vectors in V is again a vector in V.

If we repeatedly form linear combinations in (f), it follows that

(f') For any vectors $\mathbf{v}_1, \mathbf{v}_2, \ldots, \mathbf{v}_k$ in V and scalars $r_1, r_2, \ldots, r_k$,
$\mathbf{v} = r_1\mathbf{v}_1 + r_2\mathbf{v}_2 + \cdots + r_k\mathbf{v}_k$ is a vector in V.

Clearly, the various sets of 2-vectors, 3-vectors, and n-vectors that we saw in earlier chapters are vectors in this new sense and obey the four rules in part 1 in the definition above as well as scalar factoring. However, part 2 of our vector space definition—that V must be closed under linear combinations—does not apply to some collections of vectors that we considered in Chapter 2. Further, the zero vector $\mathbf{0}$ did not belong to some of those collections. The set of all 2-vectors will be a vector space, but the set of 2-vectors that lies on the line with equation $2x_1 + 3x_2 = 6$ does not form a vector space, since (1) $\mathbf{0}$ is not on this line, and (2) [3, 0] and [0, 2] lie on that line, but [2, 3] = [3, 0] + [0, 2] does not.

Vector spaces composed of vectors in $\mathbf{R}^k$, for some fixed k, are called *Euclidean vector spaces*.

EXAMPLE 1.
Examples of
Euclidean Vector
Spaces

In the following examples, all scalars are real numbers. Recall that rule (e) of vector spaces, closure under scalar multiples, was shown to be a special case of rule (f), closure under linear combinations.

(a) For any given n, the set $\mathbf{R}^n$ of all n-vectors forms a Euclidean vector space using vector addition and scalar multiplication. The rules in parts 1 and 3 in the definition of a vector space were shown to apply to $\mathbf{R}^n$ in Theorem 1 of Section 2.1. In $\mathbf{R}^n$, the "zero vector" is indeed the zero vector $\mathbf{0} = [0, 0, \ldots, 0]$. Finally, any linear combination of vectors in $\mathbf{R}^n$ is necessarily a vector in $\mathbf{R}^n$.

(b) The set $\{\alpha\mathbf{u}\}$ of all scalar multiples of a given n-vector $\mathbf{u}$ is a Euclidean vector space. The rules in parts 1 and 3 hold, since they hold for all vectors in $\mathbf{R}^n$. The n-vector $\mathbf{0} = [0, 0, \ldots, 0]$ is in this vector space, since $0\mathbf{u} = \mathbf{0}$. We show that the set $\{\alpha\mathbf{u}\}$ is closed under linear combinations as follows. The linear combination $r_1(\alpha_1\mathbf{u}) + r_2(\alpha_2\mathbf{u})$ of scalar multiples of $\mathbf{u}$ can be rewritten, using scalar factoring rules, as $(r_1\alpha_1 + r_2\alpha_2)\mathbf{u}$, a scalar multiple of $\mathbf{u}$.

(c) The set of 2-vectors lying on the line $a_1x_1 + a_2x_2 = 0$, for any choice of a_1 and a_2, is a Euclidean vector space. In Section 2.1 we saw that the set of 2-vectors on this line consists of all multiples of $\mathbf{u} = [a_2, -a_1]$. We have seen that the set of all such multiples of $\mathbf{u}$ forms a vector space.

(d) The set $E(\lambda, \mathbf{A})$ of all the eigenvectors associated with a given (real) eigenvalue λ of a matrix $\mathbf{A}$, together with the zero vector $\mathbf{0}$, form a Euclidean vector space. Parts 1 and 3 hold for $E(\lambda, \mathbf{A})$, since they hold for all vectors in $\mathbf{R}^n$. To check that linear combinations of eigenvectors for λ are eigenvectors for λ, let $\mathbf{u}_1$ and $\mathbf{u}_2$ be two eigenvectors of $\mathbf{A}$ associated with λ; then for the linear combination $r_1\mathbf{u}_1 + r_2\mathbf{u}_2$, we compute

$$\mathbf{A}(r_1\mathbf{u}_1 + r_2\mathbf{u}_2) = \mathbf{A}(r_1\mathbf{u}_1) + \mathbf{A}(r_2\mathbf{u}_2) = r_1(\mathbf{A}\mathbf{u}_1) + r_2(\mathbf{A}\mathbf{u}_2)$$
$$= r_1(\lambda\mathbf{u}_1) + r_2(\lambda\mathbf{u}_2) = \lambda(r_1\mathbf{u}_1 + r_2\mathbf{u}_2)$$

(e) The set of all vectors $\mathbf{b}$ for which a given system of equations $\mathbf{Ax} = \mathbf{b}$ has a solution is a Euclidean vector space. (We discuss this vector space in detail in Section 4.4.)

(f) For a given homogeneous system of equations $\mathbf{Ax} = \mathbf{0}$, the set of all solution vectors $\mathbf{x}$ is a Euclidean vector space. (We discuss this vector space in detail in Section 5.2.)

(g) For any n, the set consisting of just the n-vector $\mathbf{0}$ forms a one-element vector space. The reader should check that all rules of a vector space hold for this vector space. This vector space is called the *trivial vector space*. ■

EXAMPLE 2.
Matrices Forming
Vector Spaces

For any given m and n, the set $\mathcal{M}_{m,n}$ of all $m \times n$ matrices (with real-valued components) forms a vector space using matrix addition and scalar multiplication. The negation of matrix $\mathbf{A}$ is obtained by multiplying all entries of $\mathbf{A}$ by -1. The zero vector will be the matrix $\mathbf{O}$ of all 0's. When treated as a "vector," an $m \times n$ matrix is equivalent to the mn-vector obtained by taking the first row, followed by the second row, followed by the third row, and so on, to form one long vector. Then by Example 1(a), this set of vectors is a vector space. For example, $\begin{bmatrix} 1 & 2 \\ 3 & 4 \end{bmatrix}$ is equivalent to $[1, 2, 3, 4]$. And the linear combination

$$2\begin{bmatrix} 1 & 2 \\ 3 & 4 \end{bmatrix} + 3\begin{bmatrix} 2 & 0 \\ 1 & 5 \end{bmatrix} = \begin{bmatrix} 8 & 4 \\ 9 & 23 \end{bmatrix}$$

is equivalent to $2[1, 2, 3, 4] + 3[2, 0, 1, 5] = [8, 4, 9, 23]$. ■

EXAMPLE 3.
Sets of Functions
As Vector Spaces

(a) The set $\mathcal{P}$ of all polynomials (with real coefficients) forms a vector space using polynomial addition and scalar multiplication. Rules in parts 1 and 3 of the definition of a vector space are immediately seen to hold. (Again scalars are assumed to be the real numbers.) The reader should check that the polynomial 0 or $0(x)$, which equals 0 for all x, is the $\mathbf{0}$ element. For example, $f(x) + g(x) = g(x) + f(x)$, $f(x) - f(x) = 0$, and $r(g(x) + f(x)) = rg(x) + rf(x)$. It is obvious that any linear combination of two polynomials is again a polynomial; for example, $5(8x^5 + 3x^4 + 7x^3 - 2x^2 + x - 3) + 3(x^3 + 3x^2 - 2x + 5)$ equals the polynomial $40x^5 + 15x^4 + 38x^3 - x^2 - x$. Also, the set $\mathcal{P}_k$ of all polynomials with a maximum degree $\leq k$ forms a vector space.

(b) The set of all continuous real-valued functions in one variable forms a vector space. We define the addition of two functions as follows: Given $f(x)$ and $g(x)$, $f(x) + g(x)$ is the function whose value for any x is the sum of $f(x)$ and $g(x)$. As with polynomials, the rules in parts 1 and 3 of a vector space are seen to hold for continuous functions. It is a basic, and intuitive, fact of the theory of functions that the linear combination of two continuous functions is again a continuous function.

Similarly, the set of differentiable single-variable functions forms a vector space. The set of all real-valued functions in one variable, continuous or not, forms a vector space. Also, the set of continuous functions or the set of differentiable functions in two variables (or any fixed number of variables) forms a vector space. Many other classes of functions also form a vector space. ■

We discuss vector spaces of functions in some detail in Section 4.6.

EXAMPLE 4.
Examples of Sets
That Are Not
Vector Spaces

(a) The set of vectors lying on a line or on a plane cannot be a vector space if the line or plane does not contain the zero vector $\mathbf{0}$. For example, the set of 2-vectors x satisfying $3x_1 + 2x_2 = 5$ does not contain $\mathbf{0}$ and hence does not form a vector space.

(b) The set of vectors with positive entries is not a vector space, since this collection is not closed under scalar multiplication—multiplying a positive vector by -1 yields a vector with negative entries.

(c) The set of 3-vectors whose Euclidean norm is a given value, say 1, is not a vector space (this is true no matter what type of norm is used). This set is not closed under scalar multiplication. For example, $i_1 = [1, 0, 0]$ has norm 1, while the scalar multiple $3i_1$ does not.

(d) For any given n, the set $\mathcal{M}_{m,n}$ of $n \times n$ matrices that are invertible is not a vector space. In particular, it does not contain the zero matrix $\mathbf{O}$, the $\mathbf{0}$ element for this vector space (see Example 2).

(e) The set consisting of the single n-vector $\mathbf{1} = [1, 1, \ldots, 1]$ is not a vector space, because it is not closed under scalar multiplication and does not contain the zero vector.

(f) The set of 2×2 matrices with determinant equal to 0 is not a vector space, since the matrices $\begin{bmatrix} 1 & 0 \\ 0 & 0 \end{bmatrix}$, $\begin{bmatrix} 0 & 1 \\ 0 & 0 \end{bmatrix}$, $\begin{bmatrix} 0 & 0 \\ 1 & 0 \end{bmatrix}$, and $\begin{bmatrix} 0 & 0 \\ 0 & 1 \end{bmatrix}$ all have determinants of 0, but linear combinations of these matrices will produce all 2×2 matrices, most of which have nonzero determinant.

(g) The set of real-valued functions in one variable which never equal 0 for any value of x is not a vector space, since multiplying any such function $f(x)$ by the scalar 0 yields a function $0 \cdot f(x)$ that is 0 for all values of x.

(h) For any given scalar q, the set of 2×2 matrices with q as one of its eigenvalues is not a vector space. (Verification is left to the reader.) ■

One basic consequence of being closed under linear combinations is that when a vector space V contains the two vectors $\mathbf{u}$, $\mathbf{v}$, it also contains the line through $\mathbf{u}$ and $\mathbf{v}$.

Theorem 1. If a vector space V contains the two distinct vectors $\mathbf{u}$ and $\mathbf{v}$, then V also contains all vectors $\mathbf{w}$ on the line $L = \{\mathbf{w} : \mathbf{w} = \mathbf{u} + r(\mathbf{v} - \mathbf{u})$, for some scalar $r\}$ through $\mathbf{u}$ and $\mathbf{v}$.

Proof: The line L has been expressed in parametric form (see Section 2.1). Observe that $\mathbf{w} = \mathbf{u} + r(\mathbf{v} - \mathbf{u})$, which may be rewritten as $(1 - r)\mathbf{u} + r\mathbf{v}$, is a linear combination of $\mathbf{u}$ and $\mathbf{v}$. Since vector spaces are closed under linear combinations and since $\mathbf{u}$ and $\mathbf{v}$ are in V, then for any scalar r, $\mathbf{u} + r(\mathbf{v} - \mathbf{u})$ is in V. (While the parametric form of a line was defined in Section 2.1 for Euclidean vector spaces, it can also be used to define lines in non-Euclidean vector spaces.) ■

A ***vector subspace*** is a vector space that is a subset of another vector space (with the same addition operation). Any collection S of elements in a vector space already obeys the eight laws in parts 1 and 3 of the definition of a vector space. To form a vector subspace, S must contain the zero vector $\mathbf{0}$ and be closed under linear combinations (recall that being closed under scalar multiplication was shown to be a special case of closure under linear combinations). However, if S is closed under linear combinations, then for a vector $\mathbf{v}$ in S, $1\mathbf{v} + (-1)\mathbf{v} = \mathbf{v} + (-\mathbf{v}) = \mathbf{0}$ must be in S. Thus for S to be a subspace, we only need to require that S be closed under linear combinations.

Theorem 2. A nonempty collection S of elements in a vector space V is a vector space if S is closed under linear combinations.

Since $\mathbf{R}^n$, for any given n, is a vector space, then by Theorem 2, any collection S of n-vectors is a vector subspace if S is closed under linear combinations. All the vector spaces in Example 1 are vector subspaces of $\mathbf{R}^n$ for some n.

Theorem 3. Any plane P in $\mathbf{R}^3$ through the origin is a vector space.

Proof: The equation of a plane P in $\mathbf{R}^3$ is $\mathbf{a} \cdot \mathbf{x} = b$ or $a_1x_1 + a_2x_2 + a_3x_3 = b$ (see Section 2.1). Since we are told that the origin $\mathbf{0}$ is in P and $\mathbf{a} \cdot \mathbf{0} = 0$, it follows that $b = 0$. Let $\mathbf{x}_1$ and $\mathbf{x}_2$ be two vectors on P, so that $\mathbf{a} \cdot \mathbf{x}_1 = 0$ and $\mathbf{a} \cdot \mathbf{x}_2 = 0$. We must show by Theorem 2 that for any scalars r_1, r_2, the vector $\mathbf{x}^* = r_1\mathbf{x}_1 + r_2\mathbf{x}_2$ lies in P; that is, $\mathbf{a} \cdot \mathbf{x}^* = 0$.

$$\mathbf{a} \cdot \mathbf{x}^* = \mathbf{a} \cdot (r_1\mathbf{x}_1 + r_2\mathbf{x}_2) = \mathbf{a} \cdot (r_1\mathbf{x}_1) + \mathbf{a} \cdot (r_2\mathbf{x}_2)$$
$$= r_1(\mathbf{a} \cdot \mathbf{x}_1) + r_2(\mathbf{a} \cdot \mathbf{x}_2) = r_1(0) + r_1(0) = 0 \qquad (1)$$

An argument similar to the proof in Theorem 3 shows that any line passing through the origin in $\mathbf{R}^n$, for some n, is a vector subspace.

EXAMPLE 5.
Examples of
Vector Subspaces

(a) The vector space of all solutions to $a_1x_1 + a_2x_2 = 0$, for given a_1, a_2 [see Example 1(c)], is a subspace of $\mathbf{R}^2$.

(b) The set of 3-vectors of the form $[a, a, b]$ in which the first two entries are equal is a vector subspace of $\mathbf{R}^3$. To verify this assertion using Theorem 2, we need to show that any linear combination of such vectors will produce a vector in which the first two entries are equal. If $\mathbf{u} = [u, u, u^*]$ and $\mathbf{v} = [v, v, v^*]$, the linear combination $r_1\mathbf{u} + r_2\mathbf{v}$ equals $r_1[u, u, u^*] + r_2[v, v, v^*] = [r_1u + r_2v, r_1u + r_2v, r_1u^* + r_2v^*]$, a vector in which the first two entries are equal, as required.

(c) In the vector space $\mathcal{M}_{2,2}$ of 2×2 matrices, the subset of all 2×2 matrices with 0 in the upper, left entry [entry (1, 1)] forms a vector subspace by Theorem 2, since any linear combination of such matrices will still have entry (1, 1) equal to 0.

(d) The vector space $\mathcal{P}$ of all polynomials is a vector subspace of the vector space of continuous real-valued functions in one variable. Also, the vector space $\mathcal{P}_h$ of all polynomials with a maximum degree $\le h$ is a subspace of the vector space $\mathcal{P}_k$ of all polynomials with a maximum degree $\le k$, for $h < k$.

One can create a new vector space from existing vector spaces. The following theorem is an example of this process.

Theorem 4. Let W_1 and W_2 be subspaces of the vector space V. Then $W_1 \cap W_2$ is also a subspace of V.

Proof: By Theorem 2, we need to show that $W_1 \cap W_2$ is nonempty and closed under linear combinations. Since the zero vector $\mathbf{0}$ of V must be in both W_1 and W_2, then $\mathbf{0} \in W_1 \cap W_2$ and thus $W_1 \cap W_2$ is nonempty.

To show that $W_1 \cap W_2$ is closed under linear combinations, let $\mathbf{x}_1$ and $\mathbf{x}_2$ be two vectors in $W_1 \cap W_2$. This means that $\mathbf{x}_1$ and $\mathbf{x}_2$ are both contained in W_1 and both contained in W_2. Since W_1 is a vector space, then for scalars r_1, r_2, the linear combination $r_1\mathbf{x}_1 + r_2\mathbf{x}_2$ is a vector in W_1. By the same reasoning, $r_1\mathbf{x}_1 + r_2\mathbf{x}_2$ is also a vector in W_2. Then $r_1\mathbf{x}_1 + r_2\mathbf{x}_2$ is a vector in $W_1 \cap W_2$, as required. ■

EXAMPLE 6.
Intersection of
Two Vector
Spaces

Show that the set S of 3-vectors $\mathbf{x}$ in the plane $x_1 - 3x_2 + x_3 = 0$ such that $x_1 = x_2$ is a vector space. By Theorem 3, the set of 3-vectors in a plane through the origin forms a vector space. By Example 5(b), the set of 3-vectors in which the first two entries are equal forms a vector space. By Theorem 4, the intersection of these two vector spaces is a vector space. This intersection is exactly S. Hence S is a vector space. To find S, we set $x_1 = x_2$ in $x_1 - 3x_2 + x_3 = 0$ to obtain $x_1 - 3x_1 + x_3 = 0$ or $-2x_1 + x_3 = 0$. Solving for x_3, we have $x_3 = 2x_1$. Thus S consists of vectors of the form $[x_1, x_1, 2x_1]$, that is, all multiples of $[1, 1, 2]$. ■

Another way to create a vector space is to form the set of all possible linear combinations of some specified small number of vectors. For example, $\mathbf{R}^3$ is generated by all linear combinations of the three coordinate vectors $\mathbf{i}_1 = [1, 0, 0]$, $\mathbf{i}_2 = [0, 1, 0]$, and $\mathbf{i}_3 = [0, 0, 1]$. In Section 2.1 we showed that any plane P in $\mathbf{R}^3$ through the origin can be generated by all linear combinations of two noncollinear vectors $\mathbf{u}$, $\mathbf{v}$ that lie in P. (*Noncollinear* means the two vectors are not multiples of one another.)

We shall now generalize these examples. Given the collection of vectors $G = \{\mathbf{v}_1, \mathbf{v}_2, \ldots, \mathbf{v}_k\}$, define the set $V(G)$ to be

$$V(G) = \{\mathbf{w} : \mathbf{w} = r_1\mathbf{v}_1 + r_2\mathbf{v}_2 + \cdots + r_k\mathbf{v}_k, \text{ for some scalars } r_1, r_2, \ldots, r_k\} \quad (2)$$

Theorem 5. Let G be a set of k vectors $\mathbf{v}_1, \mathbf{v}_2, \ldots, \mathbf{v}_k$ in some vector space V. Then $V(G)$ is a vector space. Any vector space W containing $\mathbf{v}_1, \mathbf{v}_2, \ldots, \mathbf{v}_k$ contains $V(G)$.

Proof: By Theorem 2, we must show that if $\mathbf{w}_1$ and $\mathbf{w}_2$ are in $V(G)$, then for any scalars p, q, the vector $p\mathbf{v}_1 + q\mathbf{w}_2$ is in $V(G)$. Suppose that

$$\mathbf{w}_1 = r_1\mathbf{v}_1 + r_2\mathbf{v}_2 + \cdots + r_k\mathbf{v}_k \quad \text{and} \quad \mathbf{w}_2 = s_1\mathbf{v}_1 + s_2\mathbf{v}_2 + \cdots + s_k\mathbf{v}_k$$

Then

$$p\mathbf{v}_1 + q\mathbf{w}_2 = p(r_1\mathbf{v}_1 + r_2\mathbf{v}_2 + \cdots + r_k\mathbf{v}_k) + q(s_1\mathbf{v}_1 + s_2\mathbf{v}_2 + \cdots + s_k\mathbf{v}_k)$$
$$= (pr_1\mathbf{v}_1 + pr_2\mathbf{v}_2 + \cdots + pr_k\mathbf{v}_k) + (qs_1\mathbf{v}_1 + qs_2\mathbf{v}_2 + \cdots + qs_k\mathbf{v}_k) \quad (3)$$
$$= (pr_1 + qs_1)\mathbf{v}_1 + (pr_2 + qs_2)\mathbf{v}_2 + \cdots + (pr_k + qs_k)\mathbf{v}_k$$

So $p\mathbf{v}_1 + q\mathbf{w}_2$ is a linear combination of the vectors in G, as required.

The last sentence in the theorem follows from the fact that if W contains $\mathbf{v}_1, \mathbf{v}_2, \ldots, \mathbf{v}_k$, then by the definition of a vector space, W contains all linear combinations of the $\mathbf{v}_i$'s. ■

We write $V(\mathbf{v}_1, \mathbf{v}_2, \ldots, \mathbf{v}_k)$ to denote the vector space spanned by $\mathbf{v}_1$, $\mathbf{v}_2$, ... , $\mathbf{v}_k$.

Observe that Theorem 5 yields an alternative way of proving Theorem 3—that a plane passing though the origin in $\mathbf{R}^3$ is a vector space. As noted above, it was shown in Section 2.1 that a plane P passing through the origin in $\mathbf{R}^3$ consists of all linear combinations of any two nonzero vectors $\mathbf{u}$, $\mathbf{v}$ in P that are not collinear. Thus P = $V(\mathbf{u}, \mathbf{v})$, and by Theorem 5, $V(\mathbf{u}, \mathbf{v})$ is a vector space.

Definition

A set of vectors G in a vector space V are said to **span** V if every vector in V can be expressed as a linear combination of vectors in G. That is, $V = V(G)$. We say that V is **spanned by G**.

EXAMPLE 7.
Spaces Spanned
by Sets of Vectors

In this example we give sets of vectors that span many of the vector spaces presented in Examples 1, 2, and 3.

(a) For any given n, the set $\mathbf{R}^n$ of all n-vectors is a vector space. The most natural set to choose to span $\mathbf{R}^n$ is I = $\{\mathbf{i}_1, \mathbf{i}_2, \mathbf{i}_3, \ldots, \mathbf{i}_n\}$, where $\mathbf{i}_k$ is the coordinate vector whose kth entry is 1 and other entries 0. Thus $\mathbf{R}^n = V(\text{I})$.

(b) The set $\{r\mathbf{u}\}$ of all scalar multiples of a given vector $\mathbf{u}$ is a vector space. Obviously, $\mathbf{u}$ spans this space. Any (nonzero) multiple $r\mathbf{u}$ also spans this vector space.

(c) A spanning set for the collection of 2-vectors lying on the line $a_1x_1 + a_2x_2 = 0$ is $\mathbf{u} = [a_2, -a_1]$ [see Example 1(c)].

(d) The set of all $\mathbf{b}$ for which a given system of equations $\mathbf{Ax} = \mathbf{b}$ has a solution is a vector space. A spanning set is the set of columns $\mathbf{a}_j$ of $\mathbf{A}$, since $\mathbf{Ax} = \mathbf{b}$ is equivalent to $x_1\mathbf{a}_1 + x_2\mathbf{a}_2 + \cdots + x_n\mathbf{a}_n = \mathbf{b}$. The "solution" $\mathbf{x}$ is the vector of scalar multiples x_i needed to express $\mathbf{b}$ as a linear combination of the columns. (We discuss this example extensively in Section 4.4.)

For example, if $\mathbf{A} = \begin{bmatrix} 2 & 1 \\ 3 & 4 \end{bmatrix}$ and $\mathbf{b} = \begin{bmatrix} 7 \\ 8 \end{bmatrix}$, then $\mathbf{Ax} = \mathbf{b}$ has the solution $\mathbf{x} = \begin{bmatrix} 4 \\ -1 \end{bmatrix}$. Thus

$$x_1\mathbf{a}_1 + x_2\mathbf{a}_2 = \mathbf{b} \quad \text{or} \quad 4\begin{bmatrix} 2 \\ 3 \end{bmatrix} - \begin{bmatrix} 1 \\ 4 \end{bmatrix} = \begin{bmatrix} 7 \\ 8 \end{bmatrix}$$

So $\mathbf{b}$ is a linear combination of the columns $\mathbf{a}_1$, $\mathbf{a}_{(2v)}$.

(e) For any given m and n, the set $\mathcal{M}_{m,n}$ of all $m \times n$ matrices forms a vector space. One natural spanning set G is the $m \times n$ matrices $\mathbf{M}_{ij}$ with entry $(i, j) = 1$ and all other entries 0. For 2×2 matrices, the spanning set is

$$\mathbf{M}_{11} = \begin{bmatrix} 1 & 0 \\ 0 & 0 \end{bmatrix}, \quad \mathbf{M}_{12} = \begin{bmatrix} 0 & 1 \\ 0 & 0 \end{bmatrix}, \quad \mathbf{M}_{21} = \begin{bmatrix} 0 & 0 \\ 1 & 0 \end{bmatrix}, \quad \mathbf{M}_{22} = \begin{bmatrix} 0 & 0 \\ 0 & 1 \end{bmatrix}$$

Any $m \times n$ matrix $\mathbf{A}$ with entries a_{ij} can be expressed as the linear combination $\Sigma\, a_{ij}\mathbf{M}_{ij}$. For example, $\begin{bmatrix} 5 & 1 \\ 0 & 2 \end{bmatrix} = 5\mathbf{M}_{11} + \mathbf{M}_{12} + 2\mathbf{M}_{22}$.

(f) The set $\mathcal{P}$ of all polynomials (with real coefficients) form a vector space. An obvious spanning set is $1, x, x^2, x^3, \ldots, x^k, \ldots$. Any polynomial, such as $5x^6 + 3x^4 - 2x + 5$ is obviously a linear combination of the different powers of x. The subspace $\mathcal{P}_k$ of polynomials with maximum degree k has the $k + 1$ generators $1, x, x^2, x^3, \ldots, x^k$. ■

There are four vector spaces from Examples 1 and 3 for which we did not give spanning sets:

1. The vector space of eigenvectors for a given eigenvalue.
2. The vector space of solutions $\mathbf{x}$ to the homogeneous system $\mathbf{A}\mathbf{x} = \mathbf{0}$.
3. The vector space of all continuous functions (of a single variable).
4. The one-element vector space consisting of the zero n-vector $\mathbf{0}$.

The reader should be able to appreciate that there is no simple way to find a relatively small, but still infinite, set of continuous functions to span the space of all continuous functions (a reader interested in this problem should take a course in real analysis). The vector space consisting of just the zero vector is the vector space we get when the spanning set is the zero vector. Ways to determine spanning sets for the other two vector spaces are developed later in the book.

In part (a) of Example 5, we gave the set of n coordinate n-vectors $\{\mathbf{i}_1, \mathbf{i}_2, \mathbf{i}_3, \ldots, \mathbf{i}_n\}$ as a spanning set for $\mathbf{R}^n$. Clearly, there are many other possible spanning sets for $\mathbf{R}^n$. The set of all n-vectors is one extreme possibility. Obviously, any set that contains the n coordinate vectors will span $\mathbf{R}^n$.

We can use matrices to get useful spanning sets for $\mathbf{R}^n$. Let $\mathbf{A}$ be an invertible $n \times n$ matrix. Then $\mathbf{A}\mathbf{x} = \mathbf{b}$ has a solution for any n-vector $\mathbf{b}$, so the vector space of all such $\mathbf{b}$ equals $\mathbf{R}^n$. By Example 7(d), a spanning set of this vector space is the n columns of $\mathbf{A}$. Thus spanning sets of $\mathbf{R}^n$ and invertible matrices are closely related. This relation is developed further in the next two sections.

Optional

We close this section with an example of a vector space that is very different from the Euclidean or function vector spaces seen thus far.

**EXAMPLE 8.
Vector Space of
Circuits in a
Graph**

In Section 3.1 we introduced graphs that consist of vertices and edges. A *circuit* in a graph is a set of edges forming a closed path (a path that ends at the vertex where it starts). In the graph in Figure 4.1, the set of edges $\{e_1, e_2, e_3\}$ forms a circuit, as does the set of edges $\{e_1, e_4, e_6, e_9\}$.

We now shall define a vector space of circuits. The elements in this space are the sets of edges that form a circuit or an edge-disjoint union of circuits. For example, $\{e_1, e_2, e_3, e_7, e_8, e_9\}$ is such an edge-disjoint union of two triangular circuits. The

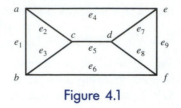

Figure 4.1

associated scalars for this vector space will not be the real numbers. Instead, we use the field with elements 0 and 1. This field uses regular multiplication of these scalars (i.e., $0 \cdot 1 = 1 \cdot 0 = 0 \cdot 0 = 0$, and $1 \cdot 1 = 1$) but addition will be modulo 2 (i.e., $0 + 0 = 1 + 1 = 0$ and $0 + 1 = 1 + 0 = 1$).

The ''sum'' $C_1 + C_2$ of two circuit vectors will be what is called the *Boolean sum* of sets. The Boolean sum of two sets consists of the elements in one but not both of the sets. The sum $C_1 + C_2$ of circuits $C_1 = \{e_1, e_2, e_3\}$ and $C_2 = \{e_1, e_4, e_6, e_9\}$ equals the circuit $\{e_2, e_3, e_4, e_6, e_9\}$. The edge e_1 disappears in this sum because we have e_1 occurring twice and $2e_1 = 0e_1$, by our addition operation for scalars. It is a theorem in graph theory that the Boolean sum of two circuits or disjoint union of circuits is again a circuit or disjoint union of circuits (see Exercise 25). The reader should pick some pairs of circuits in Figure 4.1 and confirm that their Boolean sums are circuits or disjoint unions of circuits.

The zero vector in this vector space is the empty set, the *null circuit*. The negation of a set is the set itself, since the Boolean sum of any set with itself yields the empty set. It is not hard to verify that all the laws of a vector space hold for our circuit space. But this certainly is a very different-looking vector space!

There is a lovely theory of vector spaces in graphs. This theory plays a central role in the analysis of electric networks (see Section 6.4). Besides circuits, one can define a similar type of vector space for cutsets. A *cutset* is a collection of edges whose removal disconnects the graph into two pieces. In Figure 4.1, the set of edges $\{e_4, e_5, e_6\}$ is a cutset. (A cutset is also required to have no unnecessary edges, so that no proper subset of edges in a cutset is a disconnecting set; for example, $\{e_4, e_5, e_6, e_9\}$ is not a cutset.)

One can also define a scalar product of two sets, which is defined to be the number of elements (modulo 2) that the two sets have in common. That is, the scalar product of two sets is 0 if the two sets have an even number of elements in common (0 is considered an even number) and is 1 if they have an odd number of elements in common. One of the interesting results involving such scalar products is that the scalar product of a circuit and a cutset is always 0—circuits and cutsets are orthogonal vectors! The reader should verify this result with some examples in Figure 4.1. ■

Section 4.1 Exercises

1. Show that the following collections of Euclidean vectors form a vector space with real scalars (actually, vector subspaces in $\mathbf{R}^k$, for the appropriate k). Addition is the standard vector addition.
   (a) All 2-vectors $\mathbf{x} = [x_1, x_2]$ such that $x_2 = -x_1$
   (b) All 3-vectors $\mathbf{x} = [x_1, x_2, x_3]$ such that $x_1 = 0$

    (c)   All 3-vectors on the line: $x_1 = 2t$, $x_2 = 3t$, and $x_3 = -2t$

    (d)   All 4-vectors $\mathbf{x} = [x_1, x_2, x_3, x_4]$ such that $x_1 + 5x_2 - 2x_3 + 3x_4 = 0$

    (e)   All 5-vectors $\mathbf{x} = [x_1, x_2, x_3, x_4, x_5]$ such that $x_1 = 5x_2$ and $x_4 = 3x_5$

2.   Show that the following collections of matrices form a vector space with real scalars (actually, vector subspaces in $\mathcal{M}_{m,n}$ for appropriate m, n). Addition is the standard matrix addition.

    (a)   All 2×2 matrices in which all entries have the same value

    (b)   All 3×3 symmetric matrices ($\mathbf{M}$'s with $m_{ij} = m_{ji}$)

    (c)   All 3×3 matrices $\mathbf{M}$ in which entries in diametrically opposite positions are equal, that is, $m_{12} = m_{32}$, $m_{11} = m_{33}$, $m_{13} = m_{31}$, and $m_{21} = m_{23}$

    (d)   All 4×4 diagonal matrices (where all entries not on the main diagonal are zero)

    (e)   All 4×4 matrices $\mathbf{M}$ in which entries double as you move to the right or down (i.e., $m_{i,j+1} = 2m_{ij}$ and $m_{i+1,j} = 2m_{ij}$)

3.   Show that the following collections of functions form a vector space with real scalars. Addition is the standard addition of functions (as in Example 3).

    (a)   All real-valued functions $f(x)$ on the real line for which $f(0) = 0$

    (b)   All real-valued symmetric functions $f(x)$ on the real line [i.e., $f(x) = f(-x)$]

    (c)   All real-valued functions $f(x)$ that are lines through the origin [i.e., $f(x) = ax$], for some scalar a

    (d)   All real-valued functions $f(x)$ whose first and second derivatives exist

4.   Show that the following collections of vectors form a vector space. Scalars are real numbers unless otherwise stated.

    (a)   All 1-vectors (i.e., the real scalars), where addition is the standard addition of scalars

    (b)   All 5-vectors whose entries are 0 or 1, where addition adds corresponding entries modulo 2 ($0 + 1 = 0 + 1 = 1$ while $0 + 0 = 1 + 1 = 0$). The scalars are just 0 and 1

    (c)   All real numbers of the form $a + \sqrt{5}b$, where a and b are rational numbers. Addition is the standard addition of scalars

5.   Show that the following collections of Euclidean vectors do not form a vector space (with real scalars).

    (a)   All 2-vectors $\mathbf{x} = [x_1, x_2]$ such that either $x_1 = 0$ or $x_2 = 0$, with the standard vector addition

    (b)   All 2-vectors $\mathbf{x} = [x_1, x_2]$, where the addition of $\mathbf{x}$ and $\mathbf{x}'$ equals $(\mathbf{x} \cdot \mathbf{x}')\mathbf{1}$ (the 1's vector multiplied by the scalar product of $\mathbf{x}$ and $\mathbf{x}'$)

    (c)   For a given 3×3 matrix $\mathbf{A}$, the set of 3-vectors $\mathbf{x}$ such that $\mathbf{Ax} \neq \mathbf{0}$, with standard vector addition

6.   Show that the following collections of non-Euclidean vectors do not form a vector space (with real scalars).

    (a)   All 3×3 matrices with positive entries, with standard matrix addition

    (b)   For any given scalar q, the set of 2×2 matrices with eigenvalue q [this was Example 4(h)]

    (c)   All 3×3 matrices with the addition of $\mathbf{M}_1 + \mathbf{M}_2$ defined to be $\mathbf{M}_1\mathbf{M}_2$ (the matrix product of $\mathbf{M}_1$ and $\mathbf{M}_2$)

    (d)   All real-valued functions $f(x)$ that are monotonically increasing [i.e., $x' < x$ implies that $f(x') \leq f(x)$]

    (e)   All real-valued functions on the real line that are not continuous, with standard addition of functions

7. Prove by induction from property f, that property f′ of a vector space holds: For any vectors $\mathbf{v}_1$, $\mathbf{v}_2$, . . . , $\mathbf{v}_k$ in V and scalars r_1, r_2, . . . , r_k, $\mathbf{v} = r_1\mathbf{v}_1 + r_2\mathbf{v}_2 + \cdots + r_k\mathbf{v}_k$ is a vector in V.

8. Verify that rules 1 and 3 hold for the vector space of polynomials $\mathscr{P}$ in Example 3(a).

9. Verify that the function $0(x) = 0$ is the zero vector in the vector space $\mathscr{C}$ of continuous functions.

10. Verify that in any given vector space, the zero vector is unique.

11. Verify that the collection consisting of just the zero n-vector $\mathbf{0}$ is a vector space (a vector subspace of $\mathbf{R}^n$).

12. Use Theorem 4 to show that the following sets are vector spaces.
    (a) For given a_1, a_2, a_3, the set of vectors $\mathbf{x}$ in the plane $a_1x_1 + a_2x_2 + a_3x_3 = 0$ with $x_1 = 0$
    (b) The set of 4-vectors in which the first two entries are equal and also the last two entries are equal
    (c) The set of 3×3 matrices $\mathbf{M}$ in which $m_{11} = 0$ and $m_{32} = m_{33}$

13. Verify that the given vector can be written as a linear combination of the given spanning set.

    (a) $\begin{bmatrix} 3 \\ 6 \end{bmatrix}$ in $\mathbf{R}^2$; $\left\{ \begin{bmatrix} 1 \\ 2 \end{bmatrix}, \begin{bmatrix} 2 \\ 0 \end{bmatrix} \right\}$ (b) $\begin{bmatrix} -2 \\ -1 \end{bmatrix}$ in $\mathbf{R}^2$; $\left\{ \begin{bmatrix} 1 \\ 2 \end{bmatrix}, \begin{bmatrix} 1 \\ 1 \end{bmatrix} \right\}$

    (c) $\begin{bmatrix} 3 & 0 \\ 0 & 4 \end{bmatrix}$ in the diagonal matrix subspace of $M_{2,2}$ [see Exercise 2(d)];

    $\left\{ \begin{bmatrix} 1 & 0 \\ 0 & 0 \end{bmatrix}, \begin{bmatrix} 0 & 0 \\ 0 & 1 \end{bmatrix} \right\}$

    (d) $\begin{bmatrix} 2 & 5 \\ 1 & 0 \end{bmatrix}$ in $M_{2,2}$; $\left\{ \begin{bmatrix} 1 & 1 \\ 0 & 0 \end{bmatrix}, \begin{bmatrix} 0 & 0 \\ 1 & 1 \end{bmatrix}, \begin{bmatrix} 1 & 0 \\ 1 & 0 \end{bmatrix}, \begin{bmatrix} 0 & 0 \\ 0 & 1 \end{bmatrix} \right\}$.

    (e) $3x^3 + 2x^2 - 4x + 5$ in $\mathscr{P}_3$ [see Example 3(a)]; $\{x^3, x^2, x, 1\}$ (Here 1 denotes the function that is 1 for all values of x.)
    (f) $2x^2 + 3x - 6$ in $\mathscr{P}_2$ [see Example 3(a)]; $\{x^2 + x, x^2 + 1, x + 1\}$

14. Find a spanning set of two vectors for the following vector spaces.
    (a) The set of 3-vectors $\mathbf{v}$ such that $v_3 = 0$
    (b) The set of vectors $\mathbf{x}$ lying in the plane $x_1 + x_2 + x_3 = 0$ (Hint: See the parametric definition of a plane in Section 2.1.)
    (c) The set of vectors lying in the plane $2x_1 + 4x_2 - 3x_3 = 0$ (Hint: See the parametric definition of a plane in Section 2.1.)
    (d) The set of 4×2 matrices such that all entries in the first column are equal and all entries in the second column are equal

15. Find a spanning set of three vectors for the following vector spaces.
    (a) The set of 5-vectors $\mathbf{v}$ such that $v_2 = 2v_1$ and $v_4 = 3v_3$
    (b) The set of 2×2 matrices $\mathbf{M}$ such that $m_{11} = m_{22}$
    (c) The set of polynomials $p(x)$ of greatest degree ≤ 3 such that $p(0) = 0$

16. (a) Show that two 3-vectors cannot span $\mathbf{R}^3$.
    (b) Show that three polynomials cannot span $\mathscr{P}_3$.
    (c) Show that three 2×2 matrices cannot span $M_{2,2}$.

17. Show that no finite set of polynomials can span $\mathscr{P}$, the vector space of all polynomials.

18. If a set of k 2×2 matrices spans $\mathcal{M}_{2,2}$, show that $k \geq 4$.

19. If $\mathbf{u}$ is orthogonal to each vector in the set $G = \{\mathbf{v}_1, \mathbf{v}_2, \ldots, \mathbf{v}_k\}$, show that $\mathbf{u}$ is orthogonal to all vectors in the vector space $V(G)$ spanned by G.

20. Let the vectors $\mathbf{v}_i$ in the set $\{\mathbf{v}_1, \mathbf{v}_2, \ldots, \mathbf{v}_k\}$ be mutually orthogonal and let $\mathbf{u} = a_1\mathbf{v}_1 + a_2\mathbf{v}_2 + \cdots + a_k\mathbf{v}_k$. Then show that $|\mathbf{u}|^2$ $(= \mathbf{u} \cdot \mathbf{u}) = |a_1|^2|\mathbf{v}_1|^2 + |a_2|^2|\mathbf{v}_2|^2 + \cdots + |a_k|^2|\mathbf{v}_k|^2$.

Problems for Optional Material

21. Find two circuits in Figure 4.1 with five edges, both containing e_1 and e_5. Check that the Boolean sum of these two circuits is a circuit.

22. Find the three circuits with six edges in Figure 4.1. Find the Boolean sum of each pair of these three circuits and confirm that the resulting set is a circuit.

23. Find a cutset of three edges and a cutset of four edges, both containing edges e_1 and e_3. Check that the Boolean sum of these two cutsets is a cutset.

24. Defining the scalar product of two sets to be the number of elements (modulo 2) that the two sets have in common, verify that the scalar product of the following circuit and cutset equals 0.
   (a) Circuit $\{e_1, e_4, e_6, e_9\}$ and cutset $\{e_4, e_5, e_6\}$
   (b) Circuit $\{e_1, e_2, e_3\}$ and cutset $\{e_4, e_7, e_9\}$

25. It is a well-known fact of graph theory that the following two statements are equivalent: (1) C is a set of edges forming a circuit or an edge-disjoint union of circuits; and (2) every vertex in the graph is incident to an even number (possibly zero) edges in C. Use this fact to show that the Boolean sum $C_1 + C_2$ of two such C's is again a circuit or edge-disjoint union of circuits. [Hint: Determine the number of edges of $C_1 + C_2$ incident to a given vertex in terms of the number of edges in C_1 and in C_2 incident to that vertex.]

4.2 Linear Independence

In this section we continue developing the theory of vector spaces started in Section 4.1. We begin with the question of whether a particular vector is a member of a given vector space. This question is central to understanding when a system of equations $\mathbf{Ax} = \mathbf{b}$ will have a solution.

In Section 4.1 we showed that all linear combinations of a given set $G = \{\mathbf{a}_1, \mathbf{a}_2, \ldots, \mathbf{a}_k\}$ of vectors form a vector space which we denoted $V(G)$:

$$V(G) = \{\mathbf{w} : \mathbf{w} = r_1\mathbf{a}_1 + r_2\mathbf{a}_2 + \cdots + r_k\mathbf{a}_k, \text{ for some scalars } r_1, r_2, \ldots, r_k\} \quad (1)$$

We sometimes write $V(G)$ as $V(\mathbf{a}_1, \mathbf{a}_2, \ldots, \mathbf{a}_k)$. Recall that a set of vectors G in a vector space V is said to **span** V if every vector in V can be expressed as a linear combination of vectors in G, that is, if $V = V(G)$.

At the end of Section 4.1 we noted a relation between spanning sets of $\mathbf{R}^n$ and solving a system of equations $\mathbf{Ax} = \mathbf{b}$. We now give a theorem that amplifies that relation.

Theorem 1. The following statements about a vector **b** and a vector space $V(\mathbf{a}_1, \mathbf{a}_2, \ldots, \mathbf{a}_k)$ are equivalent:

(i) The vector **b** is in the vector space $V(\mathbf{a}_1, \mathbf{a}_2, \ldots, \mathbf{a}_k)$.

(ii) The vector **b** can be expressed as a linear combination of vectors $\mathbf{a}_1, \mathbf{a}_2, \ldots, \mathbf{a}_k$.

(iii) There exist scalars $x_1, x_2, \ldots, x_k$ such that $x_1\mathbf{a}_1 + x_2\mathbf{a}_2 + \cdots + x_k\mathbf{a}_k = \mathbf{b}$.

If the vectors under consideration are n-vectors in a Euclidean vector space, statements (i), (ii), and (iii) are equivalent to:

(iv) The system $\mathbf{Ax} = \mathbf{b}$ has a solution, where **A** is an $n \times k$ matrix with jth column $\mathbf{a}_j$.

Proof: Statements (i) and (ii) are equivalent, since $V(\mathbf{a}_1, \mathbf{a}_2, \ldots, \mathbf{a}_k)$ is defined to be the set of all linear combinations of the $\mathbf{a}_i$'s; see (1). Statement (iii) is the same as the defining equation (1) for $V(G)$, with the difference that x_i's instead of r_i's are used. Statements (iii) and (iv) are equivalent, since when $\mathbf{Ax} = \mathbf{b}$ is written out in column vectors, it becomes $x_1\mathbf{a}_1 + x_2\mathbf{a}_2 + \cdots + x_k\mathbf{a}_k = \mathbf{b}$. ■

**EXAMPLE 1.
Connection
Between
Membership in a
Vector Space and
Solving the System
of Equations
$\mathbf{Ax} = \mathbf{b}$**

Consider the vector space $V(\mathbf{a}_1, \mathbf{a}_2)$, where $\mathbf{a}_1 = \begin{bmatrix} 1 \\ 1 \\ 2 \end{bmatrix}$, $\mathbf{a}_2 = \begin{bmatrix} 0 \\ 3 \\ 1 \end{bmatrix}$. The vector $\mathbf{b} = \begin{bmatrix} 2 \\ 5 \\ 5 \end{bmatrix}$ is in $V(\mathbf{a}_1, \mathbf{a}_2)$ since $2\mathbf{a}_1 + \mathbf{a}_2 = \mathbf{b}$, that is,

$$2\begin{bmatrix} 1 \\ 1 \\ 2 \end{bmatrix} + \begin{bmatrix} 0 \\ 3 \\ 1 \end{bmatrix} = \begin{bmatrix} 2 \\ 5 \\ 5 \end{bmatrix}$$

If $\mathbf{A} = \begin{bmatrix} 1 & 0 \\ 1 & 3 \\ 2 & 1 \end{bmatrix}$ is the matrix with columns $\mathbf{a}_1$ and $\mathbf{a}_2$, then $2\mathbf{a}_1 + \mathbf{a}_2 = \mathbf{b}$

is equivalent to $\mathbf{Ax} = \mathbf{b}$ having the solution $\mathbf{x} = \begin{bmatrix} 2 \\ 1 \end{bmatrix}$. ■

While the reader may still be thinking of a vector space as a collection of Euclidean vectors in $\mathbf{R}^n$, Theorem 1 [except part (iv)] applies to vector spaces of functions or whatever. A goal of this chapter is to extend the familiar picture of $\mathbf{R}^n$ and the associated useful theory and computational methods to other, more general settings.

Part (iv) of Theorem 1 cannot currently be applied to a vector space of functions, because then the columns of matrix **A** would have to be functions, not vectors of coordinate values. By the end of the next section, we shall have developed the theoretical framework for constructing coordinate systems for function spaces. Then part (iv) will be applicable to vector spaces of functions.

Theorem 1 is very important for two reasons. First, it allows us to use systems of equations $\mathbf{Ax} = \mathbf{b}$ to answer questions about whether a vector **b** is in the Euclidean vector space spanned by a given set of vectors. (The reader should not be surprised at this point by the fact that most questions in this book are answered by solving a system of linear equations.) Second, the theorem helps explain why a system of equations $\mathbf{Ax} = \mathbf{b}$ has a solution, namely, because **b** lies in the vector space spanned by the columns of **A**.

EXAMPLE 2.
Expressing a
Euclidean Vector
As a Linear
Combination of
Other Vectors

(a) Can vector $\mathbf{a}_3 = \begin{bmatrix} 1 \\ 8 \\ 3 \end{bmatrix}$ be expressed as a linear combination of $\mathbf{a}_1 = \begin{bmatrix} 1 \\ 0 \\ 1 \end{bmatrix}$

and $\mathbf{a}_2 = \begin{bmatrix} 2 \\ 4 \\ 3 \end{bmatrix}$? That is, is $\mathbf{a}_3$ in the vector space $V(\mathbf{a}_1, \mathbf{a}_2)$? By part

(iv) of Theorem 1, such a linear combination is equivalent to finding a solution to $\mathbf{Ax} = \mathbf{a}_3$, where $\mathbf{A}$ has $\mathbf{a}_1$ and $\mathbf{a}_2$ as its columns. Let us use Gauss–Jordan elimination to reduce the augmented matrix $[\mathbf{A}|\mathbf{b}]$ to reduced row echelon form.

$$\begin{bmatrix} 1 & 2 & | & 1 \\ 0 & 4 & | & 8 \\ 1 & 3 & | & 3 \end{bmatrix} \Longrightarrow \begin{bmatrix} 1 & 2 & | & 1 \\ 0 & 4 & | & 8 \\ 0 & 1 & | & 2 \end{bmatrix} \Longrightarrow \begin{bmatrix} 1 & 0 & | & -3 \\ 0 & 1 & | & 2 \\ 0 & 0 & | & 0 \end{bmatrix}$$

Then $x_1 = -3$ and $x_2 = 2$, so $\begin{bmatrix} 1 \\ 8 \\ 3 \end{bmatrix} = -3 \begin{bmatrix} 1 \\ 0 \\ 1 \end{bmatrix} + 2 \begin{bmatrix} 2 \\ 4 \\ 3 \end{bmatrix}$. Thus $\mathbf{a}_3$ is in $V(\mathbf{a}_1, \mathbf{a}_2)$.

(b) Can $\mathbf{i}_2 = \begin{bmatrix} 0 \\ 1 \\ 0 \end{bmatrix}$ be expressed as a linear combination of

$$\mathbf{a}_1 = \begin{bmatrix} 1 \\ 0 \\ 1 \end{bmatrix} \quad \text{and} \quad \mathbf{a}_2 = \begin{bmatrix} 2 \\ 4 \\ 3 \end{bmatrix}?$$

Following the same approach as in part (a), we apply Gauss–Jordan elimination to $[\mathbf{A} \mid \mathbf{i}_2]$:

$$\begin{bmatrix} 1 & 2 & | & 1 \\ 0 & 4 & | & 1 \\ 1 & 3 & | & 0 \end{bmatrix} \Longrightarrow \begin{bmatrix} 1 & 2 & | & 0 \\ 0 & 4 & | & 1 \\ 0 & 1 & | & 0 \end{bmatrix} \Longrightarrow \begin{bmatrix} 1 & 0 & | & -\frac{1}{2} \\ 0 & 1 & | & \frac{1}{4} \\ 0 & 0 & | & -\frac{1}{4} \end{bmatrix}$$

The last row of the reduced augmented matrix is the impossible equation $0x_1 + 0x_2 = -\frac{1}{4}$. Thus $\mathbf{Ax} = \mathbf{i}_2$ has no solution, and $\mathbf{i}_2$ cannot be expressed as a linear combination of $\mathbf{a}_1$ and $\mathbf{a}_2$. Then $\mathbf{i}_2$ is not in $V(\mathbf{a}_1, \mathbf{a}_2)$. ∎

EXAMPLE 3.
Expressing a Non-
Euclidean Vector
As a Linear
Combination of
Other Vectors

(a) Can the polynomial $g(x) = 3x^3 + 5x^2 + 1$ be written as a linearly combination of polynomials $p_1(x) = 2x^3 - 3x^2 + 2x + 4$ and $p_2(x) = x^3 - 11x^2 + 4x + 7$? That is, are there scalars r and s so that $g(x)$ can be expressed as $g(x) = rp_1(x) + sp_2(x)$?

Equating powers of x in $g(x) = rp_1(x) + sp_2(x)$, we see for x^3 that $3x^3 = r(2x^3) + s(x^3)$ or $3 = 2r + s$. Similarly, for x^2 we obtain $5 = -3r - 11s$. We can also get equations from the x and constant terms, but only two equations are needed to solve for r and s. Solving these two equations for r and s, we find that $r = 2$ and $s = -1$. The reader should check that

$$3x^3 + 5x^2 + 1 = 2(2x^3 - 3x^2 + 2x + 4) - (x^3 - 11x^2 + 4x + 7) \quad (2)$$

So $g(x)$ is expressible as a linear combination of $p_1(x)$ and $p_2(x)$, and thus $g(x)$ is in the vector space $V(p_1(x), p_2(x))$.

(b) Can matrix $\mathbf{A} = \begin{bmatrix} 0 & 2 \\ 4 & 3 \end{bmatrix}$ be written as a linear combination of matrices

$\mathbf{M}_1 = \begin{bmatrix} 1 & 3 \\ 2 & 1 \end{bmatrix}$ and $\mathbf{M}_2 = \begin{bmatrix} 2 & 5 \\ 2 & 3 \end{bmatrix}$? That is, do there exist scalars r and

s so that $\mathbf{A} = r\mathbf{M}_1 + s\mathbf{M}_2$?

Equating entries, we see in entry $(1, 1)$, $0 = r + 2s$, and in entry $(1, 2)$, $2 = 3r + 5s$. Solving for r and s, we get $r = 4$ and $s = -2$. But $4\mathbf{M}_1 -$

$2\mathbf{M}_2 = \begin{bmatrix} 0 & 2 \\ 4 & -2 \end{bmatrix} \neq \mathbf{A}$. Thus $\mathbf{A}$ is not expressible as a linear combination

of $\mathbf{M}_1$ and $\mathbf{M}_2$, so $\mathbf{A}$ is not a member of $V(\mathbf{M}_1, \mathbf{M}_2)$. ▪

An important question about spanning sets of a vector space is: What is the minimum set of vectors that will span a given vector space? For example, let vector space V be generated by vectors $\mathbf{a}$, $\mathbf{b}$, $\mathbf{c}$, so $V = V(\mathbf{a}, \mathbf{b}, \mathbf{c})$. Suppose that $\mathbf{c}$ is a linear combination of $\mathbf{a}$ and $\mathbf{b}$: $\mathbf{c} = r\mathbf{a} + s\mathbf{b}$, for scalars r, s. We claim that V is generated by just $\mathbf{a}$ and $\mathbf{b}$. For if $\mathbf{d}$ is a vector in V, then for some scalars t_1, t_2, t_3,

$$\mathbf{d} = t_1\mathbf{a} + t_2\mathbf{b} + t_3\mathbf{c} = t_1\mathbf{a} + t_2\mathbf{b} + t_3(r\mathbf{a} + s\mathbf{b})$$
$$= (t_1 + t_3 r)\mathbf{a} + (t_2 + t_3 r)\mathbf{b} \tag{3}$$

Thus $\mathbf{d}$ is a linear combination of $\mathbf{a}$ and $\mathbf{b}$, so $V = V(\mathbf{a}, \mathbf{b})$.

In Example 2, the vector space $V(\mathbf{a}_1, \mathbf{a}_2, \mathbf{a}_3)$ equals $V(\mathbf{a}_1, \mathbf{a}_2)$, since the third

vector $\mathbf{a}_3 = \begin{bmatrix} 1 \\ 8 \\ 3 \end{bmatrix}$ is a linear combination of $\mathbf{a}_1 = \begin{bmatrix} 1 \\ 0 \\ 1 \end{bmatrix}$ and $\mathbf{a}_2 = \begin{bmatrix} 2 \\ 4 \\ 3 \end{bmatrix}$. If the

3×3 matrix $\mathbf{A}'$ had as its columns the three vectors $\mathbf{a}_1$, $\mathbf{a}_2$, $\mathbf{a}_3$ of Example 2, then the vector space of $\mathbf{b}$'s for which $\mathbf{A}'\mathbf{x} = \mathbf{b}$ has a solution is the same as the vector space of $\mathbf{b}$'s for which $\mathbf{A}\mathbf{x} = \mathbf{b}$ has a solution, where $\mathbf{A}$ is the 3×2 matrix with columns $\mathbf{a}_1$, $\mathbf{a}_2$. In many practical problems giving rise to a system of linear equations, it will be useful to know if a smaller system (fewer columns) could be used. This concern helps motivate the following definition.

Definition

A set of n-vectors $\mathbf{a}_1$, $\mathbf{a}_2$, . . . , $\mathbf{a}_k$, $k \geq 2$, is said to be *linearly independent* if no vector in the set can be expressed as a linear combination of the other vectors. If such a linear combination exists, the set is said to be *linearly dependent*.

Observe that no vector in a linearly independent set of vectors can be the zero vector $\mathbf{0}$, since $\mathbf{0}$ equals 0 times any other vector.

It is common to say **b** is *linearly dependent on* $\mathbf{a}_1, \mathbf{a}_2, \ldots, \mathbf{a}_k$ if **b** can be expressed as a linear combination of $\mathbf{a}_1, \mathbf{a}_2, \ldots, \mathbf{a}_k$. In particular, if $\mathbf{Ax} = \mathbf{b}$ has a solution, then **b** is linearly dependent on the columns of **A**.

EXAMPLE 4.
Linearly
Independent Sets
of Vectors

(a) The set of n coordinate n-vectors $\{\mathbf{i}_1, \mathbf{i}_2, \mathbf{i}_3, \ldots, \mathbf{i}_n\}$ are linearly independent (recall that $\mathbf{i}_k$ has a 1 in the kth position and 0's in the other $n - 1$ positions). For example, $\mathbf{i}_1$ cannot be a linear combination of the other coordinate vectors because all the other coordinate vectors have a 0 in their first entry; similarly for the other coordinate vectors.

(b) In the vector space of polynomials, the monomials $x^k, x^{k-1}, \ldots, x, 1$ are linearly independent. (We shall not give a proof of this assertion; it is a problem in analysis, not linear algebra.)

(c) In the vector space $\mathcal{M}_{2,2}$ of all 2×2 matrices, the matrices $\begin{bmatrix} 1 & 0 \\ 0 & 0 \end{bmatrix}, \begin{bmatrix} 0 & 1 \\ 0 & 0 \end{bmatrix},$ $\begin{bmatrix} 0 & 0 \\ 1 & 0 \end{bmatrix},$ and $\begin{bmatrix} 0 & 0 \\ 0 & 1 \end{bmatrix}$ are linearly independent, for the same reason that the coordinate vectors in part (a) are linearly independent. Recall that in Example 7(e) of Section 4.1, we noted that these four matrices span $\mathcal{M}_{2,2}$.

(d) We claim that any set of mutually orthogonal (nonzero) vectors $\mathbf{b}_1, \mathbf{b}_2, \ldots, \mathbf{b}_k$ in $\mathbf{R}^n$ is linearly independent. Recall that **a** and **b** are orthogonal when $\mathbf{a} \cdot \mathbf{b} = 0$. Suppose that

$$\mathbf{b}_1 = r_2\mathbf{b}_2 + r_3\mathbf{b}_3 + \cdots + r_k\mathbf{b}_k \tag{4}$$

If we take the scalar product of $\mathbf{b}_1$ with the left and right sides of (4), we get 0 on the right side since orthogonal vectors have scalar products of 0, while on the left we get $\mathbf{b}_1 \cdot \mathbf{b}_1 = |\mathbf{b}_1|^2 > 0$. Thus (4) is impossible—$\mathbf{b}_1$ is not a linear combination of the other $\mathbf{b}_i$. The same argument applies for any other $\mathbf{b}_i$, so the $\mathbf{b}_i$'s are linearly independent. ■

The natural question to ask now is: How can we tell if a set of n-vectors is linearly independent or linearly dependent? If we knew which vector in a set is likely to be a linear combination of others, say, vector **d** is likely to be linearly dependent on the other vectors, then we can use Theorem 1 (as in Example 2) and seek to solve the system $\mathbf{Ax} = \mathbf{d}$, where the columns of matrix **A** are the other vectors in the set. But what do we do when we do not know which n-vector is the linear combination of the others? As usual, the answer involves solving a system of linear equations. We start by giving an alternative definition of linear independence and dependence.

Alternative Definition

A set of n-vectors $\mathbf{a}_1, \mathbf{a}_2, \ldots, \mathbf{a}_k, k \geq 2$, is said to be *linearly independent* when $x_1 = x_2 = \cdots = x_k = 0$ is the only solution to the vector equation

$$x_1\mathbf{a}_1 + x_2\mathbf{a}_2 + x_3\mathbf{a}_3 + \cdots + x_k\mathbf{a}_k = \mathbf{0} \tag{5}$$

If (5) has other solutions, the set of $\mathbf{a}_i$'s is said to be *linearly dependent*.

The next theorem shows the equivalence of our two definitions of linear independence and their connection with the solution of an associated system of equations.

Theorem 2. The following are equivalent conditions on a set of n-vectors $\mathbf{a}_1$, $\mathbf{a}_2$, $\ldots$, $\mathbf{a}_k$.

(i) The $\mathbf{a}_i$'s are linearly independent— no one of them can be written as a linear combination of the others.

(ii) The only solution to the vector equation (5) is $x_1 = x_2 = \cdots = x_k = 0$.

(iii) The only solution to the homogeneous system $\mathbf{Ax} = \mathbf{0}$ is $\mathbf{x} = \mathbf{0}$, where $\mathbf{A}$ is the matrix whose k columns are the $\mathbf{a}_i$'s.

Proof: Clearly, (ii) and (iii) are equivalent, since $\mathbf{Ax} = \mathbf{0}$ is just the matrix algebra version of the vector equation $x_1\mathbf{a}_1 + x_2\mathbf{a}_2 + x_3\mathbf{a}_3 + \cdots + x_k\mathbf{a}_k = \mathbf{0}$. We shall prove the equivalence of (i) and (ii) in terms of linear dependence. That is, we shall show that the $\mathbf{a}_i$'s are linearly dependent if and only if (5) has another solution besides $x_1 = x_2 = \cdots = x_k = 0$. Suppose that the $\mathbf{a}_i$'s are linearly dependent so that there exist scalars $k_2, k_3, \ldots, k_k$ such that

$$\mathbf{a}_1 = k_2\mathbf{a}_2 + k_3\mathbf{a}_3 + \cdots + k_k\mathbf{a}_k$$

A similar argument applies if any other $\mathbf{a}_i$ is a linear combination of the other vectors. Bringing $\mathbf{a}_1$ over to the other side, we have

$$-1\mathbf{a}_1 + k_2\mathbf{a}_2 + k_3\mathbf{a}_3 + \cdots + k_k\mathbf{a}_k = \mathbf{0} \tag{6}$$

Then (6) represents a nonzero solution to (5) (since the coefficient of $\mathbf{a}_1$ is nonzero).

Conversely, suppose that there is a nonzero solution $x_1, x_2, x_3, \ldots, x_k$ to

$$x_1\mathbf{a}_1 + x_2\mathbf{a}_2 + x_3\mathbf{a}_3 + \cdots + x_k\mathbf{a}_k = \mathbf{0} \tag{7}$$

For simplicity, suppose that $x_1 \neq 0$. Then dividing (7) by x_1, we get

$$\mathbf{a}_1 + \frac{x_2}{x_1}\mathbf{a}_2 + \frac{x_3}{x_1}\mathbf{a}_3 + \cdots + \frac{x_k}{x_1}\mathbf{a}_k = \mathbf{0} \tag{8}$$

or

$$\mathbf{a}_1 = -\frac{x_2}{x_1}\mathbf{a}_2 - \frac{x_3}{x_1}\mathbf{a}_3 - \cdots - \frac{x_k}{x_1}\mathbf{a}_k \tag{9}$$

From (9), we have shown that $\mathbf{a}_1$ is a linear combination of the other vectors, so the $\mathbf{a}_i$'s are linearly dependent. ■

EXAMPLE 5.
Linear
Dependence and
Homogeneous
Systems

(a) Are the vectors $\mathbf{a}_1 = \begin{bmatrix} 1 \\ 2 \end{bmatrix}$ and $\mathbf{a}_2 = \begin{bmatrix} 4 \\ 1 \end{bmatrix}$ linearly independent or linearly dependent? By Theorem 2, $\mathbf{a}_1$ and $\mathbf{a}_2$ are linearly independent if and only if the only scalars x_1, x_2 satisfying the vector equation $x_1\mathbf{a}_1 + x_2\mathbf{a}_2 = \mathbf{0}$, or

$$x_1 = \begin{bmatrix} 1 \\ 2 \end{bmatrix} + x_2 \begin{bmatrix} 4 \\ 1 \end{bmatrix} = \begin{bmatrix} 0 \\ 0 \end{bmatrix},$$ are $x_1 = x_2 = 0$. This vector equation is equivalent to the system of linear equations:

$$\begin{aligned} x_1 + 4x_2 &= 0 \\ 2x_1 + x_2 &= 0 \end{aligned} \quad \text{or} \quad \mathbf{Ax} = \mathbf{0}, \quad \text{where} \quad \mathbf{A} = \begin{bmatrix} 1 & 4 \\ 2 & 1 \end{bmatrix} \tag{10}$$

We can solve (10) by Gaussian elimination or Gauss–Jordan elimination. The determinant of $\mathbf{A}$ is nonzero, so we can also use Cramer's rule. Since $\det(\mathbf{A}) \neq 0$, then the inverse $\mathbf{A}^{-1}$ of $\mathbf{A}$ exists and we can solve (10) with inverses—$\mathbf{x} = \mathbf{A}^{-1}\mathbf{b}$. Any of these methods yields the result that $\mathbf{x} = \mathbf{0}$ is the unique solution to (10). Thus $\mathbf{a}_1$ and $\mathbf{a}_2$ are linearly independent.

(b) Are the vectors $\mathbf{a}_1 = \begin{bmatrix} 1 \\ 2 \end{bmatrix}$, $\mathbf{a}_2 = \begin{bmatrix} 4 \\ 1 \end{bmatrix}$, and $\mathbf{a}_3 = \begin{bmatrix} -1 \\ 5 \end{bmatrix}$ linearly independent?

By Theorem 2, $\mathbf{a}_1$, $\mathbf{a}_2$, $\mathbf{a}_3$ are linearly independent if and only if $\mathbf{x} = \mathbf{0}$ is the only solution to

$$x_1 \begin{bmatrix} 1 \\ 2 \end{bmatrix} + x_2 \begin{bmatrix} 4 \\ 1 \end{bmatrix} + x_3 \begin{bmatrix} -1 \\ 5 \end{bmatrix} = \begin{bmatrix} 0 \\ 0 \end{bmatrix} \quad \text{or} \quad \mathbf{Ax} = \mathbf{0}, \quad \text{where} \quad \mathbf{A} = \begin{bmatrix} 1 & 4 & -1 \\ 2 & 1 & 5 \end{bmatrix} \tag{11}$$

Applying Gauss–Jordan elimination to the augmented matrix $[\mathbf{A}|\mathbf{0}]$, we obtain

$$\begin{bmatrix} 1 & 4 & -1 & | & 0 \\ 2 & 1 & 5 & | & 0 \end{bmatrix} \Longrightarrow \begin{bmatrix} 1 & 4 & -1 & | & 0 \\ 0 & -7 & 7 & | & 0 \end{bmatrix} \Longrightarrow \begin{bmatrix} 1 & 0 & 3 & | & 0 \\ 0 & 1 & -1 & | & 0 \end{bmatrix}$$

or

$$\begin{aligned} x_1 \quad + 3x_3 &= 0 \\ x_2 - x_3 &= 0 \end{aligned} \tag{12}$$

From (12), we see that any vector $\mathbf{x}$ with $x_1 = -3x_3$ and $x_2 = x_3$ is a solution to $\mathbf{Ax} = \mathbf{0}$. For example, if $x_3 = 1$, then $x_1 = -3x_3 = -3$ and $x_2 = x_3 = 1$. So

$$-3 \begin{bmatrix} 1 \\ 2 \end{bmatrix} + \begin{bmatrix} 4 \\ 1 \end{bmatrix} + \begin{bmatrix} -1 \\ 5 \end{bmatrix} = \begin{bmatrix} 0 \\ 0 \end{bmatrix}.$$ Thus $\mathbf{a}_1$, $\mathbf{a}_2$, and $\mathbf{a}_3$ are linearly dependent. ■

Before giving other examples of linear independence calculations, we first expand on the observation in Example 5(a) that there are several ways to solve the system of equations (10). The following result is one of the central theorems in vector space theory. It links linear independence with the theory of solving systems of linear equations developed in Chapter 1.

Theorem 3. Let S be a set of n-vectors $\mathbf{a}_1, \mathbf{a}_2, \ldots, \mathbf{a}_n$, and let $\mathbf{A}$ be the $n \times n$ matrix whose columns are $\mathbf{a}_1, \mathbf{a}_2, \ldots, \mathbf{a}_n$. Then the following statements are equivalent.

(i) The set S of n-vectors $\mathbf{a}_1, \mathbf{a}_2, \ldots, \mathbf{a}_n$ is linearly independent.

(ii) For any n-vector $\mathbf{b}$, the system $\mathbf{Ax} = \mathbf{b}$ has one and only one solution.

(iii) $\det(\mathbf{A}) \neq 0$.

(iv) $\mathbf{A}$ is invertible.

Proof: Theorem 7 in Section 1.6 says that (ii), (iii), and (iv) are equivalent. We finish by showing that (i) and (iii) are equivalent. Restating this equivalence in terms of Theorem 2, we want to show that $\mathbf{Ax} = \mathbf{0}$ has no nonzero solution if and only if $\det(\mathbf{A}) \neq 0$. Note that $\mathbf{x} = \mathbf{0}$ is always a solution of $\mathbf{Ax} = \mathbf{0}$. The desired equivalence follows from Theorem 8 of Section 1.6, which states that $\mathbf{Ax} = \mathbf{0}$ has a unique solution (namely, $\mathbf{x} = \mathbf{0}$) if and only if $\det(\mathbf{A}) \neq 0$. ■

Corollary. If S is any set of n linearly independent n-vectors, then S spans $\mathbf{R}^n$.

Proof: This result follows immediately from the equivalence of (i) and (ii) in Theorem 3. That is, if the n n-vectors in S are linearly independent, there is a (unique) solution to $\mathbf{Ax} = \mathbf{b}$ for any $\mathbf{b}$. Thus all n-vectors $\mathbf{b}$ can be expressed as a linear combination of vectors in S. ■

Theorem 3 is an example of how a nice body of theory works. Theorem 8 from Section 1.6 was sitting around just waiting for situations where multiple solutions of homogeneous equations might arise. We have obtained so many results in the preceding three chapters that most of the questions we now face are likely to have useful theorems that may apply to them.

EXAMPLE 6.
Determining
Linear
Dependence

Are the following three vectors from Example 2 linearly independent:

$$\begin{bmatrix} 1 \\ 0 \\ 1 \end{bmatrix}, \quad \begin{bmatrix} 2 \\ 4 \\ 3 \end{bmatrix}, \quad \begin{bmatrix} 1 \\ 8 \\ 3 \end{bmatrix}?$$

To answer this question using Theorem 3, we either solve $\mathbf{Ax} = \mathbf{0}$ where the 3×3 matrix $\mathbf{A}$ consists of the three vectors, or we compute $\det(\mathbf{A})$ and see if it is nonzero. We shall use both approaches. First let us compute the determinant of $\mathbf{A}$.

$$\det(\mathbf{A}) = \begin{vmatrix} 1 & 2 & 1 \\ 0 & 4 & 8 \\ 1 & 3 & 3 \end{vmatrix}$$

$$= 1 \cdot 4 \cdot 3 + 2 \cdot 8 \cdot 1 + 1 \cdot 3 \cdot 0 - 1 \cdot 4 \cdot 1 - 8 \cdot 3 \cdot 1 - 3 \cdot 0 \cdot 2 \qquad (13)$$

$$= 12 + 16 + 0 - 4 - 24 - 0 = 0$$

Thus the vectors are linearly dependent (not linearly independent).

Equivalently, we could solve $\mathbf{Ax} = \mathbf{0}$ by reducing the augmented matrix $[\mathbf{A} \mid \mathbf{0}]$ with Gauss–Jordan elimination to reduced row echelon form.

$$\begin{bmatrix} 1 & 2 & 1 & | & 0 \\ 0 & 4 & 8 & | & 0 \\ 1 & 3 & 3 & | & 0 \end{bmatrix} \Longrightarrow \begin{bmatrix} 1 & 2 & 1 & | & 0 \\ 0 & 4 & 8 & | & 0 \\ 0 & 1 & 2 & | & 0 \end{bmatrix} \Longrightarrow \begin{bmatrix} 1 & 0 & -3 & | & 0 \\ 0 & 1 & 2 & | & 0 \\ 0 & 0 & 0 & | & 0 \end{bmatrix} \qquad (14)$$

[Compare (14) with the elimination computation in Example 2.] From the rightmost matrix in (14), we see that

$$
\begin{array}{rl}
x_1 \quad - 3x_3 = 0 & \quad x_1 = \quad 3x_3 \\
x_2 + 2x_3 = 0 & \quad x_2 = -2x_3
\end{array}
\quad \text{or} \quad
\tag{15}
$$

Thus $\mathbf{x} = \begin{bmatrix} 3r \\ -2r \\ r \end{bmatrix}$, so (with $r = 1$)

$$
3\begin{bmatrix} 1 \\ 0 \\ 1 \end{bmatrix} - 2\begin{bmatrix} 2 \\ 4 \\ 3 \end{bmatrix} + \begin{bmatrix} 1 \\ 8 \\ 3 \end{bmatrix} = 0 \quad \text{or} \quad \begin{bmatrix} 1 \\ 8 \\ 3 \end{bmatrix} = -3\begin{bmatrix} 1 \\ 0 \\ 1 \end{bmatrix} + 2\begin{bmatrix} 2 \\ 4 \\ 3 \end{bmatrix}
$$
■

EXAMPLE 7.
Determining
Linear
Independence

Are the following vectors linearly independent:

$$
\begin{bmatrix} 1 \\ 2 \\ 1 \end{bmatrix}, \quad \begin{bmatrix} 1 \\ 4 \\ 2 \end{bmatrix}, \quad \begin{bmatrix} 3 \\ 6 \\ 4 \end{bmatrix} ?
$$

We either solve $\mathbf{Ax} = \mathbf{0}$ where the 3×3 matrix $\mathbf{A}$ consists of the three vectors, or we compute $\det(\mathbf{A})$ and see if it is nonzero. We shall again use both approaches. First let us compute the determinant of $\mathbf{A}$.

$$
\begin{aligned}
\det(\mathbf{A}) &= \begin{vmatrix} 1 & 1 & 3 \\ 2 & 4 & 6 \\ 1 & 2 & 4 \end{vmatrix} \\
&= 1 \cdot 4 \cdot 4 + 1 \cdot 6 \cdot 1 + 3 \cdot 2 \cdot 2 - 3 \cdot 4 \cdot 1 - 6 \cdot 2 \cdot 1 - 4 \cdot 2 \cdot 1 \\
&= 16 + 6 + 12 - 12 - 12 - 8 = 2
\end{aligned}
\tag{16}
$$

Thus the vectors are linearly independent.

Next we solve $\mathbf{Ax} = \mathbf{0}$ by reducing the augmented matrix $[\mathbf{A} \mid \mathbf{0}]$ with Gauss–Jordan elimination to reduced row echelon form.

$$
\begin{bmatrix} 1 & 1 & 3 & \big| & 0 \\ 2 & 4 & 6 & \big| & 0 \\ 1 & 2 & 4 & \big| & 0 \end{bmatrix}
\Longrightarrow
\begin{bmatrix} 1 & 1 & 3 & \big| & 0 \\ 0 & 2 & 0 & \big| & 0 \\ 0 & 1 & 1 & \big| & 0 \end{bmatrix}
$$

$$
\Longrightarrow
\begin{bmatrix} 1 & 0 & 3 & \big| & 0 \\ 0 & 1 & 0 & \big| & 0 \\ 0 & 0 & 1 & \big| & 0 \end{bmatrix}
\Longrightarrow
\begin{bmatrix} 1 & 0 & 0 & \big| & 0 \\ 0 & 1 & 0 & \big| & 0 \\ 0 & 0 & 1 & \big| & 0 \end{bmatrix}
\tag{17}
$$

From the rightmost matrix in (17), we see that all $x_i = 0$, and $\mathbf{x} = \mathbf{0}$ is the only solution. ■

Theorem 3 lets us apply previously developed theory about square $n \times n$ matrices to the problem of determining linear independence of n n-vectors. This is very practical theory, since finding the determinant of a matrix is a very simple task that any scientific software for a computer can do.

If the matrix formed by a set of vectors is not square, then there seems to be no hope for testing for linear independence by computing a determinant. However, we actually can do this. Recall Theorem 2 of Section 3.1: If $\mathbf{A}$ is an $m \times n$ matrix $\mathbf{A}$ $(m > n)$ with associated $n \times n$ matrix $\mathbf{A}^T\mathbf{A}$, then $(\mathbf{A}^T\mathbf{A})\mathbf{x} = \mathbf{0}$ has the unique solution $\mathbf{x} = \mathbf{0}$ if and only if $\mathbf{A}\mathbf{x} = \mathbf{0}$ has the unique solution $\mathbf{x} = \mathbf{0}$. Since $\mathbf{A}^T\mathbf{A}$ is a square matrix, our theory about square matrices applies and we have (following the same reasoning in the proof of Theorem 3):

Theorem 4. Let S be a set of n m-vectors $\mathbf{a}_1, \mathbf{a}_2, \ldots, \mathbf{a}_m$ $(m > n)$, and let $\mathbf{A}$ be the $m \times n$ matrix whose columns are $\mathbf{a}_1, \mathbf{a}_2, \ldots, \mathbf{a}_n$. Then the following statements are equivalent.
  (i) The set S of n-vectors $\mathbf{a}_1, \mathbf{a}_2, \ldots, \mathbf{a}_n$ is linearly independent.
  (ii) $\det(\mathbf{A}^T\mathbf{A}) \neq 0$.
  (iii) $\mathbf{A}^T\mathbf{A}$ is invertible.

EXAMPLE 8.
Determining
Linear
Independence

Are

$$\begin{bmatrix} 1 \\ 0 \\ 1 \\ 2 \end{bmatrix}, \quad \begin{bmatrix} 2 \\ 4 \\ 3 \\ 0 \end{bmatrix}, \quad \begin{bmatrix} 1 \\ 8 \\ 3 \\ 1 \end{bmatrix}$$

linearly independent? We form the matrix $\mathbf{A} = \begin{bmatrix} 1 & 2 & 1 \\ 0 & 4 & 8 \\ 1 & 3 & 3 \\ 2 & 0 & 1 \end{bmatrix}$. Following Theorem 4, we compute

$$\mathbf{A}^T\mathbf{A} = \begin{bmatrix} 1 & 0 & 1 & 2 \\ 2 & 4 & 3 & 0 \\ 1 & 8 & 3 & 1 \end{bmatrix} \begin{bmatrix} 1 & 2 & 1 \\ 0 & 4 & 8 \\ 1 & 3 & 3 \\ 2 & 0 & 1 \end{bmatrix} = \begin{bmatrix} 6 & 5 & 6 \\ 5 & 29 & 43 \\ 6 & 43 & 75 \end{bmatrix} \quad \text{and} \quad \begin{vmatrix} 6 & 5 & 6 \\ 5 & 29 & 43 \\ 6 & 43 & 75 \end{vmatrix} = 1617$$

By Theorem 4, since $\det(\mathbf{A}^T\mathbf{A}) \neq 0$, the three vectors are linearly independent. ■

There is one situation in which linear dependence always occurs.

Theorem 5. If $k > n$, any set of k n-vectors $\{\mathbf{a}_1, \mathbf{a}_2, \ldots, \mathbf{a}_k\}$ is linearly dependent.
  Proof: Suppose that the first n $\mathbf{a}_i$ are linearly independent; otherwise, we are finished. If the $n \times n$ matrix $\mathbf{A}$ formed by these first $n\mathbf{a}_i$ is invertible, then $\mathbf{A}\mathbf{x} = \mathbf{a}_{n+1}$ has a solution by Theorem 3. Thus $\mathbf{a}_{n+1}$ is linearly dependent on earlier $\mathbf{a}_i$'s. ■

Note that to determine whether or not a *column* is a linear combination of other *columns*, we undertook computations in (12) and (14) using Gaussian elimination, which involved linear combinations of *rows*. When linear dependence was found in (12), the actual computational result was the zeroing out of the last row of the matrix by subtracting multiples of the first two rows from the last row. This zeroing out means that *the last row equals a linear combination of the first two rows*. That is, the rows are linearly dependent! We shall explore the relationship between row dependence and column dependence further in Section 4.4.

Section 4.2 Exercises

1. For each of the following sets of Euclidean vectors, express the first vector as a linear combination of the remaining vectors, if possible.

   (a) $\begin{bmatrix} 1 \\ 1 \end{bmatrix} : \begin{bmatrix} 2 \\ 1 \end{bmatrix}, \begin{bmatrix} 2 \\ -1 \end{bmatrix}$ (b) $\begin{bmatrix} 3 \\ 2 \end{bmatrix} : \begin{bmatrix} 2 \\ -3 \end{bmatrix}, \begin{bmatrix} -3 \\ 6 \end{bmatrix}$ (c) $\begin{bmatrix} -4 \\ 2 \end{bmatrix} : \begin{bmatrix} 1 \\ -3 \end{bmatrix}, \begin{bmatrix} 1 \\ -5 \end{bmatrix}$

2. For each of the following sets of Euclidean vectors, express the first vector as a linear combination of the remaining vectors.

   (a) $\begin{bmatrix} 3 \\ -1 \end{bmatrix} : \begin{bmatrix} 1 \\ 3 \end{bmatrix}, \begin{bmatrix} -2 \\ 3 \end{bmatrix}$ (b) $\begin{bmatrix} 1 \\ 1 \\ 1 \end{bmatrix} : \begin{bmatrix} 2 \\ 1 \\ 0 \end{bmatrix}, \begin{bmatrix} 0 \\ 1 \\ 2 \end{bmatrix}, \begin{bmatrix} 2 \\ 2 \\ 1 \end{bmatrix}$ (c) $\begin{bmatrix} 1 \\ 0 \\ 0 \end{bmatrix} : \begin{bmatrix} 1 \\ 1 \\ 0 \end{bmatrix}, \begin{bmatrix} 1 \\ 0 \\ 1 \end{bmatrix}, \begin{bmatrix} 1 \\ 1 \\ 2 \end{bmatrix}$

3. Express each of the following matrices as a linear combination of the matrices $\begin{bmatrix} 1 & 1 \\ 1 & 1 \end{bmatrix}$, $\begin{bmatrix} 1 & 1 \\ 0 & 1 \end{bmatrix}$, $\begin{bmatrix} 1 & 0 \\ 1 & 1 \end{bmatrix}$, if possible.

   (a) $\begin{bmatrix} 1 & 2 \\ 2 & 1 \end{bmatrix}$ (b) $\begin{bmatrix} 1 & 0 \\ 0 & 0 \end{bmatrix}$ (c) $\begin{bmatrix} 1 & 3 \\ -5 & 1 \end{bmatrix}$

4. For each of the following sets of polynomials, express the first polynomial as a linear combination of the remaining polynomials, if possible.
   (a) $x^2 + 3x + 1$: $x^2 + x$, $x^2 + 1$, $x + 1$
   (b) 2: $x^2 + x$, $x^2 + 1$, $x + 1$
   (c) x^3: x^2, x, 1

5. For each of the following pairs of a matrix and a vector, express the vector as a linear combination of the columns of the matrix.

   (a) $\begin{bmatrix} 4 & 0 \\ 0 & 3 \end{bmatrix}, \begin{bmatrix} 2 \\ 1 \end{bmatrix}$ (b) $\begin{bmatrix} 1 & -1 \\ 2 & 1 \end{bmatrix}, \begin{bmatrix} 3 \\ 0 \end{bmatrix}$ (c) $\begin{bmatrix} 2 & 3 \\ 5 & 8 \end{bmatrix}, \begin{bmatrix} -2 \\ 3 \end{bmatrix}$

6. Tell which of the following sets of vectors are linearly independent. If linearly dependent, express one vector as a linear combination of the others.

   (a) $\begin{bmatrix} 1 \\ 2 \end{bmatrix}, \begin{bmatrix} -2 \\ 4 \end{bmatrix}$ (b) $\begin{bmatrix} 1 \\ 3 \end{bmatrix}, \begin{bmatrix} 3 \\ -1 \end{bmatrix}$ (c) $\begin{bmatrix} 2 \\ 1 \end{bmatrix}, \begin{bmatrix} 2 \\ 3 \end{bmatrix}, \begin{bmatrix} 2 \\ 8 \end{bmatrix}$ (d) $\begin{bmatrix} 1 \\ -1 \end{bmatrix}, \begin{bmatrix} 2 \\ -1 \end{bmatrix}, \begin{bmatrix} 1 \\ 3 \end{bmatrix}$

7. Use determinants to test which of the following sets of vectors are linearly independent. If linearly dependent, express one vector as a linear combination of the others.

   (a) $\begin{bmatrix} 1 \\ 1 \\ 1 \end{bmatrix}, \begin{bmatrix} -2 \\ 0 \\ -2 \end{bmatrix}, \begin{bmatrix} 2 \\ 1 \\ 2 \end{bmatrix}$ (b) $\begin{bmatrix} 2 \\ 1 \\ 0 \end{bmatrix}, \begin{bmatrix} 1 \\ 1 \\ 3 \end{bmatrix}, \begin{bmatrix} 0 \\ 2 \\ 1 \end{bmatrix}$

   (c) $\begin{bmatrix} 1 \\ 0 \\ 1 \end{bmatrix}, \begin{bmatrix} 1 \\ 1 \\ 0 \end{bmatrix}, \begin{bmatrix} 1 \\ -2 \\ 3 \end{bmatrix}$ (d) $\begin{bmatrix} 2 \\ 1 \\ 3 \end{bmatrix}, \begin{bmatrix} 1 \\ 2 \\ 3 \end{bmatrix}, \begin{bmatrix} 1 \\ 3 \\ -2 \end{bmatrix}$

8. Tell which of the following sets of matrices are linearly independent. If linearly dependent, express one matrix as a linear combination of the others.

(a) $\begin{bmatrix} 2 & 1 \\ 1 & 1 \end{bmatrix}, \begin{bmatrix} 1 & 1 \\ 0 & 1 \end{bmatrix}, \begin{bmatrix} 1 & 0 \\ 1 & 1 \end{bmatrix}$ (b) $\begin{bmatrix} 1 & 1 \\ -1 & 1 \end{bmatrix}, \begin{bmatrix} 1 & 1 \\ 0 & 1 \end{bmatrix}, \begin{bmatrix} 1 & 0 \\ 1 & 1 \end{bmatrix}$

9. Tell which of the following sets of polynomials are linearly independent. If linearly dependent, express one polynomial as a linear combination of the others.
   (a) $x^2 + 2x$, $x^2 + 2$, $x + 1$
   (b) $3x^2 + 4x + 7$, $x^2 + 1$, $x + 1$
   (c) $x^2 - x + 3$, $2x^2 + 3x + 1$, $2x - 1$

10. Use the determinant of A^TA, as specified in Theorem 4, to test which of the following sets of vectors are linearly independent.

(a) $\begin{bmatrix} 2 \\ 1 \\ 1 \\ 1 \end{bmatrix}, \begin{bmatrix} 2 \\ 0 \\ 0 \\ -2 \end{bmatrix}, \begin{bmatrix} 3 \\ 2 \\ 1 \\ 3 \end{bmatrix}$ (b) $\begin{bmatrix} -1 \\ 2 \\ 1 \\ 0 \end{bmatrix}, \begin{bmatrix} 2 \\ 0 \\ 1 \\ 3 \end{bmatrix}, \begin{bmatrix} 0 \\ 4 \\ 3 \\ 3 \end{bmatrix}$ (c) $\begin{bmatrix} 0 \\ 1 \\ 0 \\ 1 \end{bmatrix}, \begin{bmatrix} 0 \\ 1 \\ 1 \\ 0 \end{bmatrix}, \begin{bmatrix} 1 \\ 1 \\ 1 \\ 1 \end{bmatrix}$

11. Use Theorem 5 to show that the following collections are linearly dependent.
    (a) Any four vectors in $\mathbf{R}^3$
    (b) Any five matrices in $\mathcal{M}_{2,2}$

12. If S is a linearly independent set of vectors in vector space V, and if $\mathbf{v}$ is a vector of V not contained in $V(S)$, the vector subspace spanned by S, then show that $S \cup \mathbf{v}$ is a linearly independent set.

13. If $\{\mathbf{v}_1, \mathbf{v}_2, \ldots, \mathbf{v}_n\}$ is a set of linearly dependent nonzero vectors in vector space V, show that for some k, $\mathbf{v}_k$ can be expressed as a linear combination of vectors with smaller subscripts.

14. If $\{\mathbf{v}_1, \mathbf{v}_2, \ldots, \mathbf{v}_n\}$ is a set of linearly independent vectors in vector space V, show that $\{\mathbf{v}_1, \mathbf{v}_1 + \mathbf{v}_2, \mathbf{v}_1 + \mathbf{v}_2 + \mathbf{v}_3, \ldots, \mathbf{v}_1 + \mathbf{v}_2 + \cdots + \mathbf{v}_n\}$ is a set of linearly independent vectors.

4.3 Bases, Dimension, and Change of Basis

In this section we present the concept of a basis for a vector space. A basis will be a minimum set of vectors that span a vector space. A basis provides a coordinate system for representing other vectors in a vector space, in the way that the Euclidean coordinate vectors $\mathbf{i}_1, \mathbf{i}_2, \ldots, \mathbf{i}_n$ provide a coordinate system for n-vectors in $\mathbf{R}^n$. The size of a basis is defined to be the dimension of the vector space. This definition of dimension corresponds to the reader's intuitive use of the word *dimension*. Finally, in this section, we discuss the procedure of changing from one coordinate basis to another.

Definition
   A linearly independent set of vectors that span a vector space V is called a *basis* for V.

EXAMPLE 1.
Examples of Bases

(a) The coordinate vectors $\mathbf{i}_1 = [1, 0, 0]$, $\mathbf{i}_2 = [0, 1, 0]$, $\mathbf{i}_3 = [0, 0, 1]$ form a basis for $\mathbf{R}^3$, since they span $\mathbf{R}^3$ and are linearly independent [see Example 4(a) in Section 4.2]. More generally, the n coordinate n-vectors are a basis for $\mathbf{R}^n$. The coordinate vectors are called the *standard basis for $\mathbf{R}^n$*.

(b) The three columns of any invertible 3×3 matrix $\mathbf{A}$ are a basis for $\mathbf{R}^3$. The columns $\mathbf{a}_1$, $\mathbf{a}_2$, $\mathbf{a}_3$ of $\mathbf{A}$ span $\mathbf{R}^3$, since $\mathbf{Ax} = \mathbf{b}$, or equivalently, $x_1\mathbf{a}_1 + x_2\mathbf{a}_2 + x_3\mathbf{a}_3 = \mathbf{b}$ has a solution $\mathbf{x}$ for any $\mathbf{b}$ (the solution is $\mathbf{x} = \mathbf{A}^{-1}\mathbf{b}$). Thus linear combinations of the columns yield all vectors in $\mathbf{R}^3$. Further, the columns are linearly independent when $\mathbf{A}$ is invertible, by Theorem 3 in Section 4.2. More generally, by the corollary to that theorem, *any set of n linearly independent n-vectors forms a basis for $\mathbf{R}^n$*.

(c) For the vector subspace of $\mathbf{R}^3$ formed by 3-vectors in which the first two entries are equal [see Example 5(b) in Section 4.1], a basis is formed by the 3-vectors $[1, 1, 0]$ and $[0, 0, 1]$ (verification is left to the reader).

(d) The vector space $\mathcal{M}_{m,n}$ of all $m \times n$ matrices has a basis consisting of the $m \times n$ matrices $\mathbf{M}_{ij}$ with entry $(i, j) = 1$ and all other entries 0. For 2×2 matrices, the basis is $\mathbf{M}_{11} = \begin{bmatrix} 1 & 0 \\ 0 & 0 \end{bmatrix}$, $\mathbf{M}_{21} = \begin{bmatrix} 0 & 1 \\ 0 & 0 \end{bmatrix}$, $\mathbf{M}_{22} = \begin{bmatrix} 0 & 0 \\ 1 & 0 \end{bmatrix}$, and $\mathbf{M}_{22} = \begin{bmatrix} 0 & 0 \\ 0 & 1 \end{bmatrix}$.

These matrices were observed to be linearly independent in Example 4(c) in Section 4.2. The $\mathbf{M}_{ij}$ are called the *standard basis for $\mathcal{M}_{m,n}$*.

(e) The monomials $1, x, x^2, x^3, \ldots, x^k, \ldots$ form a basis for the vector space $\mathcal{P}$ of all polynomials. They span $\mathcal{P}$ and are linearly independent [see Example 4(b) in Section 4.2]. Note that they are an infinite basis. These monomials are called the *standard basis for $\mathcal{P}$*. ■

If we have a set G of vectors that span a vector space V, but the vectors in G are not linearly independent, how does one find a subset G' of G that is a basis, that is, a linearly independent set that spans V? If the vectors are n-vectors in $\mathbf{R}^n$ (for some n), we can convert this problem to a matrix problem. The problem now becomes: When the columns of matrix $\mathbf{A}$ are not linearly independent, find a subset of columns of $\mathbf{A}$ that are linearly independent and such that all other columns are linearly dependent on this subset of columns.

In Examples 5, 6, and 7 of Section 4.2, we tested a set of vectors for linear dependence by putting the vectors in a matrix and applying Gauss–Jordan elimination to reduce the matrix to reduced row echelon form. This approach can also be used to find a linearly independent subset of columns that forms a basis.

Let us review how we were using the reduced row echelon form. Recall that a matrix is in *reduced row echelon form* if:

1. The first nonzero number in each row is a 1, called the *leading 1* of the row.
2. Any row(s) of all 0's appear at the bottom of the matrix.
3. The leading 1 in row i, for $i \geq 2$, is to the right of the leading 1 in earlier rows.
4. A column that contains a leading 1 for some row has 0's as its other entries; that is, a column with a leading 1 is a coordinate vector.

EXAMPLE 2.
Reduced Row
Echelon Form and
Bases

The following are examples of matrices in reduced row echelon form.

$$\begin{bmatrix} 1 & 0 \\ 0 & 1 \end{bmatrix} \quad \begin{bmatrix} 1 & 0 & 3 & 0 \\ 0 & 1 & 5 & 0 \\ 0 & 0 & 0 & 1 \end{bmatrix} \quad \begin{bmatrix} 1 & 0 & 4 & 0 & 6 \\ 0 & 1 & 1 & 0 & 0 \\ 0 & 0 & 0 & 1 & 3 \\ 0 & 0 & 0 & 0 & 0 \end{bmatrix}$$

In each of these matrices, the columns of the leading 1's, that is, the columns that are coordinate vectors, are linearly independent (since the coordinate vectors are linearly independent). The other columns are clearly linear combinations of these coordinate-vector columns. Thus for each of the matrices above, the columns containing leading 1's form a basis for the vector space spanned by all the columns. ■

This easy way of finding a basis in a matrix in reduced row echelon form can be extended, using Gauss–Jordan elimination, to any $m \times n$ matrix.

Theorem 1. Let an $m \times n$ matrix $\mathbf{A}$ be reduced by Gauss–Jordan elimination into reduced row echelon form. Let the reduced matrix be called $\mathbf{A}^*$.

(i) Then the set I of columns of $\mathbf{A}$ that reduce to columns with leading 1's in $\mathbf{A}^*$ is a linearly independent set, and the other columns of $\mathbf{A}$ are linearly dependent on this set.

(ii) The set I of columns of $\mathbf{A}$ is a basis for the vector space spanned by the columns of $\mathbf{A}$.

Proof: Let $\mathbf{A}_I$ denote the submatrix of $\mathbf{A}$ formed by the set I (the set of columns of $\mathbf{A}$ that reduce to columns with leading 1's in $\mathbf{A}^*$), and let $\mathbf{A}_I^*$ denote the submatrix of $\mathbf{A}^*$ consisting of columns with leading 1's. To show that the columns of I are linearly independent, we need by Theorem 2 of Section 4.2 to prove that the only solution to $\mathbf{A}_I\mathbf{x} = \mathbf{0}$ is $\mathbf{x} = \mathbf{0}$. Applying Gauss–Jordan elimination reduces $\mathbf{A}_I\mathbf{x} = \mathbf{0}$ to $\mathbf{A}_I^*\mathbf{x} = \mathbf{0}$. Since the columns of $\mathbf{A}_I^*$ are distinct coordinate vectors, the only solution is $\mathbf{x} = \mathbf{0}$, as required.

Looking, for concreteness, at the matrices in Example 2 in reduced row echelon form, we see that any column $\mathbf{a}_i^*$ without leading 1's is linearly dependent on the columns with leading 1's (the coordinate vectors), in the way that any n-vector is linearly dependent on the n coordinate n-vectors. The linear dependence of $\mathbf{a}_i^*$ in $\mathbf{A}^*$ can be expressed as a nonzero solution $\mathbf{x}_o$ to $\mathbf{A}^*\mathbf{x} = \mathbf{0}$. For example, for the matrix $\mathbf{A}^*$ in Example 2(b), $\mathbf{a}_3^* = 3\mathbf{a}_1^* + 5\mathbf{a}_2^*$ or $3\mathbf{a}_1^* + 5\mathbf{a}_2^* -$

$\mathbf{a}_3^* = \mathbf{0}$, implying $\mathbf{A}^*\mathbf{x}_o = \mathbf{0}$, where $\mathbf{x}_o = \begin{bmatrix} 3 \\ 5 \\ -1 \\ 0 \end{bmatrix}$. But if $\mathbf{x}_o$ is a nonzero

solution to $\mathbf{A}^*\mathbf{x} = \mathbf{0}$, then $\mathbf{x}_o$ is also a nonzero solution to $\mathbf{A}\mathbf{x} = \mathbf{0}$. Thus a linear dependence among columns of $\mathbf{A}^*$ is equivalent to the same linear dependence in $\mathbf{A}$.

Part (ii) of the theorem follows immediately from part (i). ■

EXAMPLE 3.
Finding a Basis
from a Spanning
Set

(a) Consider the vector space V spanned by the vectors $\mathbf{a}_1 = \begin{bmatrix} 1 \\ -2 \end{bmatrix}$, $\mathbf{a}_2 = \begin{bmatrix} 1 \\ 3 \end{bmatrix}$, $\mathbf{a}_3 = \begin{bmatrix} -2 \\ 4 \end{bmatrix}$. We want to find a subset of the $\mathbf{a}_i$'s that forms a basis for V. We form a 2×3 matrix $\mathbf{A}$ with the $\mathbf{a}_i$'s as columns and we perform Gauss–Jordan elimination on $\mathbf{A}$ to reduce it to a reduced row echelon form matrix $\mathbf{A}^*$. By Theorem 1, the set of columns with leading 1's in $\mathbf{A}^*$ will correspond to columns of $\mathbf{A}$ which are linearly independent and form a basis for V.

$$\mathbf{A} = \begin{bmatrix} 1 & 1 & -2 \\ -2 & 3 & 4 \end{bmatrix} \Longrightarrow \begin{bmatrix} 1 & 1 & -2 \\ 0 & 5 & 0 \end{bmatrix} \Longrightarrow \begin{bmatrix} 1 & 0 & -2 \\ 0 & 1 & 0 \end{bmatrix} = \mathbf{A}^*$$

Since leading 1's occur in the first and second columns of $\mathbf{A}^*$, then $\mathbf{a}_1$ and $\mathbf{a}_2$ are linearly independent and form a basis for V.

(b) Consider the vector space V spanned by vectors $\mathbf{a}_1 = \begin{bmatrix} 1 \\ 1 \\ 2 \end{bmatrix}$, $\mathbf{a}_2 = \begin{bmatrix} 1 \\ 2 \\ 0 \end{bmatrix}$, $\mathbf{a}_3 \begin{bmatrix} 1 \\ 0 \\ 4 \end{bmatrix}$, $\mathbf{a}_4 = \begin{bmatrix} 0 \\ 1 \\ 1 \end{bmatrix}$. We want to to find a subset of the $\mathbf{a}_i$'s that form a basis for V. We form a 3×4 matrix $\mathbf{A}$ with the $\mathbf{a}_i$'s as columns and we perform Gauss–Jordan elimination on $\mathbf{A}$ to reduce it to a reduced row echelon form matrix $\mathbf{A}^*$.

$$\mathbf{A} = \begin{bmatrix} 1 & 1 & 1 & 0 \\ 1 & 2 & 0 & 1 \\ 2 & 0 & 4 & 1 \end{bmatrix} \Longrightarrow \begin{bmatrix} 1 & 1 & 1 & 0 \\ 0 & 1 & -1 & 1 \\ 0 & -2 & 2 & 1 \end{bmatrix}$$

$$\Longrightarrow \begin{bmatrix} 1 & 0 & 2 & -1 \\ 0 & 1 & -1 & 1 \\ 0 & 0 & 0 & 3 \end{bmatrix} \Longrightarrow \begin{bmatrix} 1 & 0 & 2 & 0 \\ 0 & 1 & -1 & 0 \\ 0 & 0 & 0 & 1 \end{bmatrix} = \mathbf{A}^*$$

Since leading 1's occur in the first, second, and fourth columns of $\mathbf{A}^*$, then $\mathbf{a}_1$, $\mathbf{a}_2$, $\mathbf{a}_4$ are linearly independent and form a basis for V. ∎

The reader should also review Examples 5, 6, and 7 in Section 4.2 to find the bases found by the reduced row echelon form.

EXAMPLE 4.
Finding a Basis
for Vectors in a
Plane

Find a basis for the vector space of 3-vectors in the plane P: $3x_1 + 2x_2 - x_3 = 0$. As shown in Section 2.1, any two nonzero noncollinear vectors $\mathbf{u}$, $\mathbf{v}$ in P span P. If $\mathbf{u}$ and $\mathbf{v}$ are not collinear (not multiples of one another), they are linearly independent. To find two such vectors, pick values for x_1 and x_2, and choose x_3 to satisfy the equation of the plane, that is, $x_3 = 3x_1 + 2x_2$. Choosing $x_1 = 0$ and $x_2 = 1$, we have $x_3 = 3x_1 + 2x_2 = 3(0) + 2(1) = 2$, yielding the vector $\mathbf{u} = [0, 1, 2]$. Choosing $x_1 = 1$ and $x_2 = 0$, we have $x_3 = 3(1) + 2(0) = 3$, yielding the vector $\mathbf{v} = [1, 0, 3]$. ∎

Theorem 2. If vectors $\mathbf{a}_1$, $\mathbf{a}_2$, . . . , $\mathbf{a}_k$ are a basis for vector space V, then any vector $\mathbf{b}$ in V has a *unique* representation as a linear combination of the $\mathbf{a}_i$'s.

Proof: By the definition of a basis, we know that any $\mathbf{b}$ can be expressed as a linear combination of the $\mathbf{a}_i$'s. The problem is to show that the representation is unique. Suppose that there are two representations:

$$r_1\mathbf{a}_1 + r_2\mathbf{a}_2 + \cdots + r_k\mathbf{a}_k = \mathbf{b} \quad \text{and} \quad s_1\mathbf{a}_1 + s_2\mathbf{a}_2 + \cdots + s_k\mathbf{a}_k = \mathbf{b} \quad (1)$$

Then

$$(r_1 - s_1)\mathbf{a}_1 + (r_2 - s_2)\mathbf{a}_2 + \cdots + (r_k - s_k)\mathbf{a}_k = \mathbf{b} - \mathbf{b} = \mathbf{0} \quad (2)$$

But a linear combination (with not all scalars equal to 0) of the $\mathbf{a}_i$'s can only equal $\mathbf{0}$ in (2) if the $\mathbf{a}_i$'s are linearly dependent, by Theorem 2 in Section 4.2. Since the $\mathbf{a}_i$'s are linearly independent, then $(r_i - s_i) = 0$ for all i, and thus the representation of $\mathbf{b}$ is unique. ■

Corollary. If vectors $\mathbf{a}_1$, $\mathbf{a}_2$, . . . , $\mathbf{a}_k$ in some vector space W are linearly independent and if $\mathbf{b}$ can be written as an linear combination of the $\mathbf{a}_i$'s, this linear combination is unique.

Proof: Use the same argument as in the proof of Theorem 2. Or, one can apply Theorem 2 to the subspace $V = V(\mathbf{a}_1, \mathbf{a}_2, \ldots, \mathbf{a}_k)$ of W spanned by the $\mathbf{a}_i$'s (the $\mathbf{a}_i$'s are a basis for V). ■

Theorem 2 confirms that any basis in any vector space V acts very much like the basis of coordinate vectors in a Euclidean space. That is, one can express any vector in V in terms of a unique set of "coordinates" with respect to that basis. We do this automatically with polynomials. That is, in $\mathcal{P}_3$ (polynomials with greatest degree ≤ 3), a polynomial such as $g(x) = 4x^3 - 2x^2 + 6x + 1$ is written as a unique linear combination of the standard basis for polynomials, $x^3, x^2, x, 1$. The "coordinates" for $g(x)$ are $[4, -2, 6, 1]$.

There is one further basic question about bases we need to address. If sets B_1 and B_2 are different bases for the vector space V, do B_1 and B_2 have the same number of vectors? Intuitively, the answer should be yes.

Theorem 3. If $B_1 = \{\mathbf{u}_1, \mathbf{u}_2, \ldots, \mathbf{u}_m\}$ and $B_2 = \{\mathbf{v}_1, \mathbf{v}_2, \ldots, \mathbf{v}_n\}$ are both bases for the vector space V, then $m = n$.

Proof: Assume that $m \leq n$. Possibly B_1 and B_2 have several vectors in common. Consider a vector $\mathbf{v}_h$ in B_2 that is not in B_1 (if there is no such vector, $B_1 = B_2$). Since B_1 is a basis of V, then $\mathbf{v}_h$ can be expressed as a linear combination of vectors in B_1.

$$\mathbf{v}_h = r_1\mathbf{u}_1 + r_2\mathbf{u}_2 + \cdots + r_m\mathbf{u}_m \quad (3)$$

If each of the $\mathbf{u}_i$ for which $r_i \neq 0$ in (3) is also a member of B_2, we would have a linear dependence among vectors in B_2—impossible since B_2 is a basis. For simplicity, suppose that $r_1 \neq 0$ and $\mathbf{u}_1$ is not in B_2. Then we can bring $\mathbf{u}_1$ over to the left side of (3) and $\mathbf{v}_h$ over to the right side.

$$-r_1\mathbf{u}_1 = -\mathbf{v}_h + r_2\mathbf{u}_2 + \cdots + r_m\mathbf{u}_m \quad \text{or} \quad \mathbf{u}_1 = -\frac{1}{r_1}\mathbf{v}_h - \frac{r_2}{r_1}\mathbf{u}_2 - \cdots - \frac{r_m}{r_1}\mathbf{u}_m$$

(4)

Then $B_1' = \{\mathbf{v}_h, \mathbf{u}_2, \ldots, \mathbf{u}_m\}$ is also a basis, since by (4), the role of $\mathbf{u}_1$ in generating V can be replaced by $\mathbf{v}_h$ and the other $\mathbf{u}_i$. B_1' has one more vector of B_2 than B_1 does. Let us repeat the preceding reasoning with B_1' and B_2. Again we produce a new basis of m vectors with one more member of B_2. Eventually, all vectors in B_1' will also be in B_2. Then the relation in (3) will be a linear dependence among vectors in B_2. This is impossible, so we must have that $m = n$. ■

The proof of Theorem 3 is one of the most sophisticated proofs given in this book and takes time to understand (it must be read several times).

Corollary A. If vector space V has a basis of k vectors, any set S of k linearly independent vectors of V is also a basis for V.

Corollary B. If vector space V has a basis of k vectors, any set S of h vectors, $h > k$, in V is linearly dependent.

Proofs of the corollaries are left as exercises. Recall that Corollary B was proved easily for Euclidean vector spaces in Theorem 5 of Section 4.2. For non-Euclidean spaces, we need Theorem 3.

With Theorem 3, we know that the size of any basis for a particular vector space is a fixed number. That number has a very familiar name.

Definition

The *dimension* dim(V) of a vector space V is the size of any basis for V.

This definition of dimension corresponds to one's natural concept of dimension. For example, the dimension of $\mathbf{R}^n$ is n, since the n coordinate n-vectors are a basis for $\mathbf{R}^n$. If G is a linearly independent set of k vectors in a k-dimensional vector space V, then G will be a basis for V (see Exercise 20).

EXAMPLE 5.
Dimension of
Vector Spaces

(a) The dimension of a vector space consisting of a line through the origin is 1, since the line consists of all multiples of any (nonzero) vector $\mathbf{u}$ lying on the line.

(b) The dimension of a vector space that is a plane through the origin in $\mathbf{R}^3$ is 2, since we saw in Example 4 that such a plane has a basis of size 2.

(c) The dimension of $\mathcal{M}_{2,2}$ of all 2×2 matrices is 4. The basis for this vector space was given in Example 1(d). More generally, the dimension of $\mathcal{M}_{m,n}$ is mn.

(d) The dimension of the vector space of 3-vectors whose first two entries are equal is 2. Two basis vectors for this vector space were given in Example 1(c).

(e) The dimension of the vector space of all polynomials $\mathcal{P}$ is infinite. The monomials $1, x, x^2, x^3, \ldots$ are a basis for $\mathcal{P}$. The dimension of $\mathcal{P}_k$, the vector space of polynomials whose largest degree is $\leq k$, is $k + 1$. ■

We are now in a position to describe all the possible types of vector subspaces of $\mathbf{R}^2$ and $\mathbf{R}^3$. First we consider $\mathbf{R}^2$. Any two linearly independent vectors in $\mathbf{R}^2$ form a basis for $\mathbf{R}^2$ (see the sentence before Example 5). Thus a subspace of $\mathbf{R}^2$ (other than $\mathbf{R}^2$ itself) must have dimension 1 or 0. A basis of size 0 gives the trivial vector subspace consisting of just the $\mathbf{0}$ vector [see Example 1(g) in Section 4.1]. A basis of size 1, say the 2-vector $\mathbf{u}$, yields the vector subspace of all scalar multiples of $\mathbf{u}$. Recall that these multiples form a line through the origin in direction $\mathbf{u}$. Thus the proper subspaces of $\mathbf{R}^2$ consist of all lines through the origin and the trivial subspace of $\mathbf{0}$.

For $\mathbf{R}^3$, all subspaces (other than $\mathbf{R}^3$ itself) will have dimension 0, 1, or 2. The subspaces of dimension size 0 or 1 are the same as for $\mathbf{R}^2$: lines through the origin and the trivial subspace of $\mathbf{0}$. Bases of size 2 generate planes through the origin since, as shown in Section 2.1, all planes through the origin are characterized as all linear combinations of two nonzero noncollinear vectors in the plane. Summarizing our result for $\mathbf{R}^3$, we have:

Theorem 4. The vector subspaces of $\mathbf{R}^3$, other than $\mathbf{R}^3$ itself, are all planes through the origin, all lines through the origin, and the trivial subspace of $\mathbf{0}$.

Many of the theorems in this section are true for non-Euclidean vector spaces, such as the vector space of polynomials. However, for most readers, the Euclidean vector spaces of $\mathbf{R}^2$, $\mathbf{R}^3$, and more generally $\mathbf{R}^n$, are what come to mind when thinking about vector spaces. We now prove that this mental image is valid.

Two vector spaces V and V' are called *isomorphic* if there exists a one-to-one correspondence between V and V' such that if $\mathbf{u}, \mathbf{v}$ are vectors in V and $\mathbf{u}', \mathbf{v}'$ are the corresponding vectors in V', then for any scalars r, s:

1. The scalar multiple $r\mathbf{u}$ in V corresponds to the scalar multiple $r\mathbf{u}'$ in V'.
2. The linear combination $r\mathbf{u} + s\mathbf{v}$ in V corresponds to the linear combination $r\mathbf{u}' + s\mathbf{v}'$ in V'.

Intuitively, if two vector spaces are isomorphic, they are the same or equivalent, just with different names for their vectors.

Theorem 5. All n-dimensional vector spaces with real scalars, for a given n, are isomorphic.
    Proof: Let V and V' be two n-dimensional vector spaces. Let $\mathbf{b}_1, \mathbf{b}_2, \ldots, \mathbf{b}_n$ and $\mathbf{d}_1, \mathbf{d}_2, \ldots, \mathbf{d}_n$ be bases for V and V', respectively. We construct the isomorphism by making $\mathbf{b}_1$ correspond to $\mathbf{d}_1$, $\mathbf{b}_2$ correspond to $\mathbf{d}_2$, $\ldots$, and $\mathbf{b}_n$

correspond to d_n. As noted in Theorem 2, any u vector in V has a unique representation in terms of the basis of b_i's:

$$u = c_1 b_1 + c_2 b_2 + \cdots + c_n b_n \tag{5}$$

We can make a coordinate vector $[c_1, c_2, \ldots, c_n]$ of the "coordinates" c_i in this basis. Each vector is described completely by this coordinate vector. There is a vector u' in V' with the same coordinates as u, but here the coordinates are in terms of $d_1, d_2, \ldots, d_n$.

$$u' = c_1 d_1 + c_2 d_2 + \cdots + c_n d_n \tag{6}$$

Our correspondence will be: *Vectors in V and V' with the same coordinate vector correspond*. This correspondence is consistent with the correspondence that we defined for the basis vectors; for example, b_1 and d_1 have the same coordinate vector of $[1, 0, 0, \ldots, 0]$.

Let us check that for any scalar r and any vector u in V, the scalar multiple ru in V has the same coordinate vector as the corresponding scalar multiple ru' in V'. Since u in V and u' in V' have the same coordinate vector $[c_1, c_2, \ldots, c_n]$, then obviously both ru and ru' have the coordinate vector $[rc_1, rc_2, \ldots, rc_n]$. If v in V and the corresponding vector v' in V' have the same coordinate vector $[g_1, g_2, \ldots, g_n]$, then the linear combinations $ru + sv$ in V and $ru' + sv'$ in V' both have the coordinate vector $[rc_1 + sg_1, rc_2 + sg_2, \ldots, rc_n + sg_n]$.

We have verified that our correspondence is an isomorphism. ■

Corollary. All vector subspaces of a finite-dimensional vector space with real scalars are isomorphic to R^k, for some k.

**EXAMPLE 6.
Isomorphic Vector
Spaces**

(a) Any plane through the origin, such as $3x_1 - 4x_2 + x_3 = 0$, in R^3 is isomorphic to R^2 by Theorem 5. Following Theorem 5, we can construct the isomorphism by pairing off coordinate vectors $[c_1, c_2]$ with respect to bases for the plane and for R^2. For R^2, we use the standard basis $b_1 = \begin{bmatrix} 1 \\ 0 \end{bmatrix}$, $b_2 = \begin{bmatrix} 0 \\ 1 \end{bmatrix}$. For the plane $3x_1 - 4x_2 + x_3 = 0$, we find two linearly independent vectors in the plane, say,

$$d_1 = \begin{bmatrix} 1 \\ 0 \\ -3 \end{bmatrix} \quad \text{and} \quad d_2 = \begin{bmatrix} 0 \\ 1 \\ 4 \end{bmatrix}$$

to form a basis. Then the isomorphism is $c_1 b_1 + c_2 b_2 \leftrightarrow c_1 d_1 + c_2 d_2$, that is

$$c_1 \begin{bmatrix} 1 \\ 0 \end{bmatrix} + c_2 \begin{bmatrix} 0 \\ 1 \end{bmatrix} \leftrightarrow c_1 \begin{bmatrix} 1 \\ 0 \\ -3 \end{bmatrix} + c_2 \begin{bmatrix} 0 \\ 1 \\ 4 \end{bmatrix}$$

(b) The vector space $\mathcal{M}_{m,n}$ of 2×2 matrices and the vector space $\mathcal{P}_3$ of polynomials of highest degree ≤ 3 are four-dimensional vector spaces with real scalars, and hence, by Theorem 5, both are isomorphic to $\mathbf{R}^4$. In this case we can express an isomorphism between $\mathcal{M}_{m,n}$ and $\mathcal{P}_3$ directly (without using bases):

$$\begin{bmatrix} c_1 & c_2 \\ c_3 & c_4 \end{bmatrix} \leftrightarrow c_1 x^3 + c_2 x^2 + c_3 x + c_4 \qquad ■$$

Change of Basis

In the proof of Theorem 5 above, when $\mathbf{u} = c_1 \mathbf{b}_1 + c_2 \mathbf{b}_2 + \cdots + c_n \mathbf{b}_n$, we associated a vector $\mathbf{u}$ in the vector space V with a coordinate vector $[c_1, c_2, \ldots, c_n]$ of the "coordinates" c_i of $\mathbf{u}$ with respect to the basis $B = \{\mathbf{b}_1, \mathbf{b}_2, \ldots, \mathbf{b}_n\}$. Using another basis $D = \{\mathbf{d}_1, \mathbf{d}_2, \ldots, \mathbf{d}_n\}$ for V would produce a different vector of coordinates, say, $[c_1', c_2', \ldots, c_n']$, corresponding to $\mathbf{u} = c_1' \mathbf{d}_1 + c_2' \mathbf{d}_2 + \cdots + c_n' \mathbf{d}_n$.

We now introduce some notation to indicate which basis we are using to generate the coordinate vector for $\mathbf{u}$.

Notation

For a vector $\mathbf{u}$ in vector space V, let $[\mathbf{u}]_B$ denote the vector of coordinates of $\mathbf{u}$ with respect to the basis $B = \{\mathbf{b}_1, \mathbf{b}_2, \ldots, \mathbf{b}_n\}$. Thus, if $\mathbf{u} = c_1 \mathbf{b}_1 + c_2 \mathbf{b}_2 + \cdots + c_n \mathbf{b}_n$, then $[\mathbf{u}]_B = [c_1, c_2, \ldots, c_n]$.

In the preceding discussion, $[\mathbf{u}]_D = [c_1', c_2', \ldots, c_n']$.

Depending on the situation, sometimes one basis may be more convenient for computations or geometric insights, in other problems another basis may be more convenient. For example, in Section 3.3 we wanted to represent vectors in eigenvector coordinates rather than the standard basis coordinates. In the following discussion, we shall show how to convert a coordinate vector for $\mathbf{u}$ from one basis to another.

Consider two bases for a three-dimensional vector space V: basis $B = \{\mathbf{b}_1, \mathbf{b}_2, \mathbf{b}_3\}$ and basis $D = \{\mathbf{d}_1, \mathbf{d}_2, \mathbf{d}_3\}$. The key to converting from B-coordinates to D-coordinates is to represent each $\mathbf{b}_i$ in D-coordinates. We want to determine $[\mathbf{b}_1]_D$, $[\mathbf{b}_2]_D$, and $[\mathbf{b}_3]_D$. For example, in $\mathbf{R}^3$, let

$$B = \left\{ \mathbf{b}_1 = \begin{bmatrix} 1 \\ 1 \\ 0 \end{bmatrix}, \quad \mathbf{b}_2 = \begin{bmatrix} 1 \\ 0 \\ 1 \end{bmatrix}, \quad \mathbf{b}_3 = \begin{bmatrix} 0 \\ 1 \\ 1 \end{bmatrix} \right\}$$

and

$$D = \left\{ \mathbf{d}_1 = \begin{bmatrix} 2 \\ 1 \\ 1 \end{bmatrix}, \quad \mathbf{d}_2 = \begin{bmatrix} 1 \\ 2 \\ 1 \end{bmatrix}, \quad \mathbf{d}_3 = \begin{bmatrix} 1 \\ 1 \\ 2 \end{bmatrix} \right\}$$

Then $[\mathbf{b}_1]_D = \begin{bmatrix} \frac{1}{2} \\ \frac{1}{2} \\ -\frac{1}{2} \end{bmatrix}$, that is,

$$\begin{bmatrix} 1 \\ 1 \\ 0 \end{bmatrix} = \frac{1}{2}\begin{bmatrix} 2 \\ 1 \\ 1 \end{bmatrix} + \frac{1}{2}\begin{bmatrix} 1 \\ 2 \\ 1 \end{bmatrix} - \frac{1}{2}\begin{bmatrix} 1 \\ 1 \\ 2 \end{bmatrix}$$

also,

$$[\mathbf{b}_2]_D = \begin{bmatrix} \frac{1}{2} \\ -\frac{1}{2} \\ \frac{1}{2} \end{bmatrix} \qquad \text{and} \qquad [\mathbf{b}_3]_D = \begin{bmatrix} -\frac{1}{2} \\ \frac{1}{2} \\ \frac{1}{2} \end{bmatrix}$$

(We show how to determine these vectors shortly.) We extend the conversion to all 3-vectors by the following method. If $[\mathbf{u}]_B = \begin{bmatrix} c_1 \\ c_2 \\ c_3 \end{bmatrix}$, that is, $\mathbf{u} = c_1\mathbf{b}_1 + c_2\mathbf{b}_2 + c_3\mathbf{b}_3$, then

$$[\mathbf{u}]_D = c_1[\mathbf{b}_1]_D + c_2[\mathbf{b}_2]_D + c_3[\mathbf{b}_3]_D \tag{7}$$

Equation (7) applies to any three-dimensional vector space, not just $\mathbf{R}^3$, and it extends to any finite-dimensional vector space. Equation (7) represents a strategy that arises over and over in problems involving vectors. Solve a problem for the basis vectors and then extend the solution to all vectors in the vector space by representing other vectors as linear combinations of the basis solutions.

As an illustration of (7) using the bases B and D given above, let $\mathbf{u}$ be the 3-vector $\begin{bmatrix} 3 \\ 1 \\ 0 \end{bmatrix}$. Since $\begin{bmatrix} 3 \\ 1 \\ 0 \end{bmatrix} = 2\begin{bmatrix} 1 \\ 1 \\ 0 \end{bmatrix} + \begin{bmatrix} 1 \\ 0 \\ 1 \end{bmatrix} - \begin{bmatrix} 0 \\ 1 \\ 1 \end{bmatrix}$, then $[\mathbf{u}]_B = \begin{bmatrix} 2 \\ 1 \\ -1 \end{bmatrix}$. As noted above, $[\mathbf{b}_1]_D = \begin{bmatrix} \frac{1}{2} \\ \frac{1}{2} \\ -\frac{1}{2} \end{bmatrix}$, $[\mathbf{b}_2]_D = \begin{bmatrix} \frac{1}{2} \\ -\frac{1}{2} \\ \frac{1}{2} \end{bmatrix}$, $[\mathbf{b}_3]_D = \begin{bmatrix} -\frac{1}{2} \\ \frac{1}{2} \\ \frac{1}{2} \end{bmatrix}$. Then, by (7),

$$[\mathbf{u}]_D = 2[\mathbf{b}_1]_D + 1[\mathbf{b}_2]_D + (-1)[\mathbf{b}_3]_D = 2\begin{bmatrix} \frac{1}{2} \\ \frac{1}{2} \\ -\frac{1}{2} \end{bmatrix} + \begin{bmatrix} \frac{1}{2} \\ -\frac{1}{2} \\ \frac{1}{2} \end{bmatrix} - \begin{bmatrix} -\frac{1}{2} \\ \frac{1}{2} \\ \frac{1}{2} \end{bmatrix} = \begin{bmatrix} 2 \\ 0 \\ -1 \end{bmatrix}$$

The key step in changing bases is finding the D-coordinates $[\mathbf{b}_1]_D$, $[\mathbf{b}_2]_D$, and $[\mathbf{b}_3]_D$ for the basis vectors $\mathbf{b}_i$. We want to express each $\mathbf{b}_i$ as a linear combination of $\mathbf{d}_j$'s. (Note that this representation is unique by Theorem 2.)

For $\mathbf{b}_1$, we seek "coordinates" a_{11}, a_{21}, a_{31} such that

$$\mathbf{b}_1 = a_{11}\mathbf{d}_1 + a_{21}\mathbf{d}_2 + a_{31}\mathbf{d}_3 \quad \text{or} \quad \begin{bmatrix} 1 \\ 1 \\ 0 \end{bmatrix} = a_{11}\begin{bmatrix} 2 \\ 1 \\ 1 \end{bmatrix} + a_{21}\begin{bmatrix} 1 \\ 2 \\ 1 \end{bmatrix} + a_{31}\begin{bmatrix} 1 \\ 1 \\ 2 \end{bmatrix} \tag{8}$$

The reason for the double subscripts will be apparent shortly. The vector equation in (8) can be rewritten as a system of three equations in three unknowns.

$$
\begin{aligned}
2a_{11} + a_{21} + a_{31} &= 1 \\
a_{11} + 2a_{21} + a_{31} &= 1 \\
a_{11} + a_{21} + 2a_{31} &= 0
\end{aligned}
\tag{9}
$$

By Gaussian elimination, we find that the solution to (9) is

$$
a_{11} = \tfrac{1}{2}, \quad a_{21} = \tfrac{1}{2}, \quad a_{31} = -\tfrac{1}{2}; \quad \text{so} \quad \mathbf{b}_1 = \tfrac{1}{2}\mathbf{d}_1 + \tfrac{1}{2}\mathbf{d}_2 - \tfrac{1}{2}\mathbf{d}_3
$$

Observe that if $\mathbf{D}$ is a 3×3 matrix whose columns are the basis vectors $\mathbf{d}_i$, then (9) and (8) become the matrix equation

$$
\mathbf{D}\mathbf{a}_1 = \mathbf{b}_1, \quad \text{where} \quad \mathbf{a}_1 = \begin{bmatrix} a_{11} \\ a_{21} \\ a_{31} \end{bmatrix}
\tag{10}
$$

Similarly, to find the D-coordinates of $\mathbf{b}_2$, we obtain the system of equations

$$
\begin{aligned}
2a_{12} + a_{22} + a_{32} &= 1 \\
a_{12} + 2a_{22} + a_{32} &= 0 \\
a_{12} + a_{22} + 2a_{32} &= 1
\end{aligned}
\tag{11}
$$

whose solution is $a_{12} = \tfrac{1}{2}$, $a_{22} = -\tfrac{1}{2}$, $a_{32} = \tfrac{1}{2}$. For $\mathbf{b}_3$, we get $a_{13} = -\tfrac{1}{2}$, $a_{23} = \tfrac{1}{2}$, $a_{33} = \tfrac{1}{2}$.

If $[\mathbf{u}]_B = \begin{bmatrix} c_1 \\ c_2 \\ c_3 \end{bmatrix}$, then (7) becomes

$$
[\mathbf{u}]_D = c_1[\mathbf{b}_1]_D + c_2[\mathbf{b}_2]_D + c_3[\mathbf{b}_3]_D = c_1 \begin{bmatrix} \tfrac{1}{2} \\ \tfrac{1}{2} \\ -\tfrac{1}{2} \end{bmatrix} + c_2 \begin{bmatrix} \tfrac{1}{2} \\ -\tfrac{1}{2} \\ \tfrac{1}{2} \end{bmatrix} + c_3 \begin{bmatrix} -\tfrac{1}{2} \\ \tfrac{1}{2} \\ \tfrac{1}{2} \end{bmatrix}
\tag{12}
$$

If we define the matrix $\mathbf{A}$ to have entries a_{ij} (so that the jth column of $\mathbf{A}$ is $[\mathbf{b}_j]_D$), (12) becomes the matrix equation

$$
[\mathbf{u}]_D = \mathbf{A}([\mathbf{u}]_B), \quad \text{where} \quad \mathbf{A} = \begin{bmatrix} \tfrac{1}{2} & \tfrac{1}{2} & -\tfrac{1}{2} \\ \tfrac{1}{2} & -\tfrac{1}{2} & \tfrac{1}{2} \\ -\tfrac{1}{2} & \tfrac{1}{2} & \tfrac{1}{2} \end{bmatrix}
\tag{13}
$$

We return to the change of basis for $\mathbf{u} = \begin{bmatrix} 3 \\ 1 \\ 0 \end{bmatrix}$. Here $[\mathbf{u}]_B = \begin{bmatrix} 2 \\ 1 \\ -1 \end{bmatrix}$, since

$$
\begin{bmatrix} 3 \\ 1 \\ 0 \end{bmatrix} = 2\begin{bmatrix} 1 \\ 1 \\ 0 \end{bmatrix} + \begin{bmatrix} 1 \\ 0 \\ 1 \end{bmatrix} - \begin{bmatrix} 0 \\ 1 \\ 1 \end{bmatrix}
$$

Similarly,y, since $\begin{bmatrix} 3 \\ 1 \\ 0 \end{bmatrix} = 2\begin{bmatrix} 2 \\ 1 \\ 1 \end{bmatrix} - \begin{bmatrix} 1 \\ 1 \\ 2 \end{bmatrix}$, then $[\mathbf{u}]_D = \begin{bmatrix} 2 \\ 0 \\ -1 \end{bmatrix}$. We confirm that (13)

gives the proper change of coordinates:

$$[\mathbf{u}]_D = \mathbf{A}([\mathbf{u}]_B): \quad \begin{bmatrix} 2 \\ 0 \\ -1 \end{bmatrix} = \begin{bmatrix} \frac{1}{2} & \frac{1}{2} & -\frac{1}{2} \\ \frac{1}{2} & -\frac{1}{2} & \frac{1}{2} \\ -\frac{1}{2} & \frac{1}{2} & \frac{1}{2} \end{bmatrix}\begin{bmatrix} 2 \\ 1 \\ -1 \end{bmatrix} \tag{14}$$

The matrix $\mathbf{A}$ in (13) is called the **transition matrix from basis B to basis D**. This analysis applies for any conversion, even for non-Euclidean vector spaces. Thus we obtain the following theorem.

Theorem 6. Let $B = \{\mathbf{b}_1, \mathbf{b}_2, \ldots, \mathbf{b}_n\}$ and $D = \{\mathbf{d}_1, \mathbf{d}_2, \ldots, \mathbf{d}_n\}$ be two bases for the n-dimensional vector space V. Let $\mathbf{A}$ be the $n \times n$ transition matrix whose jth column is $\mathbf{a}_1 = [\mathbf{b}_j]_D$. Then we convert a vector $\mathbf{u}$ in V from B-coordinates $[\mathbf{u}]_B$ to D-coordinates $[\mathbf{u}]_D$ by the matrix equation $[\mathbf{u}]_D = \mathbf{A}([\mathbf{u}]_B)$.

Note that if V is a Euclidean vector space, the columns $\mathbf{a}_j$ of $\mathbf{A}$ are found by solving the system of equations $\mathbf{Da}_j = \mathbf{b}_j$, as illustrated in (9) and (11). If V is not a Euclidean vector space, there is no one simple way in all cases to determine the columns of $\mathbf{A}$.

EXAMPLE 7.
Change of Basis in a Euclidean Vector Space

For the vector space $\mathbf{R}^2$, consider the bases $B = \left\{ \mathbf{b}_1 = \begin{bmatrix} 1 \\ 0 \end{bmatrix}, \mathbf{b}_2 = \begin{bmatrix} 0 \\ 1 \end{bmatrix} \right\}$ and $D = \left\{ \mathbf{d}_1 = \begin{bmatrix} 1 \\ 1 \end{bmatrix}, \mathbf{d}_2 = \begin{bmatrix} 1 \\ -1 \end{bmatrix} \right\}$. Note that $\mathbf{B}$ is the standard coordinate basis for $\mathbf{R}^2$. To change a 2-vector $\mathbf{u}$ from the standard B-coordinates $[\mathbf{u}]_B$ $(= \mathbf{u})$ to the D-coordinates $[\mathbf{u}]_D$, we must determine the transition matrix $\mathbf{A}$ whose columns are $\mathbf{a}_1 = [\mathbf{b}_1]_D$ and $\mathbf{a}_2 = [\mathbf{b}_2]_D$. As noted above, $\mathbf{a}_1$ is found by solving the following matrix equation [see (9)]:

$$\mathbf{Da}_1 = \mathbf{b}_1: \quad \begin{aligned} 1a_{11} + 1a_{21} &= 1 \\ 1a_{11} - 1a_{21} &= 0 \end{aligned} \tag{15}$$

Similarly, to find $\mathbf{a}_2$, we solve

$$\mathbf{Da}_1 = \mathbf{b}_1: \quad \begin{aligned} 1a_{12} + 1a_{22} &= 0 \\ 1a_{12} - 1a_{22} &= 1 \end{aligned} \tag{16}$$

The reader may recall from Section 1.5 that (15) and (16) are the systems of equations satisfied by the entries a_{ij} of $\mathbf{A}$ when $\mathbf{A}$ is the inverse of $\mathbf{D}$, that is, $\mathbf{A} = \mathbf{D}^{-1}$. (The reason for this appearance of the inverse will be explained shortly.) By Gauss–Jordan elimination, we determine the transition matrix $\mathbf{A}$ whose entries are the solutions of (15) and (16) to be $\mathbf{A} = \begin{bmatrix} \frac{1}{2} & \frac{1}{2} \\ \frac{1}{2} & -\frac{1}{2} \end{bmatrix}$. ■

EXAMPLE 8.
Change of Basis in
a Non-Euclidean
Vector Space

For the vector space $\mathcal{P}_2$ of polynomials of greatest degree ≤ 2, the standard basis is $B = \{x^2, x, 1\}$. Let us give another basis $D = \{x^2 + x, x^2 + 1, x + 1\}$. To change a polynomial $p(x)$ in $\mathcal{P}_2$ from standard B-coordinates $[p(x)]_B$ to D-coordinates $[p(x)]_D$, we must determine the 3×3 transition matrix $\mathbf{A}$ with columns $\mathbf{a}_1 = [x^2]_D$, $\mathbf{a}_2 = [x]_D$, and $\mathbf{a}_3 = [1]_D$.

To represent x^2 as a linear combination of polynomials in the D basis, that is,

$$x^2 = a_{11}(x^2 + x) + a_{21}(x^2 + 1) + a_{31}(x + 1) \tag{17}$$

we require that the coefficients of x^2 terms, the coefficients of x, and the constant terms are equal on each side of (17). Equating these coefficients on each side of (17), we have

$$\begin{aligned} \text{Coefficient of } x^2\text{: } & 1a_{11} + 1a_{21} && = 1 \\ \text{Coefficient of } x\text{: } & 1a_{11} && + 1a_{31} = 0 \\ \text{Constant term: } & & 1a_{21} + 1a_{31} = 0 \end{aligned} \tag{18}$$

Solving (18) by Gaussian elimination yields $a_{11} = \frac{1}{2}$, $a_{21} = \frac{1}{2}$, $a_{31} = -\frac{1}{2}$.

Next we repeat this process to represent x as a linear combination of polynomials in the D basis:

$$x = a_{12}(x^2 + x) + a_{22}(x^2 + 1) + a_{32}(x + 1) \tag{19}$$

Equating coefficients on each side of (19), we have

$$\begin{aligned} \text{Coefficient of } x^2\text{: } & 1a_{12} + 1a_{22} && = 0 \\ \text{Coefficient of } x\text{: } & 1a_{12} && + 1a_{32} = 1 \\ \text{Constant term: } & & 1a_{22} + 1a_{32} = 0 \end{aligned} \tag{20}$$

Solving (20) by Gaussian elimination yields $a_{12} = \frac{1}{2}$, $a_{22} = -\frac{1}{2}$, $a_{32} = \frac{1}{2}$.

Repeating this process to represent 1 as a linear combination of polynomials in the D basis, we have

$$\begin{aligned} \text{Coefficient of } x^2\text{: } & 1a_{13} + 1a_{23} && = 0 \\ \text{Coefficient of } x\text{: } & 1a_{13} && + 1a_{33} = 0 \\ \text{Constant term: } & & 1a_{23} + 1a_{33} = 1 \end{aligned} \tag{21}$$

Solving (21) by Gaussian elimination yields $a_{13} = -\frac{1}{2}$, $a_{23} = \frac{1}{2}$, $a_{33} = \frac{1}{2}$. Then the required transition matrix $\mathbf{A}$ is

$$\mathbf{A} = \begin{bmatrix} \frac{1}{2} & \frac{1}{2} & -\frac{1}{2} \\ \frac{1}{2} & -\frac{1}{2} & \frac{1}{2} \\ -\frac{1}{2} & \frac{1}{2} & \frac{1}{2} \end{bmatrix} \tag{22}$$

■

Again we note that the systems of equations (18), (20), (21) are those defining the inverse of the matrix of coefficients in these equations.

We finish our discussion of change of basis by examining the relationship between the matrix $\mathbf{A}$ for changing from a basis B to a basis D and the matrix $\mathbf{A}^*$ for changing from basis D back to basis B. That is,

$$[\mathbf{u}]_D = \mathbf{A}([\mathbf{u}]_B) \quad \text{and} \quad [\mathbf{u}]_B = \mathbf{A}^*([\mathbf{u}]_D)$$

Combining the change from B to D and the change from D to B, we have

$$[\mathbf{u}]_B = \mathbf{A}^*([\mathbf{u}]_D) = \mathbf{A}^*(\mathbf{A}([\mathbf{u}]_B)) \quad \text{or} \quad [\mathbf{u}]_B = \mathbf{A}^*\mathbf{A}([\mathbf{u}]_B) \tag{23}$$

So $\mathbf{A}\mathbf{A}^*$ is a matrix that converts from B-coordinates to D-coordinates and then back to B-coordinates—back to the original coordinates. The net effect of $\mathbf{A}^*\mathbf{A}$ is to leave the vector $[\mathbf{u}]_B$ unchanged. That is, $\mathbf{A}\mathbf{A}^*$ acts just like the identity matrix $\mathbf{I}$.

Theorem 7. Let V be an n-dimensional vector space with bases B and D. Let $\mathbf{A}$ be the transition matrix from basis B to basis D, and let $\mathbf{A}^*$ be the transition matrix from basis D to basis B. Then $\mathbf{A}^* = \mathbf{A}^{-1}$.

Proof: We have just shown that $\mathbf{A}^*\mathbf{A}$ has the same effect on all B-coordinate vectors $[\mathbf{u}]_B$ as the identity matrix $\mathbf{I}$. If $\mathbf{A}^*\mathbf{A} = \mathbf{I}$, then by the definition of the inverse, $\mathbf{A}^* = \mathbf{A}^{-1}$. The issue is: If $\mathbf{A}^*\mathbf{A}$ and $\mathbf{I}$ act the same on all vectors of V, then are $\mathbf{A}^*\mathbf{A}$ and $\mathbf{I}$ equal? We shall now give a proof that indeed, $\mathbf{A}^*\mathbf{A} = \mathbf{I}$.

In the B-coordinates, $[\mathbf{b}_1]_B = \mathbf{i}_1$; that is, $\mathbf{b}_1 = 1\mathbf{b}_1 + 0\mathbf{b}_2 + \cdots + 0\mathbf{b}_n$. Substituting $[\mathbf{b}_1]_B = \mathbf{i}_1$ in the right side of (23), we see

$$[\mathbf{b}_1]_B = \mathbf{A}^*\mathbf{A}([\mathbf{b}_1]_B) \implies \mathbf{i}_1 = \mathbf{A}^*\mathbf{A}(\mathbf{i}_1) \tag{24}$$

But one of the basic properties of matrix algebra is that $\mathbf{C}\mathbf{i}_1 = \mathbf{c}_1$, where $\mathbf{c}_1$ denotes the first column of matrix $\mathbf{C}$. So (24) says that the first column of $\mathbf{A}^*\mathbf{A}$ is $\mathbf{i}_1$. Using the other basis vectors, we obtain $\mathbf{A}^*\mathbf{A}(\mathbf{i}_k) = \mathbf{i}_k$, for all k. So the columns of $\mathbf{A}^*\mathbf{A}$ are the columns of the identity matrix $\mathbf{I}$. Thus $\mathbf{A}^*\mathbf{A} = \mathbf{I}$. ■

In Section 5.1, we shall prove a much more general form of the claim that when a matrix behaves like the identity matrix, it is the identity matrix. Politicians are famous for saying, "If it walks like a duck and talks like a duck, it is a duck." Mathematicians, however, require rigorous proofs of such assertions!

**EXAMPLE 9.
Reverse Change
of Basis**

(a) Find the reverse transition matrix $\mathbf{A}^*$ in Example 7 for changing from $D = \left\{ \mathbf{d}_1 = \begin{bmatrix} 1 \\ 1 \end{bmatrix}, \mathbf{d}_2 = \begin{bmatrix} 1 \\ -1 \end{bmatrix} \right\}$ to $B = \left\{ \mathbf{b}_1 = \begin{bmatrix} 1 \\ 0 \end{bmatrix}, \mathbf{b}_2 = \begin{bmatrix} 0 \\ 1 \end{bmatrix} \right\}$. Recall that the transition matrix $\mathbf{A}$ from basis B to basis D was $\mathbf{A} = \begin{bmatrix} \frac{1}{2} & \frac{1}{2} \\ \frac{1}{2} & -\frac{1}{2} \end{bmatrix}$. Note that B is the standard basis: $\mathbf{b}_2 = \mathbf{i}_1$ and $\mathbf{b}_2 = \mathbf{i}_2$, and thus writing the vectors of D in terms of B is trivial:

$$\mathbf{d}_1 = \begin{bmatrix} 1 \\ 1 \end{bmatrix} = 1\mathbf{i}_1 + 1\mathbf{i}_2 \quad \text{and} \quad \mathbf{d}_2 = \begin{bmatrix} 1 \\ -1 \end{bmatrix} = 1\mathbf{i}_1 - 1\mathbf{i}_2 \tag{25}$$

Then $\mathbf{A}^* = \mathbf{D} = \begin{bmatrix} 1 & 1 \\ 1 & -1 \end{bmatrix}$, where the $\mathbf{D}$'s columns are $\mathbf{d}_1$, $\mathbf{d}_2$ in (25). By Theorem 7, $\mathbf{A}$ and $\mathbf{A}^*$ are inverses of each other, that is, $\mathbf{A} = \mathbf{A}^{*-1}$. Thus if $\mathbf{A}^* = \mathbf{D}$, then $\mathbf{A} = \mathbf{D}^{-1}$. (The reader should verify that $\mathbf{A} = \mathbf{D}^{-1}$.)

(b) Find the reverse transition matrix $\mathbf{A}^*$ in Example 8 from $D = \{x^2 + x, x^2 + 1, x + 1\}$ to $B = \{x^2, x, 1\}$. Expressing the polynomials in D in terms of the standard monomial basis B is trivial: $x^2 + x = 1(x^2) + 1(x)$, $x^2 + 1 = 1(x^2) + 1(1)$,

and $x + 1 = 1(x) + 1(1)$. Therefore, $\mathbf{A}^* = \mathbf{D} = \begin{bmatrix} 1 & 1 & 0 \\ 1 & 0 & 1 \\ 0 & 1 & 1 \end{bmatrix}$, whose columns

are these "coordinates" (just given) of D's vectors with respect to B. The reader

should verify that the transition matrix $\mathbf{A} = \begin{bmatrix} \frac{1}{2} & \frac{1}{2} & -\frac{1}{2} \\ \frac{1}{2} & -\frac{1}{2} & \frac{1}{2} \\ -\frac{1}{2} & \frac{1}{2} & \frac{1}{2} \end{bmatrix}$ found in Exam-

ple 8 is the inverse of $\mathbf{A}^*$, as asserted in Theorem 7. ■

The situation in Example 9 in which $\mathbf{A}^* = \mathbf{D}$ and $\mathbf{A} = \mathbf{D}^{-1}$ depends only on the fact that B is the standard coordinate basis.

Theorem 8. Let B be the standard coordinate basis $\{\mathbf{i}_1, \mathbf{i}_2, \ldots, \mathbf{i}_n\}$ for $\mathbf{R}^n$, and let $D = \{\mathbf{d}_1, \mathbf{d}_2, \ldots, \mathbf{d}_n\}$ be any other basis for $\mathbf{R}^n$. Then the transition matrix $\mathbf{A}^*$ from the basis D to the basis B is $\mathbf{A}^* = \mathbf{D}$, where the $\mathbf{D}$ is the matrix whose jth column is $\mathbf{d}_j$. Thus the transition matrix $\mathbf{A}$ from the basis B to the basis D is $\mathbf{A} = \mathbf{D}^{-1}$.

The details of the proof of Theorem 8 are left as an exercise.

This completes our development of the foundational theory of vector spaces. This theory is one of the most powerful in all mathematics and a large part of advanced mathematics casts problems in a vector space framework in order to take advantage of this theory. The following sections in this chapter examine particular vector spaces and special types of bases.

Section 4.3 Exercises

1. Give another basis, other than the standard basis given in Example 1, for:
   (a) $\mathbf{R}^2$ (b) $\mathbf{R}^3$ (c) $\mathbf{R}^4$
   (d) $\mathcal{M}_{2,2}$ (e) $\mathcal{M}_{3,3}$ (f) $\mathcal{P}_3$

2. Confirm that $[1, 1, 0, 0]$, $[0, 0, 1, 0]$, and $[0, 0, 0, 1]$ form a basis for the vector space of 4-vectors in which the first two entries are equal.

3. Find a subset of vectors in the following spanning sets G that form a basis for the vector space $V(G)$ spanned by G.

   (a) $G = \left\{ \begin{bmatrix} 1 \\ 2 \end{bmatrix}, \begin{bmatrix} 1 \\ 3 \end{bmatrix}, \begin{bmatrix} 1 \\ 4 \end{bmatrix} \right\}$ (b) $G = \left\{ \begin{bmatrix} 2 \\ -1 \end{bmatrix}, \begin{bmatrix} -2 \\ 4 \end{bmatrix} \right\}$

   (c) $G = \left\{ \begin{bmatrix} 1 \\ -3 \end{bmatrix}, \begin{bmatrix} -2 \\ 6 \end{bmatrix}, \begin{bmatrix} 3 \\ 4 \end{bmatrix} \right\}$

4. Find a subset of vectors in the following spanning sets G that form a basis for the vector space $V(G)$ spanned by G.

(a) $G = \left\{ \begin{bmatrix} 1 \\ 1 \\ 0 \end{bmatrix}, \begin{bmatrix} 1 \\ 1 \\ 3 \end{bmatrix}, \begin{bmatrix} 1 \\ 1 \\ 4 \end{bmatrix}, \begin{bmatrix} 0 \\ 1 \\ 3 \end{bmatrix} \right\}$ (b) $G = \left\{ \begin{bmatrix} 1 \\ 0 \\ -3 \end{bmatrix}, \begin{bmatrix} -1 \\ 2 \\ 6 \end{bmatrix}, \begin{bmatrix} 3 \\ 4 \\ 0 \end{bmatrix} \right\}$

(c) $G = \left\{ \begin{bmatrix} 2 \\ 1 \\ 0 \end{bmatrix}, \begin{bmatrix} 1 \\ 1 \\ 1 \end{bmatrix}, \begin{bmatrix} 1 \\ 0 \\ 3 \end{bmatrix}, \begin{bmatrix} 0 \\ 1 \\ 3 \end{bmatrix} \right\}$ (d) $G = \left\{ \begin{bmatrix} 2 \\ 1 \\ 3 \end{bmatrix}, \begin{bmatrix} -2 \\ 3 \\ 0 \end{bmatrix}, \begin{bmatrix} 0 \\ 3 \\ 4 \end{bmatrix} \right\}$

5. Find a basis for the vector space of:
   (a) All 3-vectors $\mathbf{v} = [v_1, v_2, v_3]$ with $v_2 = 0$
   (b) All 3-vectors $\mathbf{v} = [v_1, v_2, v_3]$ with $v_1 = v_2 = v_3$
   (c) All 3-vectors $\mathbf{v} = [v_1, v_2, v_3]$ with $v_1 = 2v_3$

6. Find a basis for the vector space of:
   (a) All 2×2 matrices $\mathbf{M}$ with $m_{11} = 0$
   (b) All 2×2 matrices with $m_{11} = m_{22}$
   (c) All symmetric 3×3 matrices (Recall that symmetric means $m_{ij} = m_{ji}$.)

7. Find a basis for the vector space of:
   (a) All polynomials with greatest degree ≤ 3
   (b) All polynomials $p(x)$ with greatest degree ≤ 3 and $p(0) = 0$

8. Find a basis for the vector space of the following planes in $\mathbf{R}^3$.
   (a) $x_1 + 2x_2 - x_3 = 0$
   (b) $2x_1 - x_2 + 2x_3 = 0$
   (c) $5x_1 + x_2 - 4x_3 = 0$
   (d) $-3x_1 - 4x_2 + x_3 = 0$
   (e) $x_1 - 4x_2 - 3x_3 = 0$

9. Which of the following sets of vectors form a basis for $\mathbf{R}^3$?
   (a) $\{[1, 0, 2], [2, 0, 1], [1, 1, 1]\}$
   (b) $\{[1, 3, 5], [2, 6, 1]\}$
   (c) $\{[1, 1, 0], [2, 1, 1], [0, 1, -1]\}$

10. Which of the following sets of vectors form a basis for $\mathcal{P}_2$, polynomials of greatest degree ≤ 2?
    (a) $\{x^2, x^2 + 2x, x^2 + x + 1\}$
    (b) $\{x^2 + x, x^2 + 1, x + 1\}$
    (c) $\{x^2 + x, x^2 + 3x - 1, x^2 + x + 1\}$

11. What is the dimension of each of the following Euclidean vector spaces?
    (a) 4-vectors whose first two entries are equal
    (b) 3-vectors $\mathbf{v} = [v_1, v_2, v_3]$ with $v_1 = 2v_3$
    (c) 3-vectors $\mathbf{x}$ on a plane through the origin and such that $x_1 = 0$
    (d) 5-vectors $\mathbf{x}$ such that $\mathbf{a} \cdot \mathbf{x} = 0$ and $\mathbf{b} \cdot \mathbf{x} = 0$, for two 5-vectors $\mathbf{a}$ and $\mathbf{b}$ (where $\mathbf{b}$ is not a multiple of $\mathbf{a}$)

12. What is the dimension of each of the following non-Euclidean vector spaces?
    (a) 2×2 matrices $\mathbf{M}$ with $m_{11} = 0$
    (b) Symmetric 3×3 matrices
    (c) Polynomials $p(x)$ in $\mathcal{P}_3$ with $p(0) = 0$

13. For what k is $\mathbf{R}^k$ isomorphic to the following vector spaces?
    (a) The vector space of 4-vectors in which the first two entries are equal
    (b) The vector space $\mathcal{M}_{3,3}$

   (c)   The vector space of symmetric 3×3 matrices
   (d)   The vector space $\mathcal{P}_3$ of polynomials of greatest degree ≤ 3
   (e)   The vector subspace of polynomials $p(x)$ in $\mathcal{P}_3$ for which $p(0) = 0$
   (f)   The vector space $\mathcal{P}$ of all polynomials

14.   Give an isomorphism (in the spirit of the proof of Theorem 5) between $\mathbf{R}^k$, for the appropriate k, and the following vector spaces.
   (a)   The vector space of 4-vectors in which the first two entries are equal
   (b)   The plane $x_1 + 2x_2 - x_3 = 0$
   (c)   The vector space of symmetric 3×3 matrices
   (d)   The vector space $\mathcal{P}_3$ of polynomials of greatest degree ≤ 3

15.   Give an isomorphism (in the spirit of the proof of Theorem 5) between the following pairs of vector spaces.
   (a)   The vector space of 4-vectors in which the first two entries are equal and the vector space of symmetric 2×2 matrices
   (b)   The plane $x_1 + 2x_2 - x_3 = 0$ and the vector space of 4-vectors in which the first two entries are equal and the last two entries are equal
   (c)   The vector space of symmetric 2×2 matrices and the vector subspace of polynomials in $\mathcal{P}_3$ (whose greatest degree ≤ 3) for which $p(0) = 0$

16.   (a)   Prove Corollary A of Theorem 3.
   (b)   Prove Corollary B of Theorem 3.

17.   Give the transition matrix for the following changes of basis.
   (a)   From the standard basis for $\mathbf{R}^2$ to the basis $\begin{bmatrix} 1 \\ 2 \end{bmatrix}, \begin{bmatrix} 3 \\ 2 \end{bmatrix}$

   (b)   From the standard basis for $\mathbf{R}^2$ to the basis $\begin{bmatrix} -2 \\ 3 \end{bmatrix}, \begin{bmatrix} 1 \\ 5 \end{bmatrix}$

   (c)   From the standard basis for $\mathbf{R}^3$ to the basis $\begin{bmatrix} 1 \\ 1 \\ 0 \end{bmatrix}, \begin{bmatrix} 1 \\ 1 \\ 3 \end{bmatrix}, \begin{bmatrix} 0 \\ 1 \\ 4 \end{bmatrix}$

   (d)   From the standard basis for $\mathbf{R}^3$ to the basis $\begin{bmatrix} 0 \\ 1 \\ 0 \end{bmatrix}, \begin{bmatrix} 1 \\ -1 \\ 3 \end{bmatrix}, \begin{bmatrix} 2 \\ 0 \\ 1 \end{bmatrix}$

   (e)   From the standard basis for $\mathbf{R}^3$ to the basis $\begin{bmatrix} 2 \\ 1 \\ 1 \end{bmatrix}, \begin{bmatrix} 0 \\ -1 \\ 1 \end{bmatrix}, \begin{bmatrix} 2 \\ 1 \\ 4 \end{bmatrix}$

18.   Give the transition matrix for the following changes of basis.
   (a)   From the standard basis for $\mathcal{M}_{2,2}$ to the basis $\begin{bmatrix} 1 & 1 \\ 0 & 0 \end{bmatrix}, \begin{bmatrix} 1 & 0 \\ 1 & 0 \end{bmatrix}, \begin{bmatrix} 0 & 1 \\ 0 & 1 \end{bmatrix}, \begin{bmatrix} 0 & 0 \\ 0 & 1 \end{bmatrix}$
   (b)   From the standard basis for $\mathcal{P}_2$ to the basis $\{x^2 - 1, x^2 - 2x, x - 1\}$
   (c)   From the standard basis for $\mathcal{P}_2$ to the basis $\{x^2 + x - 1, x^2 - 2x + 2, 2x^2 - x\}$

19.   Find the transition matrix $\mathbf{A}$ for changing:

   (a)   From basis $B = \left\{ \begin{bmatrix} 1 \\ 1 \end{bmatrix}, \begin{bmatrix} 1 \\ -1 \end{bmatrix} \right\}$ to basis $D = \left\{ \begin{bmatrix} 1 \\ 2 \end{bmatrix}, \begin{bmatrix} 2 \\ 3 \end{bmatrix} \right\}$

(b) From basis $B = \left\{ \begin{bmatrix} 1 \\ 2 \end{bmatrix}, \begin{bmatrix} 3 \\ 1 \end{bmatrix} \right\}$ to basis $D = \left\{ \begin{bmatrix} 2 \\ -1 \end{bmatrix}, \begin{bmatrix} 1 \\ -2 \end{bmatrix} \right\}$

(c) From basis $B = \left\{ \begin{bmatrix} 1 \\ 1 \\ 0 \end{bmatrix}, \begin{bmatrix} 1 \\ 1 \\ 3 \end{bmatrix}, \begin{bmatrix} 0 \\ 1 \\ 4 \end{bmatrix} \right\}$ to basis $D = \left\{ \begin{bmatrix} 0 \\ 1 \\ 0 \end{bmatrix}, \begin{bmatrix} 1 \\ -1 \\ 3 \end{bmatrix}, \begin{bmatrix} 2 \\ 0 \\ 1 \end{bmatrix} \right\}$

20. If G is a linearly independent set of k vectors in a k-dimensional vector space V, show that G is a basis for V.

21. **(a)** Let W and W' be subspaces of the Euclidean vector space V with $W \cap W' = \{0\}$. Then define the collection $W + W' = \{\mathbf{w} + \mathbf{w}' : \mathbf{w} \in W, \mathbf{w} \in W'\}$. Show that $W + W'$ is a subspace of V.

(b) If $B = \{\mathbf{b}_1, \mathbf{b}_2, \ldots, \mathbf{b}_n\}$ and $B' = \{\mathbf{b}_1', \mathbf{b}_2', \ldots, \mathbf{b}_n'\}$ are bases for W and W', respectively, use them to find a basis for $W + W'$. What is the dimension of $W + W'$?

22. Show that any set G of k linearly independent n-vectors, $k < n$, can be extended to a basis for $\mathbf{R}^n$. [Hint: Form an $n \times (k + n)$ matrix $\mathbf{A}$ whose first k columns come from G and whose last n columns are the identity matrix (thus the columns of $\mathbf{A}$ span $\mathbf{R}^n$).]

23. Let D and E be nonstandard bases for $\mathbf{R}^k$, let $\mathbf{A}^*$ be the transition matrix from basis D to the standard basis B, and let $\mathbf{A}'$ be the transition matrix from B to basis E. Then show that the transition matrix $\mathbf{A}$ from D to E is $\mathbf{A} = \mathbf{A}'\mathbf{A}^*$.

24. Prove Theorem 8. Show that $\mathbf{A}^* = \mathbf{D}$, a matrix whose jth column is $\mathbf{d}_j$, and that $\mathbf{A} = \mathbf{D}^{-1}$.

4.4 Column Space, Row Space, and Rank of a Matrix

In this section we examine two important vector spaces associated with a matrix $\mathbf{A}$: the column space of $\mathbf{A}$ and the row space of $\mathbf{A}$.

Definition

Let $\mathbf{A}$ be any matrix.

1. The column space of $\mathbf{A}$, denoted $\text{Col}(\mathbf{A})$, is the vector space spanned by the columns of $\mathbf{A}$.
2. The row space of $\mathbf{A}$, denoted $\text{Row}(\mathbf{A})$, is the vector space spanned by the rows of $\mathbf{A}$.

Column Space of a Matrix

On several occasions in previous sections, we took a set S of vectors that spanned a Euclidean vector space V and let S be the columns of a matrix $\mathbf{A}$. We then can determine whether a given vector $\mathbf{b}$ is in V, since $\mathbf{b}$ is in V if and only if $\mathbf{Ax} = \mathbf{b}$ has a solution (Theorem 1 in Section 4.2). Using the language of column spaces, we constructed a matrix $\mathbf{A}$ with the property that the column space $\text{Col}(\mathbf{A}) = V$ and then used the fact that $\mathbf{b}$ is in $\text{Col}(\mathbf{A})$ if and only if $\mathbf{A}\mathbf{x} = \mathbf{b}$ has a solution. If $\mathbf{A}$ happens to be an invertible $n \times n$ matrix, the columns of $\mathbf{A}$ are linearly independent (Theorem 3 of Section 4.2) and form a basis for $\text{Col}(\mathbf{A})$, and thus for V. Then $\text{Col}(\mathbf{A})$ has dimension n and equals $\mathbf{R}^n$.

EXAMPLE 1.
Refinery Problem
As a Column
Space Problem

The refinery problem introduced in Section 1.1 involved three refineries each producing different amounts of heating oil, diesel oil and gasoline from a barrel of crude oil. Production levels of each refinery were sought to satisfy a vector of demands. The resulting system of equations is

$$\begin{aligned} \text{Heating oil:} \quad 4x_1 + \quad 2x_2 + 2x_3 &= 600 \\ \text{Diesel oil:} \quad 2x_1 + \quad 5x_2 + 2x_3 &= 800 \\ \text{Gasoline:} \quad 1x_1 + 2.5x_2 + 5x_3 &= 1000 \end{aligned}$$

This system $\mathbf{A}\mathbf{x} = \mathbf{b}$ is just seeking to express the demand vector $\mathbf{b} = \begin{bmatrix} 600 \\ 800 \\ 1000 \end{bmatrix}$ as a linear combination of the three refinery production vectors, which form the basis for this column space. That is, we seek the "coordinates" x_1, x_2, x_3 of $\mathbf{b}$ with respect to the columns of $\mathbf{A}$:

$$x_1 \begin{bmatrix} 4 \\ 2 \\ 1 \end{bmatrix} + x_2 \begin{bmatrix} 2 \\ 5 \\ 2.5 \end{bmatrix} + x_3 \begin{bmatrix} 2 \\ 2 \\ 5 \end{bmatrix} = \begin{bmatrix} 600 \\ 800 \\ 1000 \end{bmatrix} \tag{1}$$

▪

In Chapter 1, solving a system $\mathbf{A}\mathbf{x} = \mathbf{b}$ was viewed as a problem about the rows of $\mathbf{A}$, that is, about the equations specified by the rows. The solution is found with Gaussian elimination by forming linear combinations of the equations (rows of $\mathbf{A}$) to obtain a reduced system that can easily be solved. As illustrated in Example 1, solving $\mathbf{A}\mathbf{x} = \mathbf{b}$ can equally be viewed as a problem about a linear combination of the columns of $\mathbf{A}$. This vector approach to solving $\mathbf{A}\mathbf{x} = \mathbf{b}$ has a natural geometric picture, illustrated by the next example.

EXAMPLE 2.
Geometric Picture
of Solution to a
System of
Equations

Consider the system of equations

$$\begin{aligned} x_1 + \quad x_2 &= 4 \\ x_1 - 2x_2 &= 1 \end{aligned} \quad \text{or} \quad x_1 \begin{bmatrix} 1 \\ 1 \end{bmatrix} + x_2 \begin{bmatrix} 1 \\ -2 \end{bmatrix} = \begin{bmatrix} 4 \\ 1 \end{bmatrix}$$

Solving by elimination, we find that $x_1 = 3$ and $x_2 = 1$. Figure 4.2 graphs this solution in vector space terms, showing the right-side vector $\begin{bmatrix} 4 \\ 1 \end{bmatrix}$ as a linear combination of the column vectors $\begin{bmatrix} 1 \\ 1 \end{bmatrix}$ and $\begin{bmatrix} 1 \\ -2 \end{bmatrix}$. Note that the picture gives no insight as to why $x_1 = 3$, $x_2 = 1$ is the solution.

▪

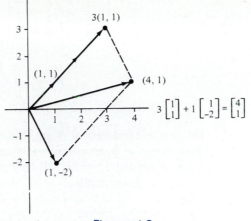

Figure 4.2

The problem of finding a basis from a given spanning set S for a Euclidean vector space V was also answered in terms of an associated matrix $\mathbf{A}$. Again $\mathbf{A}$'s columns would consist of the spanning set S, so that $V = \mathrm{Col}(\mathbf{A})$. Then Theorem 1 of Section 4.3 allows us to use Gauss–Jordan elimination to find a basis for the column space of $\mathbf{A}$. We restate that theorem here.

Theorem 1.   Let an $m \times n$ matrix $\mathbf{A}$ be reduced by Gauss–Jordan elimination into reduced row echelon form. Let the reduced matrix be called $\mathbf{A}^*$. Then the set of columns of $\mathbf{A}$ that reduce to columns with leading 1's in $\mathbf{A}^*$ is a linearly independent set and forms a basis for $\mathrm{Col}(\mathbf{A})$.

Note that by rearranging the order of the columns in $\mathbf{A}$, we may obtain a different basis.

EXAMPLE 3.
Finding a Basis of
the Column Space
of a Matrix

(a) Find a basis for the column space of the refinery problem matrix $\mathbf{A} = \begin{bmatrix} 4 & 2 & 2 \\ 2 & 5 & 2 \\ 1 & 2.5 & 5 \end{bmatrix}$. By Theorem 1, we reduce this matrix by Gauss–Jordan elimination.

$$\begin{bmatrix} 4 & 2 & 2 \\ 2 & 5 & 2 \\ 1 & 2.5 & 5 \end{bmatrix} \Longrightarrow \begin{bmatrix} 1 & 0.5 & 0.5 \\ 0 & 4 & 1 \\ 0 & 2 & 5.5 \end{bmatrix} \Longrightarrow \begin{bmatrix} 1 & 0 & \frac{3}{8} \\ 0 & 1 & \frac{1}{4} \\ 0 & 0 & 4 \end{bmatrix} \Longrightarrow \begin{bmatrix} 1 & 0 & 0 \\ 0 & 1 & 0 \\ 0 & 0 & 1 \end{bmatrix}$$

(2)

All the columns of $\mathbf{A}$ reduce to columns with leading 1's, so all the columns of $\mathbf{A}$ are in the basis of the column space of $\mathbf{A}$ (the columns are linearly independent). This means that the columns of $\mathbf{A}$ span all of $\mathbf{R}^3$ and are a basis for $\mathbf{R}^3$.

(b) Find a basis for the column space of the matrix $\mathbf{A} = \begin{bmatrix} -2 & 0 & 2 & 4 \\ 1 & 2 & 5 & 2 \\ 6 & 3 & 3 & -6 \end{bmatrix}$. Again we reduce $\mathbf{A}$ by Gauss–Jordan elimination.

$$
\begin{bmatrix} -2 & 0 & 2 & 4 \\ 1 & 2 & 5 & 2 \\ 6 & 3 & 3 & -6 \end{bmatrix} \implies \begin{bmatrix} 1 & 0 & -1 & -2 \\ 0 & 2 & 6 & 4 \\ 0 & 3 & 9 & 6 \end{bmatrix}
$$

$$
\implies \begin{bmatrix} 1 & 0 & -1 & -2 \\ 0 & 1 & 3 & 2 \\ 0 & 0 & 0 & 0 \end{bmatrix}
$$

(3)

The first two columns of **A** are reduced to columns with leading 1's, so the first two columns of **A** are a basis for the column space of **A**. Thus the column space of **A** has dimension 2, which will be a plane in $\mathbf{R}^3$ through the origin. Since no two columns of **A** are multiples of one another, any pair of columns of **A** are linearly independent and will be a basis for Col(**A**). ■

Row Space of a Matrix

The row space Row(**A**) of a matrix **A** is the vector space spanned by the rows of **A**. Gaussian elimination and Gauss–Jordan elimination utilize elementary row operations, which involve linear combinations of rows of **A**. Thus the rows of the reduced matrix (after elimination) are linear combinations of the original rows, and reversing the elimination steps, the original rows of **A** are linear combinations of the rows in the reduced matrix **A***. Then any vector that is expressible as a linear combination of the rows of **A** is expressible as a linear combination of the rows of **A***, and vice versa. The following lemma is an immediate consequence of this reasoning.

Lemma. Let **A** be an $m \times n$ matrix. Let **A** be reduced by Gauss–Jordan elimination to reduced row echelon form, and call the reduced matrix **A***. Then Row(**A**) = Row(**A***).

EXAMPLE 4.
Row Space of a
Matrix A and of
Its Reduced
Matrix A*

The matrix

$$
\mathbf{A} = \begin{bmatrix} 1 & 2 & 3 \\ 2 & 6 & 4 \\ 2 & 8 & 2 \end{bmatrix}
$$

has the reduced row echelon form $\mathbf{A}^* = \begin{bmatrix} 1 & 0 & 5 \\ 0 & 1 & -1 \\ 0 & 0 & 0 \end{bmatrix}$. By the above lemma, Row(**A**) = Row(**A***); that is, a 3-vector **b** can be expressed as a linear combination of the rows of **A** if and only if it can be expressed as a linear combination of **A***. For example, $[1, -2, 7] = 5[1, 2, 3] - 2[2, 6, 4]$ and so $[1, -2, 7] \in$ Row(**A**). Then $[1, -2, 7] \in$ Row(**A***). Using the leading 1's in **A*** (only the first row of **A*** has a nonzero first entry; similarly for the second row), we find $[1, -2, 7] = 1[1, 0, 5] - 2[0, 1, -1]$. ■

As with questions about the column space Col(**A**) of a matrix **A**, so with questions about the row space Row(**A**), the key is reducing **A** with Gauss–Jordan elimination to the reduced row echelon form matrix **A***. In **A***, each row either is all 0's or has a leading 1. Because the leading 1's are in different columns—with 0's in the other entries in any column with a leading 1—the nonzero rows of **A*** are linearly independent. These nonzero rows of **A*** must span Row(**A***), since the other rows of **A*** are all 0's. So the nonzero rows form a basis of Row(**A***), and dim(Row(**A***)) equals the number of nonzero rows in **A***. This number of nonzero rows in **A*** is the number of pivots performed during elimination. Since Row(**A**) = Row(**A***) by the lemma, then dim(Row(**A**)) also equals the number of nonzero rows in **A***, or equivalently, the number of pivots performed. In Example 4, dim(Row(**A***)) is clearly two, so dim-(Row(**A**)) must also be two.

Let us next find a basis for Row(**A**). If the (nonzero) kth row $\mathbf{a}°$ in the original matrix **A** becomes a zero row of **A***, that means the multiples of pivot rows that were subtracted from row $\mathbf{a}°$ zeroed out that row. In other words, $\mathbf{a}°$ equals a linear combination of pivot rows. Hence all the rows of **A** that are zeroed out in elimination are linearly dependent on the rows of **A** where pivots are performed. Thus the pivot rows of **A** span Row(**A**). For example, in Example 4, the first two rows of **A** (where pivots were made to get leading 1's in **A***) will span Row(**A**). Since we just noted that dim(Row(**A**)) equals the number of pivots, all the pivot rows of **A** are needed to form a basis. (The fact that the pivot rows of **A** are linearly independent can also be deduced from the fact that the corresponding rows with leading 1's in **A*** are clearly linearly independent.)

Theorem 2. Let **A** be an $m \times n$ matrix. Let **A** be reduced by Gauss–Jordan elimination to reduced row echelon form. The rows where pivots are performed during elimination form a basis of the row space Row(**A**), and the other rows of **A** are linearly dependent on the pivot rows.

EXAMPLE 5.
Finding the Basis for the Row Space of a Matrix

(a) Find a basis for the row space for the matrix $\mathbf{A} = \begin{bmatrix} 1 & 0 & 4 & 0 & 6 \\ 0 & 1 & 1 & 0 & 0 \\ 0 & 0 & 0 & 1 & 3 \\ 0 & 0 & 0 & 0 & 0 \end{bmatrix}$. **A** is

already in reduced row echelon form. Then by Theorem 2, the nonzero rows of **A** form a basis for the row space of **A**.

(b) Find a basis for the row space for the matrix $\mathbf{A} = \begin{bmatrix} -2 & 0 & 2 & 4 \\ 1 & 2 & 5 & 2 \\ 6 & 3 & 3 & -6 \end{bmatrix}$ in Exam-

ple 3(b). By Theorem 2, we find a basis by performing Gauss–Jordan elimination to reduce **A** to a matrix **A*** in reduced row echelon form. The pivot rows of **A**, which are the rows of **A** that are reduced to nonzero rows of **A***, will be a basis for Row(**A**).

$$\begin{bmatrix} -2 & 0 & 2 & 4 \\ 1 & 2 & 5 & 2 \\ 6 & 3 & 3 & -6 \end{bmatrix} \Longrightarrow \begin{bmatrix} 1 & 0 & -1 & -2 \\ 0 & 2 & 6 & 4 \\ 0 & 3 & 9 & 6 \end{bmatrix}$$

$$\Longrightarrow \begin{bmatrix} 1 & 0 & -1 & -2 \\ 0 & 1 & 3 & 2 \\ 0 & 0 & 0 & 0 \end{bmatrix}$$

(4)

The first two rows of **A** are then a basis for the row space of **A**. So Row(**A**) has dimension 2. Since no two rows of **A** are multiples of one another, any two rows of **A** will form a basis of Row(**A**).

(c) Find a basis for the row space for the matrix $\mathbf{A} = \begin{bmatrix} 2 & 6 & 0 \\ 4 & 15 & 3 \\ 1 & 4 & 1 \\ 0 & 3 & 5 \end{bmatrix}$. Performing

Gauss–Jordan elimination on **A**, we obtain

$$\begin{bmatrix} 2 & 6 & 0 \\ 4 & 15 & 3 \\ 1 & 5 & 2 \\ 0 & 3 & 5 \end{bmatrix} \Longrightarrow \begin{bmatrix} 1 & 3 & 0 \\ 0 & 3 & 3 \\ 0 & 2 & 2 \\ 0 & 3 & 5 \end{bmatrix} \Longrightarrow \begin{bmatrix} 1 & 0 & -3 \\ 0 & 1 & 1 \\ 0 & 0 & 0 \\ 0 & 0 & 2 \end{bmatrix} \Longrightarrow \begin{bmatrix} 1 & 0 & 0 \\ 0 & 1 & 0 \\ 0 & 0 & 0 \\ 0 & 0 & 1 \end{bmatrix}$$

(5)

The first, second, and fourth rows of **A** are thus a basis for the row space of **A**.

■

Rank of a Matrix

Our methods for determining whether a vector is in the column space of a matrix and determining a basis for the column space have all involved reducing the matrix to reduced row echelon form. As noted in the preceding section, these *column* results were answered with computations involving *row* operations, not column operations. We now examine the relation between the column space and row space of a matrix. The relationship between these two spaces is in many ways more interesting than properties of either individual vector space. A critical link between the row space and the column space is the concept of the rank of a matrix.

Definition

The *rank* of an $m \times n$ matrix **A**, rank(**A**), is the number of pivots performed during Gauss–Jordan (or Gaussian) elimination on **A**.

The rank of a matrix is usually defined to be the dimension of the column space (or range), but we use this elimination-based definition of rank in keeping with the central role given to Gauss–Jordan (and Gaussian) elimination throughout this book.

Note that a 3×3 identity matrix $\mathbf{I}$ has rank 3: Although in elimination there is no work to do in each pivot for $\mathbf{I}$, a pivot formally occurs at each diagonal element of $\mathbf{I}$.

Theorem 3. The rank of a matrix $\mathbf{A}$ is uniquely defined. Further,

$$\text{rank}(\mathbf{A}) = \dim(\text{Col}(\mathbf{A})) = \dim(\text{Row}(\mathbf{A}))$$

Proof: Theorem 2 states that the rows where pivots are performed form the basis of $\text{Row}(\mathbf{A})$. That is, $\dim(\text{Row}(\mathbf{A}))$ equals the number of pivots. Similarly, Theorem 1 states that $\dim(\text{Col}(\mathbf{A}))$ equals the number of leading 1's in the reduced row echelon form. But the number of leading 1's in the reduced row echelon form is also equal to the number of pivots. The number of pivots, which is the rank of $\mathbf{A}$, is this common dimension. ■

Corollary. For any matrix, the number of linearly independent columns equals the number of linearly independent rows.

To illustrate Theorem 3, look at the matrices in Example 5 in reduced row echelon form. The number of leading 1's in the reduced row echelon form matrix tells us that the column space and row space each have dimension 3 in Example 5(a) and (c) and each have dimension 2 in Example 5(b). Note that the number of pivots in Gaussian elimination (where we only subtract the pivot row from later rows) can be used instead of pivots in Gauss–Jordan elimination. It is an exercise (Exercise 12) to show that the nonzero rows of the reduced matrix in Gaussian elimination are also linearly independent and form a basis of the row space.

While Theorem 3 appears to be a very interesting theoretical result, it also has a very important practical consequence, namely, that *no matter what way the rows and columns of a matrix may be permuted and what sequence of pivots are used during elimination, the same number of pivots will always be performed*. We also now have some additional conditions that guarantee the unique solution to a system of equations. Collecting results from Chapter 1 and Section 4.2, we have the following list.

Theorem 4. Let $\mathbf{A}$ be an $n \times n$ matrix. Then the following are equivalent statements.
  (i) For any n-vector $\mathbf{b}$, the system $\mathbf{Ax} = \mathbf{b}$ has one and only one solution.
  (ii) The columns of $\mathbf{A}$ are linearly independent.
  (iii) The rows of $\mathbf{A}$ are linearly independent.
  (iv) $\det(\mathbf{A}) \neq 0$.
  (v) $\mathbf{A}$ is invertible.
  (vi) $\text{rank}(\mathbf{A}) = n$.
  (vii) $\dim(\text{Col}(\mathbf{A})) = n$.
  (viii) $\dim(\text{Row}(\mathbf{A})) = n$.
Proof: By Theorem 3 of Section 4.2, (i), (ii), (iv), and (v) are equivalent. By Theorem 3 above and its corollary, (ii) is equivalent to (iii), (vi), (vii), and (viii). ■

Theorem 3 is a surprising result, since it might seem likely that an $m \times n$ matrix (with $m \neq n$) would have different dimensions for its column space and its row space.

The explanation of why these two spaces have the same dimension can be found in the simple matrix representation of Gaussian elimination given in Section 3.2.

Recall that given a *column m*-vector $\mathbf{c}$ and *row n*-vector $\mathbf{d}$, a ***simple matrix*** $\mathbf{S}$ is the $m \times n$ matrix $\mathbf{c} * \mathbf{d}$ formed by the matrix multiplication $\mathbf{cd}$, in which $\mathbf{c}$ is treated as an $m \times 1$ matrix and $\mathbf{d}$ as a $1 \times n$ matrix. For example, the following matrix product of a column vector times a row vector yields a simple matrix.

$$
\begin{bmatrix} 3 \\ -1 \end{bmatrix} * [1, 2, 3] = \begin{bmatrix} 3 \cdot 1 & 3 \cdot 2 & 3 \cdot 3 \\ -1 \cdot 1 & -1 \cdot 2 & -1 \cdot 3 \end{bmatrix} = \begin{bmatrix} 3 & 6 & 9 \\ -1 & -2 & -3 \end{bmatrix} \quad (6)
$$

All rows (and columns) are multiples of each other, so a simple matrix has rank one.

Gaussian elimination on a matrix $\mathbf{A}$ was interpreted at the end of Section 3.2 as the process of repeatedly subtracting simple matrices from $\mathbf{A}$ until the resulting matrix is all 0's. For example, the first step of Gaussian elimination for the matrix

$$
\mathbf{A} = \begin{bmatrix} 4 & 2 & 2 \\ 2 & 5 & 2 \\ 1 & 2.5 & 5 \end{bmatrix}
$$

in the refinery problem (Example 1), where we want to subtract $\frac{1}{2}$ the first row from the second row and $\frac{1}{4}$ the first row from the third row, can be achieved by subtracting the simple matrix $\mathbf{S}_1$:

$$
\mathbf{A} - \mathbf{S}_1 = \mathbf{A} - \mathbf{l}_1 * \mathbf{u}_1' = \begin{bmatrix} 4 & 2 & 2 \\ 2 & 5 & 2 \\ 1 & 2.5 & 5 \end{bmatrix} - \begin{bmatrix} \frac{1}{2} \\ \frac{1}{2} \\ \frac{1}{4} \end{bmatrix} * [4, 2, 2]
$$

$$
= \begin{bmatrix} 4 & 2 & 2 \\ 2 & 5 & 2 \\ 1 & 2.5 & 5 \end{bmatrix} - \begin{bmatrix} 4 & 2 & 2 \\ 2 & 1 & 1 \\ 1 & 0.5 & 0.5 \end{bmatrix} = \begin{bmatrix} 0 & 0 & 0 \\ 0 & 4 & 1 \\ 0 & 2 & 4.5 \end{bmatrix} \quad (7)
$$

where $\mathbf{u}_1'$ is the first row of $\mathbf{U}$, the upper triangular reduced matrix resulting from Gaussian elimination on $\mathbf{A}$, and $\mathbf{l}_1$ is the first column in $\mathbf{L}$, the matrix of the multipliers used in Gaussian elimination on $\mathbf{A}$ (see Section 3.2 for details). Recall that $\mathbf{A} = \mathbf{LU}$ (the LU decomposition of a matrix).

If, instead of subtracting these simple matrices from $\mathbf{A}$, we add the subtracted simple matrices together, we see that $\mathbf{A}$ is the sum of simple matrices obtained from the columns of $\mathbf{L}$ and the rows of $\mathbf{U}$. For the refinery matrix $\mathbf{A}$, the sum of simple matrices is

$$
\mathbf{A} = \mathbf{l}_1 * \mathbf{u}_1' \quad + \quad \mathbf{l}_2 * \mathbf{u}_2' \quad + \quad \mathbf{l}_3 * \mathbf{u}_3'
$$

$$
= \begin{bmatrix} 1 \\ \frac{1}{2} \\ \frac{1}{4} \end{bmatrix} * [4, 2, 2] + \begin{bmatrix} 0 \\ 1 \\ \frac{1}{2} \end{bmatrix} * [0, 4, 1] + \begin{bmatrix} 0 \\ 0 \\ 1 \end{bmatrix} * [0, 0, 4] \quad (8a)
$$

$$
= \begin{bmatrix} 4 & 2 & 2 \\ 2 & 1 & 1 \\ 1 & 0.5 & 0.5 \end{bmatrix} + \begin{bmatrix} 0 & 0 & 0 \\ 0 & 4 & 1 \\ 0 & 2 & 0.5 \end{bmatrix} + \begin{bmatrix} 0 & 0 & 0 \\ 0 & 0 & 0 \\ 0 & 0 & 4 \end{bmatrix} = \begin{bmatrix} 4 & 2 & 2 \\ 2 & 5 & 5 \\ 1 & 2.5 & 5 \end{bmatrix} \quad (8b)
$$

Let us think about the column space of A in terms of the simple matrix decomposition in (8). The three column vectors l_1, l_2, l_3 of L will be a basis for the column space of A, since the columns of the simple matrices in (8b) that sum to A are multiples of these l_i's, so the columns of A are linear combinations of the three l_i's. [An algebraic proof that l_i's are a basis for $Col(A)$ is given in Exercise 22.]

The same argument applies to the row space, since the simple matrices in (8b) are formed from multiples of the row vectors u_i' (also see Exercise 22). Thus the u_i''s are a basis of $Row(A)$. The number of l_i's equals the number of u_i''s, and both equal the number of pivots, that is, the rank of A. Thus *the simple matrix decomposition in (8) gives a concrete demonstration of how the dimension of the column space and the dimension of the row space of a matrix have a common origin in Gaussian elimination.* A generalization of the preceding argument yields the following theorem.

Theorem 5

(i) If A is written as the sum of k simple matrices S_i, $A = S_1 + S_2 + \cdots + S_k$, where $S_i = c_i * d_i'$, the columns c_k span $Col(A)$ and the rows d_i' span $Row(A)$.

(ii) In the LU decomposition of a matrix A, the nonzero columns of L are a basis of $Col(A)$ and the nonzero rows of U are a basis of $Row(A)$.

(iii) The rank of a matrix A equals the minimum number of simple matrices whose sum is A.

Proof: Part (i) was proved above for a sum of three simple matrices (for a general proof, see Exercise 22).

Part (ii) is the special case of part (i) illustrated in (8). It arises from the simple-matrix interpretation of the LU decomposition given in Theorem 4 of Section 3.2. The columns of L and rows of U are bases, not just spanning sets, because their number equals the number of pivots performed during Gaussian elimination, which by Theorem 4 is the dimension of $Col(A)$ and $Row(A)$.

To prove part (iii), we observe that by part (ii) the simple matrices arising from the LU decomposition sum to A and the number of these simple matrices is $rank(A)$, the number of pivots performed during Gaussian elimination. Thus part (iii) follows if we can show that fewer simple matrices cannot sum to A. However, if k simple matrices $c_i * d_i'$, with $k < rank(A)$, did sum to A, their k columns c_i would span $Col(A)$ and the c_i's (or a subset of them) would form a basis of $Col(A)$ of size $< rank(A)$. This contradicts the fact that $dim(Col(A)) = rank(A)$. ■

**EXAMPLE 6.
LU Simple-Matrix
Decomposition of
Nonsquare Matrix**

In Example 3 of Section 3.2, we obtained L and U for the following 3×4 matrix A.

$$A = \begin{bmatrix} -2 & 2 & 0 & 4 \\ 1 & 5 & 2 & 2 \\ 6 & 3 & 3 & -6 \end{bmatrix} \quad L = \begin{bmatrix} 1 & 0 & 0 \\ -\frac{1}{2} & 1 & 0 \\ -3 & \frac{3}{2} & 0 \end{bmatrix} \quad U = \begin{bmatrix} -2 & 2 & 0 & 4 \\ 0 & 6 & 2 & 4 \\ 0 & 0 & 0 & 0 \end{bmatrix}$$

(9)

Since there are two pivots, the column and row spaces of A have dimension 2, with bases given by the nonzero columns of L and nonzero rows of U, respectively.

$$A = l_1 * u_1' + l_2 * u_2' = \begin{bmatrix} 1 \\ -\frac{1}{2} \\ -3 \end{bmatrix} * [-2, 2, 0, 4] + \begin{bmatrix} 0 \\ 1 \\ \frac{3}{2} \end{bmatrix} * [0, 6, 2, 4]$$

$$= \begin{bmatrix} -2 & 2 & 0 & 4 \\ 1 & -1 & 0 & -2 \\ 6 & -6 & 0 & -12 \end{bmatrix} + \begin{bmatrix} 0 & 0 & 0 & 0 \\ 0 & 6 & 2 & 4 \\ 0 & 9 & 3 & 6 \end{bmatrix}$$

(10)

To illustrate Theorem 5, we express the vector $\mathbf{b} = \begin{bmatrix} 100 \\ 350 \\ 300 \end{bmatrix}$, which is in Col($\mathbf{A}$), as a linear combination of the columns of $\mathbf{L}$. Solving $\mathbf{Ax} = \mathbf{b}$ yields $\mathbf{x} = [0, 50, 50, 0]^{\mathrm{T}}$ (there are other possible solutions). Using (10), we have

$$\begin{aligned}
\mathbf{b} = \mathbf{Ax} &= (l_1 * u_1' + l_2 * u_2')\mathbf{x} \\
&= l_1(u_1' \cdot \mathbf{x}) + l_2(u_2' \cdot \mathbf{x}) \\
&= \begin{bmatrix} 1 \\ -\frac{1}{2} \\ -3 \end{bmatrix}\left([-2, 2, 0, 4]\begin{bmatrix} 0 \\ 50 \\ 50 \\ 0 \end{bmatrix}\right) + \begin{bmatrix} 0 \\ 1 \\ \frac{3}{2} \end{bmatrix}\left([0, 6, 2, 4]\begin{bmatrix} 0 \\ 50 \\ 50 \\ 0 \end{bmatrix}\right) \\
&= \begin{bmatrix} 1 \\ -\frac{1}{2} \\ -3 \end{bmatrix}(100) + \begin{bmatrix} 0 \\ 1 \\ \frac{3}{2} \end{bmatrix}(400)
\end{aligned}$$

(11)

■

Section 4.4 Exercises

1. Find a set of columns that forms a basis for the column space of each of the following matrices. Give the dimension of the column space.

   (a) $\begin{bmatrix} 1 & -2 \\ -2 & 4 \end{bmatrix}$ (b) $\begin{bmatrix} 1 & -2 \\ 2 & 4 \end{bmatrix}$ (c) $\begin{bmatrix} 1 & 3 \\ -1 & 2 \end{bmatrix}$

2. Find a set of columns that forms a basis for the column space of each of the following matrices. Give the dimension of the column space and the rank of the matrix.

   (a) $\begin{bmatrix} 1 & 1 & 3 \\ 1 & 0 & 3 \\ 1 & 2 & 3 \end{bmatrix}$ (b) $\begin{bmatrix} 4 & -1 & 2 \\ 2 & 5 & 1 \\ -2 & 3 & -1 \end{bmatrix}$

   (c) $\begin{bmatrix} -1 & 3 & 1 \\ 5 & -1 & 3 \\ 2 & 1 & 2 \end{bmatrix}$ (d) $\begin{bmatrix} 2 & 1 & -1 \\ 1 & -2 & 2 \\ 3 & 4 & -4 \end{bmatrix}$

3. For each matrix in Exercise 2, find all sets of columns that form a basis for the column space.

4. Consider a coefficient matrix in a refinery problem, similar to the refinery coefficient matrix as in Example 1, with each column representing the production vector of a refinery.

Explain the practical significance of having one column be a linear combination of the others. What constraints or what freedom does this permit the manager of the refineries?

5. The first column in the inverse A^{-1} of a 2×2 matrix A gives the weights in a linear combination of A's columns that equals $i_1 = \begin{bmatrix} 1 \\ 0 \end{bmatrix}$. The second column in A^{-1} gives the weights in a linear combination of A's columns that equals $i_2 = \begin{bmatrix} 0 \\ 1 \end{bmatrix}$. Find these weights for expressing i_1 and i_2 as linear combinations of the columns and plot the linear combination (as in Figure 4.2) for the following matrices.

   (a) $\begin{bmatrix} 4 & 0 \\ 0 & 3 \end{bmatrix}$ (b) $\begin{bmatrix} 1 & -1 \\ 2 & 1 \end{bmatrix}$ (c) $\begin{bmatrix} 2 & 3 \\ 5 & 8 \end{bmatrix}$

6. Find a set of rows that forms a basis for the row space of each of the following matrices. Give the dimension of the row space.

   (a) $\begin{bmatrix} 1 & -2 \\ -2 & 4 \end{bmatrix}$ (b) $\begin{bmatrix} 1 & -2 \\ 2 & 4 \end{bmatrix}$ (c) $\begin{bmatrix} 1 & 3 \\ -1 & 2 \end{bmatrix}$

7. Find a set of rows that forms a basis for the row space of each of the following matrices. Give the dimension of the row space and the rank of the matrix.

   (a) $\begin{bmatrix} 1 & 1 & 3 \\ 1 & 0 & 3 \\ 1 & 2 & 3 \end{bmatrix}$ (b) $\begin{bmatrix} 4 & -2 & 2 \\ 2 & 5 & 1 \\ -2 & 1 & -1 \end{bmatrix}$

   (c) $\begin{bmatrix} -1 & 3 & 1 \\ 5 & -1 & 3 \\ 2 & 1 & 2 \end{bmatrix}$ (d) $\begin{bmatrix} 2 & 1 & -1 \\ 1 & -2 & 2 \\ 3 & 4 & -4 \end{bmatrix}$

8. In this exercise the reader should try to find by inspection a linear dependence among the rows of each matrix.

   (a) $\begin{bmatrix} 5 & 2 & 3 \\ 0 & 1 & 2 \\ 5 & 3 & 5 \end{bmatrix}$ (b) $\begin{bmatrix} 1 & 2 & 0 \\ 1 & 1 & 1 \\ 3 & 4 & 2 \end{bmatrix}$ (c) $\begin{bmatrix} 5 & 7 & 9 \\ 4 & 5 & 6 \\ 1 & 2 & 3 \end{bmatrix}$

9. Determine the rank of matrix A, if possible, from the information given.
   (a) A is an $n \times n$ matrix with linearly independent columns.
   (b) A is a 3×3 matrix and $\det(A) = 17$.
   (c) A is a 5×5 matrix and $\dim(\text{Row}(A)) = 3$.
   (d) A is an invertible 4×4 matrix.
   (e) A is a 4×3 matrix and $Ax = b$ has either a unique solution or else no solution.
   (f) A is an 8×8 matrix and $\dim(\text{Row}(A^T)) = 6$.

10. Let A be an 3×3 matrix. Which of the following statements guarantees that $Ax = b$ has a unique solution? Which guarantee that $Ax = b$ does not have a unique solution?
    (a) $\text{rank}(A) = 2$.
    (b) $\text{rank}(A^{-1}) = 3$.
    (c) $\dim(\text{Row}(A)) = \dim(\text{Col}(A))$.
    (d) Two rows of A are equal.
    (e) The columns of A form a basis for $\mathbf{R}^3$.

11. Let A be an $n \times n$ matrix. Which of the following statements guarantees that $Ax = b$ has a unique solution? Which guarantee that $Ax = b$ does not have a unique solution?
    (a) $\text{rank}(A) = n - 1$.

(b) $\dim(\text{Col}(\mathbf{A})) = n$.

(c) $\text{rank}(\mathbf{A}^T) = n$.

(d) $\det(\mathbf{A}) = n$.

(e) The rows of $\mathbf{A}$ form a basis for $\mathbf{R}^n$.

(f) $\text{Col}(\mathbf{A}) = \text{Row}(\mathbf{A})$.

(g) $\mathbf{A}$ is not the zero matrix and $\dim(\text{Col}(\mathbf{A})) = \det(\mathbf{A})$.

12. Show that the nonzero rows in the upper triangular matrix resulting from Gaussian elimination are linearly independent and form a basis for the row space.

13. Show that $\text{Row}(\mathbf{A}^T)$ ($\mathbf{A}^T$ is the transpose of $\mathbf{A}$) equals $\text{Col}(\mathbf{A})$ and that $\text{Col}(\mathbf{A}^T)$ equals $\text{Row}(\mathbf{A})$. Show then that $\text{rank}(\mathbf{A}) = \text{rank}(\mathbf{A}^T)$.

14. **(a)** Show that the rank of a matrix does not change when a multiple of one row is subtracted from another row.

(b) Show that the rank of a matrix does not change when a multiple of one column is subtracted from another column. (Hint: Use $\mathbf{A}^T$ and Exercise 13.)

15. **(a)** Show that if $\mathbf{A}$ is an $n \times n$ matrix and $\mathbf{b}$ an m-vector, then $\mathbf{b}$ is in $\text{Col}(\mathbf{A})$ if and only if $\text{rank}([\mathbf{A} \quad \mathbf{b}]) = \text{rank}(\mathbf{A})$, where $[\mathbf{A} \quad \mathbf{b}]$ denotes the augmented $m \times (n + 1)$ matrix with $\mathbf{b}$ added as an extra column to $\mathbf{A}$.

(b) If $\mathbf{A}\mathbf{x} = \mathbf{b}$ has no solution, show that $\text{rank}([\mathbf{A} \quad \mathbf{b}])$ must be $\text{rank}(\mathbf{A}) + 1$.

16. Prove that if a matrix $\mathbf{A}$ is not square, either the rows of $\mathbf{A}$ are linearly dependent or the columns of $\mathbf{A}$ are linearly dependent.

17. Use the LU decomposition of the following matrices $\mathbf{A}$ to obtain bases for the column and row spaces as well as a simple-matrix decomposition of $\mathbf{A}$.

(a) $\begin{bmatrix} 1 & 2 \\ 2 & 3 \end{bmatrix}$ **(b)** $\begin{bmatrix} 1 & 2 & 3 \\ 3 & 4 & 5 \\ 7 & 8 & 9 \end{bmatrix}$ **(c)** $\begin{bmatrix} 1 & 4 & 6 \\ 0 & 1 & 3 \\ 4 & 9 & 3 \end{bmatrix}$ **(d)** $\begin{bmatrix} 2 & 1 & 5 \\ 0 & 3 & -2 \\ 4 & 5 & 7 \end{bmatrix}$

18. Use the LU decomposition of the following matrices $\mathbf{A}$ to obtain bases for the column and row spaces as well as a simple-matrix decomposition of $\mathbf{A}$.

(a) $\begin{bmatrix} 1 & 2 \\ 3 & 4 \\ 5 & 6 \end{bmatrix}$ **(b)** $\begin{bmatrix} 2 & 5 & 2 \\ 1 & 3 & 6 \end{bmatrix}$

(c) $\begin{bmatrix} 2 & 1 & 0 & 1 \\ 3 & 0 & 1 & 2 \\ 3 & -3 & -1 & 1 \end{bmatrix}$ **(d)** $\begin{bmatrix} 1 & 3 & 1 & 1 \\ 2 & 2 & 0 & 2 \\ 1 & -1 & -2 & 1 \end{bmatrix}$

19. **(a)** Express the vector $\begin{bmatrix} 1 \\ 1 \\ 1 \end{bmatrix}$ as a linear combination of the columns of $\mathbf{L}$ in the LU decomposition for the matrices in Exercise 18(a) and (d).

(b) Express the vector $[10, 10]$ as a linear combination of the rows of $\mathbf{U}$ in the LU decomposition of the matrix in Exercise 18(a).

20. Prove part (ii) of Theorem 5.

21. If $\mathbf{A}$ and $\mathbf{B}$ are $m \times n$ matrices, show that $\text{rank}(\mathbf{A} + \mathbf{B}) \leq \text{rank}(\mathbf{A}) + \text{rank}(\mathbf{B})$.

22. **(a)** This exercise proves that if $\mathbf{A} = \mathbf{l}_1 * \mathbf{u}_1' + \mathbf{l}_2 * \mathbf{u}_2' + \cdots + \mathbf{l}_k * \mathbf{u}_k'$ [see equation (8) in this section] and $\text{rank}(\mathbf{A}) = k$, the $\mathbf{l}_i$'s form a basis for $\mathbf{A}$. One must show that any linear combination of the columns of $\mathbf{A}$, $\Sigma\ c_i\mathbf{a}_i$, can be expressed as a linear combination of the $\mathbf{l}_i$'s. The key to this proof is the following algebraic manipulations. (Explain the steps involved.)

$$\Sigma\ c_i\mathbf{a}_i = \mathbf{Ac} = (\mathbf{l}_1 * \mathbf{u}_1' + \mathbf{l}_2 * \mathbf{u}_2' + \mathbf{l}_3 * \mathbf{u}_3')\mathbf{c}$$
$$= (\mathbf{l}_1 * \mathbf{u}_1')\mathbf{c} + (\mathbf{l}_2 * \mathbf{u}_2')\mathbf{c} + (\mathbf{l}_3 * \mathbf{u}_3')\mathbf{c}$$
$$= \mathbf{l}_1(\mathbf{u}_1' \cdot \mathbf{c}) + \mathbf{l}_2(\mathbf{u}_2' \cdot \mathbf{c}) + \mathbf{l}_3(\mathbf{u}_3' \cdot \mathbf{c})$$

(b) Use similar reasoning with $\mathbf{dA}$ to show that $\mathbf{u}_1', \mathbf{u}_2', \ldots, \mathbf{u}_k'$ form a basis for the row space of $\mathbf{A}$.

23. If $\mathbf{A} = \mathbf{BC}$, where $\mathbf{B}$ and $\mathbf{C}$ are $n \times n$ matrices of rank n, show that $\mathbf{A}$ has rank n. (Hint: Use Theorem 4.)

24. If $\mathbf{A} = \mathbf{BC}$, show that the columns of $\mathbf{B}$ span $\mathrm{Col}(\mathbf{A})$ and the rows of $\mathbf{C}$ span $\mathrm{Row}(\mathbf{A})$. (Hint: Use Theorem 5 plus Theorem 3 of Section 3.2.)

4.5 Orthogonal Bases

We started this chapter with the concept of a vector space and spanning sets, and then we introduced linear independence and bases. All these concepts and associated theory were generalizing properties of vectors in $\mathbf{R}^2$, $\mathbf{R}^3$, and $\mathbf{R}^n$. In this section we complete this vector space development with a discussion of orthogonal bases, bases that capture critical properties of the standard coordinate vector basis for $\mathbf{R}^n$. Projections, introduced in Section 2.4, play a central role in this discussion. We shall also show how to transform a given basis into an orthogonal basis.

A basis is called *orthogonal* if its vectors are mutually orthogonal, and is called *orthonormal* if it is orthogonal and has unit-length vectors. Length here, as in the discussions about orthogonality in Chapter 2, will be measured using the Euclidean norm $|\mathbf{v}| = |\mathbf{v}|_e = \sqrt{v_1^2 + v_2^2 + \cdots + v_n^2}$.

The standard coordinate system for Euclidean n-space $\mathbf{R}^n$ uses the orthonormal basis of $I = \{\mathbf{i}_1, \mathbf{i}_2, \ldots, \mathbf{i}_n\}$, where $\mathbf{i}_j$ is the ith coordinate vector (with a 1 in the jth entry and 0's elsewhere). Thus when we write a 3-vector $\mathbf{c}$ as $\mathbf{c} = [c_1, c_2, c_3]$, we really mean $\mathbf{c} = c_1\mathbf{i}_1 + c_2\mathbf{i}_2 + c_3\mathbf{i}_3$. (In this case, the $\mathbf{i}_j$'s are 3-vectors.)

Orthonormal bases have many nice properties. Computations with orthonormal bases are simpler and more accurate. Thus one should try to use an orthonormal basis for any vector space in which one is working. This computational simplicity and accuracy will be illustrated dramatically in the next section, where we consider bases for vector spaces of functions. In this section we give some less complicated but still convincing examples as well as some of the basic theory of orthogonal and orthonormal bases.

Suppose that we have found an orthogonal basis Q of k vectors $\mathbf{q}_i$ for a k-dimensional Euclidean vector space V in $\mathbf{R}^n$. To express a vector $\mathbf{u} \in V$ in Q-coordinates as $\mathbf{u}^* = [\mathbf{u}]_Q$, one must solve the following system of equations for $\mathbf{u}^*$:

$$\mathbf{Qu}^* = u_1^*\mathbf{q}_1 + u_2^*\mathbf{q}_2 + \cdots + u_k^*\mathbf{q}_k = \mathbf{u} \quad \text{or, equivalently,} \quad \mathbf{Qu}^* = \mathbf{u} \quad (1)$$

Here $\mathbf{Q}$ is the matrix whose columns are the $\mathbf{q}_i$'s. Recall from Theorem 2 of Section 2.4 that when the columns of a matrix $\mathbf{Q}$ are orthogonal, the ith component $u_i^*\mathbf{q}_i$ in (1) is simply the projection of $\mathbf{u}$ onto $\mathbf{q}_i$, where $u_i^* = (\mathbf{q}_i \cdot \mathbf{u})/(\mathbf{q}_i \cdot \mathbf{q}_i)$. If, in addition, the $\mathbf{q}_i$ are of unit length so that $\mathbf{q}_i \cdot \mathbf{q}_i = 1$, the projection formula simplifies to $u_i^* = \mathbf{q}_i \cdot \mathbf{u}$ (see Figure 4.3 below). Summarizing, we have:

Theorem 1 (Orthogonal and Orthonormal Bases)

Let $Q = \{\mathbf{q}_i\}$ be an orthogonal basis for a k-dimensional vector space V in $\mathbf{R}^n$.

(i) To express an n-vector $\mathbf{u}$ in the basis Q, $\mathbf{u} = u_1^*\mathbf{q}_1 + u_2^*\mathbf{q}_2 + \cdots + u_k^*\mathbf{q}_k$, the Q-coordinates $u_i^{*'}$ are simply the projection coefficients of $\mathbf{u}$ onto the $\mathbf{q}_i$'s: $u_i^* = (\mathbf{q}_i \cdot \mathbf{u})/(\mathbf{q}_i \cdot \mathbf{q}_i)$.

(ii) If the $\mathbf{q}_i$'s are an orthonormal basis (they are of unit length), $u_i^* = \mathbf{q}_i \cdot \mathbf{u}$.

EXAMPLE 1.
Expressing a Vector in Terms of an Orthonormal Basis in $\mathbf{R}^2$

Consider the orthonormal basis

$$Q = \left\{ \mathbf{q}_1 = \begin{bmatrix} 0.8 \\ 0.6 \end{bmatrix}, \ \mathbf{q}_2 = \begin{bmatrix} -0.6 \\ 0.8 \end{bmatrix} \right\}$$

for $\mathbf{R}^2$. We use Theorem 1 to express the vector $\mathbf{u} = \begin{bmatrix} 1 \\ 2 \end{bmatrix}$ in the Q basis. We compute

$$u_1^* = \mathbf{q}_1 \cdot \mathbf{u} = 0.8 \cdot 1 + 0.6 \cdot 2 = 2 \qquad u_2^* = \mathbf{q}_2 \cdot \mathbf{u} = -0.6 \cdot 1 + 0.8 \cdot 2 = 1$$

Thus $[\mathbf{u}]_Q = \begin{bmatrix} 2 \\ 1 \end{bmatrix}$, or

$$\mathbf{u} = u_1^*\mathbf{q}_1 + u_2^*\mathbf{q}_2 = 2\begin{bmatrix} 0.8 \\ 0.6 \end{bmatrix} + 1\begin{bmatrix} -0.6 \\ 0.8 \end{bmatrix} \qquad (2)$$

A geometric picture of (2) is given in Figure 4.3, where $\mathbf{u}$ is depicted as the sum of its projections onto $\mathbf{q}_1$ and $\mathbf{q}_2$. ■

EXAMPLE 2.
Expressing a Vector in Terms of an Orthonormal Basis in $\mathbf{R}^3$

Express $\mathbf{u} = \begin{bmatrix} 2 \\ 1 \\ 4 \end{bmatrix}$ in the orthonormal basis

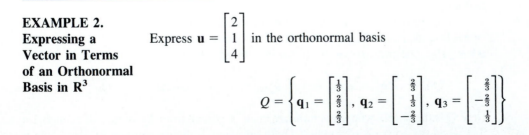

$$Q = \left\{ \mathbf{q}_1 = \begin{bmatrix} \frac{1}{3} \\ \frac{2}{3} \\ \frac{2}{3} \end{bmatrix}, \ \mathbf{q}_2 = \begin{bmatrix} \frac{2}{3} \\ \frac{1}{3} \\ -\frac{2}{3} \end{bmatrix}, \ \mathbf{q}_3 = \begin{bmatrix} \frac{2}{3} \\ -\frac{2}{3} \\ \frac{1}{3} \end{bmatrix} \right\}$$

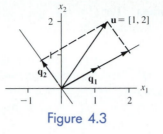

Figure 4.3

By Theorem 1, we compute the new Q-coordinates:

$$u_2^* = \mathbf{q}_2 \cdot \mathbf{u} = \tfrac{1}{3} \cdot 2 + \tfrac{2}{3} \cdot 1 + \tfrac{2}{3} \cdot 4 = 4$$

$$u_2^* = \mathbf{q}_2 \cdot \mathbf{u} = \tfrac{2}{3} \cdot 2 + \tfrac{1}{3} \cdot 1 + (-\tfrac{2}{3}) \cdot 4 = -1$$

$$u_3^* = \mathbf{q}_3 \cdot \mathbf{u} = \tfrac{2}{3} \cdot 2 + (-\tfrac{2}{3}) \cdot 1 + \tfrac{1}{3} \cdot 4 = 2$$

Thus

$$[\mathbf{u}]_Q = \begin{bmatrix} 4 \\ -1 \\ 2 \end{bmatrix} \quad \text{or} \quad \mathbf{u} = 4 \begin{bmatrix} \tfrac{1}{3} \\ \tfrac{2}{3} \\ \tfrac{2}{3} \end{bmatrix} - \begin{bmatrix} \tfrac{2}{3} \\ \tfrac{1}{3} \\ -\tfrac{2}{3} \end{bmatrix} + 2 \begin{bmatrix} \tfrac{2}{3} \\ -\tfrac{2}{3} \\ \tfrac{1}{3} \end{bmatrix} \tag{3}$$

■

Recall that if the basis $\{\mathbf{q}_i\}$ were not orthogonal, then to express a vector $\mathbf{u}$ as a linear combination of the $\mathbf{q}_i$'s, one would have to use Gaussian elimination to solve the system of equations $\mathbf{Qu^*} = \mathbf{u}$.

Expressing a vector $\mathbf{u}$ in terms of an orthonormal basis as in Example 2 is just a change of basis from the standard coordinate basis $I = \{\mathbf{i}_1, \mathbf{i}_2, \mathbf{i}_3\}$ to the basis Q. The transition matrix $\mathbf{A}$ for this change from basis I to basis Q is, by Theorem 7 in Section 4.3, $\mathbf{A} = \mathbf{Q}^{-1}$, where $\mathbf{Q}$ is the matrix whose columns form the orthonormal basis Q.

The inverse of a matrix such as $\mathbf{Q}$ whose columns form an orthonormal basis has a very simple form.

Theorem 2. Let $\mathbf{Q}$ be an $n \times n$ matrix with orthonormal columns $\mathbf{q}_i$. Then the inverse $\mathbf{Q}^{-1}$ is simply $\mathbf{Q}^T$, the transpose of $\mathbf{Q}$.

Proof: This theorem can be deduced from Theorem 3 in Section 2.3, which describes the inverse of matrices with orthogonal columns. Instead, we shall show directly that $\mathbf{Q}^T\mathbf{Q} = \mathbf{I}$. Observe that the matrix product $\mathbf{Q}^T\mathbf{Q}$ involves the scalar products of rows of $\mathbf{Q}^T$ times columns of $\mathbf{Q}$. But the rows of $\mathbf{Q}^T$ are the columns of $\mathbf{Q}$. Thus entry (i, j) in $\mathbf{Q}^T\mathbf{Q}$ equals $\mathbf{q}_i \cdot \mathbf{q}_j$. By orthogonality, this scalar product is 0, unless $i = j$ when the scalar product is 1. ■

We note that the pseudoinverse of a nonsquare matrix $\mathbf{Q}$ with orthonormal columns is also $\mathbf{Q}^T$ (see Exercise 9).

Returning to our change-of-basis discussion, the transition matrix $\mathbf{A} = \mathbf{Q}^{-1}$ for a change of basis from the standard basis I to orthonormal basis Q is, by Theorem 2, simply $\mathbf{A} = \mathbf{Q}^T$. With this fact, we can show that several important properties of vectors are not altered by an orthonormal change of bases.

Lemma. Let $Q = \{\mathbf{q}_1, \mathbf{q}_2, \ldots, \mathbf{q}_n\}$ be an orthonormal basis for $\mathbf{R}^n$. Let $\mathbf{u}$ and $\mathbf{v}$ be n-vectors whose coordinate vectors in the basis Q are $[\mathbf{u}]_Q$ and $[\mathbf{v}]_Q$, respectively. Then $([\mathbf{u}]_Q) \cdot ([\mathbf{v}]_Q) = \mathbf{u} \cdot \mathbf{v}$.

Proof: As just noted, the transition matrix from the standard basis I to basis Q is $\mathbf{Q}^T$. That is, multiplying $\mathbf{u}$ by $\mathbf{Q}^T$ converts $\mathbf{u}$ into the Q-coordinate vector $[\mathbf{u}]_Q$:

$$[\mathbf{u}]_Q = \mathbf{Q}^T\mathbf{u} \quad \text{and} \quad [\mathbf{v}]_Q = \mathbf{Q}^T\mathbf{v} \tag{4}$$

By (4), proving that $[\mathbf{u}]_Q \cdot [\mathbf{v}]_Q = \mathbf{u} \cdot \mathbf{v}$ can be restated as proving that

$$(\mathbf{Q}^T\mathbf{u}) \cdot (\mathbf{Q}^T\mathbf{v}) = \mathbf{u} \cdot \mathbf{v} \tag{5}$$

The right side of (5) can be rewritten using the following result about transposes, given in the corollary to Theorem 1 in Section 3.1:

$$(\mathbf{Q}^T\mathbf{u}) \cdot (\mathbf{Q}^T\mathbf{v}) = \mathbf{u}^T\mathbf{Q}\mathbf{Q}^T\mathbf{v} \tag{6}$$

Since $\mathbf{Q}^T = \mathbf{Q}^{-1}$, then $\mathbf{Q}\mathbf{Q}^T = \mathbf{Q}\mathbf{Q}^{-1} = \mathbf{I}$. Thus the right side of (6), $\mathbf{u}^T\mathbf{Q}\mathbf{Q}^T\mathbf{v}$, becomes $\mathbf{u}^T\mathbf{v}$ or $\mathbf{u} \cdot \mathbf{v}$, as required. ■

Note that the conversion in the lemma could be from Q to I instead of from I to Q. For if the scalar product does not change in the forward conversion, it will not change in the reverse conversion.

Theorem 3. An orthonormal change of basis in $\mathbf{R}^n$ does not alter the Euclidean norm of a vector (its length) or the angle between two vectors.

Proof: It is sufficient to prove this result for changing from the standard coordinate basis I for $\mathbf{R}^n$ to an orthonormal basis Q. If we were converting from another orthonormal basis Q' to Q, we could think of converting from Q' to I and then from I to Q.

If $\mathbf{u} = \mathbf{v}$ in the lemma, then $|[\mathbf{u}]_Q|^2 = ([\mathbf{u}]_Q) \cdot ([\mathbf{u}]_Q) = $ (by the lemma) $\mathbf{u} \cdot \mathbf{u} = |\mathbf{u}|^2$. Thus lengths of vectors do not change. The cosine of the angle θ between two vectors $\mathbf{u}$, $\mathbf{v}$ is $\cos \theta = (\mathbf{u} \cdot \mathbf{v})/(|\mathbf{u}||\mathbf{v}|)$ (see Section 2.3). Since scalar products and (Euclidean) norms do not change with an orthonormal change of basis, the formula for the cosine of their angle will not change. ■

EXAMPLE 3.
Preserving Length and Angles with Orthonormal Change of Basis

We illustrate the effect of an orthonormal change of basis on length and angles using the orthonormal basis

$$Q = \left\{ \mathbf{q}_1 = \begin{bmatrix} 0.8 \\ 0.6 \end{bmatrix}, \ \mathbf{q}_2 = \begin{bmatrix} -0.6 \\ 0.8 \end{bmatrix} \right\}$$

for $\mathbf{R}^2$ from Example 1. Consider the vectors $\mathbf{u} = \begin{bmatrix} 1 \\ 2 \end{bmatrix}$ and $\mathbf{v} = \begin{bmatrix} 5 \\ 12 \end{bmatrix}$. We compute

$$|\mathbf{u}| = \sqrt{\mathbf{u} \cdot \mathbf{u}} = \sqrt{1 \cdot 1 + 2 \cdot 2} = \sqrt{5} \tag{7a}$$

and

$$|\mathbf{v}| = \sqrt{\mathbf{v} \cdot \mathbf{v}} = \sqrt{5 \cdot 5 + 12 \cdot 12} = \sqrt{169} = 13 \qquad (7b)$$

Further,

$$\cos \theta(\mathbf{u}, \mathbf{v}) = \frac{\mathbf{u} \cdot \mathbf{v}}{|\mathbf{u}||\mathbf{v}|} = \frac{1 \cdot 5 + 2 \cdot 12}{\sqrt{5} \cdot 13} = \frac{29}{13\sqrt{5}} \qquad (8)$$

Now convert $\mathbf{u}$ and $\mathbf{v}$ to Q-coordinates. From Example 1, $[\mathbf{u}]_Q = \begin{bmatrix} 2 \\ 1 \end{bmatrix}$. As noted following Theorem 2, the transition matrix for converting $\mathbf{v}$ into Q-coordinates is $\mathbf{Q}^T$. Thus $[\mathbf{v}]_Q = \mathbf{Q}^T \mathbf{v} = \begin{bmatrix} 0.8 & 0.6 \\ -0.6 & 0.8 \end{bmatrix} \begin{bmatrix} 5 \\ 12 \end{bmatrix} = \begin{bmatrix} 11.2 \\ 6.6 \end{bmatrix}$. The length of $[\mathbf{u}]_Q$ is readily seen to be $\sqrt{5}$. Further, we obtain

$$|[\mathbf{v}]_Q| = \sqrt{[\mathbf{v}]_Q \cdot [\mathbf{v}]_Q} = \sqrt{11.2 \cdot 11.2 + 6.6 \cdot 6.6} = \sqrt{169} = 13 \qquad (9)$$

and

$$\cos \theta([\mathbf{u}]_Q, [\mathbf{v}]_Q) = \frac{[\mathbf{u}]_Q \cdot [\mathbf{v}]_Q}{|[\mathbf{u}]_Q||[\mathbf{v}]_Q|} = \frac{2 \cdot 11.2 + 1 \cdot 6.6}{\sqrt{5} \cdot 13} = \frac{29 \cdot 1}{13\sqrt{5}} \qquad (10)$$

confirming that lengths and angles are unchanged by an orthonormal change of basis. ■

The following theorem, whose proof we omit (the two-dimensional case is worked out in the exercises), gives a geometric interpretation of an orthonormal change of bases.

Theorem 4. An orthonormal change of basis $\mathbf{b} \rightarrow [\mathbf{b}]_Q = \mathbf{Q}^T \mathbf{b}$ (where $\mathbf{Q}$ is a matrix whose columns form an orthonormal basis) is simply a rotation of the coordinate axes, a permutation of the axes, or a combination of both. The entries in $\mathbf{Q}$ can be expressed in terms of the sines and cosines of the angles of this rotation.

The rotation of axes in the $\mathbf{R}^2$ plane by $\theta°$ is a change of basis:

$$\begin{array}{l} x' = x \cos \theta + y \sin \theta \\ y' = -x \sin \theta + y \cos \theta \end{array} \quad \text{or} \quad [\mathbf{v}]_Q = \mathbf{Q}^T \mathbf{v}, \quad \text{where} \quad \mathbf{Q} = \begin{bmatrix} \cos \theta & -\sin \theta \\ \sin \theta & \cos \theta \end{bmatrix}$$

$$(11)$$

It is easy to check that this $\mathbf{Q}$ has orthonormal columns. Conversely, any change of orthonormal bases in $\mathbf{R}^2$ has the form of $\mathbf{Q}$, with possibly a permutation of the rows. For example, in Figure 4.3, the new orthonormal basis for $\mathbf{Q}$ in Example 1 represents a rotation of $\cos^{-1}(0.8) \approx 37°$.

Orthogonal columns have another important advantage besides easy computation. They lead to more accurate calculations. A highly nonorthogonal basis—that is, basis vectors that are almost parallel–can result in unstable (inaccurate) computations.

EXAMPLE 4.
Bad
Nonorthogonal
Basis

Consider the following basis: $\mathbf{u} = \begin{bmatrix} 3 \\ 4 \end{bmatrix}$, $\mathbf{v} = \begin{bmatrix} 4 \\ 5 \end{bmatrix}$ for $\mathbf{R}^2$. The cosine of their angle is

$$\cos \theta(\mathbf{u}, \mathbf{v}) = \frac{\mathbf{u} \cdot \mathbf{v}}{|\mathbf{u}||\mathbf{v}|} = \frac{3 \cdot 4 + 4 \cdot 5}{\sqrt{9 + 16}\sqrt{16 + 25}} = \frac{32}{5\sqrt{41}} = 0.9995 \qquad (12)$$

The angle with a cosine of 0.9995 is about 2°. Thus $\mathbf{u}$ and $\mathbf{v}$ are almost parallel (almost the same vector). Representing any 2-vector $\mathbf{b}$ as a linear combination of two vectors that are almost the same is tricky, that is, unstable. For example, to express the vector $\mathbf{i}_1 = \begin{bmatrix} 0 \\ 1 \end{bmatrix}$ in terms of the basis $\mathbf{u}$, $\mathbf{v}$, we must find weights r, s such that (see Figure 4.4; compare with Figure 4.3)

$$r\begin{bmatrix} 4 \\ 3 \end{bmatrix} + s\begin{bmatrix} 5 \\ 4 \end{bmatrix} = \begin{bmatrix} 0 \\ 1 \end{bmatrix} \qquad (13)$$

The condition number of the matrix $\mathbf{A}$ whose columns are $\mathbf{u}$ and $\mathbf{v}$ is 81 (in sum or max norm). [Recall that the condition number $c(\mathbf{A}) = \|\mathbf{A}\| \cdot \|\mathbf{A}^{-1}\|$ measures how much a relative error in the entries of $\mathbf{A}$ or $\mathbf{b}$ could affect the relative error in $\mathbf{x}$ when solving $\mathbf{Ax} = \mathbf{b}$.] So an error of 1% in $\mathbf{A}$ or $\mathbf{b}$ could cause an error of around 80% in $\mathbf{x}$. ■

Applying the results of Example 4 in reverse, it can be shown that when errors arise in solving an ill-conditioned system of equations $\mathbf{Ax} = \mathbf{b}$ (in which $\mathbf{A}$ has a large condition number), the problem will be that some column vector (or a linear combination of them) forms a small angle with another column vector. This means that the columns are almost linearly dependent. If the columns were close to mutually orthogonal, the system $\mathbf{Ax} = \mathbf{b}$ would be well conditioned.

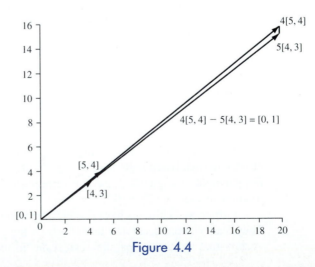

Figure 4.4

Gram–Schmidt Orthogonalization and the QR Decomposition

We shall now show how to convert a given basis $\mathbf{a}_1$, $\mathbf{a}_2$, . . . , $\mathbf{a}_m$ for an m-dimensional vector space V into a new orthonormal basis $\mathbf{q}_1$, $\mathbf{q}_2$, . . . , $\mathbf{q}_m$ for V. The method is called **Gram–Schmidt orthogonalization**. Currently, this method can only be used for Euclidean vector spaces. In the next section we extend it to non-Euclidean vector spaces.

This process starts by setting $\mathbf{q}_1$ equal to a multiple of $\mathbf{a}_1$. To make $\mathbf{q}_1$ have norm 1, we set $\mathbf{q}_1 = \mathbf{a}_1/|\mathbf{a}_1|$. Next we must construct from $\mathbf{a}_2$ a second unit vector $\mathbf{q}_2$ orthogonal to $\mathbf{q}_1$. We divide $\mathbf{a}_2$ into two "parts": the part of $\mathbf{a}_2$ parallel to $\mathbf{q}_1$ and the part of $\mathbf{a}_2$ orthogonal (perpendicular) to $\mathbf{q}_1$ (see Figure 4.5). The component of $\mathbf{a}_2$ in $\mathbf{q}_1$'s direction is simply the projection of $\mathbf{a}_2$ onto $\mathbf{q}_1$. This projection is $s\mathbf{q}_1$, where the length s of the projection is

$$s = \mathbf{a}_2 \cdot \mathbf{q}_1 \tag{14}$$

[We do not have to divide by $\mathbf{q}_1 \cdot \mathbf{q}_1$ in (14) since $\mathbf{q}_1 \cdot \mathbf{q}_1 = 1$.] The rest of $\mathbf{a}_2$, the vector $\mathbf{a}_2 - s\mathbf{q}_1$, is orthogonal to the projection $s\mathbf{q}_1$, and hence orthogonal to $\mathbf{q}_1$. So $\mathbf{a}_2 - s\mathbf{q}_1$ is the orthogonal vector we want for $\mathbf{q}_2$. To have unit norm, we set $\mathbf{q}_2 = (\mathbf{a}_2 - s\mathbf{q}_1)/|\mathbf{a}_2 - s\mathbf{q}_1|$.

Let us show how the procedure works thus far.

**EXAMPLE 5.
Gram–Schmidt
Orthogonalization
in Two
Dimensions**

Suppose that $\mathbf{a}_1 = \begin{bmatrix} 3 \\ 4 \end{bmatrix}$ and $\mathbf{a}_2 = \begin{bmatrix} 2 \\ 1 \end{bmatrix}$ (see Figure 4.5). We set

$$\mathbf{q}_1 = \frac{1}{|\mathbf{a}_1|}\mathbf{a}_1 = \frac{1}{5}\begin{bmatrix} 3 \\ 4 \end{bmatrix} = \begin{bmatrix} \frac{3}{5} \\ \frac{4}{5} \end{bmatrix}$$

We project $\mathbf{a}_2$ onto $\mathbf{q}_1$ to get the part of $\mathbf{a}_2$ parallel to $\mathbf{q}_1$. From (14), the length of the projection is

$$s = \mathbf{a}_2 \cdot \mathbf{q}_1 = 2 \cdot \tfrac{3}{5} + 1 \cdot \tfrac{4}{5} = 2$$

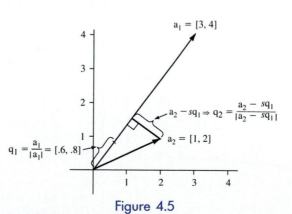

Figure 4.5

and the projection is $s\mathbf{q}_1 = 2\begin{bmatrix} \frac{3}{5} \\ \frac{4}{5} \end{bmatrix} = \begin{bmatrix} \frac{6}{5} \\ \frac{8}{5} \end{bmatrix}$. We determine the remaining part of $\mathbf{a}_2$, the part orthogonal to $s\mathbf{q}_1$, to be

$$\mathbf{a}_2 - s\mathbf{q}_1 = \begin{bmatrix} 2 \\ 1 \end{bmatrix} - \begin{bmatrix} \frac{6}{5} \\ \frac{8}{5} \end{bmatrix} = \begin{bmatrix} \frac{4}{5} \\ -\frac{3}{5} \end{bmatrix} \tag{15}$$

Since the norm of $\begin{bmatrix} \frac{4}{5} \\ -\frac{3}{5} \end{bmatrix}$ is already 1,

$$\mathbf{q}_2 = \frac{\mathbf{a}_2 - s\mathbf{q}_1}{|\mathbf{a}_2 - s\mathbf{q}_1|} = \frac{\begin{bmatrix} \frac{4}{5} \\ -\frac{3}{5} \end{bmatrix}}{1} = \begin{bmatrix} \frac{4}{5} \\ -\frac{3}{5} \end{bmatrix} \tag{16}$$

■

We extend the previous construction by finding the projections of $\mathbf{a}_3$ onto $\mathbf{q}_1$ and $\mathbf{q}_2$. Then the vector $\mathbf{a}_3 - s_1\mathbf{q}_1 - s_2\mathbf{q}_2$, which is orthogonal to $\mathbf{q}_1$ and $\mathbf{q}_2$, will yield $\mathbf{q}_3$. As before, we divide $\mathbf{a}_3 - s_1\mathbf{q}_1 - s_2\mathbf{q}_2$ by its norm to make $\mathbf{q}_3$ unit length. We continue this process to find $\mathbf{q}_4$, $\mathbf{q}_5$, and so on.

Note that the computation of the new orthonormal vectors involves just linear combinations of the original vectors. For example, the second orthonormal vector $\mathbf{q}_2 = (\mathbf{a}_2 - s\mathbf{q}_1)/|\mathbf{a}_2 - s\mathbf{q}_1|$ is a linear combination of $\mathbf{a}_2$ and $\mathbf{q}_1$ (dividing by the norm $t = |\mathbf{a}_2 - s\mathbf{q}_1|$ is just multiplying by scalar $1/t$), but $\mathbf{q}_1$ is a multiple of $\mathbf{a}_1$.

EXAMPLE 6.
Gram–Schmidt
Orthogonalization
in Three
Dimensions

Let us perform orthogonalization on the basis

$$\mathbf{a}_1 = \begin{bmatrix} 0 \\ 3 \\ 4 \end{bmatrix} \qquad \mathbf{a}_2 = \begin{bmatrix} 3 \\ 5 \\ 0 \end{bmatrix} \qquad \mathbf{a}_3 = \begin{bmatrix} 2 \\ 5 \\ 5 \end{bmatrix} \tag{17}$$

First $\mathbf{q}_1 = \mathbf{a}_1/|\mathbf{a}_1| = \frac{1}{5}\begin{bmatrix} 0 \\ 3 \\ 4 \end{bmatrix} = \begin{bmatrix} 0 \\ \frac{3}{5} \\ \frac{4}{5} \end{bmatrix}$. The length of the projection $\mathbf{a}_2$ onto $\mathbf{q}_1$ is

$$s = \mathbf{a}_2 \cdot \mathbf{q}_1 = 3 \cdot 0 + 5 \cdot \tfrac{3}{5} + 0 \cdot \tfrac{4}{5} = 3 \tag{18a}$$

So the projection of $\mathbf{a}_2$ onto $\mathbf{q}_1$ is $s\mathbf{q}_1 = 3\begin{bmatrix} 0 \\ \frac{3}{5} \\ \frac{4}{5} \end{bmatrix} = \begin{bmatrix} 0 \\ \frac{9}{5} \\ \frac{12}{5} \end{bmatrix}$.

Next we compute

$$\mathbf{a}_2 - s\mathbf{q}_1 = \begin{bmatrix} 3 \\ 5 \\ 0 \end{bmatrix} - \begin{bmatrix} 0 \\ \frac{9}{5} \\ \frac{12}{5} \end{bmatrix} = \begin{bmatrix} 3 \\ \frac{16}{5} \\ -\frac{12}{5} \end{bmatrix} \quad \text{and} \quad |\mathbf{a}_2 - s\mathbf{q}_1| = \sqrt{9 + \frac{256}{25} + \frac{144}{25}} = 5$$

Then

$$\mathbf{q}_2 = \frac{\mathbf{a}_2 - s\mathbf{q}_1}{|\mathbf{a}_2 - s\mathbf{q}_1|} = \frac{1}{5}\begin{bmatrix} 3 \\ \frac{16}{5} \\ -\frac{12}{5} \end{bmatrix} = \begin{bmatrix} \frac{3}{5} \\ \frac{16}{25} \\ -\frac{12}{25} \end{bmatrix}$$

We now compute the length of the projections of $\mathbf{a}_3$ onto $\mathbf{q}_1$ and $\mathbf{q}_2$.

$$s_1 = \mathbf{a}_3 \cdot \mathbf{q}_1 = 2 \cdot 0 + 5 \cdot \tfrac{3}{5} + 5 \cdot \tfrac{4}{5} = 3 + 4 = 7$$

$$s_2 = \mathbf{a}_3 \cdot \mathbf{q}_2 = 2 \cdot \tfrac{3}{5} + 5 \cdot \tfrac{16}{25} + 5 \cdot -\tfrac{12}{25} = \tfrac{6}{5} + \tfrac{16}{5} - \tfrac{12}{5} = 2 \qquad (18b)$$

Then

$$s_1\mathbf{a}_1 = 7 \begin{bmatrix} 0 \\ \tfrac{3}{5} \\ \tfrac{4}{5} \end{bmatrix} = \begin{bmatrix} 0 \\ \tfrac{21}{5} \\ \tfrac{28}{5} \end{bmatrix} \quad \text{and} \quad s_2\mathbf{a}_2 = 2 \begin{bmatrix} \tfrac{3}{5} \\ \tfrac{16}{25} \\ -\tfrac{12}{25} \end{bmatrix} = \begin{bmatrix} \tfrac{6}{5} \\ \tfrac{32}{25} \\ -\tfrac{24}{25} \end{bmatrix}$$

Thus

$$\mathbf{a}_3 - s_1\mathbf{q}_1 - s_2\mathbf{q}_2 = \begin{bmatrix} 2 \\ 5 \\ 5 \end{bmatrix} - \begin{bmatrix} 0 \\ \tfrac{21}{5} \\ \tfrac{28}{5} \end{bmatrix} - \begin{bmatrix} \tfrac{6}{5} \\ \tfrac{32}{25} \\ -\tfrac{24}{25} \end{bmatrix} = \begin{bmatrix} \tfrac{4}{5} \\ -\tfrac{12}{25} \\ \tfrac{9}{25} \end{bmatrix}$$

A computation reveals that $|\mathbf{a}_3 - s_1\mathbf{q}_1 - s_2\mathbf{q}_2| = 1$, so

$$\mathbf{q}_3 = (\mathbf{a}_3 - s_1\mathbf{q}_1 - s_2\mathbf{q}_2) = \begin{bmatrix} \tfrac{4}{5} \\ -\tfrac{12}{25} \\ \tfrac{9}{25} \end{bmatrix} \qquad (19)$$

Collectively, the new basis is

$$\mathbf{q}_1 = \begin{bmatrix} 0 \\ \tfrac{3}{5} \\ \tfrac{4}{5} \end{bmatrix} \qquad \mathbf{q}_2 = \begin{bmatrix} \tfrac{3}{5} \\ \tfrac{16}{25} \\ -\tfrac{12}{25} \end{bmatrix} \qquad \mathbf{q}_3 = \begin{bmatrix} \tfrac{4}{5} \\ -\tfrac{12}{25} \\ \tfrac{9}{25} \end{bmatrix} \qquad (20) \quad ■$$

We observed earlier that computations with highly nonorthogonal vectors can lead to erroneous results. This assertion, unfortunately, applies to this Gram–Schmidt orthogonalization. If some $\mathbf{a}_j$ or a linear combination of $\mathbf{a}_i$'s form a small angle with another basis vector $\mathbf{a}_k$, the new orthonormal basis may have errors, making it not exactly orthogonal. However, more stable orthogonalization methods are available using advanced techniques such as *Householder transformations*.

Suppose that the $\mathbf{a}_i$'s are not linearly independent. If, say, $\mathbf{a}_3$ is a linear combination of $\mathbf{a}_1$ and $\mathbf{a}_2$, then in the Gram–Schmidt procedure the error vector $\mathbf{a}_3 - s_1\mathbf{q}_1 - s_2\mathbf{q}_2$ will be 0. In this case we skip $\mathbf{a}_3$ and use $\mathbf{a}_4 - s_1\mathbf{q}_1 - s_2\mathbf{q}_2$ to define $\mathbf{q}_3$. The number of vectors $\mathbf{q}_i$ formed will equal the dimension of vector space spanned by the $\mathbf{a}_i$'s.

Suppose that we form a matrix $\mathbf{A}$ of the vectors $\mathbf{a}_i$ in the discussion above. Then the Gram–Schmidt orthogonalization process can be represented as the following matrix factorization.

Theorem 5. Any $m \times n$ matrix $\mathbf{A}$ with columns $\mathbf{a}_i$ can be factored in the form

$$\mathbf{A} = \mathbf{QR} \qquad (21)$$

where $\mathbf{Q}$ is the $m \times r$ orthogonal matrix whose r columns $\mathbf{q}_i$ are obtained by Gram–Schmidt orthogonalization, and $\mathbf{R}$ is an upper triangular matrix of size $r \times n$ (described below).

For $i \leq j$, entry r_{ij} of $\mathbf{R}$ is $\mathbf{a}_j \cdot \mathbf{q}_i$, the length of the projection of $\mathbf{a}_j$ onto $\mathbf{q}_i$. The diagonal entries in $\mathbf{R}$ are the lengths, before normalization, of the new columns: $r_{11} = |\mathbf{a}_1|$, $r_{22} = |\mathbf{a}_2 - s\mathbf{q}_1|$, $r_{33} = |\mathbf{a}_3 - s_1\mathbf{q}_1 - s_2\mathbf{q}_2|$, and so on. The diagonal entry r_{ii} also has the interpretation of being the length of the projection of $\mathbf{a}_j$ on $\mathbf{q}_j$. The entries of $\mathbf{R}$ below the main diagonal are all 0.

EXAMPLE 7.
QR Decomposition

Give the **QR** decomposition for the matrix $\mathbf{A} = \begin{bmatrix} 0 & 3 & 2 \\ 3 & 5 & 5 \\ 4 & 0 & 5 \end{bmatrix}$ in Example 6.

The orthogonal matrix $\mathbf{Q}$ has the columns given in (20). We form $\mathbf{R}$ from the information about the sizes of new columns and the projections in Example 6 as described in the preceding paragraph. Here, $r_{12} = s = 3$ in (18a), and $r_{13} = s_1 = 7$, $r_{23} = s_2 = 2$ in (18b). Then

$$\mathbf{QR} = \begin{bmatrix} 0 & \frac{3}{4} & \frac{4}{5} \\ \frac{3}{5} & \frac{16}{25} & -\frac{12}{25} \\ \frac{4}{5} & -\frac{12}{25} & \frac{9}{25} \end{bmatrix} \begin{bmatrix} 5 & 3 & 7 \\ 0 & 5 & 2 \\ 0 & 0 & 1 \end{bmatrix}$$

To verify this factorization, let us compute the second column of **QR**—multiplying $\mathbf{Q}$ by $\mathbf{r}_2$, the second column of $\mathbf{R}$—and show that the result is $\mathbf{a}_2$, the second column of $\mathbf{A}$.

$$\mathbf{Qr}_2 = \begin{bmatrix} 0 & \frac{3}{4} & \frac{4}{5} \\ \frac{3}{5} & \frac{16}{25} & -\frac{12}{25} \\ \frac{4}{5} & -\frac{12}{25} & \frac{9}{25} \end{bmatrix} \begin{bmatrix} 3 \\ 5 \\ 0 \end{bmatrix} \tag{22a}$$

$$= 3 \begin{bmatrix} 0 \\ \frac{3}{5} \\ \frac{4}{5} \end{bmatrix} + 5 \begin{bmatrix} \frac{3}{5} \\ \frac{16}{25} \\ -\frac{12}{25} \end{bmatrix} + 0 \begin{bmatrix} \frac{4}{5} \\ -\frac{12}{25} \\ \frac{9}{25} \end{bmatrix} = \begin{bmatrix} 3 \\ 5 \\ 0 \end{bmatrix} = \mathbf{a}_2 \tag{22b}$$

■

Proof of Theorem 5: The columns of $\mathbf{Q}$, which are the new orthonormal basis, are obtained from linear combinations of the columns of $\mathbf{A}$. Reversing this procedure yields the columns of $\mathbf{A}$ as linear combinations of the columns of $\mathbf{Q}$. This reversal is what is accomplished by the matrix product **QR**. Consider the computation in Example 7 of the second column in $\mathbf{A} = \mathbf{QR}$, illustrated in (22a)–(22b). In terms of its columns $\mathbf{q}_i$, $\mathbf{Qr}_2 = \mathbf{a}_2$ is [see (22b)]

$$3\mathbf{q}_1 + 5\mathbf{q}_2 + 0\mathbf{q}_3 = \mathbf{a}_2 \tag{23}$$

or, replacing the numbers in (23) by their entries of $\mathbf{R}$ (and deleting $0\mathbf{q}_3$),

$$\mathbf{Qr}_2 = r_{12}\mathbf{q}_1 + r_{22}\mathbf{q}_2 = \mathbf{a}_2 \tag{24}$$

As noted above, the r_{i2}'s are the lengths of projections of $\mathbf{a}_2$ onto the $\mathbf{q}_i$'s. So (24) says, in geometric terms, that $\mathbf{a}_2$ equals the sum of its projection onto $\mathbf{q}_1$ and its projection onto $\mathbf{q}_2$.

The fact that $\mathbf{Q}\mathbf{r}_2 = \mathbf{a}_2$ can also be proved algebraically. Consider the formula for $\mathbf{q}_2$:

$$\mathbf{q}_2 = \frac{\mathbf{a}_2 - s\mathbf{q}_1}{|\mathbf{a}_2 - s\mathbf{q}_1|} \tag{25a}$$

$$= \frac{\mathbf{a}_2 - r_{12}\mathbf{q}_1}{r_{22}} \tag{25b}$$

since $r_{22} = |\mathbf{a}_2 - s\mathbf{q}_1|$ and $r_{12} = s$. Solving for $\mathbf{a}_2$ in (25b), we obtain (24):

$$\mathbf{q}_2 = \frac{\mathbf{a}_2 - r_{12}\mathbf{q}_1}{r_{22}} \implies r_{12}\mathbf{q}_1 + r_{22}\mathbf{q}_2 = \mathbf{a}_2 \tag{26}$$

A similar analysis shows that the jth column in the product $\mathbf{Q}\mathbf{R}$ equals $\mathbf{a}_j$. ■

The matrix $\mathbf{R}$ is upper triangular because column $\mathbf{a}_i$ is only involved in building columns $\mathbf{q}_i, \mathbf{q}_{i+1}, \ldots, \mathbf{q}_n$ of $\mathbf{Q}$. Gram–Schmidt orthogonalization involves the columns of $\mathbf{A}$, and $\mathbf{A} = \mathbf{Q}\mathbf{R}$ says that the columns of $\mathbf{A}$ are linear combinations of the columns of $\mathbf{Q}$. This is why the matrix of lengths $\mathbf{R}$ is the second matrix in the product $\mathbf{Q}\mathbf{R}$. [Postmultiplying by $\mathbf{R}$ forms linear combinations of the columns of $\mathbf{Q}$; see (24).] Recall that the LU decomposition, $\mathbf{A} = \mathbf{L}\mathbf{U}$, represents the *rows* of $\mathbf{A}$ as linear combinations of the rows of $\mathbf{U}$ and there the matrix of multipliers $\mathbf{L}$ comes first in the product. [Premultiplying by $\mathbf{L}$ forms linear combinations of the rows of $\mathbf{U}$.]

The $\mathbf{Q}\mathbf{R}$ decomposition is used frequently in numerical procedures to improve the accuracy of computations. One of its most frequent uses is finding the inverse or pseudoinverse of an ill-conditioned matrix. If $\mathbf{A}$ is an $n \times n$ matrix with linearly independent columns, the decomposition $\mathbf{A} = \mathbf{Q}\mathbf{R}$ yields

$$\mathbf{A}^{-1} = (\mathbf{Q}\mathbf{R})^{-1} = \mathbf{R}^{-1}\mathbf{Q}^{-1} = \mathbf{R}^{-1}\mathbf{Q}^{\mathrm{T}} \tag{27}$$

(The fact that $\mathbf{Q}^{-1} = \mathbf{Q}^{\mathrm{T}}$ was given in Theorem 2.) Given the $\mathbf{Q}\mathbf{R}$ decomposition of $\mathbf{A}$, (27) says that to get $\mathbf{A}^{-1}$, we only need to determine $\mathbf{R}^{-1}$, and since $\mathbf{R}$ is an upper triangular matrix, its inverse is obtained quickly by back substitution (see Exercise 19 in Section 1.5). When $\mathbf{A}$ is very ill-conditioned, one should compute $\mathbf{A}^{-1}$ via (27): first, determining the $\mathbf{Q}\mathbf{R}$ decomposition of $\mathbf{A}$, using advanced (more stable) variations of the Gram–Schmidt procedure, then determining $\mathbf{R}^{-1}$, and thus obtaining $\mathbf{A}^{-1} = \mathbf{R}^{-1}\mathbf{Q}^{\mathrm{T}}$.

Equation (27) extends to pseudoinverses. That is, if $\mathbf{A}$ is an $m \times n$ matrix with linearly independent columns and $m > n$, its pseudoinverse $\mathbf{A}^{+}$ can be shown to equal

$$\mathbf{A}^{+} = \mathbf{R}^{-1}\mathbf{Q}^{\mathrm{T}} \tag{28}$$

See the exercises for instructions on how to verify (28) and examples of its use. This formula for the pseudoinverse is the standard way that pseudoinverses are computed in practice.

Section 4.5 Exercises

1. Express the vectors $\mathbf{b} = \begin{bmatrix} 2 \\ 3 \end{bmatrix}$ and $\mathbf{c} = \begin{bmatrix} -1 \\ 4 \end{bmatrix}$ in terms of the following orthogonal bases.

   (a) $\begin{bmatrix} 2 \\ 1 \end{bmatrix}, \begin{bmatrix} -2 \\ 4 \end{bmatrix}$ (b) $\begin{bmatrix} 2 \\ 0 \end{bmatrix}, \begin{bmatrix} 0 \\ 3 \end{bmatrix}$ (c) $\begin{bmatrix} 3 \\ 4 \end{bmatrix}, \begin{bmatrix} 4 \\ -3 \end{bmatrix}$

2. Express the vectors $\mathbf{b} = \begin{bmatrix} 1 \\ 1 \\ 0 \end{bmatrix}$ and $\mathbf{c} = \begin{bmatrix} 2 \\ 1 \\ 3 \end{bmatrix}$ in terms of the following orthogonal bases.

   (a) $\begin{bmatrix} 2 \\ -6 \\ 3 \end{bmatrix}, \begin{bmatrix} -3 \\ 2 \\ 6 \end{bmatrix}, \begin{bmatrix} 6 \\ 3 \\ 2 \end{bmatrix}$ (b) $\begin{bmatrix} 1 \\ -1 \\ 2 \end{bmatrix}, \begin{bmatrix} 2 \\ 2 \\ 0 \end{bmatrix}, \begin{bmatrix} -1 \\ 1 \\ 1 \end{bmatrix}$ (c) $\begin{bmatrix} -1 \\ 2 \\ 1 \end{bmatrix}, \begin{bmatrix} 4 \\ 1 \\ 2 \end{bmatrix}, \begin{bmatrix} -1 \\ -2 \\ 3 \end{bmatrix}$

3. Express the vectors $\mathbf{b} = \begin{bmatrix} 0 \\ 2 \end{bmatrix}$ and $\mathbf{c} = \begin{bmatrix} 2 \\ -1 \end{bmatrix}$ in terms of the following orthonormal bases.

   (a) $\begin{bmatrix} 0.6 \\ 0.8 \end{bmatrix}, \begin{bmatrix} 0.8 \\ -0.6 \end{bmatrix}$ (b) $\dfrac{1}{\sqrt{2}}\begin{bmatrix} 1 \\ 1 \end{bmatrix}, \dfrac{1}{\sqrt{2}}\begin{bmatrix} -1 \\ 1 \end{bmatrix}$

4. Express the vectors $\mathbf{b} = \begin{bmatrix} 1 \\ 0 \\ 2 \end{bmatrix}$ and $\mathbf{c} = \begin{bmatrix} 0 \\ -1 \\ 4 \end{bmatrix}$ in terms of the following orthonormal bases.

   (a) $\begin{bmatrix} \frac{1}{3} \\ -\frac{2}{3} \\ \frac{2}{3} \end{bmatrix}, \begin{bmatrix} -\frac{2}{3} \\ \frac{1}{3} \\ \frac{2}{3} \end{bmatrix}, \begin{bmatrix} \frac{2}{3} \\ \frac{2}{3} \\ \frac{1}{3} \end{bmatrix}$ (b) $\begin{bmatrix} \frac{4}{5} \\ \frac{3}{5} \\ 0 \end{bmatrix}, \begin{bmatrix} \frac{9}{25} \\ -\frac{12}{25} \\ \frac{4}{5} \end{bmatrix}, \begin{bmatrix} -\frac{12}{25} \\ \frac{16}{25} \\ \frac{3}{5} \end{bmatrix}$

5. Show that if $\mathbf{A}$ is an $n \times n$ upper triangular matrix with orthonormal columns, $\mathbf{A}$ is a diagonal matrix with diagonal entries ± 1.

6. Let $\mathbf{Q}$ be an $n \times n$ matrix with orthonormal columns and let $\mathbf{u}$ be any n-vector. Show that $|\mathbf{u}|^2 = (\mathbf{Qu}) \cdot (\mathbf{Qu})$. (Hint: Use the corollary to Theorem 1 in Section 3.1.)

7. Prove that if $\mathbf{Q}$ is an $n \times n$ matrix with orthonormal columns, then $\det(\mathbf{Q}) = \pm 1$. [Hint: Use the facts that $\det(\mathbf{A}^{-1}) = 1/\det(\mathbf{A})$ and that $\det(\mathbf{A}^{\mathrm{T}}) = \det(\mathbf{A})$.]

8. Prove that if $\mathbf{Q}$ and $\mathbf{Q}'$ are $n \times n$ matrices with orthonormal columns, the product matrix $\mathbf{QQ}'$ has orthonormal columns. (Hint: Use Theorem 4.)

9. Show that if $\mathbf{Q}$ is an $m \times n$ matrix ($m > n$) with orthonormal columns having a pseudo-inverse $\mathbf{Q}^+$, then $\mathbf{Q}^+ = \mathbf{Q}^{\mathrm{T}}$. (Use the fact that $\mathbf{Q}^+$ is characterized by the equation $\mathbf{Q}^+\mathbf{Q} = \mathbf{I}$.)

10. Compute the length of $\mathbf{b} = \begin{bmatrix} -1 \\ 3 \end{bmatrix}$ and the cosine of the angle between $\mathbf{b}$ and $\mathbf{c} = \begin{bmatrix} 2 \\ 1 \end{bmatrix}$. Then perform an orthonormal change of basis to the basis $Q = \left\{ \begin{bmatrix} 0.6 \\ 0.8 \end{bmatrix}, \begin{bmatrix} 0.8 \\ -0.6 \end{bmatrix} \right\}$ and compute the length of $[\mathbf{b}]_Q$ and the cosine of the angle between $[\mathbf{b}]_Q$ and $[\mathbf{c}]_Q$. Confirm that the two lengths and cosines are the same.

11. Compute the length of $\mathbf{b} = \begin{bmatrix} 1 \\ 1 \\ 0 \end{bmatrix}$ and the cosine of the angle between $\mathbf{b}$ and

$\mathbf{c} = \begin{bmatrix} 1 \\ 2 \\ 2 \end{bmatrix}$. Then perform an orthonormal change of basis to the basis

$Q = \left\{ \begin{bmatrix} \frac{1}{3} \\ -\frac{2}{3} \\ \frac{2}{3} \end{bmatrix}, \begin{bmatrix} -\frac{2}{3} \\ \frac{1}{3} \\ \frac{2}{3} \end{bmatrix}, \begin{bmatrix} \frac{2}{3} \\ \frac{2}{3} \\ \frac{1}{3} \end{bmatrix} \right\}$ and compute the length of $[\mathbf{b}]_Q$ and the cosine of the angle

between $[\mathbf{b}]_Q$ and $[\mathbf{c}]_Q$. Confirm that the two lengths and cosines are the same.

12. Determine the angle of rotation involved (and whether there was a reflection) in the change of basis from the standard coordinate basis for $\mathbf{R}^2$ to each of the orthonormal bases in Exercise 3.

13. Verify that Theorem 4 is true in two dimensions: namely, that a change from the standard $\{\mathbf{i}_1, \mathbf{i}_2\}$ basis to some other orthonormal basis $\{\mathbf{q}_1, \mathbf{q}_2\}$ corresponds to a rotation (around the origin) and possibly a reflection. Note that since $\mathbf{q}_1, \mathbf{q}_2$ have unit length, they are completely determined by knowing the (counterclockwise) angles θ_1, θ_2 that they make with the positive $\mathbf{i}_1$ axis; also, since $\mathbf{q}_1, \mathbf{q}_2$ are orthogonal, $|\theta_1 - \theta_2| = 90°$.

14. Compute the angle between the following pairs of nonorthogonal vectors. Which are close to orthogonal?

(a) $\begin{bmatrix} 3 \\ 2 \end{bmatrix}, \begin{bmatrix} -3 \\ 4 \end{bmatrix}$ (b) $\begin{bmatrix} 1 \\ 2 \\ 5 \end{bmatrix}, \begin{bmatrix} 2 \\ 5 \\ 3 \end{bmatrix}$ (c) $\begin{bmatrix} 1 \\ -3 \\ 2 \end{bmatrix}, \begin{bmatrix} -2 \\ 4 \\ -3 \end{bmatrix}$

15. Use the Gram–Schmidt orthogonalization to find an orthonormal basis that generates the same vector space as the following sets in $\mathbf{R}^2$.

(a) $\begin{bmatrix} 1 \\ 1 \end{bmatrix}, \begin{bmatrix} 2 \\ -1 \end{bmatrix}$ (b) $\begin{bmatrix} 1 \\ -2 \end{bmatrix}, \begin{bmatrix} -2 \\ 4 \end{bmatrix}$ (c) $\begin{bmatrix} 3 \\ 1 \end{bmatrix}, \begin{bmatrix} 1 \\ 2 \end{bmatrix}$ (d) $\begin{bmatrix} 3 \\ 4 \end{bmatrix}, \begin{bmatrix} 4 \\ 5 \end{bmatrix}$

16. Use the Gram–Schmidt orthogonalization to find an orthonormal basis that generates the same vector space as the following sets in $\mathbf{R}^3$.

(a) $\begin{bmatrix} 1 \\ 1 \\ 0 \end{bmatrix}, \begin{bmatrix} 0 \\ 1 \\ 1 \end{bmatrix}, \begin{bmatrix} 1 \\ 0 \\ 1 \end{bmatrix}$ (b) $\begin{bmatrix} 1 \\ 0 \\ 0 \end{bmatrix}, \begin{bmatrix} 1 \\ 1 \\ 0 \end{bmatrix}, \begin{bmatrix} 1 \\ 1 \\ 1 \end{bmatrix}$

(c) $\begin{bmatrix} 2 \\ 1 \\ 2 \end{bmatrix}, \begin{bmatrix} 4 \\ 1 \\ 1 \end{bmatrix}$ (d) $\begin{bmatrix} 1 \\ -1 \\ 1 \end{bmatrix}, \begin{bmatrix} 1 \\ 2 \\ 1 \end{bmatrix}, \begin{bmatrix} 2 \\ 1 \\ 2 \end{bmatrix}$

17. Find the $\mathbf{QR}$ decomposition of the following matrices.

(a) $\begin{bmatrix} 3 & -1 \\ 4 & 1 \end{bmatrix}$ (b) $\begin{bmatrix} 2 & 1 \\ 1 & 1 \\ 2 & 3 \end{bmatrix}$ (c) $\begin{bmatrix} 1 & 0 & 1 \\ 1 & 1 & 0 \\ 0 & 1 & 1 \end{bmatrix}$ (d) $\begin{bmatrix} 1 & 1 & 2 \\ -1 & 2 & 1 \\ 1 & 1 & 2 \end{bmatrix}$

18. Find an orthonormal basis for the vector spaces defined by the following planes.
(a) $x_1 - x_2 + x_3 = 0$
(b) $2x_1 + x_2 - 2x_3 = 0$

19. Compute the inverse $\mathbf{A}^{-1}$ of the following matrices $\mathbf{A}$ by first finding the $\mathbf{QR}$ decomposition of $\mathbf{A}$ and then using (27) to get the inverse. (See Exercise 19 in Section 1.5 for

instructions on computing $\mathbf{R}^{-1}$.) Check your answer by computing the inverse by Gauss–Jordan elimination.

(a) $\begin{bmatrix} 0 & 3 & 2 \\ 3 & 5 & 5 \\ 4 & 0 & 5 \end{bmatrix}$ (see Example 7)

(b) $\begin{bmatrix} 1 & 0 & 1 \\ 1 & 1 & 0 \\ 0 & 1 & 1 \end{bmatrix}$ [see Exercise 17(c)]

20. (a) Find the pseudoinverse $\mathbf{A}^+$ of the matrix $\mathbf{A} = \begin{bmatrix} 2 & 1 \\ 1 & 1 \\ 2 & 3 \end{bmatrix}$ by using the $\mathbf{QR}$ decomposition of $\mathbf{A}$ [asked for in Exercise 17(b)] and computing $\mathbf{A}^+$ as $\mathbf{A}^+ = \mathbf{R}^{-1}\mathbf{Q}^\mathrm{T}$.

(b) Check your answer by finding the pseudoinverse from the formula $\mathbf{A}^+ = (\mathbf{A}^\mathrm{T}\mathbf{A})^{-1}\mathbf{A}^\mathrm{T}$. Note that this is a very poorly conditioned matrix; compute the condition number of $(\mathbf{A}^\mathrm{T}\mathbf{A})$.

21. Verify (28), $\mathbf{A}^+ = \mathbf{R}^{-1}\mathbf{Q}^\mathrm{T}$, by substituting $\mathbf{QR}$ for $\mathbf{A}$ (and $\mathbf{R}^\mathrm{T}\mathbf{Q}^\mathrm{T}$ for $\mathbf{A}^\mathrm{T}$) in the pseudoinverse formula $\mathbf{A}^+ = (\mathbf{A}^\mathrm{T}\mathbf{A})^{-1}\mathbf{A}^\mathrm{T}$ and simplifying. (We assume that the columns of $\mathbf{A}$ are linearly independent, so $\mathbf{R}$ is invertible by Exercise 22.)

22. Show that if the columns of the $m \times n$ matrix $\mathbf{A}$ are linearly independent, the $m \times n$ matrix $\mathbf{R}$ of the $\mathbf{QR}$ decomposition must be invertible. (Hint: Show that the main diagonal entries of $\mathbf{R}$ are all nonzero and from this prove that $\mathbf{R}$'s columns are linearly independent.)

23. Show that any set H of k orthonormal n-vectors can be extended to an orthonormal basis for n-dimensional space. [Hint: Form an $n \times (k + n)$ matrix whose first k columns come from H and whose remaining n columns form the identity matrix; now apply the Gram–Schmidt orthogonalization to this matrix.]

4.6 Inner Product Spaces

In the preceding sections we have presented examples using function spaces to illustrate vector space concepts in non-Euclidean settings. However, we were limited in these examples by the fact that there was no equivalent of the matrix computations that existed for Euclidean vectors. Matrix multiplication as well as orthogonality, projections, and the Euclidean norm are all based on the scalar product. If we had a scalar product for function spaces, we could construct orthogonal bases, or even better, orthonormal bases. For polynomials $p(x)$, powers of x (i.e., $1, x, x^2, \ldots, x^k, \ldots$) form a basis, but do they form an orthogonal basis? We could convert this basis into an orthonormal basis by Gram–Schmidt orthogonalization if there were a scalar product operation for functions.

In motivating how we define a scalar product for functions, *let us initially restrict our attention to functions on the interval* [0, 1]. If we wanted to approximate the function $f(x) = x^2$ on the interval [0, 1], we could make a vector of its values at many points along the interval. For example, if we use increments of $\frac{1}{10}$, we get (see Figure 4.6) the vector

$$\mathbf{f}_{10} = [0, 0.01, 0.04, 0.09, 0.16, 0.25, 0.36, 0.49, 0.64, 0.81, 1] \tag{1}$$

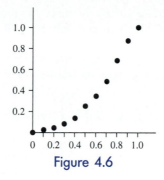

Figure 4.6

Or we could use increments of 1/100 to get $\mathbf{f}_{100}$ or increments of 1/1,000,000 to get $\mathbf{f}_{1,000,000}$, and so on. As the size of the increments approaches 0, the "vector" we get is the continuous function x^2 on the interval [0, 1]. This is the way to think of a function $f(x)$ as the "vector." It is the infinite vector that gives the value of the function at every point on the interval [0, 1].

With this image of a function as a continuous vector on the interval [0, 1], we can now define the function generalization of the scalar product. For functions $f(x)$ and $g(x)$, consider approximations $\mathbf{f}_{10}$ and $\mathbf{g}_{10}$, as in (1). The scalar product $\mathbf{f}_{10} \cdot \mathbf{g}_{10}$ involves multiplying corresponding entries and summing the products; similarly for $\mathbf{f}_{100} \cdot \mathbf{g}_{100}$. In the limit, as these vectors become infinite, we want to multiply the values at all points and sum. Calculus has an operation that does exactly that! The integral $\int f(x)g(x)\,dx$ over the interval [0, 1] is the "continuous" sum of these term-by-term products. This generalization of the scalar product is called the *inner product* of functions.

Definition

The ***inner product*** (f, g) ***of two functions*** $f(x)$ and $g(x)$ on the interval [0, 1] is defined as

$$\langle f, g \rangle = \int_0^1 f(x)g(x)\,dx \tag{2}$$

Note that we can use any interval, not just [0, 1]. We are assuming that the functions we use are integrable in the interval.

Back in Section 2.1 we presented the basic laws obeyed by scalar products (see Theorem 1 of Section 2.1). Now we state the essential laws for defining an inner product:

1. Inner products are commutative: $\langle f, g \rangle = \langle g, f \rangle$.
2. Inner products are distributive: $\langle f, (g + h) \rangle = \langle f, g \rangle + \langle f, h \rangle$.
3. Inner products have scalar factoring: $\langle f, rg \rangle = \langle rf, g \rangle = r\langle f, g \rangle$.

4. Nonnegativity of inner product of a vector with itself: $\langle f, f \rangle \geq 0$, and further, $\langle f, f \rangle = 0$ if and only if $f = \mathbf{0}$ (the zero vector).

It is easy to check that the inner product defined by (2) satisfies these four laws, because integration satisfies these laws. For example, for law 2,

$$\langle f, (g + h) \rangle = \int_0^1 f(x)[g(x) + h(x)] \ dx = \int_0^1 [\ f(x)g(x) + f(x)h(x)] \ dx \tag{3}$$

$$= \int_0^1 f(x)g(x) \ dx + \int_0^1 f(x)h(x) \ dx = \langle f, g \rangle + \langle f, h \rangle$$

Definition

A vector space V with real scalars and with an inner product $\langle \ , \ \rangle$ defined for any two vectors in V and satisfying laws 1 to 4 is called an **inner product space**.

We note that real scalars are not essential. Complex numbers can be used also.

EXAMPLE 1.
Examples of Inner Product Spaces

(a) Any Euclidean vector space is an inner product space using the standard scalar product: $\langle \mathbf{u}, \mathbf{v} \rangle = \mathbf{u} \cdot \mathbf{v}$.

(b) Any vector space whose vectors are continuous functions defined on an interval $[a, b]$ is an inner product space with the inner product defined in (2) (now the limits of integration are from a to b). The vectors could be the polynomials.

(c) Any vector space whose vectors are continuous functions $f(x, y)$ in two variables defined in the region $0 \leq x \leq 1, 0 \leq y \leq 1$ becomes an inner product space with the following generalization of the inner product in (2):

$$\langle f, g \rangle = \int_0^1 \int_0^1 f(x, y)g(x, y) \ dx \ dy$$

(d) For any vector space V whose vectors are continuous functions defined on an interval $[0, 1]$ and for any function $h(x)$ in V which is never zero in the interval $[0, 1]$, V becomes an inner product space with the following inner product $\langle \ , \ \rangle_h$:

$$\langle f, g \rangle_h = \int_0^1 f(x)g(x)h(x) \ dx \tag{4}$$

(e) For *any* finite-dimensional vector space V with a given basis $B = \mathbf{b}_1, \mathbf{b}_2, \ldots,$ $\mathbf{b}_n$ (the $\mathbf{b}_i$'s could be functions or matrices, etc.), let vectors $\mathbf{u}$ and $\mathbf{v}$ in V be expressed as coordinate vectors in basis B, $[\mathbf{u}]_B = [u_1, u_2, \ldots, u_n]$ and $[\mathbf{v}]_B = [v_1, v_2, \ldots, v_n]$, that is,

$$\mathbf{u} = u_1\mathbf{b}_1 + u_2\mathbf{b}_2 + \cdots + u_n\mathbf{b}_n \quad \text{and} \quad \mathbf{v} = v_1\mathbf{b}_1 + v_2\mathbf{b}_2 + \cdots + v_n\mathbf{b}_n$$

Define an inner product as follows (verification that this is an inner product is an exercise):

$$\langle \mathbf{u}, \mathbf{v} \rangle_B = [\mathbf{u}]_B \cdot [\mathbf{v}]_B = u_1 v_1 + u_2 v_2 + \cdots + u_n v_n \qquad (5)$$

Note that the value of this inner product is dependent on the particular basis B used.

■

The reader should be warned that our use of integration in defining inner products raises several interesting questions about integration. The collection of all possible real-valued functions on the interval $[0, 1]$ is a vector space. It includes such weird functions as the function $w(x)$ defined to be 1 if x is a rational number and 0 if x is irrational. Then what is the inner product $\langle w, w \rangle$ using the definition in (2)? The integral, as presented in standard calculus courses, is not defined in this case. Moreover, if it were defined, it would probably be 0, but then law 4 for inner products would be violated, since by law 4, $\langle w, w \rangle = 0$ should imply that $w(x)$ is the zero function (everywhere zero). But $w(x)$ is not zero everywhere.

The functional equivalent of the Euclidean norm can be defined with the inner product using the square of the norm, just as for Euclidean vectors we defined $|\mathbf{v}|_e^2 = \mathbf{v} \cdot \mathbf{v}$.

$$|f(x)|_e^2 = \langle f, f \rangle = \int_0^1 f(x)^2 \, dx \quad \text{or} \quad |f(x)|_e = \sqrt{\int_0^1 f(x)^2 \, dx} \qquad (6)$$

The counterpart of the sum norm $|\mathbf{c}|_s = \Sigma |c_i|$ for functions is $|f(x)|_s = \int_0^1 |f(x)| \, dx$. The max norm is unchanged: $|f(x)|_m = \max_{0 \le x \le 1} |f(x)|$. It is left as exercises for the reader to check that these sum and max norms satisfy the definition for a norm given in Section 2.2.

Most of the theory for Euclidean vector spaces associated with scalar products can be extended to inner product spaces. In particular, we can define orthogonal vectors. Two vectors in an inner product space are **orthogonal** if their inner product is zero. We can also define the "angle" between two vectors in an inner product space. Recall that the cosine of the angle $\theta(\mathbf{a}, \mathbf{b})$ between two Euclidean vectors $\mathbf{a}, \mathbf{b}$ is $(\mathbf{a} \cdot \mathbf{b})/(|\mathbf{a}||\mathbf{b}|)$. The inner product generalization is

$$\cos \theta(f, g) = \frac{\langle f, g \rangle}{|f||g|} \qquad (7)$$

Examples of inner product norms and angles are given in the exercises.

The following is an example of a basic theorem about orthogonality which was discussed for Euclidean vector spaces earlier [see Example 4(d) in Section 4.2] but can now be extended to inner product spaces.

Theorem 1. In an inner product space, any set $\{f_i\}$ of k mutually orthogonal (non-zero) vectors is linearly independent.

Proof: Suppose that the vectors f_i are linearly dependent. Then one can be expressed as a linear combination of the others, such as

$$f_1 = r_2 f_2 + r_3 f_3 + \cdots + r_k f_k \qquad (8)$$

If we take the inner product of each side of (8) with f_1, we obtain

$$\langle f_1, f_1 \rangle = \langle f_1, (r_2 f_2 + r_3 f_3 + \cdots + r_k f_k) \rangle$$
$$= r_2 \langle f_1, f_2 \rangle + r_3 \langle f_1, f_3 \rangle + \cdots + r_k \langle f_1, f_k \rangle \qquad (9)$$

The left side of (9) is positive, since $\langle f_i, f_i \rangle = |f_i|^2 > 0$, and all the inner products $\langle f_1, f_i \rangle$ on the right side of (9) are 0, since the f_i's are orthogonal. This contradiction proves that the f_i's are not linearly dependent. ■

Projections and Gram–Schmidt orthogonalization are based on scalar products. Using inner products in place of scalar products, projections can now be defined, and Gram–Schmidt orthogonalization can be applied to any basis in an inner product space V to convert it into an orthonormal basis. Once we have an orthonormal basis $\{q_i\}$ for V, then we can easily represent any vector f in V in terms of "coordinates" in this basis, that is, as a linear combination of the q_i's. Recall from Theorem 1 in Section 4.5 that the q_i-coordinate of f is simply the scalar product of f with q_i—now the inner product $\langle f, q_i \rangle$.

There are many bases of orthogonal and orthonormal functions that have been developed over the years. We shall discuss two, *Legendre polynomials* and *Fourier trigonometric functions*. We shall work now with the inner product space of continuous functions on the interval $[-1, 1]$.

An orthonormal basis for continuous functions on $[-1, 1]$ will be a set of functions $\{q_i(x)\}$ which are orthogonal—$\int_{-1}^{1} q_i(x) q_j(x) \, dx = 0$, for all $i \neq j$—and whose norms are 1—$\int_{-1}^{1} q_i(x)^2 \, dx = 1$. Given such an orthonormal basis $\{q_i(x)\}$, a function $f(x)$ can be written:

$$f(x) = \langle f, q_1 \rangle q_1(x) + \langle f, q_2 \rangle q_2(x) + \cdots \quad \text{or} \quad f(x) = f_1 q_1(x) + f_2 q_2(x) + \cdots \qquad (10)$$

where the coordinates $f_i = \langle f, q_i \rangle$ of the $q_i(x)$'s are the inner product versions of the projection formula for orthonormal vectors (see Theorem 1 in Section 4.5).

How do we develop such an orthonormal basis for continuous functions? One candidate might be the set of powers of x: $1, x, x^2, x^3, \ldots$. These are linearly independent; that is, x^k cannot be expressed as a linear combination of smaller powers of x. Unfortunately, there is no interval on which 1, x, and x^2 are mutually orthogonal. On $[-1, 1]$, $\langle 1, x \rangle = \int_{-1}^{1} x \, dx = 0$ and $\langle x, x^2 \rangle = \int_{-1}^{1} x^3 \, dx = 0$, but $\langle 1, x^2 \rangle = \int_{-1}^{1} x^2 \, dx = \frac{2}{3} \neq 0$.

We can construct an orthonormal basis from a nonorthogonal basis using the Gram–Schmidt orthogonalization procedure. The calculations in this procedure use scalar products, now inner products, and hence this procedure can be applied to the powers of x (which are linearly independent but, as we just said, far from orthogonal) to find an orthonormal set of polynomials.

When the interval is $[-1, 1]$, the polynomials obtained from Gram–Schmidt orthogonalization applied to $1, x, x^2, x^3, \ldots$ are called *Legendre polynomials* $L_i(x)$. Actually, we shall not worry about making their norms equal to 1. Then by Theorem 1 in Section 4.5, the $L_i(x)$-coordinate of $f(x)$ will be $\langle f, L_i \rangle / \langle L_i, L_i \rangle$. As noted above, the functions $x^0 = 1$ and x are orthogonal on $[-1, 1]$. Then we set $L_0(x) = 1$ and $L_1(x) = x$. Further, x^2 is orthogonal to x but not to 1 on $[-1, 1]$. Following the Gram–Schmidt process, we must subtract off the projection of x^2 onto 1.

$$L_2(x) = x^2 - \frac{\langle 1, x^2 \rangle}{\langle 1, 1 \rangle} 1 = x^2 - \frac{\int_0^1 x^2 \, dx}{\int_0^1 1 \, dx} = x^2 - \frac{1}{3} \tag{11}$$

A similar orthogonalization computation shows that $L_3(x) = x^3 - \frac{3}{5}x$.

EXAMPLE 2.
Approximating e^x
by Legendre
Polynomials

Let us use the first four Legendre polynomials $L_0(x) = 1$, $L_1(x) = x$, $L_2(x) = x^2 - \frac{1}{3}$, $L_3(x) = x^3 - \frac{3}{5}x$ to approximate e^x on the interval $[-1, 1]$. That is, we want the first four terms in (10) when $f(x) = e^x$:

$$\begin{aligned} e^x &\approx w_0 L_0(x) + w_1 L_1(x) + w_2 L_2(x) + w_3 L_3(x) \\ &\approx w_0 + w_1 x + w_2(x^2 - \frac{1}{3}) + w_3(x^3 - \frac{3}{5}x) \end{aligned} \tag{12}$$

where

$$w_i = \frac{\langle e^x, L_i \rangle}{\langle L_i, L_i \rangle} = \frac{\int_0^1 e^x L_i(x) \, dx}{\int_0^1 L_i(x)^2 \, dx}$$

For example,

$$w_2 = \int_{-1}^1 e^x (x^2 - \frac{1}{3}) \, dx \Big/ \int_{-1}^1 (x^2 - \frac{1}{3})^2 \, dx$$

With a little calculus, we compute the w_i to be (approximately)

$$w_0 = \frac{2.35}{2} = 1.18 \qquad w_1 = \frac{0.736}{0.667} = 1.10$$

$$w_2 = \frac{0.096}{0.178} = 0.53 \qquad w_3 = \frac{0.008}{0.046} = 0.18$$

Then (12) becomes

$$e^x \approx 1.18 + 1.10x + 0.53(x^2 - \frac{1}{3}) + 0.18(x^3 - \frac{3}{5}x) \tag{13}$$

If we collect like powers of x together on the right side, (13) simplifies to

$$e^x \approx 1 + x + 0.53x^2 + 0.18x^3 \tag{14}$$

Comparing our approximation against the real values of e^x at the points -1, -0.5, 0, 0.5, 1, we find

x	-1	-0.5	0	0.5	1
e^x	0.37	0.61	1	1.64	2.72
Legendre approximation	0.37	0.61	1	1.65	2.71

A pretty good fit. In particular, it is a better fit on $[-1, 1]$ than simply using the first terms of the standard power series for e^x, namely, $1 + x + \frac{1}{2}x^2 + \frac{1}{6}x^3$. The approximation gets more accurate as more Legendre polynomials are used. ■

Over the interval $[0, 2\pi]$, the trigonometric functions $(1/\sqrt{\pi}) \sin kx$ and $(1/\sqrt{\pi}) \cos kx$, for $k = 1, 2, \ldots$, plus the constant function $1/\sqrt{2\pi}$ are an orthonormal basis. To verify that they are orthogonal requires showing that

$$\left\langle \frac{1}{\sqrt{\pi}} \sin jx, \frac{1}{\sqrt{\pi}} \cos kx \right\rangle = \frac{1}{\pi} \int_0^{2\pi} \sin jx \cos kx \, dx = 0 \qquad \text{for all } j, k$$

$$\left\langle \frac{1}{\sqrt{\pi}} \sin jx, \frac{1}{\sqrt{\pi}} \sin kx \right\rangle = \frac{1}{\pi} \int_0^{2\pi} \sin jx \sin kx \, dx = 0 \qquad \text{for all } j, k, j \neq k$$

$$\left\langle \frac{1}{\sqrt{\pi}} \cos jx, \frac{1}{\sqrt{\pi}} \cos kx \right\rangle = \frac{1}{\pi} \int_0^{2\pi} \cos jx \cos kx \, dx = 0 \qquad \text{for all } j, k, j \neq k$$

(15)

plus showing that these trigonometric functions are orthogonal to the constant function $1/\sqrt{2\pi}$. To verify that these trigonometric functions have unit length requires showing that

$$\left\langle \frac{1}{\sqrt{\pi}} \sin kx, \frac{1}{\sqrt{\pi}} \sin kx \right\rangle = \frac{1}{\pi} \int_0^{2\pi} \sin kx^2 \, dx = 1 \qquad \text{for all } k$$

$$\left\langle \frac{1}{\sqrt{\pi}} \cos kx, \frac{1}{\sqrt{\pi}} \cos kx \right\rangle = \frac{1}{\pi} \int_0^{2\pi} \cos kx^2 \, dx = 1 \qquad \text{for all } k$$

(16)

When $q_{2k-1}(x) = (1/\sqrt{\pi}) \sin kx$ and $q_{2k}(x) = (1/\sqrt{\pi}) \cos kx$, $k = 1, 2, \ldots$ and $q_0(x) = 1/\sqrt{2\pi}$ are used as a basis, a representation of $f(x)$ is called a **Fourier series**, and the projection coefficients $\langle f, q_i \rangle$ are called **Fourier coefficients**. Using the Fourier series, we see that any piecewise continuous function can be expressed as a linear combination of sine and cosine waves. One important physical interpretation of this fact is that any complex electrical signal can be expressed as a sum of simple sinusoidal signals.

**EXAMPLE 3.
Fourier Series
Representation of
a Jump Function**

Let us determine the Fourier series representation of the discontinuous function $f(x) = 1$ for $0 < x \leq \pi$ and $f(x) = 0$ for $\pi < x \leq 2\pi$ (see Figure 4.7). The Fourier coefficients $\langle f, q_i \rangle$ are

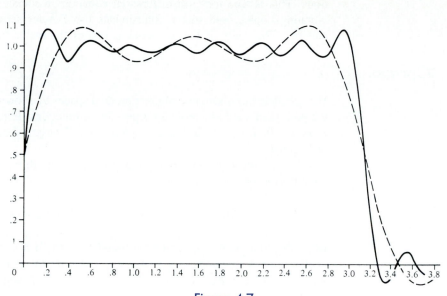

Figure 4.7

$$\langle f, q_{2k-1} \rangle = \left\langle f, \frac{1}{\sqrt{\pi}} \sin kx \right\rangle = \frac{1}{\sqrt{\pi}} \int_0^\pi \sin kx \, dx$$

$$= \frac{1}{k\sqrt{\pi}} [-\cos kx] \, \Big|_0^\pi = \begin{cases} 2/k\sqrt{\pi} & k \text{ odd} \\ 0 & k \text{ even} \end{cases} \tag{17}$$

$$\langle f, q_{2k} \rangle = \left\langle f, \frac{1}{\sqrt{\pi}} \cos kx \right\rangle = \frac{1}{\sqrt{\pi}} \int_0^\pi \cos kx \, dx$$

$$= \frac{1}{k\sqrt{\pi}} [\sin kx]_0^\pi = 0$$

Further, we calculate $\langle f, 1/\sqrt{2\pi} \rangle = \sqrt{\pi}/2$, so the constant term of the Fourier series for this $f(x)$ is $(\sqrt{\pi}/2)(1/\sqrt{2\pi}) = \frac{1}{2}$. By (17), only the odd sine terms occur. Letting an odd k be written as $2n - 1$, we obtain the Fourier series.

$$f(x) = \frac{1}{2} + \sum_{n=1}^\infty \frac{2}{(2n-1)\sqrt{\pi}} \sin[(2n-1)x] \tag{18}$$

Figure 4.7 shows the approximation to $f(x)$ obtained when the first three sine terms in (18) are used (dashed line) and when the first eight sine terms are used. The fit is impressive. ▪

Representing a function in terms of an orthonormal set of functions has a virtually unlimited number of applications in the physical sciences and elsewhere. If one can solve a physical problem for the orthonormal basis functions, one can typically obtain a solution for any function as a linear combination of the solutions for the basis functions. This is true for most differential equations associated with electrical circuits, vibrating bodies, and so on. Statisticians use Fourier series to analyze time-series patterns. The study of Fourier series is one of the major fields of mathematics.

Optional

We complete our discussion of inner product spaces by showing how badly conditioned the powers of x are as a basis for representing functions. Remember that the powers of x, x^i, $i = 0, 1, \ldots$, are linearly independent. The problem is that they are far from orthogonal.

Let us consider how we might approximate an arbitrary function $f(x)$ as a linear combination of, say, the powers of x up to x^5:

$$f(x) \approx w_0 + w_1 x + w_2 x^2 + w_3 x^3 + w_4 x^4 + w_5 x^5 \tag{19}$$

using the continuous version of the pseudoinverse. If $f(x)$ and the powers of x were vectors $\mathbf{a}_i$, not functions, our problem would have the familiar matrix form $w_0 \mathbf{a}_0 + w_1 \mathbf{a}_1 + w_2 \mathbf{a}_2 + w_3 \mathbf{a}_3 + w_4 \mathbf{a}_4 + w_5 \mathbf{a}_5 = \mathbf{f}$ or $\mathbf{Aw} = \mathbf{f}$ and the approximate (least squares) solution $\mathbf{w}$ would be given by $\mathbf{w} = \mathbf{A}^+ \mathbf{f}$, where $\mathbf{A}^+ = (\mathbf{A}^T \mathbf{A})^{-1} \mathbf{A}^T$ (see Section 3.5). Because matrix multiplication is built on scalar products, we can extend the formula for the pseudoinverse to function spaces by using inner products.

Let us generalize $\mathbf{Aw} = \mathbf{f}$ to functions by letting the columns of a matrix be functions. We define the functional "matrix" $\mathbf{A}(x)$:

$$\mathbf{A}(x) = [1, x, x^2, x^3, x^4, x^5]$$

Now $\mathbf{Aw} = \mathbf{f}$ becomes

$$\mathbf{A}(x)\mathbf{w} = f(x) \quad \text{or} \quad \mathbf{A}(x)\mathbf{w} = w_0 + w_1 x + w_2 x^2 + w_3 x^3 + w_4 x^4 + w_5 x^5 = f(x) \tag{20}$$

To find the approximate solution to (20), we need to compute the functional version of the pseudoinverse $\mathbf{A}(x)^+$:

$$\mathbf{A}(x)^+ = (\mathbf{A}(x)^T \mathbf{A}(x))^{-1} \mathbf{A}(x)^T$$

and then find the vector $\mathbf{w} = [w_0, w_1, w_2, w_3, w_4, w_5]^T$ of coefficients in (19):

$$\mathbf{w} = \mathbf{A}(x)^+ f(x) = (\mathbf{A}(x)^T \mathbf{A}(x))^{-1} (\mathbf{A}(x)^T f(x)) \tag{21}$$

The ith "row" of matrix $\mathbf{A}(x)^T$ is x^i [the same as the ith column of $\mathbf{A}(x)$], so the matrix product $\mathbf{A}(x)^T \mathbf{A}(x)$ involves computing the following inner products of each "row" of $\mathbf{A}(x)^T$ with each "column" of $\mathbf{A}(x)$:

$$\text{entry } (i, j) \text{ in } \mathbf{A}(x)^T \mathbf{A}(x) = \langle x^i, x^j \rangle \left(= \int x^i x^j \, dx \right)$$

Similarly, the matrix–"vector" product $\mathbf{A}(x)^T f(x)$ in (21) is the vector of inner products $\langle x^i, f \rangle$. The computations are simplest if we use the interval $[0, 1]$. Then entry (i, j) of $\mathbf{A}(x)^T\mathbf{A}(x)$ is

$$\langle x^i, x^j \rangle = \int_0^1 x^{i+j}\, dx = \left[\frac{x^{i+j+1}}{i+j+1} \right]_0^1 = \frac{1}{i+j+1} \tag{22}$$

For example, entry $(1, 2)$ is $\int_0^1 xx^2\, dx = \int_0^1 x^3\, dx = \frac{1}{4}$. Note that we consider the constant function $1\ (= x^0)$ to be the zeroth row of $\mathbf{A}(x)$.

Computing all the inner products for $\mathbf{A}(x)^T\mathbf{A}(x)$ yields

$$\mathbf{A}(x)^T\mathbf{A}(x) = \begin{bmatrix} 1 & \frac{1}{2} & \frac{1}{3} & \frac{1}{4} & \frac{1}{5} & \frac{1}{6} \\ \frac{1}{2} & \frac{1}{3} & \frac{1}{4} & \frac{1}{5} & \frac{1}{6} & \frac{1}{7} \\ \frac{1}{3} & \frac{1}{4} & \frac{1}{5} & \frac{1}{6} & \frac{1}{7} & \frac{1}{8} \\ \frac{1}{4} & \frac{1}{5} & \frac{1}{6} & \frac{1}{7} & \frac{1}{8} & \frac{1}{9} \\ \frac{1}{5} & \frac{1}{6} & \frac{1}{7} & \frac{1}{8} & \frac{1}{9} & \frac{1}{10} \\ \frac{1}{6} & \frac{1}{7} & \frac{1}{8} & \frac{1}{9} & \frac{1}{10} & \frac{1}{11} \end{bmatrix} \tag{23}$$

This matrix is very ill-conditioned, since the columns are all similar to each other. When the fractions in (23) are expressed to six decimal places, such as $\frac{1}{3} = 0.333333$, the inverse given by the author's microcomputer was (with entries rounded to integer values)

$$(\mathbf{A}(x)^T\mathbf{A}(x))^{-1} = \begin{bmatrix} 17 & -116 & -47 & 1{,}180 & -1{,}986 & 958 \\ -116 & 342 & 7{,}584 & -34{,}881 & 49{,}482 & -22{,}548 \\ -47 & 7{,}584 & -76{,}499 & 242{,}494 & -301{,}846 & 129{,}004 \\ 1{,}180 & -34{,}881 & 242{,}494 & 644{,}439 & 723{,}636 & -289{,}134 \\ -1{,}986 & 49{,}482 & -301{,}846 & 723{,}636 & -747{,}725 & 278{,}975 \\ 958 & -22{,}548 & 129{,}004 & -289{,}134 & 278{,}975 & -97{,}180 \end{bmatrix} \tag{24}$$

The (absolute) sum of the fifth column in (24) is about 2,000,000. The first column in (23) sums to about 2.5. So the condition number of $\mathbf{A}(x)^T\mathbf{A}(x)$, in the sum norm, is about $2{,}000{,}000 \cdot 2.5 = 5{,}000{,}000$. Now that is an ill-conditioned matrix! We rounded fractions to six significant digits, but our condition number tells us that without a seventh significant digit, our numbers in (24) could be off by 500% error. [A relative error of 0.000001 in $\mathbf{A}(x)^T\mathbf{A}(x)$ could yield answers off by a factor of 5 in pseudoinverse calculations.] Thus the numbers in (24) may be worthless. To compute the inverse accurately would require double-precision computation.

It is only fair to note that the ill-conditioned matrix (23) is infamously bad. It is called a 6×6 **Hilbert matrix**. In general, an $n \times n$ Hilbert matrix has $1/(i + j - 1)$ in entry (i, j).

Suppose that we try using the numbers in (24) for $(\mathbf{A}(x)^T\mathbf{A}(x))^{-1}$ in computing the pseudoinverse. Let us proceed to calculate $\mathbf{A}(x)^+ f(x)$ and obtain the coefficients w_i

in the fifth-degree polynomial approximating $f(x)$. As in Example 2, let us choose $f(x) = e^x$. Observe that

$$\mathbf{A}(x)^+ e^x = (\mathbf{A}(x)^\mathrm{T}\mathbf{A}(x))^{-1}\mathbf{A}(x)^\mathrm{T} e^x = (\mathbf{A}(x)^\mathrm{T}\mathbf{A}(x))^{-1} \cdot (\mathbf{A}(x)^\mathrm{T} e^x) \qquad (25)$$

In the rightmost expression in (25), we have broken $\mathbf{A}(x)^+ e^x$ into the scalar product of the matrix $(\mathbf{A}(x)^\mathrm{T}\mathbf{A}(x))^{-1}$ with $(\mathbf{A}(x)^\mathrm{T} e^x)$. Here $\mathbf{A}(x)^\mathrm{T} e^x$ is the vector of inner products of the rows, x^i, of $\mathbf{A}(x)^\mathrm{T}$ with e^x, namely $\langle x^i, e^x \rangle = \int_0^1 x^i e^x \, dx$, $i = 0, 1, \ldots, 5$. Some calculus yields $\mathbf{A}(x)^\mathrm{T} e^x = [2.718, 1, 0.718, 0.563, 0.465, 0.396]^\mathrm{T}$ (expressed to three significant digits).

Using our values for $(\mathbf{A}(x)^\mathrm{T}\mathbf{A}(x))^{-1}$ in (24) and $\mathbf{A}^\mathrm{T} e^x$, we obtain

$$\mathbf{w} = (\mathbf{A}(x)^\mathrm{T}\mathbf{A}(x))^{-1} \cdot (\mathbf{A}(x)^\mathrm{T} e^x)$$

$$= \begin{bmatrix} 17 & -116 & -47 & 1{,}180 & -1{,}986 & 958 \\ -116 & 342 & 7{,}584 & -34{,}881 & 49{,}482 & -22{,}548 \\ -47 & 7{,}584 & -76{,}499 & 242{,}494 & -301{,}846 & 129{,}004 \\ 1{,}180 & -34{,}881 & 242{,}494 & 644{,}439 & 723{,}636 & -289{,}134 \\ -1{,}986 & 49{,}482 & -301{,}846 & 723{,}636 & -747{,}725 & 278{,}975 \\ 958 & -22{,}548 & 129{,}004 & -289{,}134 & 278{,}975 & -97{,}180 \end{bmatrix} \begin{bmatrix} 2.718 \\ 1 \\ 0.718 \\ 0.563 \\ 0.465 \\ 0.396 \end{bmatrix}$$

$$= \begin{bmatrix} 17 \\ -87 \\ -219 \\ 1{,}611 \\ 2{,}449 \\ 1{,}135 \end{bmatrix}$$

$$(26)$$

Thus our fifth-degree polynomial approximation of e^x on the interval $[0, 1]$ is

$$e^x \approx 17 - 87x - 219x^2 + 1611x^3 + 2449x^4 + 1135x^5 \qquad (27)$$

Setting $x = 1$ in (26), we have $e^1 = 17 - 86 - 219 + 1611 + 2449 + 1135 = 4907$. Pretty bad estimate of e^1 (≈ 2.72). Since our computed values in $(\mathbf{A}(x)^\mathrm{T}\mathbf{A}(x))^{-1}$ are meaningless, such a bad approximation of e^x was to be expected. Compare (27) with the Legendre polynomial approximation in Example 2. Now the reader should appreciate the importance of orthogonal polynomials, such as Legendre polynomials.

If we wanted to make a more systematic study of the accuracy of an approximation of a function such as e^x by a linear approximation $p(x)$ of basis functions, the natural measure is the norm of their difference $|e^x - p(x)|$ (an inner product norm $|f| = \sqrt{\langle f \cdot f \rangle}$ is usually used). For example, it can be shown that with enough terms, a (finite) Fourier series approximation of any continuous function on the interval $[0, 2\pi]$ can be made arbitrarily close, as measured by the inner product norm. The study of such functional approximations was a major focus of mathematical research in the first half of the twentieth century.

Section 4.6 Exercises

1. Over the interval $[0, 1]$, compute the following inner products.
   (a) $\langle x, x \rangle$ (b) $\langle x^3, x^2 \rangle$ (c) $\langle x, x^3 \rangle$ (d) $\langle x^2 + x + 1, 2x + 2 \rangle$

2. (a) Verify that the inner product defined by (2) is distributive and has scalar factoring by checking that $\langle x, 2x^3 + 3x - 4 \rangle = 2\langle x, x^3 \rangle + 3\langle x, x \rangle - 4\langle x, 1 \rangle$ on the interval $[0, 1]$.
   (b) Prove that the inner product defined by (2) satisfies the law of scalar factoring.

3. Find the inner product $\langle \ , \ \rangle$ as defined in Example 1(c) for:
   (a) $\langle xy, x - y \rangle$ (b) $\langle x + y, x^2 - y \rangle$

4. Find the inner product $\langle \ , \ \rangle$ as defined in Example 1(d) with, $h(x) = 1/x$, for:
   (a) $\langle x, x \rangle$ (b) $\langle x, x^3 \rangle$

5. Find the Euclidean norm with the inner product on $[0, 1]$ for:
   (a) x (b) x^2 (c) x^3 (d) $x^2 + x - 2$

6. (a) Verify that the inner product version of the sum norm satisfies the properties of a norm (see Theorems 2 and 3 in Section 2.2).
   (b) Verify that the inner product version of the max norm satisfies the properties of a norm.

7. For the inner product defined on the interval $[-1, 1]$, show that the following pairs of functions are orthogonal.
   (a) x^2 and x^3 (b) $2x^2 + 3$ and x
   (c) x^{2m} and x^{2n+1}, for any nonnegative integers m, n

8. Find the angle between the following functions using (7) with the inner product defined on the interval $[0, 1]$.
   (a) 1 and x (b) x and x^2 (c) $x + 1$ and $x^2 - 2x$

9. Verify that the third Legendre polynomial is $x^3 - \frac{3}{5}x$.

10. Verify the values found for the weights w_0, w_1, w_2, w_3 in Example 2. (**Note:** You must use integration by parts or symbolic algebra software or a table of integrals.)

11. Approximate the following functions as a linear combination of the first four Legendre polynomials over the interval $[-1, 1]$: $L_0(x) = 1$, $L_1(x) = x$, $L_2(x) = x^2 - \frac{1}{3}$, $L_3(x) = x^3 - \frac{3}{5}x$.
    (a) $f(x) = x^4$ (b) $f(x) = |x|$ (c) $f(x) = x^5$ (d) $f(x) = \{-1 : x < 0, +1 : x \geq 0\}$

12. Approximate $x^3 + 2x - 1$ as a linear combination of the first four Legendre polynomials over the interval $[-1, 1]$: $L_0(x) = 1$, $L_1(x) = x$, $L_2(x) = x^2 - \frac{1}{3}$, $L_3(x) = x^3 - \frac{3}{5}x$. Your "approximation" should equal $x^3 + 2x - 1$ since this polynomial is a linear combination of the functions 1, x, x^2, x^3, from which the Legendre polynomials were derived by orthogonalization.

13. (a) Find the Legendre polynomial of degree 4 on the interval $[-1, 1]$.
    (b) Find the Legendre polynomial of degree 5 on the interval $[-1, 1]$.

14. (a) Using the interval $[0, 1]$, instead of $[-1, 1]$, find three orthogonal polynomials of the form $K_0(x) = a$, $K_1(x) = bx + c$, and $K_2(x) = dx^2 + ex + f$.
    (b) Approximate x^4 on the interval $[0, 1]$ using your three polynomials in part (a).
    (c) Approximate $\sqrt{x}$ on the interval $[0, 1]$ using your three polynomials in part (a).

(d) Approximate $x^2 - 2x + 1$ on the interval $[0, 1]$ using your three polynomials in part (a). Hopefully, your approximation will equal $x^2 - 2x + 1$, since this polynomial is a linear combination of 1, x, x^2, the functions used to build your set of orthogonal polynomials.

15. **(a)** Find a fourth polynomial $K_3(x)$ of order 3 orthogonal on $[0, 1]$ to the three polynomials in Exercise 14(a).

    (b) Approximate x^4 on the interval $[0, 1]$ using your four orthogonal polynomials.

16. Find the Fourier series for the following functions $f(x)$ on the interval form $-\pi$ to π.

    (a) $f(x) = -1$ for $-\pi \leq x \leq 0$, $f(x) = 1$ for $0 < x \leq \pi$

    (b) $f(x) = 0$ for $-\pi \leq x \leq -\pi/2$ and $-\pi/2 < x \leq \pi$, $f(x) = 1$ for $-\pi/2 < x \leq \pi/2$

    (c) $f(x) = x$ **(d)** $f(x) = x^2$ **(e)** $f(x) = e^x$

    [Hint: You may use computer algebra systems or a table of integrals to evaluate the integrals in parts (c), (d), and (f).]

17. Compute the inverse and find the condition number (in sum norm) of the following Hilbert matrices.

    (a) $\begin{bmatrix} \frac{1}{8} & \frac{1}{9} \\ \frac{1}{9} & \frac{1}{10} \end{bmatrix}$ **(b)** $\begin{bmatrix} \frac{1}{4} & \frac{1}{5} & \frac{1}{6} \\ \frac{1}{5} & \frac{1}{6} & \frac{1}{7} \\ \frac{1}{6} & \frac{1}{7} & \frac{1}{8} \end{bmatrix}$ **(c)** $\begin{bmatrix} \frac{1}{8} & \frac{1}{9} & \frac{1}{10} \\ \frac{1}{9} & \frac{1}{10} & \frac{1}{11} \\ \frac{1}{10} & \frac{1}{11} & \frac{1}{12} \end{bmatrix}$

5 Linear Transformations

5.1 Introduction to Linear Transformations

In this chapter we discuss a class of mappings on vector spaces, called *linear transformations*, that generalize the matrix-based mapping $\mathbf{u} \to \mathbf{w} = \mathbf{Au}$. These matrix mappings arose in various linear models, such as population growth, $\mathbf{x}' = \mathbf{Ax}$, and Markov chains, $\mathbf{p}' = \mathbf{Ap}$ and were implicitly used in writing systems of linear equations, $\mathbf{Ax} = \mathbf{b}$.

In Chapter 4, Euclidean spaces $\mathbf{R}^n$ were generalized to arbitrary vector spaces. In this chapter we develop a parallel generalization of matrix mappings to linear transformations for general vector spaces. Concepts associated with matrix systems, such as inverses, eigenvalues and eigenvectors, and matrix norms, also generalize for linear transformations. In this setting we can develop a theory of solutions to linear systems that applies to any vector space, not just Euclidean spaces.

The primary focus of a linear algebra course is systems of equations, matrices, and vector spaces. However, linear transformations include a broad range of other important mathematical operations, such as differentiation and integration, and provide a general framework for many fields of advanced mathematics that grow out of calculus. In this chapter we only scratch the surface of this broader role of linear transformations. However, insights gained here should be very helpful in further study of calculus-based mathematics.

Definition

Given vector spaces V and W, a function $T: V \to W$ mapping any vector $\mathbf{u}$ in V to a vector $\mathbf{w} = T(\mathbf{u})$ in W is a **linear transformation** if

$$T(r\mathbf{u} + s\mathbf{v}) = rT(\mathbf{u}) + sT(\mathbf{v}) \quad \text{for all } \mathbf{u}, \mathbf{v} \in V \text{ and } r, s \text{ scalars} \quad (1)$$

Note that (1) generalizes to

$$T(r_1\mathbf{u}_1 + r_2\mathbf{u}_2 + \cdots + r_n\mathbf{u}_n) = r_1T(\mathbf{u}_1) + r_2T(\mathbf{u}_2) + \cdots + r_nT(\mathbf{u}_n) \quad (1a)$$

Also, if $r = 1$ and $s = -1$ in (1), we get

$$T(\mathbf{u} - \mathbf{v}) = T(\mathbf{u}) - T(\mathbf{v}) \quad (1b)$$

And if $\mathbf{v} = \mathbf{0}$ or $s = 0$, then (1) becomes

$$T(r\mathbf{u}) = rT(\mathbf{u}) \quad (1c)$$

As noted above, linear transformations are a generalization of the matrix-based mapping $\mathbf{u} \to \mathbf{w} = \mathbf{A}\mathbf{u}$.

Theorem 1. If $\mathbf{A}$ is an $m \times n$ matrix, the mapping $T_\mathbf{A}: \mathbf{R}^n \to \mathbf{R}^m$, where $T_\mathbf{A}(\mathbf{u}) = \mathbf{A}\mathbf{u}$, is a linear transformation.

Proof: We must show that $T_\mathbf{A}$ satisfies (1).

$$\begin{aligned} T_\mathbf{A}(r\mathbf{u} + s\mathbf{v}) &= \mathbf{A}(r\mathbf{u} + s\mathbf{v}) \\ &= r\mathbf{A}\mathbf{u} + s\mathbf{A}\mathbf{v} = rT_\mathbf{A}(\mathbf{u}) + sT_\mathbf{A}(\mathbf{v}) \end{aligned} \quad (2)$$

■

**EXAMPLE 1.
Linear
Transformations
in Euclidean
Vector Spaces**

(a) Let V be any Euclidean vector space (a subspace of $\mathbf{R}^n$, for some n). We claim that if k is some given (nonzero) scalar, then $T_k: V \to V$ defined by $T_k(\mathbf{u}) = k\mathbf{u}$ is a linear transformation. Note that by the closure property of vector spaces, if $\mathbf{u}$ is in V, then $k\mathbf{u}$ must be in V. Property (1) for T_k becomes $k(r\mathbf{u} + s\mathbf{v}) = rk\mathbf{u} + sk\mathbf{v}$, an equation that follows from the laws of scalar factoring for vector spaces. (See the definition of a vector space at the beginning of Section 4.1.)

(b) Define $T_\mathbf{A}: \mathbf{R}^2 \to \mathbf{R}^2$ by $T_\mathbf{A}(\mathbf{u}) = \mathbf{A}\mathbf{u}$, where $\mathbf{A} = \begin{bmatrix} 2 & 3 \\ 1 & -5 \end{bmatrix}$. Therefore, $T_\mathbf{A}\left(\begin{bmatrix} x_1 \\ x_2 \end{bmatrix}\right) = \begin{bmatrix} 2x_1 + 3x_2 \\ x_1 - 5x_2 \end{bmatrix}$. By Theorem 1, $T_\mathbf{A}$ is a linear transformation.

(c) Define $T: \mathbf{R}^3 \to \mathbf{R}$ by $T(\mathbf{u}) = u_1$. For example, $T\left(\begin{bmatrix} 3 \\ 5 \\ 1 \end{bmatrix}\right) = 3$. The reader should check that $T(\mathbf{v}) = \mathbf{B}\mathbf{v}$, where $\mathbf{B}$ is the 1×3 matrix $[1 \quad 0 \quad 0]$. Thus T is a linear transformation by Theorem 1.

(d) Let V be some m-dimensional Euclidean vector space with basis $B = \{b_1, b_2, \ldots, b_m\}$. Using the notation developed in Section 4.3, for a vector u in V, we write $[u]_B$ to denote the vector of u's coordinates, $[u]_B = [c_1, c_2, \ldots, c_n]$, with respect to basis B, that is, $u = c_1 b_1 + c_2 b_2 + \cdots + c_n b_n$. If v is another vector in V, then $u + v$ has a coordinate vector $[u + v]_B$, which will be equal to $[u]_B + [v]_B$. Also, it can be checked that $[ru + sv]_B = r[u]_B + s[v]_B$.

Define $T: V \to R^m$ by $T(u) = [u]_B$. From the preceding discussion it follows that T is a linear transformation.

(e) Let V be any Euclidean vector space and let W be a subspace of V. We define the projection transformation $P: V \to W$ by $P(u)$ is the projection of u onto W. If q_1, $q_2, \ldots, q_k$ is an orthonormal basis for W, the projection of u into W is the vector $P(u) = (u \cdot q_1)q_1 + (u \cdot q_2)q_2 + \cdots + (u \cdot q_k)q_k$. (See Section 4.5 for how to build an orthonormal basis for a vector space and how to compute projections with an orthonormal basis.)

As a simple example, let V be R^3 and let W be the subspace spanned by the orthonormal basis $q_1 = \begin{bmatrix} \frac{1}{3} \\ \frac{2}{3} \\ \frac{2}{3} \end{bmatrix}$, $q_2 = \begin{bmatrix} \frac{2}{3} \\ -\frac{2}{3} \\ \frac{1}{3} \end{bmatrix}$. For the vector $u = \begin{bmatrix} 2 \\ 2 \\ 6 \end{bmatrix}$,

$$P(u = (u \cdot q_1)q_1 + (u \cdot q_2)q_2$$
$$= (2 \cdot \tfrac{1}{3} + 2 \cdot \tfrac{2}{3} + 6 \cdot \tfrac{2}{3})q_1 + (2 \cdot \tfrac{2}{3} + 2 \cdot -\tfrac{2}{3} + 6 \cdot \tfrac{1}{3})q_2$$
$$= 4q_1 + 2q_2 = 6\begin{bmatrix} \frac{1}{3} \\ \frac{2}{3} \\ \frac{2}{3} \end{bmatrix} + 2\begin{bmatrix} \frac{2}{3} \\ -\frac{2}{3} \\ \frac{1}{3} \end{bmatrix} = \begin{bmatrix} \frac{10}{3} \\ \frac{8}{3} \\ \frac{14}{3} \end{bmatrix}$$

To see that this projection is a linear transformation, observe that for the linear combination $ru + sv$, the q_i-component of $P(ru + sv)$ is $(ru + sv) \cdot q_i$, which equals $r(u \cdot q_i) + s(v \cdot q_i)$. Thus equation (1) holds for each q_i-component. Collecting the q_i-components together, we have $P(ru + rv) = rP(u) + sP(v)$. This projection transformation can be defined for any inner product space. (See the discussion of projections in inner product spaces in Section 4.6.) ■

**EXAMPLE 2.
Geometric Effects
of Linear
Transformations
in R^2 and R^3**

The following four linear transformations map R^2 (or R^3) into R^2 so as to achieve a certain geometric effect. Repeated transformations of the plane, usually linear transformations, are used to create the special graphic effects one sees on television where letters and images are fancifully manipulated. We shall give only simple graphic effects here. To illustrate the effects of the first three transformations, they are each applied to the unit square in Figure 5.1(a). (Also see Example 6 in Section 1.2.)

(a) Let $T_1: R^2 \to R^2$ be defined by the matrix mapping $\begin{bmatrix} x_1' \\ x_2' \end{bmatrix} = \begin{bmatrix} 1 & 1 \\ 0 & 1 \end{bmatrix}\begin{bmatrix} x_1 \\ x_2 \end{bmatrix}$. The transformation T_1 slants vertical lines of the square [in Figure 5.1(a)] by 45° [see Figure 5.1(b)].

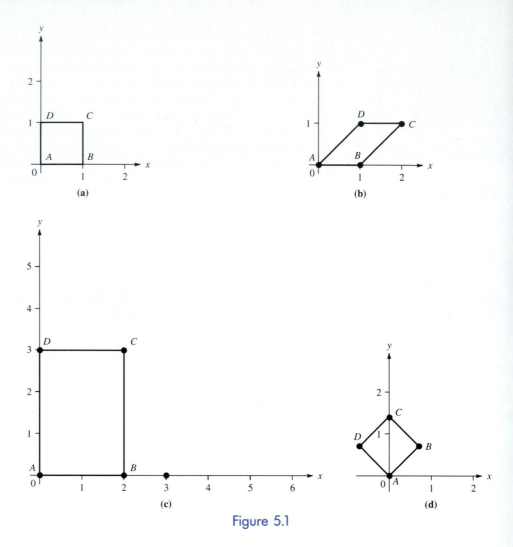

Figure 5.1

(b) Let T_1: $\mathbf{R}^2 \rightarrow \mathbf{R}^2$ be defined by the matrix mapping $\begin{bmatrix} x'_1 \\ x'_2 \end{bmatrix} = \begin{bmatrix} 2 & 0 \\ 0 & 3 \end{bmatrix}\begin{bmatrix} x_1 \\ x_2 \end{bmatrix}$. The transformation T_2 doubles the width and triples the height of the square [see Figure 5.1(c)].

(c) The rotational transformation R_θ rotates any vector $\mathbf{u}$ in $\mathbf{R}^2$ θ degrees in the counterclockwise direction around the origin. The length of $\mathbf{u}$ is unchanged. See Figure 5.1(d), in which $\theta = 45°$. Geometrically, it is clear that if we form the linear combination vector $\mathbf{z} = r\mathbf{u} + s\mathbf{v}$ and rotate it θ degrees to get $\mathbf{z}'$, then $\mathbf{z}'$ is the same vector we would get by rotating $\mathbf{u}$ and $\mathbf{v}$ each θ degrees to get $\mathbf{u}'$ and $\mathbf{v}'$ and then forming the linear combination $r\mathbf{u}' + s\mathbf{v}'$. Thus R_θ is a linear transformation.

We can also show that this rotation is a linear transformation by producing a matrix $\mathbf{A}$ that performs the transformation R_θ, so that $\mathbf{u}' = R_\theta(\mathbf{u}) = \mathbf{A}\mathbf{u}$. The following is the desired matrix: $\mathbf{A} = \begin{bmatrix} \cos\theta & -\sin\theta \\ \sin\theta & \cos\theta \end{bmatrix}$. For $\theta = 45°$, $\mathbf{A} =$

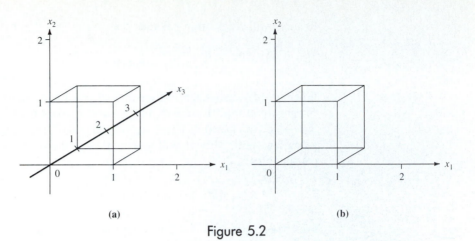

Figure 5.2

$$\begin{bmatrix} 1/\sqrt{2} & -1/\sqrt{2} \\ 1/\sqrt{2} & 1/\sqrt{2} \end{bmatrix} = \begin{bmatrix} 0.707 & -0.707 \\ 0.707 & 0.707 \end{bmatrix}.$$ Recall that in Section 4.5, we stated that such a rotational transformation is equivalent to an orthonormal change of basis. That is, **A** is a transition matrix for an orthonormal change of basis.

(d) Devise a linear transformation T that projects the three-dimensional unit cube shown in Figure 5.2(a) into the (x_1, x_2)-plane so that the cube looks just the way it is drawn on the (two-dimensional) page of this book in Figure 5.2(a). (Note that we have interchanged the positions of the x_2- and x_3-axes here compared to how they were drawn in Chapter 2.)

In Figure 5.2(a), the x_3-axis is represented as a line at a 30° angle to the x_1-axis. Moreover, distances along the x_3-axis in Figure 5.2(a) are drawn with half the length of distances along the x_1- or x_2-axes. So the mapping we want acts on a point (x_1, x_2, x_3) as follows: The x_3-coordinate should alter the x_1, x_2 coordinates in the direction of a 30° angle above the x_1-axis and the distance of the displacement should be half the value of the x_3-coordinate.

$$\begin{aligned} T: x_1' &= x_1 + \tfrac{1}{2}(\cos 30°)x_3 = x_1 + 0.433x_3 \\ x_2' &= x_2 + \tfrac{1}{2}(\sin 30°)x_3 = x_2 + 0.25x_3 \end{aligned} \tag{3}$$

[see Figure 5.2(b)]. ■

**EXAMPLE 3.
Linear
Transformations
in Non-Euclidean
Vector Spaces**

(a) *Reflection transformation R.* In the vector space F of all (real-valued) functions on the real line, define the reflection transformation $R: F \to F$ by $R(f(x)) = f(-x)$. For example, $R(2x^2 + 3x) = 2x^2 - 3x$. The reader should check that R satisfies (1).

(b) *Differentiation transformation $\mathscr{D}$.* Consider the vector space C' of functions $f(x)$ with a continuous derivative (C' is a subspace of the vector space C of continuous real-valued functions; recall that verifying C' is a subspace of C requires showing that C' is closed under linear combinations.) Define the differentiation transformation $\mathscr{D}: C' \to C$ by $\mathscr{D}(f(x)) = df(x)/dx$. That is, $\mathscr{D}(f(x))$ is the derivative of $f(x)$.

Readers will recall from calculus that

$$\frac{d}{dx}\{af(x) + bg(x)\} = a\frac{d}{dx}f(x) + b\frac{d}{dx}g(x) \qquad (4)$$

Equation (4) verifies that differentiation is a linear transformation.

(c) *Integration transformation* $\mathcal{I}$. For the vector space C of continuous functions, let $\mathcal{I}: C \rightarrow C$ by $\mathcal{I}(f(x)) = \int f(x)\, dx$, the integral of $f(x)$ (integration actually requires a constant term; here we will assume the constant is 0). Readers will recall from calculus that

$$\int [af(x) + bg(x)]\, dx = a\int f(x)\, dx + b\int g(x)\, dx \qquad (5)$$

Equation (5) verifies that integration is a linear transformation.

(d) In the vector space F of all functions on the real line, define the *shift transformation* $S_a: C \rightarrow C$ by $S_a(f(x)) = f(x + a)$. That is, S_5 shifts the values of a function $f(x)$ 5 units to the left. For example, if $f(x) = x^2 - 2x$, then $S_5(f(x)) = (x + 5)^2 - 2(x + 5) = x^2 + 8x + 15$. The reader should check that the shift operation is indeed a linear transformation. ■

EXAMPLE 4.
Linear
Transformations
of Matrices

(a) For given m and n (possibly $m = n$), the transpose operation $T: \mathcal{M}_{m,n} \rightarrow \mathcal{M}_{n,m}$ defined by $T(\mathbf{A}) = \mathbf{A}^{\mathsf{T}}$ is a linear transformation.

(b) For given m, m', n, and n', where $m' \leq m$ and $n' \leq n$, the reduction operation $T: \mathcal{M}_{m,n} \rightarrow \mathcal{M}_{m',n'}$ defined by $T(\mathbf{A}) = \mathbf{A}'$, where $\mathbf{A}'$ is the $m' \times n'$ matrix formed by the first m' rows and n' columns of $\mathbf{A}$, is a linear transformation. ■

EXAMPLE 5.
Linear
Transformations
Between Vector
Spaces

(a) The mapping $T_{\mathbf{a}}: \mathbf{R}^n \rightarrow \mathbf{R}$ defined by $T_{\mathbf{a}}(\mathbf{v}) = \mathbf{a} \cdot \mathbf{v}$, for a given n-vector $\mathbf{a}$, is a linear transformation.

(b) The mapping $T: \mathcal{M}_{n,n} \rightarrow \mathbf{R}$ defined by $T(\mathbf{M}) = m_{1,1} + m_{2,2} + \cdots + m_{n,n}$ (the sum of the main diagonal entries is called the trace of a square matrix) is a linear transformation.

(c) The mapping $T_2: \mathcal{C} \rightarrow \mathbf{R}$ defined by $T_2(f(x)) = f(2)$ is a linear transformation.

(d) The mapping $T: \mathcal{P}_3 \rightarrow \mathbf{R}^4$ that is defined by $T(a_3x^3 + a_2x^2 + a_1x + a_0) = [a_3, a_2, a_1, a_0]$ is a linear transformation.

EXAMPLE 6.
Special Linear
Transformations

There are two special linear transformations defined in any vector space V.

(a) The identity transformation $I: V \rightarrow V$ is defined as $I(\mathbf{v}) = \mathbf{v}$, for all $\mathbf{v}$ in V. For Euclidean vector spaces, $I(\mathbf{v}) = \mathbf{I}\mathbf{v}$, where $\mathbf{I}$ is the identity matrix.

(b) The zero transformation $O: V \rightarrow \mathbf{0}$ is defined as $O(\mathbf{v}) = \mathbf{0}$, for all $\mathbf{v}$ in V, where $\mathbf{0}$ is the zero vector in V. For Euclidean vector spaces, $O(\mathbf{v}) = \mathbf{O}\mathbf{v}$, where $\mathbf{O}$ is the matrix of all 0's. ■

EXAMPLE 7.
Mappings That
Are Not Linear
Transformations

(a) The mapping $T: \mathbf{R}^n \to \mathbf{R}^n$ defined by $T(\mathbf{v}) = 2\mathbf{v} + 3$ is not a linear transformation, since $T(\mathbf{u} + \mathbf{v}) = 2(\mathbf{u} + \mathbf{v}) + 3 = 2\mathbf{u} + 2\mathbf{v} + 3$, whereas $T(\mathbf{u}) + T(\mathbf{v}) = (2\mathbf{u} + 3) + (2\mathbf{v} + 3) = 2\mathbf{u} + 2\mathbf{v} + 6$. Thus $T(\mathbf{u} + \mathbf{v}) \neq T(\mathbf{u}) + T(\mathbf{v})$.

(b) The mapping $T: \mathbf{R}^n \to \mathbf{R}$ defined $T(\mathbf{v}) = \mathbf{v} \cdot \mathbf{v}$ is not a linear transformation. As in (a), $T(\mathbf{u} + \mathbf{v}) \neq T(\mathbf{u}) + T(\mathbf{v})$ (details are left to the reader).

(c) For any vector space V, the mapping $T: V \to \mathbf{R}$ defined by $T(\mathbf{v}) = 1$, for all $\mathbf{v}$ in V, is not a linear transformation because $T(\mathbf{u} + \mathbf{v}) = 1$, but $T(\mathbf{u}) + T(\mathbf{v}) = 1 + 1 = 2$. ■

In the case of vector spaces of finite dimension, such as $\mathbf{R}^n$ or $\mathcal{P}_k$ (the space of all polynomials of degree $\leq k$), any linear transformation can be defined in terms of a matrix.

Theorem 2. Any linear transformation $\mathbf{w} = T(\mathbf{v})$ from an n-dimensional vector space V to an m-dimensional vector space W, where n and m are finite, can be represented by matrix multiplication: $\mathbf{w} = \mathbf{A}\mathbf{v}$, where $\mathbf{A}$ is an $m \times n$ matrix.

Proof: Let $B = \{\mathbf{b}_1, \mathbf{b}_2, \ldots, \mathbf{b}_n\}$ be a basis for V and let $D = \{\mathbf{d}_1, \mathbf{d}_2, \ldots, \mathbf{d}_m\}$ be a basis for W. We represent a vector $\mathbf{v}$ in V by its n-vector $[\mathbf{v}]_B$ of coordinates in the basis B for V. That is, if

$$\mathbf{v} = v_1\mathbf{b}_1 + v_2\mathbf{b}_2 + \cdots + v_n\mathbf{b}_n \tag{6}$$

then $\mathbf{v}$ is represented as the n-vector $[\mathbf{v}]_B = [v_1, v_2, \ldots, v_n]$. Similarly, a vector $\mathbf{w}$ in W is represented by $[\mathbf{w}]_C$ its m-vector of coordinates in the basis D for W. The linear transformation T now can be viewed as mapping an n-vector into an m-vector.

We define an $m \times n$ matrix $\mathbf{A}$ as follows. The jth column $\mathbf{a}_j$ of $\mathbf{A}$ equals the m-vector $[T(\mathbf{b}_j)]_D$ in W, the image of $\mathbf{b}_j$ under T.

$$\mathbf{a}_j = [T(\mathbf{b}_j)]_D \tag{7}$$

Then by the linearity of linear transformations [equation (1a)] and by (7), we have

$$\begin{aligned}
[T(\mathbf{v})]_D &= [T(v_1\mathbf{b}_1 + v_2\mathbf{b}_2 + \cdots + v_n\mathbf{b}_n)]_D \\
&= v_1[T(\mathbf{b}_1)]_D + v_2[T(\mathbf{b}_2)]_D + \cdots + v_n[T(\mathbf{b}_n)]_D \\
&= v_1\mathbf{a}_1 + v_2\mathbf{a}_2 + \cdots + v_n\mathbf{a}_n = \mathbf{A}\mathbf{v}
\end{aligned} \tag{8}$$

The linear combination of the columns of $\mathbf{A}$ in the third line of (8) is exactly the definition of $\mathbf{A}\mathbf{v}$. ■

Note that Theorem 2 is the linear transformation counterpart to Theorem 5 in Section 4.3 that showed all n-dimensional vector spaces with real scalars are isomorphic to $\mathbf{R}^n$. That result was proved by using coordinate vectors with respect to a basis, as in the proof of Theorem 2.

EXAMPLE 8.
Linear
Transformations
Represented by a
Matrix

Consider the four-dimensional vector space $\mathcal{P}_3$ of polynomials of degree ≤ 3. The natural basis of four vectors for $\mathcal{P}_3$ is $B = \{x^3, x^2, x, 1\}$. Let us express two of the linear transformations in Example 3 in terms of matrices, using the construction in the proof of Theorem 2.

(a) The differentiation transformation $\mathcal{D}: \mathcal{P}_3 \to \mathcal{P}_3$ [see Example 3(b)]. Following the proof of Theorem 2, we want to express $\mathcal{D}(x^j)$, for each j, as a "vector" in terms of the basis B. The vector is $\mathcal{D}(x^j) = jx^{j-1}$. For $j = 2$, we have $\mathcal{D}(x^2) = 2x$, or in the basis B, $[2x]_B = [0, 0, 2, 0]$. The desired matrix for the differentiation transformation is

$$\mathbf{D} = \begin{bmatrix} 0 & 0 & 0 & 0 \\ 3 & 0 & 0 & 0 \\ 0 & 2 & 0 & 0 \\ 0 & 0 & 1 & 0 \end{bmatrix} \tag{9}$$

For the polynomial $p(x) = 7x^3 - 4x + 5x + 2$, differentiation yields $21x^2 - 8x + 5$. In vector form,

$$\mathbf{p} = [P(x)]_B = \begin{bmatrix} 7 \\ -4 \\ 5 \\ 2 \end{bmatrix}$$

is mapped by $\mathcal{D}$ to

$$\mathbf{Dp} = \begin{bmatrix} 0 & 0 & 0 & 0 \\ 3 & 0 & 0 & 0 \\ 0 & 2 & 0 & 0 \\ 0 & 0 & 1 & 0 \end{bmatrix} \begin{bmatrix} 7 \\ -4 \\ 5 \\ 2 \end{bmatrix} = \begin{bmatrix} 0 \\ 21 \\ -8 \\ 5 \end{bmatrix}$$

(b) The shift transformation S_5, which maps a function $f(x)$ into the function $f(x + 5)$ maps x^3 into $(x + 5)^3 = x^3 + 15x^2 + 75x + 125$, or in vector form,

$$S_5 \quad \text{maps} \quad \begin{bmatrix} 1 \\ 0 \\ 0 \\ 0 \end{bmatrix} \quad \text{to} \quad \begin{bmatrix} 1 \\ 15 \\ 75 \\ 125 \end{bmatrix}$$

Related shifts occur for x^2, x, and 1. [Note that $S_5(1) = 1$, since the constant function has the same value for all x.] The desired matrix thus is

$$\mathbf{S} = \begin{bmatrix} 1 & 0 & 0 & 0 \\ 15 & 1 & 0 & 0 \\ 75 & 10 & 1 & 0 \\ 125 & 25 & 5 & 1 \end{bmatrix} \tag{10}$$

■

An interesting question that mathematicians like to ask is: When is an operation uniquely defined? In our own current setting, when is a linear transformation uniquely defined? Intuitively, if $T_A: \mathbf{R}^n \to \mathbf{R}^n$ is defined by $T_A(\mathbf{v}) = A\mathbf{v}$ and $T_B: \mathbf{R}^n \to \mathbf{R}^n$ is defined by $T_B(\mathbf{v}) = \mathbf{Bv}$, then T_A and T_B are the same linear transformation if and only if $\mathbf{A} = \mathbf{B}$. What about linear transformations in non-Euclidean vector spaces? The following theorem tells when a linear transformation is unique.

Theorem 3. Let V and W be finite-dimensional vector spaces with bases $B = \{\mathbf{b}_1, \mathbf{b}_2, \ldots, \mathbf{b}_n\}$ and $D = \{\mathbf{d}_1, \mathbf{d}_2, \ldots, \mathbf{d}_n\}$, respectively. Let $T: V \rightarrow W$ be a linear transformation with $[T(\mathbf{b}_i)]_D = \mathbf{c}_i$, for $i = 1, 2, \ldots, n$. If $T^*: V \rightarrow W$ is another linear transformation, then $T = T^*$ if and only if $[T^*(\mathbf{b}_i)]_D = \mathbf{c}_i$, for $i = 1, 2, \ldots, n$.

Proof: If $T = T^*$, then obviously $[T(\mathbf{b}_i)]_D = [T^*(\mathbf{b}_i)]_D = \mathbf{c}_i$. Suppose that $[T^*(\mathbf{b}_i)]_D = \mathbf{c}_i$, for $i = 1, 2, \ldots, n$. Then the matrices $\mathbf{A}$ and $\mathbf{A}^*$, in the proof of Theorem 2, for the transformations T and T^*, respectively, will be the same, since the jth column of $\mathbf{A}$ will be $[T(\mathbf{b}_j)]_D$ ($= \mathbf{c}_i$) and the jth column in $\mathbf{A}^*$ will be $[T^*(\mathbf{b}_i)]_D$ ($= \mathbf{c}_i$). Since $\mathbf{A} = \mathbf{A}^*$, then T and T^* are the same linear transformation. ■

One of the important properties of linear transformations is that they map lines into lines. In 2-space, this basic result for matrix transformations was given in Theorem 2 of Section 1.2 and was used in drawing the transformation square in Example 2 above. A line L in a general vector space is defined with the same parametric form used in $\mathbf{R}^n$: $L = \{\mathbf{v} + r\mathbf{d} : r \text{ a scalar}\}$, for given vectors $\mathbf{v}$ and $\mathbf{d}$.

Theorem 4. A linear transformation $T: V \rightarrow W$ maps lines into lines.

Proof: Consider the line L in parametric form, $L = \{\mathbf{v} + r\mathbf{d} : r \text{ a scalar}\}$, where $\mathbf{v}$ and $\mathbf{d}$ are vectors in V. Let $\mathbf{w} = T(\mathbf{v})$ and $\mathbf{f} = T(\mathbf{d})$. We need to show that T maps $\mathbf{v} + r\mathbf{d}$ to $\mathbf{w} + r\mathbf{f}$.

$$T(\mathbf{v} + r\mathbf{d}) = T(\mathbf{v}) + rT(\mathbf{d}) = \mathbf{w} + r\mathbf{f} \tag{11}$$

Thus the image under T of the line L is the line $L' = \{\mathbf{w} + r\mathbf{f} : r \text{ a scalar}\}$. ■

Most of the theory of matrices extends to linear transformations of functions. For example, we can talk about inverse transformations T^{-1}, for which $T^{-1}T(\mathbf{v}) = \mathbf{v}$, and about eigenfunctions $u(x)$ of a linear transformation, for which $T(u(x)) = \lambda u(x)$.

EXAMPLE 9.
Inverse
Transformations
of Linear
Transformations

(a) For the reflection transformation $R(f(x)) = f(-x)$, the inverse R^{-1} is simply the reflection transformation itself. So $R^{-1} = R$.

(b) For differentiation, there is no (unique) inverse $\mathscr{D}^{-1}$. If two functions differ by a constant, say, $x^2 + x + 2$ and $x^2 + x + 5$, they have the same derivative, $2x + 1$. So $\mathscr{D}^{-1}(2x + 1)$ cannot be uniquely defined.

(c) For integration, the inverse $\mathscr{I}^{-1}$ is differentiation: $\mathscr{I}^{-1} = \mathscr{D}$. That is,

$$\frac{d}{dx}\left\{\int f(x)\ dx\right\} = f(x)$$

This is the fundamental theorem of calculus!

(d) For the shift transformation $S_a(f(x)) = f(x + a)$, the inverse transformation S^{-1} is S_{-a}, since $S_{-a}[S_a(f(x))] = f(x + a - a) = f(x)$.

(e) For the transpose operation T on square matrices, the transpose is its own inverse. ■

Note that if a linear transformation is defined on a finite-dimensional vector space, then Theorem 2 applies and the transformation can be represented as a matrix. The transformation will be invertible if and only if the matrix is invertible. The inverse transformation, if it exists, can be found by inverting its matrix. See the exercises for examples of matrix inverses associated with inverse transformations.

EXAMPLE 10.
Eigenfunctions of
Linear
Transformations

(a) For the reflection transformation $R(f(x)) = f(-x)$, an eigenfunction $u(x)$ associated with $\lambda = 1$ is any symmetric function $u(x)$, that is, $u(x) = u(-x)$.

(b) For differentiation, $e^{\lambda x}$ is the eigenfunction of D associated with eigenvalue $\lambda = k$, since $\dfrac{d}{dx}e^{kx} = ke^{kx}$. (*Note*: This property of e^{kx} is why e is called the "natural" base for the exponential and logarithmic functions.)

(c) For integration, $e^{(1/\lambda)x}$ is the eigenfunction of $\mathcal{I}$ associated with eigenvalue $\lambda = k$, since $\int e^{(1/\lambda)x}\,dx = \lambda e^{(1/\lambda)x}$.

(d) For the shift transformation $S_a(f(x)) = f(x + a)$, an eigenfunction $u(x)$ must have the property that $u(x + a) = u(x)$. For example, when $a = 2\pi$, the trigonometric functions, such as $\sin x$ and $\cos x$, are eigenfunctions of $S_{2\pi}$, with eigenvalue 1.

(e) For the transpose operation on matrices, the eigenvectors are symmetric matrices with associated eigenvalue of 1. ■

There is one very important extension of differentiation that bears special mention, namely, differential equations. The differential equation

$$y''(x) - 2y'(x) + 3y(x) = f(x) \tag{12}$$

can be viewed as a linear transformation DE of $y(x)$ to $f(x)$, that is, $DE(y(x)) = f(x)$. It is left to the reader to check that DE satisfies the defining property (1) for a linear transformation. Any differential equation whose left side is a linear combination of derivatives [and $y(x)$] will be a linear transformation. More generally, any linear combination of linear transformations is again a linear transformation. Indeed, one can define vector spaces of linear transformations as a natural extension of the vector spaces of matrices introduced in Chapter 4. The advanced theory of differential equations is based heavily on eigenfunctions for these linear transformations.

Section 5.1 Exercises

1. Construct linear transformations to do the following to the square in Figure 5.1(a). (Give the matrix of the transformation.)
   (a) Rotate the square 180° counterclockwise around the origin (in the plane).
   (b) Make the height 4 and width 5.

(c) Make the horizontal lines of the square slant upward at a 45° angle (width unchanged).

(d) Reflect the square about the x_2-axis.

2. Compute the new coordinates of the corners of the square in Figure 5.1(a) for each of the transformations in Exercise 1.

3. If T is the triangle with corners at $(-1, 1)$, $(1, 1)$, $(1, -1)$, then draw T after it is transformed by each of the transformations in Exercise 1.

4. Construct a linear transformation that rotates the plane the specified angle in the counterclockwise direction and apply this transformation to the triangle in Exercise 3.

(a) 45° (b) 135° (c) 60°

(d) 45° plus a reflection (so that the image of the positive x_1-axis is 90° counterclockwise of the image of the positive x_2-axis)

5. Which of the following transformations are linear transformations? Give the matrix for those that are linear transformations.

(a) $T\left(\begin{bmatrix} x \\ y \end{bmatrix}\right) = \begin{bmatrix} 2x \\ 3y \end{bmatrix}$ (b) $T\left(\begin{bmatrix} x \\ y \end{bmatrix}\right) = \begin{bmatrix} 2x + 2 \\ 3x + y + 1 \end{bmatrix}$

(c) $T\left(\begin{bmatrix} x \\ y \end{bmatrix}\right) = \begin{bmatrix} 2y \\ x \end{bmatrix}$ (d) $T\left(\begin{bmatrix} x \\ y \end{bmatrix}\right) = \begin{bmatrix} 1 \\ 1 \end{bmatrix}$

6. Show that the mapping in Example 5(a) is a linear transformation.

7. Which of the following mappings from R^2 to R are linear transformations?

(a) $T(\mathbf{v}) = v_1$

(b) $T(\mathbf{v}) = 2v_1 - 3v_2$

(c) $T(\mathbf{v}) = |\mathbf{v}|_s$

(d) $T(\mathbf{v}) = |\mathbf{v}|_m$

(e) $T(\mathbf{v}) = v_1 v_2$

8. Construct linear transformations to do the following to the square in Figure 5.1(a) and plot the square after the transformation is performed.

(a) Double the width of the square (double x_1 coordinates) and rotate it 90° (around the origin).

(b) Reflect the square about the x_1-axis and then reflect about the x_2-axis.

(c) Reflect the square about the x_2-axis and then double the x_1-coordinate and then rotate 45°.

9. This exercise verifies which types of linear transformations are commutative. Let

T_a double the x_1-coordinate: $x_1' = 2x_1$, $x_2' = x_2$.
T_b double the x_2-coordinate: $x_1' = x_1$, $x_2' = 2x_2$.
T_c reflect about the x_1-axis: $x_1' = -x_1$, $x_2' = x_2$.
T_d reflect about the x_2-axis: $x_1' = x_1$, $x_2' = -x_2$.
T_e rotate plane 45° counterclockwise.

We write $T_x T_y$ to denote T_x followed by T_y. Note that $T_x T_y(\mathbf{v}) = T_y(T_x(\mathbf{v}))$.

(a) Determine the matrix for $T_a T_b$ and $T_b T_a$—do T_a and T_b commute?

(b) Determine the matrix for $T_a T_c$ and $T_c T_a$—do T_a and T_c commute?

(c) Determine the matrix for $T_a T_e$ and $T_e T_a$—do T_a and T_e commute?

(d) Determine the matrix for $T_c T_d$ and $T_d T_c$—do T_c and T_d commute?

(e) Determine the matrix for $T_c T_e$ and $T_e T_c$—do T_c and T_e commute?

10. Find a linear transformation that is a projection of R^2 onto the following lines.
(a) $x_2 = x_1$ (b) $x_2 = 3x_1$ (c) $x_2 = -2x_1$

11. Find a linear transformation that is a projection of R^3 onto the following planes. (Hint: Tricky; use pseudoinverse.)
(a) $x_1 + x_2 + x_3 = 0$ (b) $3x_1 - x_2 + 2x_3 = 0$

12. (a) Verify that $\mathbf{x}' = T(\mathbf{x})$: $x_1' = 2x_1 - x_2$, $x_2' = 2x_1 - x_2$ maps any vector $\mathbf{x} = (x_1, x_2)$ onto the line $x_2' = x_1'$.
(b) Show that the line segment connecting the vector $\mathbf{x}$ with the vector $T(\mathbf{x})$ has a slope 2.
(c) Construct a linear transformation T' that maps any vector $\mathbf{x} = (x_1, x_2)$ onto the line $x_2' = 2x_1'$ so that the line segment between $\mathbf{x}$ and $T'(\mathbf{x})$ has a slope 1.

13. Find a linear transformation to rotate three-dimensional space $45°$ around the x_3-axis. (The x_3-axis stays fixed; the x_1-axis moves $45°$ counterclockwise around the x_3-axis.)

14. Which of the following mappings are linear transformations on $\mathcal{P}_k$, the vector space of polynomials? [Here $p(x) = a_k x^k + a_{k-1} x^{k-1} + \cdots + a_1 x + a_0$.]
(a) $T(p(x)) = xp(x)$ (b) $T(p(x)) = p(x)^2$ (c) $T(p(x)) = a_k x^k$
(d) $T(p(x)) = a_0 x^k + a_1 x^{k-1} + \cdots + a_{k-1} x + a_k$

15. (a) Verify that the reflection transformation $R(f(x)) = f(-x)$ in Example 3(a) satisfies the property of a linear transformation.
(b) In $\mathcal{P}_3$, give the matrix for the reflection transformation.

16. (a) Verify that the mapping in Example 4(a) is a linear transformation.
(b) Verify that the mapping in Example 4(b) is a linear transformation.

17. Verify that the mapping in Example 5(b) is a linear transformation.

18. (a) Verify that the mapping in Example 5(c) is a linear transformation.
(b) Verify that the mapping in Example 5(d) is a linear transformation.

19. The integration transformation $\mathcal{I}$ in Example 3(c) maps $\mathcal{P}_k$ into $\mathcal{P}_{k+1}$. Give the matrix for the integration transformation from $\mathcal{P}_3$ into $\mathcal{P}_4$ (assuming the constant term is 0).

20. (a) Consider the transformation $T_{a,b}$: $\mathcal{P} \to \mathcal{P}$ induced by multiplying a polynomial $p(x)$ times the linear function $ax + b$. Show that $T_{a,b}$ is a linear transformation.
(b) For finite-dimensional spaces of polynomials, we have $T_{a,b}$: $\mathcal{P}_k \to \mathcal{P}_{k+1}$. Give the matrix for $T_{2,5}$: $\mathcal{P}_3 \to \mathcal{P}_4$.
(c) Give the 5×4 matrix for the more general transformation $T_{a,b}$: $\mathcal{P}_3 \to \mathcal{P}_4$.
(d) Give the 5×3 matrix for the related transformation $T_{a,b,c}$: $\mathcal{P}_2 \to \mathcal{P}_4$ of multiplying a polynomial by the quadratic $ax^2 + bx + c$.

21. Give the matrix for the linear transformation from $\mathcal{P}_3$ to $\mathcal{P}_3$ of taking the second derivative of a polynomial $p(x)$ in $\mathcal{P}_3$. How is this matrix related to the matrix in (9) for the first derivative?

22. Show that T: $R^n \to R$ defined by $T(\mathbf{v}) = \mathbf{v} \cdot \mathbf{v}$ is not a linear transformation.

23. Which of the following mappings T: $\mathcal{P}_3 \to R$ are linear transformations? Give the matrix for those that are linear transformations (in terms of the standard basis for $\mathcal{P}_3$). [Here $p(x) = a_3 x^3 + a_2 x^2 + a_1 x + a_0$.]
(a) $T(p(x)) = a_3$ (b) $T(p(x)) = p(1) - p(0)$ (c) $T(p(x)) = \int_0^1 p(x)\, dx$

24. Which of the following mappings are linear transformations on $M_{2,2}$? Give the matrix for those that are linear transformations (in terms of the standard basis for $M_{2,2}$). Let **A** be a given invertible 2×2 matrix.
(a) $T(\mathbf{M}) = 3\mathbf{M}$ (b) $T(\mathbf{M}) = \mathbf{M} + \mathbf{A}$ (c) $T(\mathbf{M}) = \mathbf{AM}$

25. Which of the following mappings $T: M_{2,2} \to R$ are linear transformations? Give the matrix for those that are linear transformations (in terms of the standard basis for $M_{2,2}$).
(a) $T(\mathbf{M}) = 2m_{11}$ (b) $T(\mathbf{M}) = 2m_{12} - m_{21}$ (c) $T(\mathbf{M}) = \det(\mathbf{M})$

(d) $T(\mathbf{M}) = [1, \ 1]\mathbf{M}\begin{bmatrix} 1 \\ 1 \end{bmatrix}$

26. Verify that the identity transformation in Example 6(a) is a linear transformation.

27. Verify that the zero transformation in Example 6(b) is a linear transformation.

28. The collection $\mathcal{S}$ of all infinite sequences of real numbers $\{r_1, r_2, r_3, \ldots\}$ forms a vector space. Are the following mappings from $\mathcal{S}$ to $\mathcal{S}$ linear transformations?
(a) Double each element in a sequence.
(b) Add 2 to each element in a sequence.
(c) Shift all elements one position to the left, that is, $\{r_1, r_2, r_3, \ldots\} \to \{r_2, r_3, r_4, \ldots\}$.

29. Find the inverse transformation, if possible, of each linear transformation in Exercise 1; give the matrix of the inverse transformation.

30. Find the inverse transformation of the linear transformation in Exercise 24(a) and (c); give the matrix of the inverse transformation.

31. Find the eigenvalue(s) and eigenvectors for the linear transformation in Exercise 14(c).

32. Find the eigenvalue(s) and eigenvectors for the linear transformation in Exercise 24(a).

33. (a) Show that the left side of the differential equation in (12), $y''(x) - 2y'(x) + 3y(x)$, is a linear transformation. [Hint: You need to use the identity transformation I in Example 6(a).]
(b) Give the matrix for the linear transformation of $y''(x) - 2y'(x) + 3y(x)$ on $\mathcal{P}_3$.

34. Consider the vector space V with a basis consisting of the functions, $e^x, xe^x, x^2e^x, x^3e^x$. Give the 4×4 matrix for the linear transformation of differentiation on V.

35. Show that if T and T' are linear transformations from V to W, any linear combination of T and T' is a linear transformation from V to W. (The collection of all linear transformations from V to W can be shown to form a vector space.)

36. A linear transformation $T: V \to W$ is called an *isomorphism* if $R(T) = W$ (T maps V onto all elements of W) and $\mathbf{K}(T) = \mathbf{0}$ (implying that distinct elements of V are mapped by T to distinct elements of W). Show that T is an isomorphism if and only if there is a map S: $W \to V$ such that ST is the identity transformation on V and TS is the identity transformation on W.

5.2 Kernel and Range of a Linear Transformation

In this section we look at two fundamental vector spaces associated with any linear transformation $\mathbf{w} = T(\mathbf{v})$. These two vector spaces are especially helpful in analyzing solutions to a system $\mathbf{Ax} = \mathbf{b}$ of linear equations.

Fundamental Vector Spaces of a Linear Transformation

Let V and W be vector spaces (possibly $V = W$).

1. The **kernel** $K(T)$ of a linear transformation $T: V \to W$ is the vector space of vectors $\mathbf{v}$ in V that are mapped to the zero vector $\mathbf{0}$ in W. That is, $K(T) = \{\mathbf{v} \in V: T(\mathbf{v}) = \mathbf{0}\}$. The dimension of $K(T)$ is called the **nullity** of T.
2. The **range** $R(T)$ of a linear transformation $T: V \to W$ is the set of vectors $\mathbf{w}$ in W such that $T(\mathbf{v}) = \mathbf{w}$ for some $\mathbf{v}$ in V. That is, $R(T) = \{\mathbf{w} \in W: T(\mathbf{v}) = \mathbf{w}, \text{ for some } \mathbf{v} \in V\}$. The dimension of $R(T)$ is called the **rank** of T.

Another name that is sometimes used for $K(T)$ is the **null space** of T.

When the linear transformation is generated by a matrix $\mathbf{A}$, that is, $\mathbf{w} = T(\mathbf{v}) = \mathbf{A}\mathbf{v}$, it is common to speak of the *kernel of* $\mathbf{A}$ and the *range of* $\mathbf{A}$. Conversely, recall from Theorem 2 of Section 5.1 that any linear transformation $T: V \to W$ between finite-dimensional vector spaces V and W can be represented as a matrix mapping $T(\mathbf{v}) = \mathbf{A}\mathbf{v}$.

For convenience, we restate the definition of the kernel and range for a matrix.

Fundamental Vector Spaces of a Matrix

1. The **kernel** $K(\mathbf{A})$ of a matrix $\mathbf{A}$ is the set of vectors $\mathbf{x}$ that are solutions to the homogeneous system of equations $\mathbf{A}\mathbf{x} = \mathbf{0}$.
2. The **range** $R(\mathbf{A})$ of a matrix $\mathbf{A}$ is the set of vectors $\mathbf{b}$ such that $\mathbf{A}\mathbf{x} = \mathbf{b}$ has a solution.

Let us show that $K(T)$ and $R(T)$ do indeed satisfy the definition of a vector space.

If T is a linear transformation $T: V \to W$ from vector space V to vector space W, $K(T)$ will be a collection of vectors in V and $R(T)$ will be a collection of vectors in W. (We shall assume that the scalars for V and W are the real numbers.) Recall from Theorem 2 of Section 4.1 that if S is a collection of vectors in a vector space V, then S is a vector space (a vector subspace of V) if

$$\text{for any vectors } \mathbf{u}, \mathbf{v} \in S \text{ and any scalars } r, s, \text{ then } r\mathbf{u} + s\mathbf{v} \text{ is in } S \qquad (1)$$

We shall show that $K(T)$ satisfies (1). If $T(\mathbf{u}) = \mathbf{0}$ and $T(\mathbf{v}) = \mathbf{0}$, then by the linearity of linear transformations, we have

$$T(r\mathbf{u} + s\mathbf{v}) = rT(\mathbf{u}) + sT(\mathbf{v}) = r(\mathbf{0}) + s(\mathbf{0}) = \mathbf{0} \qquad (2)$$

Thus $K(T)$ is a vector space.

Next we show that $R(T)$ satisfies (1). Let $\mathbf{u}$ and $\mathbf{v}$ be two vectors in $R(T)$. Suppose that $T(\mathbf{x}) = \mathbf{u}$ and $T(\mathbf{y}) = \mathbf{v}$. We need to prove that $r\mathbf{u} + s\mathbf{v}$ is in $R(T)$. That is, we must find a vector $\mathbf{z}$ such that $T(\mathbf{z}) = r\mathbf{u} + s\mathbf{v}$. We claim that $\mathbf{z} = r\mathbf{x} + s\mathbf{y}$ is such a vector:

$$T(\mathbf{z}) = T(r\mathbf{x} + s\mathbf{y}) = rT(\mathbf{x}) + sT(\mathbf{y}) = r\mathbf{u} + s\mathbf{v} \tag{3}$$

Thus $R(T)$ is a vector space.

In the case of matrices, the range of $\mathbf{A}$ is a vector space we encountered in Section 4.4. In the beginning of Section 4.4 we noted that $\mathbf{Ax} = \mathbf{b}$ has a solution if and only if $\mathbf{b}$ is in the column space $\text{Col}(\mathbf{A})$, that is, $\mathbf{b}$ can be expressed as a linear combination of the columns $\mathbf{a}_i$ of $\mathbf{A}$:

$$\mathbf{Ax} = \mathbf{b} \;\leftrightarrow\; x_1\mathbf{a}_1 + x_2\mathbf{a}_2 + \cdots + x_n\mathbf{a}_n = \mathbf{b} \tag{4}$$

Thus a vector $\mathbf{b}$ is in $R(\mathbf{A})$ if and only if $\mathbf{b}$ is in the column space $\text{Col}(\mathbf{A})$.

Theorem 1. The range $R(\mathbf{A})$ equals the column space $\text{Col}(\mathbf{A})$ of $\mathbf{A}$.

Note that if $\mathbf{b}$ is not in the range of $\mathbf{A}$, then $\mathbf{Ax} = \mathbf{b}$ has no solution.

Recall that $\text{rank}(\mathbf{A})$ of a matrix $\mathbf{A}$ was shown to equal the dimension of $\text{Col}(\mathbf{A})$ (see Theorem 3 of Section 4.4). By Theorem 1, $\text{rank}(\mathbf{A})$ equals the dimension of $R(\mathbf{A})$. Thus $\text{rank}(\mathbf{A}) = \text{rank}(T)$, where T is the associated linear transformation $T(\mathbf{v}) = \mathbf{Av}$.

Suppose that $\mathbf{A}$ is an $n \times n$ matrix for which the system $\mathbf{Ax} = \mathbf{b}$ has a unique solution for every $\mathbf{b}$ (so $\mathbf{A}$ has an inverse). Then $R(\mathbf{A})$ equals $\mathbf{R}^n$, that is, $R(\mathbf{A})$ contains all possible n-vectors. And $K(\mathbf{A})$ is just the zero vector $\mathbf{0}$, since $\mathbf{0}$ is always a solution to $\mathbf{Ax} = \mathbf{0}$ and by assumption there can only be one solution to $\mathbf{Ax} = \mathbf{0}$. Building on Theorem 4 in Section 4.4, we have:

Theorem 2. Let $\mathbf{A}$ be an $n \times n$ matrix. Then the following statements are equivalent.
  (i) For any n-vector $\mathbf{b}$, the system $\mathbf{Ax} = \mathbf{b}$ has one and only one solution.
 (ii) $\mathbf{A}$ is invertible.
(iii) $R(\mathbf{A}) = \mathbf{R}^n$, or equivalently, $\text{rank}(\mathbf{A}) = n$.
(iv) $K(\mathbf{A}) = \{\mathbf{0}\}$, or equivalently, $\dim(K(\mathbf{A})) = 0$ [or nullity$(T) = 0$ for associated transformation T].

**EXAMPLE 1.
Kernel and Range
of a Projection
Mapping**

Let us determine the range $R(\mathbf{A})$ and kernel $K(\mathbf{A})$ of the matrix $\mathbf{A} = \begin{bmatrix} \frac{1}{2} & \frac{1}{2} \\ \frac{1}{2} & \frac{1}{2} \end{bmatrix}$.

$$\begin{aligned} \tfrac{1}{2}x_1 + \tfrac{1}{2}x_2 &= b_1 \\ \tfrac{1}{2}x_1 + \tfrac{1}{2}x_2 &= b_2 \end{aligned} \tag{5}$$

The linear transformation $T_\mathbf{A}: \mathbf{x} \rightarrow \mathbf{b} = \mathbf{Ax}$ projects all 2-vectors onto the line $b_1 = b_2$. That is, $T_\mathbf{A}$ maps any $[x_1, x_2]$ to the point $[(x_1 + x_2)/2, (x_1 + x_2)/2]$. So the range is

$$R(\mathbf{A}) = \left\{ r\begin{bmatrix} 1 \\ 1 \end{bmatrix} : r \text{ any scalar} \right\}$$

To find $K(\mathbf{A})$, we solve (5) by Gauss–Jordan elimination when $b_1 = b_2 = 0$. We obtain

$$x_1 + x_2 = 0 \tag{6}$$
$$0 = 0$$

The first equation in (6) defines vectors $\mathbf{x}$ in $K(\mathbf{A})$. It may be rewritten $x_1 = -x_2$. Thus the kernel is

$$K(\mathbf{A}) = \left\{ r \begin{bmatrix} 1 \\ -1 \end{bmatrix} : r \text{ any scalar} \right\}$$

Before leaving this example, let us return to the range $R(\mathbf{A})$ and derive the equation $b_1 = b_2$ defining range vectors directly from $\mathbf{A}$. We rewrite $\mathbf{Ax} = \mathbf{b}$ in (6) as $\mathbf{Ax} = \mathbf{Ib}$:

$$
\begin{aligned}
\tfrac{1}{2}x_1 + \tfrac{1}{2}x_2 &= 1b_1 + 0b_2 \\
\tfrac{1}{2}x_1 + \tfrac{1}{2}x_2 &= 0b_1 + 1b_2
\end{aligned}
\qquad
\begin{bmatrix} \tfrac{1}{2} & \tfrac{1}{2} & 1 & 0 \\ \tfrac{1}{2} & \tfrac{1}{2} & 0 & 1 \end{bmatrix}
\tag{7}
$$

Now let us perform Gauss–Jordan elimination, trying to convert $\mathbf{Ax} = \mathbf{Ib}$ into the form $\mathbf{Ix} = \mathbf{A}^{-1}\mathbf{b}$, as we do when computing the inverse of $\mathbf{A}$. The first pivot step yields

$$
\begin{aligned}
x_1 + x_2 &= 2b_1 \\
0 &= -1b_1 + 1b_2
\end{aligned}
\qquad
\begin{bmatrix} 1 & 1 & 2 & 0 \\ 0 & 0 & -1 & 1 \end{bmatrix}
\tag{8}
$$

The second equation in (8), which can be rewritten $b_1 = b_2$, gives the range constraint. ■

**EXAMPLE 2.
Kernel and Range
of a Refinery
Production
Problem**

Consider the following variation on our refinery production problem with the third refinery removed.

$$
\begin{aligned}
\text{Heating oil:} \quad & 4x_1 + 2x_2 = b_1 \\
\text{Diesel oil:} \quad & 2x_1 + 5x_2 = b_2 \\
\text{Gasoline:} \quad & 1x_1 + 2.5x_2 = b_3
\end{aligned}
\tag{9}
$$

We first seek an equation describing the possible output vectors $\mathbf{b}$. That is, if $\mathbf{A}$ is the coefficient matrix of (9), we seek a defining constraint on the range $R(\mathbf{A})$. Note that the dimension of $R(\mathbf{A})$, which is the same as the dimension of $\text{Col}(\mathbf{A})$ by Theorem 1, equals 2, since the columns of $\mathbf{A}$ are not multiples of one another [the two columns form a basis for $R(\mathbf{A})$].

We use the same technique introduced in Example 1. We write the system $\mathbf{Ax} = \mathbf{b}$ in (9) as $\mathbf{Ax} = \mathbf{Ib}$ and perform Gauss–Jordan elimination on the augmented matrix $[\mathbf{A} \quad \mathbf{I}]$:

$$\begin{bmatrix} 4 & 2 & | & 1 & 0 & 0 \\ 2 & 5 & | & 0 & 1 & 0 \\ 1 & 2.5 & | & 0 & 0 & 1 \end{bmatrix} \longrightarrow \begin{bmatrix} 1 & 0.5 & | & \frac{1}{4} & 0 & 0 \\ 0 & 4 & | & -\frac{1}{2} & 1 & 0 \\ 0 & 2 & | & -\frac{1}{4} & 0 & 1 \end{bmatrix} \longrightarrow \begin{bmatrix} 1 & 0 & | & \frac{5}{16} & -\frac{1}{8} & 0 \\ 0 & 1 & | & -\frac{1}{8} & \frac{1}{4} & 0 \\ 0 & 0 & | & 0 & -\frac{1}{2} & 1 \end{bmatrix}$$

(10)

The reduced augmented matrix in (10) corresponds to the system of equations

$$
\begin{aligned}
x_1 &= \tfrac{5}{16}b_1 - \tfrac{1}{8}b_2 \\
x_2 &= -\tfrac{1}{8}b_1 + \tfrac{1}{4}b_2 \\
0 &= -\tfrac{1}{2}b_2 + b_3
\end{aligned}
$$

(11)

The last equation in (11) can be rewritten as

$$\tfrac{1}{2}b_2 = b_3 \quad \text{or} \quad b_2 = 2b_3 \tag{12}$$

This is the range constraint we were looking for. In terms of refinery production, it means we can achieve any production vector **b** in which the diesel oil output (b_2) equals twice the gasoline output (b_3).

The kernel $K(\mathbf{A})$ of **A** in this example is $K(\mathbf{A}) = \{\mathbf{0}\}$, since setting $b_1 = b_2 = b_3 = 0$ in (11) yields the unique solution $x_1 = x_2 = 0$.

Suppose that heating oil (b_1) and diesel oil (b_2) are the outputs of primary interest and we want $b_1 = 400$ and $b_2 = 600$. Then we pick b_3 using (12) to get a vector in the range. We set $b_3 = \tfrac{1}{2}b_2 = \tfrac{1}{2}(600) = 300$. We can determine the appropriate production levels x_1, x_2 from the first two equations in (11) which only involve b_1 and b_2:

$$
\begin{aligned}
x_1 &= \tfrac{5}{16}(400) - \tfrac{1}{8}(600) = 125 - 75 = 50 \\
x_2 &= -\tfrac{1}{8}(400) + \tfrac{1}{4}(600) = -50 + 150 = 100
\end{aligned}
$$ ■

In function spaces, the range of a linear transformation is often difficult to define and requires advanced knowledge of real analysis. However, the kernel is frequently easy to determine, as illustrated by the next example.

EXAMPLE 3.
Kernel of the
Differentiation
Transformation

In Example 3(b) of Section 5.1 we noted that the operation $\mathscr{D}(f(x))$ of taking the derivative of a function $f(x)$ was a linear transformation which is defined on the vector space of all continuously differentiable functions. Let us determine the kernel $K(\mathscr{D})$ of the differentiation transformation.

The kernel $K(\mathscr{D})$ is the set of functions $f(x)$ for which $\mathscr{D}(f(x)) = 0$; that is, $\dfrac{d}{dx}f(x) = 0$. These are just the constant functions, $f(x) = c$, for some constant c. So

$$K(\mathscr{D}) = \{\text{all constant functions}\}$$ ■

We now develop some theory about the kernel of a linear transformation. The following two theorems are particularly important for the kernel $K(\mathbf{A})$ of a matrix, and its relation of solutions to a system $\mathbf{Ax} = \mathbf{b}$ of linear equations. For this reason we state these theorems for matrices rather than for general linear transformations. (All the results are easily extended to general linear transformations; verifying this is left to the exercises.)

Theorem 3. Let $\mathbf{A}$ be any $m \times n$ matrix.
  (i) If $K(\mathbf{A})$ contains at least one nonzero vector $\mathbf{x}^\circ$, then $K(\mathbf{A})$ contains an infinite number of vectors.
  (ii) If $\mathbf{x}^\circ \in K(\mathbf{A})$ and $\mathbf{x}^*$ is a solution to $\mathbf{Ax} = \mathbf{b}$, then $\mathbf{x}^* + \mathbf{x}^\circ$ is also a solution to $\mathbf{Ax} = \mathbf{b}$.
  (iii) If $\mathbf{x}'$, $\mathbf{x}''$ are two different solutions to $\mathbf{Ax} = \mathbf{b}$, for some given $\mathbf{b}$, their difference $\mathbf{x}' - \mathbf{x}''$ is a vector in $K(\mathbf{A})$.
  (iv) Given a solution $\mathbf{x}^*$ to $\mathbf{Ax} = \mathbf{b}$, any other solution $\mathbf{x}'$ to this matrix equation can be written as

$$\mathbf{x}' = \mathbf{x}^* + \mathbf{x}^\circ \qquad \text{for some } \mathbf{x}^\circ \in K(\mathbf{A})$$

  (v) If $K(\mathbf{A}) = \{\mathbf{0}\}$, then $\mathbf{Ax} = \mathbf{b}$ has at most one solution.
  Proof: (i) If $\mathbf{x}^\circ \in K(\mathbf{A})$, then $\mathbf{A}(r\mathbf{x}^\circ) = r(\mathbf{Ax}^\circ) = r(\mathbf{0}) = \mathbf{0}$, so $r\mathbf{x}^\circ \in K(\mathbf{A})$, for any scalar r. Thus $K(\mathbf{A})$ is infinite.
  (ii) Let $\mathbf{x}^\circ \in K(\mathbf{A})$, so $\mathbf{Ax}^\circ = \mathbf{0}$. Since $\mathbf{Ax}^* = \mathbf{b}$, then

$$\mathbf{A}(\mathbf{x}^* + \mathbf{x}^\circ) = \mathbf{Ax}^* + \mathbf{Ax}^\circ$$
$$= \mathbf{b} + \mathbf{0} = \mathbf{b}$$

Thus $\mathbf{x}^* + \mathbf{x}^\circ$ is a solution to $\mathbf{Ax} = \mathbf{b}$, as claimed.
  (iii) Given solutions $\mathbf{x}'$, $\mathbf{x}''$, then $\mathbf{A}(\mathbf{x}' - \mathbf{x}'') = \mathbf{Ax}' - \mathbf{Ax}'' = \mathbf{b} - \mathbf{b} = \mathbf{0}$. Thus $\mathbf{x}' - \mathbf{x}'' \in K(\mathbf{A})$.
  (iv) Given solutions $\mathbf{x}^*$, $\mathbf{x}'$, let $\mathbf{x}^\circ = \mathbf{x}' - \mathbf{x}^*$. Then $\mathbf{x}' = \mathbf{x}^* + \mathbf{x}^\circ$, and by part (iii), $\mathbf{x}^\circ \in K(\mathbf{A})$.
  (v) Suppose that one solution $\mathbf{x}^*$ to $\mathbf{Ax} = \mathbf{b}$ is known. By (iv), all solutions $\mathbf{x}'$ of $\mathbf{Ax} = \mathbf{b}$ can be expressed in the form $\mathbf{x}' = \mathbf{x}^* + \mathbf{x}^\circ$, where $\mathbf{x}^\circ \in K(\mathbf{A})$. If $K(\mathbf{A}) = \{\mathbf{0}\}$, then $\mathbf{x}' = \mathbf{x}^* + \mathbf{0}$—there is only the one solution $\mathbf{x}^*$ to $\mathbf{Ax} = \mathbf{b}$. ■

Theorem 4. Let $\mathbf{A}$ be an $m \times n$ matrix. If $\mathbf{Ax} = \mathbf{b}'$ has two solutions for some $\mathbf{b}'$, then for any other $\mathbf{b}$, either $\mathbf{Ax} = \mathbf{b}$ has no solution or an infinite number of solutions.
  Proof: By Theorem 3(iii), the difference of two solutions yields a nonzero vector $\mathbf{x}^\circ$ in $K(\mathbf{A})$. Then, by Theorem 3(i), $K(\mathbf{A})$ contains an infinite number of vectors.
  Suppose that $\mathbf{Ax} = \mathbf{b}$ has a solution $\mathbf{x}^*$. From Theorem 3(ii), $\mathbf{x}^* + \mathbf{x}^\circ$ is also a solution, for any $\mathbf{x}^\circ \in K(\mathbf{A})$. Since we just showed that there are an infinite number of vectors in $K(\mathbf{A})$, it follows that $\mathbf{Ax} = \mathbf{b}$ has an infinite number of solutions. ■

We illustrate with some examples how kernel vectors can be used to find the "right" solution to a system of equations.

EXAMPLE 4.
Multiple Solutions
in an Oil Refinery
Problem

Consider a variation on our oil refinery problem in which the three refineries produce only the first two products.

$$\text{Heating oil: } 4x_1 + 2x_2 + 2x_3 = 600$$
$$\text{Diesel oil: } 2x_1 + 5x_2 + 2x_3 = 800 \tag{13}$$

Suppose that we are given one solution $\mathbf{x}^* = \begin{bmatrix} 50 \\ 100 \\ 100 \end{bmatrix}$. We want to find another

solution with $x_3 = 220$. Let us find the kernel of this coefficient matrix and then, using Theorem 3(iv), add an appropriate kernel vector to $\mathbf{x}^*$ to get a solution with $x_3 = 220$.

The kernel for this system of equations is all solutions to the associated homogeneous system

$$4x_1 + 2x_2 + 2x_3 = 0$$
$$2x_1 + 5x_2 + 2x_3 = 0 \tag{14}$$

Solving with Gauss–Jordan elimination, we obtain

$$\begin{array}{ll} x_1 \quad + \tfrac{3}{8}x_3 = 0 \\ \quad x_2 + \tfrac{1}{4}x_3 = 0 \end{array} \quad \text{or} \quad \begin{bmatrix} 1 & 0 & \tfrac{3}{8} & | & 0 \\ 0 & 1 & \tfrac{1}{4} & | & 0 \end{bmatrix}$$

Thus $x_1 = -\tfrac{3}{8}x_3$ and $x_2 = -\tfrac{1}{4}x_3$. So vectors in the kernel have the form

$$\begin{bmatrix} -\tfrac{3}{8}x_3 \\ -\tfrac{1}{4}x_3 \\ x_3 \end{bmatrix} \quad \text{or} \quad r\begin{bmatrix} -\tfrac{3}{8} \\ -\tfrac{1}{4} \\ 1 \end{bmatrix} \tag{15}$$

Theorem 3(iv) says that any solution $\mathbf{x}'$ to the system (14) can be expressed as $\mathbf{x}' = \mathbf{x}^* + \mathbf{x}^\circ$, where $\mathbf{x}^\circ$ is a kernel vector. In this case $\mathbf{x}^\circ$ is a vector of the form (15). Then

$$\mathbf{x}' = \mathbf{x}^* + r\mathbf{x}^\circ: \quad \begin{bmatrix} x_1' \\ x_2' \\ 220 \end{bmatrix} = \begin{bmatrix} 50 \\ 100 \\ 100 \end{bmatrix} + r\begin{bmatrix} -\tfrac{3}{8} \\ -\tfrac{1}{4} \\ 1 \end{bmatrix} \tag{16}$$

We want $\mathbf{x}'$ to have $x_3' = 220$. Matching the third entry on each side of (16), we have $220 = 100 + r$. So $r = 120$ and the desired solution is

$$\mathbf{x}' = \mathbf{x} + \mathbf{x}^\circ = \begin{bmatrix} 50 \\ 100 \\ 100 \end{bmatrix} + 120\begin{bmatrix} -\tfrac{3}{8} \\ -\tfrac{1}{4} \\ 1 \end{bmatrix} = \begin{bmatrix} 5 \\ 70 \\ 220 \end{bmatrix} \tag{17}$$

■

We next give an example of the generalization of Theorem 3 to general linear transformations.

EXAMPLE 5.
Multiple Solutions
to the Differential
Transformation

In Example 3 we showed that the kernel $K(\mathscr{D})$ of the differential transformation $\mathscr{D}$ is the set of constant functions $f(x) = c$, for some constant c. That is, these are the functions for which $\dfrac{d}{dx} f(x) = 0$. Suppose that we want to find all solutions to the differential equation

$$\frac{d}{dx}(f(x)) = 6x^2 + 4x \qquad [\text{find } f(x) \text{ such that } f'(x) = 6x^2 + 4x] \qquad (18)$$

By Theorem 3(iv), if $f^*(x)$ is one solution to (18), say, $f^*(x) = 2x^3 + 2x^2$, every solution to (18) can be written as the sum of $f^*(x)$ plus a member of $K(\mathscr{D})$ (i.e., plus some constant function). So every solution has the form $2x^3 + 2x^2 + c$. We have just derived a well-known result in calculus about the form of the integral of a function. ■

We now give two applications where kernels play a central role, one in chemistry and one in management science.

EXAMPLE 6.
Balancing
Chemical
Equations

Consider the chemical reaction in which permanganate (MnO_4) and hydrogen (H) ions combine to form manganese (Mn) and water (H_2O):

$$MnO_4 + H \rightarrow Mn + H_2O \qquad (19)$$

where H represents hydrogen and O represents oxygen. Let x_1 be the number of permanganate ions (MnO_4), x_2 the number of hydrogen ions (H), x_3 the number of manganese atoms (Mn), and x_4 the number of water molecules (H_2O). Laws of chemistry require that the right-side products have the same number of atoms of Mn, of H, and of O as there are in the left-side inputs. Thus we obtain the "balancing" system of equations.

$$MnO_4 + H \rightarrow Mn + H_2O$$

$$
\begin{array}{llll}
\text{H:} & x_2 = & 2x_4 & \\
\text{Mn:} & x_1 & = x_3 & \qquad (20a)\\
\text{O:} & 4x_1 & = & x_4
\end{array}
$$

or

$$
\begin{array}{llll}
x_2 & & -2x_4 = 0 & \\
x_1 & -x_3 & = 0 & \qquad (20b)\\
4x_1 & & -x_4 = 0 &
\end{array}
$$

Notice that we have four unknowns but only three equations. Solving the homogeneous system (20b) by Gauss–Jordan elimination [pivoting on entries (2, 1), (1, 2), (3, 3)], one obtains

$$
\begin{array}{lll}
x_2 & - 2x_4 = 0 & x_2 = 2x_4 \\
x_1 & - \tfrac{1}{4}x_4 = 0 \quad \text{or} \quad & x_1 = \tfrac{1}{4}x_4 \\
& x_3 - \tfrac{1}{4}x_4 = 0 & x_3 = \tfrac{1}{4}x_4
\end{array}
\tag{21}
$$

As vectors, the solutions in (21) have the form

$$
\begin{bmatrix} \tfrac{1}{4}x_4 \\ 2x_4 \\ \tfrac{1}{4}x_4 \\ x_4 \end{bmatrix}
\quad \text{or} \quad
r\begin{bmatrix} \tfrac{1}{4} \\ 2 \\ \tfrac{1}{4} \\ 1 \end{bmatrix}
\tag{22}
$$

These vectors form the kernel for the system (20b). For example, if $r = x_4 = 4$, then $x_2 = 8$, $x_1 = x_3 = 1$, and the reaction equation becomes

$$
MnO_4 + 8H \rightarrow Mn + 4H_2O \qquad ■
$$

EXAMPLE 7.
Two-Variable
Kernel

Consider the following transportation problem that occurs in many guises in management science. Warehouses 1, 2, 3 have 20, 30, and 15 tons, respectively, of chicken wings. Colleges A and B need 25 and 40 tons, respectively, of chicken wings (to serve to students). The following table indicates the *amount* and *cost* of shipping a ton of chicken wings from a given warehouse to a given college (where x_1 is the number of tons of chicken wings shipped from warehouse 1 to college A, x_2 is the number of tons of chicken wings shipped from warehouse 1 to college B, etc.).

$$
\begin{array}{c}
\text{College} \\
\begin{array}{cc}
\text{A (25)} & \text{B (40)}
\end{array} \\
\begin{array}{r}
1\ (20) \\
\text{Warehouse } 2\ (30) \\
3\ (15)
\end{array}
\begin{bmatrix}
x_1/\$80 & x_2/\$45 \\
x_3/\$60 & x_4/\$55 \\
x_5/\$40 & x_6/\$65
\end{bmatrix}
\end{array}
\tag{23}
$$

Note that since the overall demand at the two colleges equals the warehouses' supply—65—all the supplies of each warehouse must be used. The constraints are that the total amount of chicken wings shipped from warehouse 1 equals 20 tons, from warehouse 2 equals 30 tons, from warehouse 3 equals 15 tons, and the total amount shipped to college A equals 25 tons and to college B equals 40 tons.

$$
\begin{aligned}
x_1 + x_2 \qquad\qquad\qquad\qquad &= 20 \\
x_3 + x_4 \qquad\qquad &= 30 \\
x_5 + x_6 &= 15 \\
x_1 \qquad + x_3 \qquad + x_5 \qquad &= 25 \\
x_2 \qquad + x_4 \qquad + x_6 &= 40
\end{aligned} \tag{24}
$$

The objective is to find the solution to (24) that is the cheapest, in terms of the costs in (23). That is, we want to minimize, subject to (24) and $x_i \geq 0$, the expression

$$
80x_1 + 45x_2 + 60x_3 + 55x_4 + 40x_5 + 65x_6 \tag{25}
$$

We shall restrict ourselves here to the issue of how to change from one solution $\mathbf{x}^*$ of (24) to another (cheaper) solution $\mathbf{x}^{**}$. To do so, we would add some kernel vector to $\mathbf{x}^*$, according to Theorem 3. To find the kernel, we solve the associated homogeneous system.

$$
\begin{aligned}
x_1 + x_2 \qquad\qquad\qquad\qquad &= 0 \\
x_3 + x_4 \qquad\qquad &= 0 \\
x_5 + x_6 &= 0 \\
x_1 \qquad + x_3 \qquad + x_5 \qquad &= 0 \\
x_2 \qquad + x_4 \qquad + x_6 &= 0
\end{aligned} \tag{26}
$$

Pivoting on entries (1, 1), (4, 2), (2, 3), (3, 5), we obtain

$$
\begin{aligned}
x_1 \qquad\qquad - x_4 \qquad - x_6 &= 0 \\
x_3 + x_4 \qquad\qquad &= 0 \\
x_5 + x_6 &= 0 \\
x_2 \qquad\quad x_4 \qquad x_6 &= 0 \\
0 &= 0
\end{aligned} \quad \text{or} \quad
\begin{aligned}
x_1 &= \quad x_4 + x_6 \\
x_3 &= -x_4 \\
x_5 &= -x_6 \\
x_2 &= -x_4 - x_6
\end{aligned} \tag{27}
$$

The solutions in (27) produce the set of kernel vectors

$$
\begin{bmatrix} x_4 + x_6 \\ -x_4 - x_6 \\ -x_4 \\ x_4 \\ -x_6 \\ x_6 \end{bmatrix} = x_4 \begin{bmatrix} 1 \\ -1 \\ -1 \\ 1 \\ 0 \\ 0 \end{bmatrix} + x_6 \begin{bmatrix} 1 \\ -1 \\ 0 \\ 0 \\ -1 \\ 1 \end{bmatrix} \tag{28}
$$

Then the kernel is all linear combinations of the two vectors on the right in (28). That is, the two vectors on the right in (28) span the kernel. The dimension of the kernel is 2.

Suppose that we are given a solution to (24): $\mathbf{x}^* = [20, 0, 5, 25, 0, 15]^T$, but we want another solution in which $x_1 = 10$ and $x_5 = 5$. We can achieve this by adding the right linear combination of kernel vectors $\mathbf{k}_1 = [1, -1, -1, 1, 0, 0]^T$ and $\mathbf{k}_2 = [1, -1, 0, 0, -1, 1]^T$. To make $x_5 = 5$ (now 0 in $\mathbf{x}^*$), we can add to $\mathbf{x}^*$ the vector $(-5)\mathbf{k}_2 = [-5, 5, 0, 0, 5, -5]^T$. To make $x_1 = 10$ (it is 15 in $\mathbf{x}^* - 5\mathbf{k}_2$), we add to $\mathbf{x}^*$ the vector $(-5)\mathbf{k}_1 = [-5, 5, 5, -5, 0, 0]^T$. So our desired solution is

$$\mathbf{x}^* + (-5)\mathbf{k}_1 + (-5)\mathbf{k}_2 = \begin{bmatrix} 20 \\ 0 \\ 5 \\ 25 \\ 0 \\ 15 \end{bmatrix} + \begin{bmatrix} -5 \\ 5 \\ 5 \\ -5 \\ 0 \\ 0 \end{bmatrix} + \begin{bmatrix} -5 \\ 5 \\ 0 \\ 0 \\ 5 \\ -5 \end{bmatrix} = \begin{bmatrix} 10 \\ 10 \\ 10 \\ 20 \\ 5 \\ 10 \end{bmatrix} \qquad (29)$$

    ■

The kernel in Example 7 is a little more complicated. If we had performed a different pivot sequence in (27), the two vectors that generated the kernel would be different from the two vectors in (28). By the theory developed in Section 4.3, we know that we would always obtain the kernel with two vectors, since all bases of this kernel space have the same size.

The kernel of a matrix is closely linked with the property of linear dependence. If a nonzero vector $\mathbf{x}^\circ$ is in $K(\mathbf{A})$, then $\mathbf{A}\mathbf{x}^\circ = \mathbf{0}$ says that there is a linear dependence among the columns $\mathbf{a}_i$ of $\mathbf{A}$, since

$$\mathbf{A}\mathbf{x}^\circ = \mathbf{0} \iff x_1^\circ \mathbf{a}_1 + x_2^\circ \mathbf{a}_2 + \cdots + x_m^\circ \mathbf{a}_m = \mathbf{0}$$

The reader should review Examples 1, 4, 6, and 7 to check this. We now show how to derive a basis for the kernel of $\mathbf{A}$, $K(\mathbf{A})$, in terms of such linear dependence among the columns of $\mathbf{A}$. For simplicity, we first consider matrices that are in reduced row echelon form. A basis for the range is also easy to find for such matrices.

EXAMPLE 8.
Bases for the Range and Kernel of a Matrix in Reduced Row Echelon Form

Consider the following two matrices in reduced row echelon form.

$$\mathbf{B} = \begin{bmatrix} 1 & 0 & 3 & 0 \\ 0 & 1 & 5 & 0 \\ 0 & 0 & 0 & 1 \end{bmatrix} \qquad \mathbf{C} = \begin{bmatrix} 1 & 0 & 4 & 0 & 6 \\ 0 & 1 & 1 & 0 & 0 \\ 0 & 0 & 0 & 1 & 3 \\ 0 & 0 & 0 & 0 & 0 \end{bmatrix}$$

As noted in Section 4.4, the columns with leading 1's (which are coordinate vectors) form a basis for the vector space spanned by the columns, that is, for the range of the matrix. For $\mathbf{B}$, this basis is $\{\mathbf{b}_1, \mathbf{b}_2, \mathbf{b}_4\}$, and for $\mathbf{C}$, this basis is $\{\mathbf{c}_1, \mathbf{c}_2, \mathbf{c}_4\}$.

In matrix $\mathbf{B}$, the third column $\mathbf{b}_3$ is clearly equal to $3\mathbf{b}_1 + 5\mathbf{b}_2$, or

$$3\mathbf{b}_1 + 5\mathbf{b}_2 - \mathbf{b}_3 = 0 \implies \mathbf{k} = \begin{bmatrix} 3 \\ 5 \\ -1 \\ 0 \end{bmatrix} \in K(\mathbf{B}) \qquad (30)$$

There is no other linear dependence among columns in $\mathbf{B}$, so the only nonzero solutions to $\mathbf{Bx} = \mathbf{0}$ are (nonzero) multiples of $\mathbf{k}$. Thus $\mathbf{k}$ spans $K(\mathbf{B})$. (Observe that the first two entries, 3, 5, in $\mathbf{k}$ are equal to the nonzero entries in the dependent column $\mathbf{b}_3$.)

Similarly, in $\mathbf{C}$, $\mathbf{c}_3 = 4\mathbf{c}_1 + 1\mathbf{c}_2$ and $\mathbf{c}_5 = 6\mathbf{c}_1 + 3\mathbf{c}_4$, so

$$
4\mathbf{c}_1 + 1\mathbf{c}_2 - \mathbf{c}_3 = 0 \implies \mathbf{k}' = \begin{bmatrix} 4 \\ 1 \\ -1 \\ 0 \\ 0 \end{bmatrix} \in K(\mathbf{C})
$$

$$
6\mathbf{c}_1 + 3\mathbf{c}_4 - \mathbf{c}_5 = 0 \implies \mathbf{k}'' = \begin{bmatrix} 4 \\ 0 \\ 0 \\ 3 \\ -1 \end{bmatrix} \in K(\mathbf{C})
$$

(31)

We note that $\mathbf{k}'$ and $\mathbf{k}''$ are linearly independent. To see this, look at the third and fifth entries in $\mathbf{k}'$ and $\mathbf{k}''$, corresponding to the (dependent) columns $\mathbf{c}_3$ and $\mathbf{c}_5$ of $\mathbf{C}$. Since $\mathbf{k}'$ came from the dependency of $\mathbf{c}_3$, then $k'_3 = -1$ and $k'_5 = 0$. Whereas in $\mathbf{k}''$, $k''_5 = -1$ and $k''_3 = 0$.

Any linear dependence in $\mathbf{C}$ involves $\mathbf{c}_3$ or $\mathbf{c}_5$ (or both). Hence it is reasonable to expect that any solution of $\mathbf{Cx} = \mathbf{0}$ will be a linear combination of $\mathbf{k}'$ and $\mathbf{k}''$, so $\mathbf{k}'$ and $\mathbf{k}''$ should form a basis for $K(\mathbf{C})$. ■

Lemma 1. Let $\mathbf{A}$ be an $m \times n$ matrix. Let $\mathbf{A}$ be reduced by Gauss–Jordan elimination to reduced row echelon form, and call the reduced matrix $\mathbf{A}^*$. Then $\mathbf{x}'$ is a solution to $\mathbf{Ax} = \mathbf{0}$ if and only if $\mathbf{x}'$ is a solution to $\mathbf{A}^*\mathbf{x} = \mathbf{0}$, so $K(\mathbf{A}) = K(\mathbf{A}^*)$.
Proof: The elementary row operations in Gauss–Jordan (or Gaussian) elimination preserve solutions (by Theorem 1 of Section 1.4); that is, $\mathbf{x}'$ is a solution to $\mathbf{Ax} = \mathbf{0}$ if and only if $\mathbf{x}'$ is a solution to $\mathbf{A}^*\mathbf{x} = \mathbf{0}$. ■

Theorem 5. Let $\mathbf{A}$ be an $m \times n$ matrix. Let $\mathbf{A}^*$ be the matrix in reduced row echelon form obtained from $\mathbf{A}$ by Gauss–Jordan elimination.
  (i) The columns of $\mathbf{A}$ corresponding to columns of $\mathbf{A}^*$ with leading 1's form a basis for the range $R(\mathbf{A})$.
  (ii) The columns of $\mathbf{A}$ corresponding to columns of $\mathbf{A}^*$ *without* leading 1's have linear dependencies. The dependencies generate a linearly independent collection of kernel vectors [which will be shown in Theorem 10 below to form a basis for $K(\mathbf{A})$].

Proof: Part (i) is a restatement of Theorem 1 in Section 4.4.
   Part (ii) is a generalization of Examples 7 and 8. By Lemma 1 it suffices to consider linear dependency of columns and the associated kernel for the reduced row echelon matrix. As shown in Examples 7 and 8, the columns in a reduced row echelon matrix without leading 1's are dependent on the columns with leading columns (which form a coordinate–vector basis). Then, as shown in equations

(30) and (31) of Example 8, the dependence relations (showing how a non-leading-1 column is linearly dependent on leading-1 columns) yield kernel vectors. Finally, these kernel vectors are linearly independent, because if $\mathbf{k}'$ is a kernel vector arising from a dependency of column i of $\mathbf{A}^*$ [e.g., in (31), $\mathbf{k}'$ comes from a dependency of column 3 in $\mathbf{A}^*$], then $\mathbf{k}'$ has a -1 in entry i, while entry i is 0 in other kernel vectors (associated with other column dependencies).

■

We have a little more work to do to show that the kernel vectors in Theorem 5(ii) form a basis for $K(\mathbf{A})$. Exercise 45 gives a direct proof of the fact that the dimension of $K(\mathbf{A})$ is $n - \dim(R(\mathbf{A}))$. We shall instead obtain this result as a consequence of some general properties of orthogonal vector spaces.

Kernels and Orthogonal Vector Spaces

Since $\mathbf{A}\mathbf{x} = \mathbf{0}$ if $\mathbf{x}$ is in the kernel of $\mathbf{A}$, it follows that the scalar product of a kernel vector $\mathbf{x}$ with each row $\mathbf{a}'_i$ of $\mathbf{A}$ is zero. That is, kernel vectors and row vectors of $\mathbf{A}$ are orthogonal. In the rest of this section we assume that all vector spaces have a scalar product, or more generally, an inner product, so that orthogonality can be defined. We say that two vector spaces V, W are **orthogonal** if every vector in V is orthogonal to every vector in W. By the preceding reasoning, we have proved:

Theorem 6. For any matrix $\mathbf{A}$, the row space Row($\mathbf{A}$) of $\mathbf{A}$ and the kernel $K(\mathbf{A})$ of $\mathbf{A}$ are orthogonal vector spaces.

We now give a lemma and the two theorems about orthogonal vector spaces. The proofs of the first two results are straightforward and are left as exercises.

Lemma 2. Let V_1 and V_2 be orthogonal vector subspaces of a vector space W. Then the only vector in the subspace $V_1 \cap V_2$ is $\mathbf{0}$ (the zero vector).

Theorem 7. Let W be any vector space and let V be a vector subspace of W. The set of vectors in W that are orthogonal to every vector in V forms a vector subspace.

Theorem 8. Let W be a vector space of dimension d. Let V_1 be any vector subspace of W and let V_2 be the subspace of vectors of W orthogonal to vectors in V_1. Then $\dim(V_1) + \dim(V_2) = d$. Further, any vector $\mathbf{b}$ in W has a unique decomposition

$$\mathbf{b} = \mathbf{b}_1 + \mathbf{b}_2, \qquad \text{where} \quad \mathbf{b}_1 \in V_1 \text{ and } \mathbf{b}_2 \in V_2 \tag{32}$$

Proof: Let B be a given basis for W. To show that $\dim(V_1) + \dim(V_2) = d$, we shall apply the Gram–Schmidt orthogonalization process to the following set S of $d_1 + d$ vectors: The first d_1 vectors of S are some basis for V_1 and the next d vectors are the basis B. The first d_1 vectors will be turned into an orthonormal basis B_1 for V_1. In applying the Gram–Schmidt process, the remaining d vectors will either become a **0** vector (and dropped from the basis) or be orthonormal vectors which are, by the Gram–Schmidt process, orthogonal to the first d_1 orthonormal vectors (a basis for V_1). So the set of d_2 (nonzero) remaining orthonormal vectors is a basis B_2 for V_2. Altogether the $d_1 + d_2$ orthonormal vectors of B_1 and B_2 are a basis for W. So $d_1 + d_2 = d$.

Expressing any vector **b** in W in terms of this basis $B_1 \cup B_2$ decomposes **b** into a vector $\mathbf{b}_1$ formed from the basis B_1 and a vector $\mathbf{b}_2$ formed from the basis B_2. This is the desired orthogonal decomposition $\mathbf{b} = \mathbf{b}_1 + \mathbf{b}_2$. The representation is unique because any vector has a unique representation as a linear combination of vectors in a basis (by Theorem 2 of Section 4.3). ■

EXAMPLE 9.
Orthogonal Vector Spaces

Consider the vector space V_1 of $\mathbf{R}^3$ formed by 3-vectors **x** in the plane P: $x_1 + x_2 - x_3 = 0$. We seek the vector space V_2 in $\mathbf{R}^3$ orthogonal to V_1. A basis of V_1 can be found by picking vectors in P with $x_2 = 0$, $\begin{bmatrix} 1 \\ 0 \\ 1 \end{bmatrix}$, and with $x_1 = 0$, $\begin{bmatrix} 0 \\ 1 \\ 1 \end{bmatrix}$. To find a basis for V_2, we use the approach in the proof of Theorem 8 of applying Gram–Schmidt orthogonalization to the set S of five vectors consisting of our basis for V_1 plus the standard basis for $\mathbf{R}^3$.

$$S = \left\{ \begin{bmatrix} 1 \\ 0 \\ 1 \end{bmatrix}, \begin{bmatrix} 0 \\ 1 \\ 1 \end{bmatrix}, \begin{bmatrix} 1 \\ 0 \\ 0 \end{bmatrix}, \begin{bmatrix} 0 \\ 1 \\ 0 \end{bmatrix}, \begin{bmatrix} 0 \\ 0 \\ 1 \end{bmatrix} \right\}$$

$$\longrightarrow \left\{ \begin{bmatrix} 1/\sqrt{2} \\ 0 \\ 1/\sqrt{2} \end{bmatrix}, \begin{bmatrix} -1/\sqrt{6} \\ 2/\sqrt{6} \\ 1/\sqrt{6} \end{bmatrix}, \begin{bmatrix} 1/\sqrt{3} \\ 1/\sqrt{3} \\ -1/\sqrt{3} \end{bmatrix}, \begin{bmatrix} 0 \\ 0 \\ 0 \end{bmatrix}, \begin{bmatrix} 0 \\ 0 \\ 0 \end{bmatrix} \right\}$$

(33)

The first two vectors resulting from the orthogonalization [the right side in (33)] are an orthogonal basis for V_1. The third vector is a basis for V_2. So V_2 consists of all multiples of $\begin{bmatrix} 1/\sqrt{3} \\ 1/\sqrt{3} \\ -1/\sqrt{3} \end{bmatrix}$, or more simply, multiples of $\begin{bmatrix} 1 \\ 1 \\ -1 \end{bmatrix}$.

This result can be checked by recalling the normal form of a plane: $\mathbf{a} \cdot \mathbf{x} = 0$ (Section 2.3), in which **a** is a vector orthogonal to the plane. For the plane $x_1 + x_2 - x_3 = 0$ in this example, $\mathbf{a} = \begin{bmatrix} 1 \\ 1 \\ -1 \end{bmatrix}$. ■

Corollary A. Let **A** be an $m \times n$ matrix. Then $\dim(\text{Row}(\mathbf{A})) + \dim(K(\mathbf{A})) = n$. Further, any n-vector **b** can be uniquely decomposed to $\mathbf{b} = \mathbf{b}_1 + \mathbf{b}_2$, where $\mathbf{b}_1 \in \text{Row}(\mathbf{A})$ and $\mathbf{b}_2 \in K(\mathbf{A})$.

Proof: From the discussion above, we know that the row space $\text{Row}(\mathbf{A})$ and kernel $K(\mathbf{A})$ are orthogonal. The corollary then follows from Theorem 8. ■

Corollary B. Let $\mathbf{x}'$ be a solution to the system $\mathbf{Ax} = \mathbf{b}$. Then $\mathbf{x}'$ has the unique orthogonal decomposition $\mathbf{x}' = \mathbf{x}^* + \mathbf{x}^\circ$, where $\mathbf{x}^* \in \text{Row}(\mathbf{A})$ and $\mathbf{x}^\circ \in K(\mathbf{A})$. Further, $\mathbf{Ax} = \mathbf{b}$ has a unique solution $\mathbf{x}^*$ that lies in the row space of $\mathbf{A}$.

Proof: The unique decomposition $\mathbf{x}' = \mathbf{x}^* + \mathbf{x}^\circ$ is just a restatement of Corollary A. Suppose that $\mathbf{x}^*$ and $\mathbf{x}^{**}$ are two solutions of $\mathbf{Ax} = \mathbf{b}$, and both $\mathbf{x}^*$ and $\mathbf{x}^{**}$ are in $\text{Row}(\mathbf{A})$. Since $\text{Row}(\mathbf{A})$ is a vector space, $\mathbf{x}^* - \mathbf{x}^{**}$ is a vector in $\text{Row}(\mathbf{A})$. However,

$$\mathbf{A}(\mathbf{x}^* - \mathbf{x}^{**}) = \mathbf{Ax}^* - \mathbf{Ax}^{**} = \mathbf{b} - \mathbf{b} = \mathbf{0}$$

proving that $\mathbf{x}^* - \mathbf{x}^{**}$ is also in $K(\mathbf{A})$, actually in $K(\mathbf{A}) \cap \text{Row}(\mathbf{A})$. Since $\text{Row}(\mathbf{A})$ and $K(\mathbf{A})$ are orthogonal, by Lemma 2, $\mathbf{x}^* - \mathbf{x}^{**} = \mathbf{0}$. Thus $\mathbf{x}^* = \mathbf{x}^{**}$. ▪

Corollary B has the surprising result that when $\mathbf{Ax} = \mathbf{b}$ has an infinite number of solutions, there is a natural way to select one special solution, namely, the unique solution that lies in the row space of $\mathbf{A}$. The following example illustrates the result in Corollary B.

EXAMPLE 10.
Solution in the
Row Space

Consider the following system of two equations in three unknowns.

$$\begin{array}{r} x_1 + x_2 + x_3 = 2 \\ 2x_1 \quad\quad - x_3 = 3 \end{array} \quad \text{or} \quad \mathbf{Ax} = \mathbf{b}, \text{ where } \mathbf{A} = \begin{bmatrix} 1 & 1 & 1 \\ 2 & 0 & -1 \end{bmatrix} \text{ and } \mathbf{b} = \begin{bmatrix} 2 \\ 3 \end{bmatrix}$$
(34)

We want to find a solution $\mathbf{x}^*$ to $\mathbf{Ax} = \mathbf{b}$ with $\mathbf{x}^*$ in $\text{Row}(\mathbf{A})$. Suppose that we are given the solution $\mathbf{x}' = \begin{bmatrix} 3 \\ -4 \\ 3 \end{bmatrix}$. We find a kernel vector $\mathbf{k} = \begin{bmatrix} 1 \\ -3 \\ 2 \end{bmatrix}$ of $\mathbf{A}$ using the method described in Theorem 5(ii). The rows $\mathbf{a}_1'$, $\mathbf{a}_2'$ of $\mathbf{A}$ are a basis for $\text{Row}(\mathbf{A})$. To find Corollary B's unique orthogonal decomposition $\mathbf{x}' = \mathbf{x}^* + \mathbf{x}^\circ$, where $\mathbf{x}^* \in \text{Row}(\mathbf{A})$ and $\mathbf{x}^\circ \in K(\mathbf{A})$, we form a matrix $\mathbf{D}$ whose three columns are $\mathbf{a}_1'^{\mathrm{T}}$, $\mathbf{a}_2'^{\mathrm{T}}$, $\mathbf{k}$ and solve $\mathbf{Dz} = \mathbf{x}'$.

$$\mathbf{Dz} = \mathbf{x}': \begin{bmatrix} 1 & 2 & 1 \\ 1 & 0 & -3 \\ 1 & -1 & 2 \end{bmatrix} \begin{bmatrix} z_1 \\ z_2 \\ z_3 \end{bmatrix} = \begin{bmatrix} 3 \\ -4 \\ 3 \end{bmatrix}$$
(35)

By Gaussian elimination, we find the solution is $z_1 = \tfrac{1}{2}$, $z_2 = \tfrac{1}{2}$, $z_3 = \tfrac{3}{2}$. Then

$$\mathbf{x}^* = \tfrac{1}{2}\mathbf{a}_1'^{\mathrm{T}} + \tfrac{1}{2}\mathbf{a}_2'^{\mathrm{T}} = \tfrac{1}{2}\begin{bmatrix} 1 \\ 1 \\ 1 \end{bmatrix} + \tfrac{1}{2}\begin{bmatrix} 2 \\ 0 \\ -1 \end{bmatrix} = \begin{bmatrix} \tfrac{3}{2} \\ \tfrac{1}{2} \\ 0 \end{bmatrix} \quad \text{and} \quad \mathbf{x}^\circ = \tfrac{3}{2}\mathbf{k} = \tfrac{3}{2}\begin{bmatrix} 1 \\ -3 \\ 2 \end{bmatrix} = \begin{bmatrix} \tfrac{3}{2} \\ -\tfrac{9}{2} \\ 3 \end{bmatrix}$$
(36)

Here $\mathbf{x}^*$ is the desired solution in $\text{Row}(\mathbf{A})$. ▪

With Theorem 8, we can now complete our unfinished agenda of finding a basis for $K(\mathbf{A})$. In the process we obtain the second fundamental theorem about the dimension of vector spaces associated with a matrix. The first fundamental dimension theorem (Theorem 3 of Section 4.4) is $\mathrm{rank}(\mathbf{A}) = \dim(\mathrm{Col}(\mathbf{A})) = \dim(\mathrm{Row}(\mathbf{A}))$.

Theorem 9. Let $\mathbf{A}$ be an $m \times n$ matrix. Then $\dim(R(\mathbf{A})) + \dim(K(\mathbf{A})) = n$.
Proof: From Corollary A,

$$\dim(\mathrm{Row}(\mathbf{A})) + \dim(K(\mathbf{A})) = n$$

Then $\dim(\mathrm{Col}(\mathbf{A})) + \dim(K(\mathbf{A})) = n$, since $\dim(\mathrm{Col}(\mathbf{A})) = \dim(\mathrm{Row}(\mathbf{A}))$ (from the first fundamental dimension theorem just cited). Since $R(\mathbf{A}) = \mathrm{Col}(\mathbf{A})$) by Theorem 1, the theorem is proved. ■

Restating this theorem in terms of linear transformations, we have:

Theorem 9′. Let $T: V \to W$ be a linear transformation between finite-dimensional vector spaces V and W. Then $\mathrm{rank}(T) + \mathrm{nullity}(T) = n$.

Theorem 10. Let $\mathbf{A}$ be an $m \times n$ matrix. Let $\mathbf{A}^*$ be the matrix in reduced row echelon form obtained from $\mathbf{A}$ by Gauss–Jordan elimination. The kernel vectors generated from the dependencies of columns of $\mathbf{A}^*$ *without* leading 1's form a basis for $K(\mathbf{A})$.
Proof: In Example 8 and Theorem 5, we showed that each column of $\mathbf{A}^*$ *without* leading 1's gave rise to a kernel vector and the set S of these kernel vectors was linearly independent. From Theorem 5(i), $\dim(R(\mathbf{A}))$ equals the number of columns of $\mathbf{A}^*$ with leading 1's. Then the number of columns of $\mathbf{A}^*$ without leading 1's, and hence the size of S, is $n - \dim(R(\mathbf{A}))$. By Theorem 9, $\dim(K(\mathbf{A})) = n - \dim(R(\mathbf{A}))$. Since S has $n - \dim(R(\mathbf{A}))$ linearly independent kernel vectors, it is a basis for $K(\mathbf{A})$ (by Corollary A of Theorem 3 in Section 4.3). ■

A direct proof of Theorem 9, without using orthogonal vector spaces, is given in Exercise 45.

Remember that everything said in this section about matrix mappings applies to a general linear transformation $T: V \to W$ (between finite-dimensional vector spaces V and W), since by Theorem 2 of Section 5.1, any such linear transformation T can be represented by matrix multiplication; that is, $T(\mathbf{v}) = \mathbf{A}\mathbf{v}$.

Section 5.2 Exercises

1. Find a set of columns that form a basis for the range of each of the following matrices.

(a) $\begin{bmatrix} -3 & 5 \\ 6 & -10 \end{bmatrix}$ (b) $\begin{bmatrix} 1 & 4 \\ 2 & -2 \end{bmatrix}$ (c) $\begin{bmatrix} 3 & -4 \\ 9 & -12 \end{bmatrix}$ (d) $\begin{bmatrix} 2 & 0 \\ 4 & 1 \end{bmatrix}$

2. Which of the following vectors are in the range of the matrix in Exercise 1(a)?

(a) $\begin{bmatrix} 1 \\ 3 \end{bmatrix}$ (b) $\begin{bmatrix} 1 \\ -2 \end{bmatrix}$ (c) $\begin{bmatrix} -4 \\ 2 \end{bmatrix}$ (d) $\begin{bmatrix} 3 \\ 2 \end{bmatrix}$

3. Find a set of columns that form a basis for the range of each of the following matrices.

(a) $\begin{bmatrix} 2 & 1 & 7 \\ 1 & 2 & 5 \\ 1 & 1 & 4 \end{bmatrix}$ (b) $\begin{bmatrix} 1 & 0 & 1 \\ 2 & 3 & 5 \\ 0 & -2 & -3 \end{bmatrix}$

(c) $\begin{bmatrix} 0 & 1 & -1 \\ 1 & 0 & 1 \\ -1 & 1 & 0 \end{bmatrix}$ (d) $\begin{bmatrix} 2 & -3 & 4 \\ 1 & -2 & 3 \\ 3 & 2 & -7 \end{bmatrix}$

4. Which of the following vectors are in the range of the matrix in Exercise 3(a)?

(a) $\begin{bmatrix} 1 \\ 3 \\ 2 \end{bmatrix}$ (b) $\begin{bmatrix} 2 \\ 4 \\ 2 \end{bmatrix}$ (c) $\begin{bmatrix} 4 \\ -2 \\ 1 \end{bmatrix}$

5. Find a set of columns that form a basis for the range of each of the following matrices.

(a) $\begin{bmatrix} -3 & 4 & 1 \\ 2 & -1 & 1 \end{bmatrix}$ (b) $\begin{bmatrix} 2 & 1 & 5 & 0 \\ 1 & 2 & 4 & -3 \\ 1 & 1 & 3 & -1 \end{bmatrix}$ (c) $\begin{bmatrix} 1 & 1 & 0 & 0 & 1 \\ 1 & 0 & 1 & 0 & 1 \\ 0 & 0 & 1 & 1 & 0 \\ 0 & 1 & 0 & 1 & 0 \\ 1 & 1 & 1 & 1 & 1 \end{bmatrix}$

(d) $\begin{bmatrix} 1 & 0 & 2 & -1 & 1 \\ 1 & 1 & 1 & 1 & 2 \\ 1 & 0 & 2 & -1 & 1 \\ 0 & 1 & -1 & 2 & 1 \\ 1 & 0 & 2 & -1 & 1 \end{bmatrix}$ (e) $\begin{bmatrix} 1 & 1 & 1 & 1 & 0 & 0 \\ 0 & 1 & 0 & 1 & 0 & 1 \\ 1 & 0 & 0 & 1 & 1 & 0 \\ 0 & 1 & 1 & 0 & 1 & 0 \\ 0 & 1 & 0 & 1 & 0 & 1 \\ 1 & 0 & 1 & 0 & 0 & 1 \end{bmatrix}$

6. (a) What is the kernel and range of the identity transformation $T: V \to V$ defined by $T(\mathbf{v}) = \mathbf{v}$, for an arbitrary vector space V?

(b) What is the kernel and range of the zero transformation $T: V \to \{\mathbf{0}\}$ defined by $T(\mathbf{v}) = \mathbf{0}$, for an arbitrary vector space V?

7. Give a matrix defining a linear transformation $T: R^3 \to R^3$ with the following range.

(a) $x_2 = 0$ (b) $x_1 + x_2 + x_3 = 0$

(c) The line generated by the vector $[1, 2, 3]$

8. Give a constraint equation, if one exists, on the vectors in the range of the matrices in Exercise 1.

9. Give a constraint equation, if one exists, on the vectors in the range of the matrices in Exercise 3.

10. Give a constraint equation, if one exists, on the vectors in the range of the matrices in Exercise 5.

11. Give a (nonzero) vector, if one exists, that generates the kernel of the following systems of equations or matrices.

(a) $\begin{bmatrix} 1 & -2 \\ -2 & 4 \end{bmatrix}$ (b) $\begin{bmatrix} 4 & -1 & 2 \\ 2 & 5 & 1 \\ -2 & 3 & -1 \end{bmatrix}$

(c) $\begin{bmatrix} -1 & 3 & 1 \\ 5 & -1 & 3 \\ 2 & 1 & 2 \end{bmatrix}$ (d) $\begin{bmatrix} 2 & 1 & 1 \\ 1 & -2 & -3 \\ 3 & 4 & 5 \end{bmatrix}$

(e) $\begin{aligned} x - 2y + z &= 6 \\ -x + y - 2z &= 4 \end{aligned}$ (f) $\begin{aligned} x - y + z &= 3 \\ x + y + z &= 3 \\ 2x - 3y + z &= 7 \end{aligned}$

Consider the coefficient matrix and ignore the particular right-side values in parts (e) and (f).

12. Which of the following vectors are in the kernel of the matrix in Exercise 5(b)?

(a) $\begin{bmatrix} 2 \\ -1 \\ 1 \\ 1 \end{bmatrix}$ (b) $\begin{bmatrix} 3 \\ -1 \\ -1 \\ -1 \end{bmatrix}$ (c) $\begin{bmatrix} 2 \\ -4 \\ 0 \\ -2 \end{bmatrix}$

13. Find a basis for the kernel of each matrix in Exercise 3.

14. Find a basis for the kernel of each matrix in Exercise 5.

15. (a) For **A** in Exercise 5(a) and $\mathbf{b} = \begin{bmatrix} 10 \\ 10 \end{bmatrix}$, if $\mathbf{Ax} = \mathbf{b}$ has the given solution

$\mathbf{x}' = \begin{bmatrix} 0 \\ 0 \\ 10 \end{bmatrix}$, find the family of all solutions to $\mathbf{Ax} = \mathbf{b}$.

(b) Find a solution to $\mathbf{Ax} = \mathbf{b}$ in part (a) with $x_1 = 3$.

16. (a) For **A** in Exercise 5(b) and $\mathbf{b} = \begin{bmatrix} 30 \\ 30 \\ 20 \end{bmatrix}$, if $\mathbf{Ax} = \mathbf{b}$ has the given solution

$\mathbf{x}' = \begin{bmatrix} 10 \\ 10 \\ 0 \\ 0 \end{bmatrix}$, find the family of all solutions to $\mathbf{Ax} = \mathbf{b}$.

(b) Find a solution to $\mathbf{Ax} = \mathbf{b}$ in part (a) with $x_1 = 5$.

17. (a) For **A** in Exercise 5(c) and $\mathbf{b} = \begin{bmatrix} 10 \\ 15 \\ 5 \\ 0 \\ 15 \end{bmatrix}$, if $\mathbf{Ax} = \mathbf{b}$ has the given solution

$\mathbf{x}' = \begin{bmatrix} 5 \\ 0 \\ 5 \\ 0 \\ 5 \end{bmatrix}$, find the family of solutions to $\mathbf{Ax} = \mathbf{b}$.

(b) Find a solution to $\mathbf{Ax} = \mathbf{b}$ in part (a) with $x_4 = 10$ and $x_5 = 10$.

(c) Find a solution to $\mathbf{Ax} = \mathbf{b}$ in part (a) with $x_1 = 10$ and $x_2 = 5$.

18. Consider the modified refinery system:

$$20x_1 + 4x_2 + 4x_3 = 700$$
$$10x_1 + 14x_2 + 5x_3 = 500$$

Given the solution $x_1 = 31$, $x_2 = 10$, $x_3 = 10$, use the appropriate kernel vector to obtain a new solution in which the following is true.
(a) $x_3 = 22$ **(b)** $x_2 = 25$ **(c)** $x_1 = 28$

19. Consider the following refinery-type problem:

$$10x_1 + 5x_2 + 5x_3 = 300$$
$$5x_1 + 10x_2 + 8x_3 = 300$$

(a) Find the kernel of the system.
(b) Given the solution $x_1 = x_2 = 20$, $x_3 = 0$, find a second solution with $x_3 = 10$.
(c) Repeat part (b) to find a second solution with $x_1 = 15$.

20. Find the kernel of the following linear transformations T on $\mathcal{P}$.
(a) $T(p(x)) = xp(x)$
(b) $T(p(x)) = (2x + 3)p(x)$
(c) $T(p(x)) = p(-x) + p(x)$
(d) $T(p(x)) = p(1)$
(e) $T(p(x) = \int_0^1 p(x)\ dx$

21. Find the kernel of the following linear transformations T on $\mathcal{M}_{2,2}$.
(a) $T(\mathbf{M}) = 3\mathbf{M}$
(b) $T(\mathbf{M}) = m_{1,1}$
(c) $T(\mathbf{M}) = \mathbf{M} - \mathbf{M}^{\mathrm{T}}$

22. Write out a system of equations required to balance the following chemical reactions and solve. Here C = carbon, N = nitrogen, H = hydrogen, and O = oxygen.
(a) $N_2H_4 + N_2O_4 \rightarrow N_2 + H_2O$
(b) $C_6H_6 + O_2 \rightarrow CO_2 + H_2O$

23. Write out a system of equations required to balance the following chemical reactions and solve.
(a) $SO_2 + NO_3 + H_2O \rightarrow H + SO_4 + NO$, where S = sulfur, N = nitrogen, H = hydrogen, and O = oxygen.
(b) $PbN_6 + CrMn_2O_8 \rightarrow Cr_2O_3 + MnO_2 + Pb_3O_4 + NO$, where Pb = lead, N = nitrogen, Cr = chromium, Mn = manganese, and O = oxygen.
(c) $H_2SO_4 + MnS + As_2Cr_{10}O_{35} \rightarrow HMnO_4 + AsH_3 + CrS_3O_{12} + H_2O$, where H = hydrogen, S = sulfur, O = oxygen, Mn = manganese, As = arsenic, and Cr = chromium.

24. In the transportation problem in Example 7, add a kernel vector to the given solution $\mathbf{x}^* = [20, 0, 5, 25, 0, 15]^{\mathrm{T}}$ in order to obtain another solution with the following values.
(a) $x_1 = 17$ and $x_6 = 15$
(b) $x_2 = 8$ and $x_6 = 4$
(c) $x_3 = 10$ and $x_5 = 5$
(d) $x_3 = 5$ and $x_4 = 5$

25. Find a linear transformation $T: R^3 \rightarrow R^3$ with the following kernel.

    (a) R^3

    (b) The line formed by multiples of the vector $\begin{bmatrix} 5 \\ 1 \\ 2 \end{bmatrix}$

    (c) The plane $x_1 + x_2 + x_3 = 0$

    (d) Vectors of the form $\begin{bmatrix} -3x \\ 2x + 3y \\ x - y \end{bmatrix}$

26. Find the kernel for these Markov transition matrices.

    (a) $\begin{bmatrix} 1 & 0.5 & 0 \\ 0 & 0 & 0 \\ 0 & 0.5 & 1 \end{bmatrix}$ **(b)** $\begin{bmatrix} 0.4 & 0 & 0.2 \\ 0.3 & 0.5 & 0.4 \\ 0.3 & 0.5 & 0.4 \end{bmatrix}$ **(c)** $\begin{bmatrix} 0.5 & 0 & 0.5 \\ 0 & 1 & 0 \\ 0.5 & 0 & 0.5 \end{bmatrix}$

    (d) $\begin{bmatrix} 0.50 & 0.25 & 0 & 0 & 0 \\ 0.50 & 0.50 & 0.25 & 0 & 0 \\ 0 & 0.25 & 0.50 & 0.25 & 0 \\ 0 & 0 & 0.25 & 0.50 & 0.50 \\ 0 & 0 & 0 & 0.25 & 0.50 \end{bmatrix}$ **(e)** $\begin{bmatrix} \frac{2}{3} & \frac{1}{3} & 0 & 0 & 0 & 0 \\ \frac{1}{3} & \frac{1}{3} & \frac{1}{3} & 0 & 0 & 0 \\ 0 & \frac{1}{3} & \frac{1}{3} & \frac{1}{3} & 0 & 0 \\ 0 & 0 & \frac{1}{3} & \frac{1}{3} & \frac{1}{3} & 0 \\ 0 & 0 & 0 & \frac{1}{3} & \frac{1}{3} & \frac{1}{3} \\ 0 & 0 & 0 & 0 & \frac{1}{3} & \frac{2}{3} \end{bmatrix}$

    (f) $\begin{bmatrix} \frac{2}{3} & \frac{2}{3} & 0 & 0 & 0 & 0 \\ \frac{1}{3} & \frac{1}{6} & \frac{2}{3} & 0 & 0 & 0 \\ 0 & \frac{1}{6} & \frac{1}{6} & \frac{2}{3} & 0 & 0 \\ 0 & 0 & \frac{1}{6} & \frac{1}{6} & \frac{2}{3} & 0 \\ 0 & 0 & 0 & \frac{1}{6} & \frac{1}{6} & \frac{2}{3} \\ 0 & 0 & 0 & 0 & \frac{1}{6} & \frac{1}{3} \end{bmatrix}$ **(g)** $\begin{bmatrix} \frac{1}{2} & \frac{1}{2} & 0 & 0 & 0 & 0 \\ \frac{1}{2} & 0 & \frac{1}{2} & 0 & 0 & 0 \\ 0 & \frac{1}{2} & 0 & \frac{1}{2} & 0 & 0 \\ 0 & 0 & \frac{1}{2} & 0 & \frac{1}{2} & 0 \\ 0 & 0 & 0 & \frac{1}{2} & 0 & \frac{1}{2} \\ 0 & 0 & 0 & 0 & \frac{1}{2} & \frac{1}{2} \end{bmatrix}$

27. **(a)** For each transition matrix A in Exercise 26, find the stable probability vector p^*. (See Example 9 in Section 1.4 for an example of this process.)

    (b) For each transition matrix A in Exercise 26, find the set of probability vectors p such that $Ap = p^*$, where p^* is the stable probability vector [computed in part (a)].

28. Prove that if A is the 3×3 transition matrix of some Markov chain and the kernel of A has an infinite number of vectors, then either two columns of A are equal or else one column is a weighted average of the two other columns.

29. Determine the rank of matrix A, if possible, from the given information.
    (a) A is a 6×4 matrix and $K(A) = \{0\}$.
    (b) A is a 5×6 matrix and $\dim(K(A)) = 3$.
    (c) A is a 7×5 matrix in which $\dim(K(A^T)) = 3$.
    (d) A is a 4×5 matrix and vectors in $K(A)$ form in a plane.
    (e) A is a 4×6 matrix and $Ax = b$ has at least one solution for every 4-vector b.
    (f) A is a 3×4 matrix and vectors b in the range of A satisfy the equation $b_1 + b_2 + b_3 = 0$.

30. State and prove Theorem 3 for a general linear transformation.

31. State and prove Theorem 4 for a general linear transformation.

32. Suppose that for an $m \times n$ matrix A, the reduced matrix A^* can be written in the partitioned form $A^* = \begin{bmatrix} I & R \\ O & O \end{bmatrix}$, where I is an $r \times r$ identity matrix $[r = \text{rank}(A)]$ and R is

$r \times (n - r)$. Using the submatrix $\mathbf{R}$ and an appropriate size identity matrix $\mathbf{I}$, give a matrix $\mathbf{K}$ in partitioned form whose columns form a basis of $K(\mathbf{A})$.

33. In this exercise we examine the vectors in the range of two matrices $\mathbf{A}$ and $\mathbf{B}$, that is, vectors in $R(\mathbf{A}) \cap R(\mathbf{B})$. If $\mathbf{d}$ is such a vector, then $\mathbf{Ax}' = \mathbf{d}$ and $\mathbf{Bx}'' = \mathbf{d}$, for some $\mathbf{x}'$, $\mathbf{x}''$.

   Show that if $\mathbf{C} = [\mathbf{A} \quad -\mathbf{B}]$ and $\mathbf{x}^* = \begin{bmatrix} \mathbf{x}' \\ \mathbf{x}'' \end{bmatrix}$, then $\mathbf{d}$ is in $R(\mathbf{A}) \cap R(\mathbf{B})$ if and only if $\mathbf{x}^*$ is in $K(\mathbf{C})$.

34. Let $\mathbf{A}$, $\mathbf{B}$, $\mathbf{C}$ be $n \times n$ matrices with $\mathbf{A} = \mathbf{BC}$. If $\mathbf{A}$ has full rank [rank$(\mathbf{A}) = n$], show that $\mathbf{B}$ and $\mathbf{C}$ do also.

35. If the vectors in a basis of vector space V are mutually orthogonal to the vectors in a basis of vector space W, show that every vector in V is orthogonal to every vector in W.

36. Let $T: V \rightarrow W$ be a projection of V onto W, where $W = R(T)$ is a vector subspace of V. Describe the kernel of T.

37. For each of the following matrices $\mathbf{A}$, express the 1's vector $\mathbf{1}$ (of the appropriate size) as a unique sum, $\mathbf{1} = \mathbf{x}_1 + \mathbf{x}_2$ of a vector $\mathbf{x}_1$ in Row$(\mathbf{A})$ and a vector $\mathbf{x}_2$ in $K(\mathbf{A})$.

   (a) $\begin{bmatrix} 1 & -2 \\ -2 & 4 \end{bmatrix}$ (b) $\begin{bmatrix} 2 & 1 \\ 1 & 2 \end{bmatrix}$ (c) $\begin{bmatrix} 1 & 2 & 0 \\ 0 & 1 & 2 \end{bmatrix}$ (d) $\begin{bmatrix} 1 & 1 & 0 \\ 2 & 1 & 2 \\ 0 & 1 & 1 \end{bmatrix}$

38. Find a solution to $\mathbf{Ax} = \mathbf{1}$, for $\mathbf{A}$ the matrix in Exercise 37(c), in which $\mathbf{x}$ is in Row$(\mathbf{A})$.

39. Given an $m \times n$ matrix $\mathbf{A}$ ($m < n$) with linearly independent rows, a basis $K = \{\mathbf{k}_1, \mathbf{k}_2, \ldots, \mathbf{k}_{n-m}\}$ for $K(\mathbf{A})$, and an m-vector $\mathbf{b}$, explain how to construct an $n \times n$ matrix $\mathbf{A}'$ and n-vector $\mathbf{b}'$ such that the solution $\mathbf{x}^*$ to $\mathbf{A}'\mathbf{x} = \mathbf{b}'$ is the unique solution to $\mathbf{Ax} = \mathbf{b}$ which lies in Row$(\mathbf{A})$ (as guaranteed by Corollary B of Theorem 8).

40. Prove Lemma 2.

41. Prove Theorem 7.

42. Use Corollary B of Theorem 8 to prove that if $\mathbf{v}_1, \mathbf{v}_2, \ldots, \mathbf{v}_k$ are a linearly independent set of vectors in the row space of a matrix $\mathbf{A}$, then the vectors $\mathbf{w}_i = \mathbf{Av}_i$ are a linearly independent set of vectors in the range of $\mathbf{A}$. Thus if $\{\mathbf{v}_i\}$ are a basis for Row$(\mathbf{A})$, then $\{\mathbf{Av}_i\}$ are a basis for $R(\mathbf{A})$.

43. Let $T: V \rightarrow W$ be a linear transformation. Prove that $\dim(R(T)) \leq \dim(V)$.

44. In this exercise we provide a direct proof that for a linear transformation $T: V \rightarrow W$, $\dim(R(T)) + \dim(K(T)) = n$, where $\dim(V) = n$. Let $\mathbf{w}_1, \mathbf{w}_2, \ldots, \mathbf{w}_p$ be a basis for $R(T)$, where $p = \dim(R(T))$ and let $\mathbf{v}_i$, $i = 1, 2, \ldots, p$, be vectors in V such that $T(\mathbf{v}_i) = \mathbf{w}_i$. Further, let $\mathbf{n}_1, \mathbf{n}_2, \ldots, \mathbf{n}_q$ be a basis of $K(T)$, with $q = \dim(K(T))$. Show that $\mathbf{v}_1, \mathbf{v}_2, \ldots, \mathbf{v}_p, \mathbf{n}_1, \mathbf{n}_2, \ldots, \mathbf{n}_q$ form a basis for V.

5.3 Diagonalization and Similarity

In this section we combine our knowledge of eigenvectors and of vector space bases to simplify and better understand matrices. The bases we use are formed by eigenvectors of a matrix. Recall that in Section 3.3, we saw that it was very easy to compute expressions like $\mathbf{c}^{(k)} = \mathbf{A}^k\mathbf{c}$, when $\mathbf{c}$ is expressed as a linear combination of the eigenvectors of $\mathbf{A}$. We will see that $\mathbf{A}$ usually becomes a diagonal matrix in an eigenvector basis. We also examine the general problem of how any change of basis affects linear transformations.

Recall that an n-vector $\mathbf{e}$ is an eigenvector of the $n \times n$ matrix $\mathbf{A}$ if for some eigenvalue λ, $\mathbf{Ae} = \lambda\mathbf{e}$. Thus, multiplying $\mathbf{e}$ by a matrix $\mathbf{A}$ has the same effect as multiplying $\mathbf{e}$ by the scalar λ. For example, for $\mathbf{A} = \begin{bmatrix} 3 & 1 \\ 2 & 2 \end{bmatrix}$ with $\mathbf{e} = \begin{bmatrix} 1 \\ 1 \end{bmatrix}$, we have $\mathbf{Ae} = \begin{bmatrix} 3 & 1 \\ 2 & 2 \end{bmatrix}\begin{bmatrix} 1 \\ 1 \end{bmatrix} = \begin{bmatrix} 4 \\ 4 \end{bmatrix} = 4\mathbf{e}$. So $\mathbf{e}$ is an eigenvector of this $\mathbf{A}$ with eigenvalue $\lambda = 4$. We can exploit this nice property of matrix multiplication with eigenvectors by converting a vector to eigenvector coordinates. First we need to show that the eigenvectors can indeed form a basis for $\mathbf{R}^n$.

Theorem 1

(i) Eigenvectors associated with different eigenvalues of a square matrix $\mathbf{A}$ are linearly independent.

(ii) If the $n \times n$ matrix $\mathbf{A}$ has n distinct eigenvalues, then any set of n eigenvectors, each associated with a different eigenvalue, will form a basis for $\mathbf{R}^n$.

Proof: For explicitness, suppose that we can find just two linearly independent eigenvectors $\mathbf{e}_1$, $\mathbf{e}_2$, associated distinct eigenvalues λ_1, λ_2. Then any other eigenvector $\mathbf{e}_3$, with a different eigenvalue λ_3, will be linearly dependent on $\mathbf{e}_1$ and $\mathbf{e}_2$ (assume that $\mathbf{e}_3$ is not a multiple of just $\mathbf{e}_1$ or $\mathbf{e}_2$). Recall that $\mathbf{e}_3$ will be a unique linear combination of $\mathbf{e}_1$ and $\mathbf{e}_2$: $\mathbf{e}_3 = c_1\mathbf{e}_1 + c_2\mathbf{e}_2$. The uniqueness follows from the fact that $\mathbf{e}_1$ and $\mathbf{e}_2$ are linearly independent; see the corollary to Theorem 2 in Section 4.3. We now compute $\mathbf{Ae}_3$ in two ways.

$$\mathbf{Ae}_3 = \lambda_3\mathbf{e}_3 = \lambda_3(c_1\mathbf{e}_1 + c_2\mathbf{e}_2) = \lambda_3 c_1\mathbf{e}_1 + \lambda_3 c_2\mathbf{e}_2 \tag{1}$$

and

$$\mathbf{Ae}_3 = \mathbf{A}(c_1\mathbf{e}_1 + c_2\mathbf{e}_2) = c_1\mathbf{Ae}_1 + c_2\mathbf{Ae}_2 = \lambda_1 c_1\mathbf{e}_1 + \lambda_2 c_2\mathbf{e}_2 \tag{2}$$

The representation of $\mathbf{Ae}_3$ as a linear combination of $\mathbf{e}_1$ and $\mathbf{e}_2$ is also unique. Then the coefficients of $\mathbf{e}_1$ and $\mathbf{e}_2$ on the right in (1) and (2) must be equal: $\lambda_3 c_1 = \lambda_1 c_1$ and $\lambda_3 c_2 = \lambda_2 c_2$. Thus $\lambda_3 = \lambda_1$ and $\lambda_3 = \lambda_2$. But the eigenvalues are distinct. This contradiction proves that $\mathbf{e}_1$, $\mathbf{e}_2$, $\mathbf{e}_3$ must be linearly independent. This argument generalizes to show that no eigenvector can be linearly dependent on other eigenvectors (with different eigenvalues).

Part (ii) follows from part (i): Any set of n eigenvectors of $\mathbf{A}$, each associated with a different eigenvalue, will be linearly independent, by part (i); but any set of n linearly independent n-vectors forms a basis for $\mathbf{R}^n$ (Corollary to Theorem 3 in Section 4.3). ■

Observe how results from our vector space theory developed in Chapter 4 are critical in proving Theorem 1.

Assuming that the eigenvalues of **A** are distinct, we now show how to use an eigenvector basis to simplify multiplication with **A**. We develop our results by reviewing the motivating example from Section 3.3.

EXAMPLE 1.
Computing Powers
of a Matrix with
Eigenvectors

The computer (C) and dog (D) growth model from Section 3.3,

$$\mathbf{c}' = \mathbf{Ac}, \quad \text{where} \quad \mathbf{A} = \begin{bmatrix} 3 & 1 \\ 2 & 2 \end{bmatrix}, \quad \mathbf{c} = \begin{bmatrix} C \\ D \end{bmatrix}, \quad \mathbf{c}' = \begin{bmatrix} C' \\ D' \end{bmatrix} \qquad (3)$$

had eigenvalues and associated eigenvectors of **A**: $\lambda_1 = 4$, $\mathbf{e}_1 = \begin{bmatrix} 1 \\ 1 \end{bmatrix}$ and $\lambda_2 = 1$, $\mathbf{e}_2 = \begin{bmatrix} 1 \\ -2 \end{bmatrix}$. By Theorem 1(ii), $\mathbf{e}_1$ and $\mathbf{e}_2$ form a basis for $\mathbf{R}^2$.

Suppose that we want to determine the effects of this growth model over 20 periods with the starting vector $\mathbf{c} = \begin{bmatrix} 1 \\ 7 \end{bmatrix}$. We should convert **c** to a coordinate vector $\mathbf{c}^* = [\mathbf{c}]_E$ in the eigenvector basis $E = \{\mathbf{e}_1, \mathbf{e}_2\}$: $\mathbf{c} = c_1^*\mathbf{e}_1 + c_2^*\mathbf{e}_2$. To find $\mathbf{c}^* = \begin{bmatrix} c_1^* \\ c_2^* \end{bmatrix}$ requires solving

$$\mathbf{c} = c_1^*\mathbf{e}_1 + c_2^*\mathbf{e}_2 \quad \text{or} \quad \mathbf{c} = \mathbf{Ec}^*, \quad \text{where} \quad \mathbf{E} = [\mathbf{e}_1 \quad \mathbf{e}_2] = \begin{bmatrix} 1 & 1 \\ 1 & -2 \end{bmatrix} \qquad (4)$$

The solution to (4) is

$$\mathbf{c}^* \ (= [\mathbf{c}]_E) = \mathbf{E}^{-1}\mathbf{c} = \begin{bmatrix} \frac{2}{3} & \frac{1}{3} \\ \frac{1}{3} & -\frac{1}{3} \end{bmatrix}\begin{bmatrix} 1 \\ 7 \end{bmatrix} = \begin{bmatrix} 3 \\ -2 \end{bmatrix} \qquad (5)$$

We see that (4) is $\mathbf{c} = 3\mathbf{e}_1 - 2\mathbf{e}_2$. We have rederived here the result developed in Section 4.3 that the transition matrix from the standard coordinate basis to a new basis E equals the inverse of the matrix **E** whose columns are the basis vectors of E (expressed in the standard coordinate basis).

In E-coordinates, it is easy to multiply **c** by **A**.

$$\mathbf{Ac} = \mathbf{A}(3\mathbf{e}_1 - 2\mathbf{e}_2) = 3\mathbf{Ae}_1 - 2\mathbf{Ae}_2$$

$$= 3(4\mathbf{e}_1) - 2(1\mathbf{e}_2) = 12\mathbf{e}_1 - 2\mathbf{e}_2 = 12\begin{bmatrix} 1 \\ 1 \end{bmatrix} - 2\begin{bmatrix} 1 \\ -2 \end{bmatrix} = \begin{bmatrix} 10 \\ 16 \end{bmatrix} \qquad (6)$$

In E-coordinates, **A** becomes $\mathbf{A}^* = \begin{bmatrix} 4 & 0 \\ 0 & 1 \end{bmatrix}$ and (6) becomes

$$\mathbf{A}^*\mathbf{c}^* = \begin{bmatrix} 4 & 0 \\ 0 & 1 \end{bmatrix}\begin{bmatrix} 3 \\ -2 \end{bmatrix} = \begin{bmatrix} 12 \\ -2 \end{bmatrix} \qquad (7)$$

For 20 periods, we have

$$\mathbf{A}^{20}\mathbf{c} = \mathbf{A}^{20}(3\mathbf{e}_1 - 2\mathbf{e}_2) = 3\mathbf{A}^{20}\mathbf{e}_1 - 2\mathbf{A}^{20}\mathbf{e}_2$$

$$= 3(4^{20}\mathbf{e}_1) - 2(1^{20}\mathbf{e}_2) \tag{8}$$

$$= 3 \cdot 4^{20}\begin{bmatrix} 1 \\ 1 \end{bmatrix} - 2\begin{bmatrix} 1 \\ -2 \end{bmatrix} \approx 3 \cdot 4^{20}\begin{bmatrix} 1 \\ 1 \end{bmatrix}$$

In E-coordinates, $\mathbf{A}^{20}$ becomes $\mathbf{A}^{*20} = \begin{bmatrix} 4^{20} & 0 \\ 0 & 1 \end{bmatrix}$ and (8) becomes

$$\mathbf{A}^{*20}\mathbf{c}^* = \begin{bmatrix} 4^{20} & 0 \\ 0 & 1 \end{bmatrix}\begin{bmatrix} 3 \\ -2 \end{bmatrix} = \begin{bmatrix} 3 \cdot 4^{20} \\ -2 \end{bmatrix} \tag{9}$$

■

We now describe in matrix notation the three basic steps in computing $\mathbf{A}^{20}\mathbf{c}$ in eigenvector coordinates, as developed in Example 1.

Computing c′ = Ac in Eigenvector Coordinates

   Step 1:   Obtain a basis E of n linearly independent eigenvectors $\mathbf{e}_1$, $\mathbf{e}_2$, . . . , $\mathbf{e}_n$.

   Step 2:   Convert $\mathbf{c}$ into the E-coordinate vector $\mathbf{c}^* = [\mathbf{c}]_E$. The transition matrix is $\mathbf{E}^{-1}$, where $\mathbf{E}$ is the matrix whose columns are the eigenvectors $\mathbf{e}_i$.

$$\mathbf{c}^*(= [\mathbf{c}]_E) = \mathbf{E}^{-1}\mathbf{c} \tag{10}$$

   Step 3:   Given $\mathbf{c}^* = \begin{bmatrix} c_1^* \\ c_2^* \end{bmatrix}$, the multiplication $\mathbf{Ac}$ is performed in eigenvector coordinates simply by multiplying c_1^* by λ_1 and c_2^* by λ_2. We write this step in matrix notation as

$$\begin{bmatrix} \lambda_1 c_1^* \\ \lambda_2 c_2^* \end{bmatrix} = \begin{bmatrix} \lambda_1 & 0 \\ 0 & \lambda_2 \end{bmatrix}\begin{bmatrix} c_1^* \\ c_2^* \end{bmatrix} \quad \text{or} \quad \mathbf{c}^+ = \mathbf{D}_\lambda \mathbf{c}^* \tag{11}$$

where $\mathbf{D}_\lambda$ is the diagonal matrix of eigenvalues. *Observe that $\mathbf{A}$ in the original coordinates has been converted to a diagonal matrix in eigenvector coordinates.*

   Step 4:   Convert our result $\mathbf{c}^+$ back to a vector $\mathbf{c}'$ in the standard coordinate system using the transition matrix $\mathbf{E}$.

$$\mathbf{c}' = c_1^+ \mathbf{e}_1 + c_2^+ \mathbf{e}_2 \quad \text{or} \quad \mathbf{c}' = \mathbf{E}\mathbf{c}^+ \tag{12}$$

Combining the three products in (10), (11), and (12), we have

$$\mathbf{c}' = \mathbf{E}\mathbf{c}^+ = \mathbf{E}(\mathbf{D}_\lambda \mathbf{c}^*) = \mathbf{E}(\mathbf{D}_\lambda(\mathbf{E}^{-1}\mathbf{c})) \tag{13}$$

Thus $\mathbf{c}' = \mathbf{Ac}$ in the original coordinates becomes $\mathbf{c}' = \mathbf{ED}_\lambda\mathbf{E}^{-1}\mathbf{c}$ when we convert into and then back out of eigenvector coordinates. These two ways to compute $\mathbf{Ac}$ are true for any $\mathbf{c}$, and hence by Theorem 3 in Section 5.1, $\mathbf{A} = \mathbf{ED}_\lambda\mathbf{E}^{-1}$. We have proved the following diagonalization theorem (which is true in any dimension).

Theorem 2. Let $\mathbf{A}$ be an $n \times n$ matrix with n linearly independent eigenvectors $\mathbf{e}_1$, $\mathbf{e}_2, \ldots, \mathbf{e}_n$, let $\mathbf{E}$ be the $n \times n$ matrix whose columns are $\mathbf{e}_i$'s, and let $\mathbf{D}_\lambda$ be a diagonal matrix of the eigenvalues λ_i. Then

$$\mathbf{A} = \mathbf{ED}_\lambda\mathbf{E}^{-1} \tag{14a}$$

and

$$\mathbf{A}^k = \mathbf{ED}_\lambda^k\mathbf{E}^{-1} \tag{14b}$$

Conversely,

$$\mathbf{D}_\lambda = \mathbf{E}^{-1}\mathbf{AE} \tag{15}$$

Verification of (14b) is left as an exercise, and (15) follows directly from (14a). $\mathbf{D}_\lambda^k$ is a diagonal matrix with entry $(i, i) = (\lambda_i)^k$.

**EXAMPLE 2.
Conversion
Between A and $\mathbf{D}_\lambda$**

For the computer/dog matrix, (14a) becomes, where the inverse $\mathbf{E}^{-1} = \begin{bmatrix} \frac{2}{3} & \frac{1}{3} \\ \frac{1}{3} & -\frac{1}{3} \end{bmatrix}$,

$$\mathbf{A} = \begin{bmatrix} 3 & 1 \\ 2 & 2 \end{bmatrix} = \begin{bmatrix} 1 & 1 \\ 1 & -2 \end{bmatrix}\begin{bmatrix} 4 & 0 \\ 0 & 1 \end{bmatrix}\begin{bmatrix} \frac{2}{3} & \frac{1}{3} \\ \frac{1}{3} & -\frac{1}{3} \end{bmatrix}$$

And (14b) becomes

$$\mathbf{A}^k = \begin{bmatrix} 1 & 1 \\ 1 & -2 \end{bmatrix}\begin{bmatrix} 4^k & 0 \\ 0 & 1 \end{bmatrix}\begin{bmatrix} \frac{2}{3} & \frac{1}{3} \\ \frac{1}{3} & -\frac{1}{3} \end{bmatrix} \tag{16}$$

Multiplying out the matrix product for $\mathbf{A}^k$ in (16), we have

$$\mathbf{A}^k = \begin{bmatrix} 1 & 1 \\ 1 & -2 \end{bmatrix}\begin{bmatrix} \frac{2}{3}(4^k) & \frac{1}{3}(4^k) \\ \frac{1}{3} & -\frac{1}{3} \end{bmatrix}$$

$$= \begin{bmatrix} \frac{2}{3}(4^k) + \frac{1}{3} & \frac{1}{3}(4^k) - \frac{1}{3} \\ \frac{2}{3}(4^k) - \frac{2}{3} & \frac{1}{3}(4^k) + \frac{2}{3} \end{bmatrix}$$

Using (15), we can also go from $\mathbf{A}$ to $\mathbf{D}_\lambda$:

$$\mathbf{D}_\lambda = \mathbf{E}^{-1}\mathbf{AE} = \left(\begin{bmatrix} \frac{2}{3} & \frac{1}{3} \\ \frac{1}{3} & -\frac{1}{3} \end{bmatrix}\begin{bmatrix} 3 & 1 \\ 2 & 2 \end{bmatrix}\right)\begin{bmatrix} 1 & 1 \\ 1 & -2 \end{bmatrix}$$

$$= \left(\begin{bmatrix} \frac{8}{3} & \frac{4}{3} \\ \frac{1}{3} & -\frac{1}{3} \end{bmatrix}\right)\begin{bmatrix} 1 & 1 \\ 1 & -2 \end{bmatrix} = \begin{bmatrix} 4 & 0 \\ 0 & 1 \end{bmatrix} \tag{17}$$

■

Equations (14a) and (14b) formalize in single matrix equations our eigenvector-based computations for a growth model. They also provide us with a simple way to compute powers of an $n \times n$ matrix $\mathbf{A}$—provided that we can find n linearly independent eigenvectors of $\mathbf{A}$. The formula $\mathbf{A} = \mathbf{E}\mathbf{D}_\lambda \mathbf{E}^{-1}$ for computing $\mathbf{Ac}$ can be visualized with the diagram in Figure 5.3.

An $n \times n$ matrix $\mathbf{A}$ is said to be ***diagonalizable*** if there exist an invertible matrix $\mathbf{C}$ and a diagonal matrix $\mathbf{D}$ such that $\mathbf{A} = \mathbf{C}\mathbf{D}\mathbf{C}^{-1}$. Here one says that $\mathbf{C}$ *diagonalizes* $\mathbf{A}$. Theorem 2 says that if $\mathbf{A}$ has n linearly independent eigenvectors $\mathbf{e}_i$, then $\mathbf{A}$ is diagonalized by a matrix $\mathbf{E}$ whose columns are the $\mathbf{e}_i$'s.

Besides simplifying the computation of powers of a matrix, diagonalization has many important uses in theory and applications. Matrix diagonalization plays a critical role in solving systems of linear differential equations (developed in Section 6.3). Systems of differential equations arise frequently in many settings in the physical sciences.

EXAMPLE 3.
Diagonalization of a Matrix

The matrix $\mathbf{A} = \begin{bmatrix} 5 & 4 & 2 \\ 4 & 5 & 2 \\ 2 & 2 & 2 \end{bmatrix}$ has eigenvectors

$$\mathbf{e}_1 = \begin{bmatrix} 2 \\ 2 \\ 1 \end{bmatrix}, \quad \mathbf{e}_2 = \begin{bmatrix} 1 \\ 0 \\ -2 \end{bmatrix}, \quad \text{and} \quad \mathbf{e}_3 = \begin{bmatrix} 0 \\ 1 \\ -2 \end{bmatrix},$$

and associated eigenvalues $\lambda_1 = 10$, $\lambda_2 = 1$, and $\lambda_3 = 1$. Here

$$\mathbf{E} = \begin{bmatrix} 2 & 1 & 0 \\ 2 & 0 & 1 \\ 1 & -2 & -2 \end{bmatrix}$$

By Gauss–Jordan elimination on $[\mathbf{E} \quad \mathbf{I}]$, we find $\mathbf{E}^{-1} = \begin{bmatrix} \frac{2}{9} & \frac{2}{9} & \frac{1}{9} \\ \frac{5}{9} & -\frac{4}{9} & -\frac{2}{9} \\ -\frac{4}{9} & -\frac{5}{9} & -\frac{2}{9} \end{bmatrix}$. Then, by Theorem 2,

$$\mathbf{A} = \mathbf{E}\mathbf{D}_\lambda \mathbf{E}^{-1}: \begin{bmatrix} 5 & 4 & 2 \\ 4 & 5 & 2 \\ 2 & 2 & 2 \end{bmatrix} = \begin{bmatrix} 2 & 1 & 0 \\ 2 & 0 & 1 \\ 1 & -2 & -2 \end{bmatrix} \begin{bmatrix} 10 & 0 & 0 \\ 0 & 1 & 0 \\ 0 & 0 & 1 \end{bmatrix} \begin{bmatrix} \frac{2}{9} & \frac{2}{9} & \frac{1}{9} \\ \frac{5}{9} & -\frac{4}{9} & -\frac{2}{9} \\ -\frac{4}{9} & -\frac{5}{9} & -\frac{2}{9} \end{bmatrix} \quad (18) \quad \blacksquare$$

If we convert to eigenvector coordinates and stay in eigenvector coordinates, we have the following result.

Theorem 3. Let $\mathbf{A}$ be an $n \times n$ matrix and let $\mathbf{E}$ be an $n \times n$ matrix whose columns $\mathbf{e}_i$ are a basis for $\mathbf{R}^n$ of n linearly independent eigenvectors of $\mathbf{A}$. If $\mathbf{c} = \mathbf{Ab}$, then converting $\mathbf{b}$ and $\mathbf{c}$ into $\mathbf{e}_i$-coordinates $\mathbf{b}^*$ and $\mathbf{c}^*$, we have

$$\mathbf{c}^* = \mathbf{D}_\lambda \mathbf{b}^* \quad \text{or} \quad c_i^* = \lambda_i b_i^* \quad (19)$$

where $\mathbf{b}^* = [\mathbf{b}]_E = \mathbf{E}^{-1}\mathbf{b}$ and $\mathbf{c}^* = [\mathbf{c}]_E = \mathbf{E}^{-1}\mathbf{c}$, and $\mathbf{D}_\lambda$ is the diagonal matrix whose diagonal entries λ_i are the eigenvalues associated with the $\mathbf{e}_i$.

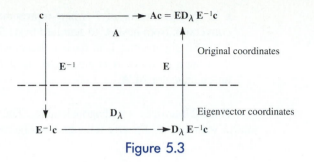

Figure 5.3

Theorem 2 tells us that if $\mathbf{A}$ has n linearly independent eigenvectors, then $\mathbf{A}$ is diagonalized by a matrix of those eigenvectors. But can there be other matrices, unconnected with eigenvectors, that diagonalize $\mathbf{A}$? The answer is no.

Theorem 4. An $n \times n$ matrix $\mathbf{A}$ is diagonalizable if and only if $\mathbf{A}$ has n linearly independent eigenvectors.

Proof: If $\mathbf{A}$ has n linearly independent eigenvectors, then by Theorem 2, $\mathbf{A}$ is diagonalizable.

If $\mathbf{A}$ is diagonalizable, there exist an $n \times n$ invertible matrix $\mathbf{C}$ and a diagonal matrix $\mathbf{D}$ such that $\mathbf{A} = \mathbf{CDC}^{-1}$. We shall show that the n columns of $\mathbf{C}$ are eigenvectors of $\mathbf{A}$. Note that the columns of $\mathbf{C}$ are linearly independent, since $\mathbf{C}$ is invertible (Theorem 3 of Section 4.2). We give two proofs that the columns of $\mathbf{C}$ are eigenvectors, one using methods of matrix algebra and one using the theory of vector spaces and linear transformations.

Matrix Algebra Proof: Consider the first column $\mathbf{c}_1$ of $\mathbf{C}$ (the argument is the same for the other columns). Let d_{11} be the first diagonal entry in the diagonal matrix $\mathbf{D}$. We shall show that $\mathbf{Ac}_1 = d_{11}\mathbf{c}_1$, proving that $\mathbf{c}_1$ is an eigenvalue of $\mathbf{A}$. The key fact we need is that $\mathbf{C}^{-1}\mathbf{c}_1 = \mathbf{i}_1$. This follows by noting that $\mathbf{C}^{-1}\mathbf{c}_1$ is the first column of the matrix product $\mathbf{C}^{-1}\mathbf{C}$, which equals identity matrix $\mathbf{I}$, whose first column is $\mathbf{i}_1$. Recall also that for any matrix $\mathbf{B}$, $\mathbf{Bi}_1 = \mathbf{b}_1$ (the first column of $\mathbf{B}$). Further, because $\mathbf{D}$ is diagonal, $\mathbf{Di}_1 (= \mathbf{d}_1) = d_{11}\mathbf{i}_1$ (a vector with d_{11} in the first entry and all other entries 0). Now we are ready for our matrix algebra calculation showing that $\mathbf{Ac}_1 = d_{11}\mathbf{c}_1$.

$$\mathbf{Ac}_1 = \mathbf{CDC}^{-1}\mathbf{c}_1 = \mathbf{CD}(\mathbf{C}^{-1}\mathbf{c}_1) = \mathbf{CD}(\mathbf{i}_1)$$
$$= \mathbf{C}(\mathbf{Di}_1) = \mathbf{C}(d_{11}\mathbf{i}_1) \qquad (20)$$
$$= d_{11}(\mathbf{Ci}_1) = d_{11}\mathbf{c}_1$$

Theoretical Proof: Let T be the linear transformation $T: \mathbf{R}^n \to \mathbf{R}^n$ associated with the matrix mapping $T([\mathbf{u}]_B) = \mathbf{A}[\mathbf{u}]_B$ in the standard basis B for $\mathbf{R}^n$. If $\mathbf{A} = \mathbf{CDC}^{-1}$, then T can also be expressed as $T([\mathbf{u}]_C) = \mathbf{D}[\mathbf{u}]_C$ in the basis C formed by the columns $\mathbf{c}_i$ of $\mathbf{C}$; see Figure 5.3 and Theorem 3 (with $\mathbf{C}$ replacing $\mathbf{E}$ and $\mathbf{D}$ replacing $\mathbf{D}_\lambda$). Observe that the coordinate vectors $\mathbf{i}_k$ are eigenvectors of T in the basis C, since $\mathbf{Di}_k = d_{kk}\mathbf{i}_k$. So there is an eigenvector $\mathbf{e}_k$ of T with $[\mathbf{e}_k]_C = \mathbf{i}_k$, for any k. Let us convert $\mathbf{i}_k = [\mathbf{e}_k]_C$ from basis C back to an eigenvector in the standard basis B [if a vector $\mathbf{e} \in \mathbf{R}^n$ is an eigenvector of T, so that $T(\mathbf{e}) = \lambda\mathbf{e}$, then

e is an eigenvector of T when represented in any coordinate basis]. For this conversion from basis C to standard basis B, the transition matrix is **C** (see Figure 5.3). So eigenvector $\mathbf{i}_k$ in basis C becomes the eigenvector $\mathbf{Ci}_k = \mathbf{c}_k$ in the standard basis B. Since **A** is the matrix for T in the standard basis, $\mathbf{c}_k$ (for each k) is an eigenvector of **A**. ■

Not all matrices are diagonalizable. The following is an example of a 2×2 matrix without two linearly independent eigenvectors. Such a matrix is called *defective*.

EXAMPLE 4.
Nondiagonalizable
Matrix

Consider the matrix $\mathbf{A} = \begin{bmatrix} 1 & 1 \\ 0 & 1 \end{bmatrix}$. Its characteristic polynomial is $\det(\mathbf{A} - \lambda\mathbf{I}) = \lambda^2 - 2\lambda + 1 = (\lambda - 1)^2$ (check this), so its eigenvalues are 1: $\lambda_1 = \lambda_2 = 1$. **A** does not have two different eigenvalues, so Theorem 1 does not apply. Any eigenvector **e** of **A** must satisfy $(\mathbf{A} - 1\mathbf{I})\mathbf{e} = \mathbf{0}$. That is,

$$(\mathbf{A} - \mathbf{I})\mathbf{e} = \mathbf{0} \quad \text{or} \quad \begin{array}{l} 0e_1 + 1e_2 = 0 \\ 0e_1 + 0e_2 = 0 \end{array} \tag{21}$$

The first equation reduces to $e_2 = 0$, and the second equation is vacuous. A solution **e** to (21) can have any value for e_1 while $e_2 = 0$. However, all such vectors $\mathbf{e} = \begin{bmatrix} e_1 \\ 0 \end{bmatrix}$ are multiples of one another. Thus **A** does not have two linearly independent eigenvectors. ■

The diagonalization process becomes simpler in the case where the eigenvectors are orthogonal, or even better orthonormal. Recall from Section 4.5 that when **E** has orthonormal columns, then $\mathbf{E}^{-1} = \mathbf{E}^T$, so $\mathbf{A} = \mathbf{ED}_\lambda\mathbf{E}^{-1}$ becomes $\mathbf{A} = \mathbf{ED}_\lambda\mathbf{E}^T$. All symmetric matrices have this property of orthogonal eigenvectors. This fact is difficult to prove and we shall only prove the following weaker result.

Theorem 5. Let **A** be a symmetric matrix. Let $\mathbf{e}_1$ and $\mathbf{e}_2$ be eigenvectors of **A** associated with distinct eigenvalues λ_1 and λ_2, respectively. Then $\mathbf{e}_1$ and $\mathbf{e}_2$ are orthogonal.

Proof: To show that $\mathbf{e}_1$ and $\mathbf{e}_2$ are orthogonal, we must show that $\mathbf{e}_1 \cdot \mathbf{e}_2 = 0$. First note that since **A** is symmetric, $\mathbf{e}_1^T\mathbf{A} = \lambda_1\mathbf{e}_1^T$ (where $\mathbf{e}_1^T$ is the transpose of $\mathbf{e}_1$—switching $\mathbf{e}_1$ into a row vector). Now we use the associative law on $\mathbf{e}_1^T\mathbf{Ae}_2$.

$$\mathbf{e}_1^T\mathbf{Ae}_2 = (\mathbf{e}_1^T\mathbf{A})\mathbf{e}_2 = (\lambda_1\mathbf{e}_1^T)\mathbf{e}_2 = \lambda_1(\mathbf{e}_1^T\mathbf{e}_2) \tag{22a}$$

$$\mathbf{e}_1^T\mathbf{Ae}_2 = \mathbf{e}_1^T(\mathbf{Ae}_2) = \mathbf{e}_1^T(\lambda_2\mathbf{e}_2) = \lambda_2(\mathbf{e}_1^T\mathbf{e}_2) \tag{22b}$$

Since $\lambda_1 \neq \lambda_2$, the only way for the expressions in (22a) and (22b) to be equal is if $\mathbf{e}_1^T\mathbf{e}_2 = 0$, or equivalently, $\mathbf{e}_1 \cdot \mathbf{e}_2 = 0$. ■

Corollary. When **A** is symmetric with distinct eigenvalues, the matrix **E** in Theorem 2 can be chosen to have orthonormal columns so that $\mathbf{E}^{-1} = \mathbf{E}^T$. Then $\mathbf{A} = \mathbf{E}\mathbf{D}_\lambda\mathbf{E}^T$.

Even when the eigenvalues are not distinct, it can be proved that there always exists a set of n orthonormal eigenvectors for any $n \times n$ symmetric matrix **A**.

EXAMPLE 5.
Diagonalization
with a Symmetric
Matrix

Suppose we know that the symmetric matrix $\mathbf{A} = \begin{bmatrix} 3 & 2 & 2 \\ 2 & 2 & 0 \\ 2 & 0 & 4 \end{bmatrix}$ has eigenvalues $\lambda_1 = 6$, $\lambda_2 = 3$, $\lambda_3 = 0$ with associated orthonormal eigenvectors

$$\mathbf{e}_1 = \begin{bmatrix} \frac{2}{3} \\ \frac{1}{3} \\ \frac{2}{3} \end{bmatrix}, \quad \mathbf{e}_2 = \begin{bmatrix} \frac{1}{3} \\ \frac{2}{3} \\ -\frac{2}{3} \end{bmatrix}, \quad \mathbf{e}_3 = \begin{bmatrix} -\frac{2}{3} \\ \frac{2}{3} \\ \frac{1}{3} \end{bmatrix}$$

Then by the preceding corollary, we have

$$\mathbf{A} = \mathbf{E}\mathbf{D}_\lambda\mathbf{E}^T: \begin{bmatrix} 6 & 0 & 0 \\ 0 & 3 & 0 \\ 0 & 0 & 0 \end{bmatrix} = \begin{bmatrix} \frac{2}{3} & \frac{1}{3} & -\frac{2}{3} \\ \frac{1}{3} & \frac{2}{3} & \frac{2}{3} \\ \frac{2}{3} & -\frac{2}{3} & \frac{1}{3} \end{bmatrix} \begin{bmatrix} 3 & 2 & 2 \\ 2 & 2 & 0 \\ 2 & 0 & 4 \end{bmatrix} \begin{bmatrix} \frac{2}{3} & \frac{1}{3} & \frac{2}{3} \\ \frac{1}{3} & \frac{2}{3} & -\frac{2}{3} \\ -\frac{2}{3} & \frac{2}{3} & \frac{1}{3} \end{bmatrix} \qquad (23)$$

■

The matrix in Example 3 was also symmetric, although the three eigenvectors given were not mutually orthogonal. Theorem 5 applies only when the eigenvalues are distinct, whereas the matrix in Example 3 has a repeated eigenvalue. The eigenvectors associated with the repeated eigenvalue can be converted into orthonormal eigenvectors using Gram–Schmidt orthogonalization.

We now apply diagonalization to a well-known problem involving symmetric matrices. ***Quadratic forms in x and y*** are expressions of the form $ax^2 + bxy + cy^2$. They were introduced briefly at the end of Section 1.2. Equations involving quadratic forms, such as

$$ax^2 + bxy + cy^2 = d \quad \text{or} \quad ax^2 + bxy + cy^2 + ex + fy = d \qquad (24)$$

define *conic sections*. Such equations form circles, ellipses, and hyperbolas centered at the origin (actually, in special degenerate cases, these equations can reduce to lines or single points). Quadratic forms can be written with matrix algebra in the form $\mathbf{z}^T\mathbf{A}\mathbf{z}$,

where $\quad \mathbf{z} = \begin{bmatrix} x \\ y \end{bmatrix}, \quad \mathbf{z}^T = [x, y], \quad \text{and} \quad \mathbf{A} = \begin{bmatrix} a & b/2 \\ b/2 & c \end{bmatrix}$

For example, if $\mathbf{A} = \begin{bmatrix} 3 & 1 \\ 1 & 3 \end{bmatrix}$, then

$$\mathbf{z}^T\mathbf{A}\mathbf{z} = [x, y]\begin{bmatrix} 3 & 1 \\ 1 & 3 \end{bmatrix}\begin{bmatrix} x \\ y \end{bmatrix} = [3x + y, x + 3y]\begin{bmatrix} x \\ y \end{bmatrix}$$

$$= (3x + y)x + (x + 3y)y \qquad (25)$$

$$= 3x^2 + 2xy + 3y^2$$

Thus, for this $\mathbf{A}$ and a right-side value of 8, we obtain the equation of the conic section

$$\mathbf{z}^T\mathbf{A}\mathbf{z} = 8 \quad \text{or} \quad 3x^2 + 2xy + 3y^2 = 8 \tag{26}$$

The graph of this equation is plotted in Figure 5.4(a); it is an ellipse. Notice that the axes of symmetry in Figure 5.4(a) for this ellipse are rotated 45° from the normal x–y axes.

By using an eigenvector basis to diagonalize $\mathbf{A}$, we can align the principal axes of the ellipse (its axes of symmetry) with the new coordinate axes, putting the conic section in what is called *standard form*. The eigenvalues of $\mathbf{A}$ are $\lambda_1 = 4$ and $\lambda_2 = 2$, since $\det(\mathbf{A} - \lambda\mathbf{I}) = \lambda^2 - 6\lambda + 8 = (\lambda - 4)(\lambda - 2)$. The orthonormal eigenvectors are $\mathbf{e}_1 = (1/\sqrt{2})\begin{bmatrix} 1 \\ 1 \end{bmatrix}$ and $\mathbf{e}_2 = (1/\sqrt{2})\begin{bmatrix} 1 \\ -1 \end{bmatrix}$ [found by solving $(\mathbf{A} - 4\mathbf{I})\mathbf{z} = \mathbf{0}$ and $(\mathbf{A} - 2\mathbf{I})\mathbf{z} = \mathbf{0}$, and then making the vectors unit length]. The change of coordinates is $\mathbf{z}' = \mathbf{E}^{-1}\mathbf{z} = \mathbf{E}^T\mathbf{z}$, where $\mathbf{E}$'s columns are $\mathbf{e}_1$, $\mathbf{e}_2$. The ellipse in (26) in $\mathbf{e}_1 - \mathbf{e}_2$ coordinates will now be

$$\mathbf{z}'^T\begin{bmatrix} 4 & 0 \\ 0 & 2 \end{bmatrix}\mathbf{z}' = 8 \quad \text{or} \quad 4x'^2 + 2y'^2 = 8 \tag{27}$$

Figure 5.4(b) shows the ellipse in standard form in the new coordinates.

There are three important observations to be made.

1. The coefficients of x'^2 and y'^2 in (27) are precisely the eigenvalues of $\mathbf{A}$.
2. If we write the quadratic form $\mathbf{z}'^T\mathbf{D}_\lambda\mathbf{z}'$ in (27) in terms of our original basis for $\mathbf{z}$, using $\mathbf{z}' = \mathbf{E}^T\mathbf{z}$, we have

$$\begin{aligned} \mathbf{z}'^T\mathbf{D}_\lambda\mathbf{z}' &= (\mathbf{E}^T\mathbf{z})^T\mathbf{D}_\lambda(\mathbf{E}^T\mathbf{z}) \\ &= \mathbf{z}^T\mathbf{E}\mathbf{D}_\lambda\mathbf{E}^T\mathbf{z} = \mathbf{z}^T\mathbf{A}\mathbf{z} \end{aligned} \tag{28}$$

Thus an orthonormal change of basis does not change the value of the quadratic form.
3. Recall Theorem 4 of Section 4.5, which stated that an orthonormal change of basis, such as $\mathbf{z}' = \mathbf{E}^T\mathbf{z}$, corresponds to a rotation (and/or reflection) of the coordinate system. In this instance, the change of coordinates was a 45° rotation [see Figure 5.4(a)].

The following theorem summarizes this analysis (a formal proof is omitted).

Theorem 6 (Principal Axes Theorem). The conic section $ax^2 + bxy + cy^2 = d$ or $\mathbf{z}^T\mathbf{A}\mathbf{z} = d$ can be rotated into standard form, so that its principal axes are the coordinate axes, by the change of basis $\mathbf{z}' = \mathbf{E}^T\mathbf{z}$, where $\mathbf{E}$'s columns are the (unit-length) eigenvectors $\mathbf{e}_1$, $\mathbf{e}_2$ of $\mathbf{A}$. The standard equation for the conic section in the new coordinates is $\lambda_1 x'^2 + \lambda_2 y'^2 = d$, where λ_1, λ_2 are the eigenvalues of $\mathbf{A}$.

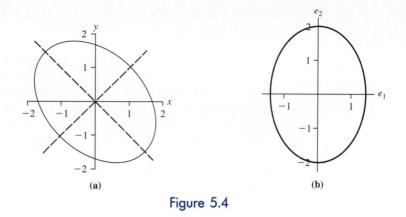

Figure 5.4

**EXAMPLE 6.
Putting a
Quadratic Form
in Standard Form**

Consider the conic section $x^2 + 4xy - 2y^2 = 10$. Here $\mathbf{A} = \begin{bmatrix} 1 & 2 \\ 2 & -2 \end{bmatrix}$. The roots of the characteristic polynomial $\det(\mathbf{A} - \lambda\mathbf{I}) = \lambda^2 + \lambda - 6 = (\lambda + 3)(\lambda - 2)$ are $\lambda_1 = -3$ and $\lambda_2 = 2$. The eigenvectors for λ_1 are the solutions to

$$(\mathbf{A} + 3\mathbf{I})\mathbf{e}_1 = \mathbf{0}: \quad \begin{matrix} 4e_{11} + 2e_{21} = 0 \\ 2e_{11} + 1e_{21} = 0 \end{matrix} \implies e_{21} = 2e_{11} \quad \text{or} \quad \mathbf{e}_1 = \frac{1}{\sqrt{5}}\begin{bmatrix} 1 \\ 2 \end{bmatrix} \qquad (29)$$

Recall that to make a vector $\mathbf{u}$ unit length, we divide it by its length $\sqrt{\mathbf{u} \cdot \mathbf{u}}$. We can avoid directly computing the second eigenvector $\mathbf{e}_2$. Since $\mathbf{A}$ is symmetric with distinct eigenvalues, $\mathbf{e}_1$ and $\mathbf{e}_2$ must be orthogonal by Theorem 5. In $\mathbf{R}^2$, the vector orthogonal to $[a, b]$ and of the same length is $[b, -a]$ (or $[-b, a]$). Then $\mathbf{e}_2 = 1/\sqrt{5}\begin{bmatrix} 2 \\ -1 \end{bmatrix}$.

By Theorem 6, the change of basis $\mathbf{z}' = \mathbf{E}^T\mathbf{z}$, where $\mathbf{E} = \frac{1}{\sqrt{5}}\begin{bmatrix} 1 & 2 \\ 2 & -1 \end{bmatrix}$ converts our conic section into the standard equation $-3x'^2 + 2y'^2 = 10$. ▪

If a conic section is not centered at the origin, a translation as well as an orthonormal transformation of coordinates will be needed to put the figure in standard form. Quadratic forms can be defined in n dimensions, and the same principal axes theorem applies to the associated $n \times n$ symmetric matrix $\mathbf{A}$.

Since the change-of-basis process outlined in (10) to (13) and in the diagram in Figure 5.3 is not dependent on the fact that the matrix in the new (eigenvector) coordinates is diagonal, Theorem 2 generalizes to any change of basis in any finite-dimensional vector space.

Theorem 7. Let $T: V \to V$ be a matrix mapping $\mathbf{x} \to T(\mathbf{x}) = \mathbf{A}\mathbf{x}$ on an n-dimensional vector space V, where $\mathbf{A}$ is the $n \times n$ matrix expressed in terms of basis $B = \{\mathbf{b}_1, \mathbf{b}_2, \ldots, \mathbf{b}_n\}$. Let $\mathbf{C}$ be the $n \times n$ matrix for the mapping T expressed in terms of a second basis $D = \{\mathbf{d}_1, \mathbf{d}_2, \ldots, \mathbf{d}_n\}$. Then there exists an invertible matrix $\mathbf{U}$ such that

$$\mathbf{C} = \mathbf{U}^{-1}\mathbf{A}\mathbf{U} \quad \text{and} \quad \mathbf{A} = \mathbf{U}\mathbf{C}\mathbf{U}^{-1} \tag{30}$$

The matrices $\mathbf{U}$ and $\mathbf{U}^{-1}$ are transition matrices for the change of basis between basis D and B: $\mathbf{U}$ goes from D to B and $\mathbf{U}^{-1}$ goes from B to D. From the change-of-basis discussion in Section 4.3, the columns of $\mathbf{U}$ are the n basis vectors $\mathbf{d}_i$ of D expressed in terms of the B basis. That is, $\mathbf{u}_i = [\mathbf{d}_i]_B$. The columns of $\mathbf{U}^{-1}$ are $[\mathbf{b}_i]_D$.

To distinguish between a linear transformation $T: V \rightarrow V$ and the different matrices that can represent T in different bases, a notation similar to the change-of-basis notation $[\cdot]_B$ is used.

Notation

We write $\mathbf{A} = [T]_B$ to indicate that $\mathbf{A}$ is the matrix for T in the basis B.

Matrices $\mathbf{A}$ and $\mathbf{C}$ are called *similar* if there is an invertible matrix $\mathbf{U}$ such that (30) is true. *Intuitively, similar matrices represent the same linear transformation expressed in different bases.*

EXAMPLE 7.
Change of Basis
for a Linear
Transformation

Consider the matrix $\mathbf{A} = \begin{bmatrix} 3 & 1 \\ 2 & 2 \end{bmatrix}$. $\mathbf{A}$ is the matrix for T: $\begin{bmatrix} x_1 \\ x_2 \end{bmatrix} \rightarrow \begin{bmatrix} 3x_1 + x_2 \\ 2x_1 + 2x_2 \end{bmatrix}$ in the standard coordinate basis $B = \{\mathbf{i}_1, \mathbf{i}_2\}$ for $\mathbf{R}^2$. Thus $\mathbf{A} = [T]_B$.

(a) Consider the basis $D = \left\{ \begin{bmatrix} 2 \\ 1 \end{bmatrix}, \begin{bmatrix} 2 \\ 0 \end{bmatrix} \right\}$. The transition matrix $\mathbf{U}$ from D to B is

$\mathbf{U} = \begin{bmatrix} 2 & 2 \\ 1 & 0 \end{bmatrix}$ and we find that $\mathbf{U}^{-1} = \begin{bmatrix} 0 & 1 \\ \frac{1}{2} & -1 \end{bmatrix}$. Converting T from matrix $\mathbf{A}$ in the basis B to a matrix $\mathbf{C}$ in basis D we obtain, by (30),

$$\mathbf{C} \ (= [T]_D) = \mathbf{U}^{-1}\mathbf{A}\mathbf{U} = \begin{bmatrix} 0 & 1 \\ \frac{1}{2} & -1 \end{bmatrix}\begin{bmatrix} 3 & 1 \\ 2 & 2 \end{bmatrix}\begin{bmatrix} 2 & 2 \\ 1 & 0 \end{bmatrix} = \begin{bmatrix} 6 & 4 \\ -\frac{5}{2} & -1 \end{bmatrix} \tag{31}$$

By (31), we see that $\mathbf{A}$ and $\mathbf{C}$ are similar matrices.

(b) Next consider the basis $E = \left\{ \begin{bmatrix} 1 \\ 1 \end{bmatrix}, \begin{bmatrix} 1 \\ -2 \end{bmatrix} \right\}$. The transition matrix from E to B is

$\mathbf{E} = \begin{bmatrix} 1 & 1 \\ 1 & -2 \end{bmatrix}$ and from B to E is $\mathbf{E}^{-1} = \begin{bmatrix} \frac{2}{3} & \frac{1}{3} \\ \frac{1}{3} & -\frac{1}{3} \end{bmatrix}$. Converting T from matrix $\mathbf{A}$ in the basis B to a matrix $\mathbf{C}$ in basis E, we obtain, by (30),

$$\mathbf{C} \ (= [T]_E) = \mathbf{E}^{-1}\mathbf{A}\mathbf{E} = \begin{bmatrix} \frac{2}{3} & \frac{1}{3} \\ \frac{1}{3} & -\frac{1}{3} \end{bmatrix}\begin{bmatrix} 3 & 1 \\ 2 & 2 \end{bmatrix}\begin{bmatrix} 1 & 1 \\ 1 & -2 \end{bmatrix} = \begin{bmatrix} 4 & 0 \\ 0 & 1 \end{bmatrix} \tag{32}$$

Note that $\mathbf{A}$ is the matrix in Examples 1 and 2, and E is the eigenvector basis. By (32), $\mathbf{A}$ is similar to the diagonal matrix $\mathbf{C}$. ▪

**EXAMPLE 8.
Change of
Basis for a
Non-Euclidean
Vector Space**

In this example we work in the non-Euclidean vector space $\mathcal{P}_2$ of polynomials of greatest degree ≤ 2. The standard basis for $\mathcal{P}_2$ is $B = \{x^2, x, 1\}$. Consider the shift S_5 that maps $p(x)$ to $p(x + 5)$. In Example 8 of Section 5.1, we determined the entries in the matrix $\mathbf{S}$ for this transformation in the standard basis B:

$$\mathbf{S} = [S_5]_B = \begin{bmatrix} 1 & 0 & 0 \\ 10 & 1 & 0 \\ 25 & 5 & 1 \end{bmatrix} \tag{33}$$

Let us convert $\mathbf{S}$ to a matrix $\mathbf{C}$ in the basis $D = \{x^2 + x, x^2 + 1, x + 1\}$. This change from basis was discussed in Example 8 of Section 4.3. The jth column of the transition matrix $\mathbf{U}$ from D to B is $[d_j(x)]_B$. Thus

$$\mathbf{U} = \begin{bmatrix} 1 & 1 & 0 \\ 1 & 0 & 1 \\ 0 & 1 & 1 \end{bmatrix} \quad \text{and we find that} \quad \mathbf{U}^{-1} = \begin{bmatrix} \frac{1}{2} & \frac{1}{2} & -\frac{1}{2} \\ \frac{1}{2} & -\frac{1}{2} & \frac{1}{2} \\ -\frac{1}{2} & \frac{1}{2} & \frac{1}{2} \end{bmatrix} \tag{34}$$

Then, by (30),

$$\mathbf{C} \,(= [S_5]_D) = \mathbf{U}^{-1}\mathbf{A}\mathbf{U} = \begin{bmatrix} \frac{1}{2} & \frac{1}{2} & -\frac{1}{2} \\ \frac{1}{2} & -\frac{1}{2} & \frac{1}{2} \\ -\frac{1}{2} & \frac{1}{2} & \frac{1}{2} \end{bmatrix} \begin{bmatrix} 1 & 0 & 0 \\ 10 & 1 & 0 \\ 25 & 5 & 1 \end{bmatrix} \begin{bmatrix} 1 & 1 & 0 \\ 1 & 0 & 1 \\ 0 & 1 & 1 \end{bmatrix}$$

$$= \begin{bmatrix} -9 & -\frac{15}{2} & -\frac{5}{2} \\ 10 & \frac{17}{2} & \frac{5}{2} \\ 20 & \frac{35}{2} & \frac{7}{2} \end{bmatrix} \tag{35}$$

▪

Theorem 7 can be extended to transformations $T: V \rightarrow W$ defined by a matrix mapping $\mathbf{x} \rightarrow T(\mathbf{x}) = \mathbf{A}\mathbf{x}$, where V has basis B and W has basis B'. If D is another basis for V and D' is another basis for W, then to convert $\mathbf{A}$ into a mapping in bases D and D' we need the transition matrix $\mathbf{U}$ from D to B with columns $\mathbf{u}_i = [\mathbf{d}_i]_B$, as in Theorem 7, and also transition matrix $\mathbf{U}'$ from D' to B' with columns $\mathbf{u}'_i = [\mathbf{d}'_i]_{B'}$. Then it can be shown that $\mathbf{C} = \mathbf{U}'^{-1}\mathbf{A}\mathbf{U}$ and $\mathbf{A} = \mathbf{U}'\mathbf{C}\mathbf{U}^{-1}$.

Section 5.3 Exercises

1. Compute the representation $\mathbf{A} = \mathbf{E}\mathbf{D}_\lambda\mathbf{E}^{-1}$ of Theorem 2 for the following matrices $\mathbf{A}$, whose eigenvalues and largest eigenvector you were asked to determine in Exercise 5 of Section 3.3.

   (a) $\begin{bmatrix} 4 & 0 \\ 2 & 2 \end{bmatrix}$ (b) $\begin{bmatrix} 1 & 2 \\ 5 & 4 \end{bmatrix}$ (c) $\begin{bmatrix} 2 & 1 \\ 2 & 3 \end{bmatrix}$ (d) $\begin{bmatrix} 4 & -1 \\ 1 & 2 \end{bmatrix}$

2. Compute the representation $A = ED_\lambda E^T$ for the following symmetric matrices A, where E consists of orthonormal eigenvectors.

(a) $\begin{bmatrix} 1 & 4 \\ 4 & 1 \end{bmatrix}$ (b) $\begin{bmatrix} 1 & 2 \\ 2 & 4 \end{bmatrix}$ (c) $\begin{bmatrix} 2 & -3 \\ -3 & 2 \end{bmatrix}$ (d) $\begin{bmatrix} -1 & -5 \\ -5 & -1 \end{bmatrix}$

(e) $\begin{bmatrix} 0 & 1 & 1 \\ 1 & 0 & 1 \\ 1 & 1 & 0 \end{bmatrix}$ (f) $\begin{bmatrix} 3 & 2 & 4 \\ 2 & 0 & 2 \\ 4 & 2 & 3 \end{bmatrix}$ (g) $\begin{bmatrix} 1 & 1 & 2 \\ 1 & 3 & 1 \\ 2 & 1 & 1 \end{bmatrix}$

3. For a starting vector of $p = \begin{bmatrix} 10 \\ 10 \end{bmatrix}$, compute $p^{(10)} = A^{10}p$ for each matrix A in Exercise 1. (Use your representation of A found in Exercise 1.)

4. Compute A^5 for each matrix in Exercise 2 using the formula $A^5 = ED_\lambda^5 E^T$.

5. (a) Given that $A = ED_\lambda E^{-1}$, prove $A^2 = ED_\lambda^2 E^{-1}$.
   (b) Use induction to prove that $A^k = ED_\lambda^k E^{-1}$.

6. (a) Obtain a formula for A^{-1} similar to $A = ED_\lambda E^{-1}$. [Hint: Only the matrix D_λ will be different.]
   (b) Verify your formula in part (a) for $A = \begin{bmatrix} 3 & 1 \\ 2 & 2 \end{bmatrix}$.

7. Which of the following matrices A are diagonalizable? If diagonalizable, give E and D_λ such that $A = ED_\lambda E^{-1}$.

(a) $\begin{bmatrix} 2 & 1 \\ 1 & 2 \end{bmatrix}$ (b) $\begin{bmatrix} 2 & 5 \\ 6 & 1 \end{bmatrix}$ (c) $\begin{bmatrix} 0 & 1 \\ 0 & 0 \end{bmatrix}$ (d) $\begin{bmatrix} \frac{1}{4} & \frac{1}{2} \\ \frac{3}{4} & \frac{1}{2} \end{bmatrix}$ (e) $\begin{bmatrix} 1 & 1 \\ 2 & 2 \end{bmatrix}$

8. Find the principal axes for the following quadratic forms and perform the change of basis to put the quadratic form into standard form. Plot the equation before and after the change of basis.
   (a) $5x^2 - 4xy + 8y^2 = 18$ (b) $3x^2 - 8xy + 3y^2 = 24$ (c) $x^2 + 8xy + y^2 = 16$
   (d) $2x^2 - 6xy + 2y^2 = 24$ (e) $x^2 + 4xy + 4y^2 = 12$

9. For the linear transformation T with associated matrix $A = \begin{bmatrix} 1 & 2 \\ 4 & 5 \end{bmatrix}$ in standard coordinates, give the matrix for T with respect to the following bases.

(a) $\begin{bmatrix} 1 \\ 1 \end{bmatrix}, \begin{bmatrix} -1 \\ 1 \end{bmatrix}$ (b) $\begin{bmatrix} 1 \\ 1 \end{bmatrix}, \begin{bmatrix} 0 \\ 1 \end{bmatrix}$ (c) $\begin{bmatrix} 2 \\ 1 \end{bmatrix}, \begin{bmatrix} 1 \\ 2 \end{bmatrix}$ (d) $\begin{bmatrix} -1 \\ 2 \end{bmatrix}, \begin{bmatrix} 1 \\ -3 \end{bmatrix}$

10. For the linear transformation T with associated matrix $A = \begin{bmatrix} 3 & 1 \\ 0 & 2 \end{bmatrix}$ in standard coordinates, give the matrix for T with respect to the following bases.

(a) $\begin{bmatrix} 0 \\ 1 \end{bmatrix}, \begin{bmatrix} -2 \\ 1 \end{bmatrix}$ (b) $\begin{bmatrix} 1 \\ -1 \end{bmatrix}, \begin{bmatrix} 3 \\ 1 \end{bmatrix}$ (c) $\begin{bmatrix} 2 \\ 3 \end{bmatrix}, \begin{bmatrix} 3 \\ 5 \end{bmatrix}$ (d) $\begin{bmatrix} -1 \\ 0 \end{bmatrix}, \begin{bmatrix} 1 \\ -3 \end{bmatrix}$

11. For the linear transformation T with associated matrix $A = \begin{bmatrix} 3 & 1 & 1 \\ 0 & 2 & 1 \\ 3 & 0 & 2 \end{bmatrix}$ in standard coordinates, give the matrix for T with respect to the following bases.

(a) $\begin{bmatrix} 0 \\ 1 \\ 1 \end{bmatrix}, \begin{bmatrix} 1 \\ -2 \\ 1 \end{bmatrix}, \begin{bmatrix} 1 \\ 0 \\ 1 \end{bmatrix}$ **(b)** $\begin{bmatrix} 1 \\ 1 \\ -1 \end{bmatrix}, \begin{bmatrix} 2 \\ 0 \\ 1 \end{bmatrix}, \begin{bmatrix} 0 \\ -2 \\ 1 \end{bmatrix}$

(c) $\begin{bmatrix} 2 \\ 0 \\ 3 \end{bmatrix}, \begin{bmatrix} 3 \\ 1 \\ 2 \end{bmatrix}, \begin{bmatrix} 1 \\ -2 \\ 1 \end{bmatrix}$ **(d)** $\begin{bmatrix} 1 \\ 2 \\ 0 \end{bmatrix}, \begin{bmatrix} 1 \\ 1 \\ -3 \end{bmatrix}, \begin{bmatrix} 1 \\ -2 \\ 1 \end{bmatrix}$

12. For the following linear transformations $T: \mathcal{P}_2 \to \mathcal{P}_2$ of polynomials of greatest degree ≤ 2, give the transformation matrix in terms of the basis $\{x^2 + x, x^2 + 1, x + 1\}$.

   (a) $T(p(x)) = R(p(x)) = (p(-x))$ **(b)** $T(p(x)) = \mathcal{D}(p(x)) = \left(\dfrac{d}{dx} p(x) \right)$

   (c) $T(p(x)) = S_3(p(x)) = (p(x + 3))$

13. If **A** is similar to **B** and **B** is similar to **C**, show that **A** is similar to **C**.

14. If **A** is similar to **B**, prove that $\mathbf{A}^n$ is similar to $\mathbf{B}^n$ for any positive integer n.

15. In the proof of Theorem 1, show that $\mathbf{u}_1$ and $\mathbf{u}_2$ are linearly independent by supposing the opposite, that $\mathbf{u}_2 = r\mathbf{u}_1$, and obtaining a contradiction when $\mathbf{Au}_2 \neq \mathbf{A}(r\mathbf{u}_1)$.

16. Show that if two $n \times n$ matrices **A** and **C** are similar, they represent the same linear transformation T on an n-dimensional vector space with respect to different bases.

17. Prove that if **A** and **B** are similar, then $\det(\mathbf{A}) = \det(\mathbf{B})$. (Hint: Use Theorems 5 and 6 in Section 1.6.)

18. Show that if a 2×2 matrix **A** has two eigenvectors that are orthogonal, **A** is symmetric.

19. If matrix **A** is similar to matrix **B**, prove that the rank of **A** equals the rank of **B**.

5.4 Eigenvector Decomposition

In Section 5.3 we used eigenvectors and the theory of vector spaces to simplify the form of matrix-based linear transformations. In this section we look directly at the problem of simplifying the information contained in matrices, whether they represent linear transformations or are just arrays of data. The simplification will be a decomposition of a matrix into simple matrices. In the process, this decomposition provides a useful scheme for finding the eigenvalues and eigenvectors of any symmetric matrix.

We started this text by introducing vectors and matrices as a convenient way to represent and manipulate sets of numbers. Now we conclude the theoretical development in this text by using our theory to extract underlying patterns in these arrays of numbers.

Our analysis in the preceding section of diagonalizing a matrix **A** yielded the eigenvector decomposition $\mathbf{A} = \mathbf{ED}_\lambda \mathbf{E}^{-1}$. We earlier saw that Gaussian elimination could be represented as the matrix decomposition $\mathbf{A} = \mathbf{LU}$ in Section 3.2, and that Gram–Schmidt orthogonalization yielded the decomposition $\mathbf{A} = \mathbf{QR}$ in Section 4.5. In Section 3.2 we showed that the **LU** decomposition in turn yielded a decomposition of **A** into a sum of simple matrices. We now show how the decomposition $\mathbf{A} = \mathbf{ED}_\lambda \mathbf{E}^{-1}$ also can be reformulated as a decomposition of **A** into simple matrices.

Recall that if $\mathbf{c} = \begin{bmatrix} 1 \\ 2 \end{bmatrix}$ and $\mathbf{d} = [3, 4, -1]$, the simple matrix $\mathbf{c} * \mathbf{d}$ equals

$$\mathbf{c} * \mathbf{d} = \begin{bmatrix} 1 \\ 2 \end{bmatrix} * [3, 4, -1] = \begin{bmatrix} 3 & 4 & -1 \\ 6 & 8 & -2 \end{bmatrix} \tag{1}$$

In general, entry (i, j) in $\mathbf{c} * \mathbf{d}$ equals $c_i d_j$. In Theorem 3 of Section 3.2 we observed that the matrix product $\mathbf{CD}$ can be decomposed into a sum of simple matrices formed by the columns $\mathbf{c}_j$ of $\mathbf{C}$ and the rows $\mathbf{d}_i'$ of $\mathbf{D}$:

$$\mathbf{CD} = \mathbf{c}_1 * \mathbf{d}_1' + \mathbf{c}_2 * \mathbf{d}_2' + \cdots + \mathbf{c}_r * \mathbf{d}_r' \tag{2}$$

Letting $\mathbf{C} = \mathbf{E}$ and $\mathbf{D} = \mathbf{D}_\lambda \mathbf{E}^{-1}$ (the ith row of $\mathbf{D}_\lambda \mathbf{E}^{-1}$ is the ith row of $\mathbf{E}^{-1}$ multiplied by λ_i), we obtain the following decomposition.

Theorem 1 (Eigenvalue Decomposition). Let $\mathbf{A}$ be an $n \times n$ matrix with n linearly independent eigenvectors $\mathbf{e}_i$ associated with eigenvalues λ_i. Let $\mathbf{e}_i^+$ denote the ith row of $\mathbf{E}^{-1}$. Then $\mathbf{A}$ is the weighted sum of simple matrices:

$$\mathbf{A} = \mathbf{E}\mathbf{D}_\lambda \mathbf{E}^{-1} = \lambda_1 \mathbf{e}_1 * \mathbf{e}_1^+ + \lambda_2 \mathbf{e}_2 * \mathbf{e}_2^+ + \cdots + \lambda_n \mathbf{e}_n * \mathbf{e}_n^+ \tag{3}$$

Theorem 1 has many uses. For typical large matrices, the eigenvalues tend to decline in size quickly. For example, if $n = 20$, perhaps $\lambda_1 = 5$, $\lambda_2 = 2$, $\lambda_3 = 0.8$, $\lambda_4 = 0.2$. Then the sum of the first two or three simple matrices in (3) would yield a very good approximation of the matrix. In a large data matrix, say, 300 by 300, which has almost 100,000 entries, if the sum of, say, the first 20 simple matrices $\lambda_i \mathbf{e}_i * \mathbf{e}_i^+$ can capture the bulk of the information in the matrix, great savings in storage space would be achieved. A striking example of this savings is given in the digitized depiction of Abraham Lincoln's face at the end of this section. This situation is similar to the way that the first few Legendre polynomials yielded a good approximation to e^x in Section 4.6.

Theorem 1 has a very nice form when $\mathbf{A}$ is a symmetric $n \times n$ matrix. Recall from the preceding section that symmetric matrices have a set of n orthogonal eigenvectors.

Theorem 2. Let $\mathbf{A}$ be a symmetric $n \times n$ matrix, so that there is a set of n orthonormal eigenvectors $\mathbf{e}_i$ and $\mathbf{E}^{-1} = \mathbf{E}^T$. Then the eigenvalue decomposition in Theorem 1 assumes the simple form

$$\mathbf{A} = \lambda_1 \mathbf{e}_1 * \mathbf{e}_1^T + \lambda_2 \mathbf{e}_2 * \mathbf{e}_2^T + \cdots + \lambda_n \mathbf{e}_n * \mathbf{e}_n^T \tag{4}$$

**EXAMPLE 1.
Eigenvalue
Decomposition of
a 2 × 2 Matrix**

We illustrate Theorem 1 with the 2×2 matrix of the computer/dog model in Example 1 of Section 5.3. From that example, we have $\lambda_1 = 4$, $\lambda_2 = 1$, and

$$\mathbf{A} = \begin{bmatrix} 3 & 1 \\ 2 & 2 \end{bmatrix} \qquad \mathbf{E} = \begin{bmatrix} 1 & 1 \\ 1 & -2 \end{bmatrix} \qquad \mathbf{E}^{-1} = \begin{bmatrix} \frac{2}{3} & \frac{1}{3} \\ \frac{1}{3} & -\frac{1}{3} \end{bmatrix}$$

Then (3) says

$$\mathbf{A} = \lambda_1 \mathbf{e}_1 * \mathbf{e}_1^+ + \lambda_2 \mathbf{e}_2 * \mathbf{e}_2^+$$

$$= 4 \begin{bmatrix} 1 \\ 1 \end{bmatrix} * [\tfrac{2}{3} \ \ \tfrac{1}{3}] + 1 \begin{bmatrix} 1 \\ -2 \end{bmatrix} * [\tfrac{1}{3} \ \ -\tfrac{1}{3}]$$

$$= 4 \begin{bmatrix} \tfrac{2}{3} & \tfrac{1}{3} \\ \tfrac{2}{3} & \tfrac{1}{3} \end{bmatrix} + 1 \begin{bmatrix} \tfrac{1}{3} & -\tfrac{1}{3} \\ -\tfrac{2}{3} & \tfrac{2}{3} \end{bmatrix}$$

$$= \begin{bmatrix} \tfrac{8}{3} & \tfrac{4}{3} \\ \tfrac{8}{3} & \tfrac{4}{3} \end{bmatrix} + \begin{bmatrix} \tfrac{1}{3} & -\tfrac{1}{3} \\ -\tfrac{2}{3} & \tfrac{2}{3} \end{bmatrix} = \begin{bmatrix} 3 & 1 \\ 2 & 2 \end{bmatrix}$$

(5)

■

EXAMPLE 2.
Eigenvector
Decomposition of
a Symmetric
Matrix

Suppose that we know that the symmetric matrix $\mathbf{A} = \begin{bmatrix} 3 & 2 & 2 \\ 2 & 2 & 0 \\ 2 & 0 & 4 \end{bmatrix}$ has eigenvalues

$\lambda_1 = 6, \lambda_2 = 3, \lambda_3 = 0$ with associated orthonormal eigenvectors

$$\mathbf{e}_1 = \begin{bmatrix} \tfrac{2}{3} \\ \tfrac{1}{3} \\ \tfrac{2}{3} \end{bmatrix}, \quad \mathbf{e}_2 = \begin{bmatrix} \tfrac{1}{3} \\ \tfrac{2}{3} \\ -\tfrac{2}{3} \end{bmatrix}, \quad \mathbf{e}_3 = \begin{bmatrix} -\tfrac{2}{3} \\ \tfrac{2}{3} \\ \tfrac{1}{3} \end{bmatrix}$$

Then by Theorem 2,

$$\mathbf{A} = \lambda_1 \mathbf{e}_1 * \mathbf{e}_1^T + \lambda_2 \mathbf{e}_2 * \mathbf{e}_2^T + \lambda_3 \mathbf{e}_3 * \mathbf{e}_3^T$$

$$= 6 \begin{bmatrix} \tfrac{2}{3} \\ \tfrac{1}{3} \\ \tfrac{2}{3} \end{bmatrix} * [\tfrac{2}{3}, \tfrac{1}{3}, \tfrac{2}{3}] + 3 \begin{bmatrix} \tfrac{1}{3} \\ \tfrac{2}{3} \\ -\tfrac{2}{3} \end{bmatrix} * [\tfrac{1}{3}, \tfrac{2}{3}, -\tfrac{2}{3}] + 0 \begin{bmatrix} -\tfrac{2}{3} \\ \tfrac{2}{3} \\ \tfrac{1}{3} \end{bmatrix} * [-\tfrac{2}{3}, \tfrac{2}{3}, \tfrac{1}{3}]$$

(6)

$$= 6 \begin{bmatrix} \tfrac{4}{9} & \tfrac{2}{9} & \tfrac{4}{9} \\ \tfrac{2}{9} & \tfrac{1}{9} & \tfrac{2}{9} \\ \tfrac{4}{9} & \tfrac{2}{9} & \tfrac{4}{9} \end{bmatrix} + 3 \begin{bmatrix} \tfrac{1}{9} & \tfrac{2}{9} & -\tfrac{2}{9} \\ \tfrac{2}{9} & \tfrac{4}{9} & -\tfrac{4}{9} \\ -\tfrac{2}{9} & -\tfrac{4}{9} & \tfrac{4}{9} \end{bmatrix} = \begin{bmatrix} 3 & 2 & 2 \\ 2 & 2 & 0 \\ 2 & 0 & 4 \end{bmatrix}$$

■

The eigenvalue decomposition (4) sheds new light on what happens when we multiply a symmetric matrix $\mathbf{A}$ times some vector $\mathbf{x}$. If we express $\mathbf{x}$ in the basis $E = \{\mathbf{e}_1, \mathbf{e}_2, \ldots, \mathbf{e}_n\}$ of orthonormal eigenvectors, then

$$\mathbf{x} = x_1^\circ \mathbf{e}_1 + x_2^\circ \mathbf{e}_2 + \cdots + x_n^\circ \mathbf{e}_n$$

(7)

where $x_i^\circ = \mathbf{e}_i \cdot \mathbf{x}$ (by Theorem 1 in Section 4.5). Using (4), the product $\mathbf{Ax}$ becomes

$$\mathbf{Ax} = (\lambda_1 \mathbf{e}_1 * \mathbf{e}_1^T + \lambda_2 \mathbf{e}_2 * \mathbf{e}_2^T + \cdots + \lambda_n \mathbf{e}_n * \mathbf{e}_n^T)\mathbf{x}$$

$$= \lambda_1(\mathbf{e}_1 * \mathbf{e}_1^T)\mathbf{x} + \lambda_2(\mathbf{e}_2 * \mathbf{e}_2^T)\mathbf{x} + \cdots + \lambda_n(\mathbf{e}_n * \mathbf{e}_n^T)\mathbf{x}$$

(8)

The terms $(\mathbf{e}_i * \mathbf{e}_i^T)\mathbf{x}$ in (8) are the projections of $\mathbf{x}$ onto each $\mathbf{e}_i$ in disguise:

$$(\mathbf{e}_i * \mathbf{e}_i^T)\mathbf{x} = \mathbf{e}_i(\mathbf{e}_i^T\mathbf{x}) = \mathbf{e}_i(\mathbf{e}_i \cdot \mathbf{x}) = x_i^\circ \mathbf{e}_i$$

so the second line in (8) says that to multiply $\mathbf{Ax}$, we take the projections of $\mathbf{x}$ onto $\mathbf{e}_i$ and multiply each projection by λ_i. Thus we are led again to the familiar result (first obtained in Section 3.3 and rederived in Section 5.3):

$$\mathbf{Ax} = \lambda_1 x_1^\circ \mathbf{e}_1 + \lambda_2 x_2^\circ \mathbf{e}_2 + \cdots + \lambda_n x_n^\circ \mathbf{e}_n \tag{9}$$

We now illustrate the usefulness of the eigenvalue decomposition of a symmetric matrix with the following problem in statistics.

EXAMPLE 3.
Principal
Components in
Statistical Analysis

Suppose that an anthropologist has collected data on 25 physical characteristics, variables $x_1, x_2, \ldots, x_{25}$, for 100 prehuman fossil remains. The researcher computes a measure of the variability V_i of each variable x_i called the *variance* of x_i. A large variance means that the x_i-variable varies substantially from fossil to fossil. The anthropologist also computes a measure of the joint variability Cov_{ij} of each pair of variables x_i, x_j called the *covariance*. The covariance is proportional to the correlation coefficient (which was discussed in Section 2.3). A positive Cov_{ij} means that variables x_i and x_j have similar values; a negative Cov_{ij} means that the variables are opposites (if the kth fossil has a large x_i-value, then the kth fossil probably has a small or negative x_j-value); and Cov_{ij} near 0 means that values of x_i and x_j are unrelated.

The anthropologist would like to find good linear combinations of the characteristics that "explain" the variability of the data. For example, let a length index L be defined as

$$L = 0.2x_1 + 0.4x_3 - 0.3x_{11} + 0.4x_{17} + 0.5x_{18} \tag{10}$$

where x_1 might be length of forearm, x_3 length of thigh, and so on. The idea is that while the lengths of forearms, thighs and other bones may individually be poor indicators of, say, whether a fossil came from town A or from town B, a *composite index*, such as possibly the length index (10), is a better way to distinguish between a fossil from town A and town B.

Among all possible indices like (10) formed by a linear combination of variables, the index I_1 that shows the greatest variability (i.e., largest variance) is called the *first principal component*. Among those other indices that are uncorrelated to I_1 (having a covariance of 0 with I_1), the index I_2 with the largest variance is called the *second principal component*, and so on. We want index I_2 uncorrelated with I_1 so that it gives us new (additional) information about variability that was not contained in I_1.

In summary, the first principal component gives an index that explains the maximum amount of variability in the data from one fossil to another. The first four principal components will typically account for over 75% of the variability in a set of 25 variables. Clearly, there are great advantages in describing each fossil with a few numbers rather than 25 numbers. The same is true for studies in psychology, finance, quality control, and any other field where people collect large amounts of data.

So how do we find these principal components? That is, how do we determine the coefficients of the x_i, as in (10)? The answer is, we form a *covariance matrix* $\mathbf{C}$ of all the covariances of the fossil data, where entry (i, j) is Cov_{ij} (and $\text{Cov}_{ii} = V_i$). Then it can be shown that the vector of weights used in the first principal component is simply the (unit-length) eigenvector $\mathbf{e}_1$ associated with the largest eigenvalue λ_1 of $\mathbf{C}$. The first simple matrix $\lambda_1(\mathbf{e}_1 * \mathbf{e}_1^T)$ in the eigenvalue decomposition of $\mathbf{C}$ of Theorem 2 shows how much of the variability in $\mathbf{C}$ is explained by the first principal component.

As an example, consider the following 4×4 covariance matrix (representing 4 of the 25 characteristics mentioned above):

$$\mathbf{C} = \begin{bmatrix} 0.86 & 1.19 & 2.02 & 1.45 \\ 1.19 & 1.68 & 2.86 & 2.06 \\ 2.02 & 2.86 & 5.05 & 3.50 \\ 1.45 & 2.06 & 3.50 & 2.53 \end{bmatrix} \tag{11}$$

All the covariances in (11) happen to be positive (this is often the case), although they can be negative.

We can determine the eigenvalues and associated eigenvectors by using some linear algebra computer software (see the References at the end of the book) or the method of deflation introduced below. We find that

$$\lambda_1 = 10 \qquad \lambda_2 \approx 0.098 \qquad \lambda_3 \approx 0.022 \qquad \lambda_4 \approx 0.0003 \tag{12}$$

The (unit-length) eigenvector associated with λ_1 is

$$\mathbf{e}_1 = [0.289, 0.408, 0.707, 0.5]^T$$

The simple matrix $\lambda_1 \mathbf{e}_1 * \mathbf{e}_1^T$ should approximate $\mathbf{C}$ well since λ_1 is much larger than the other eigenvalues.

$$\lambda_1 \mathbf{e}_1 * \mathbf{e}_1^T = \begin{bmatrix} 0.84 & 1.18 & 2.04 & 1.44 \\ 1.18 & 1.67 & 2.88 & 2.04 \\ 2.04 & 2.88 & 5.00 & 3.54 \\ 1.44 & 2.04 & 3.54 & 2.50 \end{bmatrix} \tag{13}$$

By comparing (13) with $\mathbf{C}$ in (11), it is clear that (13) accounts for virtually all the variability in the covariance matrix $\mathbf{C}$.

The first principal component index I_1 is the linear combination of variables x_1, x_2, x_3, x_4 with coefficients given by $\mathbf{e}_1$:

$$I_1 = 0.289x_1 + 0.408x_2 + 0.707x_3 + 0.5x_4 \qquad\qquad ■$$

The point of Example 3 is that not only did the simple matrix $\lambda_1 \mathbf{e}_1 * \mathbf{e}_1^T$ in (13) represent the information in the covariance matrix $\mathbf{C}$ in (11) concisely, but the first principal component derived from this simple matrix reveals an underlying structure of the model.

We conclude this section by showing how to use the eigenvector decomposition for a symmetric matrix to find all the eigenvalues and eigenvectors of a symmetric matrix. We start by reviewing the discussion in Section 3.3 on how to find the dominant (largest in absolute value) eigenvalue and associated eigenvector. Recall that expressing $\mathbf{x}$ in terms of eigenvectors, $\mathbf{x} = x_1^\circ \mathbf{e}_1 + x_2^\circ \mathbf{e}_2 + \cdots + x_n^\circ \mathbf{e}_n$, we obtained

$$\mathbf{A}^k\mathbf{x} = x_1^\circ \lambda_1^k \mathbf{e}_1 + x_2^\circ \lambda_2^k \mathbf{e}_2 + \cdots + x_n^\circ \lambda_n^k \mathbf{e}_n \tag{14}$$

Assuming that the eigenvalues are distinct and we have indexed them so that $|\lambda_1| > |\lambda_2| > |\lambda_3| > \cdots > |\lambda_n|$, then λ_1^k is going to be much larger (in absolute value) than the other λ_i^k's. Thus the first term on the right in (14) dominates the other terms, and we have

$$\mathbf{A}^k\mathbf{x} \approx x_1^\circ \lambda_1^k \mathbf{e}_1 \tag{15}$$

As noted in Section 3.3, the approximation in (15) provides a way to determine $\mathbf{e}_1$ and λ_1: Compute $\mathbf{x}^{(k)} = \mathbf{A}^k\mathbf{x}$ for a reasonably large k to get a vector $\mathbf{x}^{(k)} \approx a_1\lambda_1^k\mathbf{e}_1$, a multiple of $\mathbf{e}_1$ (this approximation will usually be very accurate), and then divide by the length (Euclidean norm) of $\mathbf{x}^{(k)}$ to get $\mathbf{e}^* = \mathbf{x}^{(k)}/|\mathbf{x}^{(k)}|$, a unit-length estimate of $\mathbf{e}_1$. Since $\mathbf{A}\mathbf{e}^* \approx \lambda_1\mathbf{e}^*$, we can also approximately determine λ_1. But how do we compute other eigenvalues and eigenvectors? (*Note*: There are simple, fast methods that use an approximate value for an eigenvalue to find the exact value of the eigenvalue.)

Suppose that $\mathbf{A}$ is a symmetric matrix and we have determined the dominant (largest in absolute value) eigenvalue λ_1 and an associated unit-length eigenvector $\mathbf{e}_1$ by some approximation method like (15). Consider the matrix

$$\mathbf{A}_2 = \mathbf{A} - \lambda_1\mathbf{e}_1 * \mathbf{e}_1^T \tag{16}$$

In words, the matrix $\mathbf{A}_2$ is $\mathbf{A}$ minus the first simple matrix in the eigenvalue decomposition of $\mathbf{A}$. By Theorem 2,

$$\mathbf{A} = \lambda_1\mathbf{e}_1 * \mathbf{e}_1^T + \lambda_2\mathbf{e}_2 * \mathbf{e}_2^T + \cdots + \lambda_n\mathbf{e}_n * \mathbf{e}_n^T \tag{17}$$

Then comparing (17) with (16), we have

$$\mathbf{A}_2 = \lambda_2\mathbf{e}_2 * \mathbf{e}_2^T + \lambda_3\mathbf{e}_3 * \mathbf{e}_3^T + \cdots + \lambda_n\mathbf{e}_n * \mathbf{e}_n^T \tag{18}$$

Since the eigenvalue decomposition of a square matrix is unique, the (nonzero) eigenvalues of $\mathbf{A}_2$ must be $\lambda_2, \lambda_3, \ldots, \lambda_n$ with associated eigenvectors $\mathbf{e}_2, \mathbf{e}_3, \ldots, \mathbf{e}_n$. In particular, λ_2 is now the dominant eigenvalue of $\mathbf{A}_2$, and applying a method like (15) to $\mathbf{A}_2$ will yield λ_2 and $\mathbf{e}_2$.

If we subtract the simple matrix $\lambda_2\mathbf{e}_2 * \mathbf{e}_2^T$ from $\mathbf{A}_2$ in (18), we will get a matrix $\mathbf{A}_3$ whose dominant eigenvalue is λ_3, and so on. This method of getting the eigenvalues and eigenvectors of $\mathbf{A}$ is called **deflation**. We note that if we have a small error in λ_1 or $\mathbf{e}_1$, the resulting $\mathbf{A}_2$ still has λ_2 as the dominant eigenvalue.

If we want the eigenvectors for use in a simple-matrix approximation of $\mathbf{A}$, as in principal component analysis, then when the entries of matrix $\mathbf{A}_i$ become sufficiently small, we can terminate this procedure for finding the eigenvalues and eigenvectors.

Deflation Method to Compute Eigenvalues and Eigenvectors of a Symmetric $n \times n$ Matrix A

0. Set $\mathbf{A}_1 = \mathbf{A}$; set $i = 1$.
1. Use a method such as (15) to determine $\mathbf{A}_i$'s dominant eigenvalue λ_i and an associated unit-length eigenvector $\mathbf{e}_i$.
2. Set $\mathbf{A}_{i+1} = \mathbf{A}_i - \lambda_i \mathbf{e}_i * \mathbf{e}_i^T$. Increase i by 1. If $i < n$, go to step 1.

We note that there are faster methods (using more advanced techniques from matrix theory) for determining all eigenvalues and eigenvectors of any $n \times n$ matrix, symmetric or not.

**EXAMPLE 4.
Finding
Eigenvalues and
Eigenvectors of a
3 × 3 Symmetric
Matrix**

Let us use the deflation method to find all eigenvalues and eigenvectors of the symmetric matrix

$$\mathbf{A} = \begin{bmatrix} 3 & 2 & 1 \\ 2 & 3 & 1 \\ 1 & 1 & 0 \end{bmatrix} \tag{19}$$

Setting $\mathbf{x} = \begin{bmatrix} 1 \\ 0 \\ 0 \end{bmatrix}$ and computing $\mathbf{A}^{25}\mathbf{x}$, we get a multiple of $\mathbf{e}_1 = \begin{bmatrix} 0.684 \\ 0.684 \\ 0.254 \end{bmatrix}$,

the unit-length eigenvector for the dominant eigenvalue $\lambda_1 \approx 5.37$.

Next we compute the deflated matrix $\mathbf{A}_2$:

$$\mathbf{A}_2 = \mathbf{A} - \lambda_1 \mathbf{e}_1 * \mathbf{e}_1^T$$

$$= \begin{bmatrix} 3 & 2 & 1 \\ 2 & 3 & 1 \\ 1 & 1 & 0 \end{bmatrix} - \begin{bmatrix} 2.51 & 2.51 & 0.93 \\ 2.51 & 2.51 & 0.93 \\ 0.93 & 0.93 & 0.35 \end{bmatrix} = \begin{bmatrix} 0.49 & -0.51 & 0.07 \\ -0.51 & 0.49 & 0.07 \\ 0.07 & 0.07 & -0.35 \end{bmatrix} \tag{20}$$

Again with $\mathbf{x} = \begin{bmatrix} 1 \\ 0 \\ 0 \end{bmatrix}$, we compute $\mathbf{A}_2^{25}\mathbf{x}$ and obtain $\begin{bmatrix} 0.5 \\ -0.5 \\ 0 \end{bmatrix}$, or $\mathbf{e}_2 = \begin{bmatrix} 0.707 \\ -0.707 \\ 0 \end{bmatrix}$

as a unit-length eigenvector, for the eigenvalue $\lambda = 1$. (The reader should verify that $\begin{bmatrix} 0.5 \\ -0.5 \\ 0 \end{bmatrix}$ is indeed an eigenvector of the original matrix $\mathbf{A}$.) Deflating again, we have

$$\mathbf{A}_3 = \mathbf{A}_2 - \lambda_2 \mathbf{e}_2 * \mathbf{e}_2^{\mathrm{T}}$$

$$= \begin{bmatrix} 0.49 & -0.51 & 0.07 \\ -0.51 & 0.49 & 0.07 \\ 0.07 & 0.07 & -0.35 \end{bmatrix} - \begin{bmatrix} 0.5 & -0.5 & 0 \\ -0.5 & 0.5 & 0 \\ 0 & 0 & 0 \end{bmatrix} = \begin{bmatrix} -0.01 & -0.01 & 0.07 \\ -0.01 & -0.01 & 0.07 \\ 0.07 & 0.07 & -0.35 \end{bmatrix}$$

$$(21)$$

Next we find that $\mathbf{e}_3 \approx \begin{bmatrix} 0.181 \\ 0.181 \\ 0.967 \end{bmatrix}$ and $\lambda_3 \approx -0.37$. Computing the last simple matrix, we obtain

$$\lambda_3 \mathbf{e}_3 * \mathbf{e}_3^{\mathrm{T}} = \begin{bmatrix} -0.01 & -0.01 & 0.07 \\ -0.01 & -0.01 & 0.07 \\ 0.07 & 0.07 & -0.35 \end{bmatrix} \tag{22}$$

This simple matrix equals $\mathbf{A}_3$, confirming the validity of our computations. ■

There is a similar eigenvector-like decomposition into simple matrices for any nonsquare matrix $\mathbf{A}$, called the ***singular value decomposition*** (the eigenvectors come from the matrix $\mathbf{A}^{\mathrm{T}}\mathbf{A}$). See *Linear Algebra and Applications* by G. Strang, which has a discussion of this important decomposition. A graphic example of singular value decomposition is given in Figure 5.5. The image at the top of the figure shows a 49-by-36 digitized image of a bust of Abe Lincoln. [Entry (i, j) is the height of the bust in that position.] The remaining figures in this set show the digitized image produced by the matrix $\mathbf{A}^{(k)}$, the sum of the first k simple matrices in the singular value decomposition of $\mathbf{A}$. These images are taken from the article "Visualization of Matrix Singular Value Decomposition" by Cliff Long in *Mathematics Magazine*, Vol. 56 (1983), pages 161–167.

Section 5.4 Exercises

1. Give the eigenvalue decomposition into a sum of two simple matrices for each of the following matrices whose eigenvalues and largest eigenvector you were asked to determine in Exercise 5 of Section 3.3. (These matrices appeared in Exercise 1 of Section 5.3.)

   (a) $\begin{bmatrix} 4 & 0 \\ 2 & 2 \end{bmatrix}$　(b) $\begin{bmatrix} 1 & 2 \\ 5 & 4 \end{bmatrix}$　(c) $\begin{bmatrix} 2 & 1 \\ 2 & 3 \end{bmatrix}$　(d) $\begin{bmatrix} 4 & -1 \\ 1 & 2 \end{bmatrix}$

2. Give the eigenvalue decomposition into a sum of two simple matrices for the following symmetric matrices. (These matrices appeared in Exercise 2 of Section 5.3.)

   (a) $\begin{bmatrix} 1 & 4 \\ 4 & 1 \end{bmatrix}$　(b) $\begin{bmatrix} 1 & 2 \\ 2 & 4 \end{bmatrix}$　(c) $\begin{bmatrix} 2 & -3 \\ -3 & 2 \end{bmatrix}$　(d) $\begin{bmatrix} -1 & 5 \\ 5 & -1 \end{bmatrix}$

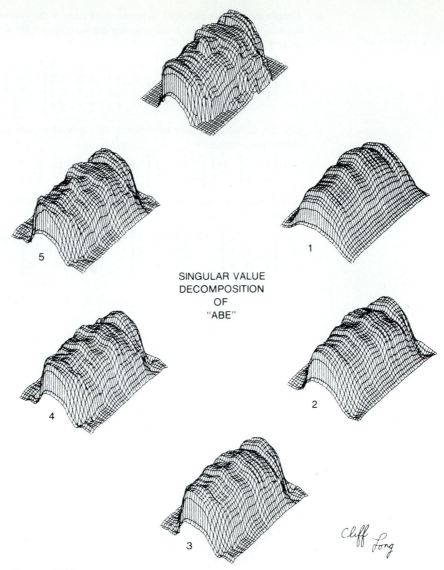

SINGULAR VALUE
DECOMPOSITION
OF
"ABE"

Figure 5.5

3. Using a software package for finding eigenvectors and eigenvalues, determine the eigenvalue decomposition into simple matrices for the following symmetric matrices.

(a) $\begin{bmatrix} 1 & 3 & 0 \\ 3 & 2 & 1 \\ 0 & 1 & 0 \end{bmatrix}$ (b) $\begin{bmatrix} 1 & 1 & 1 \\ 1 & 1 & 1 \\ 1 & 1 & 0 \end{bmatrix}$ (c) $\begin{bmatrix} 2 & -1 & 0 & 1 \\ -1 & 3 & 1 & -2 \\ 0 & 1 & 0 & 1 \\ 1 & -2 & 1 & 0 \end{bmatrix}$

4. Verify that the eigenvalues (12) and dominant eigenvector in Example 3 are correct. Determine the second principal component in Example 3 (the normalized eigenvector for λ_2).

5. Determine the first principal component for each of the following covariance matrices. In each case, tell how well it accounts for the variability. (How well does $\lambda_1 e_1 * e_1$ approximate C?)

(a) $\begin{bmatrix} 3.1 & 1.1 & 0.5 \\ 1.1 & 2.0 & 1.5 \\ 0.5 & 1.5 & 4.2 \end{bmatrix}$ (b) $\begin{bmatrix} 2.4 & 0.6 & 3.1 & 1.5 \\ 0.6 & 4.1 & 0.8 & 1.2 \\ 3.1 & 0.8 & 2.7 & 5.2 \\ 1.5 & 1.2 & 5.2 & 3.2 \end{bmatrix}$

6. Use the deflation method to compute all eigenvalues and eigenvectors of the following symmetric matrices.

(a) $\begin{bmatrix} 1 & 6 \\ 6 & 1 \end{bmatrix}$ (b) $\begin{bmatrix} 2 & -3 \\ -3 & 2 \end{bmatrix}$ (c) $\begin{bmatrix} 0 & 1 \\ 1 & 0 \end{bmatrix}$ (d) $\begin{bmatrix} 3 & -4 \\ -4 & 3 \end{bmatrix}$ (e) $\begin{bmatrix} -1 & 5 \\ 5 & -1 \end{bmatrix}$

7. Use the deflation method to compute all eigenvalues and eigenvectors of the following symmetric matrices.

(a) $\begin{bmatrix} 3 & 2 & 2 \\ 2 & 2 & 0 \\ 2 & 0 & 4 \end{bmatrix}$ (b) $\begin{bmatrix} 1 & 0 & 1 \\ 0 & 1 & 1 \\ 1 & 1 & 1 \end{bmatrix}$ (c) $\begin{bmatrix} 1 & 2 & 3 \\ 2 & 4 & -1 \\ 3 & -1 & 0 \end{bmatrix}$ (d) $\begin{bmatrix} 2 & -1 & 0 & 1 \\ -1 & 3 & 1 & -2 \\ 0 & 1 & 0 & 1 \\ 1 & -2 & 1 & 0 \end{bmatrix}$

6 Applications

6.1 Matrix Iteration and Power Series

In this section we consider an iterative approach to solving systems of equations. For many practical problems, this method rather than Gaussian elimination is preferred. The theoretical basis for iteration combines several important matrix issues, some of which we have seen before and some of which we introduce for the first time here but will see again later in this chapter. We motivate our discussion of iteration with the following widely used matrix model.

EXAMPLE 1.
Model of
Economic Supply–
Demand

We present a linear model for economic supply and demand due to W. Leontief, a Nobel Prize–winning economist. The model seeks to balance supply and demand throughout an entire economy. For each industry there will be one supply–demand equation. In practical applications, Leontief models can have hundreds or thousands of different industries. We consider an example with three industries.

The left-hand side of each equation gives the supply of the product of the ith industry. Call this quantity x_i; it is measured in dollars. On the right-hand side we have the demand for the product of the ith industry. There are two parts to the demand. The first part is demand for the product by other industries. (To produce other products, one requires this product as an input.)

For a concrete instance, let us consider an economy of three general industries: energy, construction, transportation. Suppose that the supply–demand equations are

		Supplies	Industrial demands Energy	Const.	Trans.	Consumer Demand	
Energy:	x_1	$=$	$0.4x_1$	$+\ 0.2x_2$	$+\ 0.1x_3$	$+\quad 100$	
Construction:	x_2	$=$	$0.2x_1$	$+\ 0.4x_2$	$+\ 0.1x_3$	$+\quad 50$	(1)
Transportation:	x_3	$=$		$0.2x_2$	$+\ 0.2x_3$	$+\quad 100$	

The first equation, for energy, has the supply of energy x_1 on the left. The terms on the right side of this equation are the various demands that this supply must meet. The first term on the right, $0.4x_1$, is the input of energy required to produce our x_1 dollars of energy (0.4 unit of energy input for 1 unit of energy output; this could be fuel to power tankers bringing oil). Similarly, the second term of $0.2x_2$ is the input of energy needed to make x_2 dollars of construction. And $0.1x_3$ is the energy input required for transportation. The final term of 100 is the fixed consumer demand for energy.

Each column gives the set of input demands of an industry. For example, the third column tells us that to produce the x_3 dollars of transportation requires as input $0.1x_3$ dollars of energy, $0.1x_3$ dollars of construction, and $0.2x_3$ dollars of transportation. In the Leontief model, there is an ultimate consumer demand for each output, but to meet this demand industries generate input demands on each other. Thus the demands are highly interrelated: Demand for energy depends on the production levels of other industries, these production levels depend in turn on the demand for their outputs by other industries, and so on.

When the levels of industrial output satisfy these supply–demand equations, economists say that the economy is *in equilibrium*. In matrix notation, (1) can be written

$$\mathbf{x} = \mathbf{Dx} + \mathbf{c}, \qquad \text{where} \quad \mathbf{D} = \begin{bmatrix} 0.4 & 0.2 & 0.1 \\ 0.2 & 0.4 & 0.1 \\ 0 & 0.2 & 0.2 \end{bmatrix} \quad \text{and} \quad \mathbf{c} = \begin{bmatrix} 100 \\ 50 \\ 100 \end{bmatrix} \qquad (2) \quad ■$$

Leontief's model requires an input constraint to make sense in economic terms.

Leontief Input Constraint. *The sum of the coefficients in each column of* **D** *is* < 1. The coefficients in column j of **D** represent the values of various products required as input for producing 1 unit (one dollar's worth) of product j. The economic interpretation of this constraint is that the value of inputs (of various products) to produce 1 unit of product j should be less than 1. That is, a dollar's worth of any product must cost less than a dollar to produce.

Recall from Section 3.4 that the sum norm $\|\mathbf{D}\|_s$ of **D** is the largest column sum of **D**. Then the Leontief input constraint means simply that

$$\|\mathbf{D}\|_s < 1 \qquad (3)$$

We can rewrite $\mathbf{x} = \mathbf{Dx} + \mathbf{c}$ as $\mathbf{x} - \mathbf{Dx} = \mathbf{c}$, or

$$(\mathbf{I} - \mathbf{D})\mathbf{x} = \mathbf{c} \qquad (4)$$

We can solve (4) algebraically with inverses to obtain

$$\mathbf{x} = (\mathbf{I} - \mathbf{D})^{-1}\mathbf{c} \qquad (5)$$

We shall now give an algebraic formula for computing $(\mathbf{I} - \mathbf{D})^{-1}$.

Theorem 1. Let $\mathbf{A}$ be a matrix such that $\|\mathbf{A}\| < 1$ (in the sum or max matrix norm). Then $(\mathbf{I} - \mathbf{A})^{-1}$ exists and is given by the formula

$$(\mathbf{I} - \mathbf{A})^{-1} = \sum_{k=0}^{\infty} \mathbf{A}^k = \mathbf{I} + \mathbf{A} + \mathbf{A}^2 + \mathbf{A}^3 + \cdots \tag{6}$$

Here $\mathbf{A}^0 = \mathbf{I}$ just the way for any scalar r, $r^0 = 1$. Formula (6) is the matrix version of the geometric sum for scalars.

$$\frac{1}{1 - a} = \sum_{k=0}^{\infty} a^k = 1 + a + a^2 + a^3 + \cdots \tag{7}$$

Recall that the series (7) converges only when $|a| < 1$. The verification of (7) is obtained by multiplying $1 - a$ times the infinite series and showing that the product equals 1.

Proof of Theorem 1: We follow the approach just mentioned for verifying (7). We multiply both sides of (6) by $(\mathbf{I} - \mathbf{A})$. Then (6) yields

$$\mathbf{I} \; [= (\mathbf{I} - \mathbf{A})^{-1}(\mathbf{I} - \mathbf{A})] = (\mathbf{I} + \mathbf{A} + \mathbf{A}^2 + \mathbf{A}^3 + \cdots)(\mathbf{I} - \mathbf{A}) \tag{6a}$$

The right side of (6a) is the limiting case as $k \to \infty$ of the product $(\mathbf{I} + \mathbf{A} + \mathbf{A}^2 + \cdots + \mathbf{A}^k)(\mathbf{I} - \mathbf{A})$. Using "long multiplication," we see that

$$\begin{array}{r}
\mathbf{I} + \mathbf{A} + \mathbf{A}^2 + \mathbf{A}^3 + \cdots + \mathbf{A}^k \\
\mathbf{I} - \mathbf{A} \\
\hline
\mathbf{I} + \mathbf{A} + \mathbf{A}^2 + \mathbf{A}^3 + \cdots + \mathbf{A}^k \\
\mathbf{A} - \mathbf{A}^2 - \mathbf{A}^3 - \mathbf{A}^4 - \cdots - \mathbf{A}^{k+1} \\
\hline
\mathbf{I} \qquad\qquad\qquad\qquad\qquad - \mathbf{A}^{k+1}
\end{array} \tag{8}$$

So $\mathbf{I} - \mathbf{A}^{k+1} = (\mathbf{I} + \mathbf{A} + \mathbf{A}^2 + \cdots + \mathbf{A}^k)(\mathbf{I} - \mathbf{A})$. As $k \to \infty$, we claim that $\mathbf{I} - \mathbf{A}^{k+1} \to \mathbf{I}$. Recall that $\|\mathbf{A}^{k+1}\| \leq \|\mathbf{A}\|^{k+1}$, by Theorem 3 in Section 3.4. Since $\|\mathbf{A}\| < 1$, then as $k \to \infty$, $\|\mathbf{A}\|^{k+1} \to 0$ and $\|\mathbf{A}^{k+1}\| \to 0$. Recall that the sum norm (max norm) of a matrix is the largest column (row) sum in absolute value. Then $\|\mathbf{A}^{k+1}\| \to 0$ means that the entries in $\mathbf{A}^{k+1}$ must approach 0, so in (8), $\mathbf{I} - \mathbf{A}^{k+1} \to \mathbf{I}$, as claimed. ■

Since $\|\mathbf{D}\|_s < 1$, we can use Theorem 1 to compute $(\mathbf{I} - \mathbf{D})^{-1}$. As just noted in the preceding proof, the sum norm (the largest column sum) of $\mathbf{D}^k$ approaches 0, so the individual entries of $\mathbf{D}^k$ must approach 0. Thus we only need to calculate the sum $\Sigma \, \mathbf{D}^k$ up to, say, the twentieth power of $\mathbf{D}$—the entries in the remaining powers will typically be small enough to neglect. Using the sum $\Sigma \, \mathbf{D}^k$ is not a very efficient way to compute $(\mathbf{I} - \mathbf{D})^{-1}$ for most matrices. However, the formula is simple and easy to program. Moreover, Theorem 1 guarantees that there always exists a solution to a Leontief economic model for any $\mathbf{D}$ and any $\mathbf{c}$, since the model requires $\|\mathbf{D}\|_s < 1$.

Theorem 2. Every Leontief supply–demand model $\mathbf{x} = \mathbf{Dx} + \mathbf{c}$ has a solution of nonnegative production levels for every nonnegative $\mathbf{c}$ and every nonnegative $\mathbf{D}$, provided that $\|\mathbf{D}\|_s < 1$.

The nonnegativity of production levels is very important, since a negative production level makes no sense in real-world terms. Nonnegativity follows from (6): All entries in the powers $\mathbf{D}^k$ will be ≥ 0 (since all entries in $\mathbf{D}$ are ≥ 0), so all entries in $\Sigma \, \mathbf{D}^k = (\mathbf{I} - \mathbf{D})^{-1}$ are ≥ 0. Also, $\mathbf{c} \geq 0$, so all entries in $(\mathbf{I} - \mathbf{D})^{-1}\mathbf{c}$ are ≥ 0. Hopefully, the reader can now appreciate the value of the formula in Theorem 1, even though the formula is a very tedious way to determine the inverse of $\mathbf{I} - \mathbf{A}$. (Using Gauss–Jordan elimination as shown in Section 1.5 is usually a much faster way to compute the inverse.)

EXAMPLE 2.
Solution of
Leontief Model

We solve the Leontief model in (1) by using formula (6) to obtain the inverse of $\mathbf{I} - \mathbf{D}$. We list below the powers of $\mathbf{D}$ up to $\mathbf{D}^7$ plus $\mathbf{D}^{20}$, and the sum of the right-hand side of (6) up to $\mathbf{D}^{20}$ (using a computer program).

$$\mathbf{D}^2 = \begin{bmatrix} 0.20 & 0.18 & 0.08 \\ 0.16 & 0.22 & 0.08 \\ 0.04 & 0.12 & 0.06 \end{bmatrix} \quad \mathbf{D}^3 = \begin{bmatrix} 0.116 & 0.128 & 0.054 \\ 0.108 & 0.136 & 0.054 \\ 0.04 & 0.068 & 0.028 \end{bmatrix}$$

$$\mathbf{D}^4 = \begin{bmatrix} 0.072 & 0.085 & 0.035 \\ 0.070 & 0.087 & 0.035 \\ 0.030 & 0.041 & 0.016 \end{bmatrix} \quad \mathbf{D}^5 = \begin{bmatrix} 0.046 & 0.056 & 0.023 \\ 0.046 & 0.062 & 0.023 \\ 0.020 & 0.026 & 0.010 \end{bmatrix}$$

$$\mathbf{D}^6 = \begin{bmatrix} 0.030 & 0.036 & 0.015 \\ 0.030 & 0.036 & 0.015 \\ 0.013 & 0.016 & 0.007 \end{bmatrix} \quad \mathbf{D}^7 = \begin{bmatrix} 0.019 & 0.023 & 0.009 \\ 0.019 & 0.023 & 0.009 \\ 0.008 & 0.010 & 0.004 \end{bmatrix} \tag{9}$$

$$\mathbf{D}^{20} = \begin{bmatrix} 0.00006 & 0.00008 & 0.00003 \\ 0.00006 & 0.00008 & 0.00003 \\ 0.00003 & 0.00003 & 0.00004 \end{bmatrix}$$

Summing powers of $\mathbf{D}$ from $\mathbf{I}$ ($= \mathbf{D}^0$) up through $\mathbf{D}^{20}$, we obtain

$$(\mathbf{I} - \mathbf{D})^{-1} \approx \sum_{k=0}^{20} \mathbf{D}^k \approx \begin{bmatrix} \frac{23}{12} & \frac{3}{4} & \frac{1}{3} \\ \frac{2}{3} & 2 & \frac{1}{3} \\ \frac{1}{6} & \frac{1}{2} & \frac{4}{3} \end{bmatrix} \tag{10}$$

(The entries in the inverse can be verified by computing $\mathbf{I} - \mathbf{D}$ with Gauss–Jordan elimination.) With (10) we can now solve the Leontief model in Example 1.

$$\mathbf{x} = (\mathbf{I} - \mathbf{D})^{-1}\mathbf{c} = \begin{bmatrix} \frac{23}{12} & \frac{3}{4} & \frac{1}{3} \\ \frac{2}{3} & 2 & \frac{1}{3} \\ \frac{1}{6} & \frac{1}{2} & \frac{4}{3} \end{bmatrix} \begin{bmatrix} 100 \\ 50 \\ 100 \end{bmatrix} \tag{11a}$$

$$= \begin{bmatrix} \frac{23}{12} \cdot 100 + \frac{3}{4} \cdot 50 + \frac{1}{3} \cdot 100 \\ \frac{2}{3} \cdot 100 + 2 \cdot 50 + \frac{1}{3} \cdot 100 \\ \frac{1}{6} \cdot 100 + \frac{1}{2} \cdot 50 + \frac{4}{3} \cdot 100 \end{bmatrix} \tag{11b}$$

$$= \begin{bmatrix} 191.7 + 37.5 + 33.3 \\ 66.7 + 100 + 33.3 \\ 16.7 + 25 + 133.3 \end{bmatrix}$$

$$= \begin{bmatrix} 262.5 \\ 200 \\ 175 \end{bmatrix} \begin{array}{l} \text{units of energy} \\ \text{units of construction} \\ \text{units of transportation} \end{array} \tag{11c}$$

The terms in $(\mathbf{I} - \mathbf{D})^{-1}$ allow us to see how the consumer demand affects the interindustry demands. For example, the production level for energy, the first sum in (11b), is

$$\frac{23}{12} \cdot 100 + \frac{3}{4} \cdot 50 + \frac{1}{3} \cdot 100$$

This sum says that the 100 units of consumer demand for energy requires a total of $\frac{23}{12} \cdot 100$ units of energy to be produced (100 units directly for consumers and $\frac{11}{12} \cdot 100$ units contributing to the production of energy), the 50 units of consumer demand of construction requires $\frac{3}{4} \cdot 50$ units of energy contributing to the production of construction, and the 100 units of consumer demand for transportation requires $\frac{1}{3} \cdot 100$ units of energy contributing to the production of transportation. ■

If we do not need the inverse $(\mathbf{I} - \mathbf{D})^{-1}$ (to solve the Leontief system for many different consumer vectors) but just want the solution for one specific $\mathbf{c}$, we can shorten our effort by rewriting (11a) in terms of the powers of $\mathbf{D}$.

$$\mathbf{x} = (\mathbf{I} - \mathbf{D})^{-1}\mathbf{c} = (\Sigma\, \mathbf{D}^k)\mathbf{c} = \Sigma\, \mathbf{D}^k\mathbf{c} \tag{12}$$

Computing the sum of $\mathbf{D}^k\mathbf{c}$'s is much faster than first computing the sum of $\mathbf{D}^k$'s and then multiplying by $\mathbf{c}$. First we compute the vector $\mathbf{D}\mathbf{c}$. Then we multiply $\mathbf{D}$ by the vector $\mathbf{D}\mathbf{c}$ to get $\mathbf{D}^2\mathbf{c}$, then $\mathbf{D}^3\mathbf{c}$, and so on, with each stage involving a matrix–vector product rather than matrix–matrix product. It is left as an exercise for the reader to re-solve the Leontief problem using (12) (again stop at $\mathbf{D}^{20}\mathbf{c}$).

Iteration

There is another way that we can recast the solution of $(I - D)x = c$. The method is called *solution by iteration*. Iteration was used to compute the stable probability vector of the weather Markov chain in Section 1.1. Recall that the weather Markov chain computed tomorrow's probability vector p' for sunny, cloudy, or rainy weather from today's probability vector p with the system of equations

$$p' = Ap: \begin{array}{l} p_1' = \frac{3}{4}p_1 + \frac{1}{2}p_2 + \frac{1}{4}p_3 \\ p_2' = \frac{1}{8}p_1 + \frac{1}{4}p_2 + \frac{1}{2}p_3 \\ p_3' = \frac{1}{8}p_1 + \frac{1}{4}p_2 + \frac{1}{4}p_3 \end{array}$$

The stable probability distribution for a Markov chain is a vector p^* such that $p^* = Ap^*$. In Section 1.1 we experimentally computed the next-state distribution $p^{(k)}$ using the transition equation $p^{(k)} = Ap^{(k-1)}$ and an initial distribution of $\begin{bmatrix} 0 \\ \frac{1}{2} \\ \frac{1}{2} \end{bmatrix}$.

	Sunny	Cloudy	Rainy
Today:	0	$\frac{1}{2}$	$\frac{1}{2}$
1 day ahead:	$\frac{3}{8}$	$\frac{3}{8}$	$\frac{2}{8}$
2 days ahead:	$\frac{34}{64}$	$\frac{17}{64}$	$\frac{13}{64}$
3 days ahead:	$\frac{149}{256}$	$\frac{60}{256}$	$\frac{47}{256}$
5 days ahead:	$\sim\frac{14}{23}$	$\sim\frac{5}{23}$	$\sim\frac{4}{23}$
10 days ahead:	$\sim\frac{14}{23}$	$\sim\frac{5}{23}$	$\sim\frac{4}{23}$
100 days ahead:	$\frac{14}{23}$	$\frac{5}{23}$	$\frac{4}{23}$

$$(13)$$

After many days, we observed that the probabilities stabilized at $\frac{14}{23}$, $\frac{5}{23}$, and $\frac{4}{23}$ for sunny, cloudy, and rainy weather, respectively. Thus $p^* = [\frac{14}{23}, \frac{5}{23}, \frac{4}{23}]^T$ is a solution to the equation $p^* = Ap^*$.

Let us try using iteration with the Leontief system of equations

$$x = Dx + c \tag{14}$$

We guess values for components of our initial approximate solution $x^{(0)}$, then substitute $x^{(0)}$ in the right side of (14) and compute $x^{(1)} = Dx^{(0)} + c$. We check to see if $x^{(1)}$ equals $x^{(0)}$. If not, we substitute $x^{(1)}$ on the right side of (14) and compute

$$x^{(2)} = Dx^{(1)} + c \tag{15}$$

and continue doing this.

Suppose we "guess" that $x^{(0)} = c$; that is, we produce just enough to meet consumer demand in our sample Leontief model in (1). Let us compute $Dx^{(0)} + c$ for this model using $x^{(0)} = c = \begin{bmatrix} 100 \\ 50 \\ 100 \end{bmatrix}$.

$$\mathbf{x}^{(1)} = \mathbf{D}\mathbf{x}^{(0)} + \mathbf{c} = \begin{bmatrix} 0.4 & 0.2 & 0.1 \\ 0.2 & 0.4 & 0.1 \\ 0 & 0.2 & 0.2 \end{bmatrix} \begin{bmatrix} 100 \\ 50 \\ 100 \end{bmatrix} + \begin{bmatrix} 100 \\ 50 \\ 100 \end{bmatrix}$$

$$= \begin{bmatrix} 60 \\ 50 \\ 30 \end{bmatrix} + \begin{bmatrix} 100 \\ 50 \\ 100 \end{bmatrix} = \begin{bmatrix} 160 \\ 100 \\ 130 \end{bmatrix}$$

(16)

This new estimate $\mathbf{x}^{(1)}$ of the production levels equals consumer demands $\mathbf{c}$ plus the interindustrial demands $\mathbf{D}\mathbf{c}$ to meet the consumer demand. We now compute $\mathbf{x}^{(2)} = \mathbf{D}\mathbf{x}^{(1)} + \mathbf{c}$.

$$\mathbf{x}^{(2)} = \mathbf{D}\mathbf{x}^{(1)} + \mathbf{c} = \begin{bmatrix} 0.4 & 0.2 & 0.1 \\ 0.2 & 0.4 & 0.1 \\ 0 & 0.2 & 0.2 \end{bmatrix} \begin{bmatrix} 160 \\ 100 \\ 130 \end{bmatrix} + \begin{bmatrix} 100 \\ 50 \\ 100 \end{bmatrix}$$

$$= \begin{bmatrix} 97 \\ 85 \\ 46 \end{bmatrix} + \begin{bmatrix} 100 \\ 50 \\ 100 \end{bmatrix} = \begin{bmatrix} 197 \\ 135 \\ 146 \end{bmatrix}$$

(17)

In general, we compute

$$\mathbf{x}^{(k+1)} = \mathbf{D}\mathbf{x}^{(k)} + \mathbf{c} \tag{18}$$

In terms of a computer program, we are repeatedly performing the vector assignment statement

$$\mathbf{x} \longleftarrow \mathbf{D}\mathbf{x} + \mathbf{c} \tag{19}$$

Every increase in production levels results in a further increase in the interindustry demand $\mathbf{D}\mathbf{x}$. The question is: Will this process converge to a solution $\mathbf{x}^*$ such that $\mathbf{x}^* = \mathbf{D}\mathbf{x}^* + \mathbf{c}$? We shall prove that convergence must occur in Leontief systems. Consider three successive iterates $\ldots$, $\mathbf{x}'$, $\mathbf{x}''$, $\mathbf{x}'''$, $\ldots$, where

$$\mathbf{x}'' = \mathbf{D}\mathbf{x}' + \mathbf{c} \quad \text{and} \quad \mathbf{x}''' = \mathbf{D}\mathbf{x}'' + \mathbf{c}$$

Then

$$\mathbf{x}''' - \mathbf{x}'' = (\mathbf{D}\mathbf{x}'' + \mathbf{c}) - (\mathbf{D}\mathbf{x}' + \mathbf{c}) = \mathbf{D}\mathbf{x}'' - \mathbf{D}\mathbf{x}' = \mathbf{D}(\mathbf{x}'' - \mathbf{x}') \tag{20}$$

Taking norms in (20), we have (where $\|\mathbf{D}\|_s < 1$)

$$|\mathbf{x}''' - \mathbf{x}''|_s \leq \|\mathbf{D}\|_s |\mathbf{x}'' - \mathbf{x}'|_s < |\mathbf{x}'' - \mathbf{x}'|_s \tag{21}$$

This means that $\mathbf{D}$ lessens the change in $\mathbf{x}^{(k)}$ from iteration to iteration and the change will eventually shrink to zero.

If $\mathbf{x}^*$ is the solution such that $\mathbf{x}^* = \mathbf{D}\mathbf{x}^* + \mathbf{c}$ (Theorem 1 guarantees this solution exists), then replacing $\mathbf{x}''$ by $\mathbf{x}^*$ in (20), we have

$$\mathbf{x}''' - \mathbf{x}^* = (\mathbf{D}\mathbf{x}'' + \mathbf{c}) - (\mathbf{D}\mathbf{x}^* + \mathbf{c}) = \mathbf{D}\mathbf{x}'' - \mathbf{D}\mathbf{x}^* = \mathbf{D}(\mathbf{x}'' - \mathbf{x}^*) \qquad (21a)$$

Taking norms in (21a), we have

$$|\mathbf{x}''' - \mathbf{x}^*|_s \le \|\mathbf{D}\|_s |\mathbf{x}'' - \mathbf{x}^*|_s < |\mathbf{x}'' - \mathbf{x}^*|_s \qquad (22)$$

so the iterates are getting closer and closer—that is, converging to the solution $\mathbf{x}^*$.

The following list gives the values we get in this iteration process (with the numbers rounded to integers and written as row vectors).

$$\mathbf{x}^{(0)} = [100, 50, 100]$$
$$\mathbf{x}^{(1)} = [160, 100, 130]$$
$$\mathbf{x}^{(2)} = [197, 135, 146]$$
$$\mathbf{x}^{(3)} = [220, 158, 156]$$
$$\mathbf{x}^{(4)} = [235, 173, 163]$$
$$\mathbf{x}^{(6)} = [251, 189, 170]$$
$$\mathbf{x}^{(8)} = [258, 195, 173]$$
$$\mathbf{x}^{(10)} = [261, 198, 174]$$
$$\mathbf{x}^{(12)} = [262, 199, 175]$$

$$\mathbf{x}^{(20)} = [262, 200, 175]$$

$$\mathbf{x}^{(25)} = [262, 200, 175]$$

All further $\mathbf{x}^{(n)}$, $n > 20$, equal $\mathbf{x}^{(20)}$. This is the same answer that we obtained in (11c) using the geometric series approach.

Note that we started with a very poor estimate $\mathbf{x}^{(0)} = \mathbf{c}$. Suppose that we start with the more thoughtful guess of

$$\mathbf{x}^{(0)} = [250, 200, 200]$$

Then iterating, we obtain

$$\mathbf{x}^{(1)} = [260, 200, 180]$$
$$\mathbf{x}^{(2)} = [262, 200, 176]$$
$$\mathbf{x}^{(3)} = [262.4, 256, 175.2]$$

The first iteration is already quite close to the correct solution and after three iterations, all entries are within 0.2 of the true solution. That's convergence!

Now we show how this iterative approach is actually equivalent to the previous geometric-series solution method. If $\mathbf{x}^{(0)} = \mathbf{c}$, then

$$\mathbf{x}^{(1)} = \mathbf{D}\mathbf{x}^{(0)} + \mathbf{c} = \mathbf{D}\mathbf{c} + \mathbf{c}$$

and

$$\mathbf{x}^{(2)} = \mathbf{D}\mathbf{x}^{(1)} + \mathbf{c} = \mathbf{D}(\mathbf{D}\mathbf{c} + \mathbf{c}) + \mathbf{c}$$
$$= \mathbf{D}^2\mathbf{c} + \mathbf{D}\mathbf{c} + \mathbf{c}$$

Continuing, we find that

$$\mathbf{x}^{(n)} = \mathbf{D}^n\mathbf{c} + \mathbf{D}^{n-1}\mathbf{c} + \cdots + \mathbf{D}\mathbf{c} + \mathbf{c} = \sum_{k=0}^{n} \mathbf{D}^k\mathbf{c} \qquad (23)$$

So this iterative method is just computing the partial sums in the geometric series for $(\mathbf{I} - \mathbf{D})^{-1}\mathbf{c}$ [see (12)].

A starting value $\mathbf{x}^{(0)}$ other than $\mathbf{c}$ speeds, or slows, convergence but it cannot prevent convergence. We note that most large real-world economic models are always solved by iterative methods, not by the elimination methods taught in standard mathematics books. (The reason is that iteration goes quickly in large real-world problems where the matrix $\mathbf{D}$ is mostly 0's.)

We next ask the question: Can we adapt this iterative technique to solving general systems of equations? The following theorem tells how to convert a system of equations into a form similar to a Leontief system.

Theorem 3. Given the system of equations $\mathbf{Ax} = \mathbf{b}$, let $\mathbf{D} = \mathbf{I} - \mathbf{A}$, so $\mathbf{A} = \mathbf{I} - \mathbf{D}$. Rewrite the system $\mathbf{Ax} = \mathbf{b}$ as

$$(\mathbf{I} - \mathbf{D})\mathbf{x} = \mathbf{b} \quad \text{or} \quad \mathbf{x} = \mathbf{D}\mathbf{x} + \mathbf{b} \qquad (24)$$

If $\|\mathbf{D}\| = \|\mathbf{I} - \mathbf{A}\| < 1$ (in any matrix norm), the iteration method

$$\mathbf{x}^{(k)} = \mathbf{D}\mathbf{x}^{(k-1)} + \mathbf{b} \qquad (25)$$

converges to the solution of $\mathbf{Ax} = \mathbf{b}$.

With this conversion, $\|\mathbf{D}\| < 1$ guarantees, as shown in (22), that the iteration (25) converges to the required solution $(\mathbf{I} - \mathbf{D})^{-1}\mathbf{b}$.

EXAMPLE 3.
Iteration Solution
of an Oil Refinery
Model

We turn to our oil refinery model. Recall that each refinery produces three petroleum-based products: heating oil, diesel oil, and gasoline, and x_i is the number of barrels of petroleum used by the ith refinery.

$$\begin{aligned}
4x_1 + \quad 2x_2 + 2x_3 &= 600 \\
2x_1 + \quad 5x_2 + 2x_3 &= 800 \\
1x_1 + 2.5x_2 + 5x_3 &= 1000
\end{aligned} \qquad (26)$$

Let $\mathbf{A}$ be the coefficient matrix in (26) and $\mathbf{b}$ be the vector of right-side demands. Theorem 3 does not seem to apply to the system $\mathbf{A}\mathbf{x} = \mathbf{b}$ since the sum norm of $\mathbf{D}$ $(= \mathbf{I} - \mathbf{A})$ is going to be much larger than 1 (actually, $\|\mathbf{D}\|_s = 8.5$, the second column sum). We want to rewrite the equations in (26) to make Theorem 3 apply.

To make the column sums (or row sums, for the max norm) of $\mathbf{I} - \mathbf{A}$ be less than 1, we can divide each column (or row) of $\mathbf{A}$ by its largest entry. Dividing entries in the columns this way is equivalent to changing the units of the variables. That is, dividing the first column by 4 (its largest entry) is equivalent to replacing x_1 (the number of barrels of input to refinery 1) by $x_1' = 4x_1$ (input measured in fourths of a barrel).

For all three columns we have

$$x_1' = 4x_1, \qquad x_2' = 5x_2, \qquad x_3' = 5x_3 \tag{27}$$

and hence

$$x_1 = \frac{x_1'}{4}, \qquad x_2 = \frac{x_2'}{5}, \qquad x_3 = \frac{x_3'}{5} \tag{28}$$

This change of variables divides the coefficients in column one by 4, the coefficients in column two by 5, and the coefficients in column three by 5.

$$
\begin{aligned}
x_1' + \tfrac{2}{5}x_2' + \tfrac{2}{5}x_3' &= 600 \\
\tfrac{1}{2}x_1' + x_2' + \tfrac{2}{5}x_3' &= 800 \\
\tfrac{1}{4}x_1' + \tfrac{1}{2}x_2' + x_3' &= 1000
\end{aligned}
\tag{29}
$$

It is important to note that in the new system the main diagonal entries are all 1.

Now we can try to use Theorem 3. If $\mathbf{A}'$ is the new coefficient matrix in (29), the system for iteration $\mathbf{x}' = (\mathbf{I} - \mathbf{A}')\mathbf{x}' + \mathbf{b}$ is [see (24)]

$$
\begin{aligned}
x_1' &= \qquad\quad - \tfrac{2}{5}x_2' - \tfrac{2}{5}x_3' + 600 \\
x_2' &= -\tfrac{1}{2}x_1' \qquad\quad - \tfrac{2}{5}x_3' + 800 \\
x_3' &= -\tfrac{1}{4}x_1' - \tfrac{1}{2}x_2' \qquad\quad + 1000
\end{aligned}
\tag{30}
$$

Note the nice form of (30): Each equation expresses one variable in terms of the other variables. The main diagonal entries on the right side are 0 because the main diagonal entries in (29) are 1.

In the matrix of coefficients $\mathbf{I} - \mathbf{A}'$ on the right side of (30), the sum of the (absolute values of) entries in each column is < 1. Thus $\|\mathbf{D}'\|_s = \|\mathbf{I} - \mathbf{A}'\|_s < 1$, so Theorem 3 guarantees that iteration based on (30) will converge.

For simplicity we let $\mathbf{x}'^{(0)} = \mathbf{0}$. Iterating with (30), we get (numbers are rounded to the nearest integer in this list)

$$\mathbf{x}'^{(1)} = [600, 800, 1000]$$
$$\mathbf{x}'^{(2)} = [-120, 100, 450]$$
$$\mathbf{x}'^{(3)} = [380, 680, 980]$$
$$\mathbf{x}'^{(4)} = [-64, 218, 565]$$
$$\mathbf{x}'^{(5)} = [286, 606, 907]$$
$$\vdots$$
$$\mathbf{x}'^{(10)} = [66, 374, 693] \tag{31}$$
$$\mathbf{x}'^{(11)} = [173, 490, 797]$$
$$\vdots$$
$$\mathbf{x}'^{(20)} = [117, 429, 742]$$
$$\mathbf{x}'^{(21)} = [131, 445, 756]$$
$$\vdots$$
$$\mathbf{x}'^{(40)} = [\sim125, \sim437\tfrac{1}{2}, \sim750] \text{ and no further change}$$

Observe how our iterates oscillate above and below the final solution. This is due to the minus signs in (30). The convergence is slow because the column sums (in absolute value) of (30) are all close to 1.

Converting $\mathbf{x}'^{(40)}$ back into our original variables, we have

$$x_1 = \frac{x_1'}{4} = \frac{125}{4} = 31\tfrac{1}{4}$$

$$x_2 = \frac{x_2'}{5} = \frac{437\tfrac{1}{2}}{5} = 87\tfrac{1}{2}$$

$$x_3 = \frac{x_3'}{5} = \frac{750}{5} = 150$$

■

In Example 3 we could also have rewritten the system of equations (26) by dividing each row (each equation) by its largest entry. This yields

$$x_1 + \tfrac{1}{2}x_2 + \tfrac{1}{2}x_3 = 150$$
$$\tfrac{2}{5}x_1 + x_2 + \tfrac{2}{5}x_3 = 170 \tag{32}$$
$$\tfrac{1}{5}x_1 + \tfrac{1}{2}x_2 + x_3 = 200$$

And rewriting (32) in the form $\mathbf{x} = (\mathbf{I} - \mathbf{A}^\circ)\mathbf{x} + \mathbf{b}^\circ$, we have

$$x_1 = \qquad -\tfrac{1}{2}x_2 - \tfrac{1}{2}x_3 + \quad 25$$
$$x_2 = -\tfrac{2}{5}x_1 \qquad\quad - \tfrac{2}{5}x_3 + 170 \tag{33}$$
$$x_3 = -\tfrac{1}{5}x_1 - \tfrac{1}{2}x_2 \qquad\quad + 200$$

Observe that the sum of the (absolute value of) entries in the first row of coefficients in (33), $|-\frac{1}{2}| + |-\frac{1}{2}|$ is 1, so the max norm is not < 1, as required. Also, the second column sum is 1, so the sum norm is not < 1. Then Theorem 3 does not apply to (33).

Dividing the ith row by the coefficient of x_i, as in (32) and (33), has the advantage that it does not involve a change of variables. In (33), we are simply solving the first equation of the original system for x_1 (in terms of the other variables), solving the second equation for x_2, and solving the third equation for x_3. For a general system of equations

$$
\begin{aligned}
a_{11}x_1 + a_{12}x_2 + \cdots + a_{1n}x_n &= b_1 \\
a_{21}x_1 + a_{22}x_2 + \cdots + a_{2n}x_n &= b_2 \\
\vdots \qquad \vdots \qquad\qquad \vdots \qquad \vdots \\
a_{n1}x_1 + a_{n2}x_2 + \cdots + a_{nn}x_n &= b_n
\end{aligned}
\tag{34}
$$

the row equations for iteration become

$$
\begin{aligned}
x_1 &= \frac{- a_{12}x_2 - a_{13}x_3 - \cdots - a_{1n}x_n + b_1}{a_{11}} \\
x_2 &= \frac{-a_{21}x_2 - a_{23}x_3 - \cdots - a_{2n}x_n + b_2}{a_{22}} \\
\vdots \\
x_n &= \frac{-a_{n1}x_1 - a_{n2}x_2 - a_{n3}x_3 - \cdots - a_{nn-1}x_{n-1} + b_1}{a_{nn}}
\end{aligned}
\tag{35}
$$

Iteration using this system is called **Jacobi iteration**. It can be proven that iteration scheme (35) derived from row division converges if and only if the iteration scheme derived from column division [as in (30)] converges. The following theorem states what conditions must hold for this iteration scheme to work, that is, conditions so that after row or column division the max or sum norm, respectively, will be < 1.

Theorem 4. Jacobi iteration using system (35) converges if either of the following two conditions hold:

(i) For each row i, the coefficient a_{ii} of x_i is larger than the sum of the (absolute values of) other coefficients in the row:

$$
\sum_{j,\, j \neq i} |a_{ij}| < |a_{ii}| \quad \text{each } i
\tag{36}
$$

(ii) For each column j, the coefficient a_{jj} of x_j is larger than the sum of the (absolute values of) other coefficients in the column:

$$
\sum_{i,\, i \neq j} |a_{ij}| < |a_{jj}| \quad \text{each } j
\tag{37}
$$

The reader should check that condition (37) was satisfied in the refinery problem. It is a straightforward exercise to check that (37) guarantees that after column division [as in (29) and (30)] the resulting matrix $\mathbf{D}' = \mathbf{I} - \mathbf{A}'$ will have sum norm < 1, and that (36) guarantees that after row division the max norm of $\mathbf{D}° = \mathbf{I} - \mathbf{A}°$ is < 1.

There are other iteration methods based on more advanced theory (see "Numerical Analysis" in the References). Exercise 10 mentions a simple way, called *Gauss–Seidel iteration*, to speed up the convergence of Jacobi iteration.

Optional

We conclude this section by linking the iteration, $\mathbf{x}^{(k)} = \mathbf{A}\mathbf{x}^{(k-1)}$, that normally converges to the dominant eigenvector (discussed in Section 3.3) with the iteration in Theorem 3, $\mathbf{x}^{(k)} = \mathbf{D}\mathbf{x}^{(k-1)} + \mathbf{b}$, for solving the system $(\mathbf{I} - \mathbf{D})\mathbf{x} = \mathbf{b}$. The following trick converts the latter iteration into the former.

Define the $(n + 1)$-vector $\mathbf{x}^*$ and the $(n + 1) \times (n + 1)$ matrix $\mathbf{D}^*$

$$\mathbf{x}^* = \begin{bmatrix} \mathbf{x} \\ 1 \end{bmatrix} = \begin{bmatrix} x_1 \\ x_2 \\ \vdots \\ x_n \\ 1 \end{bmatrix} \quad \text{and} \quad \mathbf{D}^* = \begin{bmatrix} \mathbf{A} & \mathbf{b} \\ \mathbf{O} & 1 \end{bmatrix}$$

For example, for the Leontief matrix $\mathbf{D}$ in (1), $\mathbf{D}^*$ is

$$\mathbf{D}^* = \begin{bmatrix} 0.4 & 0.2 & 0.1 & 100 \\ 0.2 & 0.4 & 0.1 & 50 \\ 0 & 0.2 & 0.2 & 100 \\ 0 & 0 & 0 & 1 \end{bmatrix}$$

The reader should check that

$$\mathbf{x}^{*(k)} = \mathbf{D}^*\mathbf{x}^{*(k-1)} \quad \text{is equivalent to} \quad \mathbf{x}^{(k)} = \mathbf{D}\mathbf{x}^{(k-1)} + \mathbf{b} \tag{38}$$

Then the iteration scheme $\mathbf{x}^{(k)} = \mathbf{D}\mathbf{x}^{(k-1)} + \mathbf{b}$ of Theorem 3 will converge to a solution in which

$$\text{for large } k, \quad \mathbf{x}^{(k)} \approx \mathbf{x}^{(k-1)}$$

if and only if

$$\text{for large } k, \quad \mathbf{x}^{*(k)} \approx \mathbf{x}^{*(k-1)}$$

But the latter condition for the $\mathbf{x}^*$'s means that the dominant eigenvalue of $\mathbf{D}^*$ is 1.

Theorem 5. The iteration scheme $\mathbf{x}^{(k)} = \mathbf{D}\mathbf{x}^{(k-1)} + \mathbf{b}$ for solving $(\mathbf{I} - \mathbf{D})\mathbf{x} = \mathbf{b}$ converges to a solution if and only if the augmented matrix $\mathbf{D}^*$ has a dominant eigenvalue that is equal to 1.

Section 6.1 Exercises

1. Suppose we want to solve the same Leontief system that was solved in Example 1 except that now the consumer demand vector **c** has been changed. Use the expression in equation (11a) with the following new **c**'s to determine the new vector **x** of production levels.

$$\textbf{(a) c} = \begin{bmatrix} 50 \\ 50 \\ 50 \end{bmatrix} \quad \textbf{(b) c} = \begin{bmatrix} 0 \\ 100 \\ 0 \end{bmatrix} \quad \textbf{(c) c} = \begin{bmatrix} 0 \\ 0 \\ 100 \end{bmatrix} \quad \textbf{(d) c} = \begin{bmatrix} 0 \\ 0 \\ 50 \end{bmatrix} \quad \textbf{(e) c} = \begin{bmatrix} 100 \\ 10 \\ 10 \end{bmatrix}$$

2. Use the sum of powers method in equations (10) and (11) to solve the following Leontief systems.

   (a)
   $$\begin{aligned} x_1 &= 0.1x_1 + 0.2x_2 + 0.2x_3 + 100 \\ x_2 &= 0.2x_1 + 0.1x_2 \qquad\qquad\; + 100 \\ x_3 &= 0.2x_1 \qquad\quad + 0.1x_3 + 100 \end{aligned}$$

   (b)
   $$\begin{aligned} x_1 &= 0.3x_1 + 0.1x_2 + 0.2x_3 + 100 \\ x_2 &= 0.1x_1 + 0.1x_2 + 0.1x_3 + 100 \\ x_3 &= 0.1x_1 \qquad\quad + 0.1x_3 + 100 \end{aligned}$$

   (Use computer programs for both systems; convergence is fast because the norm of **D** is small—you only need to go up to the sixth power of the coefficient matrix **D**.)

3. Find the inverse of the following matrices by writing them in the form $I - D$ and using the sum of powers method on **D**. Check the accuracy of your answer by using the determinant-based formula for the inverse of a 2×2 matrix (see Section 1.6).

   (a) $\begin{bmatrix} 0.7 & -0.2 \\ -0.4 & 0.8 \end{bmatrix}$ **(b)** $\begin{bmatrix} 0.6 & 0.3 \\ 0.2 & 0.5 \end{bmatrix}$ **(c)** $\begin{bmatrix} 0.6 & 0 \\ 0 & 0.5 \end{bmatrix}$

4. Try using the sum of powers method to solve the following system of equations. Why does the method fail?

   $$\begin{aligned} x_1 &= 0.4x_1 + 0.3x_2 + 0.4x_3 + 100 \\ x_2 &= 0.3x_1 + 0.4x_2 + 0.6x_3 + 100 \\ x_3 &= 0.7x_1 + 0.8x_2 + 0.5x_3 + 100 \end{aligned}$$

5. Use the iteration method in equation (18) to solve the Leontief systems in Exercise 2.

6. Consider systems of equations

   (i)
   $$\begin{aligned} 7x_1 + \;\; x_2 + 2x_3 &= 30 \\ x_1 + 5x_2 + 3x_3 &= 10 \\ 2x_1 + 3x_2 + 8x_3 &= 12 \end{aligned}$$

   (ii)
   $$\begin{aligned} 6x_1 + 3x_2 + \;\; x_3 &= 15 \\ 2x_1 + 5x_2 + 2x_3 &= 50 \\ x_1 + \;\; x_2 + 4x_3 &= 10 \end{aligned}$$

   (a) Mimic equations (27)–(30) to rewrite the systems in the form $\mathbf{x} = \mathbf{Dx} + \mathbf{b}$ with $\|\mathbf{D}\| < 1$.

   (b) Solve each system by iteration as described in Theorem 3, starting with $\mathbf{x}^{(0)} = \begin{bmatrix} 0 \\ 0 \\ 0 \end{bmatrix}$.

   (c) Repeat part (b) with starting vector $\bar{\mathbf{x}}^{(0)} = \begin{bmatrix} 100 \\ 100 \\ 100 \end{bmatrix}$.

7. Consider the system of equations $\mathbf{Ax} = \mathbf{b}$ where $\mathbf{A} = \begin{bmatrix} 9 & 4 & -3 \\ -3 & 3 & 10 \\ 4 & 8 & -3 \end{bmatrix}$ and $\mathbf{b} = \begin{bmatrix} 10 \\ 20 \\ 30 \end{bmatrix}$.

   (a) Rearrange the equations (rows) and divide each equation by appropriate numbers so that this system can be rewritten in the form $\mathbf{x} = \mathbf{Dx} + \mathbf{b}$ with $\|\mathbf{D}\| < 1$.
   (b) Solve this system by iteration as described in Theorem 3.

8. (a) For which of the following systems of equations does Theorem 4 apply?

   (i) $3x_1 - 4x_2 = 2$ (ii) $6x_1 + 2x_2 - x_3 = 4$ (iii) $2x_1 + x_2 = 3$

   $\quad\;\; 2x_1 + x_2 = 4$ $\qquad x_1 + 5x_2 + x_3 = 3$ $\qquad 4x_1 - x_2 = 5$

   $\qquad\qquad\qquad\qquad\;\; 2x_1 + x_2 + 4x_3 = 27$

   (b) In the systems where Theorem 4 does not apply directly, rearrange the rows and/or divide the rows or columns by the largest coefficient to make Theorem 4 apply.
   (c) Use the iterative method for solving each of the systems.

9. (a) Suppose that we use a starting vector $\mathbf{x}^{(0)} = \mathbf{w}$ in the iteration scheme $\mathbf{x}^{(k+1)} = \mathbf{Dx}^{(k)} + \mathbf{c}$. Using the same reasoning as that which led to equation (23), find formulas for $\mathbf{x}^{(1)}$, $\mathbf{x}^{(2)}$, and $\mathbf{x}^{(n)}$ in terms of $\mathbf{D}$, $\mathbf{c}$, and $\mathbf{w}$.
   (b) Use your formula for $\mathbf{x}^{(n)}$ in part (a) and the fact that $\|\mathbf{D}\| < 1$ to show that the starting vector does not influence the final values in the iteration process.

10. A well-known method to speed up the convergence of Jacobi iteration, called *Gauss–Seidel iteration*, is to use the new value of x_1 obtained from the first equation in the second and third equations (in place of the previous value of x_1); similarly, the new value for x_2 is used in the third equation. In the refinery problem, the first two equations in (30) are

$$x_1' = \qquad - \tfrac{2}{5}x_2' - \tfrac{2}{5}x_3' + 600$$
$$x_2' = -\tfrac{1}{2}x_1' \qquad - \tfrac{2}{5}x_3' + 800$$

Starting with $\mathbf{x}^{(0)} = \begin{bmatrix} 0 \\ 0 \\ 0 \end{bmatrix}$, we would compute x_1' as $-\tfrac{2}{5}(0) - \tfrac{2}{5}(0) + 600 = 600$.

Then we use this value of 600 for x_1' in the second equation to compute x_2' as $-\tfrac{1}{2}(600) - \tfrac{2}{5}(0) + 800 = 500$. The third equation would use the values for both x_1' and x_2' just computed.

   (a) Use Gauss–Seidel iteration on the refinery problem [the three equations in (30)] starting with $\mathbf{x}^{(0)} = \mathbf{0}$. Does this method give faster or slower convergence than the iteration in (31)?
   (b) Use Gauss–Seidel iteration to solve the Leontief system in Example 1. Does this method give faster or slower convergence than the standard iteration method?
   (c) Use Gauss–Seidel iteration to solve the system of equations (i) in Exercise 6.

11. Another method of iteration is to average the two previous iterates. In an iteration process such as (31), where the iterates are oscillating above and below the final solution, such an averaging will speed convergence. On the other hand, when the iterates are increasing as in the Leontief system, this averaging method will slow down the convergence.
   (a) Use this method of averaging the two previous iterates to re-solve the refinery equations [equations (30)] starting again with $\mathbf{x}^{(0)} = \mathbf{0}$. Does this method give faster or slower convergence than the iteration in (31)?

(b) Use the method of averaging to re-solve the Leontief system in Example 1. Does this method give faster or slower convergence than the standard iteration method?

12. Let **A** be an $n \times n$ matrix. How many multiplications (approximately) are required to perform each of the following operations?
    (a) Compute $\mathbf{A}^3$.
    (b) Solve $\mathbf{Ax} = \mathbf{b}$.
    (c) Iterate $\mathbf{x}^{(k+1)} = \mathbf{Ax}^{(k)}$ 10 times.
    (d) Compute $\mathbf{A}^{-1}$.

13. Find the dominant eigenvalue for the matrix **D*** preceding Theorem 5.

6.2 Markov Chains

Markov chains were introduced in Section 1.1. They are probability models for simulating the behavior of a system that randomly moves among different "states" over successive periods of time. If a Markov chain is currently in state S_j, there is a transition probability a_{ij} that 1 unit of time later it will be in state S_i. The matrix **A** of transition probabilities completely describes the Markov chain. If $\mathbf{p} = [p_1, p_2, \ldots, p_n]^T$ is the (column) vector giving the probabilities p_i that S_i is the current state of the chain and $\mathbf{p}' = [p_1', p_2', \ldots, p_n']^T$ is the vector of probabilities p_i' that S_i is the next state of the chain, we have

$$\mathbf{p}' = \mathbf{Ap} \tag{1a}$$

For a particular p_i', this is

$$p_i' = a_{i1}p_1 + a_{i2}p_2 + \cdots + a_{in}p_n \tag{1b}$$

We now review the weather Markov chain presented in Section 1.1.

EXAMPLE 1.
Weather Markov Chain Reviewed

The weather Markov chain introduced in Section 1.1 gave the probabilities that the weather would be sunny, cloudy, or rainy tomorrow based on the weather today. The transition matrix for this Markov chain was

$$
\begin{array}{c}
\phantom{\text{Tomorrow}} \quad \text{Today} \\
\begin{array}{cccc}
\text{Tomorrow} & \text{Sunny} & \text{Cloudy} & \text{Rainy}
\end{array} \\
\mathbf{A} = \begin{array}{c}
\text{Sunny} \\
\text{Cloudy} \\
\text{Rainy}
\end{array}
\begin{bmatrix}
\frac{3}{4} & \frac{1}{2} & \frac{1}{4} \\
\frac{1}{8} & \frac{1}{4} & \frac{1}{2} \\
\frac{1}{8} & \frac{1}{4} & \frac{1}{4}
\end{bmatrix}
\end{array} \tag{2}
$$

It was noted that over many periods, the probability distribution stabilized at $\frac{14}{23}$ sunny, $\frac{5}{23}$ cloudy, $\frac{4}{23}$ rainy [see (10) in Section 1.1]. Thus the probability vector $\mathbf{p}^* = [\frac{14}{23}, \frac{5}{23}, \frac{4}{23}]^T$ is a *stable probability vector* with the property that $\mathbf{p}^* = \mathbf{Ap}^*$. [This property was checked in equation (28) in Section 1.4.] That is, $\mathbf{p}^*$ satisfies the matrix equation

$$\mathbf{p}^* = \mathbf{A}\mathbf{p}^*: \quad \begin{aligned} p_1^* &= \tfrac{3}{4}p_1^* + \tfrac{1}{2}p_2^* + \tfrac{1}{4}p_3^* \\ p_2^* &= \tfrac{1}{8}p_1^* + \tfrac{1}{4}p_2^* + \tfrac{1}{2}p_3^* \\ p_3^* &= \tfrac{1}{8}p_1^* + \tfrac{1}{4}p_2^* + \tfrac{1}{4}p_3^* \end{aligned} \tag{3}$$

If we had not known the values in $\mathbf{p}^*$ from numerical computation, we could solve (3) by collecting the p^*'s on the left to get the homogeneous system of equations

$$(\mathbf{I} - \mathbf{A})\mathbf{p}^* = \mathbf{0}: \quad \begin{aligned} \tfrac{1}{4}p_1^* - \tfrac{1}{2}p_2^* - \tfrac{1}{4}p_3^* &= 0 \\ -\tfrac{1}{8}p_1^* + \tfrac{3}{4}p_2^* - \tfrac{1}{2}p_3^* &= 0 \\ -\tfrac{1}{8}p_1^* - \tfrac{1}{4}p_2^* + \tfrac{3}{4}p_3^* &= 0 \end{aligned} \tag{4}$$

To make sure we obtain a solution to (4) that is a probability vector, we needed to add the additional constraint: $p_1^* + p_2^* + p_3^* = 1$. System (4) with the added constraint $p_1^* + p_2^* + p_3^* = 1$ was solved for $\mathbf{p}^*$ in Example 9 of Section 1.4. This analytic approach to finding $\mathbf{p}^*$ can be used on any Markov chain to determine its stable probability vector, if one exists. ■

**EXAMPLE 2.
Finding a Stable
Probability Vector
in a Markov
Chain**

Consider the Markov chain given by the following transition matrix $\mathbf{A}$.

$$\mathbf{A} = \begin{array}{c} \\ 1 \\ 2 \\ 3 \\ 4 \end{array} \begin{array}{c} \begin{array}{cccc} 1 & 2 & 3 & 4 \end{array} \\ \begin{bmatrix} \tfrac{1}{2} & \tfrac{1}{4} & 0 & 0 \\ \tfrac{1}{2} & \tfrac{1}{2} & \tfrac{1}{4} & 0 \\ 0 & \tfrac{1}{4} & \tfrac{1}{2} & \tfrac{1}{2} \\ 0 & 0 & \tfrac{1}{4} & \tfrac{1}{2} \end{bmatrix} \end{array} \tag{5}$$

This could represent the random movement of a frog from minute to minute jumping among the four lanes on a highway. Column 2 says that if the frog is currently in lane 2, it has a $\tfrac{1}{2}$ chance of still being in lane 2 one minute later and $\tfrac{1}{4}$ chances of being in each of the adjacent lanes.

Let us try to determine the stable probability vector $\mathbf{p}^*$ for this Markov chain. We need to solve $(\mathbf{I} - \mathbf{A})\mathbf{p}^* = \mathbf{0}$ as in (4) above, supplemented by the constraint that the probabilities sum to 1.

$$\begin{aligned} \tfrac{1}{2}p_1^* - \tfrac{3}{4}p_2^* & & &= 0 \\ -\tfrac{1}{2}p_1^* + \tfrac{1}{2}p_2^* - \tfrac{1}{4}p_3^* & & &= 0 \\ -\tfrac{1}{4}p_2^* + \tfrac{1}{2}p_3^* - \tfrac{1}{2}p_4^* &= 0 \\ -\tfrac{1}{4}p_3^* + \tfrac{1}{2}p_4^* &= 0 \\ p_1^* + p_2^* + p_3^* + p_4^* &= 1 \end{aligned} \tag{6}$$

Solving by elimination yields $\mathbf{p}^* = [\tfrac{1}{6}, \tfrac{1}{3}, \tfrac{1}{3}, \tfrac{1}{6}]^{\mathrm{T}}$. ■

We now turn our attention to a special type of Markov chain, called an absorbing Markov chain. A state S_i in a Markov chain is called an ***absorbing state*** if $a_{ii} = 1$; that is, once you enter state S_i you never leave it. A Markov chain is an ***absorbing Markov chain*** if it has one or more absorbing states and if from every nonabsorbing state there is a positive probability (over many periods) of getting to one of the absorbing states.

EXAMPLE 3.
Gambling Model
with Absorbing
States

Consider the following Markov chain for gambling, with states representing the gambler's winnings. Each round, the gambler has a probability 0.3 of winning $1, 0.33 of losing $1, and 0.37 of staying the same. The gambler stops if he or she loses all of the money, and also stops if the winnings reach $6. So 0 and 6 will be absorbing states in this Markov chain.

$$
\mathbf{A} = \text{Next state} \begin{array}{c} 0 \\ 1 \\ 2 \\ 3 \\ 4 \\ 5 \\ 6 \end{array} \overset{\begin{array}{ccccccc} 0 & 1 & 2 & 3 & 4 & 5 & 6 \end{array}}{\left[\begin{array}{ccccccc} 1 & 0.33 & 0 & 0 & 0 & 0 & 0 \\ 0 & 0.37 & 0.33 & 0 & 0 & 0 & 0 \\ 0 & 0.30 & 0.37 & 0.33 & 0 & 0 & 0 \\ 0 & 0 & 0.30 & 0.37 & 0.33 & 0 & 0 \\ 0 & 0 & 0 & 0.30 & 0.37 & 0.33 & 0 \\ 0 & 0 & 0 & 0 & 0.30 & 0.37 & 0 \\ 0 & 0 & 0 & 0 & 0 & 0.30 & 1 \end{array} \right]}
$$

Current state

(7)

After a very long time, a person is virtually certain to have stopped playing, either having gone broke (state 0) or having won $6 (state 6). Thus after many periods, the Markov chain should converge to a stable probability vector $\mathbf{p}^*$ of the form

$$p_1^* = p, \ p_6^* = 1 - p \quad \text{and} \quad p_1^* = p_2^* = p_3^* = p_4^* = p_5^* = 0 \tag{8}$$

By looking at the transition probabilities for states 0 and 6 alone,

Current state

$$
\text{Next state} \begin{array}{c} 0 \\ 6 \end{array} \overset{\begin{array}{cc} 0 & 6 \end{array}}{\left[\begin{array}{cc} 1 & 0 \\ 0 & 1 \end{array} \right]}
$$

it is easy to see that any $\mathbf{p}^*$ of the form in (8), with $0 \le p \le 1$, is a stable vector.

Before doing any mathematical analysis of absorbing Markov chains, let us explore the behavior of (7) by letting a computer program produce a table of probability distributions over many rounds when we start with $3. Since $3 is halfway between losing and winning, we would expect the chances of losing to be close to 0.5, but a little above 0.5 since on each round there is always 0.33 versus 0.30 bias toward losing rather than winning a dollar.

Probability Distribution After k Rounds

Rounds	0	1	2	3	4	5	6
0	0	0	0	1	0	0	0
1	0	0	0.33	0.37	0.30	0	0
2	0	0.109	0.244	0.335	0.222	0.09	0
3	0.036	0.121	0.234	0.270	0.212	0.100	0.027
4	0.076	0.128	0.212	0.240	0.193	0.101	0.057
5	0.116	0.115	0.194	0.216	0.177	0.095	0.087
6	0.154	0.107	0.178	0.196	0.161	0.088	0.116
8	0.222	0.090	0.149	0.164	0.135	0.074	0.166
10	0.278	0.075	0.125	0.137	0.113	0.062	0.209
15	0.383	0.048	0.080	0.088	0.072	0.040	0.288
20	0.451	0.031	0.051	0.056	0.047	0.026	0.336
25	0.494	0.020	0.033	0.036	0.030	0.016	0.371
50	0.563	0.002	0.003	0.004	0.003	0.002	0.423
75	0.570	≈ 0	≈ 0	≈ 0	≈ 0	≈ 0	0.428
100	0.571	≈ 0	≈ 0	≈ 0	≈ 0	≈ 0	0.429

Thus if we start with \$3, the probability of eventually losing (before we reach \$6) is $p = 0.571$. Instead of repeating this computer simulation for other starting values, we shall now develop a theory that lets us calculate directly the probability of losing or winning when we start with different amounts of money. ■

The first step in our development is to divide the states of an absorbing Markov chain into two groups, the absorbing states and the nonabsorbing states. Assume that there are a absorbing states and n nonabsorbing states. If the absorbing states are listed first, the transition matrix **A** can be partitioned into the following form (partitioning of matrices was introduced in Section 3.1):

$$\mathbf{A} = \begin{array}{c} \\ \text{Absorbing} \\ \text{Nonabsorbing} \end{array} \begin{array}{cc} \text{Absorbing} & \text{Nonabsorbing} \\ \left[\begin{array}{cc} \mathbf{I} & \mathbf{R} \\ \mathbf{O} & \mathbf{Q} \end{array}\right] \end{array}$$

where **I** is an $a \times a$ identity matrix, **O** is an $n \times a$ matrix of 0's, **R** is an $a \times n$ matrix with entry r_{ij} giving the probability of going from nonabsorbing state j to absorbing state i, and **Q** is the $n \times n$ transition matrix among the nonabsorbing states. The transition matrix (7) in Example 3 becomes

$$\mathbf{A} = \text{Next state} \begin{array}{c} 0 \\ 6 \\ 1 \\ 2 \\ 3 \\ 4 \\ 5 \end{array} \begin{array}{c} \begin{array}{ccccccc} 0 & 6 & 1 & 2 & 3 & 4 & 5 \end{array} \\ \left[\begin{array}{cc|ccccc} 1 & 0 & 0.33 & 0 & 0 & 0 & 0 \\ 0 & 1 & 0 & 0 & 0 & 0 & 0.30 \\ \hline 0 & 0 & 0.37 & 0.33 & 0 & 0 & 0 \\ 0 & 0 & 0.30 & 0.37 & 0.33 & 0 & 0 \\ 0 & 0 & 0 & 0.30 & 0.37 & 0.33 & 0 \\ 0 & 0 & 0 & 0 & 0.30 & 0.37 & 0.33 \\ 0 & 0 & 0 & 0 & 0 & 0.30 & 0.37 \end{array}\right] \end{array} \quad (9)$$

Current state

Using the rule for matrix multiplication of a partitioned matrix from Section 3.1 (in which we treat the submatrices like individual entries), we have

$$\mathbf{A}^2 = \begin{bmatrix} \mathbf{I} & \mathbf{R} \\ \mathbf{O} & \mathbf{Q} \end{bmatrix} \begin{bmatrix} \mathbf{I} & \mathbf{R} \\ \mathbf{O} & \mathbf{Q} \end{bmatrix} = \begin{bmatrix} \mathbf{II} + \mathbf{RO} & \mathbf{IR} + \mathbf{RQ} \\ \mathbf{IO} + \mathbf{OQ} & \mathbf{OR} + \mathbf{QQ} \end{bmatrix} = \begin{bmatrix} \mathbf{I} & \mathbf{R} + \mathbf{RQ} \\ \mathbf{O} & \mathbf{Q}^2 \end{bmatrix} \quad (10)$$

and multiplying $\mathbf{A}$ times $\mathbf{A}^2$, we find that

$$\mathbf{A}^3 = \begin{bmatrix} \mathbf{I} & \mathbf{R} + \mathbf{RQ} + \mathbf{RQ}^2 \\ \mathbf{O} & \mathbf{Q}^3 \end{bmatrix} = \left[\begin{array}{cc|ccccc} 1 & 0 & 0.53 & 0.19 & 0.04 & 0 & 0 \\ 0 & 1 & 0 & 0 & 0.03 & 0.16 & 0.48 \\ \hline 0 & 0 & 0.16 & 0.20 & 0.12 & 0.04 & 0 \\ 0 & 0 & 0.18 & 0.27 & 0.23 & 0.12 & 0.04 \\ 0 & 0 & 0.10 & 0.21 & 0.27 & 0.23 & 0.12 \\ 0 & 0 & 0.03 & 0.10 & 0.21 & 0.27 & 0.20 \\ 0 & 0 & 0 & 0.03 & 0.10 & 0.18 & 0.16 \end{array} \right] \quad (11)$$

It is left as an exercise to check that for higher powers of $\mathbf{A}$, the partitioned form is

$$\mathbf{A}^k = \begin{bmatrix} \mathbf{I} & \mathbf{R}_k^* \\ \mathbf{O} & \mathbf{Q}^k \end{bmatrix}, \qquad \text{where} \quad \mathbf{R}_k^* = \mathbf{R} + \mathbf{RQ} + \mathbf{RQ}^2 + \cdots + \mathbf{RQ}^{k-1} \quad (12)$$

Note that $\mathbf{Q}^k$ is the standard kth power of the nonabsorbing transition matrix $\mathbf{Q}$, for play among the nonabsorbing states.

As k gets large, the entries in $\mathbf{Q}^k$ will approach $\mathbf{O}$, since over time the probability of not getting absorbed approaches 0. The important submatrix in (12) is $\mathbf{R}_k^*$. Entry (i, j) of $\mathbf{R}_k^*$ is the probability of being in absorbing state i after k rounds if we start in nonabsorbing state j. Let us explain what this probability is in detail. To go from nonabsorbing state j to absorbing state i after k rounds, we can either go immediately on the first round from j to i—with probability r_{ij}—(and then remain in absorbing state i), or we can wander among the nonabsorbing states for several rounds, ending up after w rounds in nonabsorbing state h—with probability given by entry (j, h) in $\mathbf{Q}^w$—and then go from state h to absorbing state i—with probability r_{ih} (and thereafter remain in state i). The total probability of starting in a nonabsorbing state j, wandering among nonabsorbing states for w rounds, and then going from some nonabsorbing state to absorbing state i is given by entry (i, j) in $\mathbf{RQ}^w$. Since the number w can range up to $k - 1$, we obtain the sum for $\mathbf{R}_k^*$ given in (12).

The limiting matrix $\mathbf{R}^*$ for $\mathbf{R}_k^*$ as k approaches infinity will give the probabilities r_{ij}^* that starting in nonabsorbing state j we eventually end up in absorbing state i.

$$\mathbf{R}^* = \mathbf{R} + \mathbf{RQ} + \mathbf{RQ}^2 + \cdots = \mathbf{R}(\mathbf{I} + \mathbf{Q} + \mathbf{Q}^2 + \cdots) \quad (13)$$

$\mathbf{R}^*$ is the matrix that would tell us in the gambling model the probability of eventually losing or winning when we start with different amounts.

There are two ways to compute $\mathbf{R}^*$. The first way is to compute $\mathbf{Q}^k$ for all k up to some large number, say 50, and sum these matrices and multiply by $\mathbf{R}$ to obtain $\mathbf{R}^*$ as in (13). The other way rewrites $\mathbf{R}^*$ as

$$\mathbf{R}^* = \mathbf{R}(\mathbf{I} + \mathbf{Q} + \mathbf{Q}^2 + \cdots) = \mathbf{R}(\mathbf{I} - \mathbf{Q})^{-1} \tag{14}$$

using the geometric series identity from Theorem 1 in the preceding section. We call $(\mathbf{I} - \mathbf{Q})^{-1}$ the *fundamental matrix* of an absorbing Markov chain and use the matrix $\mathbf{N}$ to denote it.

$$\mathbf{N} = \mathbf{I} + \mathbf{Q} + \mathbf{Q}^2 + \cdots = (\mathbf{I} - \mathbf{Q})^{-1} \quad \text{and} \quad \mathbf{R}^* = \mathbf{RN} \tag{15}$$

We can calculate this inverse using Gauss–Jordan elimination. For the gambling model in Example 3, $\mathbf{N}$ is given below in (17).

The geometric series identity used in (14) required that $\|\mathbf{Q}\| < 1$, for some matrix norm. The sum norm $\|\mathbf{Q}\|_s$ (largest column sum) of $\mathbf{Q}$ in (9) is 1, and the max norm is ≥ 1. However, it can be shown that $\|\mathbf{Q}\|_e < 1$ in the Euclidean norm.

The matrix $\mathbf{N}$ contains some very useful information by itself. It tells us *the expected number of times we will visit (nonabsorbing) state i if we start in (nonabsorbing) state j*. The reasoning is as follows.

The average number of times we visit state i starting from state j after exactly one round is simply $0(1 - q_{ij}) + 1q_{ij} = q_{ij}$—the weighted average of visiting state i zero times and of visiting state i one time. And the average number of times we visit state i starting from state j after exactly two rounds is $0(1 - q_{ij}^{(2)}) + 1q_{ij}^{(2)} = q_{ij}^{(2)}$, where $q_{ij}^{(2)}$ denotes entry (i, j) in $\mathbf{Q}^2$. The average number of visits after exactly k rounds is entry (i, j) in $\mathbf{Q}^k$. Probability theory says that the average number of visits from state j to state i totaled over all rounds is simply the sum of the average number of visits on each specific round. So the expected number of times we visit nonabsorbing state i starting from nonabsorbing state $\mathbf{j}$ is the sum of the (i, j) entries in $\mathbf{Q}^k$ for all $\mathbf{Q}^k$, that is, entry (i, j) in $\mathbf{N}$.

Furthermore, if we sum the entries of the jth column of $\mathbf{N}$—starting from state j, the expected number of times we visit state 1, plus the expected number of times we visit state 2, and so on—we obtain the expected number of rounds until we are absorbed. The vector–matrix product $\mathbf{1N}$ computes the sum of each column of $\mathbf{N}$.

We summarize this wealth of information about absorbing Markov chains that we can get from $\mathbf{N}$ with the following theorem. The term *absorption* is used in this theorem to mean going to an absorbing state.

Theorem 1. Let $\mathbf{N}$ be the fundamental matrix of an absorbing Markov chain. [$\mathbf{N}$ is defined in (15)]. Then the following are true.
  (i) Entry n_{ij} of $\mathbf{N}$ is the expected number of times we visit the nonabsorbing state i (before absorption) when we start in nonabsorbing state j.
  (ii) The jth entry in the vector $\mathbf{1N}$ gives the expected number of rounds before absorption when we start in nonabsorbing state j.
  (iii) Entry (i, j) in $\mathbf{RN}$ is the probability of eventually ending up in absorbing state i when we start in nonabsorbing state j.

**EXAMPLE 3
(Continued).
Gambling Model**

With Theorem 1 we can answer a variety of interesting questions about this model. We must compute the matrix $\mathbf{N}$ by finding the inverse of $\mathbf{I} - \mathbf{Q}$, where $\mathbf{Q}$ is

$$
\mathbf{Q} = \begin{array}{c} \\ 1 \\ 2 \\ 3 \\ 4 \\ 5 \end{array}
\begin{array}{c}
\begin{array}{ccccc} 1 & 2 & 3 & 4 & 5 \end{array} \\
\left[\begin{array}{ccccc}
0.37 & 0.33 & 0 & 0 & 0 \\
0.30 & 0.37 & 0.33 & 0 & 0 \\
0 & 0.30 & 0.37 & 0.33 & 0 \\
0 & 0 & 0.30 & 0.37 & 0.33 \\
0 & 0 & 0 & 0.30 & 0.37
\end{array}\right]
\end{array}
\qquad (16)
$$

Using Gauss–Jordan elimination to reduce $[(\mathbf{I} - \mathbf{Q}) \quad \mathbf{I}]$ to $[\mathbf{I} \quad (\mathbf{I} - \mathbf{Q})^{-1}]$, we obtain

$$
\mathbf{N} = (\mathbf{I} - \mathbf{Q})^{-1} = \begin{array}{c} \\ 1 \\ 2 \\ 3 \\ 4 \\ 5 \end{array}
\begin{array}{c}
\begin{array}{ccccc} 1 & 2 & 3 & 4 & 5 \end{array} \\
\left[\begin{array}{ccccc}
2.64 & 2.21 & 1.73 & 1.21 & 0.63 \\
2.01 & 4.21 & 3.30 & 2.31 & 1.21 \\
1.43 & 3.00 & 4.73 & 3.30 & 1.73 \\
0.91 & 1.91 & 3.00 & 4.21 & 2.21 \\
0.43 & 0.91 & 1.43 & 2.01 & 2.64
\end{array}\right]
\end{array}
\qquad (17)
$$

From (17) and Theorem 1 we see, for example, that if we started with \$3, there would be 3.3 rounds during an average gambling session when we would be in state 2 (when we would have \$2).

Next we sum the columns of $\mathbf{N}$:

$$
\mathbf{1N} = [7.42, 12.24, 14.19, 13.04, 8.42]
\qquad (18)
$$

The third entry in (18) tells us that if we start with \$3, we get to play about 14 rounds, on average, before the game ends.

Finally, we compute $\mathbf{R}^* = \mathbf{RN}$, where $\mathbf{R}$ is [see (9)]

$$
\mathbf{R} = \begin{array}{c} \\ 0 \\ 6 \end{array}
\begin{array}{c}
\begin{array}{ccccc} 1 & 2 & 3 & 4 & 5 \end{array} \\
\left[\begin{array}{ccccc}
0.33 & 0 & 0 & 0 & 0 \\
0 & 0 & 0 & 0 & 0.3
\end{array}\right]
\end{array}
\qquad (19)
$$

$$
\mathbf{R}^* = \mathbf{RN} = \begin{array}{c} \\ 0 \\ 6 \end{array}
\begin{array}{c}
\begin{array}{ccccc} 1 & 2 & 3 & 4 & 5 \end{array} \\
\left[\begin{array}{ccccc}
0.87 & 0.73 & 0.57 & 0.40 & 0.21 \\
0.13 & 0.27 & 0.43 & 0.60 & 0.79
\end{array}\right]
\end{array}
$$

Entry (0, 3) of $\mathbf{R}^*$ confirms our earlier simulation result that the probability of going broke when we start with \$3 is 0.57. ■

We close this discussion of absorbing Markov chains by noting that some of these results about absorbing Markov chains can be applied to other Markov chains (without absorbing states) with the following trick. Let $\mathbf{A}$ be the transition matrix of a nonabsorbing Markov chain. We convert one state, say state h, into an absorbing state by replacing the hth column $\mathbf{a}_h$ of $\mathbf{A}$ by the hth coordinate vector $\mathbf{i}_h$ (1 in the hth entry; 0 elsewhere). Now whenever we come to state h, we stay there. Our theory of absorbing Markov chains can be applied to this modified transition matrix to determine the expected number of rounds it takes to get to state h if we start from any other state j and the expected number of visits in any third state on the journey from j to h.

If we convert two states h and r into absorbing states by this method, we can compute the relative probability of reaching h or r first when we start from some state j.

EXAMPLE 4.
Ice Cream
Selection

We surveyed a group of students eating blueberry, mint, and strawberry ice cream about which flavor they would choose next time. Suppose that $\frac{3}{4}$ of those eating blueberry would choose blueberry the next time while the remaining $\frac{1}{4}$ would choose strawberry. Responses from others yielded the following transition matrix.

$$
\begin{array}{c}
\text{Current flavor} \\
\begin{array}{ccc}
\text{Blueberry} & \text{Mint} & \text{Strawberry}
\end{array} \\
\text{Next time} \quad
\begin{array}{c}
\text{Blueberry} \\ \text{Mint} \\ \text{Strawberry}
\end{array}
\left[
\begin{array}{ccc}
\frac{3}{4} & \frac{1}{2} & 0 \\
0 & \frac{1}{2} & \frac{1}{3} \\
\frac{1}{4} & 0 & \frac{2}{3}
\end{array}
\right]
\end{array}
\qquad (20)
$$

We treat the selection of flavors as a Markov chain and pose the question: How many rounds does it take on average for a person to switch from strawberry to blueberry?

To answer this question, we change the transition matrix in (20) by making blueberry an absorbing state. The modified transition matrix is

$$
\begin{array}{c}
\quad\; b \;\; m \;\; s \\
\mathbf{A} =
\left[
\begin{array}{ccc}
1 & \frac{1}{2} & 0 \\
0 & \frac{1}{2} & \frac{1}{3} \\
0 & 0 & \frac{2}{3}
\end{array}
\right]
= \left[
\begin{array}{cc}
\mathbf{I} & \mathbf{R} \\
\mathbf{O} & \mathbf{Q}
\end{array}
\right], \quad \text{with} \quad
\mathbf{Q} = \left[
\begin{array}{cc}
\frac{1}{2} & \frac{1}{3} \\
0 & \frac{2}{3}
\end{array}
\right]
\end{array}
\qquad (21)
$$

Then we find that

$$
\mathbf{N} = (\mathbf{I} - \mathbf{Q})^{-1} =
\left[
\begin{array}{cc}
\frac{1}{2} & -\frac{1}{3} \\
0 & \frac{1}{3}
\end{array}
\right]^{-1}
=
\left[
\begin{array}{cc}
2 & 2 \\
0 & 3
\end{array}
\right]
\quad \text{and} \quad \mathbf{1N} = [2,\, 5]
$$

By Theorem 1(ii), the second entry—5—in $\mathbf{1N}$ is the average number of rounds until absorption (blueberry) when starting from strawberry. Moreover, by Theorem 1(i), the second column of $\mathbf{N}$ tells us that on average a person starting with strawberry would choose strawberry three times (including the initial time) and choose mint twice before choosing blueberry. ■

Section 6.2 Exercises

1. Compute the stable distribution for the two-state weather Markov chain with transition matrix

$$
\begin{array}{cc}
& \begin{array}{cc} \text{Sunny} & \text{Cloudy} \end{array} \\
\begin{array}{c} \text{Sunny} \\ \text{Cloudy} \end{array} &
\left[\begin{array}{cc} \frac{3}{4} & \frac{1}{2} \\ \frac{1}{4} & \frac{1}{2} \end{array} \right]
\end{array}
$$

2. The printing press at a local newspaper has the following pattern of breakdowns. If it is working today, tomorrow it has a 90% chance of working (and a 10% chance of breaking down). If the press is broken today, it has a 60% chance of working tomorrow (and a 40% chance of being broken again). Compute the stable distribution for this Markov chain.

3. If the local professional basketball team, the Sneakers, wins today's game, they have a $\frac{2}{3}$ chance of winning their next game. If they lose this game, they have a $\frac{1}{2}$ chance of winning their next game. Compute the stable distribution for this Markov chain and give the approximate values of entries in $\mathbf{A}^{100}$, where $\mathbf{A}$ is this Markov chain's transition matrix.

4. Find the stable distribution for Markov chains with the following transition matrices.

(a) $\begin{bmatrix} \frac{2}{3} & \frac{1}{3} & \frac{1}{3} \\ 0 & \frac{1}{3} & 0 \\ \frac{1}{3} & \frac{1}{3} & \frac{2}{3} \end{bmatrix}$ (b) $\begin{bmatrix} \frac{1}{2} & \frac{1}{4} & \frac{1}{4} \\ \frac{1}{4} & \frac{1}{2} & \frac{1}{4} \\ \frac{1}{4} & \frac{1}{4} & \frac{1}{2} \end{bmatrix}$ (c) $\begin{bmatrix} 0 & \frac{1}{2} & \frac{1}{2} \\ \frac{1}{2} & 0 & \frac{1}{2} \\ \frac{1}{2} & \frac{1}{2} & 0 \end{bmatrix}$

5. If the stock market went up today, historical data show that it has a 60% chance of going up tomorrow, a 20% chance of staying the same, and a 20% chance of going down tomorrow. If the market is unchanged today, it has a 20% chance of being unchanged tomorrow, a 40% chance of going up, and a 40% chance of going down tomorrow. If the market goes down today, it has a 20% of going up tomorrow, a 20% chance of being unchanged, and a 60% chance of going down tomorrow. Compute the stable distribution for the stock market.

6. Write down a Markov chain to model the following situation. Assume that there are three types of voters in Texas: Republicans, Democrats, and Independents. From one (national) election to the next, 60% of Republicans remain Republican and similarly for the two other groups; among the 40% who change parties, 30% become Independent and 10% go the other major party, except that those Independents who change all become Republicans. Determine the stable distribution among the three parties.

7. Find the stable distribution for Markov chains with the following transition matrices.

(a) $\begin{bmatrix} \frac{1}{3} & \frac{1}{3} & 0 & 0 & 0 & 0 \\ \frac{2}{3} & \frac{1}{3} & \frac{1}{3} & 0 & 0 & 0 \\ 0 & \frac{1}{3} & \frac{1}{3} & \frac{1}{3} & 0 & 0 \\ 0 & 0 & \frac{1}{3} & \frac{1}{3} & \frac{1}{3} & 0 \\ 0 & 0 & 0 & \frac{1}{3} & \frac{1}{3} & \frac{2}{3} \\ 0 & 0 & 0 & 0 & \frac{1}{3} & \frac{1}{3} \end{bmatrix}$ (b) $\begin{bmatrix} \frac{2}{3} & \frac{2}{3} & 0 & 0 & 0 & 0 \\ \frac{1}{3} & \frac{1}{6} & \frac{2}{3} & 0 & 0 & 0 \\ 0 & \frac{1}{6} & \frac{1}{6} & \frac{2}{3} & 0 & 0 \\ 0 & 0 & \frac{1}{6} & \frac{1}{6} & \frac{2}{3} & 0 \\ 0 & 0 & 0 & \frac{1}{6} & \frac{1}{6} & \frac{2}{3} \\ 0 & 0 & 0 & 0 & \frac{1}{6} & \frac{1}{3} \end{bmatrix}$

(c)
$$\begin{bmatrix} \frac{2}{3} & \frac{2}{3} & 0 & 0 & 0 & 0 \\ \frac{1}{3} & \frac{1}{6} & \frac{2}{3} & 0 & 0 & 0 \\ 0 & \frac{1}{6} & \frac{1}{6} & \frac{1}{6} & 0 & 0 \\ 0 & 0 & \frac{1}{6} & \frac{1}{6} & \frac{1}{6} & 0 \\ 0 & 0 & 0 & \frac{2}{3} & \frac{1}{6} & \frac{1}{3} \\ 0 & 0 & 0 & 0 & \frac{2}{3} & \frac{2}{3} \end{bmatrix}$$
(d)
$$\begin{bmatrix} 0 & \frac{1}{2} & 0 & 0 & 0 & 0 \\ 1 & 0 & \frac{1}{2} & 0 & 0 & 0 \\ 0 & \frac{1}{2} & 0 & \frac{1}{2} & 0 & 0 \\ 0 & 0 & \frac{1}{2} & 0 & \frac{1}{2} & 0 \\ 0 & 0 & 0 & \frac{1}{2} & 0 & 1 \\ 0 & 0 & 0 & 0 & \frac{1}{2} & 0 \end{bmatrix}$$

8. (a) Make a Markov chain model for a rat wandering through the maze below if, at the end of each period, the rat is equally likely to leave its current room through any of the doorways. (It never stays where it is.)

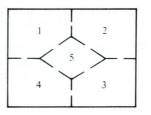

(b) What is the stable distribution?

9. Repeat the questions in Exercise 8 for the maze below.

10. Repeat Exercise 7 with the number of states expanded from six to n.

11. Determine the two eigenvalues and associated eigenvectors for the Markov chain in the following exercises. Give the distribution after six periods for the given starting distribution **p** by representing **p** as a linear combination of the eigenvectors as was done in Section 3.3.
(a) Exercise 1, starting p: sunny 0, cloudy 1
(b) Exercise 2, starting p: working 1, broken 0
(c) Exercise 3, starting p: winning $\frac{1}{2}$, losing $\frac{1}{2}$

12. Show that if **A** is a tridiagonal Markov transition matrix (like matrices in Exercise 7), then in solving $(\mathbf{A} - \mathbf{I})\mathbf{p} = \mathbf{0}$ by Gaussian elimination, the **L** in **LU** decomposition of **A** is a matrix with 1's on the main diagonal and -1 just below the diagonal entries. That is, in Gaussian elimination one always adds the current row (times 1) to the next row.

13. Re-solve the stable distribution problems in Exercise 7 with the last row of the matrix equation $(\mathbf{A} - \mathbf{I})\mathbf{p}$ replaced by the constraint $\mathbf{1} \cdot \mathbf{p} = 1$. (The last row always drops out—becomes 0—and the additional constraint that the probabilities sum to 1 can be put in its place.)

14. The following question refers to the gambling Markov chain in Example 3. If started with $4, what is the expected number of rounds when you have $3, and what is the expected number of rounds until the game ends?

15. The following model for learning a concept over a set of lessons identifies four states of learning: I = ignorance, E = exploratory thinking, S = superficial understanding, and M = mastery. If now in state I, after one lesson you have $\frac{1}{2}$ probability of still being in I and $\frac{1}{2}$ probability of being in E. If now in state E, you have $\frac{1}{4}$ probability of being in I, $\frac{1}{2}$ in E, and $\frac{1}{4}$ in S. If now in state S, you have $\frac{1}{4}$ probability of being in E, $\frac{1}{2}$ in S, and $\frac{1}{4}$ in M. If in M, you always stay in M (with probability 1).
   (a) Write out the transition matrix for this Markov chain with the absorbing state as the first state.
   (b) Compute the fundamental matrix **N**.
   (c) What is the expected number of rounds until mastery is attained if currently in the state of ignorance?

16. In the maze below, suppose that a rat has a 20% chance of going into the middle room, which is an absorbing state, and a 40% chance each of going to the room on the left or on the right. What is the expected number of times a rat starting in room 1 enters room 2? What is the expected number of rounds until the rat goes to the middle room?

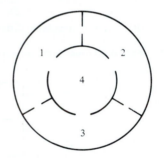

17. (a) Make a Markov chain model for a rat wandering through the maze below if, at the end of each period, the rat is equally likely to leave its current room through any of the doorways. The center room is an absorbing state. (It never stays in the same room, except for the center room.) Compute the fundamental matrix **N**.

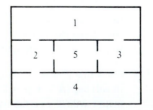

   (b) If the rat starts in room 4, what is the expected number of rounds it will be in room 2?
   (c) If the rat starts in room 4, what is the expected number of rounds until absorption?

18. Consider the game of Ping Pong with the following states:
   1: Player A hits the ball toward player B's side.
   2: Player B hits the ball toward player A's side.

3: Play is dead because A hit the ball out or in the net.

4: Play is dead because B hit the ball out or in the net.

The transition matrix is

$$
\begin{array}{c}
\begin{array}{cccc} & \;1 & \;2 & 3 & 4 \end{array} \\
\begin{array}{l}
\text{(A hits ball to B) 1} \\
\text{(B hits ball to A) 2} \\
\text{(A hits ball out) 3} \\
\text{(B hits ball out) 4}
\end{array}
\left[
\begin{array}{cccc}
0 & 0.9 & 0 & 0 \\
0.4 & 0 & 0 & 0 \\
0.6 & 0 & 1 & 0 \\
0 & 0.1 & 0 & 1
\end{array}
\right]
\end{array}
$$

If we start play with player A hitting the ball (in state 1):

(a) What is the expected number of times player A hits the ball (before the point is over)?

(b) What is the expected number of hits by A and B (before the point is over)?

(c) What is the probability that player A hits the ball out (i.e., that player B wins the point)?

19. Repeat Exercise 17 for the maze below in which rooms 1 and 5 are absorbing. Start in room 4.

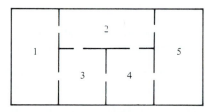

20. Repeat the Markov chain model of the gambling game given in Example 3 but now with probability $\frac{1}{3}$ that a player wins one dollar in a period, with probability $\frac{1}{3}$ that a player loses a dollar, and with probability $\frac{1}{3}$ that a player stays the same. The game ends if the player loses all his or her money OR if the player has six dollars. Compute **N**, **1N**, and **RN** for this new problem.

21. Three tanks A, B, C are engaged in a battle. Tank A, when it fires, hits its target with hit probability $\frac{1}{2}$. B hits its target with hit probability $\frac{1}{3}$, and C with hit probability $\frac{1}{6}$. Initially (in the first periods), B and C fire at A and A fires at B. Once one tank is hit, the remaining tanks aim at each other. The battle ends when there is one or no tank left. Make a Markov chain model of this battle. [**Assistance in computing probabilities:** Let the states of the Markov chain be the eight different possible subsets of tanks currently in action: ABC, AB, AC, BC, A, B, C, None. When in states A or B or C or None, the probability of staying in the current state is 1—this simulates the battle being over. One can never get to state AB (why?). So one only needs to determine the transition probabilities from states ABC, AC, and BC. From states AC and BC, the transition probabilities are products of the probability that each remaining tank hits or misses its target. For example, the probability of going form state AC to state A is the product of the probability that A hits C—$\frac{1}{2}$—times the probability that C misses A—$\frac{5}{6}$. So this probability is $(\frac{1}{2})(\frac{5}{6}) = \frac{5}{12}$. It takes some knowledge of probability to compute the transition probabilities from state ABC: From ABC there is a $\frac{5}{18}$ chance of remaining in state ABC (all tanks miss), a $\frac{5}{18}$ chance of going to state AC (A hits B but B and C miss A), a $\frac{4}{18}$ chance of going to state BC (at least one of B or C hits A and A misses B), and a $\frac{4}{18}$ chance of going to state C (at least one of B or C hits

A and A hits B).] The transition matrix for this game is (the states are the subsets of tanks surviving):

	ABC	AC	BC	A	B	C	None
ABC	$\frac{5}{18}$	0	0	0	0	0	0
AC	$\frac{5}{18}$	$\frac{5}{12}$	0	0	0	0	0
BC	$\frac{4}{18}$	0	$\frac{10}{18}$	0	0	0	0
A	0	$\frac{5}{12}$	0	1	0	0	0
B	0	0	$\frac{5}{18}$	0	1	0	0
C	$\frac{4}{18}$	$\frac{1}{12}$	$\frac{3}{18}$	0	0	1	0
None	0	$\frac{1}{12}$	$\frac{1}{18}$	0	0	0	1

(a) Determine the expected number of rounds that the battle lasts (starting from state ABC).

(b) What are the chances of the different tanks winning (being the sole surviving tank)?

22. In the weather Markov chain in Example 1, how many days does it take on average if now sunny until the first rainy day? How many days on average if now cloudy until the first rainy day?

23. In the stock market Markov chain in Exercise 5, how many days on average does it take if the market went up today until the market goes down?

24. In the Texas voters Markov chain in Exercise 6, how many national elections on average does it take for a typical Texas voter who is Republican to become Democratic?

25. In the Markov chain in Example 2, how many periods does it take on average if now in state 1 to get to state 4? How many periods on average if now in state 2 to get to state 4?

6.3 Linear Differential Equations

In this section we present a brief survey of basic solution techniques for linear differential equations culminating in the use of matrix diagonalization (developed in Section 5.3) to solve a system of linear differential equations. A common mathematical model for many dynamic systems, such as falling objects, vibrating strings, or economic growth, is a linear differential equation of the form

$$y''(t) = a_1 y'(t) + a_0 y(t) \tag{1}$$

where $y(t)$ is a function that measures the ''position'' of the quantity, $y'(t)$ denotes the first derivative with respect to t (representing time), $y''(t)$ denotes the second derivative, and a_1 and a_0 are constants. The differential equation is called *linear* because the right side is a linear combination of the function and its derivative. Solutions of (1) are known to be functions of the form

$$y(t) = Ae^{kt} \tag{2}$$

where e is Euler's constant, and A and k are constants that depend on the particular problem. This form of solution also works if higher derivatives are involved in the linear differential equation.

EXAMPLE 1.
Differential
Equation for
Instantaneous
Interest

The simple differential equation

$$y'(t) = 0.10y(t) \tag{3}$$

describes the amount of money $y(t)$ in a savings account after t years when the account earns 10% interest compounded instantaneously. Recall that $y'(t)$ is the instantaneous rate of change, or graphically, the slope, of $y(t)$. Thus (3) says that the instantaneous growth rate of the savings account is 10% of the account's current value.

Let us try setting $y(t) = Ae^{kt}$ [as suggested in (2)]. Recall that the derivative of e^{kt} is ke^{kt}. Now (3) becomes

$$kAe^{kt} = 0.10Ae^{kt} \tag{4}$$

Dividing both sides of (4) by Ae^{kt}, we obtain $k = 0.10$. So the solution is

$$y(t) = Ae^{0.10t} \tag{5}$$

The constant A is still to be determined. To know how much money we will have after t years, we must know how much we started with. Suppose that we started with 1000 dollars. The starting time is $t = 0$. Thus we have (using the fact $e^0 = 1$)

$$1000 = y(0) = Ae^0 = A \tag{6}$$

Then the solution of (3) when $y(0) = 1000$ is

$$y(t) = 1000e^{0.10t} \tag{7}$$

■

EXAMPLE 2.
Solving a Second-
Order Linear
Differential
Equation

Consider the following differential equation that might describe the height of a falling particle in a special force field.

$$y''(t) = 6y'(t) - 8y(t) \tag{8}$$

This differential equation is called a *second-order* equation because it involves the second derivative. To solve this equation, we also need to know the starting conditions—what are the initial height $y(0)$ and the initial speed $y'(0)$. Here the derivative $y'(t)$ measures speed, that is, the rate of change of the height. Suppose that in this problem

$$f(0) = 100 \quad \text{and} \quad f'(0) = -20 \tag{9}$$

We solve this problem in two stages. First we substitute $y(t) = Ae^{kt}$ in (8).

$$\frac{d^2Ae^{kt}}{dt^2} = 6\frac{dAe^{kt}}{dt} - 8Ae^{kt} \tag{10}$$

Recall that the second derivative of e^{kt} is k^2e^{kt}. So (10) becomes

$$k^2Ae^{kt} = 6kAe^{kt} - 8Ae^{kt} \tag{11}$$

Dividing by Ae^{kt}, (11) becomes

$$k^2 = 6k - 8 \quad \text{or} \quad k^2 - 6k + 8 = 0 \tag{12}$$

Equation (12) is called the **characteristic equation** of the linear differential equation (8). The roots of (12) are easily verified to be 2 and 4. So we have two possible types of solutions to (8).

$$y(t) = Ae^{4t} \quad \text{and} \quad y(t) = A'e^{2t}$$

The constants A and A' can have any value and these solutions will still satisfy (8). In fact, it can readily be checked (see Exercise 9) that any linear combination of the basic solutions e^{2t} and e^{4t} is a solution. Thus

$$y(t) = Ae^{4t} + A'e^{2t} \tag{13}$$

is the general form of a solution to (8). The constants A and A' depend on the starting values. From (9) we have

$$\begin{aligned}
100 &= y(0) = Ae^0 + A'e^0 \\
-20 &= y'(0) = 4Ae^0 + 2A'e^0
\end{aligned} \tag{14a}$$

which simplifies to

$$\begin{aligned}
A + A' &= 100 \\
4A + 2A' &= -20
\end{aligned} \tag{14b}$$

In (14a), we obtain $y'(0)$ by differentiating (13) and setting $t = 0$. Now we have our old "friend," a system of two equations in two unknowns. We solve (14b) and obtain

$$A = -110 \quad \text{and} \quad A' = 210 \tag{15}$$

Substituting these values in (13), we obtain the required solution

$$y(t) = -110e^{4t} + 210e^{2t} \tag{16}$$

■

The calculations for any other second-order linear differential equation would proceed in a similar fashion: First substitute (2) in the differential equation to obtain the characteristic equation [as in (12)] and solve for its roots; then determine A and A' from the pair of equations for starting values [as in (14)]. This method generalizes to kth-order linear differential equations. Then the characteristic equation has k roots, we need k initial values, and we have to solve k equations in k unknowns to determine the constants.

For completeness, we note that if the two roots of the characteristic equation (12) were the same, such as 2 and 2, the starting value equations cannot be solved, since the coefficients of A and A' in the two equations of (14) will be the same. In the case of identical roots of the characteristic equation, $y(t)$ instead has the form

$$y(t) = Ae^{rt} + A'te^{rt} \tag{17}$$

where r is the double root. It is an exercise to check that in the case of a multiple root (and only then), te^{rt} is also a solution to the differential equation.

**EXAMPLE 3.
System of
Differential
Equations**

Let us consider a pair of first-order differential equations that describe motion of an object in x–y space with one equation governing the x-coordinate and one the y-coordinate.

$$\begin{aligned} x'(t) &= 2x(t) - y(t) \\ y'(t) &= - x(t) + 2y(t) \end{aligned} \tag{18}$$

Suppose that the starting values are $x(0) = y(0) = 1$. Let $\mathbf{u}(t) = \begin{bmatrix} x(t) \\ y(t) \end{bmatrix}$ and $\mathbf{u}'(t) = \begin{bmatrix} x'(t) \\ y'(t) \end{bmatrix}$. Then (18) can be written in matrix notation as

$$\mathbf{u}'(t) = \mathbf{B}\mathbf{u}(t), \quad \text{where} \quad \mathbf{B} = \begin{bmatrix} 2 & -1 \\ -1 & 2 \end{bmatrix} \quad \text{and} \quad \mathbf{u}(0) = \begin{bmatrix} 1 \\ 1 \end{bmatrix} = \mathbf{1} \tag{19}$$

In Example 1 we saw that

$$y'(t) = by(t), \quad y(0) = A \quad \text{has the solution} \quad y(t) = Ae^{bt} \tag{20}$$

Substituting $\mathbf{B}$ for b and $\mathbf{1}$ for A in (20), we obtain the "formal solution" to (19):

$$\mathbf{u}(t) = \mathbf{1}e^{\mathbf{B}t} \tag{21}$$

■

A matrix in the exponent looks strange. But one definition of e^x from calculus is in terms of the power series

$$e^x = 1 + \frac{x}{1!} + \frac{x^2}{2!} + \frac{x^3}{3!} + \cdots + \frac{x^k}{k!} + \cdots \tag{22}$$

Upon extending (22), the matrix exponential $e^{\mathbf{B}}$ is defined as

$$e^{\mathbf{B}} = \mathbf{I} + \frac{\mathbf{B}}{1!} + \frac{\mathbf{B}^2}{2!} + \frac{\mathbf{B}^3}{3!} + \cdots + \frac{\mathbf{B}^k}{k!} + \cdots \qquad (23)$$

The power series (23) is well defined for all matrices. While this power series of matrices may look forbidding, it is easy to use if we work in eigenvector-based coordinates so that $\mathbf{B}$ and its powers act like scalar multipliers (as in $\mathbf{Bu} = \lambda\mathbf{u}$). Recall Theorem 2 of Section 5.3, which said that if $\mathbf{U}$ is a matrix whose columns were eigenvectors of $\mathbf{B}$ associated with different eigenvalues, and if $\mathbf{D}_\lambda$ is a diagonal matrix of the corresponding eigenvalues, then

$$\mathbf{B} = \mathbf{U}\mathbf{D}_\lambda\mathbf{U}^{-1} \qquad (24)$$

Substituting with (24) for $\mathbf{B}$ in (23), we obtain

$$e^{\mathbf{B}} = \mathbf{I} + \mathbf{U}\mathbf{D}_\lambda\mathbf{U}^{-1} + \frac{1}{2!}\mathbf{U}\mathbf{D}_\lambda\mathbf{U}^{-1} + \frac{1}{3!}\mathbf{U}\mathbf{D}_\lambda\mathbf{U}^{-1} + \cdots + \frac{1}{k!}\mathbf{U}\mathbf{D}_\lambda\mathbf{U}^{-1} + \cdots \qquad (25a)$$

$$= \mathbf{U}\left(\mathbf{I} + \mathbf{D}_\lambda + \frac{1}{2!}\mathbf{D}_\lambda + \frac{1}{3!}\mathbf{D}_\lambda + \cdots + \frac{1}{k!}\mathbf{D}_\lambda + \cdots\right)\mathbf{U}^{-1} \qquad (25b)$$

$$= \mathbf{U}e^{\mathbf{D}_\lambda}\mathbf{U}^{-1} = \mathbf{U}\begin{bmatrix} e^{\lambda_1} & 0 & 0 & \cdots \\ 0 & e^{\lambda_2} & 0 & \cdots \\ 0 & 0 & e^{\lambda_3} & \cdots \\ \cdots & \cdots & \cdots & \cdots \\ \cdots & \cdots & \cdots & \cdots \end{bmatrix}\mathbf{U}^{-1} \qquad (25c)$$

The reason that in (25c) $e^{\mathbf{D}_\lambda}$ turns out to be simply a diagonal matrix with diagonal entries $e^{\lambda_1}, e^{\lambda_2}, \ldots, e^{\lambda_n}$ is that in (25b), the matrices $\mathbf{D}_\lambda/k!$ are diagonal with entries $\lambda_1/k!, \lambda_2/k!, \ldots$, and that summing these matrices we get a matrix whose entry $(1, 1)$ is $1 + \lambda_1 + \lambda_2/2! + \lambda_3/3! + \cdots$, which equals e^{λ_1}, and similarly for the other diagonal entries. From the matrix diagonalization discussion in Section 5.3, we recognize that $e^{\lambda_1}, e^{\lambda_2}, \ldots, e^{\lambda_n}$ are the eigenvalues of the matrix $e^{\mathbf{B}t}$.

The solution to the system of differential equations in Example 3 is left as an exercise. We now work out the details of evaluating a matrix exponential in two different ways for a problem solved previously.

**EXAMPLE 4.
Converting a
Second-Order
Differential
Equation into a
Pair of First-
Order Differential
Equations**

Let us consider again the second-order equation from Example 2:

$$y''(t) = 6y'(t) - 8y(t) \qquad (26)$$

We convert (26) into a pair of first-order equations by introducing a second function $x(t)$, defined as

$$x(t) = y'(t) \quad \text{and thus} \quad x'(t) = y''(t) \qquad (27)$$

Now (26) can be written as the pair of the first-order equations

$$x'(t) = 6x(t) - 8y(t)$$
$$y'(t) = x(t)$$

(28)

With initial conditions of $\mathbf{u}(0) = \begin{bmatrix} -20 \\ 100 \end{bmatrix}$, we have

$$\mathbf{u}'(t) = \mathbf{B}\mathbf{u}(t) \quad \text{or} \quad \begin{bmatrix} x'(t) \\ y'(t) \end{bmatrix} = \begin{bmatrix} 6 & -8 \\ 1 & 0 \end{bmatrix} \begin{bmatrix} x(t) \\ y(t) \end{bmatrix}$$

(29)

As in Example 3, the solution to (29) should be

$$\mathbf{u}(t) = e^{\mathbf{B}t}\mathbf{u}(0)$$

(30)

Remember that we already know from the solution of (28) Example 2, so $\mathbf{u}(t)$ in (30) must equal

$$\mathbf{u}(t) = \begin{bmatrix} x(t) \\ y(t) \end{bmatrix} = \begin{bmatrix} -440e^{4t} + 420e^{2t} \\ -110e^{4t} + 210e^{2t} \end{bmatrix}$$

(31)

Here, $y(t)$ is obtained from (16) and $x(t) = y'(t) = -440e^{4t} + 420e^{2t}$ is obtained by differentiating the solution for $y(t)$.

To determine $\mathbf{u}(t)$ using (30), we need (25):

$$e^{\mathbf{B}t} = \mathbf{U}e^{\mathbf{D}_\lambda t}\mathbf{U}^{-1}$$

(32)

where $e^{\mathbf{D}_\lambda t}$ is a 2×2 diagonal matrix with diagonal entries $e^{\lambda_i t}$. Recall that the columns $\mathbf{u}_1$ and $\mathbf{u}_2$ of $\mathbf{U}$ are eigenvectors of $\mathbf{B}$.

In Section 3.3 we learned how to find the eigenvalues of a matrix $\mathbf{B}$—they are the roots of the characteristic polynomial $\det(\mathbf{B} - \lambda\mathbf{I})$—and from them, the associated eigenvectors. The characteristic polynomial

$$\det(\mathbf{B} - \lambda\mathbf{I}) = \begin{vmatrix} 6 - \lambda & -8 \\ 1 & 0 - \lambda \end{vmatrix} = \lambda^2 - 6\lambda + 8$$

and its roots are 4 and 2. Eigenvectors $\mathbf{u}_1$, $\mathbf{u}_2$ of $\mathbf{B}$ associated with the eigenvalues 4 and 2 are (Exercise 6)

$$\mathbf{u}_1 = \begin{bmatrix} 4 \\ 1 \end{bmatrix} \text{ for } \lambda_1 = 4 \qquad \mathbf{u}_2 = \begin{bmatrix} 2 \\ 1 \end{bmatrix} \text{ for } \lambda_2 = 2$$

Thus

$$\mathbf{U} = \begin{bmatrix} 4 & 2 \\ -\frac{1}{2} & 2 \end{bmatrix}, \quad \text{and we compute} \quad \mathbf{U}^{-1} = \begin{bmatrix} \frac{1}{2} & -1 \\ -\frac{1}{4} & 2 \end{bmatrix}$$

(33)

Using (32) to substitute for $e^{\mathbf{B}t}$ in $\mathbf{u}(t) = e^{\mathbf{B}t}\mathbf{u}(0)$, we obtain

$$
\begin{aligned}
\mathbf{u}(t) = e^{\mathbf{B}t}\mathbf{u}(0) &= \mathbf{U}e^{\mathbf{D}_\lambda t}\mathbf{U}^{-1}\mathbf{u}(0) \\[4pt]
&= \begin{bmatrix} 4 & 2 \\ 1 & 1 \end{bmatrix} \begin{bmatrix} e^{4t} & 0 \\ 0 & e^{2t} \end{bmatrix} \begin{bmatrix} \tfrac{1}{2} & -1 \\ -\tfrac{1}{2} & 2 \end{bmatrix} \begin{bmatrix} -20 \\ 100 \end{bmatrix} \\[4pt]
&= \begin{bmatrix} 4 & 2 \\ 1 & 1 \end{bmatrix} \begin{bmatrix} e^{4t} & 0 \\ 0 & e^{2t} \end{bmatrix} \begin{bmatrix} 110 \\ 210 \end{bmatrix} \\[4pt]
&= \begin{bmatrix} 4e^{4t} & 2e^{2t} \\ e^{4t} & e^{2t} \end{bmatrix} \begin{bmatrix} 110 \\ 210 \end{bmatrix} \\[4pt]
&= \begin{bmatrix} -440e^{4t} + 420e^{2t} \\ -110e^{4t} + 210e^{2t} \end{bmatrix}
\end{aligned} \tag{34}
$$

Our answer corresponds with the result (31) obtained in Example 2.

We now redo the solution of $\mathbf{u}(t) = e^{\mathbf{B}t}\mathbf{u}(0)$ without using $e^{\mathbf{B}t} = \mathbf{U}e^{\mathbf{D}_\lambda t}\mathbf{U}^{-1}$ directly. Instead, we use the eigenvector-coordinate approach from Section 5.3 to compute $e^{\mathbf{B}t}\mathbf{u}(0)$ via the power series definition of $e^{\mathbf{B}t}$ $\left(\text{with } \mathbf{B} = \begin{bmatrix} 6 & -8 \\ 1 & 0 \end{bmatrix} \right)$:

$$
e^{\mathbf{B}t} = \mathbf{I} + t\mathbf{B} + \frac{t^2\mathbf{B}^2}{2!} + \frac{t^3\mathbf{B}^3}{3!} + \cdots + \frac{t^k\mathbf{B}^k}{k!} + \cdots \tag{35}
$$

Since $e^{\mathbf{B}t}$ involves powers of $\mathbf{B}$, the computation of multiplying $e^{\mathbf{B}t}$ times $\mathbf{u}(0)$ will be simplified if we express $\mathbf{u}(0)$ in terms of $\mathbf{B}$'s eigenvectors.

Writing $\mathbf{u}(0) = \begin{bmatrix} -20 \\ 100 \end{bmatrix}$ as a linear combination of $\mathbf{u}_1$ and $\mathbf{u}_2$ (we must solve the system $\mathbf{u}(0) = a\mathbf{u}_1 + b\mathbf{u}_2$ for a and b; see Section 3.3 for details), we obtain

$$
\mathbf{u}(0) = \begin{bmatrix} -20 \\ 100 \end{bmatrix} = -110\mathbf{u}_1 + 210\mathbf{u}_2 \tag{36}
$$

We now are ready to compute $e^{\mathbf{B}t}\mathbf{u}(0)$, which we rewrite using (35) as

$$
\mathbf{u}(t) = e^{\mathbf{B}t}\mathbf{u}(0) = \mathbf{I}\mathbf{u}(0) + t\mathbf{B}\mathbf{u}(0) + \frac{t^2\mathbf{B}^2\mathbf{u}(0)}{2!} + \cdots + \frac{t^k\mathbf{B}^k\mathbf{u}(0)}{k!} + \cdots \tag{37}
$$

Substituting $\mathbf{u}(0) = -110\mathbf{u}_1 + 210\mathbf{u}_2$ in (37), we have

$$
\begin{aligned}
\mathbf{u}(t) &= e^{\mathbf{B}t}(-110\mathbf{u}_1 + 210\mathbf{u}_2) \\[4pt]
&= \mathbf{I}(-110\mathbf{u}_1 + 210\mathbf{u}_2) + t\mathbf{B}(-110\mathbf{u}_1 + 210\mathbf{u}_2) \\[4pt]
&\quad + \frac{t^2\mathbf{B}^2}{2!}(-110\mathbf{u}_1 + 210\mathbf{u}_2) + \cdots + \frac{t^k\mathbf{B}^k}{k!}(-110\mathbf{u}_1 + 210\mathbf{u}_2) + \cdots \\[4pt]
&= -110\left\{ \mathbf{I}\mathbf{u}_1 + t\mathbf{B}\mathbf{u}_1 + \frac{t^2\mathbf{B}^2}{2!}\mathbf{u}_1 + \cdots + \frac{t^k\mathbf{B}^k}{k!}\mathbf{u}_1 + \cdots \right\} \\[4pt]
&\quad + 210\left\{ \mathbf{I}\mathbf{u}_2 + t\mathbf{B}\mathbf{u}_2 + \frac{t^2\mathbf{B}^2}{2!}\mathbf{u}_2 + \cdots + \frac{t^k\mathbf{B}^k}{k!}\mathbf{u}_2 + \cdots \right\}
\end{aligned} \tag{38}
$$

Since $\mathbf{u}_1$ and $\mathbf{u}_2$ are eigenvectors, the term $\mathbf{B}^k\mathbf{u}_1$ equals $4^k\mathbf{u}_1$, and $\mathbf{B}^k\mathbf{u}_2$ equals $2^k\mathbf{u}_2$. So (38) becomes

$$\mathbf{u}(t) = -110\left\{\mathbf{u}_1 + 4t\mathbf{u}_1 + \frac{4^2t^2}{2!}\mathbf{u}_1 + \cdots + \frac{4^kt^k}{k!}\mathbf{u}_1 + \cdots\right\}$$

$$+ 210\left\{\mathbf{u}_2 + 2t\mathbf{u}_2 + \frac{2^2t^2}{2!}\mathbf{u}_2 + \cdots + \frac{2^kt^k}{k!}\mathbf{u}_2 + \cdots\right\}$$

$$= -110\left\{1 + 4t + \frac{4^2t^2}{2!} + \cdots + \frac{4^kt^k}{k!} + \cdots\right\}\mathbf{u}_1 \tag{39}$$

$$+ 210\left\{1 + 2t + \frac{2^2t^2}{2!} + \cdots + \frac{2^kt^k}{k!} + \cdots\right\}\mathbf{u}_2$$

$$= -110e^{4t}\mathbf{u}_1 + 210e^{2t}\mathbf{u}_2$$

$$= -110e^{4t}\begin{bmatrix}4\\1\end{bmatrix} + 210e^{2t}\begin{bmatrix}2\\1\end{bmatrix} = \begin{bmatrix}-440e^{4t} + 420e^{2t}\\-110e^{4t} + 210e^{2t}\end{bmatrix}$$

Comparing $\mathbf{u}(t) = e^{\mathbf{B}t}\mathbf{u}(0) = e^{\mathbf{B}t}(-110\mathbf{u}_1 + 210\mathbf{u}_2)$ from (37) and (38) with the answer $\mathbf{u}(t) = -110e^{4t}\mathbf{u}_1 + 210e^{2t}\mathbf{u}_2$ in (39), we are reminded that in eigenvector coordinates, the matrix $e^{\mathbf{B}t}$ simply multiplies the $\mathbf{u}_1$-coordinate by e^{4t} and the $\mathbf{u}_2$-coordinate by e^{2t}, since e^{4t} and e^{2t} are the eigenvalues of $e^{\mathbf{B}t}$.

Observe that for a different starting vector $\mathbf{u}^*(0)$, we would get $\mathbf{u}^*(0) = a'\mathbf{u}_1 + b'\mathbf{u}_2$, for some a', b' and then the result in (39) would be $\mathbf{u}(t) = a'e^{4t}\mathbf{u}_1 + b'e^{2t}\mathbf{u}_2$. ■

We close this section by noting that the conversion of (26) to a system (28) of first-order differential equations can be applied to any linear higher-order differential equation.

EXAMPLE 5.
Converting a
Third-Order
Differential
Equation

Consider the third-order linear differential equation

$$y'''(t) = y''(t) + 2y'(t) + 3y(t) \tag{40}$$

We introduce the two new functions $x(t)$ and $z(t)$:

$$w(t) = y'(t) \quad \text{and} \quad z(t) = w'(t) \quad [= y''(t)] \tag{41}$$

Then (40) can be written

$$z'(t) = z(t) + 2w(t) + 3y(t) \tag{42}$$

Defining the vector function $\mathbf{u}(t) = \begin{bmatrix}y(t)\\w(t)\\z(t)\end{bmatrix}$, we can rewrite (41) and (42):

$$\mathbf{u}'(t) = \mathbf{B}\mathbf{u}(t): \begin{bmatrix}z'(t)\\w'(t)\\y'(t)\end{bmatrix} = \begin{bmatrix}1 & 2 & 3\\1 & 0 & 0\\0 & 1 & 0\end{bmatrix}\begin{bmatrix}z(t)\\w(t)\\y(t)\end{bmatrix} \tag{43}$$

The solution to (43) is $\mathbf{u}(t) = e^{\mathbf{B}t}\mathbf{u}(0)$, which we would evaluate using eigenvectors, as discussed in Example 4. ■

Section 6.3 Exercises

1. Solve the following first-order differential equations, with given initial values.
   (a) $y'(t) = 0.5y(t)$, $y(0) = 100$
   (b) $y'(t) - 4y(t) = 0$, $y(0) = 10$

2. Suppose that a population of bacteria is continuously doubling its size every unit of time. Write a differential equation for $y(t)$, the size of the population.

3. Solve the following second-order differential equations, with given initial values. Use the method based on the characteristic equation.
   (a) $y''(t) = 5y'(t) - 4y(t)$, $y(0) = 20$, $y'(0) = 5$
   (b) $y''(t) = -5y'(t) + 6y(t)$, $y(0) = 1$, $y'(0) = 15$
   (c) $y''(t) = 2y'(t) + 8y(t)$, $y(0) = 2$, $y'(0) = 0$

4. Convert the following differential equations into systems of simultaneous first-order differential equations. **Do not solve**.
   (a) $y''(t) = 5y'(t) - 4y(t)$
   (b) $y''(t) = -5y'(t) - 6y(t)$
   (c) $y'''(t) = 4y''(t) + 3y'(t) - 2y(t)$
   (d) $y'''(t) = 2y'(t) + y(t)$

5. Check that 3 and 1 are the eigenvalues for $\mathbf{B} = \begin{bmatrix} 2 & -1 \\ -1 & 2 \end{bmatrix}$ in Example 3 and then find the associated eigenvectors. Solve the system of differential equations in Example 3 using the matrix exponential method.

6. Check that 4 and 2 are the eigenvalues for $\mathbf{B} = \begin{bmatrix} 6 & -8 \\ 1 & 0 \end{bmatrix}$ in Example 4 and that associated eigenvectors are $\begin{bmatrix} 4 \\ 1 \end{bmatrix}$ and $\begin{bmatrix} 2 \\ 1 \end{bmatrix}$. Also verify (36).

7. Re-solve the second-order differential equations in Exercise 3 by converting them to a pair of first-order differential equations and solving by the method in Example 4. (You must find the eigenvalues and eigenvectors.)

8. Solve the following pairs of first-order differential equations by using the solution technique in Example 4 (you must find the eigenvalues and eigenvectors). The initial condition is $x(0) = y(0) = 10$.
   (a) $x'(t) = 4x(t)$, $y'(t) = 2x(t) + 2y(t)$
   (b) $x'(t) = 2x(t) + y(t)$, $y'(t) = 2x(t) + 3y(t)$
   (c) $x'(t) = x(t) + 4y(t)$, $y'(t) = 2x(t) + 3y(t)$

9. (a) Show that any multiple $ry*(t)$ of a solution $y*(t)$ to a second-order differential equation $y''(t) = ay'(t) + by(t)$ is again a solution.
   (b) Show that any linear combination $ry*(t) + sy°(t)$ of solutions $y*(t)$, $y°(t)$ to $y''(t) = ay'(t) + by(t)$ is again a solution.

10. Suppose that $y°(t)$ is some solution to $y''(t) - ay'(t) - by(t) = f(t)$ and $y*(t)$ is a solution to $y''(t) - ay'(t) - by(t) = 0$. Then show that for any r, $y°(t) + ry*(t)$ is also a solution to $y''(t) - ay'(t) - by(t) = f(t)$.

11. Verify that $y(t) = te^{\lambda t}$ is a solution to the differential equation $y''(t) = cy'(t) + dy(t)$ whose characteristic equation $k^2 - ck - d = 0$ has λ as its double root. [**Note:** If $k^2 - ck - d = 0$ has λ as a double root, then the characteristic equation can be factored as $(k - \lambda)^2 = 0$. This means that $c = 2\lambda$ and $d = -\lambda^2$. Use these values for c and d in verifying that $te^{\lambda t}$ is a solution.]

6.4 0–1 Matrices

In Section 3.1 we represented a graph G by a 0–1 adjacency matrix $\mathbf{A}(G)$ and saw that $\mathbf{A}^k(G)$ told us how many paths of length k there were in the graph between each pair of vertices. In this section we present three other applications of 0–1 matrices. The first application concerns mathematical schemes to encode information in a fashion designed to detect, and if possible correct, errors introduced by noise when messages are transmitted.

A **binary code** is a scheme for encoding a letter or number as a binary sequence of 0's and 1's and then decoding the binary sequence back into a letter or number. The binary sequence is often transmitted over a communications channel with random noise that may change one of the digits in the binary sequence. That is, when a 1 is sent, a 0 may be received; or vice versa. We assume that the chance of two errors in one binary sequence is small enough that it can be ignored.

The examples about binary codes involve multiplication and addition mod 2. Because computers represent numbers in terms of 0's and 1's (a circuit is open or closed), it is very easy for computers to calculate mod 2. The following tables summarize the rules for addition and multiplication mod 2.

+ mod 2	0	1
0	0	1
1	1	0

· (times) mod 2	0	1
0	0	0
1	0	1

The sum of many 1's is 0 if there is an even number of 1's, and is 1 if there is an odd number of 1's. For example, the following scalar product is calculated in arithmetic mod 2:

$$[1, 1, 0, 1, 1] \cdot [1, 0, 1, 1, 1] = 1 \cdot 1 + 1 \cdot 0 + 0 \cdot 1 + 1 \cdot 1 + 1 \cdot 1$$
$$= 1 + 0 + 0 + 1 + 1 \qquad (1)$$
$$= 1 \quad (\text{mod } 2)$$

EXAMPLE 1.
Parity-Bit Code
for Error
Detection

Error-detecting binary codes are designed to make it possible to detect an error should one occur during transmission. Then the transmitter would be asked to send the binary sequence again. A standard error-detecting code used to transmit data over telephone lines between computers is a parity-bit code.

The basic unit in such communication is usually a byte, an 8-bit binary sequence. Let $\mathbf{b} = [b_1, b_2, \ldots, b_8]$ be the byte to be sent. In a parity-bit code, an additional ninth bit, p, is added to **b** to get the sequence to be transmitted $\mathbf{c} = [b_1, b_2, \ldots, b_8, p]$. This bit p is normally chosen so that the number of 1's in **c** is even. Another way to say this is: Pick p so that the sum of the bits in **c** equals 0 (mod 2):

$$b_1 + b_2 + \cdots + b_8 + p = 0 \quad \text{(mod 2)} \tag{2}$$

or, equivalently,

$$\mathbf{1} \cdot \mathbf{c} = \mathbf{0} \quad \text{(mod 2)} \tag{3}$$

For example, if the byte to be sent is $\mathbf{b} = [1, 0, 1, 0, 0, 1, 0, 0]$, which has an odd number of 1's, then $p = 1$ and we send $\mathbf{c} = [1, 0, 1, 0, 0, 1, 0, 0, 1]$. Suppose that the message we received was $\mathbf{c}' = [1, 0, 0, 0, 0, 1, 0, 0, 1]$—the third bit was erroneously changed to 0. Whenever a message $\mathbf{c}'$ is received, we compute $\mathbf{1} \cdot \mathbf{c}'$. If no errors had occurred and $\mathbf{c}' = \mathbf{c}$, we would find that $\mathbf{1} \cdot \mathbf{c}' = 0$ mod 2. On the other hand, if $\mathbf{1} \cdot \mathbf{c}' = 1$, as in this example, some digit was altered—and we ask the sender to retransmit. A simple way to compute the proper value for the parity bit p is to let

$$p = \Sigma \, b_i = \mathbf{1} \cdot \mathbf{b} \quad \text{(mod 2)} \tag{4}$$

That is, let p equal the sum mod 2 of the 1's in **b**. If $p = \mathbf{1} \cdot \mathbf{b} = 1$ (mod 2), **b** has an odd number of 1's and making p 1 will give **c** an even number of 1's. If $p = \mathbf{1} \cdot \mathbf{b} = 0$ (mod 2), **b** already has an even number of 1's and we want p to be 0. ■

EXAMPLE 2.
Hamming Code
for Error
Correction

More advanced error-correcting codes can actually correct an error and reconstruct the original binary sequence that was sent. The following scheme due to Hamming takes a 4-bit binary sequence and encodes it as a 7-bit sequence. Let $\mathbf{b} = [b_1, b_2, b_3, b_4]^{\mathrm{T}}$ be the binary message, let p_1, p_2, p_3 be the parity-check bits, and let $\mathbf{c} = [c_1, c_2, \ldots, c_7]^{\mathrm{T}}$ be the code sequence that will be transmitted. The parity-check bits are chosen to satisfy the following three parity checks:

$$
\begin{aligned}
p_1 & & + b_1 + b_2 & & + b_4 & = 0 \quad \text{(mod 2)} \\
& p_2 & + b_1 & + b_3 + b_4 & & = 0 \quad \text{(mod 2)} \\
& & p_3 & + b_2 + b_3 + b_4 & & = 0 \quad \text{(mod 2)}
\end{aligned}
\tag{5}
$$

The idea behind this scheme with three parity checks is, if an error in some bit occurs, we can "triangulate in" on which bit was altered by looking at which subsets of parity checks are violated.

Let us encode these message and parity-check bits in the code sequence **c** as follows:

$$c_1 = p_1, \quad c_2 = p_2, \quad c_3 = b_1, \quad c_4 = p_3, \quad c_5 = b_2, \quad c_6 = b_3, \; c_7 = b_4 \tag{6}$$

The reason why $c_3 = b_1$, not p_3, will be clear shortly. Now (5) is

$$
\begin{array}{llll}
c_1 & + c_3 & + c_5 & + c_7 = 0 \quad (\text{mod } 2) \\
& c_2 + c_3 & + c_6 + c_7 = 0 \quad (\text{mod } 2) \\
& c_4 + c_5 & + c_6 + c_7 = 0 \quad (\text{mod } 2)
\end{array}
\tag{7}
$$

or

$$\mathbf{Mc} = \mathbf{0} \quad (\text{mod } 2)$$

where $\mathbf{M}$ is the matrix of coefficients in (7):

$$
\mathbf{M} =
\begin{bmatrix}
1 & 0 & 1 & 0 & 1 & 0 & 1 \\
0 & 1 & 1 & 0 & 0 & 1 & 1 \\
0 & 0 & 0 & 1 & 1 & 1 & 1
\end{bmatrix}
\tag{8}
$$

Each of the c_i's is involved in a different subset of parity checks, so when (exactly) one c_i is altered in transmission, the parity checks should allow us to determine which bit was changed.

Since each of the parity bits $c_1\ (= p_1)$, $c_2\ (= p_2)$, $c_4\ (= p_3)$ is in just one of the parity equations in (7), each can be determined as the sum of the other bits in their equation (just as $p = \Sigma\, b_i$ in Example 1). For example, $p_1 + b_1 + b_2 + b_4 = 0$ (mod 2) implies that $p_1 = b_1 + b_2 + b_4$ (mod 2). Summarizing how we go from the message vector $\mathbf{b}$ to the code vector $\mathbf{c}$ yields:

$$
\begin{array}{lll}
c_1 = b_1 + b_2 + b_4, & c_2 = b_1 + b_3 + b_4, & c_3 = b_1 \\
c_4 = b_2 + b_3 + b_4, & c_5 = b_2, & c_6 = b_3, \quad c_7 = b_4
\end{array}
\tag{9}
$$

The following matrix–vector product (mod 2) does the encoding specified by (9):

$$
\mathbf{c} = \mathbf{Qb}, \qquad \text{where} \quad
\mathbf{Q} =
\begin{bmatrix}
1 & 1 & 0 & 1 \\
1 & 0 & 1 & 1 \\
1 & 0 & 0 & 0 \\
0 & 1 & 1 & 1 \\
0 & 1 & 0 & 0 \\
0 & 0 & 1 & 0 \\
0 & 0 & 0 & 1
\end{bmatrix}
\tag{10}
$$

For example, suppose that $\mathbf{b} = \begin{bmatrix} 1 \\ 0 \\ 1 \\ 0 \end{bmatrix}$. Then

$$c = Qb = \begin{bmatrix} 1 & 1 & 0 & 1 \\ 1 & 0 & 1 & 1 \\ 1 & 0 & 0 & 0 \\ 0 & 1 & 1 & 1 \\ 0 & 1 & 0 & 0 \\ 0 & 0 & 1 & 0 \\ 0 & 0 & 0 & 1 \end{bmatrix} \begin{bmatrix} 1 \\ 0 \\ 1 \\ 0 \end{bmatrix}$$

$$= \begin{bmatrix} 1 \cdot 1 + 1 \cdot 0 + 0 \cdot 1 + 1 \cdot 0 \\ 1 \cdot 1 + 0 \cdot 0 + 1 \cdot 1 + 1 \cdot 0 \\ 1 \cdot 1 + 0 \cdot 0 + 0 \cdot 1 + 0 \cdot 0 \\ 0 \cdot 1 + 1 \cdot 0 + 1 \cdot 1 + 1 \cdot 0 \\ 0 \cdot 1 + 1 \cdot 0 + 0 \cdot 1 + 0 \cdot 0 \\ 0 \cdot 1 + 0 \cdot 0 + 1 \cdot 1 + 0 \cdot 0 \\ 0 \cdot 1 + 0 \cdot 0 + 0 \cdot 1 + 1 \cdot 0 \end{bmatrix} = \begin{bmatrix} 1 \\ 0 \\ 1 \\ 1 \\ 0 \\ 1 \\ 0 \end{bmatrix} \qquad (11)$$

Suppose that the transmission received c' has a 0 instead of a 1 in the sixth position. We compute the vector e:

$$e = Mc' = \begin{bmatrix} 1 & 0 & 1 & 0 & 1 & 0 & 1 \\ 0 & 1 & 1 & 0 & 0 & 1 & 1 \\ 0 & 0 & 0 & 1 & 1 & 1 & 1 \end{bmatrix} \begin{bmatrix} 1 \\ 0 \\ 1 \\ 1 \\ 0 \\ 0 \\ 0 \end{bmatrix} = \begin{bmatrix} 0 \\ 1 \\ 1 \end{bmatrix} \qquad (12)$$

Note that Mc' is just the set of left sides in (7) with c replaced by c'. If no error had occurred and $c' = c$, then $e = Mc = 0$ as in (7). If $e \neq 0$, as in (12), an error must have occurred. Depending on which parity equations are now violated, we can figure out which bit in c' was changed in transmission. We claim that $e \, (= \, Mc')$ equals the column of M corresponding to the bit of c that was changed. This is the case in (12), where e equals the sixth column of M. The reason is that when the kth bit is altered, exactly those equations involving the kth bit (i.e., those rows of M with a 1 in the kth column) will now equal 1 (mod 2).

As the reader has probably noticed, for each i, the ith column of M is simply the binary representation of the number i. Thus the vector e "spells out" the location of the bit that was changed. To get the correct transmission, we simply change back the bit in the position spelled out by e. In this case, we would change the sixth bit from a 0 back to a 1. ■

Our second application of 0–1 matrices involves a second useful matrix associated with a graph called the incidence matrix $M(G)$.

Incidence Matrix $\mathbf{M}(G)$ of a Graph G

$\mathbf{M}(G)$ has a row for each node of G and a column for each edge of G. If the jth edge is incident to the ith node (i.e., the ith node is an endpoint of the jth edge), entry $m_{ij} = 1$; entry $m_{ii} = 2$ if there is a loop edge at i; and otherwise $m_{ij} = 0$.

The incidence matrix for the graph in Figure 6.1 is

$$\mathbf{M}(G) = \begin{array}{c} \\ a \\ b \\ c \\ d \\ e \end{array} \begin{array}{c} \begin{array}{ccccccc} e_1 & e_2 & e_3 & e_4 & e_5 & e_6 & e_7 \end{array} \\ \left[\begin{array}{ccccccc} 1 & 1 & 1 & 0 & 0 & 0 & 0 \\ 1 & 0 & 0 & 1 & 0 & 0 & 0 \\ 0 & 1 & 0 & 1 & 1 & 0 & 0 \\ 0 & 0 & 1 & 0 & 1 & 1 & 0 \\ 0 & 0 & 0 & 0 & 0 & 1 & 2 \end{array} \right] \end{array} \tag{13}$$

$\mathbf{M}(G)$ will always have exactly two 1's in each column (or one 2), since an edge has two endpoints.

An *independent set* $\mathbf{I}$ of nodes in a graph is a set of nodes with the property that there is no edge joining any two nodes in $\mathbf{I}$. In Figure 6.1, $\{a, e\}$ is an independent set of nodes (there are several other independent sets in this graph). Independent sets arise in various settings. For example, let G be a graph in which each node stands for a letter that can be transmitted over a noisy communications channel and an edge joins two nodes if the corresponding letters can be confused when transmitted (i.e., one letter is sent but another letter is received). An independent set would represent a set of letters that cannot be confused with one another.

The next example shows how, using $\mathbf{M}(G)$, one can recast the graph problem of finding a maximum-sized independent set in a graph as an optimization problem in linear algebra.

**EXAMPLE 3.
Finding a
Maximum
Independent Set**

We shall now recast the property of being an independent set in terms of a set of linear inequalities. We assume that G has n nodes. The key step is to represent a set of nodes by a membership vector. The (row) membership vector $\mathbf{x} = [x_1, x_2, \ldots, x_n]$ for a set $\mathbf{I}$ is defined to have $x_i = 1$ if the ith node is in the set $\mathbf{I}$ and $x_i = 0$ otherwise. We claim that $\mathbf{x}$ is the membership vector for an independent set if and only if the following inequality holds.

$$\mathbf{x}\mathbf{M}(G) \le 1 \tag{14}$$

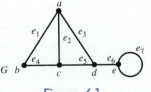

Figure 6.1

Recall that putting the vector before the matrix means that we are forming the scalar product of row vector $\mathbf{x}$ with each column of $\mathbf{M}(G)$. The columns of $\mathbf{M}(G)$ correspond to the edges of G. For the graph in Figure 6.1, (14) becomes

$$
\begin{array}{llll}
\text{edge } e_1: & x_a + x_b & & \leq 1 \\
\text{edge } e_2: & x_a + x_c & & \leq 1 \\
\text{edge } e_3: & x_a + x_d & & \leq 1 \\
\text{edge } e_4: & x_b + x_c & & \leq 1 \\
\text{edge } e_5: & x_c + x_d & & \leq 1 \\
\text{edge } e_6: & x_d + x_e \leq 1 \\
\text{edge } e_7: & 2x_e \leq 1
\end{array}
\qquad (15)
$$

Recall that each column of $\mathbf{M}(G)$ has two 1's, which correspond to the pair of nodes that form that edge. Then the left side of the first inequality in (15) refers to nodes a and b, which are the endpoints of edge e_1. The condition that $x_a + x_b \leq 1$ says that not both a and b can be in the independent set $\mathbf{I}$ (i.e., not both $x_a = 1$ and $x_b = 1$).

We are ready to pose the problem of finding a maximum independent set of G as a linear program. A *linear program* involves the maximization (or minimization) of some linear expression subject to a system of linear equations or inequalities. In the graph model for confusing letters sent over a noisy communication channel, a maximum independent set would be the largest possible set of letters that can be sent without one being confused with another.

We must restate the concept of maximizing the size of the independent set in terms of membership vectors. What we want is to maximize the number of 1's in the membership vector. Another way to say this is to maximize the sum $\sum x_i$, or in matrix algebra, $\mathbf{1} \cdot \mathbf{x}$. Combining this objective with (14), we have the linear program

$$\text{Maximize } \mathbf{1} \cdot \mathbf{x}$$
$$\text{Subject to } \mathbf{x}\mathbf{M}(G) \leq \mathbf{1} \qquad (16)$$
$$\text{and } x_i = 0 \quad \text{or} \quad 1$$

Because of the integer constraint, such a linear program is called an *integer program*. There is a large literature about solving integer programs. Virtually all optimization problems in graph theory can be posed as integer problems. ■

Our third application, also a graph problem, uses a directed version of an incidence matrix. It concerns a basic problem about electrical networks.

EXAMPLE 4.
Currents in an
Electrical Network

We want to determine the current in different parts of the electrical network in Figure 6.2(a). There are three basic laws that are used to analyze simple electrical networks. The following review of elementary physics summarizes the concepts behind these laws. A battery or other source of electrical power "forces" electricity through electrical devices, such as a light or a doorbell. The force applied to a device depends on two factors: (1) the resistance of the device, a measure of how hard it is to push electricity through the device; and (2) the current through the device, the rate at which electricity flows through the device. The fundamental law of electricity, due to Ohm, says

Ohm's Law

$$\text{force} = \text{resistance} \times \text{current}$$

Force is measured in volts, current in amperes, and resistance in ohms. In terms of these units, Ohm's law is

$$\text{volts} = \text{ohms} \times \text{amperes}$$

A battery supplies a fixed force into a network. Batteries send their electricity out from a positive terminal and receive it back at a negative terminal. All the voltage, that is, force, provided by a battery is used up by the time the electricity returns to the terminal. This property of voltage is called

Kirchhoff's Voltage Law. In any circuit (closed path) in a network, the sum of the voltages used by resistive devices equals the voltage from the battery(ies).

The final law says that current is conserved at any branch node.

Kirchhoff's Current Law. The sum of the currents flowing into any node is equal to the sum of the current flowing out of the node.

Let us use these three laws to derive a set of linear equations modeling the behavior of currents in the network in Figure 6.2. Later we will express each of Kirchhoff's laws in the form of a matrix equation.

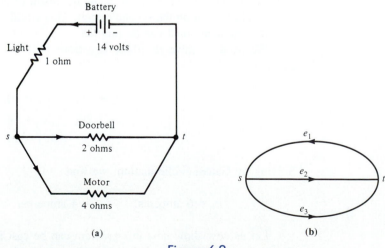

(a)   (b)

Figure 6.2

Assume that the battery delivers 14 volts of electricity. Let c_1 be the current flowing through the section of the network with the battery and the light, whose resistance is 1 ohm. Let c_2 be the current through the section with the doorbell, whose resistance is 2 ohms. And let c_3 be the current through the section with the motor, whose resistance is 4 ohms. Currents flow in the direction of the arrows along these edges. By the current law, at node s we have the equation

$$c_1 = c_2 + c_3 \tag{17}$$

Note that at node t we get the same equation. Next we use the voltage law. Following the circuit, battery to light to doorbell to battery, we have

$$\text{voltage at light} + \text{voltage at doorbell} = \text{battery voltage} \tag{18}$$

We use Ohm's law (voltage = resistance × current) to determine the voltages at the devices in (18). The battery voltage is 14. So (18) becomes

$$1c_1 + 2c_2 = 14 \tag{19}$$

Following a second circuit, battery to light to motor to battery, we have

$$1c_1 + 4c_3 = 14 \tag{20}$$

There is still a third circuit that we could use, node s to doorbell to node t to motor (going against the current flow) to node s. If we go against the current flow, the voltage is treated as negative. So the voltage law for this circuit is

$$2c_2 - 4c_3 = 0 \tag{21}$$

Note that equation (21) is simply what is obtained when we subtract (20) from (19). Intuitively, the third circuit is the net result of going forward on the first circuit and then backward on the second circuit. This circuit provides no new information—(21) is redundant—and can be omitted.

We need to solve the three equations (17), (19), (20) in the three unknown currents:

$$\begin{aligned}
c_1 - c_2 - c_3 &= 0 \\
c_1 + 2c_2 &= 14 \\
c_1 + 4c_3 &= 14
\end{aligned} \tag{22}$$

Solving by Gaussian elimination, we find

$$c_1 = 6 \text{ amperes}, \qquad c_2 = 4 \text{ amperes}, \qquad c_3 = 2 \text{ amperes} \qquad ■$$

Let us now show how this problem can be cast in matrix notation. Kirchhoff's current law can be restated in matrix form using the incidence matrix $\mathbf{M}(G)$ of the underlying graph (when batteries and resistive devices are ignored). Figure 6.2(b) shows the graph G associated with the network in Figure 6.2(a). G contains two nodes s, t and three edges e_1, e_2, e_3 joining s and t. In this graph, each edge has a direction; the directions for G are shown in Figure 6.2(b). The current c_i in edge e_i is positive if it flows in the direction of e_i and negative if it flows in the opposite direction. For an incidence matrix $\mathbf{M}(G)$ of a graph G with directed edges, entry $m_{ij} = +1$ if the jth edge is directed into the ith node, $= -1$ if the jth edge is directed out from the ith node, and $= 0$ if the jth edge does not touch the ith node. For the graph G in Figure 6.2(b), the incidence matrix $\mathbf{M}(G)$ is

$$\mathbf{M}(G) = \begin{array}{c} \\ s \\ t \end{array} \begin{array}{c} \begin{array}{ccc} e_1 & e_2 & e_3 \end{array} \\ \left[\begin{array}{ccc} +1 & -1 & -1 \\ -1 & +1 & +1 \end{array} \right] \end{array} \tag{23}$$

Kirchhoff's current law says that the flow into a node equals the flow out of the node, or in other words, the net current flow is zero. At node s, this means that

$$+c_1 - c_2 - c_3 = 0 \tag{24}$$

Observe that currents going into s are associated with edges that have $+1$ in row s of $\mathbf{M}(G)$, and currents going out are associated with edges that have -1 in row s of $\mathbf{M}(G)$. If $\mathbf{m}'_s$ denotes row s of $\mathbf{M}(G)$ and $\mathbf{c}$ is the vector of currents (c_1, c_2, c_3), then (24) can be rewritten as

$$\mathbf{m}'_s \cdot \mathbf{c} = 0 \tag{25}$$

This equation is true for all rows of $\mathbf{M}(G)$. Thus (25) extends to

Kirchhoff's Current Law

$$\mathbf{M}(G)\mathbf{c} = \mathbf{0} \tag{26}$$

This result is true for the graph G of any network.

We can also define a circuit matrix $\mathbf{K}(G)$ for G with a row for each circuit (closed path) of G and a column for each edge. Let r_j be the resistance in the jth edge. Then define entry k_{ij} of $\mathbf{K}(G) = +r_j$ if the ith circuit uses the jth edge traversing this edge in the direction of its arrow, $= -r_j$ if the ith circuit uses the jth edge in the opposite direction of its arrow, and $= 0$ otherwise. For example, $\mathbf{K}(G)$ in Example 4 would be, with circuits listed in the order they were discussed,

$$\mathbf{K}(G) = \begin{array}{c} \begin{array}{ccc} e_1 & e_2 & e_3 \end{array} \\ \left[\begin{array}{ccc} 1 & 2 & 0 \\ 1 & 0 & 4 \\ 0 & 2 & -4 \end{array} \right] \end{array}$$

Some circuits in a network are always redundant. It is sufficient to use a subset of the rows of $\mathbf{K}(G)$ that forms a basis for the row space of $\mathbf{K}(G)$. Note that at the end of

Section 4.1 we presented a vector space of circuits to illustrate an unusual type of vector space. (In Section 4.1 the circuits were represented as sets and the edges had no directions.)

If k_i' is the ith row of $K(G)$, then $k_i' \cdot c$ will be the (signed) sum of all the voltages used on the ith circuit. If f_i is the voltage force of batteries on the ith circuit, Kirchhoff's voltage law becomes

$$k_i' \cdot c = f_i \tag{27}$$

And if f is the vector of f_i's, we have

Kirchhoff's Voltage Law

$$K(G)c = f \tag{28}$$

The problem of determining currents in a network can now be stated: *Solve the system of equations given by* $M(G)c = 0$ *and* $K(G)c = f$ *to determine* c.

Section 6.4 Exercises

1. If the following bytes (8-bit binary sequences) are being sent in a parity-checking code, what should the additional parity bit be (0 or 1)?
   (a) 10101010
   (b) 11100110
   (c) 00000000

2. Suppose that the following 9-bit messages were received in a parity-checking code. Which messages are altered during transmission?
   (a) 101010101
   (b) 101101101
   (c) 000000000

3. Explain why if two errors occur during transmission, a parity-checking code of the sort in Example 1 will not detect an error.

4. Suppose that in the Hamming code in Example 2, the following 4-bit binary messages
$$b = \begin{bmatrix} b_1 \\ b_2 \\ b_3 \\ b_4 \end{bmatrix}$$
   are to be sent. What will the coded 7-bit message $c = [c_1, c_2, c_3, c_4, c_5, c_6, c_7]^T$ be?

   (a) $b = \begin{bmatrix} 1 \\ 1 \\ 0 \\ 0 \end{bmatrix}$ (b) $b = \begin{bmatrix} 1 \\ 1 \\ 1 \\ 0 \end{bmatrix}$ (c) $b = \begin{bmatrix} 1 \\ 1 \\ 1 \\ 1 \end{bmatrix}$

5. Suppose that in the Hamming code in Example 2, the following messages c' are received. In each case compute the error vector $e = Mc'$ and from it tell which bit, if any, was changed in transmission.

(a) $\begin{bmatrix} 0 \\ 0 \\ 1 \\ 1 \\ 1 \\ 1 \\ 0 \end{bmatrix}$ (b) $\begin{bmatrix} 1 \\ 1 \\ 1 \\ 1 \\ 1 \\ 0 \\ 1 \end{bmatrix}$ (c) $\begin{bmatrix} 0 \\ 1 \\ 0 \\ 0 \\ 0 \\ 0 \\ 0 \end{bmatrix}$

6. Assume that at most one error occurred in transmission of the Hamming code in Example 2. What message was originally sent if the following message was received?

(a) $\begin{bmatrix} 0 \\ 0 \\ 1 \\ 1 \\ 1 \\ 1 \\ 0 \end{bmatrix}$ (b) $\begin{bmatrix} 0 \\ 1 \\ 0 \\ 0 \\ 0 \\ 0 \\ 0 \end{bmatrix}$ (c) $\begin{bmatrix} 1 \\ 1 \\ 1 \\ 0 \\ 0 \\ 1 \\ 0 \end{bmatrix}$

7. Suppose that we let $c_1 = p_1$, $c_2 = p_2$, $c_3 = p_3$, $c_4 = b_1$, $c_5 = b_2$, $c_6 = b_3$, $c_7 = b_4$ instead of the encoding scheme in (6). With this encoding scheme, what is the new $\mathbf{Q}$ matrix in

(10)? If $\mathbf{b} = \begin{bmatrix} 1 \\ 1 \\ 0 \\ 0 \end{bmatrix}$, what would $\mathbf{c}$ be?

8. (a) Explain why the Hamming code in Example 2 will always detect two errors (i.e., if two bits in the code are changed, the error vector $\mathbf{e}$ cannot be all 0's).
   (b) Give an example to show that the Hamming code in Example 2 cannot correct two errors.
   (c) Give an example to show that the Hamming code cannot always detect three errors.

9. The Hamming code in Example 2 can be extended to a similar code for 15-bit sequences—11 message bits and 4 parity-check bits. Write out the system of parity-check equations (or equivalently, the matrix for coefficients for these equations) for a 15-bit Hamming code.

10. Another way to encode a binary sequence is by treating the sequence as the coefficients of a polynomial $p(x)$; for example, the sequence $[1, 0, 1, 1]$ yields the polynomial $p(x) = 1 + 0x + 1x^2 + 1x^3$. We encode by multiplying $p(x)$ by some other polynomial $g(x)$ to get the polynomial $p'(x) = g(x)p(x)$ whose coefficients we transmit. For example, let $g(x) = 1 + x$. Then for $p(x) = 1 + 0x + 1x^2 + 1x^3$, we compute $p'(x) = g(x)p(x) = (1 + x)(1 + 0x + 1x^2 + 1x^3) = 1 + 1x + 1x^2 + 0x^3 + 1x^4$. (Remember that arithmetic is mod 2.) So we transmit $[1, 1, 1, 0, 1]$. To decode we divide the polynomial $p'(x)$ by $g(x)$. If any error occurred in transmission, there will be a remainder—this tells us that an error occurred.
   (a) Using $g(x) = 1 + x$, perform a polynomial encoding of the sequence $[1, 1, 0, 1]$.
   (b) Suppose that the following messages are received, based on this polynomial encoding with $g(x) = 1 + x$. Which ones have errors?
      (i) $[1, 1, 0, 1, 0]$
      (ii) $[1, 0, 1, 0]$
      (iii) $[1, 0, 0, 0]$
   (c) (*Advanced*) Show that the parity of messages transmitted is always even with the polynomial encoding scheme with $g(x) = 1 + x$. [Hint: By setting $x = 1$ in $p'(x)[= g(x)p(x)]$ one can sum the coefficients of $p'(x)$ (i.e., the message bits).]

11. Draw the graphs with the following incidence matrices.

(a) $\begin{array}{c} a \\ b \\ c \\ d \end{array} \begin{bmatrix} 0 & 0 & 1 & 1 \\ 1 & 0 & 1 & 0 \\ 0 & 1 & 0 & 0 \\ 1 & 1 & 0 & 1 \end{bmatrix}$ (b) $\begin{array}{c} a \\ b \\ c \\ d \end{array} \begin{bmatrix} 0 & 1 & 0 & 1 \\ 1 & 0 & 1 & 1 \\ 1 & 0 & 0 & 0 \\ 0 & 1 & 1 & 0 \end{bmatrix}$ (c) $\begin{array}{c} a \\ b \\ c \\ d \\ e \\ f \end{array} \begin{bmatrix} 0 & 0 & 1 & 1 & 0 & 0 \\ 0 & 1 & 1 & 0 & 1 & 0 \\ 1 & 1 & 0 & 0 & 0 & 1 \\ 1 & 0 & 0 & 0 & 1 & 0 \\ 0 & 0 & 0 & 1 & 0 & 0 \\ 0 & 0 & 0 & 0 & 0 & 1 \end{bmatrix}$

12. Write the incidence matrices for the following graphs.

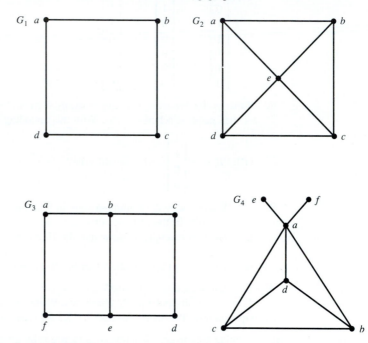

13. Let $\mathbf{A}(G)$ denote the adjacency matrix of a graph and let $\mathbf{M}(G)$ denote the incidence matrix of that graph. Show that entry (i, j) of $\mathbf{M}(G)\mathbf{M}(G)^T$ equals entry (i, j) of $\mathbf{A}(G)$, for $i \neq j$. Explain this result in words.

14. Write out the linear program for finding a maximum independent set in the following graphs in Exercise 12.
    (a) G_1 (b) G_2 (c) G_3

15. Show that $\mathbf{1} \cdot \mathbf{M}(G) = 2\mathbf{1} = [2, 2, 2, \ldots, 2]^T$ for any graph.

16. Determine the currents in each branch of the following circuit.

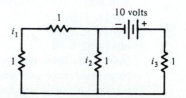

17. Determine the currents in each branch of the following circuits in which the incoming amperage (on the left) is given.

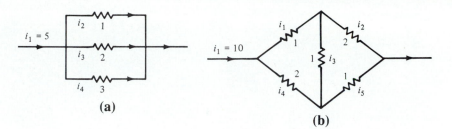

(a)   (b)

18. Determine the currents in each branch of the following circuit. The voltage in the battery is 19.

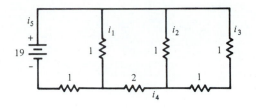

Brief History of Matrices and Linear Algebra

Matrices and linear algebra did not grow out of the study of coefficients of systems of linear equations, as one might guess. Arrays of coefficients led mathematicians to develop determinants, not matrices. Leibnitz, coinventor of calculus, used determinants in 1693, about 150 years before the study of matrices in their own right. Cramer presented his determinant-based formula for solving systems of linear equations in 1750. The first implicit use of matrices occurred in Lagrange's work on bilinear forms in the late nineteenth century. His objective was to characterize the maxima and minima of functions of several real variables. In addition to requiring that the first derivatives be zero, he needed a condition on the matrix of second derivatives: The condition was positive definiteness or negative definiteness (although he did not use matrices).

Gauss developed Gaussian elimination around 1800, to solve least squares problems in celestial computations and later in geodetic computations. It should be noted that Chinese manuscripts from several centuries earlier have been found that explain how to solve a system of three equations in three unknowns by ''Gaussian'' elimination. Gaussian elimination was for years considered part of the development of geodesy, not mathematics. Gauss–Jordan elimination's first appearance in print was in a handbook on geodesy by Wilhelm Jordan. The name Jordan in Gauss–Jordan elimination does not refer to the famous mathematician Camille Jordan, but rather to the geodesist Wilhelm Jordan. (Most linear algebra texts mistakenly identified Gauss–Jordan elimination with the mathematician Jordan, until a 1987 article in the *Mathematical Monthly* by Athloen and McLaughlin [1], motivated by a historical talk by this author's father, A. W. Tucker, set the record straight.)

For matrix algebra to develop, one needed two things: (1) the proper notation, such as a_{ij} and $\mathbf{A}$; and (2) the definition of matrix multiplication. Interestingly, both these critical factors occurred at about the same time, around 1850, and in the same country, England. Except for Newton's invention of calculus, the major mathematical advances in the seventeenth, eighteenth, and early nineteenth century were all made by continental mathematicians, names such as Bernoulli, Cauchy, Euler, Gauss, and Laplace. But in the mid-nineteenth century, English mathematicians pioneered the study of various algebraic systems. For example, Augustus DeMorgan and George Boole developed the algebra of sets (Boolean algebra) in which symbols were used for propositions and abstract elements (see Bell [2] and Cajori [3]).

The introduction of matrix notation and the invention of the word *matrix* were motivated by attempts to develop the right algebraic language for studying determinants. In 1848, J. J. Sylvester introduced the term *matrix,* the Latin word for *womb,* as a name for an array of numbers. He used ''womb'' because he viewed a matrix as a generator of determinants. That is, every subset of k rows and k columns in a matrix generated a determinant (associated with the submatrix formed by those rows and columns).

Matrix algebra grew out of work by Arthur Cayley in 1855 about linear transformations. Given transformations,

$$T_1: \quad x' = ax + by \qquad T_2: \quad x'' = \alpha x' + \beta y'$$
$$y' = cx + dy \qquad\qquad y'' = \gamma x' + \delta y'$$

he considered the transformation obtained by performing T_1 and then performing T_2.

$$x'' = (\alpha a + \beta c)x + (\alpha b + \beta d)y$$
$$y'' = (\gamma a + \delta c)x + (\gamma b + \delta d)y$$

In studying ways to represent this composite transformation, he was led to define matrix multiplication: The matrix of coefficients for the composite transformation T_2T_1 is the product of the matrix for T_2 times the matrix for T_1. Cayley went on to study the algebra of these compositions—matrix algebra—including matrix inverses. In his 1858 *Memoir on the Theory of Matrices*, Cayley gave the famous Cayley–Hamilton theorem, a square matrix satisfies its characteristic equation. The use of a single symbol **A** to represent the matrix of a transformation was essential notation of this new algebra. A link between matrix algebra and determinants was quickly established with the result: $\det(\mathbf{AB}) = \det(\mathbf{A}) \det(\mathbf{B})$. But Cayley seemed to have realized that matrix algebra might grow to overshadow the theory of determinants. He wrote, ''There would be many things to say about this theory of matrices which should, it seems to me, precede the theory of determinants.''

It is a curious sidelight to this discussion that another prominent English mathematician of this time was Charles Babbage, who built the first modern calculating machine. Abstracting the mechanics of computation as well as its algebraic structure and notation (and DeMorgan's work on the algebra of sets, which would later be crucial in computer science) seemed to be all part of the same general intellectual pattern in England in the mid-nineteenth century.

Mathematicians also tried to develop an algebra of vectors but there was no natural general definition for the product of two vectors. The first vector algebra, involving a noncommutative vector product, was proposed by Hermann Grassmann's first *Ausdehnungslehre* (1844). This text also introduced column–row products, what are now called simple matrices or rank-one matrices (formed by matrix multiplication of a column vector times a row vector). The famous treatise on vector analysis by the late-nineteenth-century American mathematical physicist J. Willard Gibbs developed vector and matrix theory further [5], including representations of general matrices, which he called *dyadics*, as a sum of simple matrices, which Gibbs called *dyads*. Later the physicist P. A. M. Dirac introduced [4] the term *bra-ket* for what we now call the scalar product of a "bra" (row) vector times a "ket" (column) vector, while the term *ket-bra* referred to the product of "ket" (column) times "bra" (row), yielding what we call here a simple matrix. (Physicists in the twentieth century developed the convention of assuming that any vector was implicitly a column vector with a row vector represented as the transpose of a column vector.)

Matrices remained closely associated with linear transformations and, from the theoretical viewpoint, were by 1900 just a finite-dimensional subcase of an emerging general theory of linear transformations. Matrices were also viewed as a powerful notation, but after an initial spurt of interest in the nineteenth century were little studied in their own right. More attention was paid to vectors, which are basic mathematical elements of physics as well as many areas of mathematics. The modern definition of a vector space was introduced by Peano in 1888. Abstract vector spaces, whose elements were functions of linear transformations, soon followed.

Interest in matrices, with emphasis on their numerical analysis, reemerged after World War II with the development of modern digital computers. John von Neumann and Herman Goldstein in 1947 introduced condition numbers in analyzing roundoff error. Alan Turing, the other giant (with von Neumann) in the development of stored-program computers, introduced the **LU** decomposition of a matrix in 1948. The usefulness of the **QR** decomposition was realized a decade later.

From Wilhelm Jordan through J. Willard Gibbs to Alan Turing, much of fundamentals of the theory of matrices have been developed by users of mathematics rather than by mathematicians themselves. On the other hand, the theory of vectors, vector spaces, and linear transformations has been developed primarily by mathematicians.

References

1. S. Athloen and R. McLaughlin, "Gauss–Jordan Reduction: A Brief History," *Mathematics Monthly*, 94 (1987), pp. 130–142.
2. E. T. Bell, *The Development of Mathematics*, McGraw-Hill, New York, 1940.
3. F. Cajori, *A History of Mathematical Notations*, Vol. II, Open Court Publishing Company, Chicago, 1929.
4. P. A. M. Dirac, *Principles of Quantum Mechanics*, 3rd ed., Clarendon Press, Oxford, 1947.
5. J. Willard Gibbs, *Collected Works*, Vol. II, Yale University Press, New Haven, Conn., 1928.

References

Introductory Linear Algebra

Anton, H., *Elementary Linear Algebra*, 6th ed., Wiley, New York, 1991.

Banchoff, T., and J. Wermer, *Linear Algebra Through Geometry*, Springer-Verlag, New York, 1983.

Curtis, C., *Linear Algebra: An Introductory Approach*, 4th ed., Springer-Verlag, New York, 1984.

Grossman, S., *Elementary Linear Algebra*, 4th ed., W. B. Saunders, Philadelphia, 1991.

Kumpel, P., and J. Thorpe, *Linear Algebra with Applications to Differential Equations*, W. B. Saunders, Philadelphia, 1983.

Nicholson, W. K., *Elementary Linear Algebra with Applications*, Prindle, Weber & Schmidt, Boston, 1986.

Strang, G., *Linear Algebra and Its Applications*, 3rd ed., Harcourt Brace Jovanovich, San Diego, 1988.

Freshman-Level Linear Algebra

Althoen, S., and R. Bumcrot, *Matrix Methods in Finite Mathematics*, W. W. Norton, New York, 1976.

Brown, J., and D. Sherbert, *Introductory Linear Algebra with Applications*, Prindle, Weber & Schmidt, Boston, 1984.

Applied Linear Algebra

Helzer, G., *Applied Linear Algebra*, Little, Brown, Boston, 1983.

Magid, A., *Applied Matrix Models*, Wiley, New York, 1985.

Noble, B., and J. Daniels, *Applied Linear Algebra*, 3rd ed., Prentice Hall, Englewood Cliffs, N.J., 1988.

Rorres, C., and H. Anton, *Applications of Linear Algebra*, 3rd ed., Wiley, New York, 1984.

Numerical Analysis

General

Burden, R., and J. Faires, *Numerical Analysis*, 4th ed., PWS-Kent, Boston, 1989.

Conte, S., and C. deBoor, *Elementary Numerical Analysis*, 3rd ed., McGraw-Hill, New York, 1980.

Hildebrand, F., *Introduction to Numerical Analysis*, McGraw-Hill, New York, 1974.

Kincaid, J., and W. Cheney, *Numerical Analysis*, Brooks/Cole, Pacific Grove, Calif., 1990.

More Advanced Numerical Linear Algebra

Golub, G., and C. VanLoan, *Matrix Computations*, 2nd ed., Johns Hopkins University Press, Baltimore, 1989.

Horn, R., and C. Johnson, *Matrix Analysis*, Cambridge University Press, New York, 1985.

Stewart, G., *Introduction to Matrix Computations*, Academic Press, New York, 1973.

Wilkinson, J., *The Algebraic Eigenvalue Problem*, Oxford University Press, New York, 1988.

Specific Applications References

Graphics

Berger, M., *Computer Graphics*, Benjamin-Cummings, Menlo Park, Calif., 1986.

Preparata, F., and M. Shamos, *Computational Geometry, An Introduction*, Springer-Verlag, New York, 1985.

Rogers, D., and J. Adams, *Mathematical Elements for Computer Graphics*, McGraw-Hill, New York, 1976.

Linear Models in Statistics

Draper, N., and H. Smith, *Applied Regression Analysis*, 2nd ed., Wiley, New York, 1981.

Graybill, F., *Theory and Application of the Linear Model*, Duxbury, Boston, 1976.

Neter, J., W. Wasserman, and M. Kutner, *Applied Linear Statistical Models*, 2nd ed., Richard D. Irwin, Homewood, Ill., 1985.

Differential Equations and Other Physical Science Applications

Boyce, W., and R. DiPrima, *Elementary Differential Equations and Boundary Value Problems*, 4th ed., Wiley, New York, 1986.

Braun, M., *Differential Equations and Their Applications: An Introduction to Applied Mathematics*, 3rd ed., Springer-Verlag, New York, 1983.

Noble, B., *Applications of Undergraduate Mathematics in Engineering*, Mathematical Association of America, Washington, D.C., 1967.

Strang, G., *Introduction to Applied Mathematics*, Wellesley-Cambridge Press, Wellesley, Mass., 1986.

Markov Chains

Hoel, P., S. Port, and C. Stone, *Introduction to Stochastic Processes*, Houghton Mifflin, Boston, 1972.

Kemeny, J., and L. Snell, *Finite Markov Chains*, Springer-Verlag, New York, 1976.

Growth Models and Recurrence Relations

Goldberg, S., *Introduction to Difference Equations*, Wiley, New York, 1958.

Kemeny, J., and L. Snell, *Mathematical Models in the Social Sciences*, MIT Press, Cambridge, Mass., 1969.

Linear Programming

Bradley, H., A. Hax, and T. Magnanti, *Applied Mathematical Programming*, Addison-Wesley, Reading, Mass., 1977.

Gass, S., *Linear Programming*, 5th ed., McGraw-Hill, New York, 1985.

Hillier, F., and G. Lieberman, *Introduction to Operations Research*, 5th ed., Holden-Day, Oakland, Calif., 1990.

For more references, see *A Basic Library List*, published by the Mathematical Association of America, Washington, D.C., 1991.

Matrix Algebra Software

General-Purpose Computer Languages and Computation Packages with Basic Matrix Operations

TRUE BASIC and other versions of the language BASIC that have matrix operations built in; for example, MATRIX 100, an enhanced BASIC for IBM PCs from Stanford Business Software.

APL. Reference: See Helzer's book listed in the References under "Applied Linear Algebra."

DERIVE, SoftWarehouse, Inc.

MACSYMA, Symbolics, Inc.

MAPLE, University of Waterloo.

Matrix Computation Packages

GAUSS (by L. Edelfsen and S. Jones for IBM PC), Applied Technical Systems.

LINEAR-KIT (by H. Anton for IBM PC and Apple), Wiley.

MAX (by E. Hermen for IBM PC), Brooks/Cole.

PC-MATLAB (C. Moler and J. Little for IBM PC), MathWorks.

The following two packages, designed for larger computers, are the best matrix computation software in existence. MAX and PC-MATLAB use parts of these packages.

LINPACK. Public domain. Reference: J. Dongarra, J. Bunch, C. Moler, and G. Stewart, *LINPACK User's Guide*, SIAM, Philadelphia, 1979.

EISPACK. Public domain. Reference: B. Smith et al., *Matrix Eigensystems Routines: EISPACK Guide*, 2nd ed., Springer-Verlag, New York, 1976.

413

Solutions to Odd-Numbered Exercises

Chapter 1

Section 1.1

1. **(a)** $[1 \quad 2 \quad 3 \quad 4]$; **(b)** $\begin{bmatrix} 2 \\ 4 \\ 5 \end{bmatrix}$; **(c)** $\begin{bmatrix} 3 \\ 6 \\ 7 \end{bmatrix}$; **(d)** 4; **(e)** 3. **3.**

7. $6x_1 + 3x_2 + 2x_3 = 280$
$\quad 4x_1 + 6x_2 + 3x_3 = 350$
$\quad 3x_1 + 2x_2 + 6x_3 = 350$

9. $10x_1 + 50x_2 + 200x_3 = 600$
$\quad 1x_1 + \quad 3x_2 + 0.2x_3 = \quad 20$
$\quad 30x_1 + 10x_2 \qquad\quad = 200$

11. $2000x_1 + 500x_2 + 5000x_3 = 280{,}000$ $5x_1 - \quad x_2 \qquad\qquad = \qquad 0$
$\qquad\qquad -x_1 \qquad\qquad 2x_3 = \qquad 0$

13. $\begin{bmatrix} \frac{1}{2} \\ \frac{1}{2} \end{bmatrix}$, $\begin{bmatrix} \frac{5}{8} \\ \frac{3}{8} \end{bmatrix}$, converges to $\begin{bmatrix} 0.6 \\ 0.4 \end{bmatrix}$; $\begin{bmatrix} \frac{3}{4} \\ \frac{1}{4} \end{bmatrix}$, $\begin{bmatrix} \frac{9}{16} \\ \frac{7}{16} \end{bmatrix}$, converges to $\begin{bmatrix} 0.6 \\ 0.4 \end{bmatrix}$.

15. **(a)** $\begin{bmatrix} 0.9 & 0.6 \\ 0.1 & 0.4 \end{bmatrix}$; **(b)** 0.75; **(c)** 0.87. **17.** **(a)** $\begin{bmatrix} 0.6 & 0.4 & 0.2 \\ 0.2 & 0.2 & 0.2 \\ 0.2 & 0.4 & 0.6 \end{bmatrix}$; **(b)** $\begin{bmatrix} 0.34 \\ 0.2 \\ 0.45 \end{bmatrix}$;

Section 1.2

1. $\begin{bmatrix} 5.8 & 7.6 & 8.8 \\ 7.6 & 6.0 & 8.8 \\ 7.8 & 7.2 & 8.0 \\ 5.4 & 5.4 & 6.2 \end{bmatrix}$. 3. (a) $\begin{bmatrix} 3 & 6 & 9 & 12 \\ 6 & 12 & 18 & 24 \\ 9 & 15 & 21 & 27 \end{bmatrix}$; (b) $\begin{bmatrix} -2 & 0 & 4 & 2 \\ 4 & -2 & -2 & 0 \\ 4 & 0 & 0 & 4 \end{bmatrix}$;

(c) $\begin{bmatrix} 3 & 0 & -6 & -3 \\ -6 & 3 & 3 & 0 \\ -6 & 0 & 0 & -6 \end{bmatrix}$; (d) $\begin{bmatrix} 0 & 2 & 5 & 5 \\ 4 & 3 & 5 & 8 \\ 5 & 5 & 7 & 11 \end{bmatrix}$; (e) $\begin{bmatrix} -1 & 4 & 12 & 11 \\ 10 & 5 & 9 & 16 \\ 12 & 10 & 14 & 24 \end{bmatrix}$;

(f) $\begin{bmatrix} 5 & 6 & 5 & 10 \\ 2 & 14 & 20 & 24 \\ 5 & 15 & 21 & 23 \end{bmatrix}$. 5. (a) $4I + 2J$; (b) $J - A$; (c) $4I + 3J - 2A$. 7. (a) 2; (b) 5;

(c) 38; (d) 14. 9. (a) -27; (b) $[50, 89, 128, 167]$; (c) not defined; (d) not defined;

(e) 64. 11. (a) $[10, 6, 3, 2]\begin{bmatrix} 5.00 & 5.00 & 4.00 \\ 1.00 & 1.50 & 0.75 \\ 0.75 & 1.00 & 1.00 \\ 8.0 & 7.00 & 10.00 \end{bmatrix}$; (b) $[74.25, 76.00, 67.50]$.

13. (a) $A = \begin{bmatrix} 3 & 4 \\ 2 & -5 \end{bmatrix}$, $b = \begin{bmatrix} 5 \\ 3 \end{bmatrix}$, $Ax = b$; (b) $A = \begin{bmatrix} 2 & 1 & -2 \\ 1 & 0 & 3 \\ 3 & -1 & 0 \end{bmatrix}$, $b = \begin{bmatrix} 0 \\ 3 \\ 5 \end{bmatrix}$, $Ax = b$;

(c) $A = \begin{bmatrix} 2 & -1 & 0 \\ 3 & 2 & 0 \\ 4 & -3 & 0 \end{bmatrix}$, $x = Ax$. 15. (a) $Ax = By + c$; (b) $Ax - By = c$.

17.

(a)   (b)   (c)

23. (a) $2x_1^2 + 8x_1x_2 + 2x_2^2$; (b) $3x_1^2 - 2x_1x_2 + 3x_2^2$;
(c) $2x_1^2 + 3x_2^2 + 5x_3^2 + 2x_1x_2 + 8x_1x_3 - 2x_2x_3$;
(d) $x_1^2 + 4x_2^2 + 5x_3^2 - 4x_1x_3 + 2x_2x_3$.

Section 1.3

1. AE 3×2 matrix, BA 2×7 matrix, BC 2×3 matrix, CA 3×7 matrix, CC 3×3 matrix, DB 2×3 matrix, DD 2×2 matrix, EB 7×3 matrix, ED 7×2 matrix.

3. (a) Not possible; (b) $\begin{bmatrix} -2 & -3 & -4 & -5 \\ -2 & -4 & -6 & -8 \\ -1 & -1 & -1 & -1 \end{bmatrix}$; (c) $\begin{bmatrix} 16 & 14 & 20 \\ 32 & 28 & 40 \\ 41 & 35 & 47 \end{bmatrix}$;

(d) $\begin{bmatrix} 16 & 31 & 46 & 61 \\ 7 & 12 & 17 & 22 \\ 10 & 19 & 28 & 37 \\ 11 & 19 & 27 & 35 \end{bmatrix}$; **(e)** $\begin{bmatrix} 13 & -7 & -6 \\ 1 & 2 & -3 \\ 7 & -3 & -4 \\ 2 & 1 & -3 \end{bmatrix}$. **5.** $\mathbf{AB} = \begin{bmatrix} 1 & 0 \\ 0 & 1 \end{bmatrix} = \mathbf{BA}$.

7. (a) $\mathbf{BA} = \begin{bmatrix} 2.30 & 3.05 \\ 1.65 & 2.10 \end{bmatrix}$; **(b)** $\mathbf{CB} = \begin{bmatrix} 7{,}000 & 12{,}500 & 5{,}500 \\ 14{,}000 & 25{,}000 & 11{,}000 \end{bmatrix}$;

(c) $(\mathbf{CB})\mathbf{A} = \begin{bmatrix} 3125 & 4100 \\ 6250 & 8200 \end{bmatrix}$. **9. (a)** [160, 155]; **(b)** DC; **(c)** CAB; **(d)** entry (1, 1) in

DCAB. **11.** $\mathbf{x}^{(k)} = \mathbf{A}^k\mathbf{x}$. **13. (a)** $\begin{bmatrix} \frac{6}{16} & \frac{5}{16} & \frac{5}{16} \\ \frac{5}{16} & \frac{6}{16} & \frac{5}{15} \\ \frac{5}{16} & \frac{5}{16} & \frac{6}{16} \end{bmatrix}$, $\begin{bmatrix} \frac{22}{64} & \frac{21}{64} & \frac{21}{64} \\ \frac{21}{64} & \frac{22}{64} & \frac{21}{64} \\ \frac{21}{64} & \frac{21}{64} & \frac{22}{64} \end{bmatrix}$, $\frac{21}{64}$; **(b)** $\begin{bmatrix} \frac{8}{16} & \frac{7}{16} & \frac{6}{16} \\ \frac{6}{16} & \frac{6}{16} & \frac{5}{16} \\ \frac{2}{16} & \frac{3}{16} & \frac{5}{16} \end{bmatrix}$,

$\begin{bmatrix} \frac{30}{64} & \frac{29}{64} & \frac{27}{64} \\ \frac{24}{64} & \frac{23}{64} & \frac{22}{64} \\ \frac{10}{64} & \frac{12}{64} & \frac{15}{64} \end{bmatrix}$, $\frac{10}{64}$. **15. (a)** $\begin{bmatrix} 6 & 0 & 0 \\ 0 & 2 & 0 \\ 0 & 0 & 15 \end{bmatrix}$; **(b)** $\begin{bmatrix} a_{11}b_{11} & 0 & 0 \\ 0 & a_{22}b_{22} & 0 \\ 0 & 0 & a_{33}b_{33} \end{bmatrix}$.

19. (a) $\begin{bmatrix} 1 & 0 & 4 \\ 0 & 1 & 0 \\ 0 & 0 & 1 \end{bmatrix}$; **(b)** $\begin{bmatrix} 1 & 0 & 0 \\ -2 & 1 & 0 \\ 3 & 0 & 1 \end{bmatrix}$. **21.** $\begin{bmatrix} 2 & 2 & 6 & 2 \\ 0 & 1 & 5 & 6 \\ 0 & 0 & 1 & 7 \\ 0 & 0 & 0 & 3 \end{bmatrix}$.

Section 1.4

1. (a) $x = \frac{14}{3}$, $y = \frac{1}{3}$; **(b)** $x = \frac{23}{13}$, $y = -\frac{2}{13}$; **(c)** $x = 2$, $y = 6$. **3. (a)** $\begin{bmatrix} 30 \\ 14 \\ -9 \end{bmatrix}$; **(b)** $\begin{bmatrix} \frac{1}{3} \\ -\frac{7}{3} \\ 0 \end{bmatrix}$;

(c) $\begin{bmatrix} -1 \\ 2 \\ \frac{3}{2} \end{bmatrix}$; **(d)** $\begin{bmatrix} \frac{37}{29} \\ \frac{15}{29} \\ \frac{9}{29} \end{bmatrix}$; **(e)** multiple solutions; **(f)** $\begin{bmatrix} -\frac{21}{11} \\ -\frac{76}{11} \\ \frac{164}{11} \end{bmatrix}$.

5. (a) Multiple solutions; **(b)** no solution; **(c)** $\begin{bmatrix} 47.82 \\ -54.35 \\ 160.87 \end{bmatrix}$; **(d)** $\begin{bmatrix} 41\frac{2}{3} \\ 41\frac{2}{3} \\ 0 \end{bmatrix}$.

11. (a) No solution; **(b)** $\begin{bmatrix} 0.5 \\ 0 \\ 0.5 \\ 0 \\ 0.5 \end{bmatrix}$. **13.** $\begin{bmatrix} 5 \\ 4 \\ 3 \\ 3 \\ 2 \\ 0 \end{bmatrix} + x_6 \begin{bmatrix} 1 \\ 1 \\ 1 \\ 1 \\ 1 \\ 1 \end{bmatrix}$.

15. $43\frac{1}{18}$ tables, $23\frac{11}{18}$ chairs, $34\frac{13}{18}$ sofas. **17.** 100,000 A's, 50,000 B's, 100,000 C's.

19. (a) $\begin{bmatrix} 1 & \frac{3}{2} \\ 0 & 1 \end{bmatrix}$; **(b)** $\begin{bmatrix} 1 & 0 & \frac{1}{2} \\ 0 & 1 & -1 \\ 0 & 0 & 1 \end{bmatrix}$; **(c)** $\begin{bmatrix} 1 & 0 & 3 \\ 0 & 1 & 4 \\ 0 & 0 & 1 \end{bmatrix}$; **(d)** $\begin{bmatrix} 1 & 2 & 3 & 0 \\ 0 & -1 & -2 & 0 \\ 0 & 0 & 1 & 1 \end{bmatrix}$.

21. (a) Unique; **(b)** no solution; **(c)** multiple solutions; **(d)** multiple solutions; **(e)** unique.

23. (a) $\begin{bmatrix} \frac{1}{2} \\ 0 \\ \frac{1}{2} \end{bmatrix}$; **(b)** $\begin{bmatrix} \frac{1}{3} \\ \frac{1}{3} \\ \frac{1}{3} \end{bmatrix}$; **(c)** $\begin{bmatrix} \frac{1}{3} \\ \frac{1}{3} \\ \frac{1}{3} \end{bmatrix}$.

Section 1.5

3. (a) $\begin{aligned} x_1 \quad\;\; + 2x_3 &= 1 \\ x_2 + 3x_3 &= 0; \end{aligned}$ **(b)** $\begin{bmatrix} 2 \\ \frac{3}{2} \\ -\frac{1}{2} \end{bmatrix}$. **5. (a)** $\dfrac{1}{5}\begin{bmatrix} -9 & 22 & -1 \\ -5 & 10 & 0 \\ 4 & -7 & 1 \end{bmatrix};$

$x_1 + \quad\;\; + 4x_3 = 0$

(b) $\dfrac{1}{6}\begin{bmatrix} -2 & 1 & 2 \\ -10 & -4 & -2 \\ -6 & -3 & 0 \end{bmatrix};$ **(c)** $\dfrac{1}{3}\begin{bmatrix} -6 & 26 & -11 \\ 3 & -16 & 7 \\ 3 & -11 & 5 \end{bmatrix};$ **(d)** $\dfrac{1}{58}\begin{bmatrix} 2 & 14 & -20 \\ 11 & -10 & 6 \\ -5 & -6 & -8 \end{bmatrix};$

(e) no solution; **(f)** $\dfrac{1}{11}\begin{bmatrix} -8 & 6 & 1 \\ -30 & 17 & 1 \\ 63 & -39 & -1 \end{bmatrix}$. **7. (a)** Second column of inverse;

(b) $-2 \times$ (third column of inverse);

(c) second column minus third column. **9. (a)** $\begin{bmatrix} 5 & 4 & 3 & 2 & 1 \\ 4 & 4 & 3 & 2 & 1 \\ 3 & 3 & 3 & 2 & 1 \\ 2 & 2 & 2 & 2 & 1 \\ 1 & 1 & 1 & 1 & 1 \end{bmatrix};$ **(b)** $\begin{bmatrix} \frac{5}{6} & \frac{2}{3} & \frac{1}{2} & \frac{1}{3} & \frac{1}{6} \\ \frac{2}{3} & \frac{4}{3} & 1 & \frac{2}{3} & \frac{1}{3} \\ \frac{1}{2} & 1 & \frac{3}{2} & 1 & \frac{1}{2} \\ \frac{1}{3} & \frac{2}{3} & 1 & \frac{4}{3} & \frac{2}{3} \\ \frac{1}{6} & \frac{1}{3} & \frac{1}{2} & \frac{2}{3} & \frac{5}{6} \end{bmatrix}$.

11. (a) $\begin{bmatrix} 0.0001 & -0.127 & 0.038 \\ -0.0004 & 0.382 & -0.013 \\ 0.005 & -0.089 & 0.001 \end{bmatrix}, \begin{bmatrix} 5.04 \\ 4.89 \\ 1.53 \end{bmatrix};$ **(b)** 0.51 less Jello; **(c)** $0.013k$ more.

13. (a) $\begin{bmatrix} \frac{1}{2} & -1 & \frac{1}{2} \\ \frac{1}{2} & 0 & -\frac{1}{2} \\ -\frac{3}{2} & 2 & \frac{1}{2} \end{bmatrix}, \begin{bmatrix} 100,000 \\ 50,000 \\ 100,000 \end{bmatrix};$ **(b)** $\begin{bmatrix} -100,000 \\ 0 \\ 200,000 \end{bmatrix};$ **(c)** $\begin{bmatrix} -5000k \\ -5000k \\ 15,000k \end{bmatrix}$.

17. $\begin{bmatrix} 1/a_{11} & 0 & 0 \\ 0 & 1/a_{22} & 0 \\ 0 & 0 & 1/a_{33} \end{bmatrix}$. **19. (a)** $\begin{bmatrix} 1 & -2 & -\frac{1}{2} & 2 \\ 0 & 1 & -\frac{1}{2} & 0 \\ 0 & 0 & \frac{1}{2} & -2 \\ 0 & 0 & 0 & 1 \end{bmatrix}$. **23. (a, c)** No inverse;

(b, d) guarantees nothing. **25. (a)** $\begin{bmatrix} 6 & 0 & 3 \\ 6 & 3 & 0 \\ 0 & 0 & 15 \end{bmatrix};$ **(b)** $\begin{bmatrix} 4 & 0 & 7 \\ 6 & 1 & 2 \\ 0 & 0 & 25 \end{bmatrix};$

(c) $\begin{bmatrix} \frac{1}{2} & 0 & -\frac{1}{10} \\ -1 & 1 & \frac{1}{5} \\ 0 & 0 & \frac{1}{5} \end{bmatrix}$. **27. (a)** $\begin{bmatrix} 2 & 0 & 0 \\ -1 & 1 & -1 \\ 0 & 0 & 2 \end{bmatrix}, \begin{bmatrix} 1 \\ -1 \\ 1 \end{bmatrix};$ **(b)** no inverse;

(c) $\begin{bmatrix} 7 & -3 & -3 \\ -3 & 7 & -3 \\ -3 & -3 & 7 \end{bmatrix}, \begin{bmatrix} 2 \\ -3 \\ 2 \end{bmatrix}$. **29. (a) (i)** $\begin{bmatrix} 18 & 5 \\ 5 & 23 \end{bmatrix},$ **(ii)** $\begin{bmatrix} 23 & 14 \\ 10 & 9 \end{bmatrix},$ **(iii)** $\begin{bmatrix} 9 & 5 \\ 3 & 14 \end{bmatrix};$

(b) (i) PG, **(ii)** QZ, **(iii)** WU.

Section 1.6

1. (a) -17; **(b)** 0; **(c)** -2. **3.** $x_1 = \dfrac{2D - G}{15}, \; x_2 = \dfrac{10G - 5D}{15a}$. **5. (a)** $\begin{bmatrix} x \\ y \end{bmatrix} = \begin{bmatrix} 0 \\ 0 \end{bmatrix};$

(b) $\begin{bmatrix} x \\ y \end{bmatrix} = r\begin{bmatrix} 1 \\ 4 \end{bmatrix};$ **(c)** $\begin{bmatrix} x \\ y \end{bmatrix} = r\begin{bmatrix} 3 \\ 1 \end{bmatrix}$. **7. (a)** -2; **(b, c, d)** 0; **(e)** -4; **(f)** $\sim 3.3 \times 10^{-8}$.

9. $x_3 = 9600/64 = 150$.   **15. (a)** 0; **(b)** -2; **(c)** -4.   **17. (a)** 9; **(b)** 81; **(c)** $\frac{1}{3}$; **(d)** 81.
21. (a, b, e) Unique; **(c, d)** infinite.

Chapter 2

Section 2.1

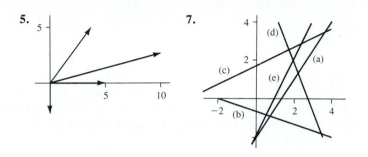

5.   **7.**

9. (a) $[1 - r, 2 - r]$; **(b)** $[-1 - 3r, 3 + 2r]$; **(c)** $[3, -2 - 5r]$; **(d)** $[-2 - r, -1 - 4r]$;
(e) $[4 + 5r, 2r]$; **(f)** $[1, 2]$.   **11. (a)** $x_1 - x_2 = -1$; **(b)** $2x_1 + 3x_2 = 7$; **(c)** $x_1 = 3$;
(d) $-4x_1 + x_2 = 7$; **(e)** $2x_1 - 5x_2 = -8$; **(f)** no standard equation.
13. (a) $\dfrac{x_1 - 1}{-1} = \dfrac{x_2 - 2}{2} = \dfrac{x_3 - 1}{-2}$; **(b)** $\dfrac{x_1 + 1}{-1} = \dfrac{x_2 - 2}{-1} = \dfrac{x_3 - 3}{2}$; **(c)** no symmetric

form; **(d)** no symmetric form; **(e)** $\dfrac{x_1 - 4}{5} = \dfrac{x_2 + 1}{1} = \dfrac{x_3 + 2}{1}$; **(f)** no symmetric form

(point $[1, 4, 3]$).   **15. (a)** $[1, 1, 2] + r_1[1, -1, -1] + r_2[-3, 0, 0]$;
(b) $[3, 0, 1] + r_1[-2, -2, 3] + r_2[5, -3, 0]$; **(c)** $[2, 0, 0] + r_1[2, -1, 0] + r_2[2, 0, -3]$;
(d) $[-2, -4, 3] + r_1[-4, 1, 4] + r_2[-3, -10, 5]$.   **17. (a)** $r_1 = 2$, $r_2 = 2$; **(b)** $r_1 = 3$,
$r_2 = -2$, **(c)** $r_1 = -3$, $r_2 = -6$; **(d)** $r_1 = 2$, $r_2 = 2$; **(e)** $r_1 = 5$, $r_2 = -4$; **(f)** $r_1 = 20$,
$r_2 = -22$.

Section 2.2

1. (a) $\sqrt{3}$; **(b)** 3; **(c)** $\sqrt{18}$; **(d)** $\sqrt{26}$; **(e)** 8.   **3. (a)** 5; **(b)** $\sqrt{5}$; **(c)** $\sqrt{24}$; **(d)** $\sqrt{84}$.
5. (a) 10, 6.

Section 2.3

1. (a) $1/\sqrt{2}$, 45°; **(b)** 7/25, ~74°; **(c)** $5/\sqrt{50}$, 45°; **(d)** 0, 90°; **(e)** $2/\sqrt{18}$, ~62°;
(f) $4/\sqrt{42}$, ~52°.   **3. (a)** Parallel; **(b)** orthogonal; **(c)** parallel; **(d)** orthogonal;

(e) neither; **(f)** neither.   **5.** 0, 0.67.   **7. (a)** 0.15; **(b)** -0.74.   **9. (a)** $\begin{bmatrix} -3 & 2 \\ 2 & -1 \end{bmatrix}$;

(b) $\begin{bmatrix} -2 & 1 \\ 1 & 0 \end{bmatrix}$; **(c)** $\begin{bmatrix} \frac{3}{2} & \frac{1}{2} \\ 2 & 1 \end{bmatrix}$.   **13. (a)** $(x_1 - 1) - 2x_2 + 2(x_3 - 2) = 0$,

$x_1 - 2x_2 + 2x_3 = 5$; **(b)** $3(x_2 - 1) + (x_3 - 4) = 0$, $3x_2 + x_3 = 7$;
(c) $(x_2 - 3) + 3(x_3 + 1) = 0$, $x_2 + 3x_3 = 0$; **(d)** $2(x_1 - 2) = 0$, $x_2 = 2$.

Section 2.4

1. (a) $\frac{19}{25}$; **(b)** $-\frac{2}{5}$; **(c)** 0; **(d)** 0; **(e)** $-\frac{37}{25}$; **(f)** 1.   **3. (a)** $\frac{19}{25}[3, 4] + \frac{1}{25}[68, -51]$;
(b) $-\frac{2}{5}[3, 1] + \frac{1}{5}[-9, 27]$; **(c)** $[0, 0] + [4, 0]$; **(d)** $[0, 0] + [-6, 2]$;

(e) $\frac{37}{25}[3, 4] + \frac{1}{25}[-36, 27]$; **(f)** $[5, 2] + [0, 0]$.   **5. (a)** $\begin{bmatrix} \frac{8}{5} \\ -\frac{3}{5} \end{bmatrix}$; **(b)** $\begin{bmatrix} -\frac{4}{5} \\ \frac{11}{20} \end{bmatrix}$; **(c)** $\begin{bmatrix} \frac{8}{5} \\ -\frac{1}{15} \end{bmatrix}$.

7. (a) 1.8; **(b)** -3; **(c)** 0; **(d)** -1.2.   **9. (a)** $\frac{3}{5}\sqrt{5}$; **(b)** $\frac{2}{5}\sqrt{5}$; **(c)** 0.   **11. (a)** $\frac{11}{14}\sqrt{14}$;
(b) $\frac{2}{14}\sqrt{14}$; **(c)** $\frac{4}{14}\sqrt{14}$.   **13.** $\hat{y} = 0.64x$.   **15.** $\hat{y} = 0.25x^2$.

Section 2.5

1. (a) $[-20, -15, 10]$; **(b)** $[8, 30, -9]$; **(c)** $[-19, 0, 0]$; **(d)** $[10, 19, -17]$;
(e) $[-20, 0, 4]$; **(f)** $[0, 0, 0]$.   **3. (a)** $[-12, 24, -30]$; **(b)** -129; **(c)** 0;
(d) $[129, 387, 258]$; **(e)** 0; **(f)** not defined.   **9. (a)** 3; **(b)** 9; **(c)** 6; **(d)** 0.   **13. (a)** Left,
right; **(b)** left, left; **(c)** right, right.   **15. (a)** Hidden; **(b)** visible.

Chapter 3

Section 3.1

1. (a) $\mathbf{I}$; **(b)** n; **(c)** 1; **(d)** $\mathbf{i}_k$; **(e)** 1; **(f)** 1; **(g)** 0; **(h)** n; **(i)** 0.   **3. (a)** $\mathbf{Qx} = \mathbf{x}$,

$\mathbf{Q} = \begin{bmatrix} 2 & -3 \\ 5 & 4 \end{bmatrix}$; **(b)** $(\mathbf{Q} - \mathbf{I})\mathbf{x} = \mathbf{0}$.   **5. (a)** $\begin{bmatrix} 1 \\ 2 \end{bmatrix}$; **(b)** $\begin{bmatrix} 1 \\ 0 \end{bmatrix}$; **(c)** $\begin{bmatrix} -3 \\ 0 \end{bmatrix}$; **(d)** $\begin{bmatrix} 7 \\ 5 \end{bmatrix}$;

(e) $\begin{bmatrix} 1 \\ 1 \end{bmatrix}$.   **9.** $\Sigma_i\, a_{ik}$.   **15. (a)** $\begin{bmatrix} 3 & 2 & 10 \\ 4 & -5 & 8 \\ 0 & 0 & 1 \end{bmatrix}$; **(b)** $\begin{bmatrix} 1 & 2 & 5 & 20 \\ 2 & -1 & -2 & -10 \\ 3 & 4 & 6 & 30 \\ 0 & 0 & 0 & 1 \end{bmatrix}$.

23. (a) $(\mathbf{AB})^2 = \mathbf{ABAB}$; **(b)** any diagonal matrices.   **27.** $\begin{bmatrix} 5 & 1 & -3 & -3 \\ - & 5 & 1 & -3 \\ - & - & 5 & 1 \\ - & - & - & 5 \end{bmatrix}$.

31. (a) $\begin{bmatrix} 0 & 1 & 0 & 1 \\ 1 & 0 & 1 & 0 \\ 0 & 1 & 0 & 1 \\ 1 & 0 & 1 & 0 \end{bmatrix}$; **(b)** $\begin{bmatrix} 0 & 1 & 0 & 1 & 1 \\ 1 & 0 & 1 & 0 & 1 \\ 0 & 1 & 0 & 1 & 1 \\ 1 & 0 & 1 & 0 & 1 \\ 1 & 1 & 1 & 1 & 0 \end{bmatrix}$; **(c)** $\begin{bmatrix} 0 & 1 & 0 & 0 & 0 \\ 1 & 0 & 1 & 0 & 0 \\ 0 & 1 & 0 & 1 & 0 \\ 0 & 0 & 1 & 0 & 1 \\ 0 & 0 & 0 & 1 & 0 \end{bmatrix}$;

(d) $\begin{bmatrix} 0 & 1 & 0 & 0 \\ 1 & 0 & 0 & 0 \\ 0 & 0 & 0 & 1 \\ 0 & 0 & 1 & 0 \end{bmatrix}$; **(e)** $\begin{bmatrix} 0 & 1 & 1 & 1 & 1 & 1 \\ 1 & 0 & 1 & 1 & 0 & 0 \\ 1 & 1 & 0 & 1 & 0 & 0 \\ 1 & 1 & 1 & 0 & 0 & 0 \\ 1 & 0 & 0 & 0 & 0 & 0 \\ 1 & 0 & 0 & 0 & 0 & 0 \end{bmatrix}$; **(f)** $\begin{bmatrix} 0 & 1 & 0 & 0 & 0 & 1 \\ 1 & 0 & 1 & 0 & 1 & 0 \\ 0 & 1 & 0 & 1 & 0 & 0 \\ 0 & 0 & 1 & 0 & 1 & 0 \\ 0 & 1 & 0 & 1 & 0 & 1 \\ 1 & 0 & 0 & 0 & 1 & 0 \end{bmatrix}$.

33. (a) $\begin{bmatrix} 0 & 4 & 0 & 4 \\ 4 & 0 & 4 & 0 \\ 0 & 4 & 0 & 4 \\ 4 & 0 & 4 & 0 \end{bmatrix}$; **(b)** $\begin{bmatrix} 4 & 8 & 4 & 8 & 8 \\ 8 & 4 & 8 & 4 & 8 \\ 4 & 8 & 4 & 8 & 8 \\ 8 & 4 & 8 & 4 & 8 \\ 8 & 8 & 8 & 8 & 8 \end{bmatrix}$; **(c)** $\begin{bmatrix} 0 & 2 & 0 & 1 & 0 \\ 2 & 0 & 3 & 0 & 1 \\ 0 & 3 & 0 & 3 & 0 \\ 1 & 0 & 3 & 0 & 2 \\ 0 & 1 & 0 & 2 & 0 \end{bmatrix}$;

(d) $\mathbf{A}^3(G_4) = \mathbf{A}(G_4)$. **35. (a)** D_1: $\begin{bmatrix} 0 & 0 & 0 & 0 \\ 1 & 0 & 0 & 0 \\ 0 & 1 & 0 & 0 \\ 1 & 0 & 1 & 0 \end{bmatrix}$, $\begin{bmatrix} 0 & 0 & 0 & 0 \\ 0 & 0 & 0 & 0 \\ 1 & 0 & 0 & 0 \\ 0 & 1 & 0 & 0 \end{bmatrix}$;

D_2: $\begin{bmatrix} 0 & 0 & 0 & 0 & 0 \\ 1 & 0 & 0 & 0 & 0 \\ 0 & 1 & 0 & 0 & 0 \\ 1 & 0 & 1 & 0 & 0 \\ 1 & 1 & 1 & 1 & 0 \end{bmatrix}$, $\begin{bmatrix} 0 & 0 & 0 & 0 & 0 \\ 0 & 0 & 0 & 0 & 0 \\ 1 & 0 & 0 & 0 & 0 \\ 0 & 1 & 0 & 0 & 0 \\ 2 & 1 & 1 & 0 & 0 \end{bmatrix}$. **39. (a)** $\begin{bmatrix} 2\mathbf{J} & \mathbf{J} \\ \mathbf{J} & 2\mathbf{J} \\ 2\mathbf{J} & \mathbf{J} \\ \mathbf{J} & 2\mathbf{J} \end{bmatrix}$; **(b)** $\begin{bmatrix} \mathbf{A} & \mathbf{J} - \mathbf{A} \\ \mathbf{A} & \mathbf{J} - \mathbf{A} \\ 2\mathbf{A} & \mathbf{J} - \mathbf{A} \\ 2\mathbf{A} & \mathbf{J} - \mathbf{A} \end{bmatrix}$,

where $\mathbf{J} = \begin{bmatrix} 1 & 1 & 1 & 1 \\ 1 & 1 & 1 & 1 \end{bmatrix}$, $\mathbf{A} = \begin{bmatrix} 1 & 0 & 1 & 0 \\ 0 & 1 & 0 & 1 \end{bmatrix}$; **(c)** $\begin{bmatrix} \mathbf{J} & 2\mathbf{J} & \mathbf{A} \\ 2\mathbf{J} & \mathbf{J} & \mathbf{A} \\ \mathbf{J} & \mathbf{J} & \end{bmatrix}$;

(d) $\begin{bmatrix} \mathbf{O} & 2\mathbf{J} & 3\mathbf{I} & 3\mathbf{I} \\ \mathbf{I} & \mathbf{I} & 3\mathbf{I} & 3\mathbf{I} \\ \mathbf{J} - \mathbf{I} & \mathbf{J} - \mathbf{I} & \mathbf{O} & \mathbf{O} \end{bmatrix}$, where $\mathbf{I}$ is 2×2 identity matrix

$\mathbf{J}$ is 2×2 1's matrix, $\mathbf{O}$ is 2×2 0's matrix
$\mathbf{A}$ is 3×3 matrix alternating 0's and 1's.

41.

$$\begin{array}{c} \\ a \\ b \\ c \\ d \\ \\ e \\ f \\ g \\ h \end{array} \begin{array}{cccccccc} a & b & c & d & e & f & g & h \\ \end{array}$$

$$\begin{bmatrix} 0 & \frac{1}{2} & \frac{1}{2} & \frac{1}{3} & 0 & 0 & 0 & \frac{1}{4} \\ \frac{1}{4} & 0 & 0 & \frac{1}{3} & 0 & 0 & 0 & 0 \\ \frac{1}{4} & 0 & 0 & \frac{1}{3} & 0 & 0 & 0 & 0 \\ \frac{1}{4} & \frac{1}{2} & \frac{1}{2} & 0 & 0 & 0 & 0 & 0 \\ 0 & 0 & 0 & 0 & 0 & \frac{1}{2} & \frac{1}{2} & \frac{1}{4} \\ 0 & 0 & 0 & 0 & \frac{1}{3} & 0 & 0 & \frac{1}{4} \\ 0 & 0 & 0 & 0 & \frac{1}{3} & 0 & 0 & \frac{1}{4} \\ \frac{1}{4} & 0 & 0 & 0 & \frac{1}{3} & \frac{1}{2} & \frac{1}{2} & 0 \end{bmatrix} = \begin{bmatrix} \mathbf{M} & \mathbf{N}^{\mathrm{T}} \\ \mathbf{N} & \mathbf{M} \end{bmatrix}.$$

43. $\begin{bmatrix} \mathbf{K} & \mathbf{L} & \mathbf{K} & \mathbf{L} \\ \mathbf{L} & \mathbf{K} & \mathbf{L} & \mathbf{K} \\ \mathbf{K} & \mathbf{L} & \mathbf{K} & \mathbf{L} \\ \mathbf{L} & \mathbf{K} & \mathbf{L} & \mathbf{K} \end{bmatrix}$, $\begin{array}{c} \mathbf{K} = 20\mathbf{J} \\ \mathbf{L} = 16\mathbf{J} \end{array}$. **45. (a)** $\begin{bmatrix} \mathbf{M}_1^2 & \mathbf{O} \\ \mathbf{O} & \mathbf{M}_2^2 \end{bmatrix}$;

(b) $\begin{bmatrix} \mathbf{M}_3^2 + \mathbf{M}_4^2 & \mathbf{M}_3\mathbf{M}_4 + \mathbf{M}_4\mathbf{M}_3 \\ \mathbf{M}_4\mathbf{M}_3 + \mathbf{M}_3\mathbf{M}_4 & \mathbf{M}_3^2 + \mathbf{M}_4^2 \end{bmatrix}$; **(c)** $\begin{bmatrix} 4\mathbf{J} & \mathbf{O} \\ \mathbf{O} & 4\mathbf{J} \end{bmatrix}$;

(d) $\begin{bmatrix} \mathbf{M}_1^2 + \mathbf{M}_5^2 & \mathbf{M}^* & \mathbf{M}_5^2 \\ \mathbf{M}^* & \mathbf{M}_1^2 + 2\mathbf{M}_5^2 & \mathbf{M}^* \\ \mathbf{M}_5^2 & \mathbf{M}^* & \mathbf{M}_1^2 + \mathbf{M}_5^2 \end{bmatrix}$, $\mathbf{M}^* = \mathbf{M}_1\mathbf{M}_5 + \mathbf{M}_5\mathbf{M}_1$.

47. $\mathbf{A}^3(G) = \begin{bmatrix} \mathbf{A}^{*3} + 3\mathbf{A}^* & 3\mathbf{A}^{*2} + \mathbf{I} \\ 3\mathbf{A}^{*2} + \mathbf{I} & \mathbf{A}^{*3} + 3\mathbf{A}^* \end{bmatrix}$.

Section 3.2

1. (a) $\begin{bmatrix} 1 & 0 & 0 \\ 0.5 & 1 & 0 \\ -0.5 & 7 & 1 \end{bmatrix} \begin{bmatrix} 2 & -3 & 2 \\ 0 & 0.5 & 0 \\ 0 & 0 & 5 \end{bmatrix}$; **(b)** $\begin{bmatrix} 1 & 0 & 0 \\ -1 & 1 & 0 \\ -2 & 0 & 1 \end{bmatrix} \begin{bmatrix} -1 & -1 & 1 \\ 0 & -3 & 4 \\ 0 & 0 & -2 \end{bmatrix}$;

(c) $\begin{bmatrix} 1 & 0 & 0 \\ -2 & 1 & 0 \\ -5 & 2.2 & 1 \end{bmatrix} \begin{bmatrix} -1 & -3 & 2 \\ 0 & -5 & 7 \\ 0 & 0 & 0.6 \end{bmatrix}$; **(d)** $\begin{bmatrix} 1 & 0 & 0 \\ 0.5 & 1 & 0 \\ -1 & -0.75 & 1 \end{bmatrix} \begin{bmatrix} 2 & 4 & -2 \\ 0 & -4 & -3 \\ 0 & 0 & -7.25 \end{bmatrix}$;

(e) $\begin{bmatrix} 1 & 0 & 0 \\ 2 & 1 & 0 \\ 5 & 3 & 1 \end{bmatrix} \begin{bmatrix} 1 & 1 & 4 \\ 0 & -1 & -5 \\ 0 & 0 & 0 \end{bmatrix}$; **(f)** $\begin{bmatrix} 1 & 0 & 0 \\ 1.5 & 1 & 0 \\ 4.5 & -39 & 1 \end{bmatrix} \begin{bmatrix} 2 & -3 & -1 \\ 0 & -0.5 & -0.5 \\ 0 & 0 & -11 \end{bmatrix}$. **5. (a)** $\begin{bmatrix} 2 \\ 0 \\ 3 \end{bmatrix}$;

(b) $\begin{bmatrix} 0 \\ -25 \\ -15 \end{bmatrix}$; **(c)** $\dfrac{1}{3} \begin{bmatrix} -40 \\ 20 \\ 25 \end{bmatrix}$; **(d)** $\dfrac{1}{29} \begin{bmatrix} -55 \\ 60 \\ -80 \end{bmatrix}$; **(f)** $\dfrac{1}{11} \begin{bmatrix} -40 \\ -205 \\ 425 \end{bmatrix}$.

7. $\dfrac{n(n+1)}{2}\left(n - \dfrac{n-1}{3}\right) \sim \dfrac{n^3}{3}$.

11. (a) $\begin{bmatrix} 3 & 6 & 9 \\ 1 & 2 & 3 \end{bmatrix}$; **(b)** $\begin{bmatrix} -1 & -2 & -3 \\ 2 & 4 & 6 \\ 1 & 2 & 3 \end{bmatrix}$; **(c)** $\begin{bmatrix} 6 & 0 \\ 2 & 0 \end{bmatrix}$; **(d)** $\begin{bmatrix} -2 & 0 \\ 4 & 0 \\ 2 & 0 \end{bmatrix}$. **13. (a)** $\begin{bmatrix} 1 \\ 2 \end{bmatrix}$ *

$[1, 2] + \begin{bmatrix} 0 \\ -1 \end{bmatrix} * [0, 1]$; **(b)** $\begin{bmatrix} 2 \\ 1 \end{bmatrix} * [2, 0] + \begin{bmatrix} 0 \\ 4 \end{bmatrix} * [0, 1]$; **(c)** $\begin{bmatrix} 2 \\ 5 \end{bmatrix}$ *

$[1, -1] + \begin{bmatrix} 0 \\ 6 \end{bmatrix} * [0, 1]$; **(d)** $\begin{bmatrix} 0 \\ 5 \end{bmatrix} * [1, 0] + \begin{bmatrix} 2 \\ 0 \end{bmatrix} * [0, 1]$. **15. (a)** $\begin{bmatrix} 1 \\ 3 \\ 5 \end{bmatrix}$ *

$[1, 2] + \begin{bmatrix} 0 \\ 2 \\ 4 \end{bmatrix} * [0, -1]$; **(b)** $\begin{bmatrix} 2 \\ 1 \end{bmatrix} * [1, 2.5, 1] + \begin{bmatrix} 0 \\ 1 \end{bmatrix} * [0, 0.5, 5]$; **(c)** $\begin{bmatrix} 1 \\ 1 \end{bmatrix}$ *

$[1, 5, 7, 2] + \begin{bmatrix} 0 \\ 1 \end{bmatrix} * [1, -4, -7, 1]$; **(d)** $\begin{bmatrix} 2 \\ -1 \end{bmatrix} * [1, -3, 2]$.

Section 3.3

1. (a) $\lambda_1 = 2, \lambda_2 = 1$; **(b)** $\lambda_1 = -1, \lambda_2 = 3$; **(c)** $\lambda_1 = -2, \lambda_2 = -3$; **(d)** $\lambda_1 = -2,$
$\lambda_2 = 2, \lambda_3 = 1$. **3. (a)** $\begin{bmatrix} 1349 \\ 645 \end{bmatrix}$; **(b)** $3 \cdot 7^n \begin{bmatrix} 1 \\ 1 \end{bmatrix}$; **(c)** $\begin{bmatrix} 2721 \\ 2017 \end{bmatrix}$. **5. (a) (i)** $\lambda_1 = 4, \lambda_2 = 2,$
(ii) $\lambda_1 = 6, \lambda_2 = -1$, **(iii)** $\lambda_1 = 4, \lambda_2 = 1$, **(iv)** $\lambda_1 = 3$ (only one), **(v)** $\lambda_1 = 3, \lambda_2 = -3,$
(vi) $\lambda_1 = 2, \lambda_2 = \lambda_3 = -1$, **(vii)** $\lambda_1 = 8, \lambda_2 = \lambda_3 = 2$, **(viii)** $\lambda_1 = 3 + \sqrt{2}, \lambda_2 = 3 - \sqrt{2},$
$\lambda_3 = -1$, **(ix)** $\lambda_1 = 6, \lambda_2 = 4, \lambda_3 = \lambda_4 = -2$; **(b) (i)** $\begin{bmatrix} 1 \\ 1 \end{bmatrix}$, **(ii)** $\begin{bmatrix} 2 \\ 5 \end{bmatrix}$, **(iii)** $\begin{bmatrix} 1 \\ 2 \end{bmatrix}$, **(iv)** $\begin{bmatrix} 1 \\ 1 \end{bmatrix}$,

(v) $\begin{bmatrix} 5 \\ 1 \end{bmatrix}$ or $\begin{bmatrix} 1 \\ -1 \end{bmatrix}$, **(vi)** $\begin{bmatrix} 1 \\ 1 \\ 1 \end{bmatrix}$, **(vii)** $\begin{bmatrix} 2 \\ 1 \\ 2 \end{bmatrix}$, **(viii)** $\begin{bmatrix} 1 \\ \sqrt{2} \\ 1 \end{bmatrix}$, **(ix)** $\begin{bmatrix} 1 \\ 1 \\ -1 \\ 1 \end{bmatrix}$. **7. (a) (i)** $\lambda_1 = 1,$

$\lambda_2 = 0.5$, (ii) $\lambda_1 = 1.2$, $\lambda_2 = 1$, (iii) $\lambda_1 = 1.24$, $\lambda_2 = 0.96$; (b) (i) $\mathbf{u} = \begin{bmatrix} 3 \\ 1 \end{bmatrix}$,

(ii) $\mathbf{u} = \begin{bmatrix} 2 \\ 1 \end{bmatrix}$, (iii) $\mathbf{u} = \begin{bmatrix} 1 \\ \sqrt{2} \end{bmatrix}$; (c) (i) $\mathbf{v} = \begin{bmatrix} 1 \\ 2 \end{bmatrix}$, (ii) $\mathbf{v} = \begin{bmatrix} 2 \\ 3 \end{bmatrix}$, (iii) $\mathbf{v} = \begin{bmatrix} 1 \\ -\sqrt{2} \end{bmatrix}$;

(d) (i) $\mathbf{x}^{(k)} = 2\mathbf{u} + 4(0.5)^k\mathbf{v}$, (ii) $\mathbf{x}^{(k)} = 2.5(1.2)^k\mathbf{u} + 2.5\mathbf{v}$,

(iii) $\mathbf{x}^{(k)} = 8.536(1.24)^k\mathbf{u} + 1.463(0.96)^k\mathbf{v}$. **9.** (a) $\lambda_1 = 3$, $\lambda_2 = 1$, $\lambda_3 = 0$, $\mathbf{u} = \begin{bmatrix} 1 \\ -2 \\ 1 \end{bmatrix}$,

$\mathbf{v} = \begin{bmatrix} 1 \\ 0 \\ -1 \end{bmatrix}$, $\mathbf{w} = \begin{bmatrix} 1 \\ 1 \\ 1 \end{bmatrix}$; (b) $\mathbf{p} = 10\mathbf{u} + 20\mathbf{v} + 20\mathbf{w}$; (c) $10 \cdot 3^{12}\begin{bmatrix} 1 \\ -2 \\ 1 \end{bmatrix}$. **19.** (a) $\lambda_1 = 1$,

$\mathbf{u} = \begin{bmatrix} 1 \\ 1 \end{bmatrix}$; (b) $\lambda_2 = -1$, $\mathbf{u} = \begin{bmatrix} 1 \\ 3 \end{bmatrix}$; (c) eigenvalues equal in absolute value.

Section 3.4

1. (a) 7, 8; (b) 6, 8; (c) 8, 9; (d) 7, 6; (e) 12, 10; (f) 19, 20; (g) 14, 13; (h) 2, 2.

3. (a) 9; (b) 27; (c) $\begin{bmatrix} 3 \\ 0 \\ 0 \end{bmatrix}$; (d) 405. **5.** $5^5 \cdot 200$, $4^5 \cdot 100$. **7.** (a) $\frac{1}{2}$, $\frac{7}{6}$; (b) $\frac{1}{5}$, $\frac{17}{2}$.

11. (a) $\begin{bmatrix} 0 & 1 & 0 & 0 & 1 & 1 \\ 1 & 0 & 1 & 0 & 1 & 0 \\ 0 & 1 & 0 & 1 & 1 & 0 \\ 0 & 0 & 1 & 0 & 1 & 0 \\ 1 & 1 & 1 & 1 & 0 & 1 \\ 1 & 0 & 0 & 0 & 1 & 0 \end{bmatrix}$; (b) 5; (c) maximum number of edges incident to any

vertex. **13.** (a) 3, 3, 3; (b) 5.601, 7, 7; (c) 6.720, 8, 8; (d) 3, 3, 3. **15.** 1.

17. (a) 96%; (b) $\begin{bmatrix} 83\frac{1}{3} \\ 66\frac{2}{3} \\ 150 \end{bmatrix}$, 27%. **19.** (a) 700%; (b) 450%. **21.** 600%. **23.** 164%.

Section 3.5

1. (a) [0.2, 0.4]; (b) $\frac{1}{14}[1, 2, 3]$; (c) $\frac{1}{11}\begin{bmatrix} 2 & 3 & 3 \\ 1 & -4 & 7 \end{bmatrix}$; (d) $\frac{1}{105}\begin{bmatrix} 20 & 10 & 5 \\ -21 & 42 & 0 \end{bmatrix}$;

(e) $\frac{1}{6}\begin{bmatrix} 2 & 1 & 0 & -1 \\ 0 & 1 & -2 & 1 \\ 1 & 0 & 1 & 2 \end{bmatrix}$; (f) $\frac{1}{1271}\begin{bmatrix} 103 & 87 & 513 & -158 \\ 124 & -31 & 124 & 217 \\ -123 & 328 & -82 & 164 \end{bmatrix}$.

3. $\hat{y} = x + 3.57$. **5.** (a) $\hat{y} = 5078x + 16{,}100$; (b) ~700; (c) $\hat{y} = 5078x + 31{,}333$.

7. (a) $\frac{1}{40}\begin{bmatrix} 12.5 & -4 & -2 \\ -5 & 8 & 4 \end{bmatrix}$, $\begin{bmatrix} 75.5 \\ 185 \end{bmatrix}$; (b) $\begin{bmatrix} 0 \\ -240 \\ 480 \end{bmatrix}$, 90°; (c) second refinery.

9. (a) A 5.64, B 13.88; (b) -0.13; (c) -0.17.

11. (a) $\begin{bmatrix} -0.352 & -0.808 & 0.596 & 0.679 & -0.114 \\ -0.995 & 0.463 & 0.898 & -2.397 & 2.031 \\ 2.534 & 2.342 & -2.829 & 0.751 & -1.798 \end{bmatrix}$, ~3390;

(b) $\hat{y} = 0.47\text{GPA}_{\text{col}} + 1.88\text{GPA}_{\text{hi}} - 0.49$; (c) $\mathbf{e} = \begin{bmatrix} -0.10 \\ 0 \\ -0.15 \\ 0.11 \\ 0.14 \end{bmatrix}$.

13. $\hat{y} = 3.916x^3 - 9.857x^2 - 1.130x + 8.5$, ~6850. 15. (a) $\begin{bmatrix} 1 \\ -2 \\ 0 \\ 2 \\ 1 \end{bmatrix}$; (b) $\begin{bmatrix} -1 \\ 3 \\ -1 \\ -2 \\ 1 \end{bmatrix}$;

(c) $\begin{bmatrix} 1 \\ -1 \\ 1 \\ -1 \\ 1 \\ -1 \end{bmatrix}$.

Chapter 4

Section 4.1

13. (a) $3\begin{bmatrix} 1 \\ 2 \end{bmatrix}$; (b) $\begin{bmatrix} 1 \\ 2 \end{bmatrix} - 3\begin{bmatrix} 1 \\ 1 \end{bmatrix}$; (c) $3\begin{bmatrix} 1 & 0 \\ 0 & 0 \end{bmatrix} + 4\begin{bmatrix} 0 & 0 \\ 0 & 1 \end{bmatrix}$;

(d) $5\begin{bmatrix} 1 & 1 \\ 0 & 0 \end{bmatrix} + 4\begin{bmatrix} 0 & 0 \\ 1 & 1 \end{bmatrix} - 3\begin{bmatrix} 1 & 0 \\ 1 & 0 \end{bmatrix} - 4\begin{bmatrix} 0 & 0 \\ 0 & 1 \end{bmatrix}$; (e) $3(x^3) + 2(x^2) - 4(x) + 5(1)$;

(f) $\frac{11}{2}(x^2 + x) - \frac{7}{2}(x^2 + 1) - \frac{5}{2}(x + 1)$. 15. (a) $\begin{bmatrix} 1 \\ 2 \\ 0 \\ 0 \\ 0 \end{bmatrix}, \begin{bmatrix} 0 \\ 0 \\ 1 \\ 3 \\ 0 \end{bmatrix}, \begin{bmatrix} 0 \\ 0 \\ 0 \\ 0 \\ 1 \end{bmatrix}$; (b) $\begin{bmatrix} 1 & 0 \\ 0 & 1 \end{bmatrix}, \begin{bmatrix} 0 & 1 \\ 0 & 0 \end{bmatrix}$,

$\begin{bmatrix} 0 & 0 \\ 1 & 0 \end{bmatrix}$; (c) x^3, x^2, x. 21. $\{e_1, e_2, e_5, e_8, e_6\}, \{e_1, e_3, e_5, e_7, e_4\}$. 23. $\{e_1, e_3, e_6\}$,

$\{e_1, e_3, e_8, e_9\}$, or $\{e_1, e_3, e_4, e_5\}$.

Section 4.2

1. (a) $\frac{3}{4}\begin{bmatrix} 2 \\ 1 \end{bmatrix} - \frac{1}{4}\begin{bmatrix} 2 \\ -1 \end{bmatrix}$; (b) $8\begin{bmatrix} 2 \\ -3 \end{bmatrix} + \frac{13}{3}\begin{bmatrix} -3 \\ 6 \end{bmatrix}$; (c) $-9\begin{bmatrix} 1 \\ -3 \end{bmatrix} + 5\begin{bmatrix} 1 \\ -5 \end{bmatrix}$.

3. (a) $3\begin{bmatrix} 1 & 1 \\ 1 & 1 \end{bmatrix} - \begin{bmatrix} 1 & 1 \\ 0 & 1 \end{bmatrix} - \begin{bmatrix} 1 & 0 \\ 1 & 1 \end{bmatrix}$; (b) not possible;

(c) $-3\begin{bmatrix} 1 & 1 \\ 1 & 1 \end{bmatrix} + 6\begin{bmatrix} 1 & 1 \\ 0 & 1 \end{bmatrix} - 2\begin{bmatrix} 1 & 0 \\ 1 & 1 \end{bmatrix}$. 5. (a) $\frac{1}{2}\begin{bmatrix} 4 \\ 0 \end{bmatrix} + \frac{1}{3}\begin{bmatrix} 0 \\ 3 \end{bmatrix}$; (b) $\begin{bmatrix} 1 \\ 2 \end{bmatrix} - 2\begin{bmatrix} -1 \\ 1 \end{bmatrix}$;

(c) $-25\begin{bmatrix} 2 \\ 5 \end{bmatrix} + 16\begin{bmatrix} 3 \\ 8 \end{bmatrix}$. **7. (a)** $\begin{bmatrix} 1 \\ 1 \\ 1 \end{bmatrix} = \frac{1}{2}\begin{bmatrix} -2 \\ 0 \\ -2 \end{bmatrix} + \begin{bmatrix} 2 \\ 1 \\ 2 \end{bmatrix}$; **(b)** independent;

(c) $\begin{bmatrix} 1 \\ -2 \\ 3 \end{bmatrix} = 3\begin{bmatrix} 1 \\ 0 \\ 1 \end{bmatrix} - 2\begin{bmatrix} 1 \\ 1 \\ 0 \end{bmatrix}$; **(d)** independent. **9. (a)** Independent;

(b) $3x^2 + 4x + 7 = 3(x^2 + 1) + 4(x + 1)$; **(c)** independent.

Section 4.3

1. Many possible answers for each part. **3. (a)** Any two of the vectors; **(b)** either vector;

(c) the last vector and one of the first two. **5. (a)** $\begin{bmatrix} 1 \\ 0 \\ 0 \end{bmatrix}, \begin{bmatrix} 0 \\ 0 \\ 1 \end{bmatrix}$; **(b)** $\begin{bmatrix} 1 \\ 1 \\ 1 \end{bmatrix}$; **(c)** $\begin{bmatrix} 2 \\ 0 \\ 1 \end{bmatrix}, \begin{bmatrix} 0 \\ 1 \\ 0 \end{bmatrix}$.

7. (a) $x^3, x^2, x, 1$; **(b)** x^3, x^2, x. **9. (a)** Basis; **(b)** no basis; **(c)** no basis. **11. (a)** 3;
(b) 2; **(c)** 1; **(d)** 3. **13. (a)** 3; **(b)** 9; **(c)** 6; **(d)** 4; **(e)** 3; **(f)** infinite k.

15. (a) $\begin{bmatrix} v_1 \\ v_1 \\ v_2 \\ v_3 \end{bmatrix} \leftrightarrow \begin{bmatrix} v_1 & v_2 \\ v_2 & v_3 \end{bmatrix}$; **(b)** $v_1\begin{bmatrix} 0 \\ 1 \\ 2 \end{bmatrix} + v_2\begin{bmatrix} 1 \\ 0 \\ 1 \end{bmatrix} \leftrightarrow \begin{bmatrix} v_1 \\ v_1 \\ v_2 \end{bmatrix}$; **(c)** $\begin{bmatrix} v_1 & v_2 \\ v_2 & v_3 \end{bmatrix} \leftrightarrow$

$v_1x^3 + v_2x^2 + v_3x$. **17. (a)** $\begin{bmatrix} -\frac{1}{2} & \frac{3}{4} \\ \frac{1}{2} & -\frac{1}{4} \end{bmatrix}$; **(b)** $\frac{1}{13}\begin{bmatrix} -5 & 1 \\ 3 & 2 \end{bmatrix}$; **(c)** $\begin{bmatrix} -\frac{1}{3} & \frac{4}{3} & -\frac{1}{3} \\ \frac{4}{3} & -\frac{4}{3} & \frac{1}{3} \\ -1 & 1 & 0 \end{bmatrix}$;

(d) $\begin{bmatrix} -\frac{1}{5} & 1 & \frac{2}{5} \\ -\frac{1}{5} & 0 & \frac{2}{5} \\ \frac{3}{5} & 0 & -\frac{1}{5} \end{bmatrix}$; **(e)** $\begin{bmatrix} \frac{5}{6} & -\frac{1}{3} & \frac{1}{3} \\ \frac{1}{2} & -1 & 0 \\ -\frac{1}{3} & \frac{1}{3} & \frac{1}{3} \end{bmatrix}$. **19. (a)** $\begin{bmatrix} -1 & -5 \\ 1 & 3 \end{bmatrix}$; **(b)** $\begin{bmatrix} \frac{4}{3} & \frac{7}{3} \\ -\frac{5}{3} & -\frac{5}{3} \end{bmatrix}$;

(c) $\begin{bmatrix} \frac{4}{5} & 2 & \frac{13}{5} \\ -\frac{1}{5} & 1 & \frac{8}{5} \\ \frac{3}{5} & 0 & -\frac{4}{5} \end{bmatrix}$.

Section 4.4

1. (a) Either column (not both); **(b)** both columns; **(c)** both columns. **3. (a)** Column 2
plus column 1 or 3; **(b)** column 2 plus column 1 or 3; **(c)** any two columns; **(d)** column 1
with column 2 or 3. **5. (a)** $\begin{bmatrix} \frac{1}{4} \\ 0 \end{bmatrix}, \begin{bmatrix} 0 \\ \frac{1}{3} \end{bmatrix}$; **(b)** $\begin{bmatrix} \frac{1}{3} \\ -\frac{2}{3} \end{bmatrix}, \begin{bmatrix} \frac{1}{3} \\ \frac{1}{3} \end{bmatrix}$; **(c)** $\begin{bmatrix} 8 \\ -5 \end{bmatrix}, \begin{bmatrix} -3 \\ 2 \end{bmatrix}$.

7. (a, c, d) Any two rows; **(b)** row 2 plus row 1 or 3. **9. (a)** n; **(b)** 3; **(c)** 3; **(d)** 4;
(e) 3; **(f)** 6. **11. (a)** Not unique; **(b)** unique; **(c)** unique; **(d)** unique; **(e)** unique;

(f) unknown; **(g)** unique. **17. (a)** $\begin{bmatrix} 1 \\ 1 \end{bmatrix}\begin{bmatrix} 0 \\ 1 \end{bmatrix}$, $[1, 2], [0, -1]$;

(b) $\begin{bmatrix} 1 \\ 3 \\ 7 \end{bmatrix}\begin{bmatrix} 0 \\ 1 \\ 3 \end{bmatrix}$, $[1, 2, 3][0, -2, -4]$; **(c)** $\begin{bmatrix} 1 \\ 0 \\ 4 \end{bmatrix}\begin{bmatrix} 0 \\ 1 \\ -7 \end{bmatrix}$, $[1, 4, 6][0, 1, 3]$; **(d)** $\begin{bmatrix} 1 \\ 0 \\ 2 \end{bmatrix}\begin{bmatrix} 0 \\ 1 \\ 1 \end{bmatrix}\begin{bmatrix} 0 \\ 0 \\ 1 \end{bmatrix}$,

$[2, 1, 5][0, 3, -2][0, 0, -1]$.   **19. (a)** $\begin{bmatrix} 1 \\ 1 \\ 1 \end{bmatrix} = \begin{bmatrix} 1 \\ 3 \\ 5 \end{bmatrix} - 2\begin{bmatrix} 0 \\ 1 \\ 2 \end{bmatrix}$, $\begin{bmatrix} 1 \\ 1 \\ 1 \end{bmatrix} = \begin{bmatrix} 1 \\ 2 \\ 1 \end{bmatrix} - \begin{bmatrix} 0 \\ 1 \\ 1 \end{bmatrix} + \begin{bmatrix} 0 \\ 0 \\ 1 \end{bmatrix}$;

(b) $[10, 10] = 10[1, .5] + 5[0, -1]$.

Section 4.5

1. (a) $\mathbf{b} = \dfrac{7}{5}\begin{bmatrix} 2 \\ 1 \end{bmatrix} + \dfrac{2}{5}\begin{bmatrix} -2 \\ 4 \end{bmatrix}$, $\mathbf{c} = \dfrac{2}{5}\begin{bmatrix} 2 \\ 1 \end{bmatrix} + \dfrac{9}{10}\begin{bmatrix} -2 \\ 4 \end{bmatrix}$; **(b)** $\mathbf{b} = \begin{bmatrix} 2 \\ 0 \end{bmatrix} + \begin{bmatrix} 0 \\ 3 \end{bmatrix}$,

$\mathbf{c} = -\dfrac{1}{2}\begin{bmatrix} 2 \\ 0 \end{bmatrix} + \dfrac{4}{3}\begin{bmatrix} 0 \\ 3 \end{bmatrix}$; **(c)** $\mathbf{b} = \dfrac{18}{25}\begin{bmatrix} 3 \\ 4 \end{bmatrix} - \dfrac{1}{25}\begin{bmatrix} 4 \\ -3 \end{bmatrix}$, $\mathbf{c} = \dfrac{13}{25}\begin{bmatrix} 3 \\ 4 \end{bmatrix} - \dfrac{16}{25}\begin{bmatrix} 4 \\ -3 \end{bmatrix}$.

3. (a) $\mathbf{b} = 1.6\begin{bmatrix} 0.6 \\ 0.8 \end{bmatrix} - 1.2\begin{bmatrix} 0.8 \\ -0.6 \end{bmatrix}$, $\mathbf{c} = 0.4\begin{bmatrix} 0.6 \\ 0.8 \end{bmatrix} + 2.2\begin{bmatrix} 0.8 \\ -0.6 \end{bmatrix}$;

(b) $\mathbf{b} = \sqrt{2}\left(\dfrac{\sqrt{2}}{2}\begin{bmatrix} 1 \\ 1 \end{bmatrix}\right) + \sqrt{2}\left(\dfrac{\sqrt{2}}{2}\begin{bmatrix} -1 \\ 1 \end{bmatrix}\right)$, $\mathbf{c} = \dfrac{\sqrt{2}}{2}\left(\dfrac{\sqrt{2}}{2}\begin{bmatrix} 1 \\ 1 \end{bmatrix}\right) -$

$\dfrac{3\sqrt{2}}{2}\left(\dfrac{\sqrt{2}}{2}\begin{bmatrix} -1 \\ 1 \end{bmatrix}\right)$.   **11.** $|\mathbf{b}| = \|[\mathbf{b}]_{\mathcal{Q}}\| = \sqrt{2}$, $\cos(\mathbf{b}, \mathbf{c}) = \cos([\mathbf{b}]_{\mathcal{Q}}, [\mathbf{c}]_{\mathcal{Q}}) = 45°$;

15. (a) $\dfrac{1}{\sqrt{2}}\begin{bmatrix} 1 \\ 1 \end{bmatrix}$, $\dfrac{1}{\sqrt{2}}\begin{bmatrix} 1 \\ -1 \end{bmatrix}$; **(b)** $\dfrac{1}{\sqrt{5}}\begin{bmatrix} 1 \\ -2 \end{bmatrix}$; **(c)** $\dfrac{1}{\sqrt{10}}\begin{bmatrix} 3 \\ 1 \end{bmatrix}$, $\dfrac{1}{\sqrt{10}}\begin{bmatrix} -1 \\ 3 \end{bmatrix}$;

(d) $\dfrac{1}{5}\begin{bmatrix} 3 \\ 4 \end{bmatrix}$, $\dfrac{1}{5}\begin{bmatrix} 4 \\ -3 \end{bmatrix}$.   **17. (a)** $\begin{bmatrix} 0.6 & -0.8 \\ 0.8 & 0.6 \end{bmatrix}\begin{bmatrix} 5 & 0.2 \\ 0 & 1.4 \end{bmatrix}$; **(b)** $\begin{bmatrix} \frac{2}{3} & -\frac{1}{\sqrt{2}} \\ \frac{1}{3} & 0 \\ \frac{2}{3} & \frac{1}{\sqrt{2}} \end{bmatrix}\begin{bmatrix} 3 & 3 \\ 0 & \sqrt{2} \end{bmatrix}$;

(c) $\begin{bmatrix} \frac{1}{\sqrt{2}} & -\frac{1}{\sqrt{6}} & \frac{1}{\sqrt{3}} \\ \frac{1}{\sqrt{2}} & \frac{1}{\sqrt{6}} & -\frac{1}{\sqrt{3}} \\ 0 & \frac{2}{\sqrt{6}} & \frac{1}{\sqrt{3}} \end{bmatrix}\begin{bmatrix} \sqrt{2} & \frac{1}{\sqrt{2}} & \frac{1}{\sqrt{2}} \\ 0 & \frac{\sqrt{6}}{2} & \frac{1}{\sqrt{6}} \\ 0 & 0 & \frac{2}{\sqrt{3}} \end{bmatrix}$; **(d)** $\begin{bmatrix} \frac{1}{\sqrt{3}} & \frac{1}{\sqrt{6}} & 0 \\ -\frac{1}{\sqrt{3}} & \frac{2}{\sqrt{6}} & 0 \\ \frac{1}{\sqrt{3}} & \frac{1}{\sqrt{6}} & 0 \end{bmatrix}\begin{bmatrix} \sqrt{3} & 0 & \sqrt{3} \\ 0 & \sqrt{6} & \sqrt{6} \\ 0 & 0 & 0 \end{bmatrix}$.

19. (a) $\mathbf{R}^{-1} = \begin{bmatrix} 0.2 & -0.12 & -1.16 \\ 0 & 0.2 & -0.4 \\ 0 & 0 & 1 \end{bmatrix}$, $\mathbf{A}^{-1} = \begin{bmatrix} -1 & 0.6 & -0.2 \\ -0.2 & 0.32 & -0.24 \\ 0.8 & -0.48 & 0.36 \end{bmatrix}$;

(b) $\mathbf{R}^{-1} = \begin{bmatrix} \frac{1}{\sqrt{2}} & -\frac{1}{\sqrt{6}} & -\frac{1}{2\sqrt{3}} \\ 0 & \frac{2}{\sqrt{6}} & -\frac{1}{2\sqrt{3}} \\ 0 & 0 & \frac{\sqrt{3}}{2} \end{bmatrix}$, $\mathbf{A}^{-1} = \begin{bmatrix} 0.5 & 0.5 & -0.5 \\ -0.5 & 0.5 & 0.5 \\ 0.5 & -0.5 & 0.5 \end{bmatrix}$.

Section 4.6

1. (a) $\frac{1}{3}$; (b) $\frac{1}{6}$; (c) $\frac{1}{5}$; (d) $\frac{35}{6}$.   3. (a) 0; (b) $-\frac{1}{6}$.   5. (a) $\sqrt{1/3}$; (b) $\sqrt{1/5}$; (c) $\sqrt{1/7}$;
(d) $\sqrt{1.7}$.   11. (a) $-\frac{3}{35} + \frac{6}{7}x^2$; (b) $0.19 + 0.94x^2$; (c) $-0.60x + 1.71x^3$;
(d) $2.8x - 2.17x^3$.   13. (a) $L_4(x) = x^4 - \frac{6}{7}x^2 + \frac{3}{35}$; (b) $L_5(x) = x^5 - \frac{10}{9}x^3 + \frac{5}{21}x$.
15. (a) $K_3(x) = x^3 - \frac{3}{2}x^2 + \frac{3}{5}x - \frac{1}{20}$; (b) $2x^3 - 1.286x^2 + 0.296x - 0.014$.

17. (a) $\begin{bmatrix} 648 & -720 \\ -720 & 810 \end{bmatrix}$, 361; (b) $\begin{bmatrix} 900 & -2520 & 1680 \\ -2520 & 7350 & -5040 \\ 1680 & -5040 & 3528 \end{bmatrix}$, 9195;

(c) $\begin{bmatrix} 16{,}201 & -39{,}603 & 23{,}762 \\ -39{,}603 & 98{,}018 & -59{,}405 \\ 23{,}762 & -59{,}405 & 36{,}303 \end{bmatrix}$, 66,222.

Chapter 5

Section 5.1

1. (a) $\begin{bmatrix} -1 & 0 \\ 0 & -1 \end{bmatrix}$; (b) $\begin{bmatrix} 5 & 0 \\ 0 & 4 \end{bmatrix}$; (c) $\begin{bmatrix} 1 & 0 \\ 1 & 1 \end{bmatrix}$; (d) $\begin{bmatrix} -1 & 0 \\ 0 & 1 \end{bmatrix}$.

3.

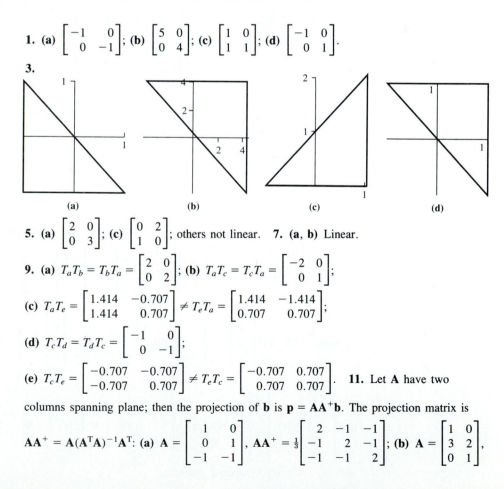

(a)   (b)   (c)   (d)

5. (a) $\begin{bmatrix} 2 & 0 \\ 0 & 3 \end{bmatrix}$; (c) $\begin{bmatrix} 0 & 2 \\ 1 & 0 \end{bmatrix}$; others not linear.   7. (a, b) Linear.

9. (a) $T_a T_b = T_b T_a = \begin{bmatrix} 2 & 0 \\ 0 & 2 \end{bmatrix}$; (b) $T_a T_c = T_c T_a = \begin{bmatrix} -2 & 0 \\ 0 & 1 \end{bmatrix}$;

(c) $T_a T_e = \begin{bmatrix} 1.414 & -0.707 \\ 1.414 & 0.707 \end{bmatrix} \neq T_e T_a = \begin{bmatrix} 1.414 & -1.414 \\ 0.707 & 0.707 \end{bmatrix}$;

(d) $T_c T_d = T_d T_c = \begin{bmatrix} -1 & 0 \\ 0 & -1 \end{bmatrix}$;

(e) $T_c T_e = \begin{bmatrix} -0.707 & -0.707 \\ -0.707 & 0.707 \end{bmatrix} \neq T_e T_c = \begin{bmatrix} -0.707 & 0.707 \\ 0.707 & 0.707 \end{bmatrix}$.   11. Let $\mathbf{A}$ have two

columns spanning plane; then the projection of $\mathbf{b}$ is $\mathbf{p} = \mathbf{A}\mathbf{A}^+\mathbf{b}$. The projection matrix is

$\mathbf{A}\mathbf{A}^+ = \mathbf{A}(\mathbf{A}^T\mathbf{A})^{-1}\mathbf{A}^T$: (a) $\mathbf{A} = \begin{bmatrix} 1 & 0 \\ 0 & 1 \\ -1 & -1 \end{bmatrix}$, $\mathbf{A}\mathbf{A}^+ = \frac{1}{3}\begin{bmatrix} 2 & -1 & -1 \\ -1 & 2 & -1 \\ -1 & -1 & 2 \end{bmatrix}$; (b) $\mathbf{A} = \begin{bmatrix} 1 & 0 \\ 3 & 2 \\ 0 & 1 \end{bmatrix}$,

$$AA^+ = \tfrac{1}{14}\begin{bmatrix} 5 & 3 & -6 \\ 3 & 13 & 2 \\ -6 & 2 & 10 \end{bmatrix}.$$ **13.** $\begin{bmatrix} 0.707 & -0.707 & 0 \\ 0.707 & 0.707 & 0 \\ 0 & 0 & 1 \end{bmatrix}.$

15. (b) $\begin{bmatrix} -1 & 0 & 0 & 0 \\ 0 & 1 & 0 & 0 \\ 0 & 0 & -1 & 0 \\ 0 & 0 & 0 & 1 \end{bmatrix}.$ **19.** $\begin{bmatrix} \frac{1}{4} & 0 & 0 & 0 \\ 0 & \frac{1}{3} & 0 & 0 \\ 0 & 0 & \frac{1}{2} & 0 \\ 0 & 0 & 0 & 1 \\ 0 & 0 & 0 & 0 \end{bmatrix}.$ **21.** $\begin{bmatrix} 0 & 0 & 0 & 0 \\ 0 & 0 & 0 & 0 \\ 6 & 0 & 0 & 0 \\ 0 & 2 & 0 & 0 \end{bmatrix}.$

23. (a) Linear $\begin{bmatrix} 1 \\ 0 \\ 0 \\ 0 \end{bmatrix}$; **(b)** linear $\begin{bmatrix} 1 \\ 1 \\ 1 \\ 0 \end{bmatrix}$; **(c)** linear $\begin{bmatrix} \frac{1}{4} \\ \frac{1}{3} \\ \frac{1}{2} \\ 1 \end{bmatrix}$. **25. (a)** Linear $\begin{bmatrix} 2 \\ 0 \\ 0 \\ 0 \end{bmatrix}$;

(b) linear, $\begin{bmatrix} 0 \\ 2 \\ -1 \\ 0 \end{bmatrix}$; **(d)** linear $\begin{bmatrix} 1 \\ 1 \\ 1 \\ 1 \end{bmatrix}$. **29. (a)** Rotate 180° clockwise; **(b)** make height $\frac{1}{4}$ and

width $\frac{1}{5}$; **(c)** make horizontal lines of square slant at $-45°$; **(d)** reflect about x_2-axis.

31. $\lambda = 1$, $p(x) = a_k x^k$. **33. (b)** $\begin{bmatrix} 3 & 0 & 0 & 0 \\ -6 & 3 & 0 & 0 \\ 6 & -4 & 3 & 0 \\ 0 & 2 & -2 & 3 \end{bmatrix}.$

Section 5.2

1. (a, c) Either column; **(b, d)** both columns. **3. (a, d)** Any two columns; **(b, c)** all columns. **5. (a, b, d)** Any two columns; **(c)** first or last column, plus two of the other three columns; **(e)** all minus one of first four columns. **7. (a)** Any matrix whose second row is **0**, other rows $\neq$ **0**; **(b)** $\begin{bmatrix} 0 & 1 & 1 \\ 1 & -1 & 0 \\ -1 & 0 & -1 \end{bmatrix}$ (or any matrix whose columns span that plane); **(c)** all columns multiples of $\begin{bmatrix} 1 \\ 2 \\ 3 \end{bmatrix}$. **9. (a)** $b_3 = \frac{1}{3}b_1 + \frac{1}{3}b_2$; **(b, c)** no constraint;

(d) $b_3 = 8b_1 - 13b_2$. **11. (a)** $\begin{bmatrix} 2 \\ 1 \end{bmatrix}$; **(b)** $\begin{bmatrix} 1 \\ 0 \\ -2 \end{bmatrix}$; **(c)** $\begin{bmatrix} 5 \\ 4 \\ -7 \end{bmatrix}$; **(d)** $\begin{bmatrix} 1 \\ -7 \\ 5 \end{bmatrix}$, **(e)** $\begin{bmatrix} 3 \\ 1 \\ -1 \end{bmatrix}$;

(f) none. **15. (a)** $\begin{bmatrix} 0 \\ 0 \\ 10 \end{bmatrix} + r\begin{bmatrix} 1 \\ 1 \\ -1 \end{bmatrix}$; **(b)** $\begin{bmatrix} 3 \\ 3 \\ 7 \end{bmatrix}$. **17. (a)** $\begin{bmatrix} 5 \\ 0 \\ 5 \\ 0 \\ 5 \end{bmatrix} + r\begin{bmatrix} 1 \\ -1 \\ -1 \\ 1 \\ 0 \end{bmatrix} + s\begin{bmatrix} 1 \\ 0 \\ 0 \\ 0 \\ -1 \end{bmatrix}$;

(b) $\begin{bmatrix} 10 \\ -10 \\ -5 \\ 10 \\ 10 \end{bmatrix}$; **(c)** $\begin{bmatrix} 10 \\ 5 \\ 10 \\ -5 \\ -5 \end{bmatrix}$. **19. (a)** $r\begin{bmatrix} 2 \\ 11 \\ -15 \end{bmatrix}$; **(b)** $\begin{bmatrix} \frac{56}{3} \\ \frac{38}{3} \\ 10 \end{bmatrix}$; **(c)** $\begin{bmatrix} 15 \\ -\frac{15}{2} \\ \frac{75}{2} \end{bmatrix}$. **21. (a)** **O**; **(b)** **M**'s with $m_{1,1} = 0$; **(c)** symmetric matrices.

23. (a) $3SO_2 + 2NO_3 + 2H_2O \rightarrow 4H + 3SO_4 + 2NO$;

(b) $15PbN_6 + 44CrMn_2O_8 \rightarrow 22Cr_2O_3 + 88MnO_2 + 5Pb_3O_4 + 90NO$;

(c) $1496H_2SO_4 + 64MnS + 52As_2Cr_{10}O_{35} \rightarrow 64HMnO_4 + 104AsH_3 + 520CrS_3O_{12} + 1308H_2O$.

25. (a) O (zero matrix); **(b)** any 3×3 matrix whose rows span the plane $5x_1 + x_2 + 2x_3 = 0$,

say, $\begin{bmatrix} 0 & 2 & -1 \\ 1 & -5 & 0 \\ 2 & 0 & -5 \end{bmatrix}$; **(c)** any 3×3 matrix with rows multiples of the 1's vector;

(d) any 3×3 matrix with rows multiples of $[5, 3, 9]$.

27. (a) (a) $\begin{bmatrix} p \\ 0 \\ 1-p \end{bmatrix}$, **(b)** $\frac{1}{7}\begin{bmatrix} 1 \\ 3 \\ 3 \end{bmatrix}$, **(c)** $\begin{bmatrix} 0.5p \\ 1-p \\ 0.5p \end{bmatrix}$, **(d)** $\frac{1}{8}\begin{bmatrix} 1 \\ 2 \\ 2 \\ 2 \\ 1 \end{bmatrix}$, **(e)** $\frac{1}{6}\begin{bmatrix} 1 \\ 1 \\ 1 \\ 1 \\ 1 \\ 1 \end{bmatrix}$, **(f)** $\frac{1}{853}\begin{bmatrix} 512 \\ 256 \\ 64 \\ 16 \\ 4 \\ 1 \end{bmatrix}$,

(g) $\frac{1}{6}\begin{bmatrix} 1 \\ 1 \\ 1 \\ 1 \\ 1 \\ 1 \end{bmatrix}$; **(b)** add kernel from Exercise 26 to answer in part (a).

29. (a) 4; **(b)** 3; **(c)** 4; **(d)** 3; **(e)** 4; **(f)** 2.

37. (a) $\mathbf{1} = \begin{bmatrix} -0.2 \\ 0.4 \end{bmatrix} + \begin{bmatrix} 1.2 \\ 0.6 \end{bmatrix}$; **(b)** $\mathbf{1} = \begin{bmatrix} 1 \\ 1 \end{bmatrix} + \begin{bmatrix} 0 \\ 0 \end{bmatrix}$; **(c)** $\mathbf{1} = \begin{bmatrix} \frac{3}{7} \\ \frac{9}{7} \\ \frac{6}{7} \end{bmatrix} + \begin{bmatrix} \frac{4}{7} \\ -\frac{2}{7} \\ \frac{1}{7} \end{bmatrix}$;

(d) $\mathbf{1} = \begin{bmatrix} 1 \\ 1 \\ 1 \end{bmatrix} + \begin{bmatrix} 0 \\ 0 \\ 0 \end{bmatrix}$. **39.** $\mathbf{A}' = \begin{bmatrix} \mathbf{A} \\ \mathbf{K} \end{bmatrix}$, $\mathbf{b}' = \begin{bmatrix} \mathbf{b} \\ \mathbf{0} \end{bmatrix}$, rows of $\mathbf{K}$ are the kernel basis $\{\mathbf{k}_i\}$.

Section 5.3

1. (a) $\begin{bmatrix} 1 & 0 \\ 1 & 1 \end{bmatrix}\begin{bmatrix} 4 & 0 \\ 0 & 2 \end{bmatrix}\begin{bmatrix} 1 & 0 \\ -1 & 1 \end{bmatrix}$; **(b)** $\begin{bmatrix} 2 & 1 \\ 5 & -1 \end{bmatrix}\begin{bmatrix} 6 & 0 \\ 0 & -1 \end{bmatrix}\begin{bmatrix} \frac{1}{7} & \frac{1}{7} \\ \frac{5}{7} & -\frac{2}{7} \end{bmatrix}$;

(c) $\begin{bmatrix} 1 & 1 \\ 2 & -1 \end{bmatrix}\begin{bmatrix} 4 & 0 \\ 0 & 1 \end{bmatrix}\begin{bmatrix} \frac{1}{3} & \frac{1}{3} \\ \frac{2}{3} & -\frac{1}{3} \end{bmatrix}$; **(d)** defective matrix (only one eigenvalue).

3. (a) $\begin{bmatrix} 10,485,760 \\ 10,485,760 \end{bmatrix}$; **(b)** $\sim \begin{bmatrix} 172,760,510 \\ 431,901,250 \end{bmatrix}$; **(c)** $\begin{bmatrix} 6,990,510 \\ 13,981,010 \end{bmatrix}$; **(d)** $\begin{bmatrix} 590,490 \\ 590,490 \end{bmatrix}$.

7. (a) $\begin{bmatrix} 1 & 1 \\ 1 & -1 \end{bmatrix}\begin{bmatrix} 3 & 0 \\ 0 & 1 \end{bmatrix}\begin{bmatrix} 0.5 & 0.5 \\ 0.5 & -0.5 \end{bmatrix}$; **(b)** $\begin{bmatrix} 1 & 5 \\ 1 & 6 \end{bmatrix}\begin{bmatrix} 7 & 0 \\ 0 & -4 \end{bmatrix}\begin{bmatrix} \frac{6}{11} & \frac{5}{11} \\ \frac{5}{11} & -\frac{5}{11} \end{bmatrix}$;

(c) not diagonalizable; **(d)** $\begin{bmatrix} 1 & 1 \\ 1.5 & -1 \end{bmatrix}\begin{bmatrix} 1 & 0 \\ 0 & -0.25 \end{bmatrix}\begin{bmatrix} 0.4 & 0.4 \\ 0.6 & -0.4 \end{bmatrix}$;

(e) $\begin{bmatrix} 1 & 1 \\ 2 & -1 \end{bmatrix}\begin{bmatrix} 3 & 0 \\ 0 & 0 \end{bmatrix}\begin{bmatrix} \frac{1}{3} & \frac{1}{3} \\ \frac{2}{3} & -\frac{1}{3} \end{bmatrix}$.

9. (a) $\begin{bmatrix} 6 & 1 \\ 3 & 0 \end{bmatrix}$; **(b)** $\begin{bmatrix} 3 & 2 \\ 6 & 3 \end{bmatrix}$; **(c)** $\begin{bmatrix} -\frac{5}{3} & -\frac{4}{3} \\ \frac{22}{3} & \frac{23}{3} \end{bmatrix}$; **(d)** $\begin{bmatrix} -15 & 26 \\ -12 & 21 \end{bmatrix}$. **11. (a)** $\begin{bmatrix} 0 & 3 & 1 \\ -1.5 & 3 & 0 \\ 3.5 & -1 & 4 \end{bmatrix}$;

$$\textbf{(b)} \begin{bmatrix} 0 & -5 & -1 \\ 1.5 & 6 & 0 \\ -0.5 & -3 & 1 \end{bmatrix}; \textbf{(c)} \tfrac{1}{11} \begin{bmatrix} 42 & 35 & 22 \\ 9 & 24 & -11 \\ -12 & -10 & 11 \end{bmatrix}; \textbf{(d)} \tfrac{1}{13} \begin{bmatrix} 50 & -8 & 13 \\ -6 & 15 & -13 \\ 21 & 6 & 26 \end{bmatrix}.$$

Section 5.4

1. (a) $\begin{bmatrix} 4 & 0 \\ 4 & 0 \end{bmatrix} + \begin{bmatrix} 0 & 0 \\ -2 & 2 \end{bmatrix}$; **(b)** $\begin{bmatrix} \frac{12}{7} & \frac{12}{7} \\ \frac{30}{7} & \frac{30}{7} \end{bmatrix} + \begin{bmatrix} -\frac{5}{7} & \frac{2}{7} \\ \frac{5}{7} & -\frac{2}{7} \end{bmatrix}$; **(c)** $\begin{bmatrix} \frac{4}{3} & \frac{4}{3} \\ \frac{8}{3} & \frac{8}{3} \end{bmatrix} + \begin{bmatrix} \frac{2}{3} & -\frac{1}{3} \\ -\frac{2}{3} & \frac{1}{3} \end{bmatrix}$;

(d) defective matrix (no decomposition). **3. (a)** $\lambda_1 = 4.67$, $\mathbf{u}_1 = \begin{bmatrix} 0.82 \\ 1 \\ 0.21 \end{bmatrix}$, $\lambda_2 = -1.79$,

$\mathbf{u}_2 = \begin{bmatrix} 1 \\ 0.93 \\ 0.52 \end{bmatrix}$, $\lambda_3 = 0.12$, $\mathbf{u}_3 = \begin{bmatrix} -0.41 \\ 0.12 \\ 1 \end{bmatrix}$, $\begin{bmatrix} 1.83 & 2.23 & 0.47 \\ 2.23 & 2.72 & 0.58 \\ 0.47 & 0.58 & 0.12 \end{bmatrix} + \begin{bmatrix} -0.84 & 0.78 & -0.44 \\ 0.78 & -0.72 & 0.41 \\ -0.44 & 0.41 & -0.22 \end{bmatrix} +$

$\begin{bmatrix} 0.01 & -0.01 & -0.03 \\ -0.01 & 0.00 & 0.01 \\ -0.03 & 0.01 & 0.10 \end{bmatrix}$; **(b)** $\lambda_1 = 2.73$, $\mathbf{u}_1 = \begin{bmatrix} 1 \\ 1 \\ 0.73 \end{bmatrix}$, $\lambda_2 = -0.73$, $\mathbf{u}_2 = \begin{bmatrix} -0.37 \\ -0.37 \\ 1 \end{bmatrix}$,

$\lambda_3 = 0$, $\mathbf{u}_3 = \begin{bmatrix} -1 \\ 1 \\ 0 \end{bmatrix} - \begin{bmatrix} 1.08 & 1.08 & 0.79 \\ 1.08 & 1.08 & 0.79 \\ 0.79 & 0.79 & 0.57 \end{bmatrix} + \begin{bmatrix} -0.08 & -0.08 & 0.21 \\ -0.08 & -0.08 & 0.21 \\ 0.21 & 0.21 & -0.57 \end{bmatrix}$;

(c) $\lambda_1 = 4.71$, $\mathbf{u}_1 = \begin{bmatrix} -0.56 \\ 1 \\ 0.1 \\ -0.52 \end{bmatrix}$, $\lambda_2 = -1.97$,

$\mathbf{u}_2 = \begin{bmatrix} -0.12 \\ 0.54 \\ -0.78 \\ 1 \end{bmatrix}$, $\lambda_3 = 1.58$, $\mathbf{u}_3 = \begin{bmatrix} 1 \\ 0.61 \\ 0.5 \\ 0.18 \end{bmatrix}$, $\lambda_4 = 0.12$, $\mathbf{u}_4 = \begin{bmatrix} -0.6 \\ -0.05 \\ 1 \\ 0.74 \end{bmatrix}$,

$\begin{bmatrix} 0.93 & -1.65 & -0.17 & 0.86 \\ -1.65 & 2.95 & 0.29 & -1.54 \\ -0.17 & 0.29 & 0.03 & -0.15 \\ 0.86 & -1.54 & -0.15 & 0.80 \end{bmatrix} + \begin{bmatrix} -0.01 & 0.07 & -0.10 & 0.12 \\ 0.07 & -0.30 & 0.43 & -0.56 \\ -0.10 & 0.43 & -0.63 & 0.80 \\ 0.12 & -0.56 & 0.80 & -1.03 \end{bmatrix} +$

$\begin{bmatrix} 0.95 & 0.58 & 0.48 & 0.17 \\ 0.58 & 0.35 & 0.29 & 0.10 \\ 0.48 & 0.29 & 0.24 & 0.08 \\ 0.17 & 0.10 & 0.08 & 0.03 \end{bmatrix} + \begin{bmatrix} 0.13 & 0.01 & -0.21 & -0.16 \\ 0.01 & 0.00 & -0.02 & -0.01 \\ -0.21 & -0.02 & 0.36 & 0.26 \\ -0.16 & -0.01 & 0.26 & 0.20 \end{bmatrix}$.

5. (a) $\begin{bmatrix} 0.9 & 1.0 & 1.7 \\ 1.0 & 1.2 & 2.0 \\ 1.7 & 2.0 & 3.2 \end{bmatrix}$, poor fit; **(b)** $\begin{bmatrix} 1.6 & 1.0 & 2.5 & 2.4 \\ 2.5 & 1.6 & 4.0 & 3.9 \\ 2.4 & 1.5 & 3.9 & 3.8 \end{bmatrix}$, fair fit.

7. (a) $\lambda_1 = 6$, $\mathbf{u}_1 = \begin{bmatrix} 2 \\ 1 \\ 2 \end{bmatrix}$, $\lambda_2 = 3$, $\mathbf{u}_2 = \begin{bmatrix} 1 \\ 2 \\ -2 \end{bmatrix}$, $\lambda_3 = 0$, $\mathbf{u}_3 = \begin{bmatrix} 2 \\ -2 \\ -1 \end{bmatrix}$; **(b)** $\lambda_1 = 2.41$,

$\mathbf{u}_1 = \begin{bmatrix} 0.71 \\ 0.71 \\ 1 \end{bmatrix}$, $\lambda_2 = 1$, $\mathbf{u}_2 = \begin{bmatrix} -1 \\ 1 \\ 0 \end{bmatrix}$, $\lambda_3 = -0.41$, $\mathbf{u}_3 = \begin{bmatrix} -0.71 \\ -0.71 \\ 1 \end{bmatrix}$; **(c)** $\lambda_1 = 5.06$,

$\mathbf{u}_1 = \begin{bmatrix} 0.62 \\ 1 \\ 0.17 \end{bmatrix}$, $\lambda_2 = -3.14$, $\mathbf{u}_2 = \begin{bmatrix} -0.92 \\ 0.4 \\ 1 \end{bmatrix}$, $\lambda_3 = 3.08$, $\mathbf{u}_3 = \begin{bmatrix} 0.81 \\ -0.66 \\ 1 \end{bmatrix}$; **(d)** $\lambda_1 = 4.71$,

$\mathbf{u}_1 = \begin{bmatrix} -0.56 \\ 1 \\ 0.1 \\ -0.52 \end{bmatrix}$, $\lambda_2 = -1.97$, $\mathbf{u}_2 = \begin{bmatrix} -0.12 \\ 0.54 \\ -0.78 \\ 1 \end{bmatrix}$, $\lambda_3 = 1.58$, $\mathbf{u}_3 = \begin{bmatrix} 1 \\ 0.61 \\ 0.5 \\ 0.18 \end{bmatrix}$, $\lambda_4 = 0.68$,

$\mathbf{u}_4 = \begin{bmatrix} -0.6 \\ -0.05 \\ 1 \\ 0.74 \end{bmatrix}$.

Chapter 6

Section 6.1

1. (a) $\begin{bmatrix} 150 \\ 150 \\ 100 \end{bmatrix}$; **(b)** $\begin{bmatrix} 75 \\ 200 \\ 50 \end{bmatrix}$; **(c)** $\begin{bmatrix} \frac{100}{3} \\ \frac{100}{3} \\ \frac{400}{3} \end{bmatrix}$; **(d)** $\begin{bmatrix} \frac{50}{3} \\ \frac{50}{3} \\ \frac{200}{3} \end{bmatrix}$; **(e)** $\begin{bmatrix} 202.5 \\ 90 \\ 35 \end{bmatrix}$.

3. (a) $\begin{bmatrix} \frac{10}{6} & \frac{5}{12} \\ \frac{5}{6} & \frac{35}{24} \end{bmatrix}$; **(b)** $\begin{bmatrix} \frac{25}{12} & -\frac{5}{4} \\ -\frac{5}{6} & \frac{5}{2} \end{bmatrix}$; **(c)** $\begin{bmatrix} \frac{5}{3} & 0 \\ 0 & 2 \end{bmatrix}$.

7. (a) $\mathbf{D} = \begin{bmatrix} 0 & -\frac{4}{9} & \frac{1}{3} \\ -\frac{1}{2} & 0 & \frac{3}{8} \\ \frac{3}{10} & -\frac{3}{10} & 0 \end{bmatrix}$; **(b)** $\begin{bmatrix} -0.60 \\ 4.25 \\ 0.54 \end{bmatrix}$.

9. (a) $\mathbf{x}^{(1)} = \mathbf{Dw} + \mathbf{c}$, $\mathbf{x}^{(2)} = \mathbf{D}^2\mathbf{w} + \mathbf{Dc} + \mathbf{c}$, $\mathbf{x}^{(n)} = \mathbf{D}^n\mathbf{w} + \mathbf{D}^{n-1}\mathbf{c} + \cdots + \mathbf{Dc} + \mathbf{c}$.
11. (a) Convergence faster; **(b)** convergence slower. **13.** 1.

Section 6.2

1. $\begin{bmatrix} \frac{2}{3} \\ \frac{1}{3} \end{bmatrix}$. **3.** $\begin{bmatrix} 0.6 \\ 0.4 \end{bmatrix}$. **5.** $\begin{bmatrix} 0.4 \\ 0.2 \\ 0.4 \end{bmatrix}$. **7. (a)** $\begin{bmatrix} 0.1 \\ 0.2 \\ 0.2 \\ 0.2 \\ 0.2 \\ 0.1 \end{bmatrix}$; **(b)** $\frac{1}{853}\begin{bmatrix} 512 \\ 256 \\ 64 \\ 16 \\ 4 \\ 1 \end{bmatrix}$; **(c)** $\frac{1}{26}\begin{bmatrix} 8 \\ 4 \\ 1 \\ 1 \\ 4 \\ 8 \end{bmatrix}$; **(d)** $\begin{bmatrix} 0.1 \\ 0.2 \\ 0.2 \\ 0.2 \\ 0.2 \\ 0.1 \end{bmatrix}$.

9. (b) Does not converge to steady state. **11. (a)** $\lambda_1 = 1$, $\mathbf{u}_1 = \begin{bmatrix} \frac{2}{3} \\ \frac{1}{3} \end{bmatrix}$, $\lambda_2 = \frac{1}{4}$,

$\mathbf{u}_2 = \begin{bmatrix} 1 \\ -1 \end{bmatrix}$, $\mathbf{p}^{(6)} = 1^6\mathbf{u}_2 - \frac{2}{3}(\frac{1}{4})^6\mathbf{u}_2 \approx \mathbf{u}_2 = \begin{bmatrix} \frac{2}{3} \\ \frac{1}{3} \end{bmatrix}$; **(b)** $\lambda_1 = 1$, $\mathbf{u}_1 = \begin{bmatrix} \frac{6}{7} \\ \frac{1}{7} \end{bmatrix}$, $\lambda_2 = 0.3$,

$\mathbf{u}_2 = \begin{bmatrix} 1 \\ -1 \end{bmatrix}$, $\mathbf{p}^{(6)} = 1^6\mathbf{u}_2 + \frac{1}{7}(0.3)^6\mathbf{u}_2 \approx \mathbf{u}_2 = \begin{bmatrix} \frac{6}{7} \\ \frac{1}{7} \end{bmatrix}$; **(c)** $\lambda_1 = 1$, $\mathbf{u}_1 = \begin{bmatrix} \frac{3}{5} \\ \frac{2}{5} \end{bmatrix}$, $\lambda_2 = \frac{1}{6}$,

$\mathbf{u}_2 = \begin{bmatrix} 1 \\ -1 \end{bmatrix}$, $\mathbf{p}^{(6)} = 1^6\mathbf{u}_2 - \frac{1}{10}(\frac{1}{6})^6\mathbf{u}_2 \approx \mathbf{u}_2 = \begin{bmatrix} \frac{3}{5} \\ \frac{2}{5} \end{bmatrix}$.

15. (b) $\begin{bmatrix} 6 & 4 & 2 \\ 8 & 8 & 4 \\ 4 & 4 & 4 \end{bmatrix}$; **(c)** 18.   **17. (a)** $\begin{bmatrix} 2 & 1 & 1 & 1 \\ 1.5 & 2 & 1 & 1.5 \\ 1.5 & 1 & 2 & 1.5 \\ 1 & 1 & 1 & 2 \end{bmatrix}$; **(b)** 1.5; **(c)** 6.

19. (a) $\begin{bmatrix} \frac{4}{3} & \frac{2}{3} & \frac{2}{3} \\ \frac{1}{2} & \frac{11}{8} & \frac{5}{8} \\ \frac{1}{2} & \frac{5}{8} & \frac{11}{8} \end{bmatrix}$; **(b)** $\frac{2}{3}$; **(c)** $\frac{8}{3}$.   **21. (a)** 2.74; **(b)** A, 0.27; B, 0.19; C, 0.44.

23. $\frac{25}{6}$.   **25.** 18, 16.

Section 6.3

1. (a) $y(t) = 100e^{0.5t}$; **(b)** $y(t) = 10e^{4t}$.   **3. (a)** $y(t) = 25e^t - 5e^{4t}$; **(b)** $y(t) = 3e^t - 2e^{-6t}$;

(c) $y(t) = \frac{2}{3}e^{4t} + \frac{4}{3}e^{-2t}$.   **5.** $\left(\begin{bmatrix} 1 & 1 \\ -1 & 1 \end{bmatrix} \begin{bmatrix} e^{3t} & 0 \\ 0 & e^t \end{bmatrix} \begin{bmatrix} 0.5 & -0.5 \\ 0.5 & 0.5 \end{bmatrix} \right) \begin{bmatrix} 1 \\ 1 \end{bmatrix} = \begin{bmatrix} e^t \\ e^t \end{bmatrix}$ [i.e.,

$x(t) = y(t) = e^t$].

Section 6.4

1. (a) 0; **(b)** 1; **(c)** 0.   **5. (a)** $\mathbf{e} = \begin{bmatrix} 0 \\ 0 \\ 1 \end{bmatrix}$, fourth bit; **(b)** $\begin{bmatrix} 0 \\ 1 \\ 1 \end{bmatrix}$, sixth bit; **(c)** $\begin{bmatrix} 0 \\ 1 \\ 0 \end{bmatrix}$, second

bit.   **7.** New $\mathbf{Q}$ obtained from old $\mathbf{Q}$ by interchanging rows 3 and 4, new $\mathbf{c}$ same as old $\mathbf{c}$.

9. M $= \begin{bmatrix} 1 & 0 & 1 & 0 & 1 & 0 & 1 & 0 & 1 & 0 & 1 & 0 & 1 & 0 & 1 \\ 0 & 1 & 1 & 0 & 0 & 1 & 1 & 0 & 0 & 1 & 1 & 0 & 0 & 1 & 1 \\ 0 & 0 & 0 & 1 & 1 & 1 & 1 & 0 & 0 & 0 & 0 & 1 & 1 & 1 & 1 \\ 0 & 0 & 0 & 0 & 0 & 0 & 0 & 1 & 1 & 1 & 1 & 1 & 1 & 1 & 1 \end{bmatrix}$;

11.

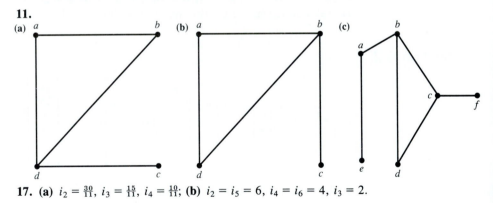

17. (a) $i_2 = \frac{30}{11}$, $i_3 = \frac{15}{11}$, $i_4 = \frac{10}{11}$; **(b)** $i_2 = i_5 = 6$, $i_4 = i_6 = 4$, $i_3 = 2$.

Index